FIELD GUIDE TO THE LICHENS OF GREAT SMOKY MOUNTAINS NATIONAL PARK

The University of Tennessee Press gratefully acknowledges the support of the following organizations in producing *Field Guide to the Lichens of Great Smoky Mountains National Park*:

Discover Life in America
Eagle Hill Institute
Highlands Biological Station
New York Botanical Gardens
Tennessee Native Plant Society
Tennessee Wildlife Resources Agency
University of Colorado Museum of Natural History

Field Guide to the Lichens of Great Smoky Mountains National Park

ERIN A. TRIPP AND JAMES C. LENDEMER

Illustrations by Bobbi Angell

The University of Tennessee Press
Knoxville

Support for this book was provided in part by the Tennessee Wildlife Resources Agency, Division of Environmental Services.

First Edition.

Library of Congress Cataloging-in-Publication Data
Names: Tripp, Erin Anne, 1979– author. | Lendemer, James C., author.
Title: Field guide to the lichens of Great Smoky Mountains National Park / Erin A. Tripp and James C. Lendemer; illustrations by Bobbi Angell.
Description: First edition. | Knoxville: The University of Tennessee Press, [2020] | Includes bibliographical references. |
Identifiers: LCCN 2019013801 (print) | LCCN 2019016520 (ebook) | ISBN 9781621905158 (pdf) | ISBN 9781621905141 (pbk.)
Subjects: LCSH: Lichens–Identification–Great Smoky Mountains National Park (N.C. and Tenn.)
Classification: LCC QK587 (ebook) | LCC QK587 .T75 2020 (print) | DDC 579.709768/89–dc23
LC record available at https://lccn.loc.gov/2019013801

Let us envision the forest as an aggregate of plant and animal life: an aggregate of trees of the canopy and lower layers, shrubs and vines, both large and small, herbaceous plants of the forest floor; mosses and lichens on the ground, on rocks, and on tree trunks, fungi, both parasitic and saprophytic, and also of the animals—grazing, burrowing, leaf-eating, seed-eating—the mammals, the birds, the insects, and the micro-organisms (plant and animal) above and below ground; an aggregate of interdependent and interacting organisms. From among this aggregate of multitudinous forms, the trees stand out pre-eminent, because of their superior size and because of their pronounced influence on all of the forest inhabitants. Yet each member of the forest community plays its part, in competition, in contribution to organic detritus, in giving to the forest its many aspects, both local and seasonal (E. Lucy Braun 1950, 3).

Contents

PREFACE

A History of Change

Between 1539 and 1543, chroniclers and painters on the De Soto expedition described an eastern North American landscape that was very different from what we see today. Canebrakes, or dense stands of our only native bamboo, *Arundinaria*, were once in great abundance along streams and floodplains and said to run for dozens of miles. They were important fodder for wildlife, habitats for birds, and Cherokee peoples used them in construction and basketry. Today, only isolated stands remain. The American Chestnut (*Castanea dentata*) was without doubt one of the most ecologically important trees in much of eastern North America. Once comprising upwards of 50% of the trees in a single forest, its tall straight trunks made it ideal for construction beams; its strength appropriate for railroad ties; its light weight suitable for ships' masts and furniture; and its rot resistance fitting for fences and barns that still dot southeastern landscapes today. Tannic acids from its bark served the leather trade, and its fruits fed whole populations of native and domesticated animals.

Like much of the rest of eastern North America, logging was once a daily routine in the area now occupied by Great Smoky Mountains National Park (GSMNP), although we are fortunate that great stretches of the park were protected from logging due their inaccessibility. Major sawmill operations could once be found from Oconoluftee to Smokemont to Tremont. That activity has ebbed since the height of the logging era in eastern North America, between the late 1800s and 1920, but air pollution increased steadily in the decades that followed.

As few as one or two centuries ago, 200-pound cats and native elk the size of small cars roamed the park. Now, the top carnivore and herbivore have been extirpated, although reintroduction experiments of the latter are underway. In the absence of these flora and fauna, non-native Wild Hogs have flourished and brought about great damage to ecosystems at all elevations. Today, one could argue that the biggest agent of environmental change in the park may be introduced species. Both the Fraser Fir and Eastern Hemlock declined rapidly after the arrival of non-native, lethal adelgids in the 20th Century. Currently, living hemlocks are nearly gone from much of southeastern North America. In the Southeast, the largest and as yet uninfected stands of hemlocks occur in protected coves of Bankhead National Forest, Alabama, where the species is disjunct from otherwise contiguous portions of its range in the Appalachian Mountains. Long-term impacts related to loss of the hemlock from eastern forests might ultimately equal or surpass that of the American Chestnut.

The ecological, biological, and cultural diversity of GSMNP is, at this point in time, inextricably intertwined, and must be thought of as a process rather than as a terminus. As plant and animal abundances and distributions have responded to pre- and post-human arrival, so, too, will the lichens, the mosses, the beetles, the tardigrades, the slime molds, and yet more inconspicuous biological organisms about which we are only beginning to learn something. Without a historical context, both evolutionary and recorded, little, if anything, makes sense. It is only fitting that we continue to document biodiversity, distributions, and change such that the present becomes an adequate account of what will become history. It is our hope that this *Field Guide to the Lichens of Great Smoky Mountains National Park* fills one much-needed niche and provides a historical foundation upon which future researchers can build.

Acknowledgments

We are indebted to the following individuals at (or formerly employed by) Great Smoky Mountains National Park: Keith Langdon, Paul Super, Susan Sachs, Emily Darling, and Becky Nichols. Our publisher, University of Tennessee Press, especially Thomas Wells, Jon Bogs, and the Board of Directors, are thanked for their extensive time, resources, expertise, and willingness to give lichens a try. We are additionally grateful to Annalisa Weaver for her very careful copy edits. Much as she has done for the prior four decades to advance understanding of the botanical world, Bobbi Angell beautifully illustrated this book, which seeks to make lichens more familiar to all. Though her illustrations, new knowledge and understanding of lichen biology is born. Richard Harris provided assistance with various lichen identifications, and we further thank him as well as Bill Buck (New York Botanical Garden) for accompanying us during fieldwork. Staff members at the NY and COLO Herbaria (Ellen Bloch, Laura Briscoe, Ana Maria Ruiz [NY]; Dina Clark, Tim Hogan, Ryan Allen [COLO]) and students are thanked for helping facilitate curation of our specimens. Other friends provided companionship and, in some cases, photography during fieldwork (Andy Moroz, Kate Deregibus) or general feedback on GSMNP (David Clarke, Jonathan Dey, Kyle Dexter, Josh Kelly). Troy McMullin is thanked for letting us "borrow" his excellent photographs of pin lichens. We are extremely grateful to two individuals for their time, expertise, and generous comments, which improved an earlier version of this manuscript: Larry St. Clair (Brigham Young University) and James Lawrey (George Mason University). The first author thanks her parents, C. D. and Linda Tripp, as well as sisters, Gretchen Lowman and Rachel Rutledge, for helping to facilitate our fieldwork and putting up with our shenanigans for so many years. Finally, we are grateful to the Eastern Bank of Cherokee who once lived much more abundantly in these mountains and today welcomed us onto their homeland.

PART 1
LICHEN BIOLOGY

FIGURE 1. CRYPTOGAMIC BIOMASS

Things grow fast in the Smokies . . . sometimes *really* fast. Here is a log that fell approximately a decade ago, and quickly became covered by *Hypnum* cf. *imponens*, a common moss of the park. Moss and lichens are often referred to as "cryptogams" by professionals, and course woody debris such as this fallen tree serves an exceptionally important role as a substrate for cryptogamic diversity in the Smokies and beyond. Photo by Erin Tripp.

WHAT ARE LICHENS?

A lichen is a symbiotic association of a minimum of two unrelated organisms. Always present are the fungal "mycobiont" and photosynthetic "photobiont" partners, the latter of which is most frequently a green alga or, less commonly, a nitrogen-fixing cyanobacterium. Think of a lichen as a peanut butter and jelly sandwich: the fungus comprises the bread (the upper and lower "cortex," see below), the alga comprises the layer directly underneath the upper layer of bread (jelly or peanut butter, depending on your mood), and the inner remaining layer (the jelly or the peanut butter) is more fungus (the "medulla," see below). The traditional concept of the lichen was a symbiosis that was mutualistic, with the fungus providing a protected environment in which the photobiont could live, and the photobiont supplying nutritional products of photosynthesis to feed the fungus. Other perspectives suggest the traditional view is not entirely accurate: the symbiosis could instead reflect a controlled parasitism (Ahmadjian & Jacobs 1981). Additionally, reductions in key genes in the fungal genome, such as those responsible for the production of energy, appear to contribute to obligate dependence of the fungus on the photobiont, which maintains a more complete set of genes needed for energy production (Pogoda et al. 2018). Recent research has also shown that the lichen symbiosis is far more complex than previously understood, with the discovery of additional partners such as endolichenic fungi and endolichenic bacteria, whose functions are for the most part not yet known (Arnold et al. 2009).

At present, taxonomy of lichens is based entirely on classification of the mycobionts. Thus, names such as *Arthonia*, *Pertusaria*, or *Pseudosagedia* reflect fungal names, but these are also the names applied to the lichen as a whole.

FIGURE 2. LICHEN LADEN SPRUCE
Northern hardwood forests are one of the most lichenologically biodiverse ecosystems in Great Smoky Mountains National Park. These forests are dominated by *Betula alleghaniensis* (Yellow Birch), *Picea rubes* (Red Spruce), American Beech (*Fagus grandifolia*), Black Cherry (*Prunus serotina*), and *Acer saccharum* (Sugar Maple). Contrary to common misconception, forests laden with lichen diversity and biomass such as that seen here are to be considered healthier ecosystems compared to forests without lichens. One need only stop the car next to any number of unhealthy, fragmented forests lining Interstate-95 to observe a near complete absence of lichens. This is not a good thing. Photo by James Lendemer.

This taxonomy reflects the widespread assumption that the fungus is the only member of the association required to be present to yield the lichen we see in nature. In contrast, it has been assumed that the photobiont can vary, with no major impact on the morphology of the lichen.

The degree to which this is entirely accurate remains to be seen, as it is now known that photobionts can and do impact lichen morphology. In any case, it is true that the same mycobiont can associate with different green algae or different cyanobacteria and still yield a comparable lichen morphologically. It is also true that photobionts have their own names and their own classification systems separate from the mycobiont and the lichen.

As is currently understood, taxonomic diversity of the mycobiont is far greater than taxonomic diversity of the photobiont. Many among the latter belong to Chlorophyta—a diverse and widespread lineage of green algae. Despite the tradition of naming lichens based on the "more diverse" partner, evidence is rapidly accumulating that photobionts also comprise a tremendous diversity of genotypes, phenotypes, and different lineages. And, of course, the alga does impact expression of the symbiotic phenotype as a whole. It is possible to envision a time in the future when symbiomes such as lichens receive formal naming based on the full set of co-evolving members (Tripp et al. 2017).

WHY LICHENS?

In 1950, the renowned plant ecologist Lucy Braun observed that every component of the forest functions in some way vital to that forest. Relative proportions of these components–plant, mammal, bird, amphibian, invertebrate, fungus, or bacterium–vary across environments, but in some habitats, lichens together with bryophytes contribute more to the total biotic diversity than do vascular plants (Kantvilas 1990; Jarman & Kantvilas 1995) (Figures 1 & 2). Each year, approximately equal numbers of new species of lichens and new species of flowering plants are reported from North America. The large number of vascular plant discoveries are primarily related to the many ongoing large-scale floristic efforts, such as *The Flora of North America* (1993+), the *New Manual of Vascular Plants of Northeastern United States and Adjacent Canada* (Naczi et al. 2017), or the revision of California's *Jepson Manual II* (Baldwin et al. 2012). New plant discoveries also reflect a large number of vascular plant specialists actively advancing taxonomic research. In contrast, there very few large-scale floristic efforts in American lichenology: Nash et al.'s *Lichen Flora of the Greater Sonoran Desert Region* (2002–2007) and McCune's *Microlichens of the Pacific Northwest* (2017) represent the two exceptions. Furthermore, professional botanists outnumber professional lichenologists in North America by a ratio that approximates 1,000 to 1. It is clear that lichens are fundamental contributors to ecological processes; however, the many gaps that remain in lichenology regarding even basic tasks such as species identification yield a clear picture of how much work remains ahead of us. A goal of *Field Guide to the Lichens of Great Smoky Mountains National Park* is to inspire new curiosity about North American lichens in hopes of recruiting new students to the field.

Whether on rock, tree, or soil, lichens influence the distribution of nutrients and water in all habitats in which they occur. Just like plants and animals, they grow and respire; and just like plants, they photosynthesize, owing to the presence of the photobiont (Figure 3). Lichens also influence nutrient distribution in another manner—as primary decomposers of parent rock material into what will ultimately become soil (Friedmann 1982). The fungal component of the lichen manufactures and secretes various chemicals that aid its penetration into rock surfaces; once below the surface, fungal hyphae grow among the crystalline structures and between rock cleavage lines, further fracturing the parent material (Figure 3). As lichens grow and decompose, they trap fine soil particles among their surfaces, encouraging saprophytic fungi and bacterial growth, furthering the process of soil formation (Jones et al. 1981).

Lichens contribute significantly to the "fixation" of nitrogen—that is, the process by which atmospheric nitrogen is converted into a molecular form useable by living organisms (Hitch & Stewart 1973; Pike 1978). In most natural landscapes, this is accomplished through nitrogen-fixing bacteria. In GSMNP, there are several genera of lichens that harbor such bacteria, including *Coccocarpia*, *Collema*, *Leptogium*, *Lobaria*, *Nephroma*, *Pannaria*, *Parmeliella*, *Peltigera*, *Pseudocyphellaria*, and *Sticta* (Figure 4). These "cyanolichens" are exceptionally diverse and abundant in GSMNP, where they contribute importantly to nitrogen cycling on both a local and regional scale.

Beyond contributions to nutrient cycling, lichens function prominently in many food webs. In boreal ecoregions, lichens serve as primary food sources for reindeer and mountain goats, and Brodo et al. (2001) noted that caribou cannot live without them. In other areas, lichens figure substantially into the diets of large- and small-bodied animals including elk, moose, deer, squirrels, birds, snails, mites, and wingless insects (Sharnoff 1994, Pettersson et al. 1995). On a microscopic level, lichens represent hubs of trophic interactions among microorganisms and invertebrates.

Lichens are used by numerous animals in nest construction (Ladd 1998); with enough patience on a calm spring day, you can watch Ruby Throated Hummingbirds use them to build beautiful nests. Lichens are also used medicinally in some human societies (Wei et al. 1982). More recently, they have been

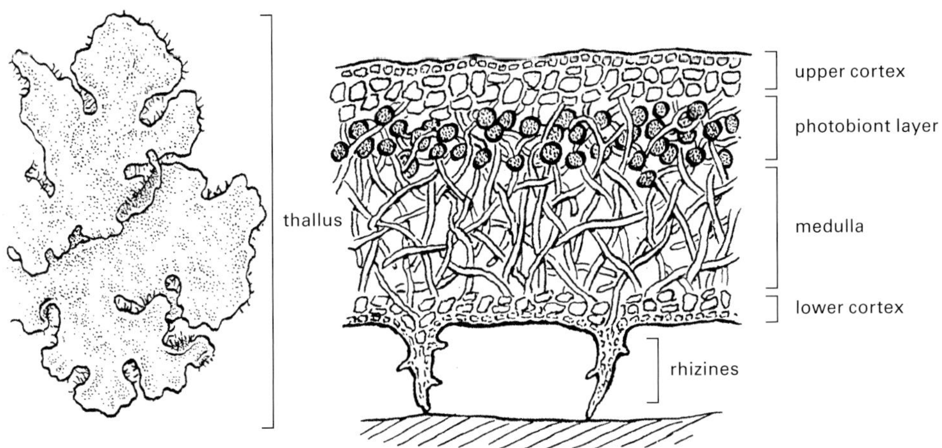

FIGURE 3. THALLUS

Thallus is the general term for the vegetative body of a lichen. Thalli (plural) are typically composed of four layers, from top to bottom: the upper cortex, photobiont layer, medulla, and lower cortex. The photobiont layer, situated immediately underneath the upper cortex, receives physical protection from the latter yet is near enough to sunlight to photosynthesize. The cortices and medulla are composed of fungal-only tissue. Illustrations by Bobbi Angell.

FIGURE 4. *COLLEMA AND LEPTOGIUM*

"Jelly lichens" refers to species of *Collema* and *Leptogium* (here, *C. subflaccidum* on left and *L. cyanescens* on right, along Goshen Prong), which are diverse and exceptionally abundant in the Smokies. When wet, these two genera of cyanolichens can be difficult to differentiate. When dry however, most *Collema* are blackish in color whereas most *Leptogium* are gray. In either case, thank these lichens for their important contributions to nitrogen cycling in the Smokies and beyond. Photo by Erin Tripp.

appreciated as important indicators of overall species richness, forest quality, landscape health, or indicators of air quality in a particular region (De Wit 1983; Nilsson et al. 1995; Hawksworth 2002; Bergamini et al. 2007). Lichen diversity (i.e., lichen species richness) both predicts the number of rare lichen species in a given landscape and serves as an indicator for old growth forests (Nordén et al. 2007; Brunialti et al. 2010). Further, lichens respond negatively to human disturbance in much the same way that flowering plants do (Jägerbrand & Alatalo 2015; Tripp et al., 2019). Air pollution, for example, negatively impacts lichen diversity as well as abundance (McCune et al. 1997; Perlmutter et al. 2017). In sum, lichens function in myriad ways important to ecosystem processes, yet are negatively impacted by disturbance in the same manner that applies to numerous other organisms.

LICHEN BIOLOGY

Lichens are miniature ecosystems that occur on a number of different substrates ranging from the bark of living plants ("corticolous") to rock ("saxicolous"), soil ("terricolous"), leaf surfaces ("foliicolous"), liverworts and mosses ("muscicolous"), rotting wood ("ligniciolous"), other lichens ("lichenicolous"), or even the exterior surfaces of animals such tortoises ("zooicolous"). A rarer number of species occurs as parasites of free-living algae ("algicolous") or ephemerally on dried tree resin ("resinicolous"). In essence, lichens grow on anything that sits still long enough, including you, when your time expires. Their morphological diversity and ecological contributions have gone unnoticed to a great number of natural history enthusiasts, owing not to a lack of interest but rather to a lack of practical and comprehensive identification manuals. Below, we equip the reader with basic information on lichen biology (morphology, reproduction, biogeography) to enable better understanding of and appreciation for the species encountered in this *Field Guide*. We cite examples of common lichens in the Park to illustrate the main messages.

LICHEN MORPHOLOGY

The vegetative, non-sexual portion of the lichen body is termed a thallus (plural: "thalli") and is generally composed of four layers (Figure 3). The upper and lower layers form the cortex and are composed of densely packed fungal filaments called hyphae that are usually elongated and somewhat irregular in shape. The upper cortex is where various accessory pigments that give lichens their outer color reside. The lower cortex is frequently white, gray, brow, or near black in color. Its absence or presence and properties such as texture and color are important in the identification of many lichens. Just underneath the upper cortex is the algal layer, which consists of algal or cyanobacterial cells that vary from light to dark green to almost blue, depending on which photobiont is present. The layer below the algae is termed the medulla, which is composed of a network of loosely arranged to densely packed hyphae. It is frequently bright white in cross section but can be brightly pigmented in some species, such as *Pseudocyphellaria aurata*. Below the lower cortex and medulla, special structures called "rhizines" can be found in some foliose lichens. These function in the attachment of a lichen to its substrate; and, although they resemble roots, rhizines are apparently not involved in the uptake of nutrients (Figure 5). Not all lichens have the four basic layers. Species of *Peltigera* and most *Heterodermia*, for example, lack a lower cortex. All species of *Lepraria* and some *Graphis* lack an upper cortex (i.e., they are "ecorticate") (Figure 6).

Lichens can be broadly characterized as having one of four different growth forms: foliose, fruticose, crustose, or squamulose. Of the four main types, foliose lichens most readily exemplify the four-layered, stratified thallus described above. Foliose species are, for the most part, the large and easily visible "macrolichens" that come to mind when one conjures an image of lichens covering tree bark. They are characterized by flattened, leaf-like thalli with a distinct upper and lower surface (Figure 7). Various aspects of both macro- and micromorphology (as well as chemistry) aid in the identification of foliose lichens. For example, the genus *Parmotrema* is predominantly characterized by its broad, rounded lobes, ruffled lobe margins, and the presence of cilia scattered along the margins of the thallus lobes (Figure 8). Among the most common foliose lichens on trees in GSMNP are species of *Heterodermia*, *Hypotrachyna*, *Lobaria*, *Punctelia*, *Parmelia*, *Parmotrema* and, on rock, *Xanthoparmelia*. A special type of foliose lichens has a unique "umbilicate" growth form wherein they form plate-like thalli that are attached to the substrate at a single, central point called a "holdfast." Such lichens are referred to as "rock tripe" and include species of *Lasallia* and *Umbilicaria* (Figure 9) that are common throughout the Park.

The squamulose growth form is a specialized type of foliose lichen (umbilicate lichens are another type of specialized foliose lichen). Squamulose species are characterized as having thalli composed of miniature, frequently overlapping, shingle-like lobes called squamules (Figure 10). Because of the small size of these lobes, they superficially resemble (and are often discussed together with) crustose lichens (see below for explanation of the crustose growth form). Other squamulose species are larger and resemble foliose macrolichens.

The primary thallus seen on many species of *Cladonia* is squamulose, and most of these species produces a fruticose portion of the thallus that grows upwardly out of the squamules (Figure 11). Here is a great lesson to learn early on: lichens do their best to defy classification into tidy categories. In the end, it is perhaps worth considering whether lichenologists should abandon the idea of the squamulose growth form altogether. Then again, without categories, nature is a blur.

Fruticose lichens are the most three-dimensional of the growth forms as they have branches rather than lobes, and thus resemble tiny shrubs. The branches, such as in *Usnea*, are typically round and cylindrical to subcylindrical in cross section. However, some species, such as members of the genus *Ramalina*, can be similar to foliose lichens in having flattened branches. Unlike foliose lichens, however, branches of fruticose lichens do not have distinct upper and lower surfaces because they grow away from (as opposed to along) the substrate. As such, the algal layer is present along the entire periphery of the branch, just below the cortex (sort of like if bread was never flat, but instead round; can you envision a peanut butter and jelly sandwich that was tubular in shape?). Fruticose lichens adhere to substrates via one or more holdfasts, are erect or pendulous, and are often seen dangling from trees and rocks. The most common fruticose lichens in GSMNP are species of *Cladonia*, *Ramalina*, and *Usnea*. Other examples of fruticose lichens in GSMNP include *Bryoria*, *Cetraria*, *Ephebe*, *Evernia*, and *Stereocaulon* (Figure 12).

Crustose lichens, often called microlichens, are by far the most understudied of the four growth forms, even though they often constitute over half of the lichen diversity in any given region, including GSMNP (Tripp & Lendemer, unpub. data [manuscript in preparation]; Figure 13). They are characterized by an indistinct or absent lower cortex, and thus are in intimate contact with their substrates. Because they lack a well-defined lower cortex, crustose lichens always lack rhizines and other specialized attachment structures. Crustose lichens can be described by various terms that define their growth patterns (Figure 14). Frequently, an initial mat of fungal hyphae called a prothallus extends as a delicate fringe along the edge of the expanding, growing thallus.

Due to the lack of a lower cortex, crustose lichens generally cannot be removed intact from their substrates; thus pieces of bark or rock containing the crust must be removed with the lichen for further study. Many species are large and easily visible to the naked eye; equally many are very small and require some form of magnification, such as a 10x hand lens. This added challenge has resulted in comparably less published information on crustose lichens than other growth forms across eastern North America. *Lecanora* and *Pertusaria* are two of the most common and most diverse genera of crustose lichens in GSMNP; they can be found abundantly on bark and, to a lesser degree, on rock. One species of crustose lichen found on rocks, *Porpidia albocaerulescens*, is among the most common lichens in all of eastern North America, including GSMNP.

It is important to note that similarity of growth form does not always indicate a close evolutionary relationship, but other times, shared growth form is part of what characterizes close evolutionary relatives. Some lichens defy placement into one category or another and a few are defined by very specific terms such as being leprose (Figure 15), such as in *Lepraria* and *Chrysothrix*) or filamentous, as in *Cystocoleus*. A bizarre example of a species that challenges placement into a well-defined growth form is *Lecanora anakeestiicola*, which consists of a crustose thallus in rock overhangs from which fruticose branches dangle downwards.

Distinctive patterns and modifications to lichen thalli can aid in the identification of many foliose, fruticose, and crustose species (Figure 16). These include cyphellae—crater-like pores typically on the lower surface (e.g., *Sticta fragilinata*); pseudocyphellae—various differentiated breaks in the cortex (e.g., *Cetrelia chicitae*, *Parmelia* spp., *Pseudocyphellaria aurata*); maculae—light-colored spots on the upper cortex (e.g., *Platismatia tuckermanii*, *Parmotrema reticulatum*); perforations—holes in the thallus (e.g., *Menegazzia subsimilis*); cilia—hair-like projects on the edge of the thallus (e.g., *Parmotrema hypotropum*); pruina—crystalline deposits on the surfaces that are often white in color (e.g., *Physconia subpallida*); tomentum—fuzzy surfaces (e.g., *Anzia americana*); veined surfaces—prominent branching patterns on the lower surfaces (e.g., *Peltigera* spp.); inflated lobes—these typically enlarged in cross section (e.g., *Hypogymnia* spp.); and fibrils—small branchlets that project from the main branches of fruticose lichens (e.g., *Usnea strigosa*).

LICHEN REPRODUCTION

Lichens reproduce both sexually and asexually (Tripp & Lendemer 2017). The fungal partner is almost always an Ascomycete fungus instead of, for example, a species of the better-known mushroom-forming Basidiomycete fungi. The fruiting body or basic unit of the sexual reproductive structure in Ascomycetes is the ascoma (plural: ascomata). Lichen ascomata generally come in two forms. Apothecia are open, disc-shaped structures that are circular in outline and may have simple margins (Figure 17) or margins that are ornamented externally (e.g., *Anaptychia palmulata*, *Lobaria ravenelii*). The discs of apothecia may be concave (e.g., *Anzia colpodes*), convex (e.g., *Buellia stillingiana*), flat (e.g., *Lecanora* or *Chaenotheca*), lirellate (lip-like) as in the "script lichens" (e.g., *Graphis scripta, Opegrapha varia*), otherwise elongate and irregular (e.g., *Arthonia anglica*), or raised on stalks or podetia, as in many *Cladonia* (also *Dibaeis baeomyces*). Perithecia are flask-shaped structures, usually partially embedded within the thallus, with pore-like openings (Figure 18). Think of perithecia as apothecia that evolutionarily have become flask-shaped through modifications (e.g., pinching) of the external margins. Some apothecia can be superficially pore-like or urn-shaped, as in *Pertusaria* or *Conotrema urceolateum*, but these should properly be considered apothecia rather than perithecia.

Ascomata are extremely diverse morphologically. They can occur with distinctive margins as in *Biatora* and *Lecidea* (Figure 19) or with thalline margins, which are derived from thallus material and therefore similarly contain algae, as in *Ochrolechia* and *Lecanora* (Figure 19). The former are termed "lecideine margins," whereas the latter are termed "lecanorine margins." Sexual reproductive structures of lichens–whether apothecia or perithecia–can range from comprising nearly the entire visible lichen body to sparsely scattered across the thallus to rare or completely unknown in many species.

Sexual structures, especially the ascospores (hereafter: "spores," for simplicity), are extremely important for lichen identification; work at this level requires up to 100x magnification via a compound microscope. (Note that this utility is no different from other groups of organisms wherein reproductive structures are more diverse than other parts of an organism, and so taxonomists have capitalized on such structures for purposes of classification and evolutionary investigation.) Inside the ascomata are asci, or sacs, that contain spores derived from the process of meiosis (Figure 20). Because sexual reproduction involves only the fungus and not the algae within the lichen thallus, dispersed fungal spores must encounter an algal counterpart in order to form a new thallus.

Among the greatest but as yet largely unanswered curiosities about lichen reproduction is just how widespread and free-living the algal partners are. Whereas some scientists think most lineages of lichen photobionts are not typically free-living, they must be—to some degree—to facilitate re-lichenization following spore dispersal. A phycologist may laugh at this assertion, shouting "Duh! . . . everyone knows they are free-living!" This uncertainty remains but one of many mysteries; other unanswered questions include: In what sorts of environments do photobionts occur (wet, dry, shaded, exposed), and to what extent do geographical distributions of these photobionts limit or facilitate re-lichenization (upon contact with a spore) and development of a new lichen thallus? Do photobionts commonly undergo sexual reproduction when free of the confines of the lichen thallus? It is otherwise assumed that they reproduce only asexually inside the lichen thallus.

Vegetative, asexual reproduction in lichens occurs through modifications of the thallus into specialized propagules. Unlike sexual reproduction, these asexual propagules are products of mitosis (not meiosis) and generally disperse the fungus and the photobiont together, as a unit. Because these propagules contain both the mycobiont and photobiont (and we presume they contain other important organisms such as bacteria), they are loosely termed "lichenized propagules" or "asexual lichenized propagules." There are numerous types of asexual lichenized propagules, the most important among these being soredia and isidia. Soredia are powdery masses of hyphae surrounding one or a few algal cells, and these rounded bundles form from the medulla when it becomes exposed in specialized breaches of the upper cortex, termed soralia (Figure 21). By definition, sorediate are "ecorticate," meaning they lack a cortex, which is precisely because they are derived from the tissue of the medulla.

Isidia are miniature, columnar outgrowths of the upper cortex and, by definition (and unlike soredia), are corticate (Figure 22). Other types of asexual lichenized propagules include phyllidia (i.e., micro-lobules that break off from the main thallus; Figure 23); schizidia (i.e., disintegrating flakes of the upper cortex that contain algae); blastidia (i.e., budding proliferations of the upper or lower cortex that contain algae); and fragments (i.e., pieces of the lichen thallus that contain both partners).

Another, perhaps less appreciated, form of asexual reproduction in lichens involves propagules that are "non-lichenized." In other words, these propagules derive from mitosis (not meiosis) but disperse only the mycobiont. The most prominent example of "asexual non-lichenized propagules" are conidia, which are microscopic cells manufactured via mitosis that bud from the ends of specialized hyphae termed conidiophores. The conidia and conidiophores are typically housed inside specialized structures termed pycnidia that resemble minute perithecia (Figure 24). Pycnidia appear as tiny black dots on the thalli of numerous species of lichens. However, pycnidia are not always dark in color and can be almost imperceptible, even with a microscope.

Much in the way that asexual lichenized propagules have diversified into numerous different types of structures, so too have the structures that produce conidia. In addition to pycnidia, other types of such structures include sporodochia, which are hemispherical structures that resemble soralia but are actually invaginated pycnidia that bear conidia (e.g., see *Tylophoron americanum*). Still other types include synnemata, which are sporodochia elevated on stipes (e.g., *Dictyocatenulata alba*). Do you think we have yet discovered all the different types of reproductive structures and modes by which lichens can reproduce? It is commonly assumed that the most important function of conidia is to serve as an agent of fertilization. Conidia are dispersed from a pycnidium, move around the environment, and eventually land on a specialized structure of receptive hyphae, termed a "trichogyne," resulting in fertilization. Cells that derive from this fertilization event are dikaryotic (i.e., they contain two, unfused nuclei, in an "n + n" configuration), which eventually gives rise to a diploid cell in the vicinity of the ascus. This diploid cell is the source for subsequent meiosis, which results in the production of spores. (Note: it is widely assumed that this microscopic area of tissue, near lichen fruiting bodies, represents the only diploid tissue across an entire lichen thallus, but recent evidence from genomic analyses suggests that patterns of haploidy and diploidy may be far more complex than previously appreciated [Tripp et al. 2017]). Thus, whereas conidia are direct products of asexual reproduction in lichens, they eventually contribute to sexual processes that give rise to spores.

It seems unfathomable that, after all of the above, conidia can also function in yet another aspect of lichen reproduction. Careful studies in *Calopadia* (Sanders 2014), for instance, have demonstrated that conidia can land on a substrate, germinate, and then make contact with algal cells to form a new lichen thallus. Thus, conidia can function both sexually (where they go on to fertilize trichogynes) or asexually (as in *Calopadia*). What other possibilities remain? How many other ways can lichens reproduce? Lucky are we that the world of lichen reproduction is still largely wide-open and ours to explore (Tripp & Lendemer 2017)!

In lichenology, mode of reproduction is tightly linked to species identification (Tripp 2016). Generally speaking, species are referred to as either sexually or asexually reproducing. Species that reproduce sexually are those that commonly produce apothecia or perithecia, and never produce asexual lichenized propagules such as soredia or isidia. Asexually reproducing species always produce isidia, soredia, or some other type of asexual lichenized propagule. Such species can, however, also occasionally produce sexual fruiting bodies under certain environmental conditions. Both sexually and asexually reproducing species can also commonly produce asexual, non-lichenized propagules (i.e., conidia).

To complicate matters, a small number of species seem to reproduce using all of the available strategies, sometimes at the same time, and on the same individual thallus. One example is *Dibaeis baeomyces,* which frequently produces pink apothecia as well as abundant white blastidia. Other examples include many species of *Cladonia* that produce apothecia and pycnidia on the tips of the fruticose podetia, and soredia on the surfaces of the podetia, or the edges of the squamules. Why wouldn't all lichens reproduce both sexually and asexually at the same time? It seems like it could be a winning strategy. Yet it happens only rarely. These are the kinds of questions that scientists, specifically ecologists and evolutionary biologists, spend whole careers attempting to answer.

Sexually reproducing lichens—those that must encounter a new photobiont prior to re-lichenization and subsequent thallus development, comprise 77% of the ~5,500 species of lichens in North American (Tripp & Lendemer, unpub. data [manuscript in preparation). This high number likely reflects the many documented advantages of sexual reproduction, including higher rates of recombination, over asexual strategies. Nonetheless, that the majority of the remaining 23% of North American lichens reproduce asexually suggests that this mode, too, is successful.

In many instances, sexual and asexual species are each other's closest relatives, indicating that transitions among reproductive mode may be a driving force in lichen evolution (Tripp 2016). This phenomenon, in which one member of a sister species pair is sexual and the other is asexual, has been termed "species pairs." Although in some cases one member of the pair may be "embedded" within the lineage of the other, in many other instances it is clear that the two entities exist as separately evolving lineages (in evolutionary biology-speak, they are "reciprocally monophyletic"; Lendemer & Harris 2014). The vast world of genetic architecture that underlies transitions in reproductive modes, including mating type loci, remains largely unexplored in lichens (but see Scherrer et al. 2006; Honegger & Zippler 2007; Singh et al. 2012) but has been more extensively surveyed in non-lichenized Ascomycetes, especially model organisms such as *Aspergillus* and *Neurospora* (e.g., Paoletti et al. 2007).

LICHEN OCCURRENCE AND BIOGEOGRAPHY

Like the modern vascular plant flora of eastern North America, some portion of the lichen biota (Yoshimura & Sharp 1968; Culberson 1972) is thought to reflect ancient North American–Asian connections made possible by earlier Bering Land Bridges (Manos & Meireles 2015). Other elements of the lichen biota clearly represent products of more recent climatic events that have helped to structure modern community assemblages. Thus, the determinants of present-day lichen occurrence and biogeography include both historical, evolutionary effects and more recent, ecological processes, as is amply evident in the lichen flora of Great Smoky Mountains National Park, which consist of ancient elements as well as derivatives of more recent speciation events (Tripp & Lendemer, in press). Below, we discuss some of processes that help determine lichen distributions, ranging from the landscape to local scale.

On a landscape scale, several factors help determine lichen presence or absence. Holding habitat type constant, lichen diversity is higher in areas that are of higher quality and are less fragmented (Allen & Lendemer 2016; Tripp et al., 2019). Studies have also shown elevation to be an important factor. For example, Wolf (2005) demonstrated that lichen diversity reaches a peak at low-to-mid elevations (2,000 m) across a 1,000–4,200 m transect in the Colombian Andes; preliminary analyses from Great Smoky Mountains National Park indicate a similar "hump" in diversity at mid elevations, much in a similar manner to mammals worldwide (McCain, Tripp, & Lendemer, unpublished data).

Distributions and abundances of host substrates also matter. In eastern North America, Schmitt and Slack (1990) compared lichen diversity and substrate affinities between the Adirondacks and southern Blue Ridge and found that individual species of lichens are rarely host specific, but communities or "groups of co-occurring species" do show host specificity. Similarly, data from Peck et al. (2002) show that lichen communities demonstrate specificity to groups of hosts rather than to individual host tree species. From our work in the Smokies, we have found strong evidence of host community specificity, as in the above, but also extreme host-specificity (Tripp & Lendemer, in review). Some species, such as *Arthonia kermesina, A. cupressina*, *Graphis sterlingiana*, and *Arthopyrenia betulicola* are only known from single host tree species.

In other instances, community impacts include not just host trees but also the presence of other lichens. From our 12+ years of research and extensive observations in the Smokies, we have learned that *Lobaria pulmonaria* and *Lobaria quercizans* co-occur on the same tree far more frequently than is explainable by chance alone; although it remains untested, it is highly likely that the two species share symbiotic partners, either photobionts, bacteria, both, or other organisms such as endolichenic fungi. Just as important are habitat quality, disturbance, habitat fragmentation, elevation, community composition, and so forth; these are but a few of the elements expected to drive patterns of lichen distributions. What remains exceptionally under-investigated is the degree to which biotic factors—such as the presence or absence of obligate symbionts (bacteria, other fungi, other algae) facilitate or limit the occurrence of a lichen in a given area.

At a local scale, habitat and microhabitat differences have significant effects on lichen presence or absence. In general, lichen species richness is affected by light, nutrient, and water availability (Becker 1980, Lücking 1998, Jüriado et al. 2003, Dyer & Letourneau 2007). Lichen occurrence depends heavily on the physical (roughness, hardness, water-holding capacity) and chemical (basic or acidic) properties of the substrate (Culberson 1955, Brodo 1973), with different classes of substrates (e.g., bark, rock, soil, or leaf) hosting different combinations of these properties.

Large-scale analyses of all ~5,500 species of North American lichens have further demonstrated that occurrence on specific substrates limits overall range size; for example, lichens that grow on bark have, on the whole, much narrower geographical distributions than do lichens that occur on decaying wood (lignum) or soil (Tripp & Lendemer, unpub. data [manuscript in preparation]). Vertical distribution across the bole of host trees also matters. For example, lichen communities of tree bases differ from those of upper portions of a tree to its canopy (Hale 1965; Hoffman & Kazmierski 1969, Lang et al. 1980, Ladd 1996; McCune 1993; McCune et al. 1997, McCune et al. 2000, Peck et al. 2002). The degree to which coarse woody debris is freshly fallen versus decayed also affects the presence and absence of lichens (Søderstrom 1988). The above factors vary among geographical regions and among different stages of forest succession (McCune 1993). Stands of large-bole trees have different lichen assemblages than stands of similar tree species with small boles (Jüriado et al. 2003). As at the landscape level, the above-mentioned factors represent only a few of the many that contribute to the presence or absence of lichens at the local level.

Finally, geographical distribution and range size are the important metrics of assessment for rare species in North America, followed by existing and future threats to the population at individual locations. How common and widely distributed a species is, is of tremendous importance to downstream initiatives aimed at conservation of threatened or endangered organisms and has widespread financial impacts. Conservation biology of lichens in the United States is essentially non-existent. Compared to larger, macroscopic organisms, a mere two species of lichens are currently recognized and protected under the US Federal Endangered Species Act. One of these, *Cetradonia linearis*, occurs in GSMNP. However, dozens or even hundreds of other lichen species in the area warrant similar, if not more, attention based on current knowledge. This is especially the case for numerous lichens of GSMNP, many of which only persist because of populations that remain in GSMNP; Examples include *Alectoria fallacina, Arthonia cupressina, Arthonia kermesina, Cladonia appalachensis, Graphis sterlingiana, Hypotrachyna virginica, Rinodina brodoana, Rinodina chrysomeleana, Rockefellera crossophylla*, and *Sticta sylvatica*.

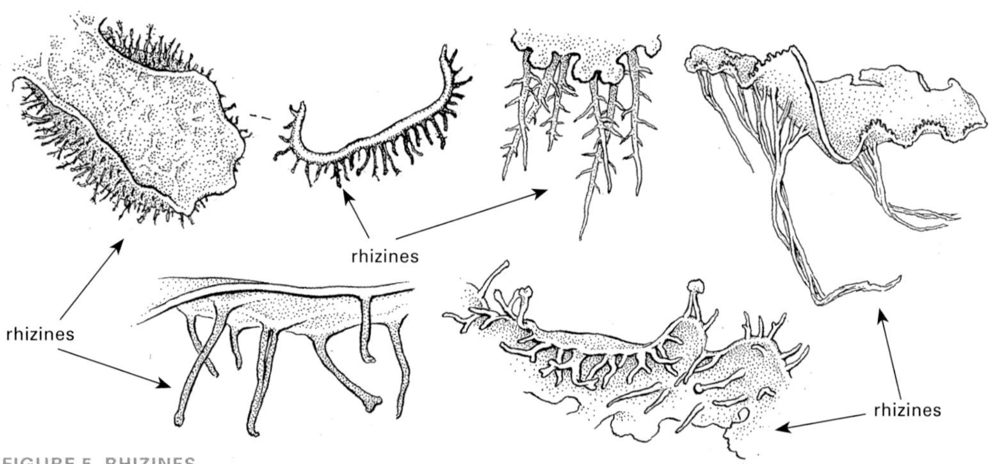

FIGURE 5. RHIZINES

Lichens do not have roots (lichens are not related to plants, and anyway, not all plants have roots!). Rhizines serve as an organ of attachment to the substrate, come in various shapes, sizes, and modifications, but are present only in some lichens. In some groups such as *Parmelia* and *Peltigera*, rhizine morphology helps inform species identification. Illustrations by Bobbi Angell.

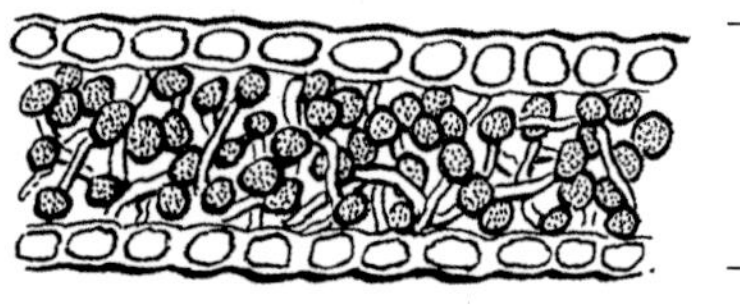

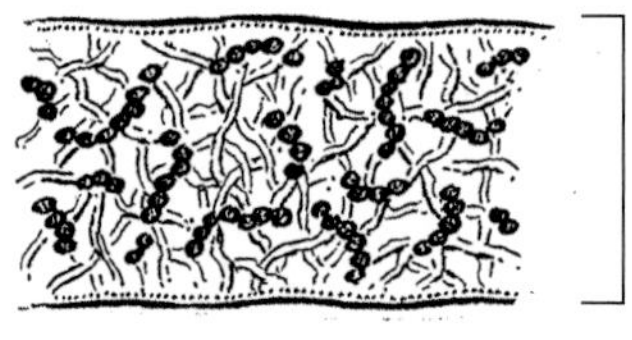

FIGURE 6. CORTICATE AND ECORTICATE THALLI

Like most other evolutionary lineages on Earth, lichens defy rules whenever you aren't watching closely. Whereas most species of lichens are corticate, either on one or both surfaces, many others are not, i.e., they are "ecorticate", either on one or both surfaces. By definition, crustose lichens are ecorticate on their lower surfaces. Some groups, such as species of *Lepraria*, have lost all signs of a cortex of any sort, period. Illustrations by Bobbi Angell.

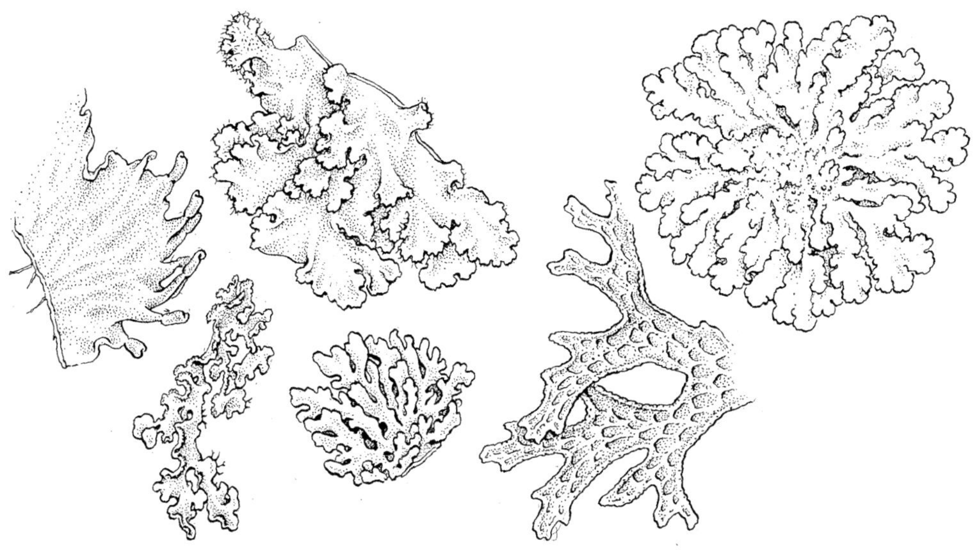

FIGURE 7. FOLIOSE LICHENS

Foliose lichens are typically the big "leafy" species that most non-lichenologists (and many lichenologists!) will first notice in the forest. They are typically large in size and have leaf-like lobes, with differentiated upper and lower surfaces. These illustrations represent but a few of the many variations on the foliose growth form. There are *a lot* of foliose lichens in the Smokies—lucky us! Illustrations by Bobbi Angell.

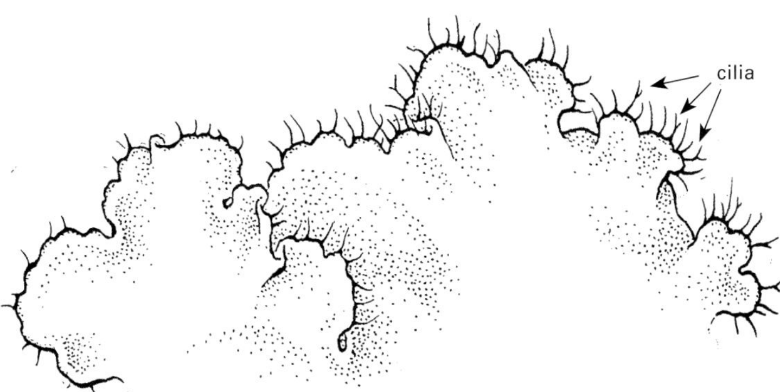

FIGURE 8. CILIA

Cilia are small, hair-like structures typically found on the margins of lobes of lichens, but sometimes elsewhere, such as on the upper cortex. Cilia are morphologically distinct from rhizines, so do not confuse the two. Genera such as *Parmotrema* are your best groups in which to learn and understand cilia. Illustration by Bobbi Angell.

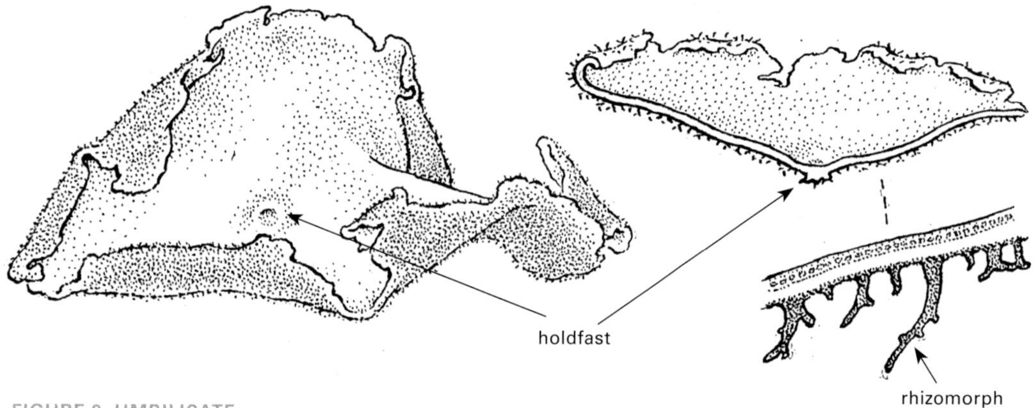

FIGURE 9. UMBILICATE

Umbilicate growth forms are attached to the substrate via a single holdfast structure. The genus *Umbilicaria* epitomizes the umbilicate morphology. Illustrations by Bobbi Angell.

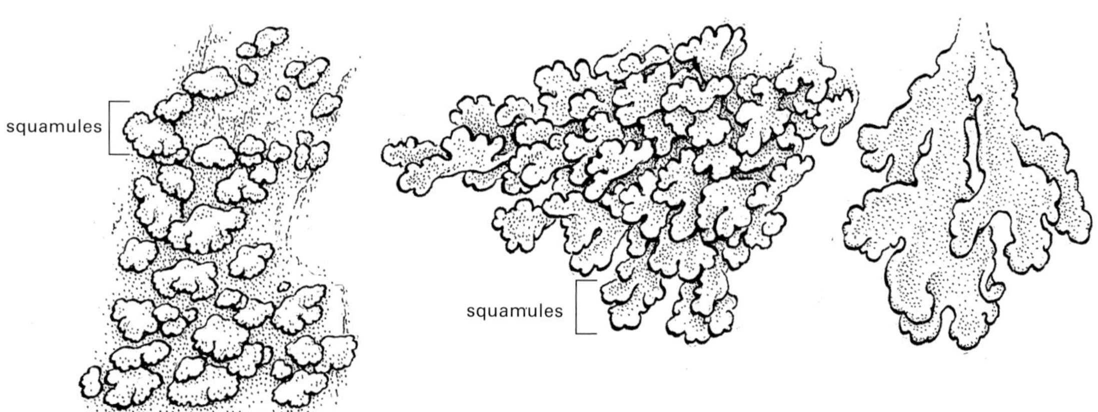

FIGURE 10. SQUAMULOSE

Squamulose refers to a growth form that lies somewhere between foliose and crustose. We are the first to admit that it is far more useful in some groups, such as *Hypocenomyce* or *Placidium*, than others. In fact, we nearly decided to abandon this term altogether for the purposes of this Field Guide, but we did not wish to shortchange anyone who was hoping for the full tour. Look for squamulose growth in *Hypocenomyce scalaris*, *Endocarpon pallidulum*, *Fuscopannaria leucosticta* and *Placidium arboreum* among many, many others. Illustrations by Bobbi Angell.

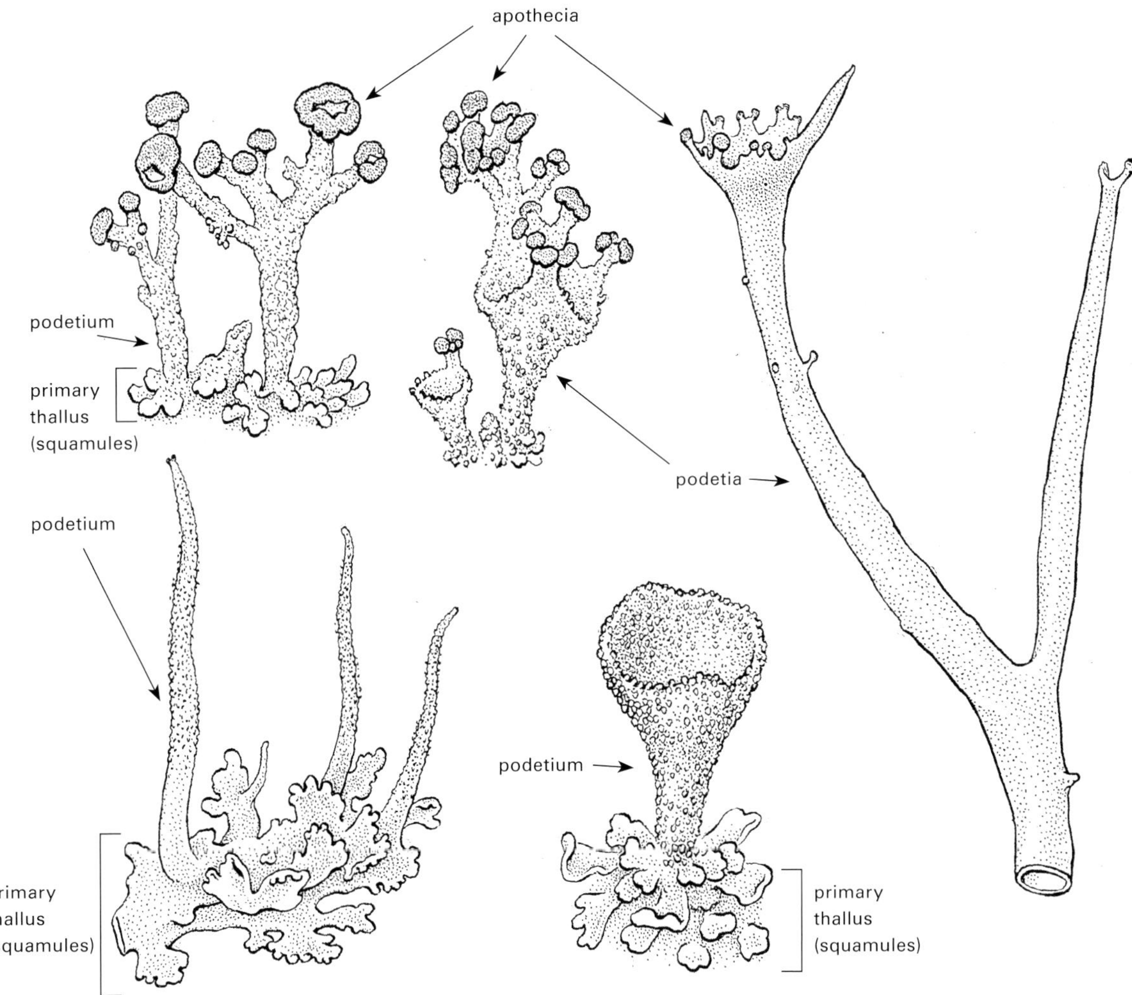

FIGURE 11. CLADONIA

Cladonia is among the top three most diverse (and abundant, and widespread . . .) genera of lichens on the planet, thus it behooves you to learn it in the Smokies, where upwards of some 50 different species can be found. Most species of *Cladonia* have a squamulose primary thallus and, from that, an ascending podetium (plural: podetia) upon which pycnidia and apothecia sit. So, they are basically squamulose or foliose while at the same time fruticose. Special. Illustrations by Bobbi Angell.

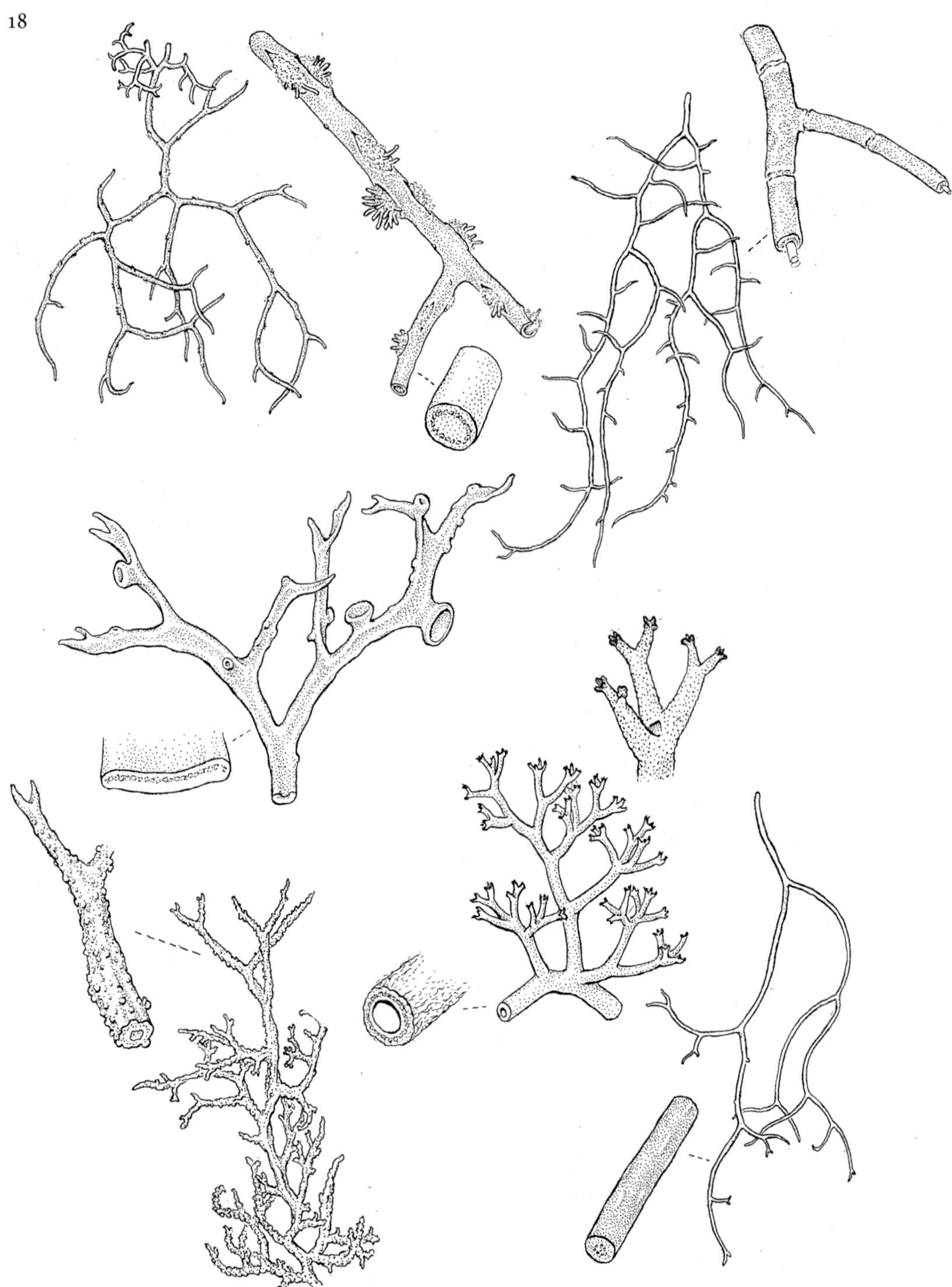

FIGURE 12. FRUTICOSE LICHENS

By definition, fruticose lichens do not have a differentiated upper and lower surface. They are simply the same color and texture all around. Branches of fruticose species can be flattened or rounded, or all shapes in between. There are numerous fruticose species in the Smokies. Think of this as the "shrubby" growth form. Illustrations by Bobbi Angell.

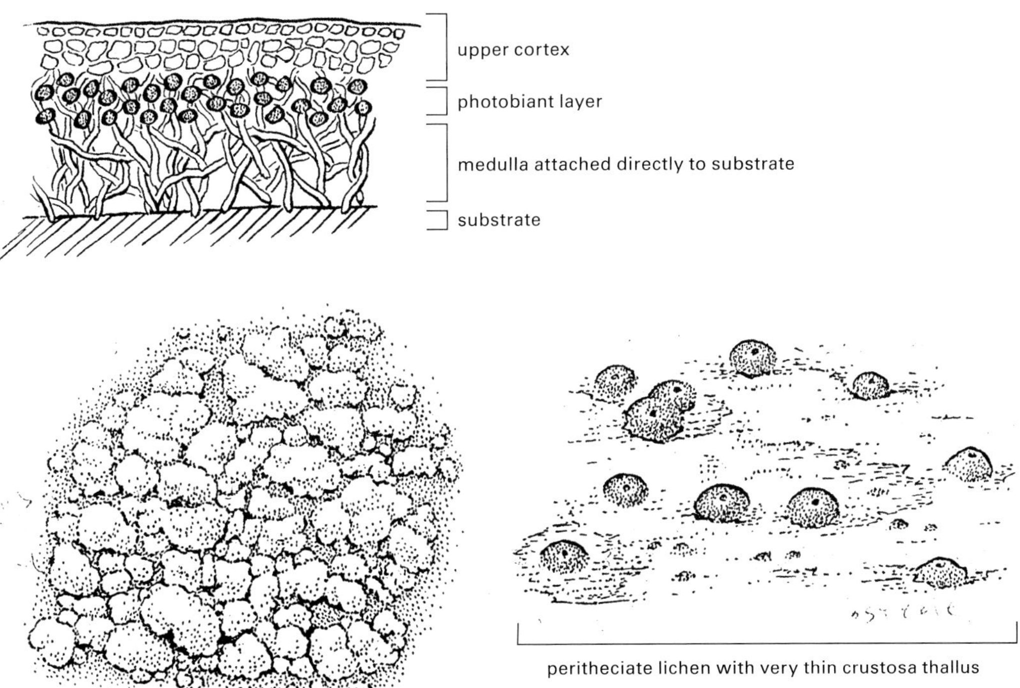

FIGURE 13. CRUSTOSE LICHENS

Even the term alone sounds somewhat demeaning . . . somewhat "lesser". But that couldn't be further from the truth. Crustose lichens are, in fact, where the party started, and where the party will end. They comprise ~70% of total lichen diversity within the Smokies, and just about any other ecosystem you walk into. Yet, they are traditionally the least studied and least understood, perhaps because they tend to be physically smaller than macrolichens (i.e., foliose and fruticose growth forms). By definition, crustose lichens lack a lower cortex. Beyond this, expect them to do just about anything and everything morphologically. They are hands down our favorite growth form, and the growth form in which the greatest numbers of discoveries are to be made. Illustrations by Bobbi Angell.

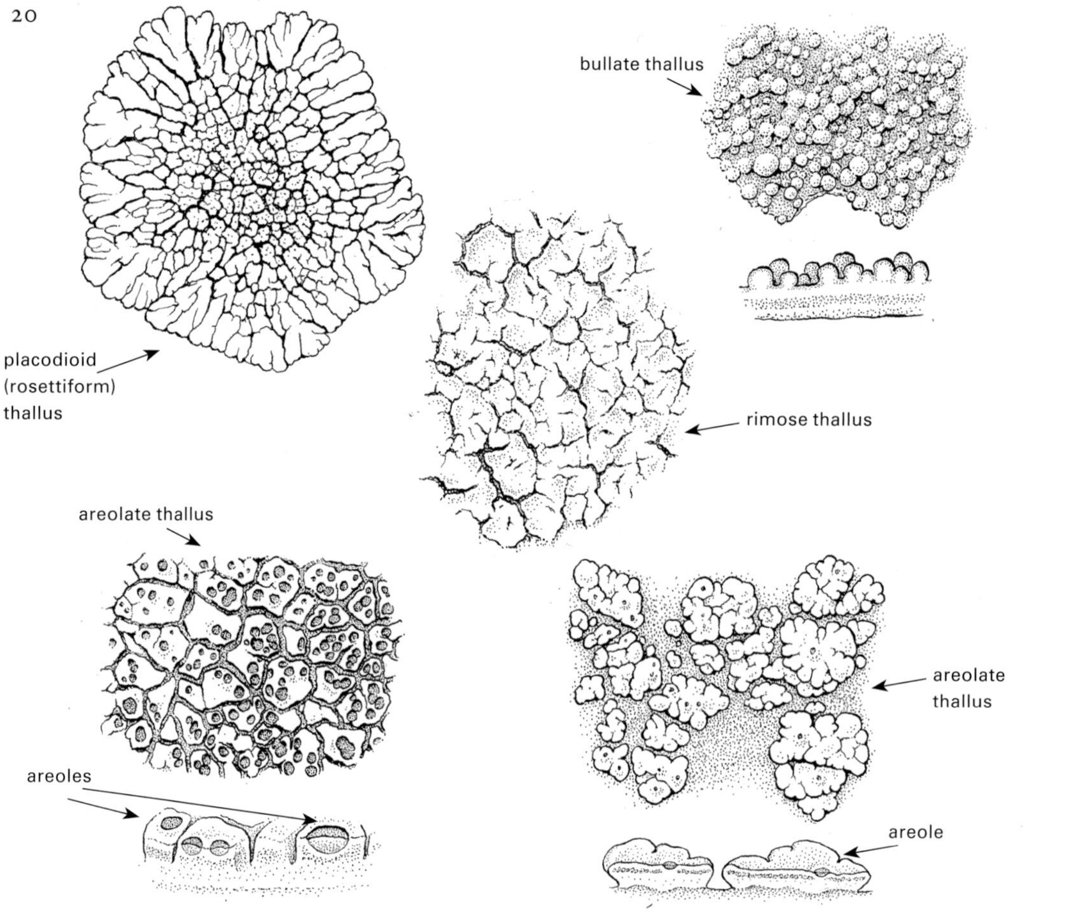

bullate thallus

placodioid (rosettiform) thallus

rimose thallus

areolate thallus

areoles

areolate thallus

areole

FIGURE 14. CRUSTOSE GROWTH FORMS

Crustose growth forms are sometimes further partitioned in types of crustose-ness. Examples include bullate (thick, bumpy, often with convex tile-like areoles), areolate (tile-like), rimose (thick, continuous), lobate (with marginal areas that become divided and elongate to resemble lobes), and rosettiform (forming circular, target-like colonies) forms. Illustrations by Bobbi Angell.

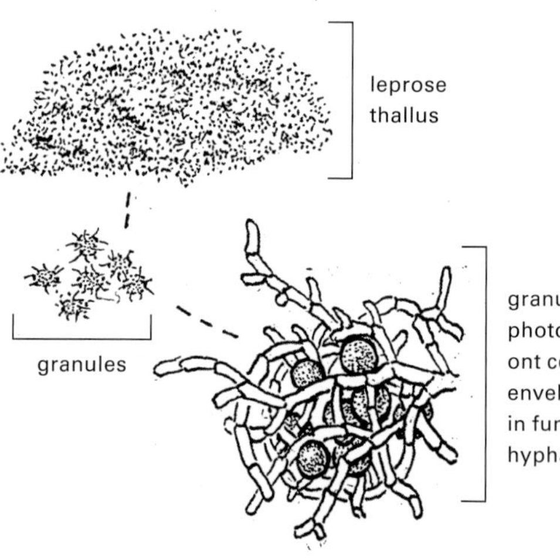

FIGURE 15. LEPROSE LICHENS

Leprose refers to a lichen growth form in which the thallus has been reduced to little more than ecorticate, dust-like granules. If you look closely at leprose lichens, you will see that each granule is made up of several photobiont cells enveloped in a mass of hyphae, basically mimicking a soredium. The genus *Lepraria* defines the leprose growth form, which nonetheless has also evolved in other, completely unrelated genera such as *Chrysothrix*. Illustrations by Bobbi Angell.

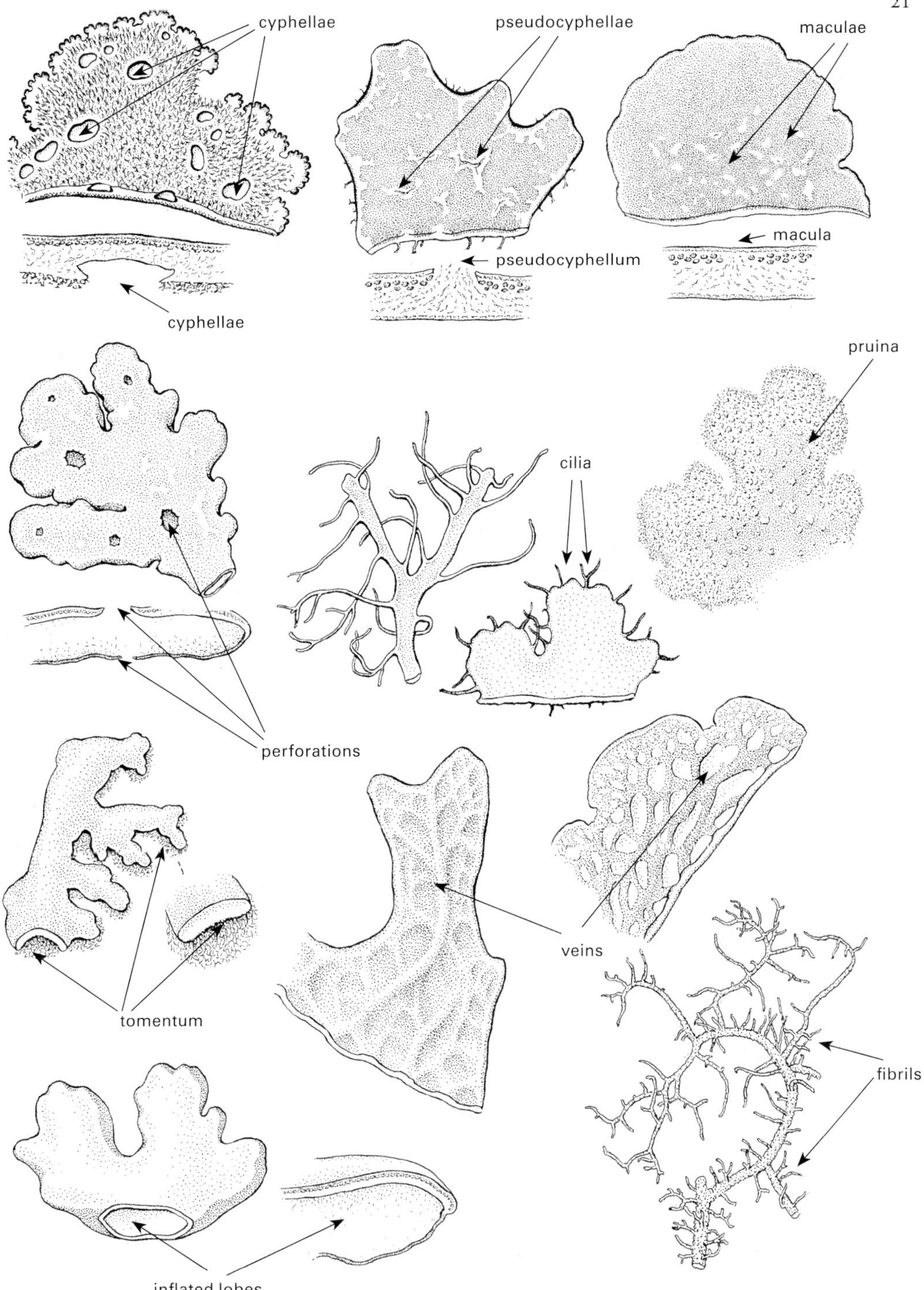

FIGURE 16. THALLUS MODIFICATIONS

Among the many joys of studying lichens is the opportunity to explore the rich and varied modifications to the thallus. Cyphellae, pseudocyphellae, maculae, perforations, cilia, pruina, tomentum, reticulae, veins, inflations, and fibrils are but a few of the examples. These modifications are highly relevant to lichen identification, both to genera and species, depending upon the structure and on the lineage. Illustrations by Bobbi Angell.

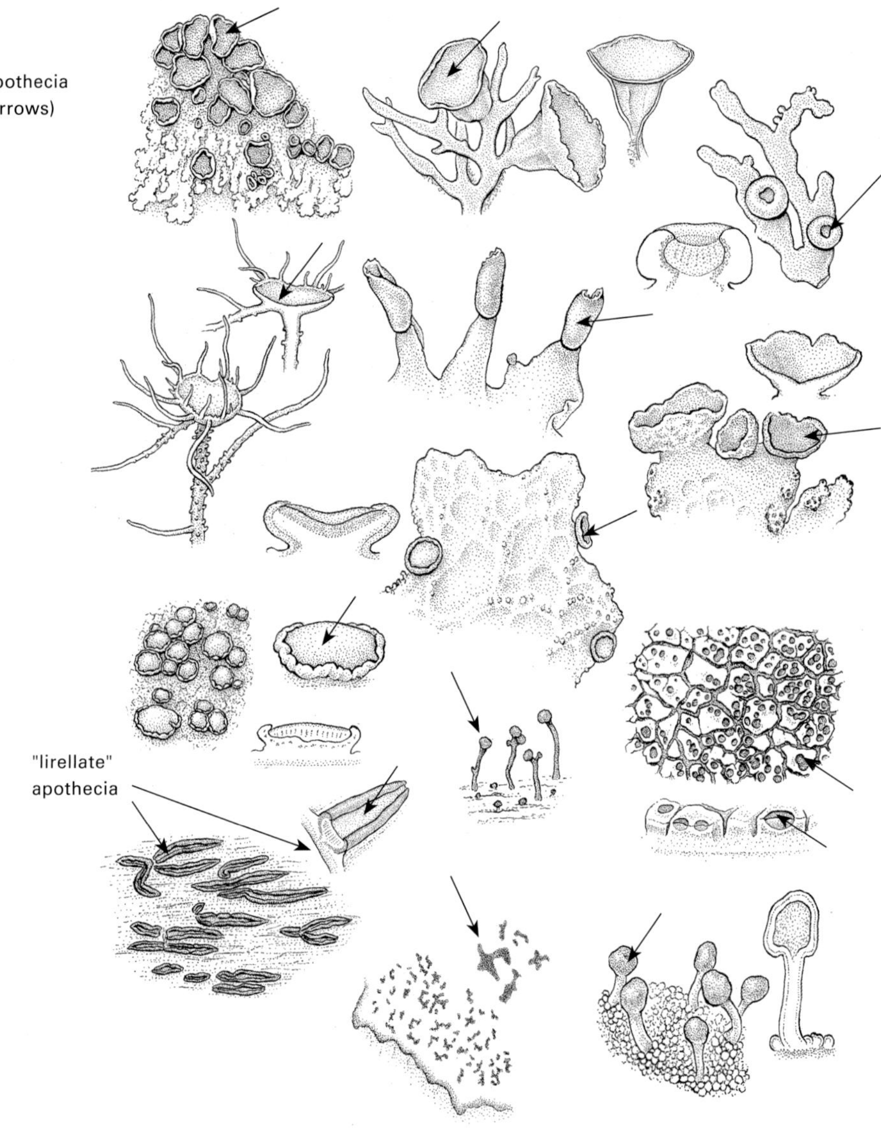

FIGURE 17. APOTHECIA

Natural selection is a process that we—as human observers of the biodiversity that surrounds us—should be extremely grateful for. It is the most important mechanism through which "endless forms most beautiful" have evolved. Selection acts particularly strongly to diversify structures relevant to reproductive processes. In lichens, apothecia (and perithecia: see Figure 18) are central to sexual reproductive biology, and not surprisingly, the evolutionary outcome is as wonderful as you can imagine. These are but a few examples of thousands of variations on apothecia, inside of which meiosis occurs (inside of asci, or sac-like structures characteristic of ascomycete fungi) to yield sexual, meiotic ascospores. There is an entire vocabulary associated with apothecia alone; but fear not, soon these terms will be rolling off the tip of your tongue. Illustrations by Bobbi Angell.

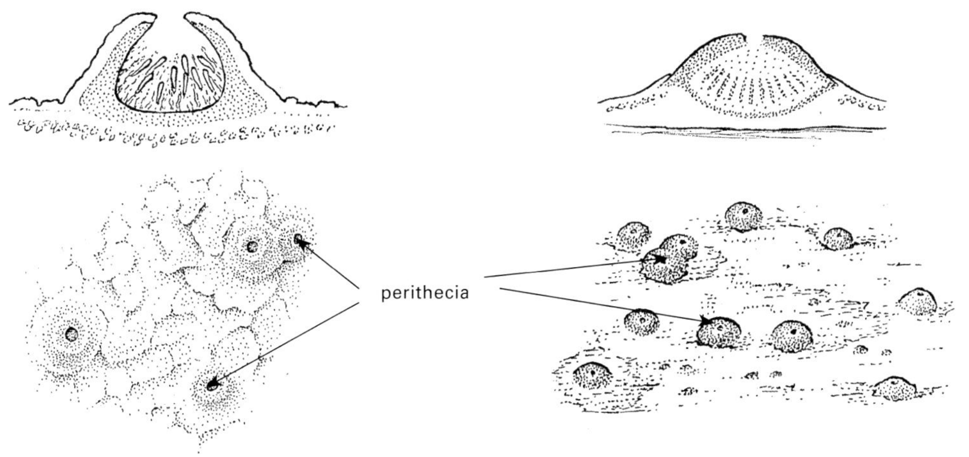

FIGURE 18. PERITHECIA

As an alternative to apothecia, the process of sexual reproduction occurs in another type of structure, termed perithecia, in some ascomycete lichens. Think of perithecia as apothecia that have become nearly closed evolutionarily—like an open clay bowl on a spinning wheel that you are able to pinch the margins to form a flask. Okay, we aren't potters, so forgive the poor imagery, but you get the idea. A given sexually reproducing ascomycete lichen makes either apothecia or perithecia, but not both. Hence, this is a good break in the dichotomous key! Peritheciate lichens sometimes get referred to as pyrenolichens. It's just the way things go. Illustrations by Bobbi Angell.

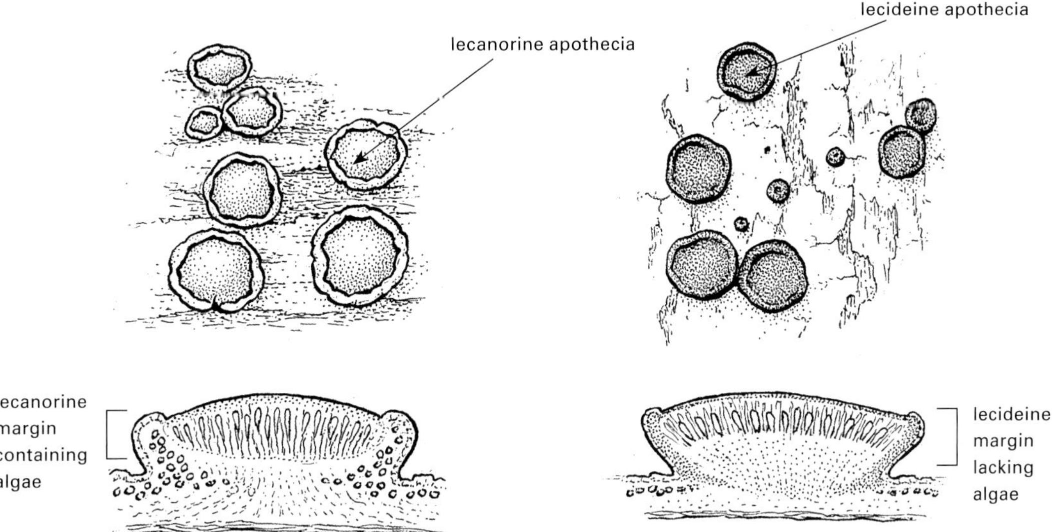

FIGURE 19. LECANORINE VS. LECIDEINE APOTHECIA

The margins of ascomata are typically termed either lecideine—when the margin lacks algae—or lecanorine—when the margin contains algae. As you might have guessed, there are gray areas in between these two states, such as pseudolecanorine margins. In this guide, we try to keep it as simple as possible. Illustrations by Bobbi Angell.

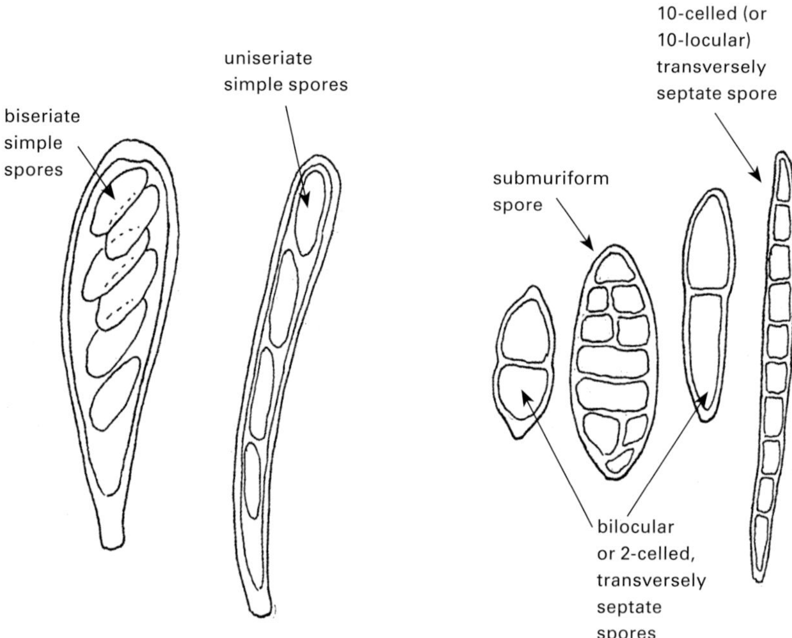

FIGURE 20. ASCI AND ASCOSPORES

Most lichens are lichenized ascomycete fungi, and in ascomycetes, sac-like structures termed asci are the locations in which meiosis occurs, a process that yields typically eight ascospores (but sometimes fewer, sometimes more!). Within an ascus, spores can be arranged in a single row (uniseriate), two rows (biseriate) or irregularly (no special term). Ascospores, which result from meiosis, are arguably *the* most important structure of sexually reproducing lichens from an identification standpoint. Most keys will ask you whether the ascospores are simple (which is not the case for any of the four illustrated on the right: simple ascospores have *no* septa of any sort) or septate in some manner. In these four, from left to right, ascospores #1, #3, and #4 are transversely septate whereas spore #2 is submuriform to muriform. Spores #1 and #3 are 1-septate (or 2-celled) whereas spore #4 is multi-septate (specifically, 9-septate or 10-celled). All four are hyaline. Other spores are distinctive in turning brown, either upon early development or later in development. Note that sexually reproducing lichens, which do not reproduce via lichenized propagules (see below), comprise ~75% of the Park's lichen biota. Illustrations by Bobbi Angell.

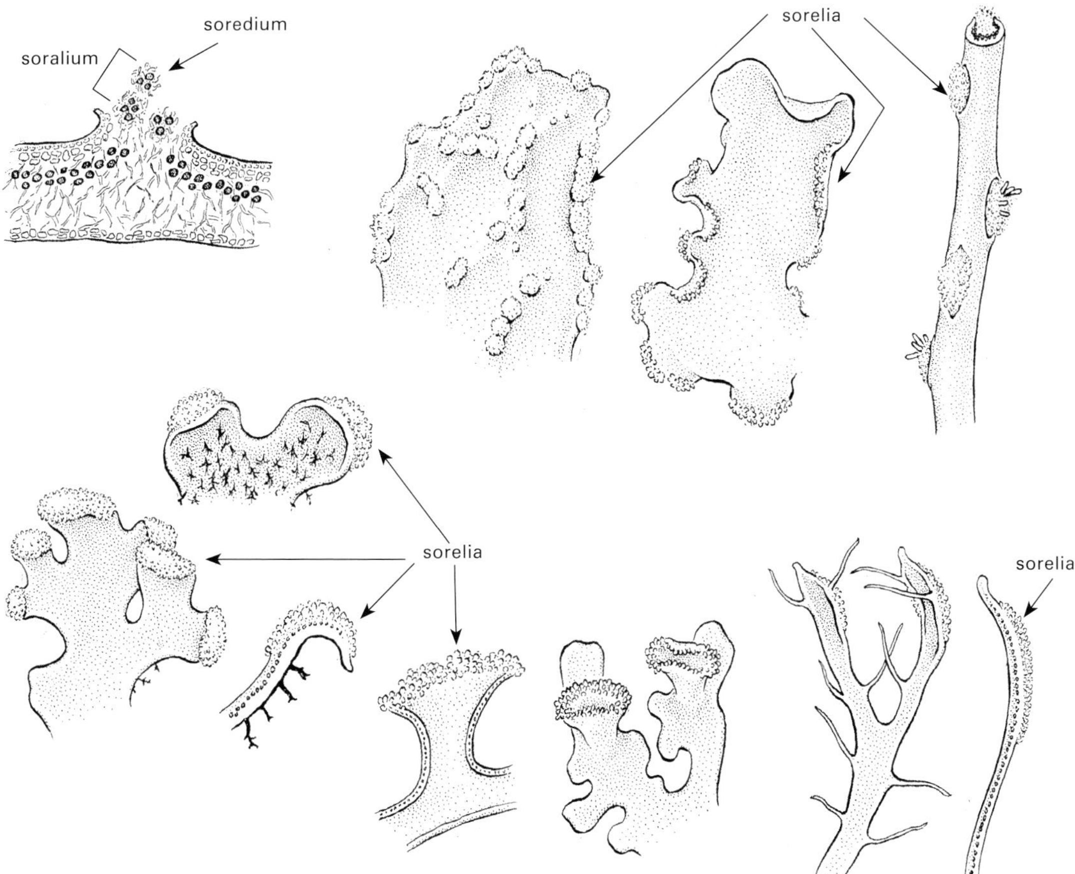

FIGURE 21. SOREDIA

There are many different types of (lichenized) asexual reproductive propagules, but soredia together with isidia (see Figure 22) are by far the most widespread and important. Think of soredia as little round packages of several photobiont cells wrapped in ecorticate fungal hyphae—sort of like leprose granules. Soredia arise from the breakdown of the thallus and are typically formed in structures termed soralia that burst through the upper surface. These can be laminal (on the surface of the thallus), marginal (on the margins), terminal (on the terminus of the lobes), or many forms in between. These little packages get swept away by the winds (or whatever) and, because they are composed of the primary photobiont and mycobiont, are presumably ready to germinate and generate a brand-new lichen thallus. Whether the habitat is right, and whether this habitat contains other necessary biotic partners (e.g., other fungi, bacteria, etc.), remains to be further understood. What we can tell you is that on the whole, lichens that reproduce via lichenized asexual propagules typically have broader geographical ranges than lichens that reproduce sexually (Tripp et. al. 2016). Why? Although lichenized asexual propagules are often physically larger than sexual ascospores, they likely have a much higher probability of establishment given both partners are co-dispersed. What this means is that lichen establishment—rather than dispersal per se—may be the primary bottleneck in explaining the next generation of lichens. In other words, a lot of things disperse, but establishment is far more limited and is furthermore made easier by co-dispersal. Finally, approximately 25% of the park's lichen biota reproduce via lichenized asexual propagules. These species can occasionally to rarely be found manufacturing sexual reproductive structures in addition to these asexual structures. As to why, how, and when—this is an area in need of a lot of research, and it promises to be very exciting research. Be the one! Illustrations by Bobbi Angell.

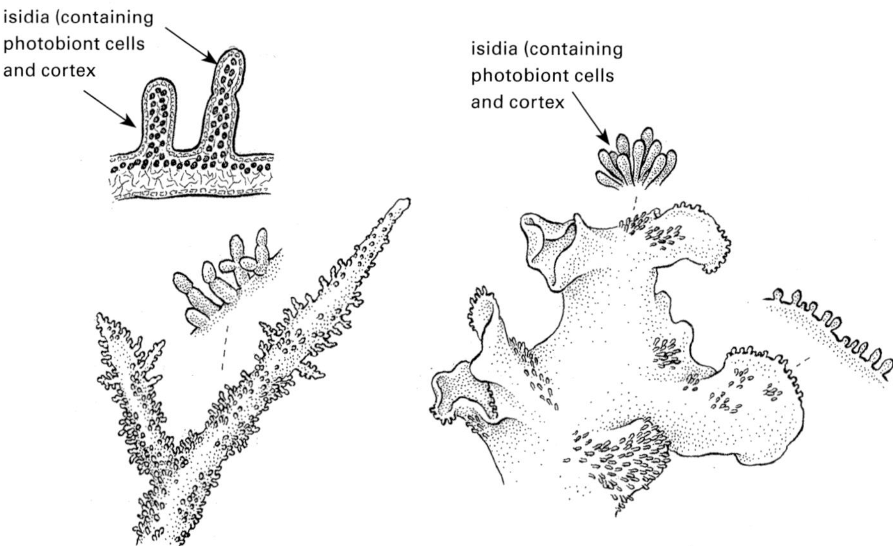

FIGURE 22. ISIDIA

Isidia are functionally like soredia in that they co-disperse both the primary photobiont and mycobiont, but morphologically, they are unlike soredia in that they are corticate. In other words, isidia are simply outgrowths of the upper cortex plus underlying photobiont layer. Like soredia, they come in numerous shapes and sizes ranging from granular to globose to cylindrical to coralloid isidia. Some of our favorite isidia in the Park are those of *Ochrolechia yasudae*. Look at them closely—it's like an entire miniature forest . . . well maybe not so miniature to the springtail crawling about these towering columns. Illustrations by Bobbi Angell.

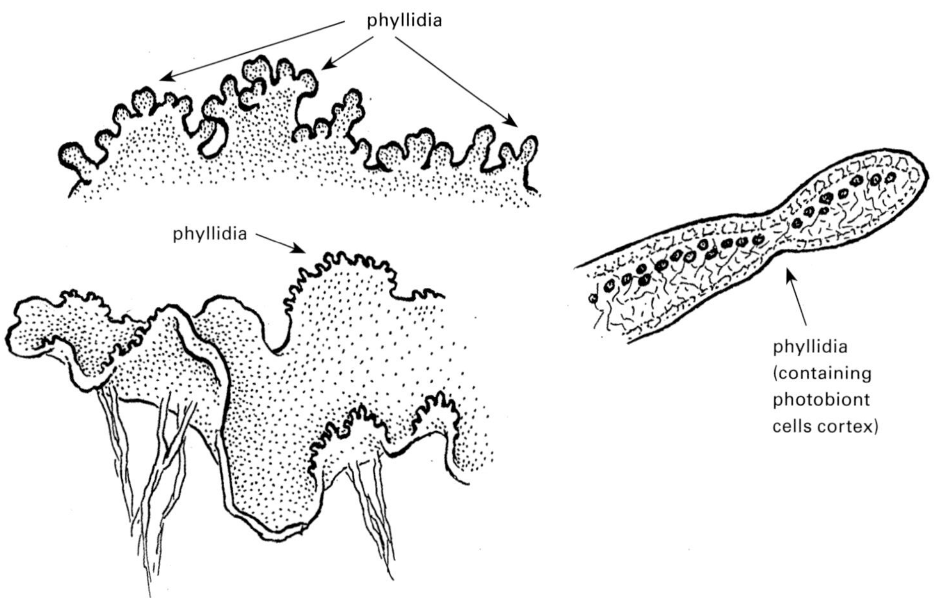

FIGURE 23. PHYLLIDIA

Think of phyllidia as isidia, but leaf-like and typically originating from the margins. Yep, that's about it. *Petligera phyllidiosa* is a good lichen in which to learn phyllidia. Illustrations by Bobbi Angell.

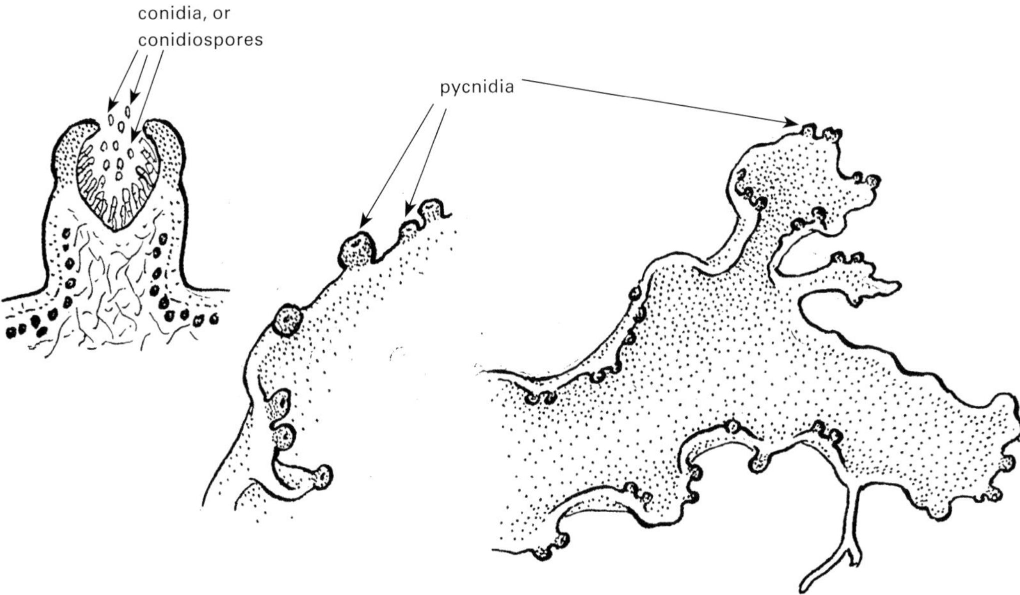

FIGURE 24. PYCNIDIA

Just when all this lichen reproduction mumbo jumbo was starting to come together in your head. . . . You've got the sexual species, the lichenized asexual species and. . . . But wait. Why do they keep saying lichenized asexual? Why not just asexual? Well you guessed it: that's because lichens (well, at least the mycobionts) reproduce asexually in isolation of their photobionts, too. Therefore: you've got asexual propagules that are lichenized (i.e., the contain both partners, such as soredia and isidia) and you've got asexual propagules that are non-lichenized (enter: pycnidia!). Pycnidia look a heckuvalot like perithecia in that they are flask shaped, but they are much smaller and the process that takes place inside of them is no different than cell division in your lower dermis: mitosis. Thus, pycnidia are flask-like structures in which mitosis occurs, yielding simple genetic clones of the parent cells. But the products of these mitotic events vary in size and shape between species and are termed conidia (or, conidiospores). We once wrote a paper (short title: Twenty-seven Modes) that detailed all the ways in which condia can contribute to lichen reproduction, and in effort to finish this Field Guide, we refer to you to that work, which was published in the journal *Brittonia* (Tripp & Lendemer 2017). Suffice to say: conidia are diverse, but they represent mitotic (asexual) processes that disperse only the mycobiont. Conidia and the pycnidia that bear them are diverse, beautiful, and worth your attention. Illustrations by Bobbi Angell.

PART 2
GREAT SMOKY MOUNTAINS NATIONAL PARK
Evolution of a Landscape and Its Culture

FIGURE 25. FALL COLOR

Great Smoky Mountains National Park is the nation's most visited Park in part because of the annual spectacle of fall color. Here is a famous view of the changing forests, from the outside. Photo by Erin Tripp.

Long before the establishment of a magnificent National Park and World Heritage Site that straddles the mountainous border regions of eastern Tennessee and western North Carolina, the land—presently home to the Great Smoky Mountains (Figure 25)—was valued by native inhabitants for its biological wealth and resources. The oldest records of human habitation in the region date back at least 12,000 years. In more recent history, members of the Eastern Band of the Cherokee Nation were the primary occupants of the area. There is ample evidence of extensive historical land alteration and management in this region, primarily for agricultural purposes, by the Cherokee and other eastern native peoples prior to European arrival (Delcourt et al. 1993).

Not until the Spanish explorer Hernando de Soto arrived in 1540 did the Cherokee encounter European members of the human species. Within two centuries, settlers were arriving in significant numbers and establishing homesteads among the lands occupied by the Eastern Band. Despite extensive Cherokee bloodshed, widespread efforts to defend their homeland, and even the invention, rapid codification, and dissemination of a new written language to deal with increasing conflict, the ousting of the Cherokee Nation proceeded with tremendous speed after gold was discovered in north Georgia in 1828. A mere two years later, President Andrew Jackson signed the Indian Removal Act—widely agreed to be among the most heinous acts ever authorized by Congress and signed by an American president—and shortly thereafter defied a Supreme Court ruling in favor of the Cherokee, setting in motion the beginning of a new era in land modification by colonists. In years to come, this series of events would have unparalleled consequences for other biological inhabitants of the eastern North America, and would eventually result in an even more magnified ecological value to the land we have dubbed our nation's favorite National park.

A handful of Cherokee continued to assert their self-governance and survived in their homeland by living among the mountains, especially between Clingman's Dome and Mt. Guyot in the Smokies. Their persistence resulted in the 1889 chartering of the 56,000-acre Qualla Indian Reservation, which today borders southern portions of the Park in Swain County, North Carolina. The Native American population of the Southeast prior to European arrival has been estimated at 1.7 million (Smith 1987, Denevan 1992). Although by no means a pristine landscape at that time, the region was far less populated than it is today: in 2009, a US Census Bureau report noted that Florida, Georgia, and North Carolina alone contained 37.5 million people. The relative proportion of native peoples to colonial descendants has reversed dramatically compared to the early nineteenth century. Indeed, the Southeast leads the nation in population growth (www.census.gov) and, with that, environmental destruction and habitat alteration, again magnifying the ecological worth of our nation's favorite national park.

Taken together, there can be little doubt that both prior and subsequent to European arrival, substantial portions of the terrain encompassed by Great Smoky Mountains National Park were burned for agricultural use, cleared for grazing, cut for resource extraction, settled, and otherwise managed by and for the benefit of the human species. But mixed among 400-year regrowth (from Cherokee use) and 90-year regrowth (from the peak of the American industrial logging regime) are the largest stands of mature forests remaining in eastern North America (Figures 26 & 27). Yet these two, often conflicting, human histories are mere threads of sentences in a narrative

FIGURE 26. BIG LIRIO AND JAMES
Low elevation forests are often dominated by *Liriodendron tulipifera* (Tulip Poplar), which is an early successional hardwood species. Great Smoky Mountains National Park is home to some of the largest and tallest tulip poplars, such as the individual seen here with James Lendemer holding our field camera, for scale. Photo by Erin Tripp.

that spans millions of pages. As detailed in subsequent sections, this landscape has been fashioned over a long history of complex evolutionary, geological, climatic, and biotic interactions that have given rise to one of the most biodiverse, celebrated, inspiring, and ecologically important landscapes in the North America—and, indeed, the world.

GEOGRAPHY

Great Smoky Mountains National Park encompasses 520,976 acres (814 square miles), making it one of the largest protected areas east of the Mississippi River. The Park is, furthermore, situated among extensive lands managed by the United States Forest Service, adding extensive forested habitats to an already sizeable protected area. This block of native habitat in the middle of the southern Appalachian Mountains is both expansive and relatively intact compared to the majority of eastern North America, and the extent of this habitat can be appreciated through study of satellite imagery. GSMNP is geographically the third largest national park in the East, following Everglades National Park (1.5 million acres) and Isle Royale National Park (572,000 acres). The Park occupies portions of Blount, Cocke, and Sevier Counties in Tennessee and Haywood and Swain Counties in North Carolina. GSMNP is bounded by the Little Tennessee and Tuckasegee Rivers to the west and south, and by the Pigeon River to the east. Situated west of the Continental Divide, rivers originating within GSMNP drain into the Ohio, Mississippi, and Tennessee River watersheds, ultimately emptying into the Gulf of Mexico. The eastern Continental Divide, which routes rivers toward the Atlantic Ocean, lies just tens of miles to the east of GSMNP, just east of Asheville, North Carolina.

The Park's geographical position at the 35th parallel enabled its escape from the *direct* effects of immense glaciers of the Pleistocene Epoch, ca. 2 million – 11,000 years ago. Further, the north-south geographical orientation of the Appalachian Mountains permitted organisms to migrate during glacial and interglacial cycles, lessening the impacts on migratory routes and thus lowering extinction rates. This orientation is in contrast to the east-west ori-

FIGURE 27. BIG TREES
High lichen biodiversity in Great Smoky Mountains National Park is in part attributable to the abundance of mature forests, which drives habitat stratification and yields numerous micro-niches for lichen diversity and other organisms. These mature forests represent a departure from much of the rest of eastern North America, where young to middle age stands are the rule, and old growth is the exception. Photo by Kate Deregibus.

entation of European mountain chains, which suffered far higher local extinction events as a result of major barriers to migration. The historical outcome was an eastern North American biota that is more species-rich than the European biota. Finally, the southern extent of the great peaks that comprise the Smokies (i.e., the presence of high peaks situated at southerly latitudes in modern times) is thought to have been (and to be) a major contributor to a biological diversity hotspot, which, perhaps more than any other single parcel of land east of the Mississippi River, characterizes the eastern deciduous forest biome (Braun 1950).

GEOLOGY AND TOPOGRAPHY

At least a billion years ago, sedimentary rocks were deposited onto the floor of an ancient ocean near the region encompassing present-day GSMNP. Millions of years of pressure compressed these layers into metamorphic gneisses, schists, slates, and quartz-rich conglomeritic sandstones. This formation has been termed the Precambrian Ocoee Series and comprises the majority of rocks in the Park. Subsequently, during the Paleozoic, depositions of sedimentary limestone and sandstone occurred. Mixed among these metamorphic and sedimentary rocks are igneous elements such as granite. The end of the Paleozoic (+/- 250 million years before present) witnessed the collision of the North American and African plates, inducing a long and violent process of mountain building known as the Allegheny Uplift. This geological event represents the origin of the Appalachian Mountains, an ancient chain that once reached upwards of 13,000 feet (4,000 m) in height (Matmon et al. 2003). Long after these three major geological events and much climate-induced weathering, GSMNP hosts, at present, 20 peaks that surpass 6,000 feet (1,830 m) in height, making the Smokies home to many of the tallest peaks in eastern North America. Among these, Clingman's Dome is the Park's highest, standing at 6,643 feet (2,025 m) above sea level. In contrast, the lowest elevations in the Park slightly exceed 800 feet (244 m). This tremendous altitudinal gradient can be traversed over very short distances, making it possible to see communities of plants and other organisms representative of habitats in central Alabama as well as those in northern New Hampshire in a single-day hike.

During the Allegheny Orogeny, older, metamorphic rocks were folded and tilted upwards and sometimes slid over the younger, sedimentary layers. Striking examples are visible today in the craggy folds of bluish gray slate outcrops of the Anakeesta formation, such as Charlies Bunion, Chimney Tops, and Alum Cave. In Cades Cove and to the north (e.g., Rich Mountain and sinkholes such as Whiteoak Sinks), erosion of metamorphic layers exposed limestone layers underneath. Other sedimentary elements such as Thunderhead Sandstone can be seen along the northern face of Sterling Ridge and among the steep northern slopes of Mount Le Conte, and Roaring Fork Sandstones are visible south of Cosby. Metamorphosed igneous formations characterize western drainages of Raven Fork and the Purchase Knob area in extreme southeastern portions of the Park, south of the Cataloochee Divide. Northeastern portions of GSMNP contain abundant boulder fields that were formed via the freezing and thawing cycles of glacial periods. Examples of these boulder fields can be readily seen north of the long, east-west running Greenbrier Ridge.

FIGURE 28. ANAKEESTA ROCK

Anakeesta rock outcrop showing a community of lichens that specializes on this type of substrate. Geologists have shown that Anakeesta is rich in graphitic and sulfidic slates and phyllites. Several lichens in the Park that occur on Anakeesta occur nowhere else on Earth, such as *Cladonia appalachensis* and *Lecanora anakeestiicola*. Photo by James Lendemer.

Other portions of the Park host unique formations such as the ultramafic amphibolites between Balsam Mountain and Beech Gap Trails in the Straight Fork drainage, or the diorites in the backcountry north of Hazel Creek Trail.

The United States Geological Survey has reported GSMNP as having one of the most complex and deformed bedrock geologies in eastern North America (Southworth et al.

FIGURE 30. COVE HARDWOOD FOREST
Great Smoky Mountains National Park hosts the best examples of eastern temperate hardwood forests in North America. We feel this point is non-negotiable and has long been appreciated by diverse visitors as fact. The iconic cove hardwood forests, seen here in the Baxter Creek watershed, are laden with hundreds of species of flowering plant spring ephemerals, such as the purple *Phacelia bipinnatifida* in the photograph. What has been far less appreciated about these cove hardwood forests is that they also host some of the highest alpha diversity of lichens anywhere in North America. So you finally made it to the Smokies not in the third week of April but in January instead. No problem! Explore the most lichenologically diverse national park in the United States anytime of year! Photo by Erin Tripp.

FIGURE 29. ROCK OUTCROP
Most rocks in the Smokies are "acidic," primarily granites and quartzites, such as the lichen-covered outcrop seen here. There are rare pockets of limestone and other calcareous rocks in northern and western portions of the Smokies, which host entirely different communities of lichens. In some instances, acidic rocks tend to show signs of being "weakly calcareous," and the lichens will be the first to let you know it. Photo by Erin Tripp.

FIGURE 31. COVE HARDWOOD FOREST
It is in large part true that mosses *do* grow on the north side of a tree, at least in the Northern Hemisphere. This is the shaded, cooler, damper surface (again, in the Northern Hemisphere; the pattern is reversed south of the Equator!). Appreciate the moss-covered north facing surfaces of the *Fraxinus americana* (American Ash) and *Aesculus flava* (Yellow Buckeye) seen in this cove hardwood forest. In contrast, lichens often (but not always) preferentially occupy south-facing surfaces of trees, where it is typically slightly more exposed, warmer, and drier (and where there is less competition from mosses). Can you imagine how different our childhoods would have been had they only taught us that "lichens grow on the south sides of trees," instead? Photo by Kate Deregibus.

2005). This geological diversity and complexity has had a striking effect on lichen occurrences and distributions in GSMNP, equal to or surpassing that of geological effects on the vascular flora, given that rocks are primary substrates for a great number of lichens and that specific species of lichens preferentially occupy one type of rock or another. Further discussion of the relationship between rocks and lichens appears below, under "Ecosystems."

CLIMATE

Rising at times more than a mile above their bases, the mountains that comprise GSMNP help to establish their own weather patterns. In general, precipitation is higher in the Park than in surrounding areas for three reasons: (1) because of this orographic effect, (2) because the Park hosts the first major peaks that weather systems encounter as they move in, primarily from the west and northwest, and (3) because the SW-NE orientation of the Park's mountains is more or less perpendicular to the direc-

FIGURE 32. MOIST WALL
Vertical rock faces are very different micro-environments from horizontal rock surfaces and rock overhangs. The orange lichen seen here (*Porpidia degelii*) covering vertical surfaces prefers rocks of this orientation. In contrast, look for species such as *Flakea papillata* or, if you are really lucky, *Rockefellera crossophylla*, deep under horizontal rock overhangs. Photo by James Lendemer.

tion of the prevailing wind. Lower elevations receive an average of +/- 55 inches of precipitation annually while some upper elevations receive up to 85 inches, making them worthy of temperate rainforest designation. Snowfall per year can range from fewer than 12 inches (30 cm) in the valleys to over eight feet (100 inches; 254 cm) at higher elevations (Perry et al. 2007). This dramatic gradient in precipitation and associated local variation in moisture and humidity yield numerous micro-niches in which lichens occur and partition themselves according to adaptive preferences and thresholds.

Temperatures within the Park also fluctuate across elevational gradients, varying as much as 20°F between a low and a high elevation site at a given time on a single day (Gaffin et al. 2001). Additionally, scientists have recorded considerable temperature variation at fine spatial scales, measured at ground level (Fridley 2009). For example, temperature has been shown to vary up to 7°F across a transect of only 1,000 linear meters (Fridley 2009). Topographic differences help to drive these great variations in local temperature regimes that have been documented in the Park. Other such topography-influenced effects include the drainage of cold air downslope at night, on average moister conditions in concave landforms such as coves (Figures 30 & 31), variation in solar radiation associated with different land covers (e.g., meadows, wetlands, slopes, ridges), different cloud covers, different slopes and aspects, and the buffering of microclimates associated with proximity to perennial streams or soils with variable moisture-holding capacities (Geiger et al. 2003).

This magnitude of microhabitats built in large part by a diverse climate fosters a tremendous diversity of lichens found in GSMNP (Figure 32). These microhabitats, spanning a broad range of altitudes, may also be key to mitigating widespread effects of climate change on populations. Organisms with dispersal limitations—species that are unable to move when the going gets tough, such as lichens–may not be able to outpace rapid, environmental transformations by migrating northwards but may, instead, be able migrate upslope, assuming there are higher elevations with suitable habitats that can accommodate migrating species.

PART 3
BIOLOGICAL DIVERSITY OF GREAT SMOKY MOUNTAINS NATIONAL PARK

FIGURE 33. WATERFALL, BAXTER

Most visitors travel to the Smokies to see a plethora of spring ephemerals in iconic cove hardwood forests such as *Trillium erectum* seen here, or the brilliant hues of autumn color. What goes less noticed are the many shades of lichen diversity present within the park, but we hope this book does something positive to change that. Areas of locally high relative humidity such as near this waterfall in the Baxter Creek watershed often host unique communities of lichens that depend on above average humidity, whether at low, middle, or high elevations. Photo by Erin Tripp.

FIGURE 34. ERIN TRIPP AND KYLE DEXTER, LONG TERM STUDY PLOT THAT MOTIVATED OUR LICHEN STUDIES
Our lichenological research in Great Smoky Mountains National Park began with a preliminary investigation of the lichens in a 1-hectare long-term plant ecology plot, established and studied continuously by one of us (Erin Tripp) together with colleague Kyle Dexter. Over the course of only a single day of collecting lichens in that plot, we documented over 110 species, 11 of which were new to science. From that moment, we recognized just how little was previously known about the lichen diversity of this remarkable tract of land and our nation's most visited national park. Photo by Erin Tripp.

"The key to an understanding of the Deciduous Forest Formation as a whole, and the spatial relations of its integral parts and the nature of these parts, is obviously to be sought in the old land areas, where ancient forest types have been able to continue in spite of vicissitudes in surrounding areas" (Braun 1950, 6). The importance of GSMNP as a refugium for biological diversity in eastern North American cannot be overstated, from glacial times to present-day land fragmentation. Although less than 5% of native habitats remain in the eastern United States, the park is 95% forested (the vast majority of this native), and upwards of 35% (> 175,000 acres) of this is mature or "old growth" forest (Wuerthner 2003). Cove hardwood forests, seen at their finest in GSMNP, are among the richest ecosystems in North America (Figure 33). The most diverse salamander fauna in the world outside of the tropics occurs in the Park.

FIGURE 35. TYPICAL BRANCHES
Lichens are incredible organisms and will grow on anything that sits still long enough—including you! Appreciate the numerous thalli of *Usnea* covering tree bark as well as thalli of a different species of *Usnea* covering the branches. Photo by James Lendemer.

GSMNP also hosts the densest black bear population in eastern North America and somewhere between 16–21 National Champion trees, celebrated for their physical size and temporal longevity. The tallest tree canopy ever recorded in eastern North America, based on work by the Eastern Native Tree Society (affectionately: "ENTS"), occurs in the lowest elevations of Baxter Creek Trail, where one of us (EAT) and colleague Kyle Dexter have a long-term plant ecology research project (Figure 34). The combination of geological, climatic, evolutionary, ecological, and other past and present effects described above creates one of the most varied habitats in the United States and, indeed, across temperate latitudes of the entire world.

Biological inhabitants of GSMNP have been and continue to be extensively documented through biodiversity inventories. Historically, study of two charismatic groups, plants and animals, has represented the vast majority of scientific inquiries into the Park's biota. Our intention is to provide only the briefest overview of these groups, for they have been discussed in detail in numerous other works. Recent initiatives, such as the All Taxa Biodiversity Inventory (ATBI) launched in 1997, seek to improve knowledge of biodiversity in GSMNP by attempting a comprehensive inventory of *all* species that occur within the Park's borders. Between 1997 and 2009, more than 800 species new to science and 6,000 species new to the Park had been recorded. A query of www.dlia.org on 7 June 2017 indicated a rise in these numbers to nearly 1,000 species new to science and nearly 9,200 species new to the Park. Within a year of publication of this *Field Guide*, the total number all species summed across lineages will have approached 20,000. Progress in documentation of "cryptic diversity" has been impressive, yet still slow compared to most macroscopic groups. Through dissemination of this *Field Guide*, we aim to bring new knowledge and enthusiasm about a historically neglected and more cryptic group of organisms—lichens—but one that comprises a large percentage of total biodiversity and biomass within the Park's borders (Figure 35).

ECOSYSTEMS

GSMNP hosts several major types of vascular plant communities. For example, Braun (1950) defined six types in the region while Whittaker's (1956) classic monograph of the Park included 15 types. More recently, 12 vegetation types have been identified (Jenkins 2007).

FIGURE 36. ANAKEESTA ROCK AND FIR DIEOFF

High elevation Anakeesta rock outcrop overlooking die-off of *Abies fraseri* (Fraser Fir) in background. Spruce-fir ecosystems are an endemic habitat type of the southern Appalachian Mountains and are highly endangered, in part owing to destruction by a non-native pest (Balsam Wooly Adelgid) introduced in the late 1950s coupled with detrimental effects of acid fog of the late twentieth century. Hundreds of species of lichens are either endemic to this endangered ecosystem (such as *Arthonia kermesina* and *Lecanora masana*) or are disjunct in southern Appalachian spruce-fir forests from much more northerly, boreal locations (such as *Rhizocarpon geographicum* and *Usnea merrillii*). It is estimated that the Smokies hosts 75% of all spruce-fir forests. Photo by James Lendemer.

Whichever concept is preferred, it is imperative to recognize that all are temporary. As new diseases are introduced and keystone species disappear and become replaced by others, or as other factors alter community composition, vegetation ensembles will undergo change and so too must our concepts of them. At present, we are nearing the latter stages of a major transformation following the demise of the American Chestnut and, at uppermost elevations, major dieback of the Balsam Fir (Figure 36). In the next century, we will find ourselves

FIGURE 37. DEAD HEMLOCKS

In the late 1990s, Great Smoky Mountains National Park looked like a very different place than it does today, less than two decades later. The non-native, now invasive Hemlock Wooly Adelgid was introduced to the USA in the 1950s and spread into the southern Appalachians in the early 2000s, where it quickly wreaked havoc on Eastern Hemlock (*Tsuga canadensis*). Once thriving stands of this important conifer are now all but extinct in the southern Appalachians, leaving behind skeletons of a once vibrant ecosystem that now look lkike this instead. Photo by Erin Tripp.

FIGURE 38. BIG CREEK

Great Smoky Mountains National Park is famous for its massive, freshwater rivers (and swimming holes) that dissect the landscape. These rivers, big and small, create persistently high-humidity microenvironments that hundreds of lichen species depend on. Big Creek, which runs through eastern portions of the Park north of Sterling Ridge, is just one of many examples. Photo by Erin Tripp.

FIGURE 39. BIG RED OAK

Quercus rubra (Red Oak) is one of the most important trees and lichen substrates of low to middle elevation hardwood forests in the Smokies. The classic gray, vertical striations typical of its bark is a great way to recognize the species, even in the dead of winter. Photo by James Lendemer.

FIGURE 40. TYPICAL BARK

Large lichens and large trees go together. This picture shows typical lichen-covered bark near the base of a Red Oak (*Quercus rubra*). *Lobaria quercizans* (gray) and *Sticta beauvoisii* (brown) can be seen in this photo and are common constituents of the community of lichens that prefer tree bases. Photo by Erin Tripp.

FIGURE 41. CHAENOTHECA FURFURACEA ON HALESIA

For any given species, bark characteristics of middle aged trees are often very different from characteristics of old growth trees. Here is an example from *Halesia carolina* (Silverbell), which becomes flakey to shaggy with old age. A given species of lichen preferentially occupies bark of old, medium, or young trees, depending on the ecology of the lichen. Just like trees, lichens run the gamut of successional strategies: some are pioneer species, others are late successional species, and as yet others are generalists and grow just about anywhere and on anything. The yellow crustose lichen seen here is *Chaenotheca furfuracea*, which typically grows in sheltered and protected overhangs, but is here likely growing on the base of a mature tree because the flaking bark mimics those conditions. Photo by James Lendemer.

needing to define new communities following the loss of the Eastern Hemlock due to irreversible damage done by the Asian wooly adelgid (Figure 37).

Whether vascular plant communities can be used to define communities of other organisms remains to be determined for a great number of lineages, particularly microorganisms. Lichen communities sometimes but not always parallel those of vascular plants (Tripp & Lendemer, in press). For example, waterfalls and riparian corridors occur in numerous different plant communities, ranging from low-elevation rich cove or acid cove forest to high-elevation northern hardwood forest (Figure 38). However, from a lichenological perspective, high humidity associated with these waterfall spray zones and riparian areas compose a unique habitat in which some foliicolous (leaf dwelling) lichens have their sole distributions (e.g., species of *Gyalectidium*, *Byssoloma subdiscordans*, *Fellhanera bouteillei*). In this sense, high-humidity spray zones and watersheds may be considered single

FIGURE 42. WOLF RIDGE TRAIL
Grassy balds such as that seen here (along Wolf Ridge Trail, near Gregory Bald) are dominated by small shrubs such as *Vaccinium pallidum* and *Rubus canadensis* and grasses such as *Danthonia compressa*. In such habitats, branches and twigs become the most important lichen substrate, hosting species such as *Everniastrum catawbiense, Lecanora masana, L. rugosella, Tuckermanopsis ciliaris, Usnea subfloridana*, and *U. subfusca*. Photo by James Lendemer.

lichen community type, even though they can occur at several different elevations. Rock outcrops, further subdivided by rock type (e.g., calcareous or non-calcareous) may serve as another community type, as many saxicolous (rock-dwelling) lichens are restricted to one or another type of rock, again regardless of elevation (e.g., *Enterographa hutchinsiae*, *Porpidia albocaerulescens,* and *Vahliella leucophaea* on non-calcareous rocks; *Leptogium lichenioides* and *Caloplaca flavorubescens* on calcareous rocks) (Figures 28 & 29).

Other habitats of lichens and their relatives, such as lichenicolous fungi that grow on top of lichens or parasitic fungi that are derived from lichen ancestors, fall squarely outside the realm of traditional plant community types, such as the saprobe *Sarea resinae*, whose entire life cycle occurs ephemerally on the sap of gymnosperms, regardless of what species of gymnosperm. Other lichens occur as epiphytes on bryophytes, most commonly mosses but also liverworts (e.g., *Gomphillus americanus* and *G. calycioides*). As yet other lichens occur only on the most distal branches of young *Picea* twigs (*Gyalideopsis piceicola, G. epicorticis*), only on upper surfaces of shelf fungi (*Chaenothecopsis balsamconensis* and *Phaeocalicium polyporaeum*), or alternatively as broad-ranging terricolous (soil-dwelling) species that occur on acidic substrates from low elevation pine-oak forests to high elevation spruce-fir forests (*Dibaeis baeomyces*). Traditional plant community definitions may be more applicable to corticolous (bark-dwelling) species that have a high affinity to a specific tree species such that said lichen species closely mirrors the presence (or absence) of this specific species of tree (Figures 39, 40 & 41).

Recognizing (1) the evolving concepts of vegetation communities, (2) the often non-specific relationship between plant community and lichen constituents, and (3) a broader need to develop and employ biogeographical standards in lichen community ecology everywhere but especially in eastern North America, we here tentatively define five plant communities highly relevant to lichen diversity plus three lichen sub-communities as occurring within GSMNP (Table 1). These five broadly defined and simplified plant communities are:

[1] mesophytic mixed hardwood forest
[2] submesic mixed hardwood forest
[3] xeric acid forest
[4] northern hardwood forest (including balds; Figure 42)
[5] sub-boreal or boreal spruce-fir forest

Within these given communities, any of three lichen sub-communities can be found:

[a] calcareous rock outcrops
[b] non-calcareous rock outcrops
[c] humid riparian / spray zone areas

PLANTS AND ANIMALS

The most recent tallies of biological diversity in GSMNP for many organisms are those in the National Park Service and ATBI retained by, or recent publications pertaining to ATBI (see in particular the special issue of *Southeastern Naturalist* in 2007). Within Park borders alone there are over 1,500 species of flowering plants, 200 birds, 66 mammals, 50 fish, and 43 amphibians. Although still largely incomplete for many taxonomic groups (such as fungi), these numbers are exceptionally high for a single national park in the greater context of the eastern North American biota.

LICHENS: HISTORY OF STUDY IN GSMNP

Compared to investigation of most vascular plants and large-bodied animals, there have been remarkably few studies devoted specifically to lichen diversity and taxonomy in GSMNP. Even across eastern North America, most modern lichen studies have not been "floristic" (i.e., including generation of checklists) in nature (but see Wetmore 1967, Brodo 1968, Moore 1968, Yoshimura & Sharp 1968, Skorepa 1972, Peck et al. 2002, Lendemer & Tripp 2008, Hodkinson 2010, Spribille et al. 2010, Lücking et al. 2011, Lendemer et al. 2016, Knudsen et al. 2017; also see Table 2 for historical accounts) but rather have focused on individual lineages (McDonald et al. 2003, Tønsberg 2007, Tripp & Lendemer 2010, Sheard 2010, Tripp et al. 2010, Spribille et al. 2014). Prior to our efforts in GSMNP, there had been only one published study that sought to develop a checklist of the Park's lichens—that of the Swedish lichenologist Gunnar Degelius (1941). In this work, Degelius (1941: 2) wrote that prior to his investigation, the lichen flora of GSMNP was "hitherto unknown, although occasional collections have been made." He then wrote (pg. 2) that he "roamed about a large part of the National Park's central and highest districts." A glimpse at his specimens reveals that he collected in only 10 different localities (Table 3), all of which were at mid to high elevations, as he indicated. Thus, Degelius's work was limited to a subset of habitats found in GSMNP. It is also evident based on his checklist that very few crustose species were collected, while more modern lichen inventories have indicated the importance of crustose taxa to overall species counts (Schmitt & Slack 1990, Peck et al. 2002; Tripp

2016 (NY WhiteRocks pub in WNAN); Tripp & Lendemer, in press). In total, Degelius documented 206 species of lichens in GSMNP.

Published studies since that of Degelius have consisted of only partial investigations of the lichen biota of GSMNP. Most notably, Dey (1978) studied "macrolichens" of the uppermost elevations (≥ 5,500 ft.) of the southern Appalachians, which included a great number of the high peaks in the Park (Table 3). Thus, like Degelius's work, Dey's work included only a subset of all habitats in the Park and excluded crustose species. Other collectors have made occasional (Evans 1947, Mozingo 1954, Thomson 1967, DePriest 1984, Ciegler et al. 2003, Keller et al. 2004, Fanning et al. 2007) or somewhat more regular (Skorepa 1972; Tønsberg 2004, 2005; Lendemer et al. 2013; Selva 2016) trips to collect or report lichens from the Park. Yet none had produced a comprehensive inventory. Since Degelius's time and prior to our work, there had been only one estimate of the total number of lichens in GSMNP: in Keller et al. (2007), Park staff biologist Becky Nichols reported 463 species (as personal communication). Following our first half decade of research within the Park's borders, we reported a total of 770 species to GSMNP (Lendemer et al. 2013). Since publication of that book, the total number of lichens and lichenicolous fungi known from GSMNP has grown to 920 species (Tripp & Lendemer, in press). To put this into a broader perspective, the continuously updated, online checklist of lichens for North America (Esslinger 2016) contains 5,561 species. Thus, a fifth of all lichens in North America north of the Mexican border occur within the confines of only 814 square miles that comprise the Smokies.

DEVELOPMENT OF THIS BOOK

We initiated our investigation of the lichen biota of the southern Appalachians over a decade ago through an inventory of the lichens of Gorges State Park (Lendemer & Tripp 2008), 45 miles SSE of GSMNP. Because this project returned numerous interesting lichenological discoveries (Lendemer & Tripp 2008, Lendemer et al. 2014), we sought to expand our studies into the adjacent, mountainous regions. An investigation of lichen diversity and ecology in GSMNP was a logical place to begin given that the Park is a known biodiversity hotspot but at the same time much in need of focused, lichenological studies. That there existed some prior work upon which to build and that there was an ongoing all-taxon biodiversity initiative provided further stimulus for prospects of a long-term research project in GSMNP. In 2006, we launched a preliminary investigation of the lichens in a 1-hectare long-term plot, established and studied continuously by one of us (E. Tripp) together with colleague Kyle Dexter (Figure 34). One day of collecting in this small area situated among north-facing rich cove forest on lower Baxter Creek yielded over 110 species, 11 of which were new to science and numerous others of which were range extensions. That moment marked the beginning of perhaps a lifelong research program on lichen diversity and ecology in GSMNP.

The present work represents one of the most important products of our first decade+ of lichenological research in GSMNP (Figure 26). As part of this work, we have traversed thousands of miles on foot, both on trail and off trail, carrying pillowcases full of rock, bark, and soil samples with lichens clinging to these substrates, targeting every corner of the Park, in summer, spring, fall, and winter (Figure 43). We have focused on all elevations, substrates, and habitats, but have given special attention to crustose and other microlichens that are often neglected in studies. We have done our best to avoid spending too much time looking at the spectacular scenery (Figure 44), staring off the porch at the Southern Appalachian Highlands Science Learning Center toward Purchase Knob, or admiring the fall color (Figure 45)—all activities we have enjoyed but that otherwise take time away from lichenizing. On nearly all of these trips, we have carried a Nikon D7100 Digital SLR camera with a 105 mm 1:1 macro lens and ring flash around our necks. We have collectively made 7,000 new herbarium vouchers as part of this fieldwork (http://sweetgum.nybg.org/science/vh/). The lichen checklist of GSMNP now includes 920 species, making the Smokies the most lichenologically diverse National Park in the United States. These 920 species also represent a large fraction of all lichens in eastern North America, making this guide highly usable across most landmasses east of the Mississippi River. We are both proud of the work we contribute in this guide as well as humble for the opportunity we have been given through the generosity of the staff at and visitors to GSMNP (Figures 46 & 47). There is little doubt that our once-modern taxonomic classifications employed in this guide will become old fashioned, and *no* doubt that future collecting efforts by lichenologists will result in additional discoveries about the Park's rich lichen biota. It is our hope that this *Field Guide to the Lichens of Great Smoky Mountains National Park* provides a historical foundation upon which future workers can build.

FIGURE 43. THE AUTHORS

The foundation of this *Field Guide* and our first decade of lichenological study in Great Smoky Mountains National Park is the cumulative 7,000 museum voucher specimens that we collected over the course of ~2,000 miles traversed on foot, both on and off trail. We have found collecting lichens in pillowcases to be both an effective means of specimen transport as well as separation of specimens into different field sites. Beyond the pillowcases, our camera, three-pound hammers, rock chisels, wood chisels, branch clippers, sharpies, #2 brown paper bags, GPS, and hand lenses are the only items essential to any given field day. Photo by Erin Tripp.

FIGURE 44. PURCHASE KNOB

Clouds forming around the summit of Purchase Knob in early autumn, as viewed from the Southern Appalachian Highlands Science Learning Center porch. Photo by Erin Tripp.

FIGURE 45. FALL COLOR

The forests comprising Great Smoky Mountains National Park record a history of change. Here is a view of the changing forest, from the inside. Photo by James Lendemer.

FIGURE 46. LICHEN DRYING

Our decade+ of lichen research in the Park has been facilitated by the staff and research facilities at the Southern Appalachian Highlands Science Learning Center (aka: Purchase Knob). This photo shows typical treatment of lichen specimens after field collection and during the early stages of curation (e.g., preliminary identifications, TLC-based chemistry assays) in preparation for final identification followed by formal depositing and accessioning collections into the herbaria at the New York Botanical Garden and The University of Colorado—Boulder. Photo by James Lendemer.

FIGURE 47. PURCHASE KNOB

Much of our work—in fact almost all of it—has been made possible by hospitality afforded to us by the National Park Service and specific staff members with the Smokies (see Acknowledgements). Old wood and lignum such as the fencepost in the foreground, just outside the porch at the Purchase Knob field station, is crucial habitat for lichen biodiversity. Species such as *Xylographa trunciseda* and, if you are having a really great day, *Cyphelium tigillare*, are more or less restricted to such surfaces. Purchase Knob is in the background, because one can never have too many photos of this mountain. Photo by James Lendemer.

FIGURE 48. SMOKY MOUNTAINS

Great Smoky Mountains National Park derives its name from "Shaconage" or "a place of blue smoke" as named by its early human inhabitants: the Eastern Band of the Cherokee. It is a fitting name, as seen in this photograph, but so too would have been Great Lichen Mountains National Park. Photo by Erin Tripp.

PART 4

LICHENS OF GREAT SMOKY MOUNTAINS NATIONAL PARK

HOW TO USE THIS FIELD GUIDE

It is intended that this *Field Guide* will facilitate the identification of nearly all of the apparent (and most of the unapparent), easy-to-distinguish species of lichens in Great Smoky Mountains National Park, both rare and common. The photographic portion of the guide is extensive but not 100% comprehensive; numerous species are separated from nearest relatives only through microscopy, chemical analyses, and sometimes DNA sequencing. Nonetheless, to facilitate utility of this guide for the biologist or enthusiastic naturalist, we include appendices and a full set of dichotomous keys to aid in the identification of *all* GSMNP lichens. Of the more than 920 species that we report occurring in the Park, more than 600 are illustrated in the main portion of this Guide. All 920 species are included in the dichotomous keys. Nearly all of these were photographed by us in the field, inside the borders of Great Smoky Mountains National Park. Each photograph is tied to a physical voucher specimen, which is housed either at the NY Herbarium, the COLO Herbarium, or both. As such, future researchers have the capacity to return to the physical specimen upon which the species entry is based. For each species entry, the collector, collection number, and photographer are cited. In a few instances, we were unable to capture reasonably good photographs of a given species in the field and instead utilized existing herbarium material collected in GSMNP by us or our colleagues. Similarly, the collector, collection number, and photographer of the collection is provided in the species entry.

The dichotomous keys represent our full understanding of the lichens of GSMNP, without any omissions, at the time of publication; however, no field guide is ever complete. We predict numerous lichen species have yet to be documented in GSMNP.

After basic taxonomy of the lichens of GSMNP is understood, we can begin to learn about the interactions between lichens and their environment. It is our goal that this *Field Guide* facilitates expanded learning of the lichen biota of the Park and broader region (i.e., much of eastern North America). Readers seeking additional, in-depth discussion of lichen biology, whether taxonomic, evolutionary, or ecological, are encouraged to refer to appropriate reference books, including *Lichens of North America* and *Lichen Flora of the Greater Sonoran Desert Region*. In particular, the topic of lichen chemistry is not covered in this guide beyond inclusion in the species pages and dichotomous keys, even though its importance to lichen taxonomy and identification cannot be exaggerated. Readers seeking additional, in-depth discussion on numerous species from the Park specifically are encouraged to refer to our earlier technical guide, *The Lichens and Allied Fungi of Great Smoky Mountains National Park* (Lendemer et al. 2013; Tripp & Lendemer, in press).

Lichens are highly complex structures with seemingly endless evolutionary modifications to morphology. In many species, this diversity can best be appreciated with magnification; we highly encourage the use of a 10x or 14x hand lens in the field.

This *Field Guide* is arranged alphabetically by genus and then by species. Each species treated includes a photograph, discussion of its habitat and ecology, as well as a range map. We also provide up-to-date discussion pertaining to recent advances in taxonomy and miscellaneous aspects of the lichen biology of some species.

LICHENS OF GREAT SMOKY MOUNTAINS NATIONAL PARK

Acarospora fuscata

Peanut Butter's Lichen

Tripp 2283 (photo: Lendemer)

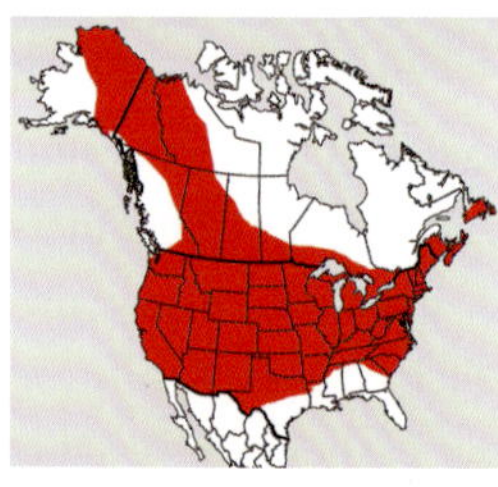

NOTES: *Acarospora fuscata* is readily identifiable by its brown areoles, each with one to several sunken apothecia that have reddish-brown discs. Like most *Acarospora*, the asci contain many tiny, ellipsoid colorless spores. *Acarospora obpallens* differs in having regularly shaped circular areoles, each with a single apothecium. *Acarospora thamnina* is a similar species that is not yet reported from the Smokies and differs in having areoles with black rather than brown lower surfaces.
CHEMISTRY: Gyrophoric acid. Spot tests. Cortex K-, KC+ pink, C+ pink, P-, UV-.
NICHE: This species occurs on both shaded and exposed non-calcareous rocks throughout the Smokies. It is here pictured on granite. *Acarospora fuscata* is common across eastern and western North America and many other temperate regions worldwide, including Europe.
KEY FEATURES: Irregularly shaped brown areoles with one to several sunken apothecia, C+ red upper surface, on both shaded and exposed non-calcareous rocks.

Acarospora obpallens

Acorn Cups

Tripp 3733 (photo: Lendemer)

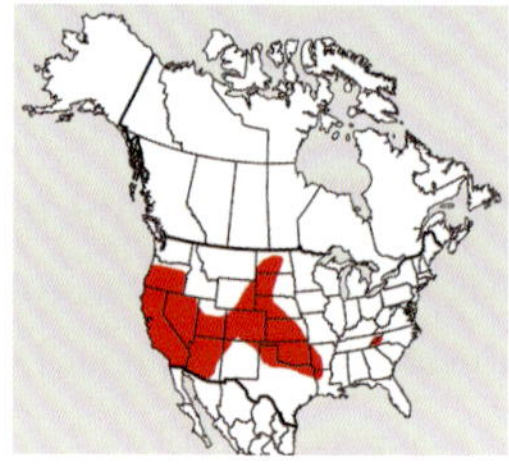

NOTES: *Acarospora obpallens* is a distinctive species by its regularly shaped brown areoles that each contain only a single, large sunken apothecium having a dark brown-to-black disc. *Acarospora* as a genus is characterized by its polysporous asci and tiny, ellipsoid, colorless spores. This species is most likely to be confused with the far commoner *A. fuscata*, but the latter has irregularly shaped areoles, each of which often contains more than one apothecium.
CHEMISTRY: Gyrophoric acid. Spot tests. Cortex K-, KC+ pink, C+ pink, P-, UV-.
NICHE: This species is rare on exposed, non-calcareous rocks in the Smokies. The above photograph was taken near the Lonesome Pine Overlook near Deep Creek. *Acarospora obpal-*

lens has also been found nearby in the southern Appalachians on exposed granitic domes. It is disjunct from more arid areas of western North America, where it occurs on exposed rocks and on soil.

KEY FEATURES: Regularly shaped brown areoles each containing only a single, sunken apothecium, C+ red upper surface, on exposed non-calcareous rocks.

Acarospora sinopica

Rusty Composure

Tripp 3697 (photo: Lendemer)

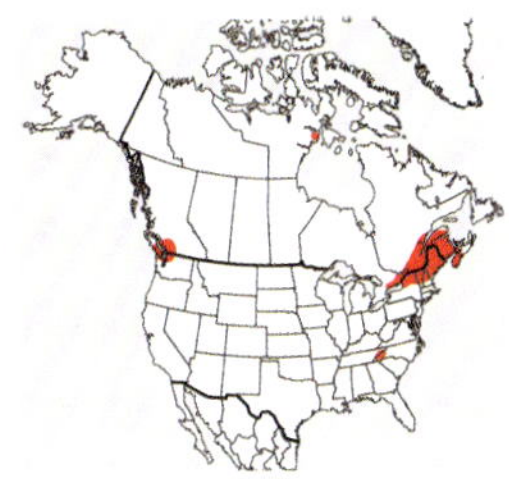

NOTES: *Acarospora sinopica* is one of the most distinctive crustose lichens in the Smokies owing to its intense, rust-orange (to sometimes reddish-orange) areolate thallus. This remarkable color derives from the uptake and accumulation of heavy metals from the unusual rock on which thalli occur. Superficially, one may first confuse *Acarospora sinopica* for a Caloplaca, but members of the latter genus typically have K+ purple thalli and polarilocular spores. This species could also be confused with the unrelated *Rhizocarpon oederi*, but the latter has black apothecia with 4-celled spores.

CHEMISTRY: No substances. Spot tests. K-, KC-, C-, P-, UV-.

NICHE: This species occurs on shaded or exposed, heavy metal-rich rocks in the Smokies, where it is very rare except on suitable substrates such as the Anakeesta Formation. *Acarospora sinopica* occurs at disjunct locations throughout the Northern Hemisphere where heavy metal-rich rocks are found.

KEY FEATURES: Rusty orange to reddish orange areoles, sunken apothecia, C- upper surface, on shaded or exposed rocks rich in heavy metals.

Agonimia opuntiella

Tiny and Spiny

Lendemer 44542 (photo: Tripp)

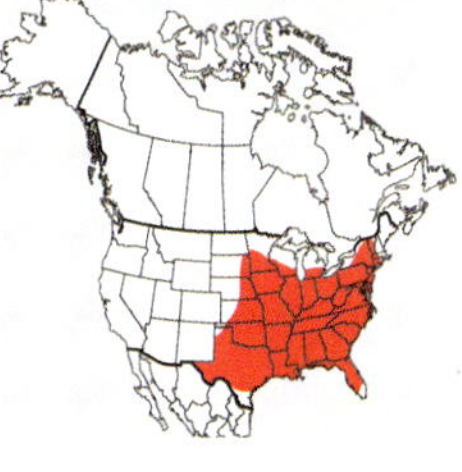

NOTES: *Agonimia opuntiella* is a highly inconspicuous but highly recognizable species by its prickly hairs (hence, the specific epithet) that terminate minute, greenish-brown squamules. It is also blastidiate, which may or may not be immediately obvious on any given thallus. If you find this lichen, then you are having a good day. If you find it fertile with perithecia, then you are having a *great* day, in which case you will want to run, not walk, to the lab to see its beautiful, muriform spores.

CHEMISTRY: No substances. Spot tests. K-, KC-, C-, P-, UV-.

NICHE: This species is most commonly found growing over mosses (as seen in the above photo) or detritus, especially on or near calcareous rocks. It is uncommon but can be found throughout the Smokies at low-to-middle elevations. Look for it near Ace Gap. *Agonimia opuntiella* also occurs throughout eastern North America, extending into Europe.

KEY FEATURES: Prickly hairs on upper surface of thallus, minute greenish-brown squamules, often blastidiate, more rarely with perithecia, commonly overgrowing mosses on calcareous rock.

Agyrium rufum

Abstention

Tripp 3593 (photo: Lendemer)

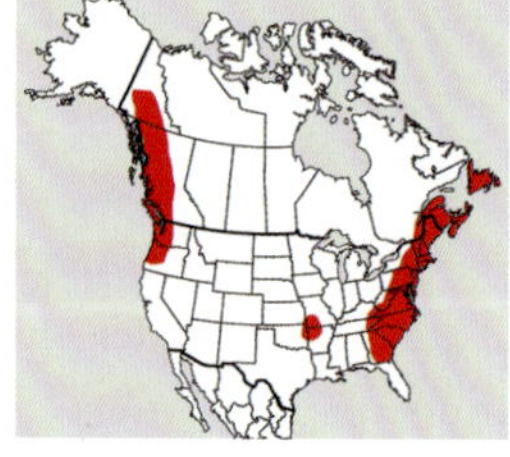

NOTES: *Agyrium rufum* is one of many inconspicuous crustose lichens in the Smokies—only the most astute observer will be fortunate enough to find it. *Agyrium rufum* is an unusual species in that it lacks a photobiont (the green in the above photo is an unassociated, free-living alga). If found, *A. rufum* can be recognized by its non-lichenized habit, its small, pale red apothecia, and its colorless simple spores. As curious as it seems, molecular studies have demonstrated that this species is closely related to the genus *Pertusaria*.

CHEMISTRY: No substances. Spot tests. K-, KC-, C-, P-, UV-.

NICHE: This species occurs exclusively on old, dry conifer wood. *Agyrium rufum* is rare in the Smokies and has a poorly understood distribution elsewhere in North America. The map presented here assumes that in the eastern United States, *A. rufum* occurs in areas between currently known locations, i.e., the Atlantic coast and the southern and northern Appalachians.

KEY FEATURES: Absence of photobiont, small, pale red apothecia, colorless simple spores, and occurrence on old, dry conifer wood.

Ahtiana aurescens

Ted's Beauty

Tripp 2627 (photo: Lendemer)

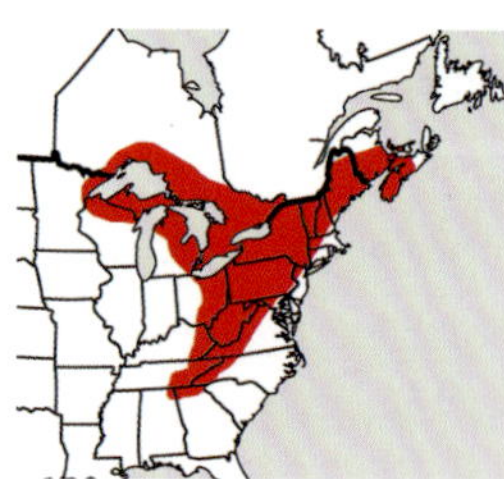

NOTES: *Ahtiana aurescens* is a beautiful macrolichen that only occurs in beautiful places, like the Smokies. It is easily recognized by its yellowish-green thallus, which reminds one of the far more common *Usnocetraria oakesiana*. The latter differs in having marginal soralia (vs. being esorediate and instead usually fertile, as in *A. aurescens*). Furthermore, the apothecia of *A. aurescens* are raised and often have slightly crenate margins. As seen in the photo, *A. aurescens* generally has large, prominent black pycnidia clustered especially along lobe margins. This genus is named after the esteemed Finnish lichenologist Teuvo (Ted) Ahti.

CHEMISTRY: Usnic acid, fatty acids. Spot tests. Cortex K-, KC+ gold, C-, P- UV-. Medulla K-, KC-, C-, P-, UV-.

NICHE: This species occurs on twigs, especially in the canopy of humid, high-quality forests in the Smokies. Look for it along the Lakeshore Trail. It is endemic to eastern North America but has been extirpated from most of its native range.

KEY FEATURES: Beautiful, yellowish-green thallus, marginal, black pycnidia, raised apothecia with crenate margins.

Alectoria fallacina

Bubbling Witches' Hair

Lendemer 32951 (photo: Tripp)

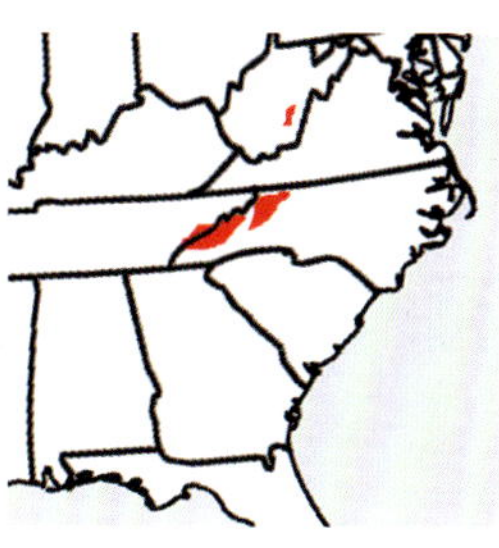

NOTES: *Alectoria fallacina* resembles *Usnea merrillii*, which occurs in similar habitats. It is easily differentiated from the latter by the lack of a central cord in the branches. *Alectoria fallacina*, if found, lacks both sexual and asexual reproductive structures, and how it reproduces is a mystery but we suspect it occurs through fragmentation of the thallus. A close inspection of the branches will reveal characteristic, clustered white pseudocyphellae that resemble soralia. This is one of several species that should be protected by the US Endangered Species Act.

CHEMISTRY: Usnic acid. Spot tests. Cortex K-, KC+ gold, C-, P- UV-. Medulla K-, KC-, C-, P-, UV-.

NICHE: This species is restricted to humid, high-elevation spruce-fir forests in the Smokies, where it is infrequent. It occurs in similar habitats elsewhere in the southern Appalachians. Disjunct populations have been reported from New England and the Canadian Maritimes, but these are actually *A. sarmentosa*, a different and much more widespread species.

KEY FEATURES: Fruticose thallus, branches without a central cord, esorediate, clustered pseudocyphellae that resemble soralia, humid high-elevation spruce-fir forests.

Amandinea punctata

Little Black Dots, and I Don't Care

Lendemer 53354 (photo: Tripp)

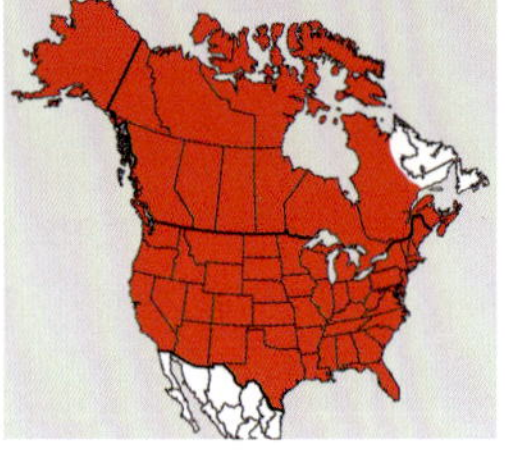

NOTES: *Amandinea punctata* is just about as close to being a weed as one can get in lichenology. Except for in the Smokies (see Niche, below). Elsewhere in North America, this is an extremely common species, marked by its thin, continuous, grayish-brown thallus and black apothecia with lecideine margins. There are of course a lot of other "little black dot" lichens in the Smokies, but hopefully its 2-celled brown spores that are 8 to the ascus helps identification of this one.

CHEMISTRY: No substances. Spot tests. K-, KC-, C-, P-, UV-.

NICHE: *Amandinea punctata* is extremely common in disturbed areas throughout North America and Europe, where it is corticolous and occurs mostly on twigs. It is relatively uncommon in the Smokies because, well, the Smokies are just too nice. Look for it in parking lots, perhaps near latrines.

KEY FEATURES: Thin, continuous thallus, black apothecia, brown, 2-celled spores, on bark (especially twigs), weedy everywhere except in the Smokies.

Anaptychia palmulata

Lobe-y McLobe Face

NOTES: *Anaptychia palmulata* is one of the most easily recognizable macrolichens in the Smokies owing to lobules produced along apothecial and lobe margins, especially near the center of the thallus, its vibrant green thallus

Lendemer 33129 (photo: Tripp)

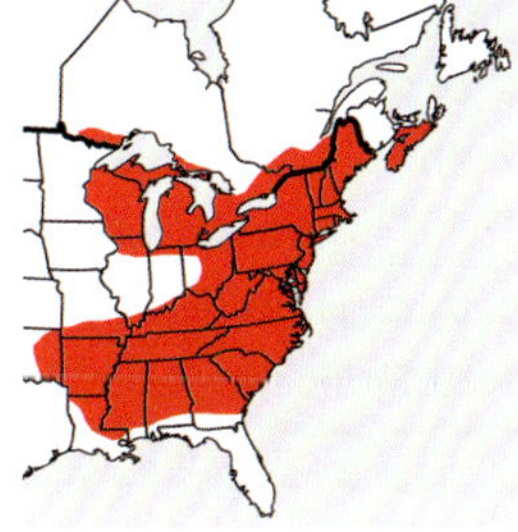

when wet (otherwise brownish gray), and its pale lower surface. *Heterodermia hypoleuca* is somewhat similar but is light gray in color and has an ecorticate lower surface. *Phaeophyscia squarrosa* is another lobulate species but differs by its dark lower surface and prominent rhizines. This is one to learn by gestalt. You will be happier for it.

CHEMISTRY: Variolaric acid, trace of atranorin. Spot tests. Cortex and medulla: K-, KC-, C-, P-, UV.

NICHE: This species is restricted to the bark of hardwoods in high-quality forests in the Smokies. Elsewhere in its range (a near-endemic to eastern North America), it is relatively rare, occurring only in similar habitats. Herbarium records indicate that this species was more common and widespread than it is now, likely owing to habitat degradation through the years.

KEY FEATURES: Abundant lobules along apothecial and thallus margins, vibrant green thallus when wet, pale and corticate lower surface. Look at the picture–this is what you will see in the field!

Andreiomyces morozianus

Andy's Dust

Tripp 5022 (photo: Lendemer)

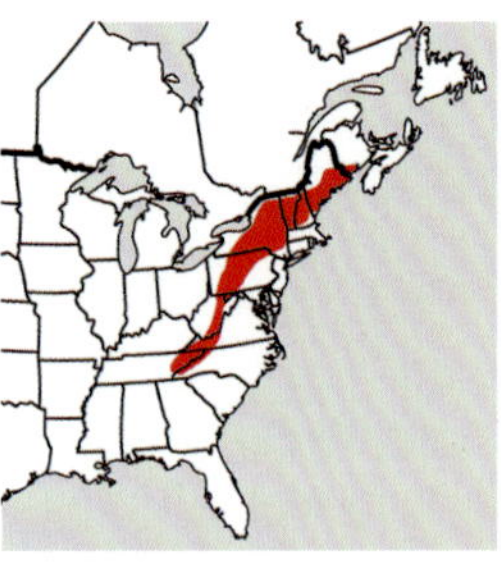

NOTES: *Andreiomyces morozianus* is a distinctive leprose lichen that can be recognized by its yellowish-green thallus with irregular patches of orange pigment. It is most likely to be confused with members of the genus *Lepraria* but can be differentiated from those species by chemistry (including its pigments). The binomial name of this taxon is formed from the first (Andrei) and last (Moroz) names of Dr. Andrei Moroz, a psychiatrist from New York City and lichen enthusiast of the Smokies.

CHEMISTRY: Isousnic acid, obtusatic acid, unknown orange pigment. Spot tests. K-, KC+ yellow, C-, P-, UV+ blue-white.

NICHE: In the Smokies, this species occurs on sheltered substrates including the bark of mature Red Spruce and Yellow Birch as well as non-calcareous rock overhangs. *Andreiomyces morozianus* occurs elsewhere throughout the Appalachians from Canada to the Smokies but is relatively rare across its range.

KEY FEATURES: Yellowish-green, leprose thallus with patches of orange pigment, in high-elevation spruce-fir forests and rocks.

Anisomeridium biforme

Moonshiner's Flask

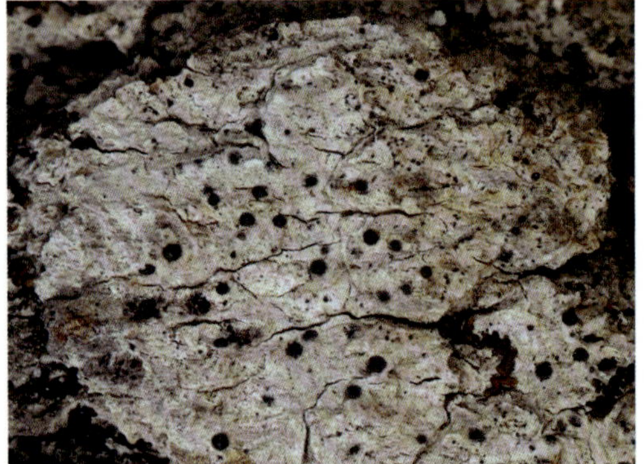

Tripp 3932 (photo: Lendemer)

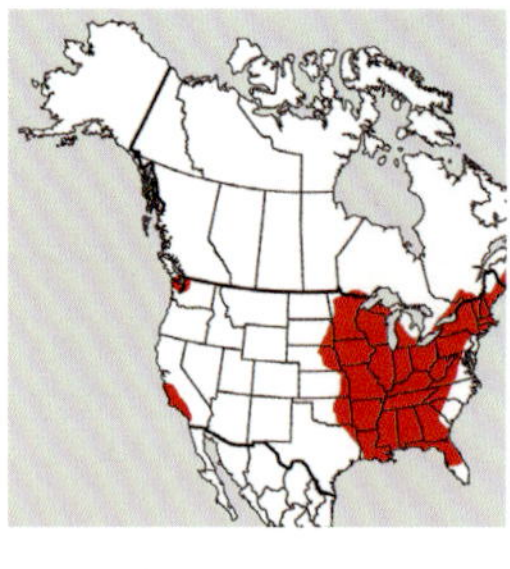

NOTES: *Anisomeridium biforme* can be recognized by its relatively large, black perithecia and its immersed white thallus. It is superficially similar to several pyrenolichens that occur in the Smokies but differs in having an inspersed hymenium, asci with spores arranged in a single row, and colorless, 1-septate spores with two equally sized cells. It is most likely to be confused with *Acrocordia megalospora*, but the latter species has much larger spores and larger perithecia with conspicuous, pointed beaks.

CHEMISTRY: No substances. Spot tests. K-, KC-, C-, P-, UV-.

NICHE: This species occurs on the bark of large hardwoods, particularly at middle-to-low elevations in the Smokies. *Anisomeridium biforme* is infrequent but widespread throughout temperate eastern North America, with disjunct populations in other temperate regions of the world. Populations in western North America are questionably conspecific.

KEY FEATURES: Corticolous, large black perithecia, inspersed hymenium, uniseriate colorless spores with two equally sized cells.

Anisomeridium polypori

Slipper Spores

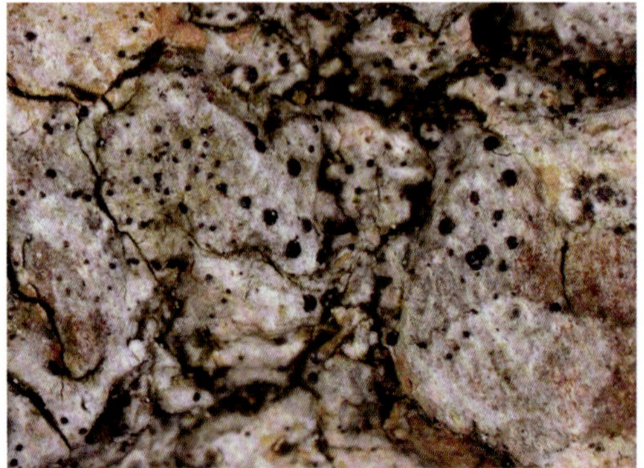

Buck 56240 (photo: Lendemer)

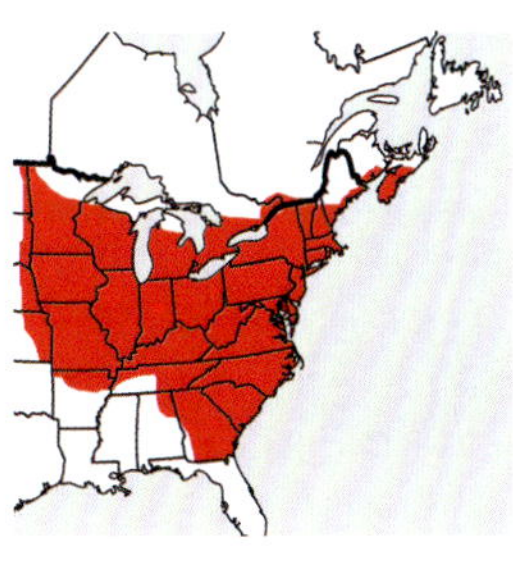

NOTES: *Anisomeridium polypori* is widespread throughout eastern North America and can easily be recognized by its thin, white thallus with a *Trentepohlia* photobiont, small black perithecia, and colorless, 2-celled spores where the submedian septum divides the cells into unequal sizes so that the spores resemble loafers. *Julella fallaciosa* also has a thin white thallus and occurs on hardwood trees; however, that species lacks a photobiont and has muriform spores. *Anisomeridium biforme* is somewhat similar, but is very rare in the Smokies, has larger perithecia, and spores with a median septum that divides the spores into cells of equal size.

CHEMISTRY: No substances. Spot tests. K-, KC-, C-, P-, UV-.

NICHE: This species occurs throughout the Smokies on the bark of trees, especially on the bases and roots. Although it is found on both hardwoods and conifers, the former are more frequent substrates.

KEY FEATURES: Corticolous, thin white thallus, small black perithecia, 2-celled, colorless spores that resemble loafers.

Anzia americana

American Spongepants

Lendemer 33236 (photo: Tripp)
Inset: Lendemer 33236 (photo: Tripp)

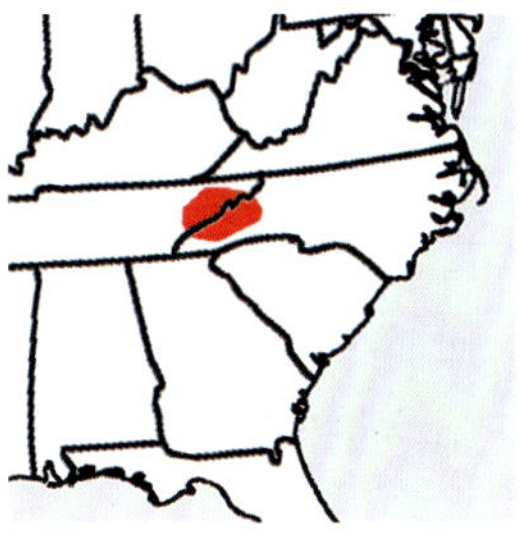

NOTES: *Anzia americana* is easily recognized, first by its thick, spongy lower surface (a feature shared only by other members of *Anzia*) and second by its elongate lobes with terminal soralia. The latter structures are difficult to discern in this photo, although a single soralium can be seen on occasion. *Anzia americana* is one of the rarest macrolichens in North America and is similarly rare in the Smokies. Unfortunately, *A. americana* cannot be considered for listing under the US Endangered Species Program because the status of populations outside of the United States is unknown.

CHEMISTRY: Atranorin, divaricatic acid. Spot tests. Medulla K-, KC+/- pale pink, C-, P-, UV+ bright blue-white.

NICHE: In the Smokies, we have seen this species only twice: once on a mature hardwood at a high-elevation site near the TN/NC border, and another on a dead mature hemlock near Cataloochee. It is not yet known from anywhere else in North America, but rare, disjunct populations occur in the mountains of South America and southeast Asia.

KEY FEATURES: Spongy lower surface, elongate lobes, terminal soralia, mature high-elevation forests.

Anzia colpodes

Fecund Spongepants

Lendemer 33070 (photo: Tripp)

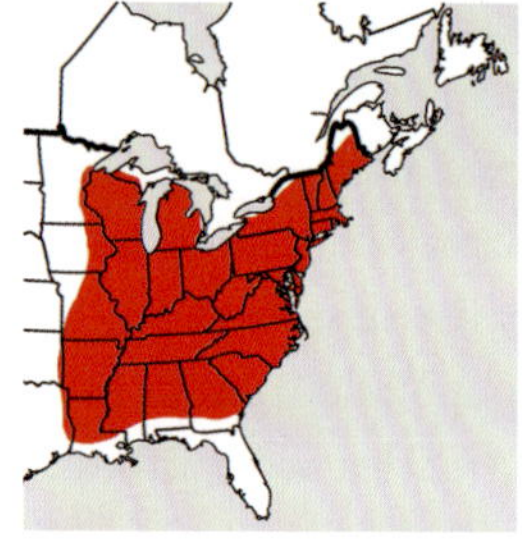

NOTES: *Anzia colpodes* is one of two species in this genus that occurs in the Smokies (three total in North America) and is easily recognized by its spongy lower surface and absence of soralia but presence of apothecia. This reproductive difference serves to easily distinguish *A. colpodes* from the far rarer *A. americana*.

CHEMISTRY: Atranorin, divaricatic acid. Spot tests. Medulla K-, KC+/- pale pink, C-, P-, UV+ bright blue-white.

NICHE: The Smokies are the last stronghold for this eastern North American endemic, which was once far more common and widespread in the region than it is today. In the Smokies, one encounters *Anzia colpodes* in middle-to-low-elevation forests on hardwoods. Rabbit Creek Trail near Abrams Creek in Tennessee, despite experiencing an awesome F5 tornado in 2011, is an excellent location in which to still find healthy populations of this species.

KEY FEATURES: Spongy lower surface, rosette forming thalli, esorediate, often (but not always) with apothecia, middle-to-low-elevation forests.

Arthonia anglica

Fish on Chips

Tripp 2542 (photo: Tripp)

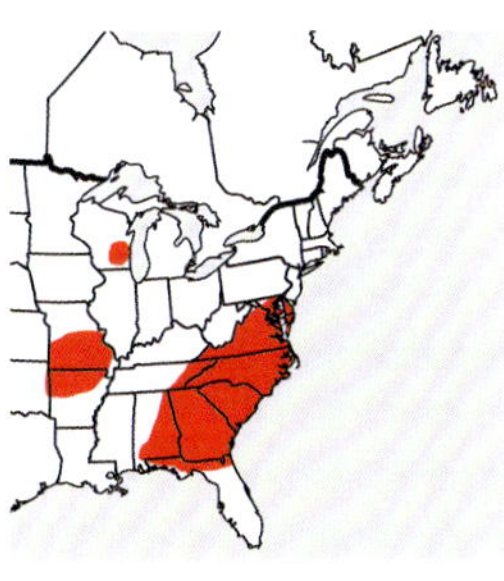

NOTES: *Arthonia anglica* is one of numerous species in this taxonomically complex genus in the Smokies. Its reddish-brown, irregularly shaped lirellae most resemble those of *A. rubella*, but the latter species has 6-celled spores with enlarged terminal cells, whereas those of *A. anglica* are 4-celled with only the uppermost cell enlarged. *Arthonia anglica* is the only member of the genus in the Smokies that has C+ pink lirellae owing to the presence of gyrophoric acid.

CHEMISTRY: Gyrophoric acid. Spot tests. *Ascomata* (in section) K-, KC+ pink, C+ pink, P-, UV-.

NICHE: This species, like most temperate *Arthonia*, is corticolous on hardwoods. In the Smokies, *A. anglica* can be found in humid, middle-to-low elevations on a diversity of corticolous substrates. The Lakeshore Trail is a superb area in which to see this species. Its distribution also includes the Ozarks and a disjunct locality in Wisconsin, but this distribution likely reflects under-collection rather than its true distribution. As the name implies, this species also occurs in Europe.

KEY FEATURES: Reddish-brown irregularly shaped ascomata, 4-celled colorless spores with the uppermost cell enlarged, C+ pink in section.

Arthonia byssacea

Cerulean Eyeshadow

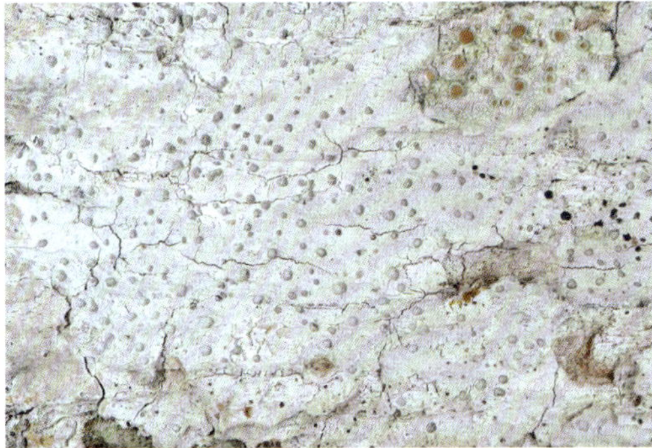

Tripp 3907 (photo: Lendemer)

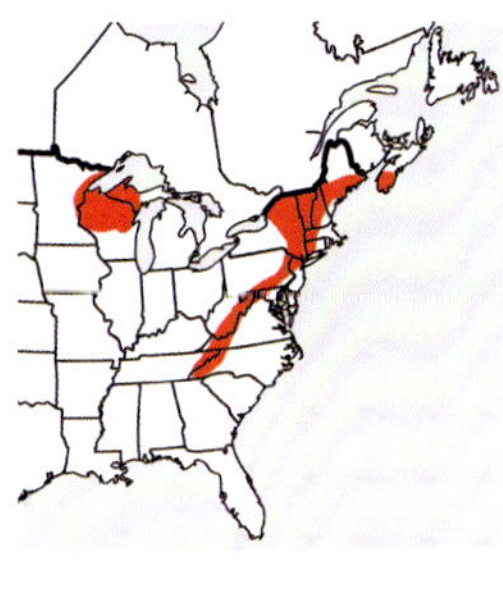

NOTES: *Arthonia byssacea* can be recognized by its thin, bluish white to near-white thallus and small apothecia marked by heavy white pruinose discs, giving these fruiting bodies from a distance the appearance of soralia. The dark hypothecium and 4–6 celled colorless spores further confirm the identification of this species. The only species with which this could be confused is *Lecanactis abietina*, but the latter has conspicuous white pruinose pycnidia that are C+ red due to the presence of lecanoric acid. *Arthonia byssacea* is not likely to be seen by the casual visitor to the Smokies, but rather only by the most careful observer.

CHEMISTRY: No substances. Spot tests. K-, KC-, C-, P-, UV-.

NICHE: This species occurs on dry faces of mature, living hardwood trees in old growth middle-to-high elevation forests of the Smokies. It occurs in similar habitats throughout the Appalachians and Great Lakes Region as well as in Europe.

KEY FEATURES: Bluish white to white thallus, heavily pruinose apothecia, darkened hypothecium, 4–6 celled spores, dry faces of mature hardwood trees.

Arthonia caesia

Mt. Cammerer Cupcakes

Tripp 3910 (photo: Lendemer)

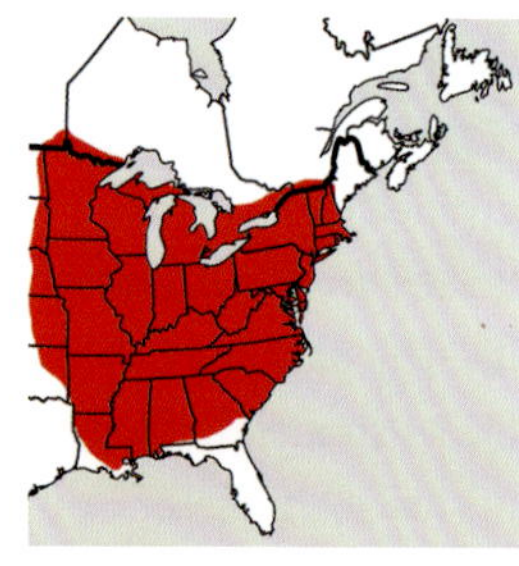

NOTES: *Arthonia caesia* is a highly distinctive crustose lichen characterized by its blue pruinose apothecia and green to golden yellow leprose thallus. This combination of features makes it unmistakable for any other crustose lichen in the Smokies. If you are lucky enough to view spores of this species, note that they are pinched near the middle septum, then further divided by two additional septa near both polar ends of the cell.

CHEMISTRY: Usnic acid, zeorin. Spot tests. K-, KC+ gold, C-, P-, UV-.

NICHE: *Arthonia caesia* is typically found in disturbed habitats and forest openings. As such, it is relatively rare in the Smokies, where overall forest quality is high and disturbance is rather minimal. A terrific place to see this lichen is on the Cammerer Trail. Elsewhere, it is very common in urban areas throughout eastern North America. This species also occurs in Europe and was originally described from Germany in the 1800s.

KEY FEATURES: Blue pruinose apothecia, green to golden yellow leprose thallus, spores pinched at the middle septum with additional septa near cell ends.

Arthonia caudata

Quite A Let Down

Tripp 2090 (photo: Lendemer)

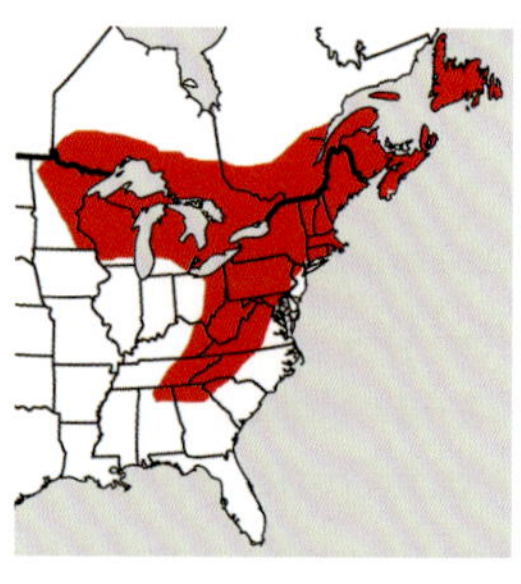

NOTES: *Arthonia caudata* is one of several non-lichenized species of *Arthonia* that occur in the Smokies. It is recognized primarily by its ecology (see below) and because it often lacks spores in its fruiting bodies. Where spores are present, look for their distinguishing long length and spindle shape. *Arthonia caudata* is most likely to be confused with *A. quintaria*, which usually occurs on

hardwoods and has 5-celled macrocephallic spores. As the name implies, *A. caudata* and the photo above are quite a letdown.
CHEMISTRY: No substances. Spot tests. K-, KC-, C-, P-, UV-.
NICHE: *Arthonia caudata* is relatively rare in the Smokies because it occurs exclusively on young branches of White Pine, a tree that is not common in the National Park. Elsewhere, this species is much more common, tracking the range of its host. The map presented here assumes that the species mirrors the distribution of its host.
KEY FEATURES: Non-lichenized thallus, irregularly shaped black ascomata, spindle-shaped spores, occurrence on White Pine, unremarkable.

Arthonia cinnabarina

Flappy Red Lips

Tripp 5437 (photo: Tripp)

NOTES: *Arthonia cinnabarina* is one of numerous secrets of the Smokies that we hope you will encounter! It is one of several species of *Arthonia* that occur within the National Park but is easily differentiated from all others by its dull reddish pruinose discs, its 6-celled spores, and its ecology (see Niche).

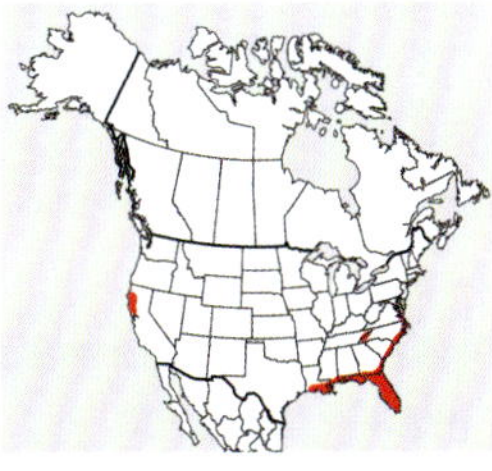

Arthonia kermesina also has red ascomata but is extremely different both morphologically and ecologically (scarlet red apothecia restricted to high-elevation old growth spruce). Other species of *Arthonia* in the Smokies have red ascomata, such as *A. helvola* and *A. vinosa*, but these are easily distinguished from *A. cinnabarina* by the number of cells per spore (3-celled in *A. helvola*, 2-celled in *A. vinosa*); furthermore, *A. helvola* and *A. vinosa* have fleck-like ascomata (vs. irregularly elongate *A. cinnabarina*).
CHEMISTRY: Anthraquinones. Spot tests. K+ red (apothecia), KC-, C-, P-, UV-.
NICHE: *Arthonia cinnabarina* is a southeastern coastal plain species and is disjunct inland in the Smokies. However, more collections throughout the Southeast are likely to yield a more continuous distribution. In the Smokies, look for it on *Ilex* twigs in very wet habitats at low elevation, such as Bone Valley. Curiously, *A. cinnabarina* also occurs in coastal forests of northern California.
KEY FEATURES: Thallus continuous and very thin, typically shiny, dull reddish apothecia erumpent and bursting above the thallus with flaps of bark attached; on twigs in wet habitats at low elevation.

Arthonia cupressina

Golden Spruce Dots

Lendemer 33042 (photo: Tripp)

NOTES: Prepare yourself: *Arthonia cupressina* is the Gunnison Sage Grouse of your lichen list. This rare and mysterious species will leave you wanting more. *Arthonia cupressina* is easily identifiable by its golden yellow pruinose apothecia, its leprose thallus, and its ecology (see Niche). *Arthonia cupressina* sometimes co-occurs in the same habitat, even on the same tree, with another rare species in the genus: *A. kermesina*.

CHEMISTRY: Usnic acid. Spot tests. K-, KC+ gold, C-, P-, UV-.

NICHE: *Arthonia cupressina* is known in the Smokies from only two localities, both of which are in high elevation, old growth spruce-fir forests. The species, like *A. kermesina*, is restricted to the bark of mature, Red Spruce trees. It is one of the rarest lichens in North America, known only from a handful of collections from the Smokies and immediate surroundings, most over century old. Its habitat and substrate are also extremely restricted in North America. This species deserves immediate protection under the US Endangered Species Program.

KEY FEATURES: Golden yellow pruinose apothecia, leprose thallus, corticolous on mature Red Spruce at high elevations, exceptionally rare.

Arthonia helvola

Covert Carmine

Lendemer 26683 (photo: Lendemer)

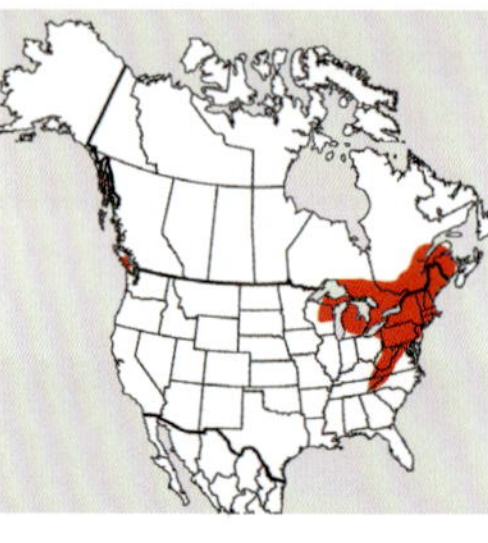

NOTES: *Arthonia helvola* is a lovely crustose lichen that will likely be first noticed by its tiny red apothecia. Like most *Arthonia*, these apothecia are irregular in shape; but, unlike most *Arthonia*, they are rather blotch-like (instead of lirellate). It's apothecia bear colorless spores that are 3-celled. There are not many species of *Arthonia* in the Smokies that have anthraquinone pigments that in the fruiting bodies, so give these a drop of K and watch the magic happen. Due to the blotch-like, red apothecia that are K+ purple, you can only confuse it for *A. cinnabarina* or *A. spadicea*, but *A. cinnabarina* differs in having 4–7-celled spores and *A. spadicea* has 2-celled spores.

CHEMISTRY: Anthraquinones. Spot tests. K+ purple (apothecia), KC-, C-, P-, UV-.

NICHE: *Arthonia helvola* prefers to grow on the roots and bases of hardwood trees in northern boreal forests, such as those near False Gap.

This ecology might tip you off: it is otherwise a species of colder climates of the northeastern United States and southeastern Canada.

KEY FEATURES: Tiny crustose lichen with tiny blotch-like apothecia, these red, K+ purple, colorless, 3-celled spores, growing at the bases of hardwoods in northern hardwood forests.

Arthonia kermesina

Hot Dots

Tripp 3834 (photo: Lendemer)

NOTES: *Arthonia kermesina* is one of the best-kept secrets of the Smokies. It is easily recognized by its small but intensely colored scarlet red apothecia and ecology (see Niche). No other species can be confused with *A. kermesina*. Hot Dots often grows with Golden Spruce Dots (*A. cupressina*), which is very similar in appearance except that the apothecia of the latter are golden-yellow instead of intense red.

CHEMISTRY: Unidentified red pigments. Spot tests. (apothecia) K+ reddish-purple, KC-, C-, P-, UV-.

NICHE: This species is restricted to high-elevation spruce-fir forests of the Smokies, where it occurs exclusively on the bark of large, mature Red Spruce trees. Despite our and colleagues' extensive searches in spruce-fir forests across the Appalachian Mountain chain, only a single population has been found outside of the Smokies—in the nearby Black Mountains of North Carolina. Thus, *A. kermesina* seems truly to be a restricted endemic species whose habitat is extremely limited. This species deserves immediate protection under the US Endangered Species Program.

KEY FEATURES: Small, intensely scarlet red apothecia, inconspicuous white thallus, restricted to large, mature Red Spruce trees in high-elevation spruce-fir forests.

Arthonia quintaria

Twig Squiggle

Tripp 5435 (photo: Tripp)

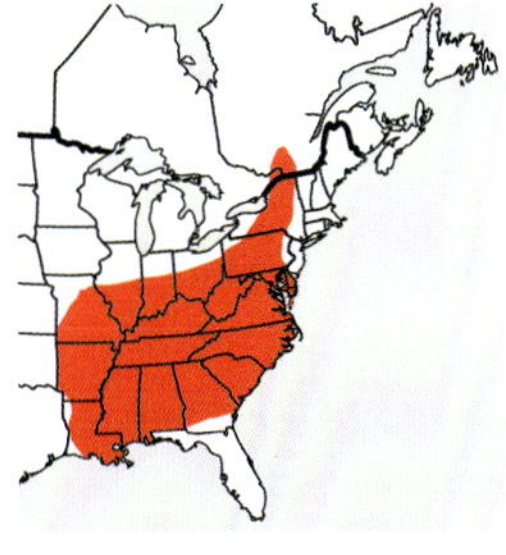

NOTES: *Arthonia quintaria* is one of several members of the genus that occur in the Smokies and has inconspicuous stain-like thalli, irregularly shaped ascomata, and lack a photobiont. Although all of these species typically lack spores,

A. quintaria can be identified by the above features combined with its ecology (see below). It is most likely to be confused with *A. caudata*, but the latter species has more rounded ascomata and always occurs on White Pine. When fertile, *A. quintaria* can easily be recognized by its 6-celled spores, with the uppermost cell enlarged.

CHEMISTRY: No substances. Spot tests. K-, KC-, C-, P-, UV-.

NICHE: This species occurs on young branches and bark of trees not yet colonized by other lichens. Though often overlooked, it is common in the Smokies along forest edges and in the canopy at middle-to-low elevations. *Arthonia quintaria* is widespread in eastern North America, abounding in less pristine habitats.

KEY FEATURES: Non-lichenized thallus, irregularly shaped black ascoma, 6-celled spores with the uppermost cell enlarged, unremarkable.

Arthonia ruana

Corpse Reviver

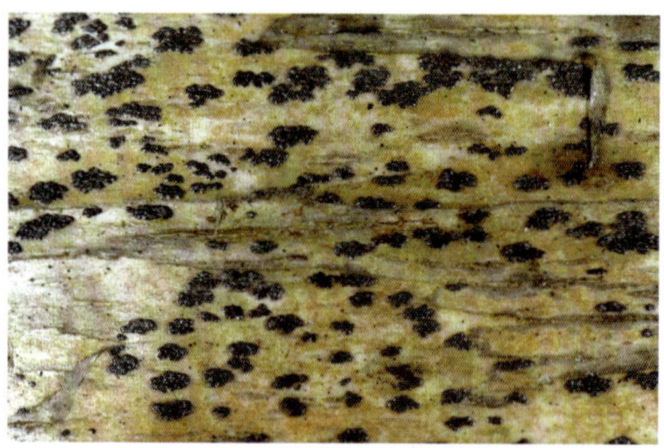

Lendemer 29281 (photo: Tripp)

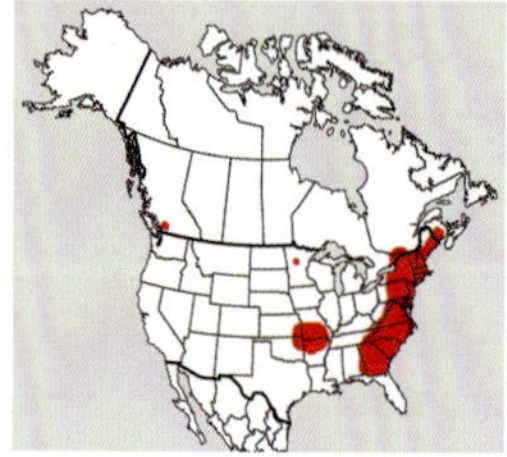

NOTES: *Arthonia ruana* is somewhat inconspicuous but nonetheless an exciting find because it associates with *Trentepohlia*, an alga with chain-forming cells that readily accumulate carotenoids, lending a rusty brown-orange hue to the thallus. *Arthonia ruana* is furthermore recognized by its small, fleck-like black ascomata with colorless muriform spores that turn dirty brown with age, just like the rest of us.

CHEMISTRY: No substances. Spot tests. K-, KC-, C-, P-, UV-.

NICHE: *Arthonia ruana* is often found on the roots and bases of hardwood trees—especially Yellow Birch and Sugar Maple—in humid low-to-middle elevation forests. It is relatively common and widespread in the Smokies as well as throughout the Appalachian Mountains. If one wants to see this species, it can be found readily at The Appalachian Highlands Learning Center (aka, Purchase Knob).

KEY FEATURES: Black, fleck-like ascomata, colorless muriform spores that turn brown with age, on hardwoods, especially roots and tree bases, in humid low-to-middle elevations.

Arthonia rubella

Rosy Contours

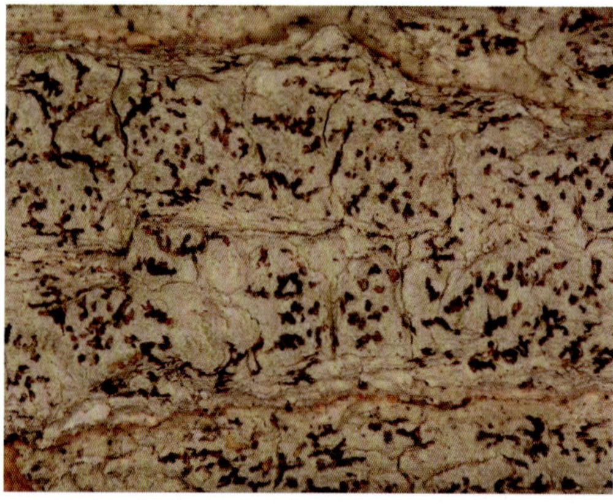

Lendemer 29489 (photo: Lendemer)

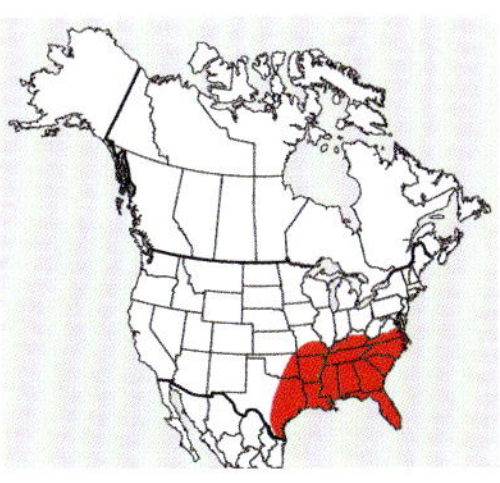

NOTES: Some lichens just have it all figured out. *Arthonia rubella* is a proud member of one of the most remarkable radiations of lichens (and related non-lichenized fungi) on the planet. *Arthonia rubella* specifically can be recognized by its white thallus, its reddish-brown apothecia that are irregularly shaped, its relatively large, 6-celled spores that bear significantly enlarged end cells, and its niche.

CHEMISTRY: No substances. Spot tests. K-, KC-, C-, P-, UV-.

NICHE: In North America, the center of distribution of *Arthonia rubella* is the southeastern United States. Thus, it is right at home with so many other species in this genus that like it low, warm, and wet. Look for it in the Smokies in similarly low, warm, and wet habitats. It grows on a variety of hardwood trees, such as the bark of hickories.

KEY FEATURES: Continuous white thallus, long, reddish-brown, irregularly shaped, lirellate apothecia, large, 6-celled spores with enlarged end cells, bark hardwoods at low elevations.

Arthonia susa

Southern Sentry

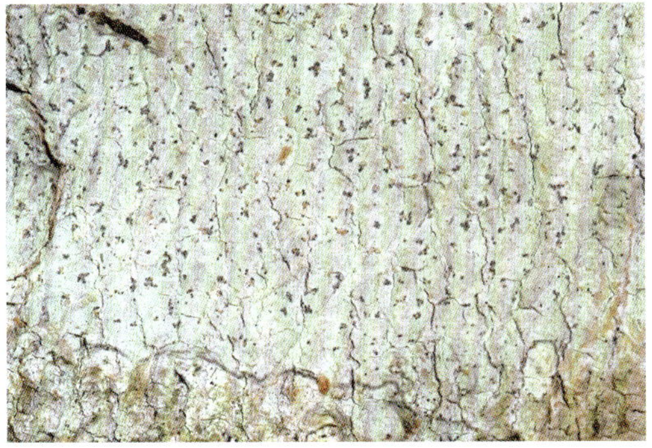

Lendemer 26993 (photo: Tripp)

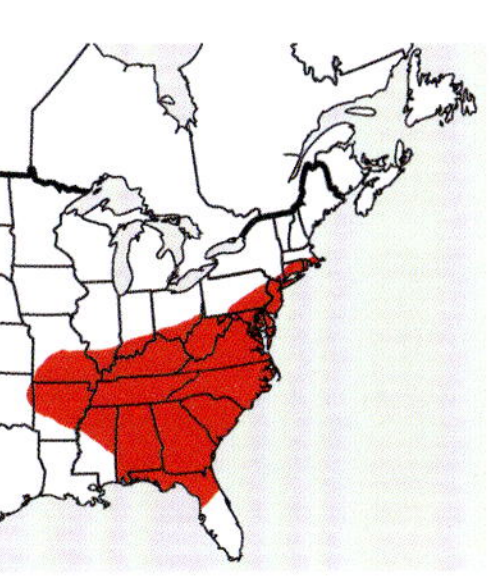

NOTES: *Arthonia susa* is a crustose lichen that, in the Smokies, can be readily recognized even in the field by its dark brownish-black irregularly shaped ascomata and pale gray to blue-gray ecorticate thallus. This species also often has a black prothallus, which is visible in the photograph above. In the laboratory, identification of this species can be confirmed by its green coccoid algal photobiont and by the presence of colorless, muriform spores. *Arthonia quintaria* might be confused with this species, but the former has a stain-like thallus, 6-celled rather than muriform spores, and lacks a photobiont. The epithet "susa" is derived from the acronym for Southeastern United States of America—the best place on Earth to be a lichenologist and eat boiled peanuts.

CHEMISTRY: No substances. Spot tests. K-, KC-, C-, P-, UV-.

NICHE: In the Smokies, *Arthonia susa* occurs on the bark of hardwoods in low-to-middle elevation forests. It is widespread in the National Park and occurs in similar habitats throughout the southeastern United States.

KEY FEATURES: Gray to blue-gray ecorticate thallus, dark brownish-black, irregularly shaped ascomata, muriform colorless spores, green coccoid photobiont, on hardwoods at low-to-middle elevations.

Arthonia vinosa

Auburn Islands

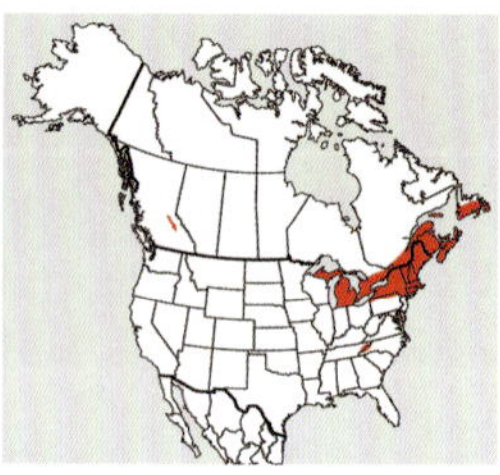

Tripp 4977 (photo: Lendemer)

NOTES: *Arthonia vinosa* is unlikely to be confused with any other species in the Smokies on account of its ecology (see Niche) and its small, fleck-like reddish ascomata that contain K+ purple pigments. *Arthonia helvola* also has reddish apothecia that contain K+ purple pigments, but that species is much smaller and has 3-celled rather than 2-celled spores. See further discussion of comparisons among the red *Arthonia* under *A. cinnabarina*. As is the case for *A. ruana,* the rusty brownish-orange color of the thallus of *A. vinosa* is due to the presence of a *Trentepohlia* photobiont.

CHEMISTRY: Red pigments. Spot tests. (apothecia) K+ purple, KC-, C-, P-, UV-.

NICHE: *Arthonia vinosa* occurs primarily in the eastern portions of the Smokies, where it is restricted to the dry, sheltered bark on the trunks of large Yellow Birch trees in mature high elevation forests. It can be found in similar habitats throughout the southern Appalachians, where it is disjunct from the boreal forests of Canada and New England. This species also occurs Europe.

KEY FEATURES: Reddish fleck-like ascomata with K+ purple pigments, colorless 3-celled spores, on dry, sheltered bark of mature Yellow Birches at high elevations.

Arthopyrenia betulicola

Birch spots

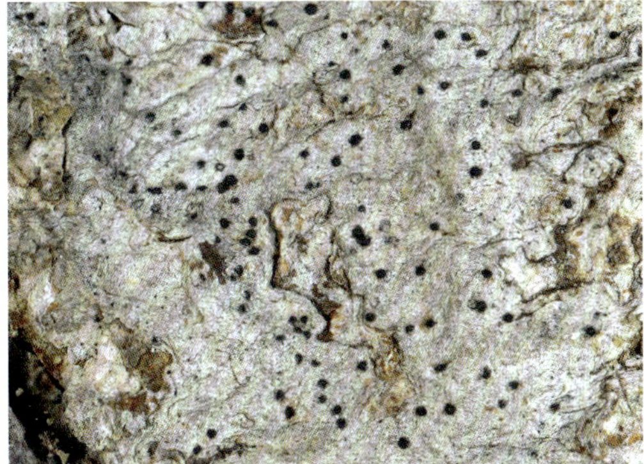

Tripp 3894 (photo: Lendemer)

NOTES: *Arthopyrenia betulicola* is an easy-to-recognize crustose lichen by its very large, black perithecia, its 2-celled colorless spores, and its ecology (see Niche), which is reflected in the specific epithet. *Arthopyrenia cinchonae* is similar but occurs at lower elevations and has wider spores. *Arthopyrenia degelii* is also similar but has much smaller perithecia and occurs only on American Witch Hazel.

CHEMISTRY: No substances. Spot tests. K-, KC-, C-, P-, UV-.

NICHE: *Arthopyrenia betulicola* occurs at middle-to-high elevations in the Smokies, where it is restricted to dry, sheltered bark of old Yellow Birch. Outside the Smokies, it has been found in a handful of nearby, similar habitats. It is a restricted endemic species with limited suitable habitat and deserves immediate protection under the US Endangered Species Program. The eastern slopes of Chiltoes Mountain is an excellent location to see populations of *A. betulicola*.

KEY FEATURES: Very large, black perithecia, 2-celled colorless spores, restricted to bark of mature Yellow Birch, at middle-to-high elevations.

Arthopyrenia cinchonae

Fire Tinder

Lendemer 44644 (photo: Tripp)

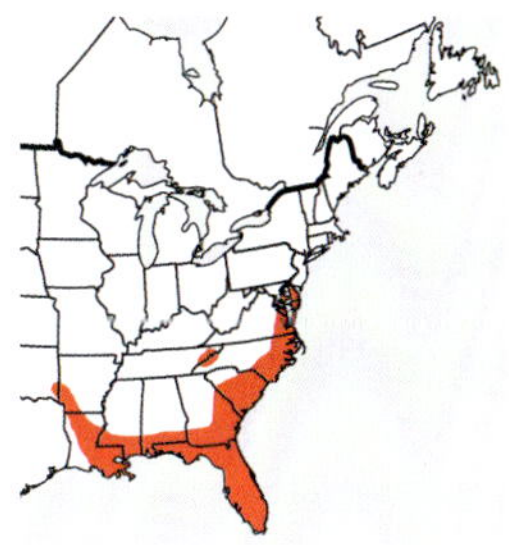

NOTES: *Arthopyrenia cinchonae* is a pyrenofungus characterized by its very large perithecia that contain fat, colorless, 2-celled spores. It is usually non-lichenized but sometimes is found associating with *Trentepohlia*. *Arthopyrenia cinchonae* is easily differentiated from *A. betulicola* (the only other species with very large perithecia in the Smokies) because the latter is always lichenized, has narrower spores, and occurs only on birches at high elevations.

CHEMISTRY: No substances. Spot tests. K-, KC-, C-, P-, UV-.

NICHE: *Arthopyrenia cinchonae* occurs on the bark of hardwoods at low elevations in the Smokies. For better or worse, it is particularly fond of disturbed forests. Hey, somebody's gotta grow there.

KEY FEATURES: White mycelial thallus, usually non-lichenized, fat, colorless, 2-celled spores, at low elevations in disturbed forests on the bark of hardwoods.

Arthopyrenia degelii

A Hopeful Encounter

Tripp 3959 (photo: Lendemer)

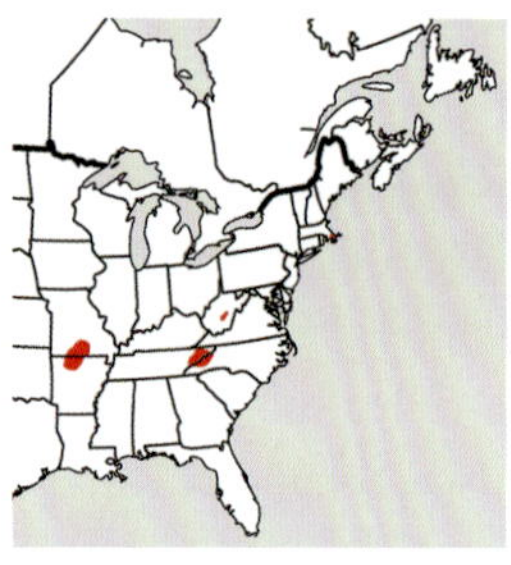

NOTES: *Arthopyrenia degelii* is a fine example of ecological specialization in lichens. It is foremost distinctive by this specialization, described under Niche below. Additionally, it is marked by a brown thallus and small black perithecia. *Arthopyrenia degelii* differs from *A. betulicola* and *A. cinchonae* by features described under the former. This is one of hundreds of crusts in the Smokies that most visitors never encounter unless they know what to look for. The epithet honors Gunnar Degelius, first lichenologist to work extensively in the Smokies.

CHEMISTRY: No substances. Spot tests. K-, KC-, C-, P-, UV-.

NICHE: This species occurs exclusively on branches and stems of American Witch Hazel. It is relatively rare in the Smokies and, interestingly, completely absent from most of the range of its host. *In situ* speciation in the Smokies? We wouldn't be surprised if this were the rule rather than the exception. *Arthopyrenia degelii* can be seen on lower portions of LeConte Creek and at the Park Headquarters at Cherokee Orchard, where it was first collected in the 1930's.

KEY FEATURES: Small black perithecia, colorless 2-celled spores, restricted to branches and stems of American Witch Hazel.

Arthothelium spectabile

Glory, Glory

Lendemer 32923 (photo: Tripp)

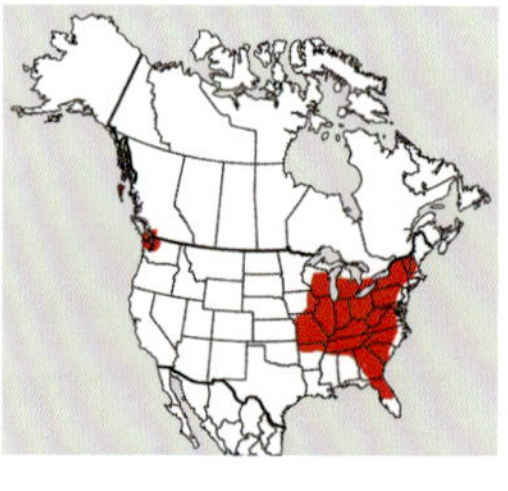

NOTES: *Arthothelium spectabile* is, as the specific epithet implies, a spectacular crustose lichen. It can be distinguished in the field from all relatives (*Arthonia*, *Arthothelium*) in the Smokies by its thick, chalky white thallus with large, dark splotchy ascomata. The identification of *A. spectabile* can further be confirmed in the laboratory by its colorless muriform spores. *Ar-*

thonia interveniens has a similar gestalt but is never quite as dramatic in appearance and has differently shaped spores that resemble bent loafers. *Arthonia ruana* has muriform spores like *A. spectabile*, but in the former they are smaller; thalli of *A. ruana* are also consistently inconspicuous.

CHEMISTRY: No substances. Spot tests. K-, KC-, C-, P-, UV-.

NICHE: *Arthothelium spectabile* occurs throughout middle and low elevations of the Smokies on hardwoods with plated bark such as buckeye and hickory. It is widespread in temperate eastern North America, with disjunct populations in Europe and the Pacific Northwest.

Key Features: Thick white thallus, large, black, fleck-like ascomata, colorless muriform spores, on hardwoods with plated bark.

Aspicilia caesiocinerea

Gray Goo

Tripp 2404 (photo: Lendemer)

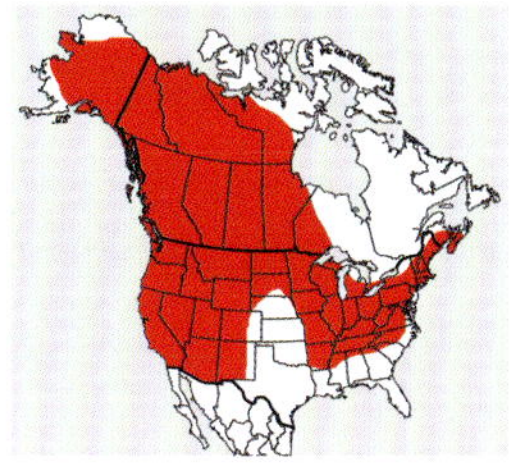

NOTES: Members of the genus *Aspicilia* are usually recognized by their gray-to-brown areolate thalli, immersed lecanorine apothecia, relatively large colorless, simple spores, and occurrence on non-calcareous rocks. *Aspicilia caesiocinerea* can be separated from other species that occur in the Smokies by the absence of secondary compounds. *Aspicilia laevata* looks similar to *A. caesiocinerea*, but has stictic acid in the thallus (and reacts P+ orange). Like all North American species of *Aspicilia*, *A. caesiocinerea* is poorly understood taxonomically, ecologically, and evolutionarily. It needs study.

CHEMISTRY: No substances. Spot tests. K-, KC-, C-, P-, UV-.

NICHE: *Aspicilia caesiocinerea* occurs on shaded non-calcareous rocks at middle and low elevations in the Smokies, particularly along waterways. It is infrequent but widespread throughout North America and, nearly everywhere it occurs, it grows on non-calcareous rock outcrops.

KEY FEATURES: Dark gray areolate thallus, immersed lecanorine apothecia with dark blue-gray discs, negative spot tests, on non-calcareous rocks.

Aspicilia cinerea

Stool Pigeon

Tripp 3947 (photo: Lendemer)

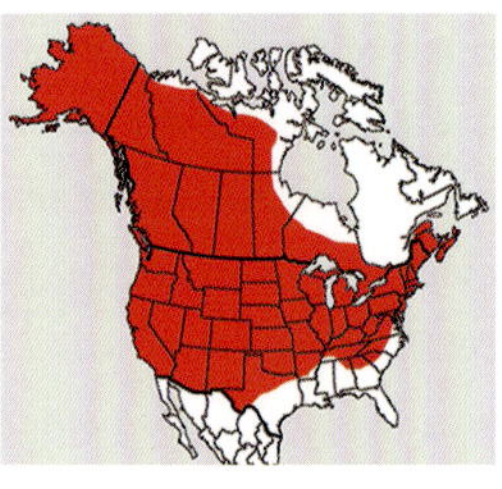

NOTES: As is the case for most members of the genus, *Aspicilia cinerea* is poorly understood in North America. Here we use the name for a species that has distinctive white-to-light-gray areolate thalli that contain norstictic acid (K+ yellow to red). Other crustose lichens with white areolate thalli such as *Lecanora oreinoides* and *Lecidea tessellata* might be confused with *A. cinerea,* but can easily be distinguished by having different secondary metabolites that do not react K+ yellow to red. Our best advice is to learn the genus first, then try to make sense of the species, second.

CHEMISTRY: Norstictic acid. Spot tests. K+ yellow turning red, KC-, C-, P+ yellow, UV-.

NICHE: This species occurs on shaded and exposed non-calcareous rocks at middle and low elevations in the Smokies and beyond.

KEY FEATURES: White-to-light-gray areolate thallus, immersed lecanorine apothecia with dark blue-gray discs, K+ yellow turning red thalli, on non-calcareous rocks.

Aspicilia laevata

Overcast Sky Lichen

Tripp 3706 (photo: Lendemer)

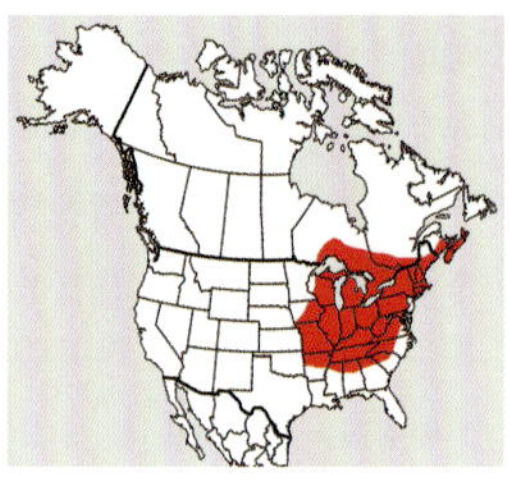

NOTES: *Aspicilia laevata* is superficially similar to *A. caesiocinerea* but can be distinguished by the presence of stictic acid in the thallus. It might also be confused with *A. cinerea*, but that species has a white-to-light-gray thallus and produces norstictic acid instead of stictic acid. Members of *Aspicilia* that occur in the Smokies are poorly understood and overlap ecologically.

CHEMISTRY: Stictic acid. Spot tests. K+ dirty yellow-brown, KC-, C-, P+ orange, UV-.

NICHE: *Aspicilia laevata* occurs on shaded and exposed non-calcareous rocks at middle and low elevations in the Smokies. Using the taxonomic concept we employ here, it is common and widespread in eastern North America.

KEY FEATURES: Light to dark gray areolate thallus, immersed lecanorine apothecia with dark blue-gray discs, K+ yellow and P+ orange thallus, on non-calcareous rocks.

Aulaxina quadrangula

Peppered Eggs

Tripp 53158 (photo: Tripp)

NOTES: Not very many leaf-dwelling lichens occur in the Smokies, and few of those are likely to be confused with *Aulaxina quadrangula*. The species can easily be recognized by its white, glossy thalli, which are composed of closely aggregated islands, each of which has a single, dark apothecium immersed near the center. Close examination with a compound microscope will reveal muriform, colorless spores. The apothecia are small and black, and thus can easily be mistaken for perithecia. Species of *Tricharia* or *Gyalideopsis* might be confused with *A. quadrangula*, but those all have specialized hair-like structures that are absent in *Aulaxina*. Sterile thalli could be confused with *Gyalectidium appendiculatum*, however the thallus in that species is dispersed rather than aggregated, and has minute, white hairs on the upper surface.

CHEMISTRY: No substances. Spot tests. K-, KC-, C-, P-, UV-.

NICHE: This is a foliicolous species that is found on the evergreen leaves of plants such as *Leucothoe* or *Rhododendron* in the Smokies, especially along streams and near waterfalls. It is rare in the southern Appalachians, where it is disjunct from the main portion of its range in the southeastern Coastal Plain and elsewhere in the tropics.

KEY FEATURES: Foliicolous, white, island-like thalli each with a single immersed black apothecium, muriform, colorless spores.

Bacidia apiahica

Tropical Emollient

Lendemer 53138 (photo: Tripp)

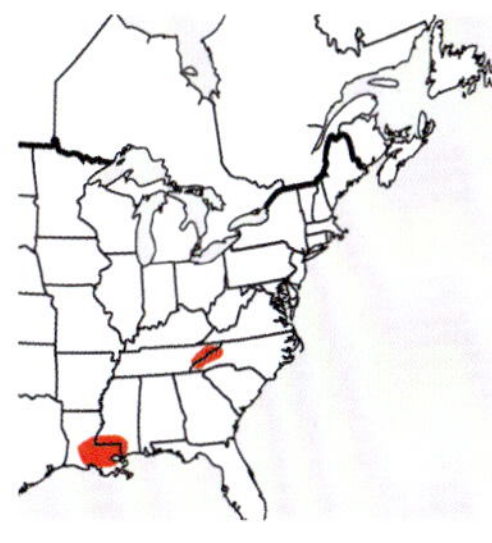

NOTES: This lichen should be added to your checklist of "easy," given it is one of the few foliicolous species present in the Park. It also has some pretty sweet needle-like spores. Together, those two traits more or less land you on this species, and nothing else. You might be tempted to confuse *Bacidia apiahica* with *Fellhanera bouteillei*, which similarly occurs on *Rhododendron* in suitable habitat (i.e., wet places), but that species has pale apothecia with 2-celled, not-at-all needle-like spores. *Gyalectidium appendiculatum* has a similar ecology, but that species bears hyphophores.

CHEMISTRY: No substances. Spot tests. K-, KC-, C-, P-, UV-.

NICHE: This species likes it wet. First, consider your conditions. On top of a grassy bald? Enjoy the view instead. Botanizing down in the laurel thickets of Mingus Mill, or the cloud-soaked ridges of Mingus Lead? Start by looking closely at the leaves of *Rhododendron maximum*. See some goop on them? Something that perhaps resembles a disease? Grab your hand lens and look for pale-to-dark apothecia (depending on state of hydration) surrounded by pale, small, irregular thalli. Elsewhere, this species is known from tropical latitudes on leaves of other plants.

KEY FEATURES: Foliicolous, pale thalli that bear pale-to-dark apothecia or pycnidia, long and skinny needle-like spores, primarily on *Rhododendron* leaves in wet places.

Bacidia coprodes

Lost for Words

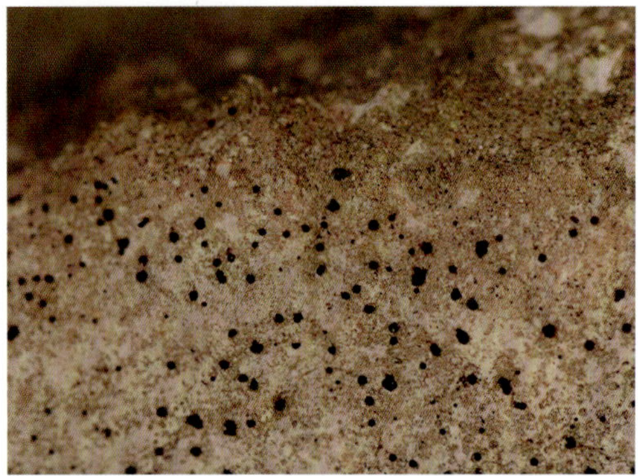

Lendemer 44565 (photo: Tripp)

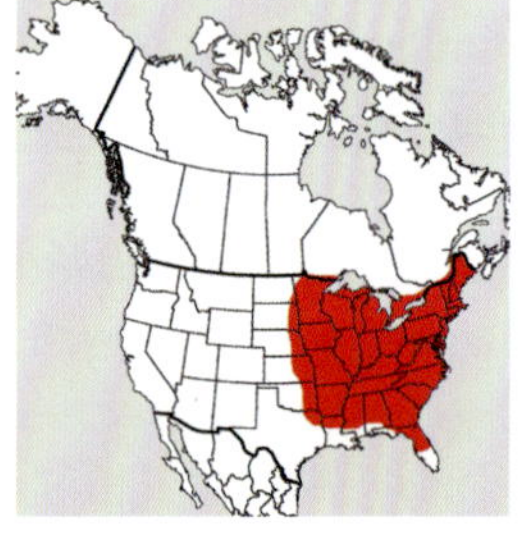

NOTES: *Bacidia coprodes* is an inconspicuous species characterized by its thin, minutely areolate greenish thallus. It is most readily recognized by its ecology (see Niche). Be sure to check out its short, 4-celled spores, atypical of the genus. *Bacidia coprodes* is not a species that you will likely encounter unless you opt to conduct a taxonomic revision of the genus for your next career move.

CHEMISTRY: No substances. Spot tests. K-, KC-, C-, P-, UV-.

NICHE: *Bacidia coprodes* occurs throughout eastern North America in the Piedmont and mountains, always on limestone or derivations thereof (e.g., concrete walls, grout).

KEY FEATURES: Minutely areolate thallus, needle-shaped spores, on calcareous rocks or rock-like surfaces (e.g., concrete).

Bacidia diffracta

Grainy Spindle-spores

Buck 56395 (photo: Lendemer)

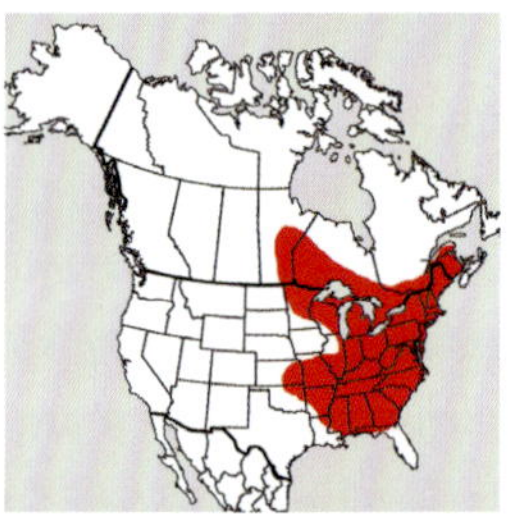

NOTES: Several species of *Bacidia* in the Smokies have frequently pruinose, reddish-brown or brown apothecia. Among these, *Bacidia diffracta* can be recognized by its granular thallus and brown hypothecium and exciple that are K+ rose. The most similar species is *B. polychroa*, which also has a K+ rose hypothecium and exciple, but differs its continuous and smooth (vs. granular) thallus. *Bacidia suffusa* is also similar, but that species has a continuous, smooth thallus, and the brown pigments are K- rather than K+ rose. All of these species appear to be quite rare and tend to be found only in mature hardwood forests in eastern North America.

CHEMISTRY: Atranorin. Spot tests: K- or K+ weak yellow, C-, KC-, P-, UV-.

NICHE: A rare species of mature hardwood forests, *Bacidia diffracta* is found throughout the Smokies at middle and low elevations. It is similarly widespread, but rare, in much of temperate eastern North America.

KEY FEATURES: Granular crustose thallus, brown hypothecium and exciple that are K+ rose, needle-shaped spores, frequently pruinose apothecia, on bark at middle and low elevations.

Bacidia helicospora

Helical and Humid

Lendemer 46017A (photo: Lendemer)

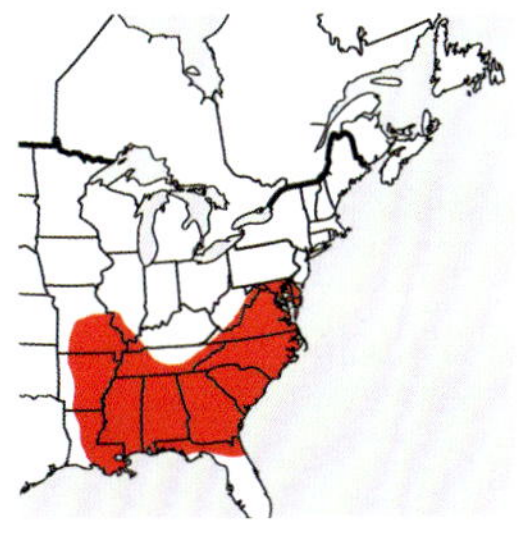

NOTES: This is an easily overlooked species that can be difficult to spot in the forest, let alone to identify. It can be recognized by its small, dark reddish-brown-to-black apothecia, colorless-to-light-yellow hypothecium, and epihymenium that contains clumps of gray-brown pigment that react K+ red-purple. Although the epithet hints that the spores are helically arranged (and these often released as a packet), this does not always seem to be the case in fresh material. In a key, the most similar species is *B. heterochroa*, which differs in having a coastal distribution, larger apothecia, and pigmented paraphyses caps. *Bacidia purpurans*, which is not yet known from the Smokies, has larger apothecia and a reddish-brown hypothecium.

CHEMISTRY: No substances. Spot tests. K-, KC-, C-, P-, UV-.

NICHE: *Bacidia helicospora* is widespread in the Smokies and throughout southeastern North America, where it is found on the bark and branches of trees, particularly hardwoods, at middle and low elevations.

KEY FEATURES: Smooth crustose thallus, inconspicuous small reddish-brown to black apothecia, needle-like spores, colorless hypothecium, K+red-purple epihymenium, on bark and branches at middle and low elevations.

Bacidia lobarica

Bright Balls

Lendemer 44524 (photo: Tripp)

NOTES: *Bacidia lobarica* is a very rare species that is known from <10 collections. It is characterized by superficially appearing sorediate, but the thallus is actually composed of minute areoles, these pale green to dirty yellow in color. We consistently find this species producing pale, biatorine apothecia. It is most likely to be confused with *Scoliciosporum pensylvanicum*, but the latter differs by having a truly leprose thallus and shorter spores.

CHEMISTRY: Lobaric acid. Spot tests. K-, KC+ fleeting pink, C-, P-, UV+ blue-white.

NICHE: *Bacidia lobarica* is restricted to the southern Appalachian Mountains of Tennessee and North Carolina, with one collection from the adjacent Piedmont of North Carolina. It has been collected at low elevations on the bark of hardwoods. Look for it on the Indian Creek Trail near Sunkota Ridge.

KEY FEATURES: Minute, light-green-to-yellow areoles that resemble soredia, pale, biatorine apothecia, lobaric acid, low elevations on the bark of hardwoods.

Bacidia polychroa

Chilled Rose

Lendemer 29486 (photo: Lendemer)

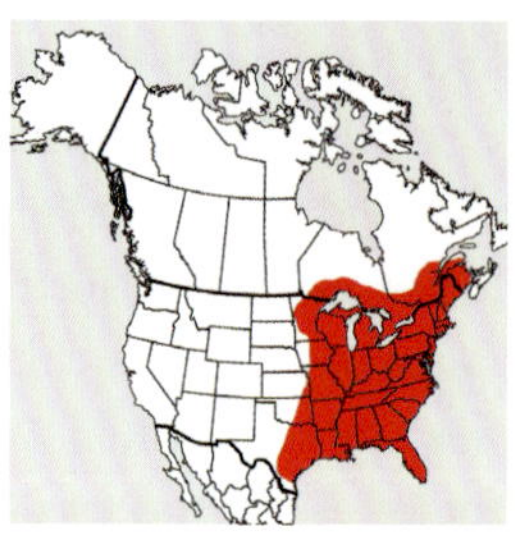

NOTES: This is another one of the uncommon *Bacidia* species that occurs in the Smokies. It can be recognized by its smooth, continuous thallus, frequently pruinose reddish-brown apothecia and brown pigmented hypothecium and exciple that react K+ rose. The most similar species is *B. diffracta*, which differs in having a granular thallus. Both species can grow together in close proximity, and, in such cases, the thallus has a surprisingly consistent and distinctive character. *Bacidia suffusa* is also similar in having frequently pruinose apothecia and a smooth thallus, but in that species the brown apothecia pigments are not K+ rose. Want to know more about *Bacidia*? It is a fascinating and beautiful genus, and one of the few with a detailed monograph in North America.

CHEMISTRY: Atranorin or no substances. Spot tests: K- or K+ weak yellow, C-, KC-, P-, UV-.

NICHE: Much like the species with which it would be confused, *B. poylchroa* is widespread throughout temperate eastern North America, where it tends to occur in mature hardwood forests. In the Smokies, and southern Appalachians generally, it is rare and found primarily at middle and low elevations.

KEY FEATURE: Smooth, continuous crustose thallus, frequently pruinose reddish-brown apothecia, needle-like spores, brown hypothecium that is K+ rose, on bark at middle and low elevations.

Bacidia schweinitzii

Champion Crust of Eastern North America

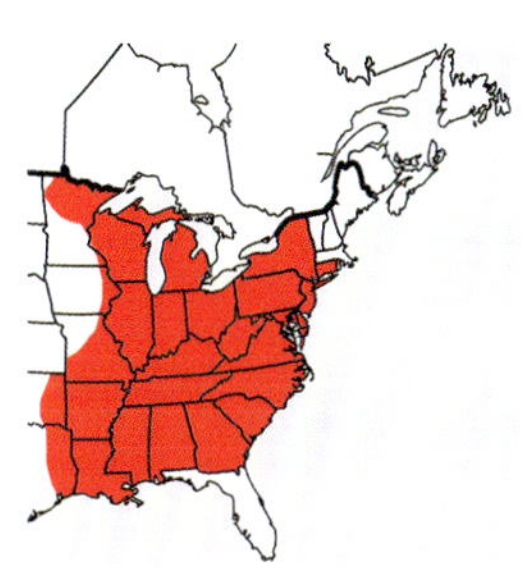

Lendemer 33319 (photo: Tripp)

NOTES: *Bacidia schweinitzii* should be the first crustose lichen you learn in the Smokies (or elsewhere east of the Mississippi River). It is absolutely one of the most abundant and easily recognizable crusts, owing to its thick green thallus, usually bordered by a conspicuous white a prothallus, its large, shiny black apothecia, and its needle-shaped, colorless spores. Occasionally, thalli are found without apothecia, but in these cases, you can still recognize it by its black pycnidia (see the small black dots in the center of the photograph).

CHEMISTRY: Traces of atranorin. Spot tests. K-, KC-, C-, P-, UV-.

NICHE: In the Smokies, this species is both widespread and common at middle-to-low elevations on the bark of hardwoods and occasionally conifers. It occurs in similar habitats throughout temperate eastern North America, but avoids Florida for some reason.

KEY FEATURES: Thick green thallus, large shiny black apothecia, needle-shaped spores, extraordinarily common and widespread on bark. This one is easy, go with your first impression!

Bacidia suffusa

Needle Knows

Tripp 2592 (photo: Lendemer)

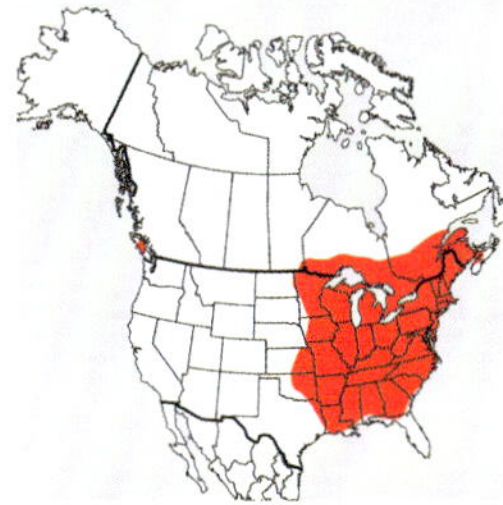

NOTES: We've reached the end of *Bacidia*, and at this point you are probably ready to go on a hunt and try to catch them all. Well, your life list wouldn't be complete without this one. *Bacidia suffusa* is similar to *B. diffracta* and *B. polychroa* in having reddish-brown-to-brown apothecia that are frequently pruinose. *Bacidia diffracta* differs in having a granular (vs. smooth) thallus and a hypothecium that reacts K+ rose (vs. K-). *Bacidia polychroa* is very similar to *B. suffusa* in having a smooth thallus and brown hypothecium. However, like *B. diffracta*, *B. polychroa* differs *from B. suffusa* in having a K+ rose hypothecium.

CHEMISTRY: Atranorin. Spot tests: K- or K+ weak yellow, C-, KC-, P-, UV-.

NICHE: A rare species of mature hardwood forests, *Bacidia suffusa* is uncommon but widespread at middle and low elevations of the southern Appalachians including the Smokies. It is similarly distributed elsewhere in temperate eastern North America.

KEY FEATURES: Smooth, continuous crustose thallus, frequently pruinose reddish-brown apothecia, needle-like spores, brown hypothecium that is K-, on bark at middle and low elevations.

Bacidina delicata

A Trivial Pursuit

Lendemer 44761 (photo: Tripp)

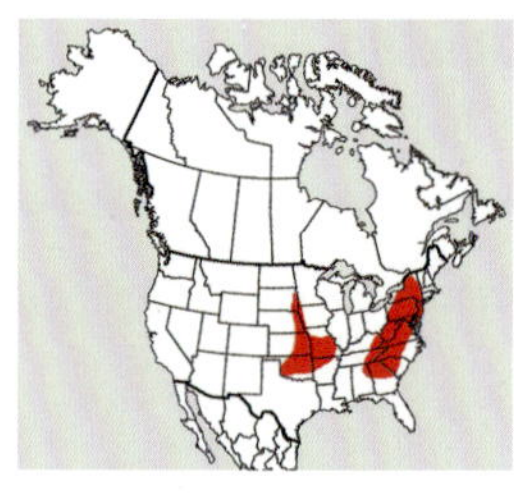

NOTES: *Bacidina delicata* is a relatively common yet often unnoticed and thus under-collected lichen. This species is characterized by a thallus consisting mostly of photobiont cells that only loosely associate with fungal hyphae (i.e., goniocysts), by its unpigmented apothecia and pycnidia (the former much rarer than the latter), and its needle-shaped spores. We don't have anything else to say about *Bacidina delicata* other than that the name is rather euphonious, isn't it?

CHEMISTRY: No substances. Spot tests. K-, KC-, C-, P-, UV-.

NICHE: This species can be found on dark, dank rocks as well as on tree bases in humid habitats at middle elevations. It is a typical element of the southern Appalachian lichen biota and also occurs in the Ozarks and surrounding areas.

KEY FEATURES: Green thallus of goniocysts, unpigmented pycnidia (less commonly with apothecia), needle-shaped spores. Look for green scuz on wet rocks.

Baeomyces rufus

Freedom Doofs

Tripp 3501 (photo: Lendemer)

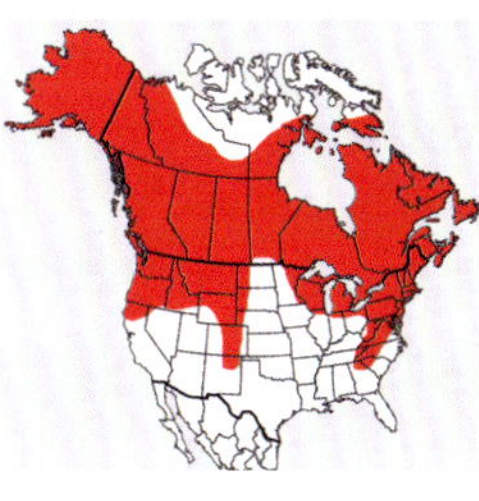

NOTES: *Baeomyces rufus* is one of several species in the Smokies with fleshy-pink-to-brown apothecia that are raised on conspicuous stalks. This species is, however, actually unrelated to others of this stalky sort and is distinctive by its greenish-to-brown thallus and occurrence on rocks. *Dibaeis baeomyces* could be confused with *B. rufus*, but can be distinguished by occurring on soil, having a thick, sordid-to-bone-white thallus, and by having bright pink apothecia on more slender stalks.
CHEMISTRY: Stictic acid. Spot tests. K+ dirty yellow-brown, KC-, C-, P+ orange, UV-.
NICHE: This species is commonly encountered on wet or damp non-calcareous rocks at all elevations throughout the Smokies. It is particularly common in eastern portions of the National Park. Cataloochee and Andrews Bald are two of many locations where *B. rufus* can be seen. Outside of the Smokies, it is widespread in the Appalachians and throughout the boreal forests of the Northern Hemisphere.
KEY FEATURES: Green-to-brown thallus, fleshy pink to brown apothecia raised on stalks, on non-calcareous rocks, widespread and common.

Baglietto baldensis

Rock Pits

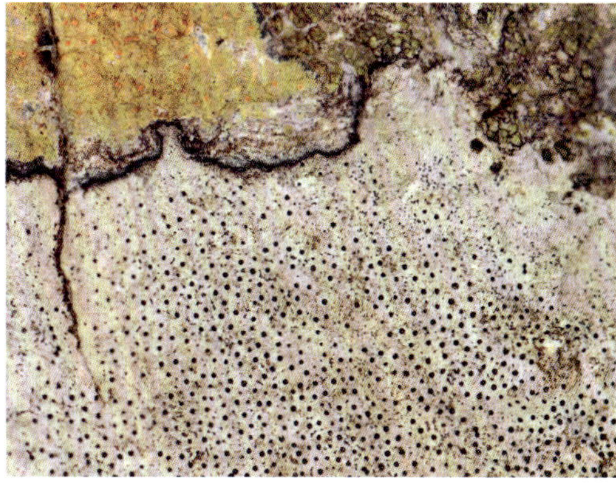

Lendemer 44544 (photo: Tripp)

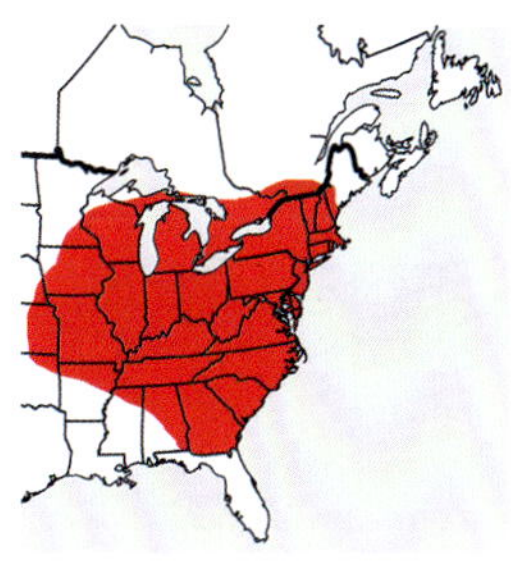

NOTES: *Baglietto baldensis!* We really like this species. It is easily recognized by its black perithecia, which are immersed in pits and often buried in (almost as if etching out) the substrate, which is invariably limestone. Curiously, the fruiting bodies are rarely fertile. Instead of containing spores, ample paraphyses may instead serve as reproductive propagules and fodder for imaginative people. The thallus of *B. baldensis* is usually dirty white in appearance.
CHEMISTRY: No substances. Spot tests. K-, KC-, C-, P-, UV-.
NICHE: This species is common throughout eastern North America in habitats with ample exposed limestone. We first reported it as new to the Smokies in 2013, and its infrequency in the Park is not surprising owing to a paucity of limestone. That being said, be sure to spend time lichenizing at White Oak Sinks and along Rich Mountain Trail near Ace Gap where *B.*

baldensis is common on exposed calcareous rocks. Look for it growing with *Caloplaca flavovirescens* on calcareous rocks.

KEY FEATURES: Sordid white thallus with small black (evenly spaced) perithecia immersed in pits and appearing to etch out rock, these usually sterile, invariably on exposed limestone, throughout eastern North America.

Bathelium carolinianum

Yellow Congregation

Tripp 2649 (photo: Lendemer)

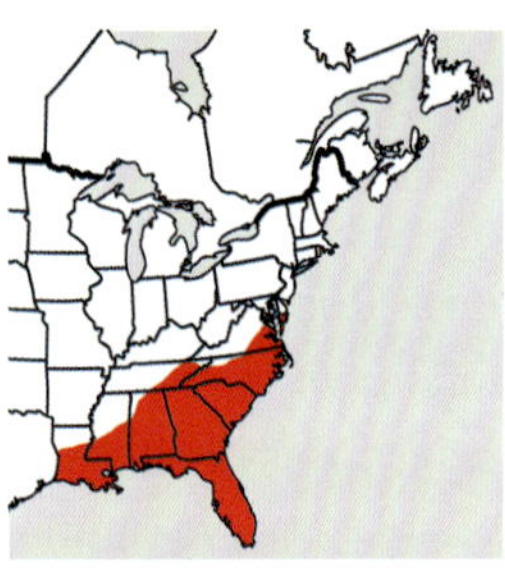

NOTES: Tropical lichens in the Smokies! *Bathelium carolinianum* is distinctive by its dark brown, clustered perithecia that are arranged in pseudostroma and filled with a yellow pigment visible in cross section under magnification. This species loves warm, sunny places everywhere, including the Smokies. *Bathelium carolinianum* is only confusable with *Trypethelium virens*, which differs in having pseudostroma that are concolorous with the thallus and lacking the distinctive yellow pigment.

CHEMISTRY: Unknown yellow pigment in the pseudostroma. Spot tests. K-, C-, KC-, P-, UV.

NICHE: You don't have to travel to South Florida to see this tropical lichen. *Bathelium carolinianum* belongs to a special club of extra-tropical lichens that, in North America, makes its home in warm areas of the southeastern Coastal Plain and make its way inland into the lowest parts of the mountains. In the Smokies, you can find large populations of *B. carolinianum* in southern portions of the Park such as along the Lakeshore Trail.

KEY FEATURES: Greenish-brown thallus, clustered, dark brown perithecia arranged in pseudostroma filled with yellow pigment, crustose on smooth-barked hardwoods such as American Holly.

Biatora appalachensis

An Asexual Dilemma

Tripp 3806 (photo: Lendemer)

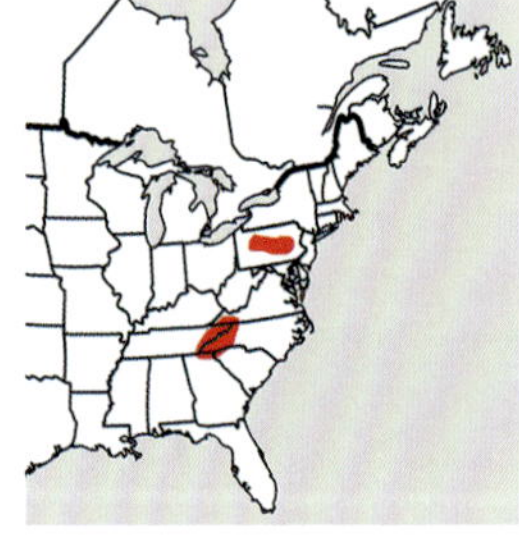

NOTES: *Biatora appalachensis* is one of several common, sterile sorediate crusts (affectionately: "SSCs") in the Smokies that make this lichen biota one of the most special, albeit challenging to novices. This species can be recognized by its lime-green soralia, areolate thallus (sometimes poorly developed), and chemistry. It is most likely to be confused with *B. printzenii*, which differs chemically (see that species, which is P+ orange-red). *Biatora chrysantha* is chemically identical to *B. appalachensis* but occurs on bryophytes and humus over rocks and has a thick, well-developed dirty-green thallus.

CHEMISTRY: Gyrophoric acid. Spot tests. K-, C+ pink, KC+ pink, P-, UV-.

NICHE: *Biatora appalachensis* is widespread in the Smokies, but only at high elevations on the bark and branches of hardwoods and conifers. Disjunct populations have been collected in Pennsylvania. We expect but have not yet confirmed that this species additionally occurs at high elevations in the central Appalachians.

KEY FEATURES: Lime-green soralia that are C+ pink and P-, weakly areolate thin thallus, on bark and branches of hardwoods and conifers at high elevations.

Biatora chrysantha

Moss Tops

Tripp 4985 (photo: Lendemer)

NOTES: *Biatora chrysantha* is chemically identical to *B. appalachensis* but occurs on bryophytes and humus over mosses and rocks instead of on the bark and branches of trees. *Biatora chrysantha* is characterized especially by its thick, well-developed seafoam-green thallus with conspicuous soralia. This species is differentiable from the similar *B. printzenii*, as the latter is P+ orange-red.

CHEMISTRY: Gyrophoric acid. Spot tests. K-, C+ pink, KC+ pink, P-, UV-.

NICHE: *Biatora chrysantha* occurs at middle-to-high elevations in the Smokies where it can form extensive colonies on bryophytes or other organic matter, especially over seepy rock faces.

KEY FEATURES: Dirty green, well-developed thallus and soralia that are C+ pink and P-, on bryophytes and organic matter over wet rocks at middle-to-high elevations.

Biatora longispora

Orange Crush

Tripp 4947 (photo: Lendemer)

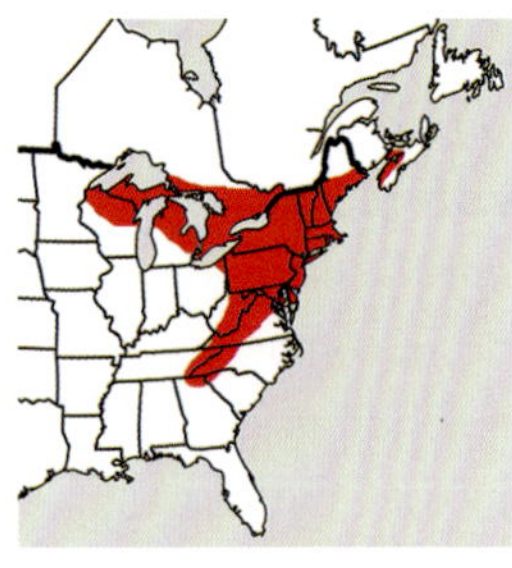

NOTES: Here is another readily identifiable and accessible crustose lichen to all natural history enthusiasts of the Smokies. *Biatora longispora* can be recognized by its relatively large, orange-to-orange-brown biatorine apothecia set against a thin, pale green thallus. In the photo, you can also see a thin white prothallus that underlays the green, primary thallus, which is typical of this species. *Biatora pycnidiata* is most similar but differs in having P+ red apothecia and shorter spores. These two species, together with *B. vernalis*, represent three sexually reproducing members of *Biatora* thus far known in the Smokies.

CHEMISTRY: No substances. Spot tests. K-, C-, KC-, P-, UV-.

NICHE: *Biatora longispora* is a common species on the bark of hardwoods at middle-to-low elevations in the Smokies and elsewhere throughout the Appalachians. This species has a classic Appalachian–Great Lakes biogeographical distribution.

KEY FEATURES: Orange-to-orange-brown biatorine apothecia, thin, pale green thalli, P- apothecia, on bark of hardwoods at middle-to-low elevations, lovely.

Biatora pontica

Lemon Lime Lichen

Tripp 4930 (photo: Lendemer)

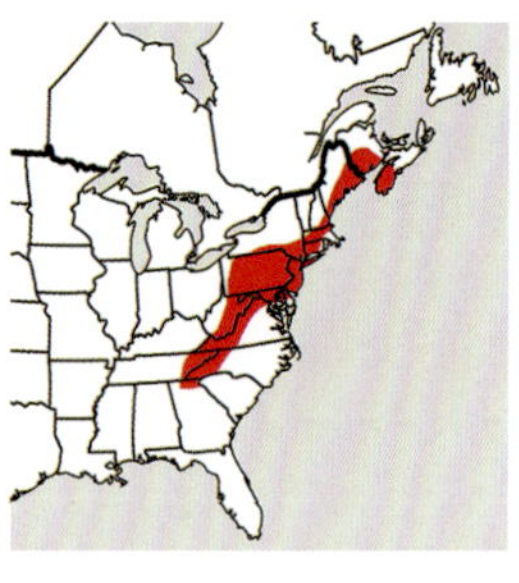

NOTES: *Biatora pontica* may cause a bit of grief because of its similarity to so many other sterile sorediate crusts in the Smokies. On a good day, it can be recognized by its very small, green soralia that react C+ orange and UV+ orange (double whew!) due to its unique chemistry: *Biatora pontica* is one of several sorediate species in the genus in the Smokies, but all of the others are C+ pink.

CHEMISTRY: Xanthones. Spot tests. K-, C+ orange, KC+ orange, P-, UV+ dull orange.

NICHE: This species is infrequent at middle-to-high elevations in the Smokies on hardwoods, especially on Yellow Birch (as in the photo). It occurs in similar habitats throughout the Appalachians.

KEY FEATURES: Minute, yellowish-green soralia that are C+ orange and UV+ orange, occurrence on hardwoods at middle-to-high elevations throughout the Smokies.

Biatora printzenii

Printzen's Lichen

Lendemer 33064 (photo: Tripp)

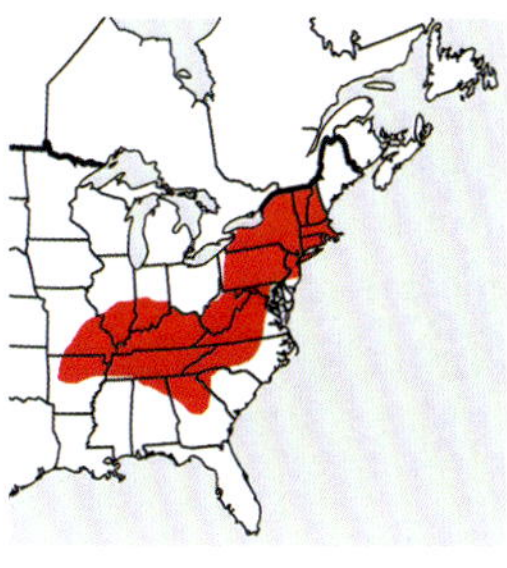

NOTES: *Biatora printzenii* is one of several asexual members of the genus in the Smokies that reproduce via soredia arranged in prominent soralia. It is easily differentiated by its chemistry, which yields a C+ pink and P+ orange reaction in the soralia. *Biatora appalachensis* has a P- thallus, and *B. pontica* has a C+ orange thallus. This species was named in honor of Christian Printzen, German lichenologist and specialist in the taxonomy of *Biatora* (die winzigen Flechten!).

CHEMISTRY: Gyrophoric acid, argopsin. Spot tests. K-, C+ pink, KC+ pink, P+ orange-red, UV-.

NICHE: *Biatora printzenii* is widespread in the Smokies, where it occurs on the bark of hardwoods and conifers at nearly all elevations. It is similarly widespread across eastern temperate broadleaf forests. We think Lucy Braun would approve.

KEY FEATURES: Lime-green soralia that are C+ pink and P+ orange-red, areolate thallus, on bark and branches of hardwoods and conifers at all elevations.

Biatora pycnidiata

P+ Orange Crust

Tripp 3889 (photo: Lendemer)

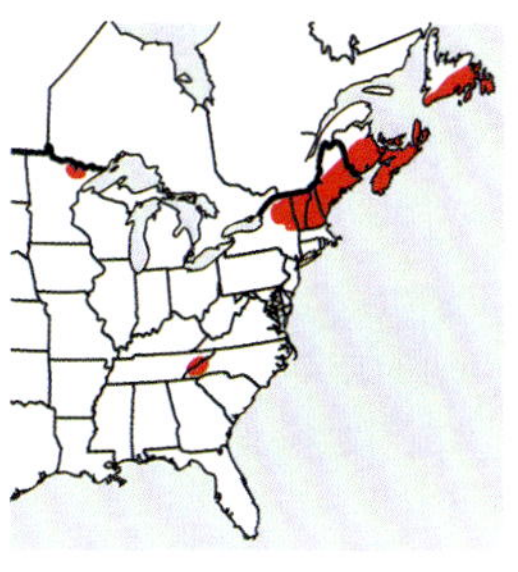

NOTES: *Biatora pycnidiata* is outwardly very similar to *B. longispora* but differs in having argopsin (and reacts P+ orange-red) and shorter spores. It also differs in its distribution (see below). *Biatora pycnidiata* is also similar to *B. vernalis*, but the latter species occurs on bryophytes and other organic matter and has much paler, flesh-colored apothecia. This photo shows the variation in color of the apothecia (e.g., pale orange to brown), typical of the species. Despite being common in habitats where it occurs, *B. pycnidiata* was described only recently.

CHEMISTRY: Argopsin. Spot tests. K-, C-, KC-, P+ orange-red, UV-.

NICHE: This species is abundant on the bark and branches of conifers and hardwoods at high elevations in the Smokies. It occurs in similar habitats throughout the southern Appalachians, with disjunct populations in spruce forests in the central and northern Appalachians as well as Great Lakes.

KEY FEATURES: Orange-to-orange-brown to reddish-brown biatorine apothecia, thin, dark green thallus, P+ orange-red apothecia, on bark and branches at high elevations.

Biatora vernalis

Spring Green

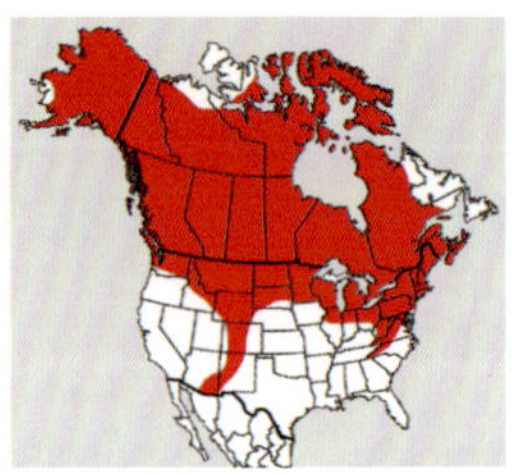

Tripp 2523 (photo: Tripp)

NOTES: *Biatora vernalis* is one three esorediate members of the genus in the Smokies that together are readily identified by their biatorine apothecia and mostly simple, ellipsoid, colorless spores. These three species are typically found abundantly fertile in the Smokies and share thick, scum-like, pale-green thalli and fleshy orange-to-brown-colored apothecia. *Biatora vernalis* is distinguished from the other two species by its spore size and ecology (see Niche).

CHEMISTRY: No substances. Spot tests. K-, C-, KC-, P-, UV-.

NICHE: *Biatora vernalis* grows exclusively on bryophytes and other organic matter at middle-to-high elevations in the Smokies, where it is relatively uncommon. It occurs in similar habitats throughout the Appalachians and is much more common in arctic boreal forests.

KEY FEATURES: Light-to-dark-flesh-colored apothecia, thick, green thallus, P- apothecia, on bryophytes and organic matter at middle-to-high elevations.

Biatoropsis usnearum

Stuck on U(snea)

Tripp 4962 (photo: Lendemer)

NOTES: This is a lichen parasite so common in the Smokies that we just had to include it in the book. Otherwise, at least one of you would have encountered it hiking down a trail and wondered what it was. As the name suggests, this species parasitizes members of the fruticose lichen genus *Usnea*. It is characterized by its large, convex, pale fruiting bodies that occur along *Usnea* branches and in some cases resemble apothecia. When this parasitic fungus forms fruiting bodies, it usually forces the lichen branches to reflex and grow in the opposite direction, as one would do if infected and trying to get away.

CHEMISTRY: No substances. Spot tests. K-, C-, KC-, P-, UV-.

NICHE: Restricted to *Usnea* thalli, particularly *U. subgracilis*, occurs throughout the Smokies but especially at higher elevations. The species is widely distributed in northern temperate regions of North America but has been overlooked because it is a lichen parasite and the lichens themselves are unfortunately deemed by some to be "hard enough already."

KEY FEATURES: Conspicuous, pale fruiting bodes formed on secondary branches of *Usnea*, especially *U. subgracilis*, throughout the Smokies but more abundant at higher elevations.

Bilimbia sabuletorum

Red Black Caps

Tripp 5006 (photo: Lendemer & Tripp)

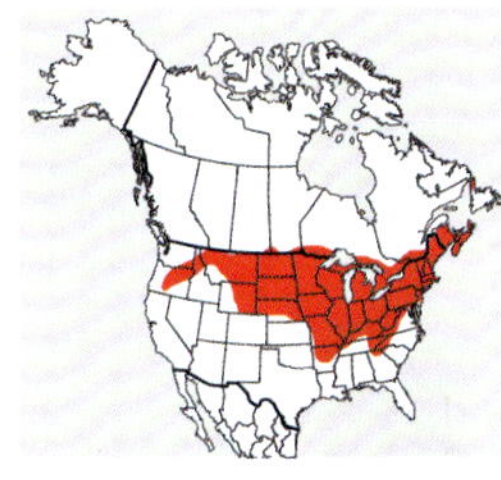

NOTES: Erin screeches with satisfaction when she encounters this species in the field, and we think you will too. It resembles some members of *Biatora* such as *B. vernalis* and may happen to be geographically proximal to them in this guide but differs in having dark, reddish-brown apothecia with longer, 6–8 celled spores.

CHEMISTRY: No substances. Spot tests. K-, C-, KC-, P-, UV-.

NICHE: *Bilimbia sabuletorum* likes to be deep and damp and grows on shaded bryophytes and other organic matter, primarily at middle elevations in eastern portions of the Smokies. It is relatively common and widespread across temperate North America and also occurs in Europe. If you are lucky, you might find it growing together with two even rarer lichens—*Flakea papillata* and *Rockefellera crossophylla*. A good place to see this trio is along Spring House Branch Trail, about half a mile above its junction with Forney Creek Trail.

KEY FEATURES: Dark reddish-brown apothecia, thick, green thallus, 6–8 celled colorless spores, on bryophytes and organic matter at middle elevations.

Botryolepraria lesdainii

Rock Candy

Tripp 3862 (photo: Lendemer)

NOTES: *Botryolepraria lesdainii* is distinguished from the true Dust Lichens of the genus *Lepraria* primarily by gestalt, which, in an era of sequencing whole genomes in a single day, remains one our favorite tools as field biologists. It also has a different algal photobiont and different chemistry from species of *Lepraria*. In the field, *Botryolepraria* can be recognized by its distinctly airy, loosely woven, cottony leprose thallus that is a characteristic seafoam green in color. Surprisingly, the genus consists of only two species (the second being Neotropical), and molecular studies have shown that *B. lesdainis* is related to *Verrucaria*. That was a surprise to many, but makes sense considering that its characteristic color is shared by both *Agonimia* and *Flakea*.

CHEMISTRY: Unknown Terpenoid. Spot tests. K-, C-, KC-, P-, UV-.

NICHE: This species is more common than you might think and can often be found lurking in cool, shaded overhangs of calcareous and weakly calcareous rocks throughout the Smokies. It is widespread but overlooked in temperate eastern North America, with disjunct populations in central California and Europe.

KEY FEATURES: Airy, loose cottony, seafoam-green thallus that is entirely leprose, all spot tests negative, on strongly to weakly calcareous rocks.

Brigantiaea leucoxantha

American Sunrise

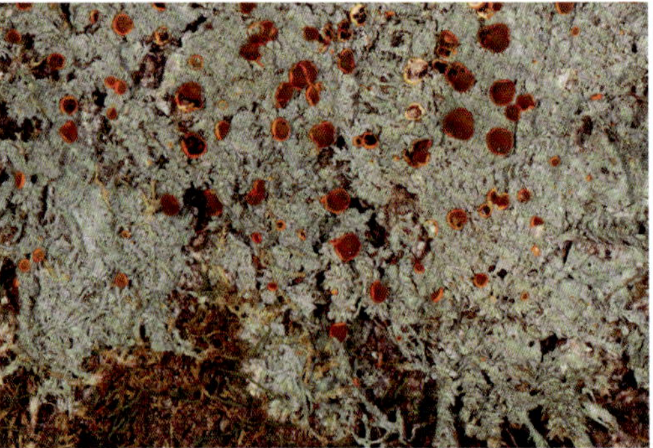

Tripp 5000 (photo: Tripp & Lendemer)

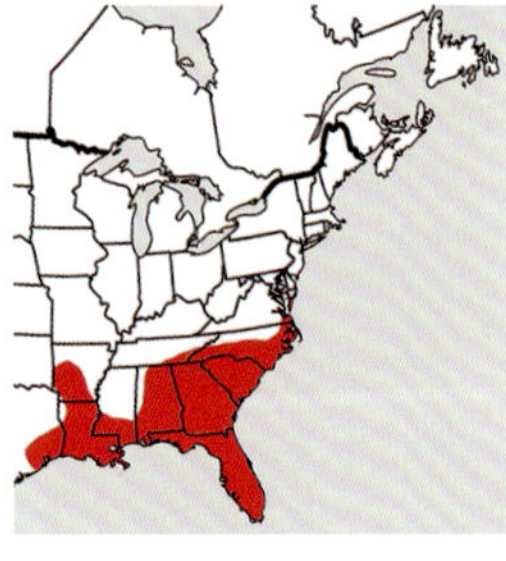

NOTES: Is it just us or is it *hot* in here? You will be the luckiest visitor in the National Park should you find *Brigantiaea leucoxantha*. This species is real show stopper, and someone should probably write a song about it. It is absolutely unmistakable for any other Smokies lichen owing to its screaming orange, large apothecia. It has a crusty blue-gray thallus as well.

CHEMISTRY: It doesn't matter. But if you must know, atranorin and zeorin. Spot tests. (apothecia) K+ purple, C-, KC-, P-, UV-.

NICHE: In the Smokies, *Brigantiaea leucoxantha* is rare, occurring only at low elevations in southern portions of the Park, where it grows on hardwoods, especially Hickory. Although rare in the Park, this species is emblematic of and common throughout the southeastern coastal plain, where it grows in both upland and lowland mesic forests.

KEY FEATURES: Large, bright-orange apothecia against a blue-gray thallus growing on bark, southern portions of Smokies at low elevations.

Bryoria americana

Unravelling Twine

Tripp 5360 (photo: Lendemer)

NOTES: *Bryoria* is an exceptionally easy-to-recognize genus by its fruticose thalli, which are brown owing to the presence of the sun-screening pigment melanin. There are five species known in the Smokies, all of which occur at high elevations in spruce-fir forest (and one of them, *B. furcellata*, extending to lower elevations). Among these, *B. nadvornikiana* is the only one that contains alectorialic acid and is sorediate–isidiate. *Bryoria furcellata* also has asexual propagules but lacks alectorialic acid. *Bryoria tenuis* and *B. bicolor* are also similar but have black branches (these brown and black in the latter) versus pure brown branches in *B. americana*.

CHEMISTRY: Fumarprotocetraric acid. Spot tests. K-, C-, KC-, P+ red, UV-.

NICHE: *Bryoria americana* can be found attached to twigs of *Abies fraseri* and *Picea rubens* at high elevations in the Smokies, where it is disjunct from the boreal forests of more northern regions. Clingman's Dome hosts the highest diversity of *Bryoria* in the Smokies.

KEY FEATURES: Fruticose thallus with brown branches, lacking asexual propagules, on twigs of spruces and firs at high elevations.

Bryoria bicolor

Two-toned Twine

Tripp 5083 (photo: Lendemer)

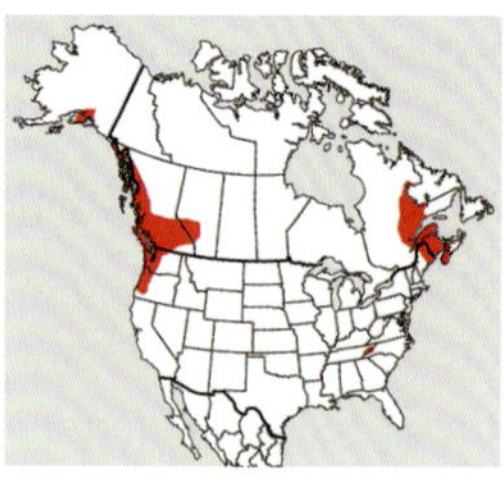

NOTES: *Bryoria bicolor* is a species that we overlooked during our early work in the Smokies because it is relatively small and inconspicuous amid an otherwise profusion of lichens at high elevations in the Park. As the specific epithet implies, this species is characterized by its bi-colored branches, the primary ones being almost black and contrasting sharply with yellowish-green secondary branches on close inspection. This species is most likely to be confused with *B. tenuis*, from which it can be distinguished by its tertiary branches that join secondary branches at ~90° angles (vs. < 90° in *B. tenuis*). It also has denser thalli than *B. tenuis*. *Bryoria bicolor* is smaller in overall size than most other Smokies *Bryoria*.
CHEMISTRY: Fumarprotocetraric acid. Spot tests. K-, C-, KC-, P+ orange-red, UV-.
NICHE: *Bryoria* bicolor is disjunct in the Smokies, where it is both uncommon and overlooked, occurring on branches (especially of Fraser Fir) only at the highest elevations. Elsewhere, this species occupies boreal forests in oceanic regions of North America.
KEY FEATURES: Small, dense thalli, bi-colored branches, right angles between tertiary and secondary branches, restricted to the highest elevations in the Smokies.

Bryoria furcellata

Ragged Ropes

Lendemer 33035 (photo: Tripp)

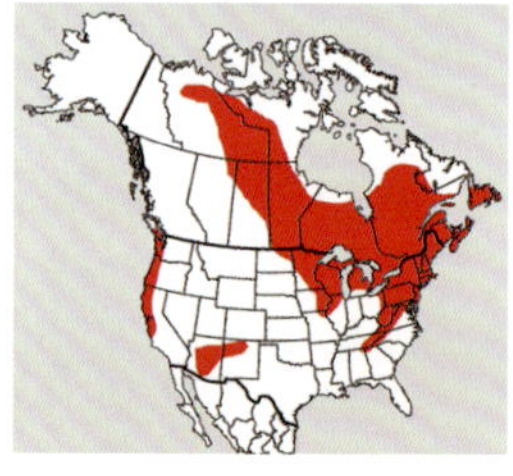

NOTES: *Bryoria furcellata* is an easy-to-identify lichen because it is the only species in the genus with spiny isidia (see photo!) growing out of slit-shaped soralia. These features make it impossible to confuse for any other *Bryoria*. Species of *Bryoria* are easily differentiable from other fruticose lichens in the Smokies such as members of *Usnea* and *Alectoria* because of their brown cortex. See further notes on the *Bryoria* of the Smokies under *B. americana*.

CHEMISTRY: Fumarprotocetraric acid. Spot tests. K-, C-, KC-, P+ orange-red, UV-.

NICHE: *Bryoria furcellata* is the most common species of *Bryoria* in the Smokies, where it grows primarily on conifers across a broad elevation range. It occurs elsewhere in similar habitats across most of eastern North America, and has disjunct populations in western North America. If you find a *Bryoria* in the field, there is a high probability that it is *B. furcellata*.

KEY FEATURES: Fruticose lichen with a brown cortex and spine-like isidia that arise from slit-shaped soralia, broadly distributed in Smokies, growing mostly on conifers.

Bryoria nadvornikiana

Ghost Ropes

Tripp 3537 (photo: Lendemer & Tripp)

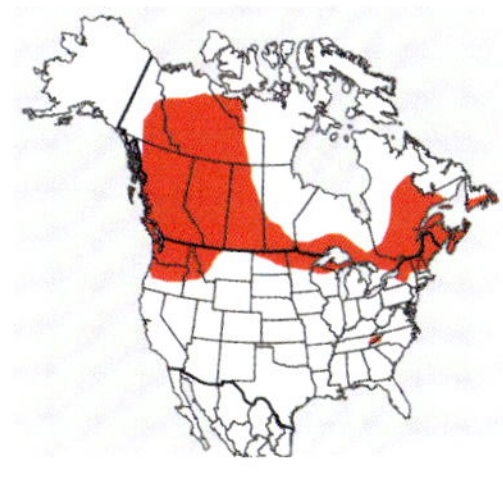

NOTES: *Bryoria nadvornikiana* is a relatively large, somewhat pendant fruticose species characterized by its pale greenish-gray thalli and large white soralia. Its gray thallus distinguishes it from all other *Bryoria* in the Smokies, which are brown owing to the presence of melanin. See further notes on the *Bryoria* of the Smokies under *B. americana*.

CHEMISTRY: Barbatolic acid plus other trace substances. Spot tests. K+ yellow, C+ pink or C-, KC+ red, PD+ yellow, UV-.

NICHE: This species occurs at high elevations, mostly on the bark of trees including Yellow Birch, Red Spruce, and Fraser Fir, but can occasionally be found on rock. A great place to find this species is on the Appalachian Trail between Newfound Gap and Mt. Kephart. Elsewhere, this species is relatively common in boreal forests across northern portions of North America.

KEY FEATURES: Gray thallus, large stature, somewhat pendant, large white soralia, high elevations on bark or occasionally on rock.

Buellia badia

Ne'er-do-wells

Tripp 3700 (photo: Lendemer)

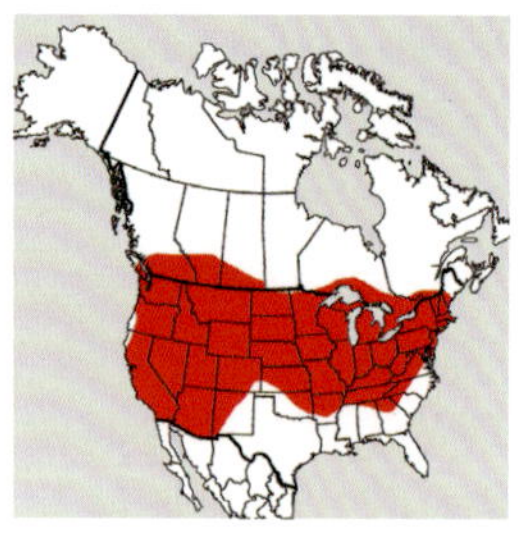

NOTES: *Buellia badia* is a bad apple. Although it might look innocent enough on the rocks where it grows, there is definitely something sinister underfoot. The species often parasitizes other lichens that grow on sun-exposed rocks. Sometimes one can find the apothecia present on the thalli of other, still-identifiable, lichens such as *Caloplaca*. However, the host seems to succumb rapidly to its infection, such that its thallus dies and only the thallus of the *Buellia*, a scattering of small gray areoles, remains. The species is only likely to be confused with forms of *Amandinea punctata*, found growing on rocks. When *A. punctata* occurs on rocks, however, it typically is found in shaded or sheltered habitats, and is never parasitic on other lichens. The spores of the two species can be similar, which unfortunately means that one should always hunt for pycnidia to determine whether the conidia are long and filiform (*Amandinea*) or short and bacilliform (*Buellia*).

CHEMISTRY: No substances. Spot tests. K-, KC-, C-, P-, UV-.

NICHE: This species is uncommon in the Smokies, where it is found on sun-exposed, non-calcareous rocks at high elevations. Look for it the next time you sit on the old chair at High Rocks. Outside of the Smokies, this species is most common in western North America, but it occurs sporadically throughout the Appalachian Mountains and elsewhere in temperate eastern North America.

KEY FEATURES: Quasi-parasitic, indistinct gray thallus, small, black, lecideine apothecia, 2-celled brown spores, 8-per ascus, filiform conidia, on sun-exposed, non-calcareous rocks at high elevations.

Buellia dialyta

Bumpy Backs

Tripp 3845 (photo: Lendemer)

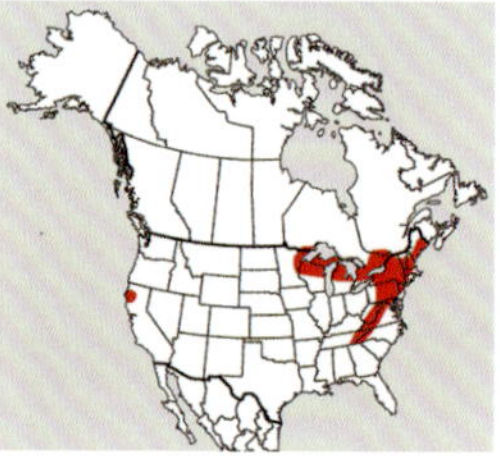

NOTES: *Buellia dialyta* is characterized by its greenish-gray areolate thallus, its convex apothecia, and its P+ orange-red thallus reaction. This and other species of *Buellia* are marked by plane, smooth apothecia and 2- to 4-celled brown spores. Species of *Rhizocarpon* have similar spores, but do not grow on trees. *Buellia dialyta* really does not look like any other species of *Buellia*, in our opinion, but if must go down that route, it is the only species in the genus in our area that produces fumarprotocetraric acid and has large spores. In the Smokies, *Amandinea punctata* has similar spores and is corticolous, but that species has much smaller apothecia and spores and no matter how much we go at it with an open mind, is never as pretty as species of *Buellia*.

CHEMISTRY: Fumarprotocetraric acid. Spot tests. K+ dirty brown, C-, KC-, P+ orange-red, UV-.

NICHE: *Buellia dialyta* is relatively common in the Smokies, where it grows noticeably well on Eastern Hemlock (this host nearly completely obliterated from the Park owing to the wooly adelgid), but also readily inhabits other corticolous substrates at low-to-middle elevations. This species has a classic Appalachian–Great Lakes distribution and is relatively common throughout this range. Interestingly, it was first described from northern California. We ourselves have not yet seen this species in the West.

KEY FEATURES: Green-gray areolate thallus, black convex apothecia, 2-celled brown spores, P+ orange-red thallus, corticolous on Hemlock and other trees at low-to-middle elevations in the Smokies.

Buellia elizae

Tuckerman's Love

Lendemer 33011 (photo: Tripp)

NOTES: *Buellia elizae* must surely be considered the most beautiful member of the genus in the

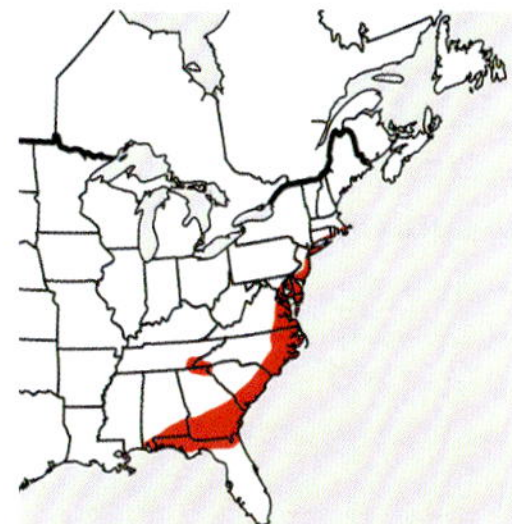

Smokies. It is easily distinguished from all congeners by its red pruinose apothecial discs. Its green thallus ranges from continuous to discontinuous and is often granular. *Buellia elizae* was named by the founder of American lichenology, Edward Tuckerman, in honor of his bride-to-be, Eliza. He collected the type specimen on their honeymoon in the pinelands of eastern Virginia.

CHEMISTRY: Barbatic acid. Spot tests. (apothecia) K+ fleeting purple, C-, KC-, P-, UV-.

NICHE: *Buellia elizae* is rare in the Smokies, where it has a strong preference for conifers. It occurs at both low and high elevations, which we really do not have a good explanation for (sorry!). A great place to see this lovely species is on Little Bottoms Trail near Abrams Creek. *Buellia elizae* is endemic to and rare in the southeastern United States. We suspect that it was once more common in Longleaf Pine and other southeastern habitats, prior to their destruction by humans.

KEY FEATURES: Crustose species with red pruinose apothecial discs, green and generally granular thallus, K+ fleeting purple reaction in the apothecia, corticolous especially on conifers, rare.

Buellia sharpiana

Evelyn's Buttons

Tripp 5020 (photo: Lendemer)

NOTES: Growing atop the highest peaks of Mount LeConte, this species is seen by upwards of a million visitors to the Smokies each year. It sits right underneath their feet on the exposed rocks of these impressive summits and yet most never realize they are stepping on a rare, narrowly endemic species. If you stoop to take a look, you can recognize *Buellia sharpiana* by its light yellow-colored areolate thallus with a prominent, dark prothallus and immersed dark apothecia. These characteristics, combined with its chemistry and brown 2-celled spores, make this species distinctive from all others in the Smokies. This species commemorats Evelyn Sharp, distinguished University of Tennessee professor and Knoxville resident.

CHEMISTRY: Arthothelin. Spot tests. K-, C+ weak orange, KC+ orange-red, P-, UV+ dull orange.

NICHE: Restricted to exposed Anakeesta rocks on the highest peaks of Mount LeConte in Tennessee. That's right, it doesn't occur anywhere else on Earth, so far as known . . .

KEY FEATURES: Areolate, dull yellow thallus, dark prothallus, immersed dark apothecia, brown 2-celled spores, restricted to Anakeesta rocks at the highest elevations of Tennessee.

Buellia spuria

Checkered Tiles

Tripp 6062 (photo: Lendemer)

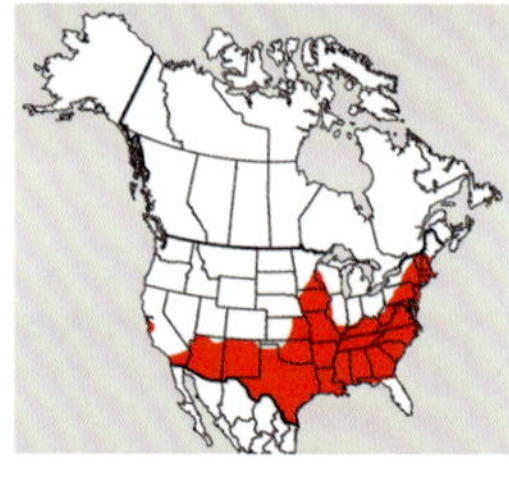

NOTES: In the Smokies, relatively few crustose lichens occur on rocks and have brown 2-celled spores. In this regard, *Buellia spuria* is quite distinctive, particularly given its tile-like areolate thallus and chemistry. The most likely culprit for confusion is actually with *Lecanora oreinoides* and *Lecidea tessellata*, two distantly related species that are externally similar to *B. spuria* in the appearance of their thalli, but completely different internally because they have colorless, simple spores. All three species can occur together in the Smokies, and one should always be careful to check the spores and the chemistry.

CHEMISTRY: (I) Atranorin and norstictic acid. Spot tests: K+ yellow turning red, KC-, C-, P+ yellow, UV-. (II) Atranorin and stictic acid. Spot tests: K+ yellow, KC-, C-, P+ orange, UV-.

NICHE: *Buellia spuria* is rare in the Smokies primarily because it occurs on sun-exposed, non-calcareous rock outcrops, which are in short supply in an area with so much precipitation and forest cover. Nonetheless it is common in such habitats throughout the southeastern United States, and becomes much more common in the arid areas of central and southeastern North America.

KEY FEATURES: Gray areolate thallus, black +/- immersed apothecia, brown 2-celled spores, K+ yellow and often K+ yellow turning red, on sun-exposed, non-calcareous rocks.

Buellia stillingiana

The Everywhere Lichen

Lendemer 26676 (photo: Tripp)

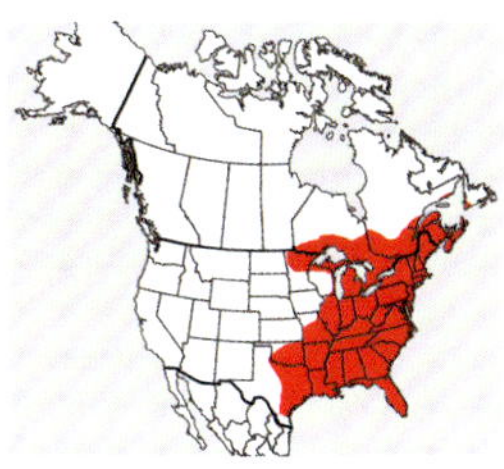

NOTES: Weedy lichen alert! *Buellia stillingiana* is one of the most common and widespread lichens in the Smokies and, indeed, much of eastern North America. Put it on your "Top 5 to learn" list. It is an easy-to-recognize member of the genus by its continuous, smooth-to-slightly-cracked white thallus with abundant black apothecia. It might be confused with three other corticolous species of *Buellia* in the Smokies, *B. vernicoma*, *B. curtisii*, and B. *dialyta*. From *B. vernicoma*, it differs by having 2-celled spores. From *B. curtisii*, it differs by having shorter spores with blunt ends. From *B. dialyta*, it differs by having a K+ yellow-to-red thallus. See the dichotomous key for specifics!

CHEMISTRY: Atranorin, norstictic acid. Spot tests. K+ yellow to red, C-, KC-, P+ yellow, UV-.

NICHE: *Buellia stillingiana* is one of the most common crustose, corticolous lichens in the Smokies. If you find black dots against a white thallus on trees, it is probably *B. stillingiana*. This species occurs throughout the Smokies, primarily at low-to-middle elevations, where it occupies the bark of hardwoods, especially smooth-bark substrates. It is widespread in similar habitats throughout eastern North America.

KEY FEATURES: Continuous white thallus, abundant black apothecia, brown, 2-celled spores with blunt ends, K+ yellow-to-red thallus.

Buellia vernicoma

Tuckerman's Green Sleepy Hair Lichen

Tripp 3636 (photo: Lendemer)

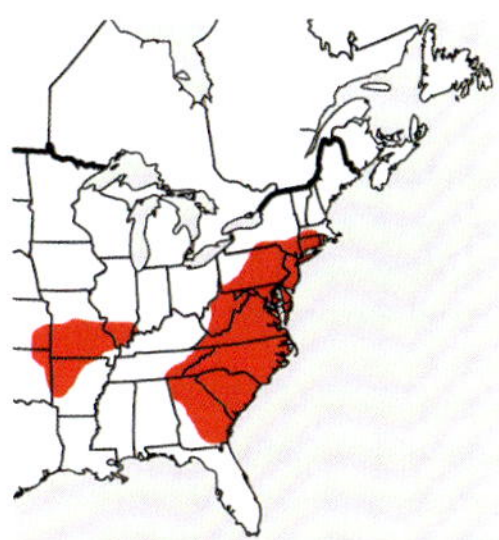

NOTES: *Buellia vernicoma* is characterized by its greenish-yellow areolate thallus, black prothallus, black apothecia, and brown, 4-celled spores. The combination of these features make it unmistakable among (and more beautiful than!) all of its closest relatives in the Smokies. The specific epithet is mysterious to us, but we like it: Green Hair? Green Sleepy?

CHEMISTRY: Unknown xanthone. Spot tests. K-, C-, KC-, P-, UV+ dull orange.

NICHE: *Buellia vernicoma* is an occasional species throughout the Smokies at most elevations except the uppermost portions of the Park. It occurs in high-quality habitats elsewhere in the eastern and mid-Atlantic United States and is disjunct in the Ozarks. It occurs primarily on trees but is here pictured on rocks, just like its type collections in Chester County, Pennsylvania, and Essex County, Massachusetts.

KEY FEATURES: Greenish-yellow areolate thallus, black prothallus, black apothecia, brown, 4-celled spores, on bark, sometimes rock, attractive!

Bulbothrix isidiza

Dainty Baubles

Dey 32659 (photo: Lendemer)

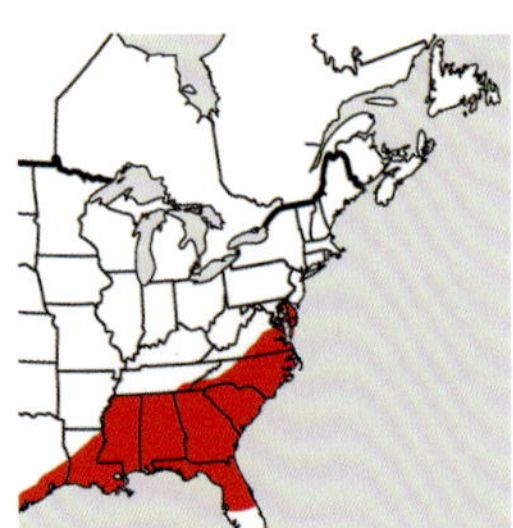

NOTES: *Bulbothrix* may be one of the most endearing genera of macrolichens, if not because of its name, then because of the characterized bulbate cilia that form a fringe around the lobe margins. Although only two species of *Bulbothrix* occur in the Smokies, this is doubtless the more elegant of the two. In contrast to *B. scortella*, it has a larger thallus and salazinic acid in the medulla (the medulla of *B. scortella* is C+ pink due to the presence of gyrophoric acid). There are not any other similar macrolichens in the Smokies, mainly because all the other corticolous species with K+ red medullas have thalli that are much larger, with ruffled or ascending lobes. In some respects, *Canoparmelia caroliniana* is most similar because it has adnate lobes and isidia. However the thallus surface of *C. caroliniana* is maculate, and the medulla is K- due to the presence of perlatolic acid instead of salazinic acid.

CHEMISTRY: Atranorin and salazinic acid. Cortex: K+ yellow, C-, KC-, P-, UV-; medulla: K+ yellow turning red, C-, KC-, P+ orange, UV-.

NICHE: In the Smokies, this species is found only at the lowest elevations in Tennessee, usually on conifer branches fallen from the canopies of trees. It is much more common southward in the Coastal Plain and can be found in many areas of southeastern North America.

KEY FEATURES: Foliose thallus, isidia, bulbate cilia on the lobe tips, K+ yellow turning red medulla, on fallen branches at the lowest elevations.

Bulbothrix scortella

Bulbil Bottoms

Lendemer 29622 (photo: Tripp)

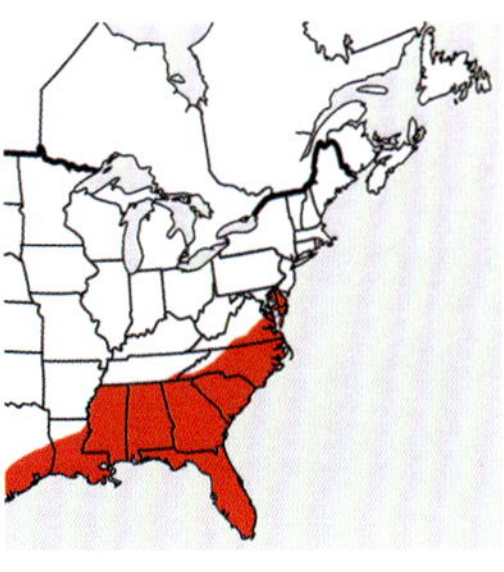

NOTES: *Bulbothrix scortella* is one of numerous foliose, gray, isidiate lichens in the Smokies. Fortunately, it can be easily differentiated from all others (except its congener, *B. isidiza*) by bulbils or inflated bases of the marginal cilia, as the generic name implies. The related *B. isidiza* grows in the canopy of pine trees and has thalli that are twice the size and therefore unlikely to be confused with *B. scortella*. Rather, *B. scortella* is more likely to be confused with *Hypotrachyna horrescens* or *H. minarum,* but differs from both by having a pale undersurface (black in *Hypotrachyna*) and by having cilia with bulbils. From *H. horrescens,* it also differs by being C-.

CHEMISTRY: Gyrophoric acid. Spot tests. (medulla) K-, C+ pink, KC+ pink, P-, UV-.

NICHE: *Bulbothrix scortella* is occasional in the Smokies at low elevations, where it occurs on the bark of hardwoods. It can be found in similar habitats in the southeastern United States. Look for this species in the Rabbit Creek area.

KEY FEATURES: Foliose, isidiate, pale lower surfaces, marginal cilia with bulbils, C+ pink medulla, bark of hardwoods, low elevations.

Byssoloma leucoblepharum

White Eyelashes

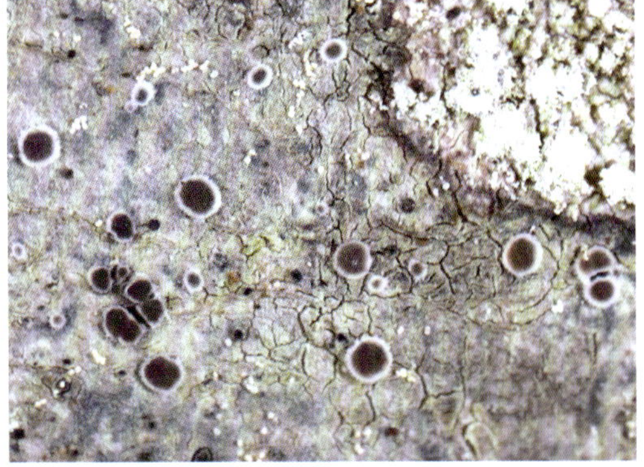

Lendemer 23659 (photo: Lendemer)

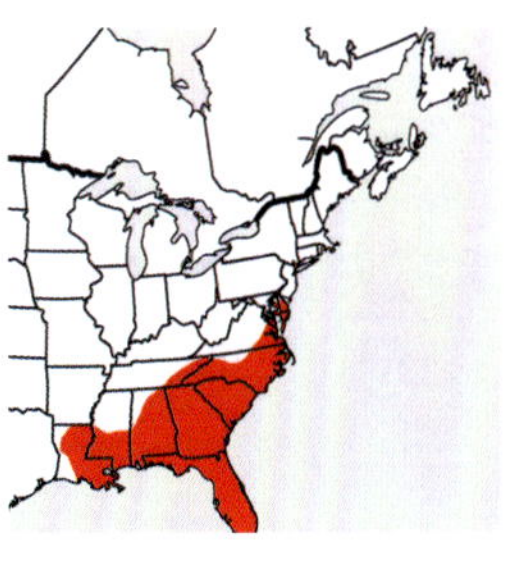

NOTES: Byssoid margins makes lichen learning easy and fun! If you see white wooly margins of an apothecium, it's only one of two species: this or *B. subdiscordans. Byssoloma leucoblepharum* is distinguished from the others by its brown apothecia and a brown-to-colorless epihymenium (black apothecia and a blue-green epihymenium in *B. subdiscordans*). *Byssoloma meadii* is another, much more similar species that is not yet known from the Smokies. It differs from *B. leucoblepharum* in having light brown apothecial discs and in producing xanthones, which make the thallus UV+ orange.

CHEMISTRY: No substances. Spot tests. K-, C-, KC-, P-, UV-.

NICHE: Look for White Eyelashes (the literal translation of "*leucoblepharum*") on branches and on *Rhododendron* leaves at low elevations in the Park, such as along Schoolhouse Gap Trail. This species has tropical affinities and, closer to home, occurs primarily in the southeastern United States. Populations in the Smokies are slightly disjunct from more Coastal Plain areas, just because they can be, and you would be too if given the option . . .

KEY FEATURES: Brown apothecia, brown-to-colorless epihymenium, occurrence at low elevations on branches and evergreen leaves, and of course, byssoid apothecia margins. And you thought *Lepidopteran* antennae were the coolest thing you'd ever seen with a magnifying lens.

Byssoloma marginatum

Four-Celled Affection

Lendemer 53164 (photo: Tripp)

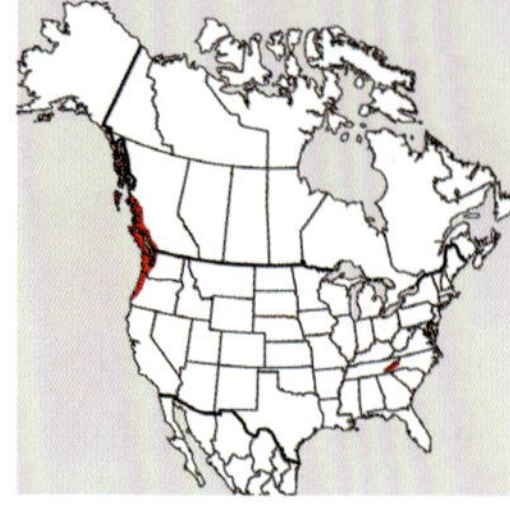

NOTES: *Byssoloma marginatum* has a funny epithet: many species in the genus have conspicuously fuzzy, often white margins. This one does not, and they are not decorated in any manner that would call attention to them by using names such as "marginatum." Instead of having those attractive margins, this species has one that has a faint, continuous thallus that bears brown-to-black discs and 3-septate, 4-celled spores. Its apothecia are typically very small, as evidenced by its size comparison to the leaves of *Frullania,* seen in this photo.

CHEMISTRY: No substances. Spot tests. K-, KC-, C-, P-, UV-.

NICHE: Four-Celled Affection takes a liking to both leaves and twigs, always in wet habitats such as riparian corridors. It occurs disjunct in the Pacific Northwest as well as in the British Isles.

KEY FEATURES: Very small, brown-to-black apothecia that bear 4-celled spores, on twigs or leaves in wet places.

Byssoloma subdiscordans

White-Ringed Full Stops

Tripp 3860 (photo: Lendemer)

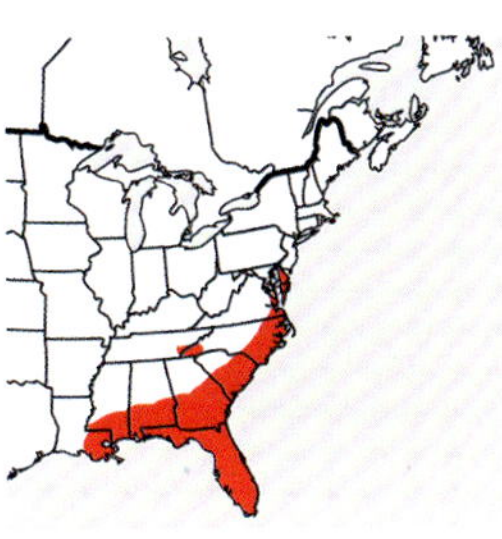

NOTES: *Byssoloma subdiscordans* is one of only two species in the Smokies with distinctive bright white, byssoid apothecial margins. *Byssoloma subdiscordans* is easily distinguished from the other species, *B. leucoblepharum*, by its large, black apothecia and its blue-green epihymenium (dirty brown apothecia and a brown or colorless epihymenium in *B. leucoblepharum*). It is possible that *B. subdiscordans* could be confused with *B. marginatum*, but the latter has margins that are compact and thus not appearing byssoid.
CHEMISTRY: No substances. Spot tests. K-, C-, KC-, P-, UV-.
NICHE: This species occurs on branches and leaves of trees in humid, middle-to-high elevations in the Smokies, where it is disjunct and uncommon but not rare. It occurs in similar, mesic habitats throughout the southeastern Coastal Plain, where it is primarily foliicolous.
KEY FEATURES: Black apothecia with bright white byssoid margins, blue-green epihymenium, high-humidity habitats at middle-to-upper elevations, spectacular.

Calicium lenticulare

White-Rimmed Twin Pins

Tripp 5077 (photo: Lendemer)

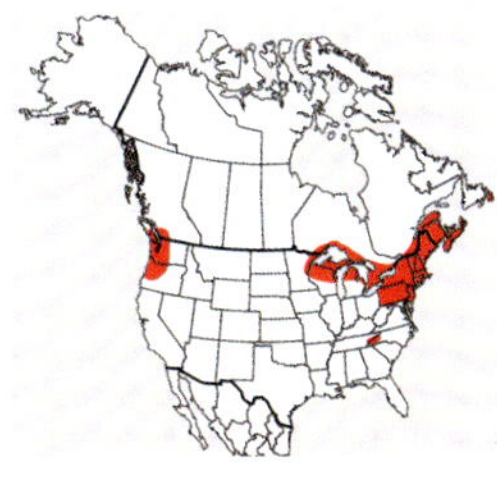

NOTES: *Calicium lenticulare* is one of many species in this genus present in the Smokies. These lichens, informally known as "the pin lichens," are minute in size by most standards and thus highly inconspicuous to all but the most discerning Park visitor, or to a snail. *Calicium lenticulare* is distinctive by its white-pruinose capitulum undersides (i.e., the portion immediately below the spore-bearing mazaedium), as seen in the photograph. This species is also characterized by apothecial stalks that are strongly I+ blue in section and its 2-celled spores. Finally, the thallus of *C. lenticulare* is almost indiscernible, more or less immersed in the substrate.
CHEMISTRY: No substances. Spot tests. K-, KC-, C-, P-, UV-.
NICHE: *Calicium lenticulare* is a relatively frequent species at high elevations in the Smokies, where it occurs on old wood, primarily standing conifer stumps. Elsewhere, it occurs in similarly mesic habitats, with an oceanic-to-Great Lakes distribution and a distant outpost in western North America.
KEY FEATURES: Minute pins, immersed thallus, densely white pruinose capitulum undersides, I+ blue apothecium stalks, 2-celled spores, high elevations, on old conifer stumps . . . a perfectly fine place to dwell.

Calicium trabinellum

Midas's Touch

Collector McMullin 19123 (photo: McMullin)

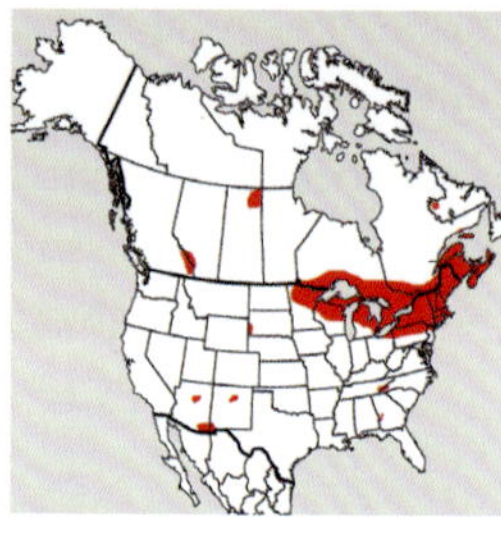

NOTES: Pin lichens, or stubbles, require a special touch to find. While you can amble the ridges and valleys casually looking for most lichens, the pins are small and often make one work for a reward. *Calicium trabinellum* is one such species, which although apparently common in the Smokies, we have rarely seen in the field. The reason is that it is rather small, and one must look carefully to find the distinctive golden yellow pruina on the undersides of the apothecial cups. Perhaps if you find it, you'll contract gold fever and spend the rest of your life looking for pins. There are not many other species in the Smokies likely to be confused with this one, especially given that the species has 2-celled, brown spores, an immersed thallus, and lacks any positive spot test reactions.

CHEMISTRY: No substances. Spot tests. K-, KC-, C-, P-, UV-.

NICHE: This species is widespread throughout the Park, where it grows on dry wood of standing, decorticate tree trunks, especially dead conifers. Although it is likely also widespread northward throughout the Appalachians, it has probably been overlooked there. In North America, it is common and widespread in northern temperate and boreal areas.

KEY FEATURES: Minute pins, immersed thallus, yellow pruinose capitulum undersides, 2-celled spores, on conifer wood throughout the Park.

Caloplaca camptidia

Backcountry Grommets

Buck 56323 (photo: Lendemer)

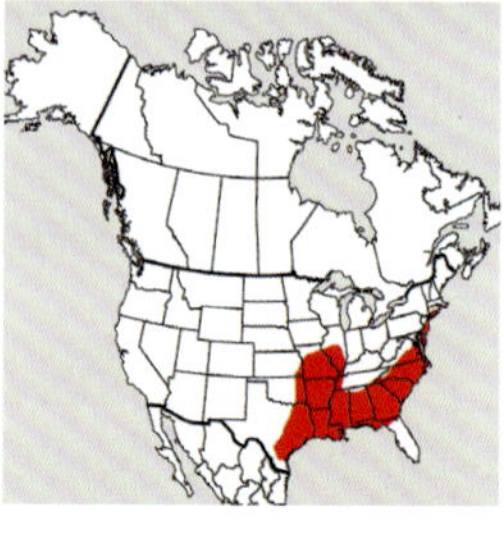

NOTES: *Caloplaca camptidia* is a small, crustose lichen with a continuous gray-to-white thallus bearing reddish brown-to-chocolate-brown apothecia. These apothecia are rimmed by a thin margin that is lighter brown in color, which is perhaps the best way to differentiate this species from *C. cerina*. The latter has orange apothecia with gray margins. Rather than confusing this species for any other *Caloplaca* (small tip: there are very few that grow on bark in the Park!), you might confuse it with one of several species of *Bacidia*. However, a quick cross section through the apothecium

will confirm the polarilocular spores typical of *Caloplaca* and that the margin is lecanorine and contains algal cells. Apothecia margins of *Bacidia* species are entirely fungal and lack algae, and the spores are long and needle like.
CHEMISTRY: No substances. Spot tests. K-, KC-, C-, P-, UV-.
NICHE: Backcountry Grommets is a species of temperate forests of the eastern United States, where it occurs widely on the bark of hardwoods. Look for it at low-to-middle elevations, such as near the walk-through tunnel on the "Road to Nowhere."
KEY FEATURES: Small crustose lichen with brown apothecia, lighter brown lecanorine margins, polarilocular spores, on the bark of hardwoods, low-to-middle elevations.

Caloplaca cerina

Twig Time

Lendemer 33197 (photo: Tripp)

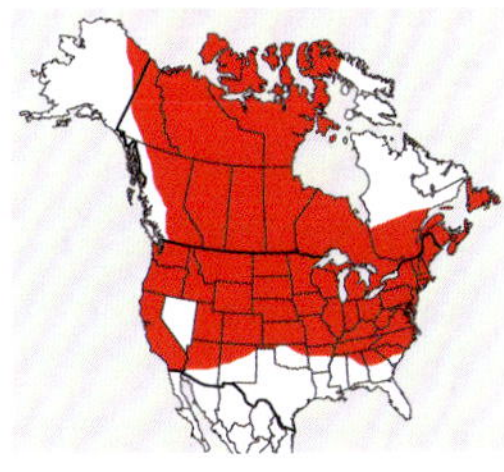

NOTES: As currently known, there are three corticolous species of *Caloplaca* in the Smokies. Among these, *Caloplaca cerina* is distinctive by having orange apothecia with gray margins. *Caloplaca flavorubescens* also has orange apothecia but its margins are +/- concolorous with the discs (vs. concolorous with the thallus, as in *C. cerina*) and its thallus is yellow instead of gray. *Caloplaca camptidia* has very different brown apothecia and cannot be confused with *C. cerina*. There are very few other crustose species (in other genera) that might be mistaken for *C. cerina*, but if questions arise, check for polarilocular spores.
CHEMISTRY: Anthraquinones. Spot tests. K+ purple (apothecia), KC-, C-, P-, UV-.
NICHE: *Caloplaca cerina* occurs sporadically at middle-to-high elevations in the Smokies, nearly always on branches, especially of hardwoods. One most commonly encounters this species on twigs that have fallen from the canopy. *Caloplaca cerina* is widespread and common in several portions of North America and Europe, where it occupies similar niches.
KEY FEATURES: Crustose greenish-gray thallus with orange apothecia that have gray margins, polarilocular spores, on twigs of hardwoods, middle-to-high elevations, uncommon.

Caloplaca chrysodeta

Dying Sun

Buck 56381 (photo: Lendemer)

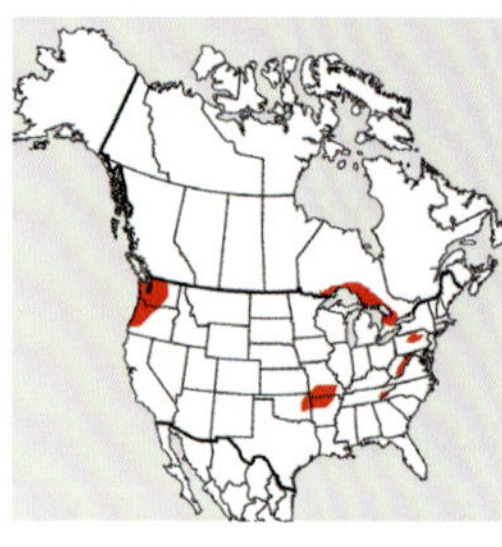

NOTES: If you like asexual reproduction (and by now, it is perfectly normal to admit that you do), the Smokies is certainly the place for your next lichen vacation. You already knew that GSMNP was the most visited National Park in the United States, right? Think of all the company you'll have there . . . admiring those beautiful eastern peaks and mounds upon mounds of soredia at the same time. Now that that has been clarified, *Caloplaca chrysodeta*. So, this species doesn't look a thing like most other *Caloplaca*. It has a distinctive ecology and leprose thallus that is continuous across its substrate, with no signs of lobes whatsoever. In fact, it looks more like a sick *Lepraria*, or a normal *L. vouauxii,* which most people think looks rather ill. Also, try a K test to separate this species from other leprose yellow things, like *Psilolechia* and species of *Chrysothrix*. We suppose it has retained some redeeming, *Caloplaca* qualities, afterall.
CHEMISTRY: Anthraquinones. Spot Tests: K+ purple-red, KC-, C-, P-, UV-.
NICHE: This species occurs on calcareous rocks in dry, shaded habitats. That is probably why it is very rare in the Smokies. More time exploring our few areas of limestone might turn up additional populations.
KEY FEATURES: Think: something that looks like a sick *Lepraria* but is K+ intense purple-red, on limestone, in shaded areas.

Caloplaca chrysophthalma

American Gemstones

Lendemer 44800 (photo: Tripp)

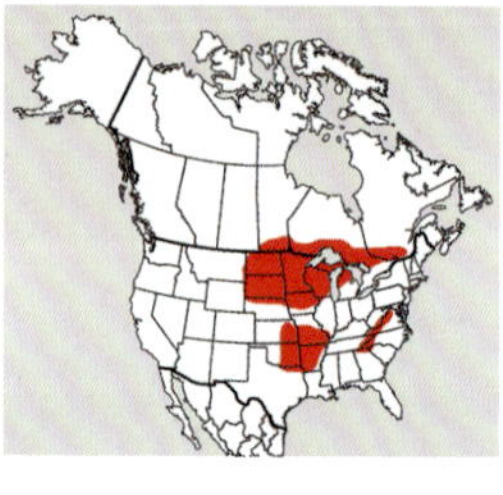

NOTES: *Caloplaca* is a species-rich (and easy to admire) genus of mostly fertile, apotheciate species. But here lies a special, asexual gem! *Caloplaca chrysophthalma* is distinctive by its golden yellow soredia that occur in discrete soralia and its well-developed but usually minute yellow thalli. You will be very lucky but *very* happy if you find this species. To date, we have seen it on only four occasions.
CHEMISTRY: Anthraquinones. Spot tests. K+ red (apothecia), KC-, C-, P-, UV-.
NICHE: *Caloplaca chrysophthalma* is a corticolous species that occurs on hardwoods throughout the western Great Lakes region and central Great Plains, with disjunct populations in the Southern Appalachians and Ozarks. In the Smokies, its center of distribution seems to be the greater Jenkins Ridge–Welch Ridge area. Maybe Kephart, who long dwelled in this area, was fond of it?
KEY FEATURES: Small, well-developed yellow thalli with conspicuous, discrete yellow soredia, rare in the Smokies but otherwise common on hardwoods in the upper midwest.

Caloplaca feracissima

Hole-in-my-Pocket

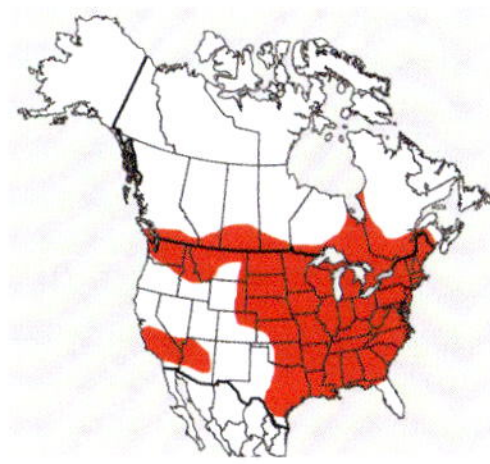

Lendemer 32801 (photo: Lendemer)

NOTES: Under the hand lens, *Caloplaca feracissima* looks like scattered little gold coins on the ground. Was it a bank heist? Or did some unfortunate hiker not notice the hole a mouse chewed through their backpack the night before? We might never know why this species ended up so commonly on concrete. Perhaps something had to grow there. But it is a pleasant surprise to find it at home in so many urban areas. It can easily be recognized by its overall lack of a thallus and the scattered apothecia that have dark yellow-orange discs and margins that are lighter yellow in color. *Caloplaca subsoluta* is similar, but differs in having an areolate thallus and less commonly growing on concrete, compared to naturally occurring calcareous rocks. *Caloplaca flavovirescens* is also similar, but has a thin, continuous yellow thallus.

CHEMISTRY: Anthraquinones. Spot Tests: K+ purple-red, KC-, C-, P-, UV-.

NICHE: This species is distributed throughout the United States, where it grows on old cement in nearly every city, town, village, and parking lot. It is very rare in the Smokies but can be found on cement and other calcareous rocks, including human-made structures.

KEY FEATURES: Indistinct crustose thallus, small yellow apothecia with darker orange-yellow discs, K+ purple, on concrete.

Caloplaca flavocitrina

Continental Firecrackers

Lendemer 26853 (photo: Lendemer)

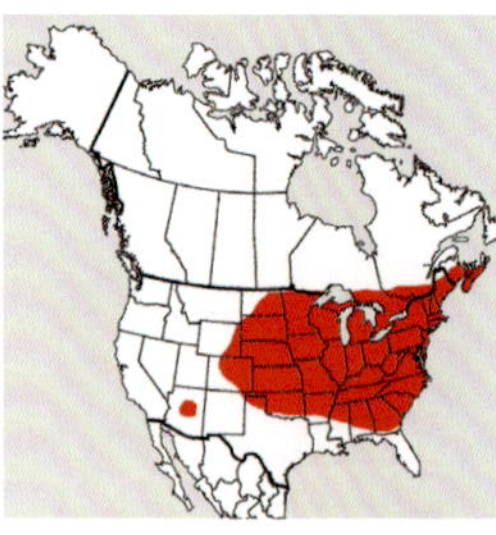

NOTES: *Caloplaca flavocitrina* is an uncommon species in the Smokies but, if you find it, you will know it by its irregularly shaped areoles that bear soredia, which emerge from the margins of the areoles, eventually dissolving them and leaving behind sad little heaps of yellow. This lichen is deep yellow-orange in color, like most of the other *Caloplacas*, but it is the only sorediate member of the genus in the Park that grows on rock and is also areolate. You might be tempted to confuse it with *C. chrysopthalma*, but that species grows on bark, has a continuous yellow thallus and soredia that are golden yellow.

CHEMISTRY: Anthraquinones. Spot Tests: K+ purple-red, KC-, C-, P-, UV-.

NICHE: Continental Firecrackers has a very wide geographical distribution across North America, primarily the United States. Its center of distribution is definitively *not* the southern Appalachian Mountains, but if you find it there or in the Smokies, it will surely be on calcareous rock in shaded overhangs or sheltered outcrops.

KEY FEATURES: Yellow-orange areolate thallus with irregular areoles that bear soredia at the margins, K+ purple, on calcareous rock, primarily low elevations.

Caloplaca flavorubescens

Full Sun

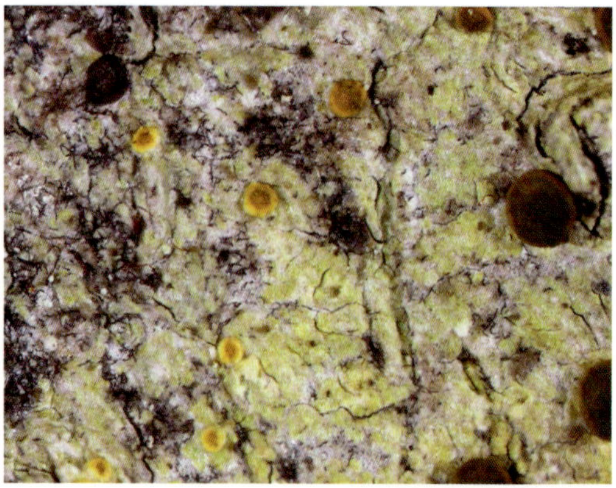

Lendemer 23353 (photo: Lendemer)

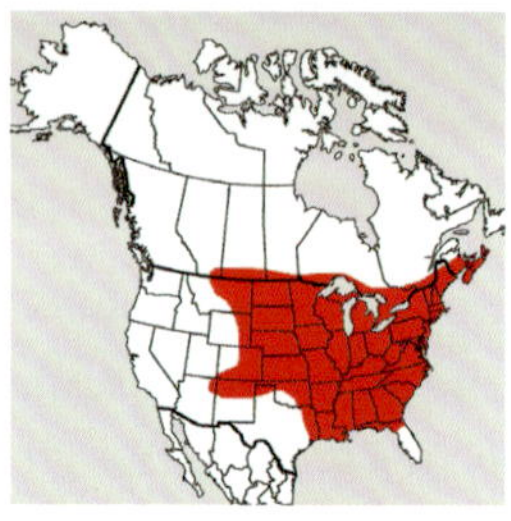

NOTES: *Caloplaca flavorubescens* looks rather similar to *C. flavovirescens*, but that species grows on rock and the former does not. You will learn Full Sun by its ecology and by its continuous, yellow thalli that bear beautiful fruiting bodies, these deep orange in color with a golden-yellow rim. On bark, there basically isn't anything else you can confuse this with. Isn't that a welcomed change?

CHEMISTRY: Anthraquinones. Spot Tests: K+ purple-red, KC-, C-, P-, UV-.

NICHE: Full Sun occurs only on bark, particularly on bark that likes to flake such as buckeyes and maples. This species was once widespread in the temperate eastern United States, but is now rare and mostly restricted to high quality forests. In the Smokies, it is known from only a few scattered locations throughout the Park

KEY FEATURES: Continuous yellow thallus, large apothecia with deep-orange discs and golden-yellow margins, K+ purple, polarilocular spores, on bark.

Caloplaca flavovirescens

Limestone Lover Lichen

Tripp 3717 (photo: Lendemer)

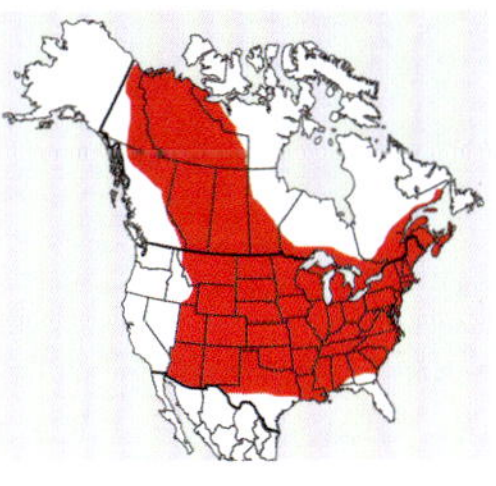

NOTES: *Caloplaca* is one of the most species-rich genera of lichens in North America. Unlike the West, the Smokies are not a center of diversity owing to comparatively less suitable substrate (i.e., rock, especially mineral-rich rock). Species of *Caloplaca* are readily recognizable by their polarilocular spores. *Caloplaca flavovirescens* is distinguished from other members of the genus in the Smokies by its striking yellow thallus, its dark orange apothecia, and its ecology (see Niche).

CHEMISTRY: Anthraquinones. Spot tests. K+ purple (apothecia discs), KC-, C-, P-, UV-.

NICHE: *Caloplaca flavovirescens* is the most common species of *Caloplaca* in the Smokies, where it occurs on rock surfaces, especially on calcareous human-built structures such as stone walls, old foundations, and mortar. The species is common and widespread in temperate North America, with a distribution extending far north into the Arctic.

KEY FEATURES: Crustose yellow thallus with orange apothecia that have orange margins, on rock, especially calcareous human-built structures, common throughout the Park.

Caloplaca reptans

Dark Creeper

Tripp 3514 (photo: Lendemer)

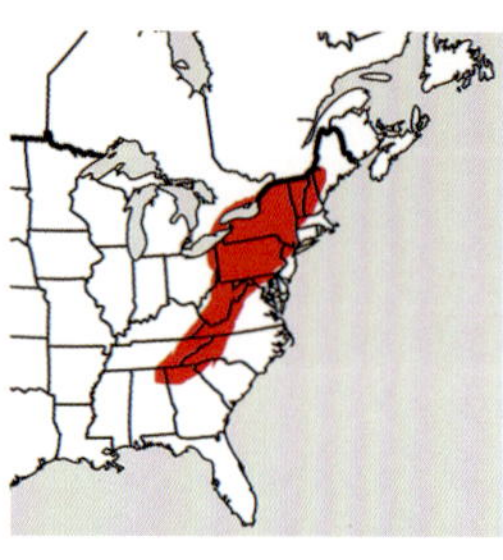

NOTES: *Caloplaca reptans* doesn't look like much of anything you will spend time pondering, but it is nonetheless a *bona fide* Appalachian treasure. Although it has a tiny thallus, it is easily recognized by its ecology (see Niche), by the areolate thallus with areoles that often become lobed or even incised, as in the above photo, and by its large soralia packed full of light green soredia. Typical of the Dark Creeper is a rosette-forming thallus, with lobes that are more or less elongated. No other species in the Smokies is similar, and we agree it certainly looks nothing like the iconic *Caloplaca* species on prior pages.

CHEMISTRY: No substances. Spot tests. K-, KC-, C-, P-, UV-.

NICHE: This species is common throughout the Smokies, where it is found on non-calcareous rocks in shaded overhangs and on vertical rock faces at all elevations. It is an Appalachian endemic that occurs in similar habitats throughout this mountain range.

KEY FEATURES: Gray-to-green-gray areolate thallus, areoles often elongate or lobed, conspicuous soralia with light-colored soredia, on non-calcareous rocks in overhangs.

Caloplaca subsoluta

Orange Atoms

Lendemer 53168 (photo: Tripp)

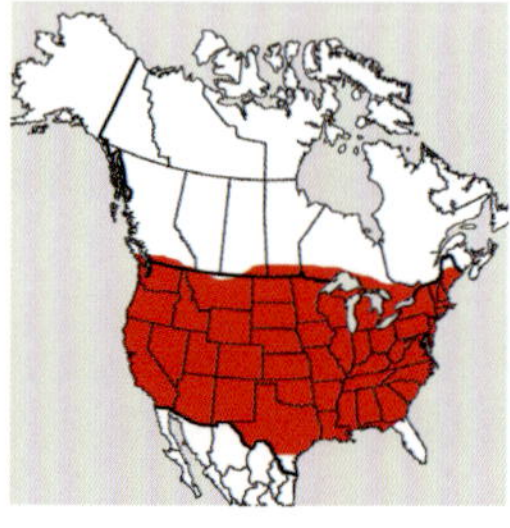

NOTES: There are a lot of *Caloplaca* species in this world. But lucky for you, you went to the Smokies for vacation rather than the Rockies. Here, your challenge is much more of a minor problem than a full-blown crisis. We have only a few species of *Caloplaca* in the Smokies. Most can be instantaneously recognized by their typically bright-orange (sometimes orangish-yellow) thalli that bear various anthraquinone pigments that react K+ purple-red (the sexual ones also have polarilocular spores). *Caloplaca subsoluta* is characterized by having typically minute, yellowish-orange areoles that surround the fruiting bodies, which have margins concolorous with the thallus and discs that are darker orange in color. In some cases, these areoles are quite tiny and the species thus appears to be lacking a thallus altogether (note variation across the thallus in this photograph). This species is also characterized by growing on rock. In our area, Orange Atoms is likely to be confused only with *C. flavovirescens* or *C. flavorubescens*, both of which are also bright orange and reproduce sexually (others in our area are primarily asexual species). *Caloplaca flavorubescens* occurs on bark and thus can be eliminated from any confusion. In contrast, *C. flavovirescens* occurs on rock, but that species nearly always has a bright, conspicuous, extensive, yellow, continuous thallus upon which the orange apothecia sit. In fact, *C. flavovirescens* is very commonly first spotted by its yellow thallus, in marked contrast to the minute, orange squamules that surround apothecia of Orange Atoms.

CHEMISTRY: Anthraquinones. Spot Tests: K+ purple-red, KC-, C-, P-, UV-.

NICHE: This species occurs on calcarerous rocks, which are rare in the Park, but more common are brick and mortar that bear lime. Look for it on person-made structures, such as the Silers Bald Shelter. Elsewhere, this species occurs throughout eastern and western North America. Look for it wherever you see calcareous rocks.

KEY FEATURES: Crustose, bright orange lichen with minute squamules that surround the fruting bodies, exclusively on calcareous rocks and cement, K+ bright purple.

Candelaria concolor

Urban Yella'

Lendemer 48562 (photo: Tripp)

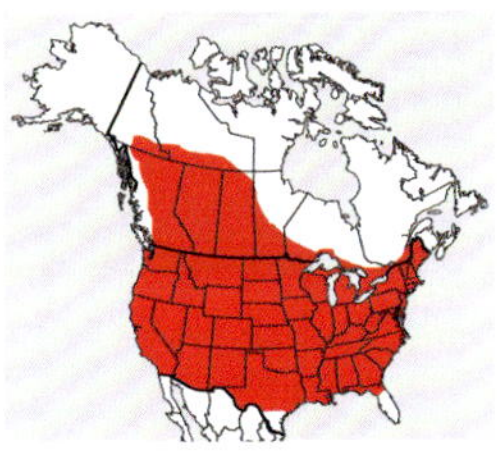

NOTES: There are not very many yellow or orange foliose lichens in the Smokies. This both means that (1) we have to make due with other lichen colors, and (2) bright colors facilitate identification. *Candelaria concolor* is the only foliose yellow lichen in the region that is K- and also sorediate. *Candelaria fibrosa* is much rarer and lacks soredia. *Xanthomendoza weberi* could be confused with *C. concolor* because both are sorediate, but *X. weberi* is orange in color and K+ magenta.

CHEMISTRY: Calycin. Spot tests. Cortex: K-, C-, KC-, P-, UV+ dull orange.

NICHE: *Candelaria concolor* belongs on a street corner in Manhattan or Atlanta and not in the Smokies. It is typical of disturbed or polluted environments and seems adapted to highly lit forest edges. Not surprisingly, it is rare in the Smokies, where it is usually found along roadsides or deforested edges, trying to make its way back to the city.

KEY FEATURES: Small foliose yellow thallus, both marginal and terminal soralia, K-, on bark of trees in forest edge habitats near development or disturbance.

Candelaria fibrosa

Sunny Disposition

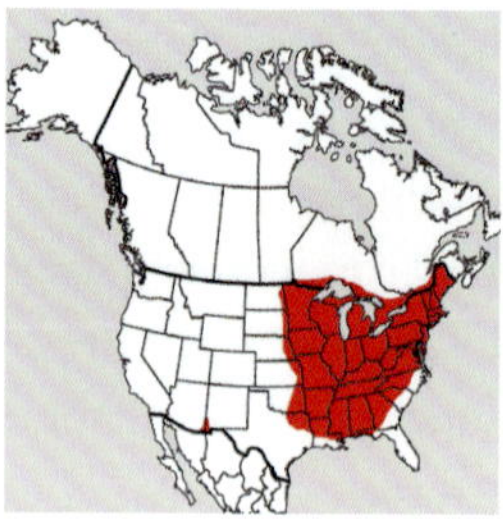

Dey 32865 (photo: Lendemer)

NOTES: *Candelaria fibrosa* is one of a number of species that was apparently common and widely distributed throughout eastern North America more than a century ago, and yet now has become rare in the region. *Teloschistes chrysophthalmus* is another such example. The rarity of *C. fibrosa* stands in contrast to its relative *C. concolor*, which seems to abound in disturbed areas and along forest edges. Maybe it really is harder to reproduce with spores than soredia. Regardless, there are very few species with which *C. fibrosa* can be confused. The K- thallus separates it from species such as *Xanthomendoza hasseana* and *X. weberi*. The greatest cause of confusion involves immature thalli of *C. concolor* that have not yet developed their characteristic asexual reproductive structure. *Candelaria fibrosa* can usually be distinguished in such cases by the presence of at least some pycnidia, which stand out as little bumps on the upper surface. It is a good lesson that not all lichens are ready for you to identify them!

CHEMISTRY: Calycin. Spot tests. Cortex: K-, C-, KC-, P-, UV+ dull orange.

NICHE: This species was once widespread in temperate eastern North America, but it is now rare, as is the case in the Smokies. It occurs on the bark and branches of hardwoods, especially in the canopies of tall trees, where it gets a lot of light. At the same time, it grows on old, isolated trees in fields and hedgerows, again where it receives all the sun it apparently needs.

KEY FEATURES: Small, yellow, foliose thallus, K-, without soredia and often with apothecia, on bark.

Candelariella efflorescens

Burn Out

Tripp 5430 (photo: Tripp)

NOTES: *Candelariella efflorescens* is a special species that you may encounter in the Smokies if the force is with you. It is easily distinguished from *C. xanthostigma* because it is sorediate (the latter is esorediate). Instead, *C. efflorescens* is most likely to be confused with *Chrysothrix*, but that genus is truly leprose, whereas *Candelariella* produces corticate areoles that bear soredia in this case. *Candelariella efflorescens* can potentially be confused with yellow species of *Caloplaca*, but the latter are always K+ purple.

CHEMISTRY: Calycin, pulvinic acid, pulvinic dilactone. Spot tests. K- to K+ pale red (fleeting), KC-, C-, P-, UV-.

NICHE: This species is relatively uncommon in the Smokies where it is corticolous on hardwoods and conifers at high elevations. Elsewhere, it is widespread in the Appalachians and northeastern North America.

KEY FEATURES: Bright yellow areolate thallus with soredia, calycin, and pulvinic acid, corticolous on hardwoods and conifers, albeit rare at high elevations in the Smokies.

Candelariella xanthostigma

Smoky Mountain Snuff

Tripp 3602 (photo: Lendemer)

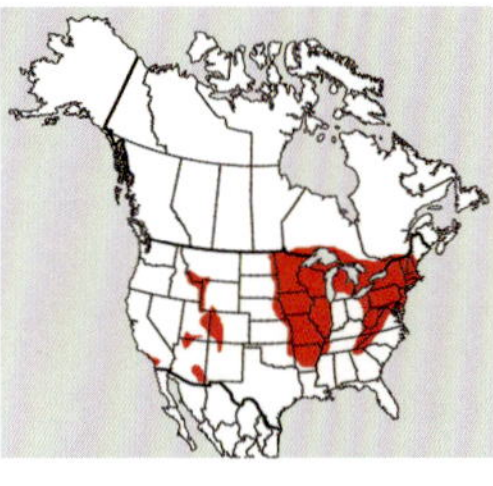

NOTES: *Candelariella xanthostigma* is one of three species in the genus in the Smokies, all of which have shiny, corticate, yellow thalli. This species is easily differentiated from the others by its granulose rather than areolate thallus with soralia. It is more likely to be confused with members of *Chrysothrix* or *Caloplaca*. However, *Chrysothrix* species have ecorticate, leprose thalli (resembling *Lepraria*) while all the yellow species of *Caloplaca* in the Smokies have K+ purple thalli. *Candelariella xanthostigma* is generally found in relatively small patch sizes, unlike *Chrysothrix* large swaths of carpet.

CHEMISTRY: Calycin, pulvinic acid, pulvinic dilactone, vulpinic acid. Spot tests. K+ pale red or K-, C-, KC-, P-, UV-.

NICHE: This species is restricted to old bark of very mature hardwoods at low-to-middle elevations in the Smokies, where it is infrequent. It is similarly infrequent elsewhere across its range.

KEY FEATURES: Shiny, corticate yellow thallus that is granulose and K-, small patch sizes, only on old, mature hardwoods, at middle-to-low elevations.

Canoparmelia caroliniana

Southern Sprawl Lichen

Lendemer 33132 (photo: Tripp)

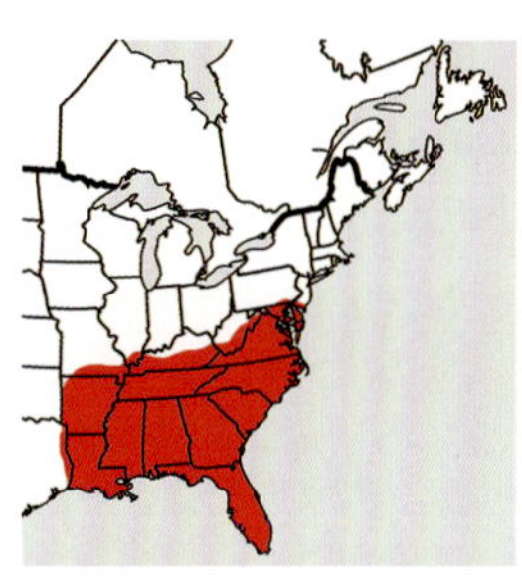

NOTES: Enter the land of gray-green foliose lichens. *Canoparmelia caroliniana* is one of numerous species in the Smokies that are large, foliose, gray-green, and isidiate. Start by learning the genus: *Canoparmelia* consistently grows tightly appressed to its substrate. In addition to isidia, *C. caroliniana* is specifically characterized by its white, maculate upper surface and presence of perlatolic acid in the medulla, which reacts UV+ blue-white. *Canoparmelia caroliniana* is most likely to be confused with *Punctelia rudecta*, which differs by having a very strong C+ red medulla reaction and white pseudocyphellae on the upper surface. Populations of *C. caroliniana* with a brown lower surface were earlier recognized under the name *C. amabilis*, but recent studies have shown these entities to be conspecific, and *C. caroliniana* has nomenclatural priority.

CHEMISTRY: Atranorin, perlatolic acid. Spot tests. (medulla) K-, C-, KC+ pinkish, P-, UV+ blue-white.

NICHE: In the Smokies, this species is occasional at low elevations, where it occurs on the bark of hardwoods and conifers, particularly in western portions of the Park. Elsewhere in its range, the southeastern United States, it is far weedier. We suspect its less frequent occurrence in the Smokies is due to higher-quality habitats in this area.

KEY FEATURES: Large, foliose, isidiate, white maculae on upper surface, UV+ blue-white medulla, low elevations, on hardwoods and conifers.

Canoparmelia texana

Creased and Crinkled

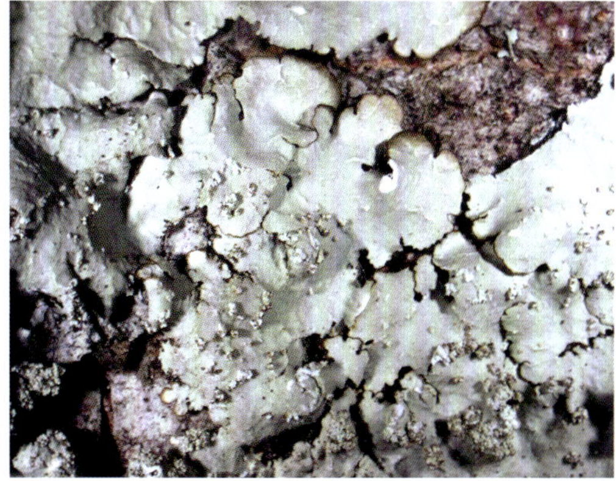

Lendemer 29616 (photo: Lendemer)

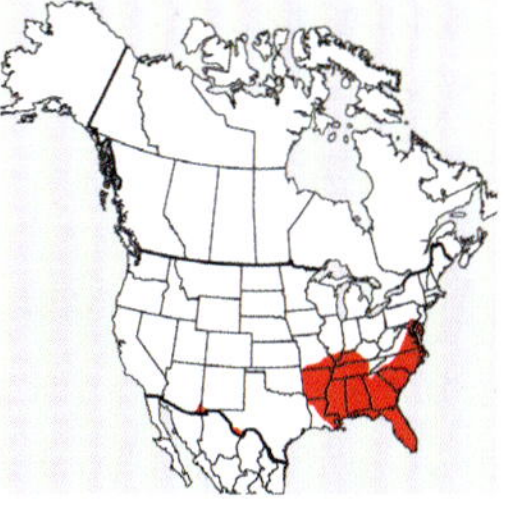

NOTES: Once you have experienced the wonder that is *Canoparmelia caroliniana*, you will be left wanting more. *Canoparmelia texana* superficially resembles *C. caroliniana* but differs in having soralia instead of isidia. Although the medulla in both species is UV+ blue-white, that reaction is due to the presence of different chemicals. Most of the time, *C. texana* can be recognized by its soralia, but occasionally one encounters thalli where the asexual propagules have gone wild and form pustules that dissolve into soredia. At first you would think it was a different species, but this does not seem to be supported by molecular data at the moment. Try not to let that get in the way of your *Canoparmelia* enjoyment. The only lichens that are likely to be confused with *C. texana* are *Punctelia missouriensis* and *P. caseana*, but both of those have a C+ red medulla and conspicuous white pseudocyphellae on the upper surface.

CHEMISTRY: Atranorin and divaricatic acid. Spot tests: (cortex) K+ yellow, C-, KC-, P-, UV-; (medulla): K-, C-, KC+ pinkish, P-, UV+ blue-white.

NICHE: In the Smokies, this species is rare and found only at low elevations, primarily on branches fallen from the canopies of trees. Its rarity is interesting given that it is common and widespread elsewhere in the southern Appalachians and the southeastern United States in general.

KEY FEATURES: Blue-gray foliose thallus, maculate upper surface, black lower surface, soredia or pustules, medulla UV+ blue-white, rare on bark at low elevations.

Catillaria lenticularis

Pass The Puck!

Lendemer 44540 (photo: Tripp)

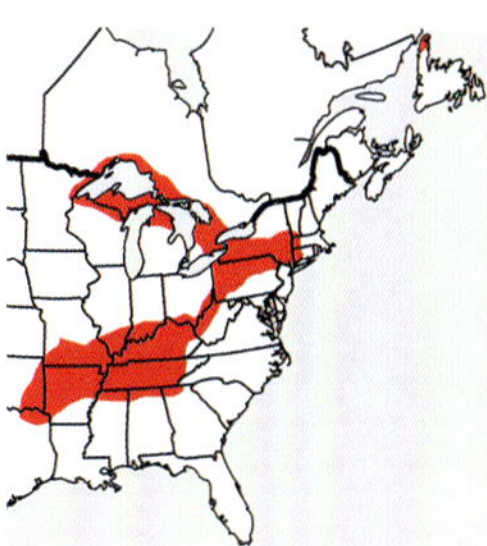

NOTES: This crustose lichen is a special find that occurs only on calcareous rocks (see Niche). It can be recognized by its tiny brown-to-black apothecia and its small, colorless, 2-celled spores. *Catillaria lenticularis* might be confused with *Bacidia coprodes*, but that species has rod-shaped spores and larger, black apothecia.

CHEMISTRY: No Substances. Spot tests. K-, KC-, C-, P-, UV-.

NICHE: *Catillaria lenticularis* is rare but locally abundant on calcareous rock outcrops that occur in the Smokies. It is found on similar substrates throughout eastern North America, specifically in the Appalachians, Ozarks, and Great Lakes. Keep an eye out for *C. lenticularis* when you hike though Ace Gap!

KEY FEATURES: Saxicolous on calcareous rocks, indistinct thallus, tiny black apothecia with 2-celled spores.

Catinaria atropurpurea

Little Brown Dots And I Don't Care

Tripp 5285 (photo: Lendemer)

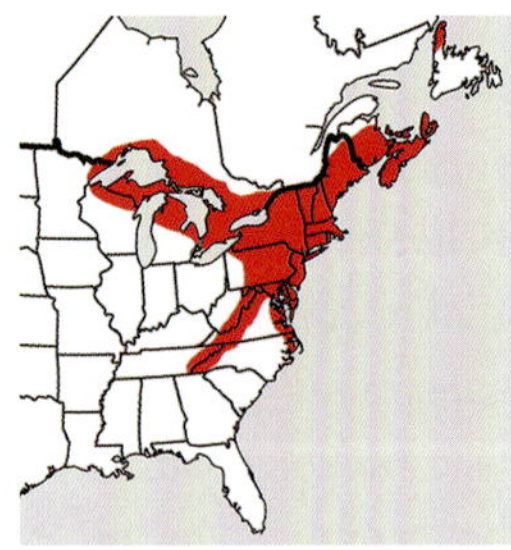

NOTES: *Catinaria atropurpurea* is a tiny treat that most Park visitors will likely never see. Nonetheless, the keenest of observers can identify it in the field by its miniscule dark apothecia set against a thin-green thallus. If you thought the apothecia of *Catillaria lenticularis* were small, better pull up a chair. In addition to its general appearance, this species has tiny, colorless 2-celled spores. It might be confused for a small *Bacidia* or *Bacidina*, but these have needle-shaped spores and are less entertaining!

CHEMISTRY: No Substances. Spot tests. K-, KC-, C-, P-, UV-.

NICHE: This species is an inconspicuous member of the lichen communities on hardwood trees at middle-to-low elevations of the Smokies. We once found it on a large fallen tree along Rough Fork Trail, and you might find it there, too. It is widespread in eastern North America but largely overlooked because of its small size.

KEY FEATURES: Tiny black dots (just like all the others) containing colorless, 2-celled spores. Look for corticolous, thin-green thalli, and trust your instincts. As we advocate in taxonomy, "Name it with confidence."

Cetradonia linearis

Rock Gnome Lichen

Not Vouchered (photo: Tripp & Lendemer)

NOTES: *Cetradonia linearis* is one of only two lichens currently protected by the US Federal Endangered Species Program. Although rare and certainly deserving conservation protection, the reality is that dozens of other Smokies lichens are in immediate need of similar or greater protection. *Cetradonia linearis* is instantly recognizable by its strap-like squamules, which are blue-green on the upper surface and pale below (see photos). This rare species might only be confused for *Cladonia petrophila*, which has a UV+ blue-white medulla and much broader squamules, which don't at all qualify as "strap-like." If you are lucky enough to see this species, take only photographs, as collecting it is strictly prohibited without special permission by the US Fish and Wildlife Service.

CHEMISTRY: Atranorin, protolichesterinic acid. Spot tests. K-, C-, KC-, P-, UV-.

NICHE: *Cetradonia linearis* is rare on seepage faces and on boulders in streams at high-to-middle elevations in the Smokies. It is endemic to such habitats in the southern Appalachians.

KEY FEATURES: Large, strap-like squamules with a blue-green upper surface, pale lower surface, terminal apothecia, on seepage rock faces and boulders in steams, high and middle elevations.

Cetraria laevigata

The Magic Hour

Tripp 5085 (photo: Lendemer)

NOTES: *Cetraria laevigata* is one of the only brown fruticose lichens that grows on soil within the Smokies. It can be recognized by its huge, flattened lobes, which are bordered by an unbroken line of white pseudocyphellae and are also conspicuously ciliate, both of which can be seen in the photograph. The P+ red medulla distinguishes this species from *Cetraria arenaria*, which occurs at lower elevations. Historically, the names Cet*raria islandica* and *C. ericetorum* have been incorrectly applied to this species in the Smokies. *Cetraria* will be familiar to natural history enthusiasts from the west but is an uncommon and exciting find in most portions of eastern North America.

CHEMISTRY: Fumarprotocetraric acid, protolichesterinic acid, and other fatty acids. Spot tests. (medulla) K-, C-, KC-, P+ orange-red, UV-.

NICHE: *Cetraria laevigata* is extremely rare in the Smokies, where it occurs only in the highest elevations in the Park, on wet, acidic soils. We have seen this species only once within Park boundaries. It is disjunct from the arctic, where it is far more common.

KEY FEATURES: Large brown fruticose lichen, flattened lobes bordered by unbroken line of white pseudocyphellae and conspicuous cilia, P+ red medulla, terricolous, highest elevations in Smokies.

Cetrelia cetrarioides

Chicita's Apprentice

Tripp 2252 (photo: Lendemer)

NOTES: There are three species of *Cetrelia* in the Smokies, which can be recognized by their large, foliose thalli, white pseudocyphellae on their upper surface, ruffled margins with abundant soralia, and black lower surfaces. *Cetrelia cetrarioides* differs from the other two by having perlatolic acid in the medulla, which reacts C- and dull UV+ blue-white. Populations of *C. cetrarioides* belong to two chemotypes that are sometimes treated as separate species. After unsuccessful attempts to distinguish these in the laboratory, we lumped them together here. It makes life easier. Besides other *Cetrelia*, this species may be confused with *Punctelia*, but none of the members of that genus has a UV+ blue-white medulla.

CHEMISTRY: Atranorin, perlatolic acid, accessories. Spot tests. (medulla): K-, KC+ pinkish, C-, P-, UV+ dull blue-white.

NICHE: *Cetrelia cetrarioides* is restricted to high elevation forests in the Smokies, where it is not uncommon. It occurs in similar habitats throughout the Appalachians and Great Lakes. Disjunct populations are found in other temperate regions including the Pacific Northwest, Europe, and eastern Asia.

KEY FEATURES: Large, blue-gray foliose thallus, white pseudocyphellae on upper surface, black lower surface, sorediate, C- and UV+ dull blue-white medulla, on bark at high elevations.

Cetrelia chicitae

Chicita's Charm

Tripp 3471 (photo: Lendemer)

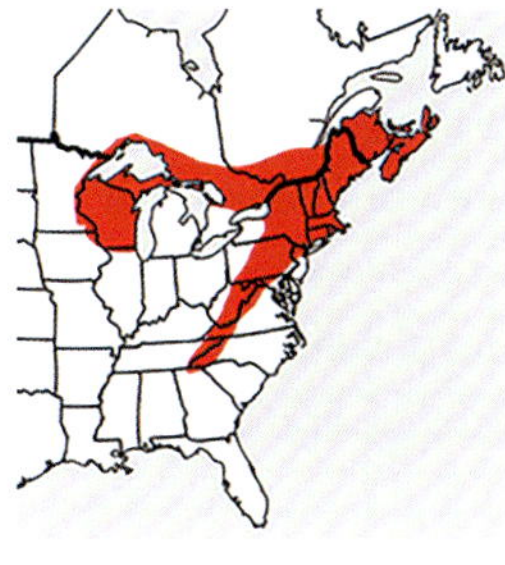

NOTES: *Cetrelia chicitae* can be distinguished from the other two congeners in the Smokies by its bright UV+ blue-white medulla that is C- and KC+ pink. *Cetrelia cetrarioides* is similar but has a less conspicuously UV+ medulla and occurs at high elevations. The specific epithet honors the pioneering lichen chemist Chicita Culberson.

CHEMISTRY: Atranorin, alectoronic acid. Spot tests. (medulla) K-, KC+ pink, C-, P-, UV+ bright blue white.

NICHE: This species is common in many areas of the Smokies, where it is found on bark or branches of hardwoods and conifers, more rarely on shaded non-calcareous rocks. Elsewhere in North America, *Cetrelia chicitae* is widespread in the Appalachian Mountains and Great Lakes Regions. Disjunct populations also occur in eastern Asia.

KEY FEATURES: Large, blue-gray foliose thalli, white pseudocyphellae on upper surface, black lower surface, sorediate, C- and UV+ bright blue-white medulla, on bark, all elevations.

Cetrelia olivetorum

A Hole-y Mess

Tripp 2296 (photo: Lendemer)

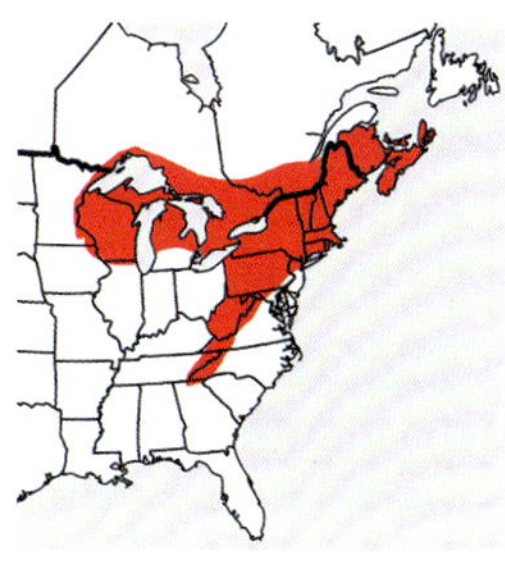

NOTES: *Cetrelia olivetorum* can easily be distinguished from the other two species of *Cetrelia* in the Smokies by a C+ red reaction in the medulla. Confusion with the very common *Punctelia rudecta* is possible, as that species also has pseudocyphellae on the upper surface and a C+ red medulla; however, the latter can easily be distinguished by its pale brown lower surface and less attractive appearance. What can we say?

CHEMISTRY: Atranorin, olivetoric acid. Spot tests. (medulla) K-, KC+ red, C+ red, P-, UV-.

NICHE: Like *Cetrelia chicitae*, *C. olivetorum* is common throughout the Smokies on bark and branches but occurs more rarely on shaded, non-calcareous rocks. It is widespread in the Appalachians and Great Lakes Region. Disjunct populations are found in many other regions including montane Central America, Europe, and eastern Asia.

KEY FEATURES: Large, blue-gray foliose thalli, white pseudocyphellae on upper surface, black lower surface, sorediate, C+ red medulla, on bark or rock.

Chaenotheca balsamconensis

Balsam Cone-Pin

Lendemer 44528 (photo: Tripp)

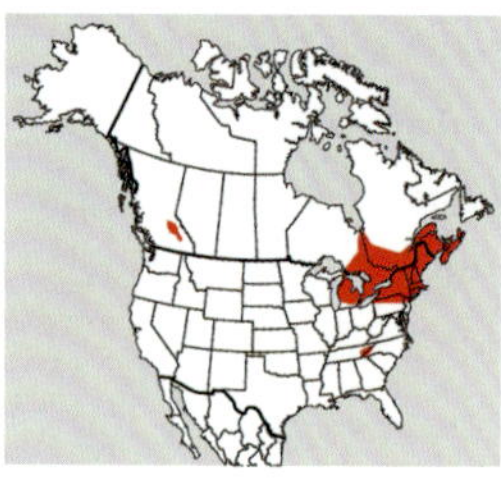

NOTES: Would you have thought there was anything that grew on a shelf fungus? We certainly didn't until Erin first found this species. We were puzzled by it until an enterprising PhD student found it again in the Black Mountains on the summit of Balsam Cone itself. There is no way to confuse this species with any other: it is the only lichen—other than *Phaeocalicium polyporaeum*—known to occur on shelf fungi. The latter differs in lacking an algal partner and in having 2-celled brown spores rather than simple globose spores, as with *Chaenotheca balsamconensis.*

CHEMISTRY: Unidentified xanthones. Spot tests. K-, KC-, C-, P-, UV+ dull orange (thallus).

NICHE: *Trichaptum abietinum* is a shelf fungus that occurs very commonly on dead or dying conifers, particularly hemlocks and firs that have been killed by invasive insects. It may be difficult to find a silver lining to the loss of our beloved conifers, but if such exists, it is the *Chaenotheca balsamconensis* that grows on *T. abietinum*, which grows on dying conifers.

KEY FEATURES: Fungicolous on shelf fungi at low, middle, and high elevations in the Smokies, lichenized, simple brown spores, restricted to *Trichaptum abietinum*, which grows on dead conifers.

Chaenotheca brunneola

Cinnamon Stalks

Tripp 4969 (photo: Lendemer)

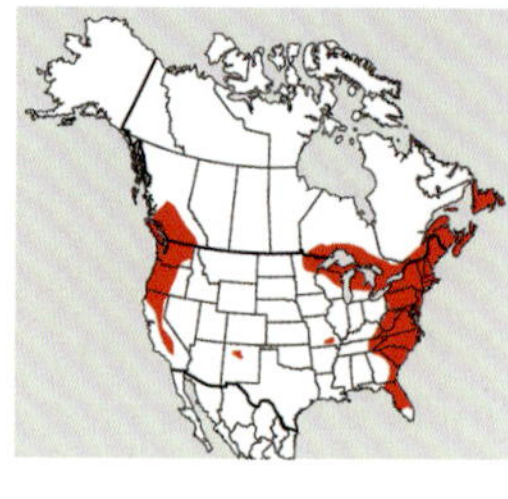

NOTES: *Chaenotheca brunneola* is a common pin lichen that can be recognized by its slender stalks, brown capitulum full of tiny, globose brown spores, epruinose undersides of the capitulum, and thin-green thallus with a coccoid green photobiont. Amateur lichenologists might confuse species of *Chaenotheca* for *Calicium*, but species in the former consistently have ornamented 2-celled spores and short, stouter pins.

CHEMISTRY: Unknown substance. Spot tests. K-, C-, KC-, P+ yellow, UV-.

NICHE: This species occurs on the hard, rotting wood of both conifers and hardwoods at all elevations in the Smokies. *Chaenotheca brunneola* is one of the most common members of the genus in the southern Appalachians and elsewhere in North America.

KEY FEATURES: Slender pins, brown, epruinose capitulum, tiny and simple globose brown spores, green thallus with coccoid photobiont, lignicolous, relatively common.

Chaenotheca chlorella

Golden Cuplets

McMullin 19095 (photo: McMullin)

NOTES: Calicoid fungi are truly one of the wonders of the lichen world, and the Smokies are unique for hosting so many species. While there are a lot of stubbles you could potentially encounter, surprisingly few of them are likely to be confused with *C. chlorella* because they almost all lack yellow pruina on the lower surfaces of the apothecia. Species of *Calicium*, like *C. trabinellum*, have such a pruina, but differ in their transversely septate spores. The most similar species are other members of *Chaenotheca* that also have brown spores, such as *C. furfuracea* and *C. chrysocephala*. However, *C. chlorella* differs from all these in having ellipsoid spores, a well-developed gray-to-greenish thallus, and a *Stichococcus* photobiont that can be recognized by its small, sausage shaped green cells.

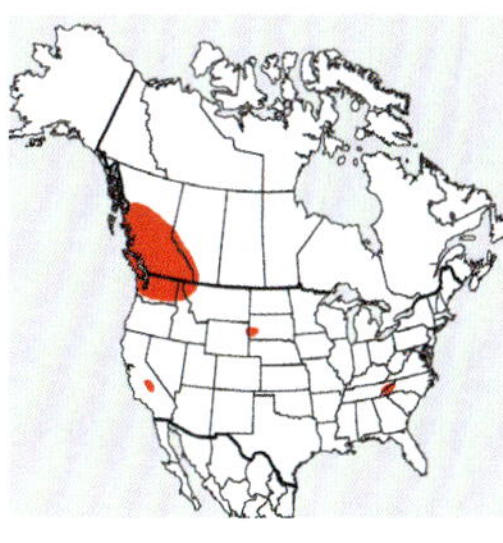

CHEMISTRY: Pulvinic acid pigments. Spot tests: K-, C-, KC-, P-, UV-.

NICHE: This is an infrequent species, restricted to mature forest stands at middle and high elevations in the Smokies and occurs on the bark and wood of mature trees. The southern Appalachian occurrences are disjunct from the primary range of the species that extends throughout much of the northern boreal forests in North America and Europe.

KEY FEATURES: Gray to greenish thallus, *Stichococcus* photobiont, tall slender stalks, lower side of cups yellow pruinose, simple brown ornamented spores.

Chaenotheca chrysocephala

Yellow All-Over Lichen

Tripp 2504 (photo: Tripp)

NOTES: *Chaenotheca chrysocephala* has, as the name implies, yellow-headed mazaedia that sit atop slender stalks as well as a yellow areolate thallus and a greenish-yellow pruinose capitulum. Yellow, yellow, yellow. But the spores are brown. These features make this species distinctive among all other pin lichens in the Smokies. *Chaenotheca chrysocephala* might be confused for *Chrysothrix xanthina*, but the latter is composed only of leprose granules (not on pins) that give the thallus a fuzzy or powdery appearance.

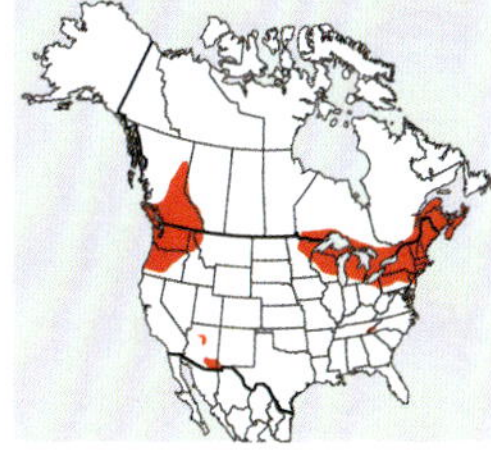

CHEMISTRY: Vulpinic acid. Spot tests. K-, C-, KC-, P-, UV-.

NICHE: This species occurs on bark and wood on large, old conifers and hardwoods. It is especially fond of mature forests at middle-to-high elevations, just like us. *Chaenotheca chrysocephala* also occurs in boreal habitats in eastern and western North America. It is special no matter where you find it.

KEY FEATURES: Slender yellow pins, yellow mazaedia, yellow areolate thallus, brown spores, vulpinic acid, corticolous and lignicolous, middle-to-high elevations.

Chaenotheca ferruginea

Ms. North Carolina!

Tripp 4921 (photo: Lendemer)

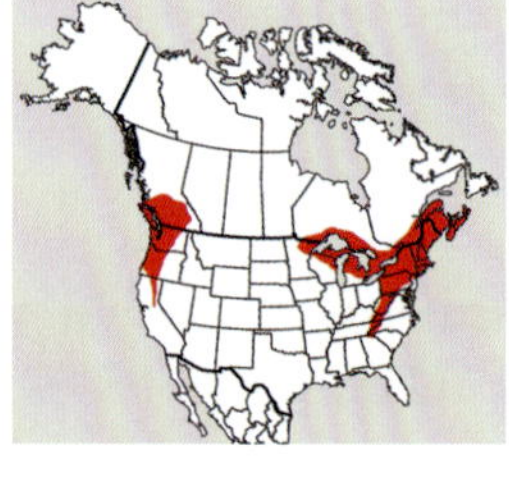

NOTES: *Chaenotheca ferruginea* is immediately recognizable by a conspicuous orange pigment that occurs in a variably developed thallus. This species also has ornamented, brown globose spores. A casual observer might first confuse this species for *C. chrysocephala*, but *C. ferruginea* is orange, not yellow, and also has an epruinose capitulum (vs. the greenish-yellow capitulum in *C. chrysocephala*). *Chaenotheca ferruginea* is pretty enough to win a beauty contest, and for that reason we like to call it Ms. North Carolina!

CHEMISTRY: Unknown orange pigment. Spot tests. K-, C-, KC-, P-, UV-.

NICHE: *Chaenotheca ferruginea* occurs at middle-to-high elevations in the Smokies, where it occupies the bark of old trees, particularly conifers as well as standing, dead stumps. Look for this species near the headwaters of Straight Creek. Elsewhere, this species occurs in similar habitats in the northern Appalachians, Great Lakes region, and the Pacific Northwest.

KEY FEATURES: Orange pigment in thallus, ornamented brown spores, epruinose capitulum, corticolous and lignicolous, middle-to-high elevations.

Chaenotheca furfuracea

The Golden Fleece

Tripp 3868 (photo: Lendemer)

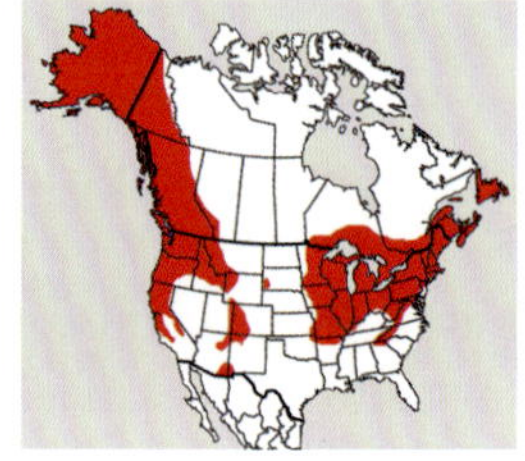

NOTES: Great things come in small packages. It is difficult to confuse *Chaenotheca furfuracea* for any congener, owing to its well-developed, fluffy green ecorticate thallus, giving it an almost powdery appearance. It also has pins that are concolorous with the thallus, and a nearly spherical capitulum. Some might mistake this species for *C. chrysocephala*, but that species has a corticate, granular thallus, a different ecology (on bark), and a different photobiont (coccoid green, instead of *Stichococcus* in *C. furfuracea*).

CHEMISTRY: Vulpinic acid. Spot tests. K-, C-, KC-, P-, UV-.

NICHE: *Chaenotheca furfuracea* occurs at middle-to-high elevations in the Park, where it grows on organic matter or rocks in cool overhangs. It occasionally grows on bark, too. Look for this species on Gunter Fork Trail, not far from Walnut Bottoms, not far from the biggest fields of Fringed Phacelia your eyes will ever see.

KEY FEATURES: Fluffy, green, ecorticate thallus, middle-to-high elevations, on organic matter or rocks.

Chaenotheca trichialis

Slender Stalks

Tripp 2112 (photo: Lendemer)

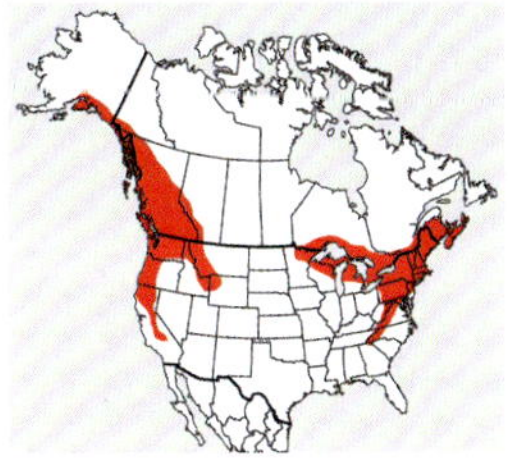

NOTES: *Chaenotheca trichialis* is an inconspicuous calicioid lichen (aka "pin lichen") that can be best identified by a combination of features including its granular, greenish-gray thallus that is P-, its weakly to densely white pruinose pins, and its *Stichococcus* photobiont. Its pins are also rather hair-like, as the specific epithet implies.

CHEMISTRY: No substances. Spot tests. K-, C-, KC-, P-, UV-.

NICHE: Just like most of the other *Chaenotheca* in the Smokies, *C. trichialis* occurs at middle-to-high elevations on the bark of hardwoods or conifers. If you are wandering around the Purchase Knob area, no doubt learning something, learn one more thing by finding this near Double Gap. This species is relatively broad ranging across the Appalachians and Great Lakes region as well as in the Pacific Northwest into Canada.

KEY FEATURES: Granular, greenish-gray thallus, +/- white pruinose pins, *Stichococcus* photobiont. Good luck!

Chaenothecopsis debilis

That's Debatable

McMullin 19111 (photo: McMullin)

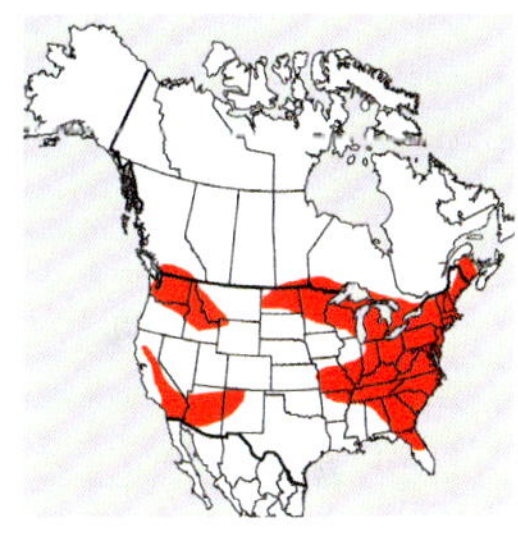

NOTES: One is likely to find only a really small number of calicioids, or stubbles, in the Smokies. So, if you come across a field of tiny black pins growing on the wood of a dead tree and the pins have 2-celled, brown spores, then there is a high degree of probability it is this species. Do not get overwhelmed by the number of other *Chaenothecopsis* species known from the area; they are not likely to be found on your daily jaunt down the trail. The reddish coloration of the stalk, when viewed under the compound microscope, makes this species distinctive, especially when combined with the transversely septate spores. Not impressed with *Chaenothecopsis*? Well it isn't even a lichen.

CHEMISTRY: No substances. Spot tests. K-, KC-, C-, P-, UV-.

NICHE: This is a common and widely distributed species throughout much of temperate, eastern North America. It is typically found on the old, dry wood of standing dead tree trunks.

KEY FEATURES: Small, slender, black pins, persistent asci, 2-celled, brown spores, reddish stalk turning more intense in K, on wood of standing stumps.

Chaenothecopsis nana

Granny Sticks

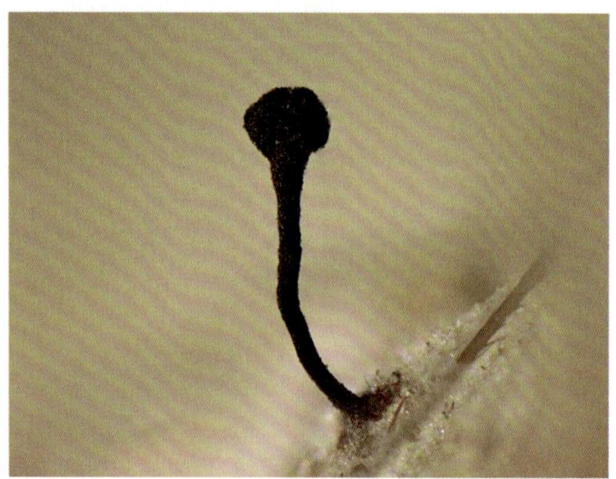

McMullin 19050 (photo: McMullin)

NOTES: Nana means "small" or "little," and the diminutive is certainly appropriate for this species. To find it, you will need to snuggle up closely with the horizon line of the substrate, look for black pins, and then look for even smaller black pins. If you happen to find *Chaenothecopsis nana*, you can distinguish it from other members of the genus by its stature, simple, ellipsoid spores, and by the fact that the stalk and cap do not react with K. Sometimes the absence of positives can be a good thing.

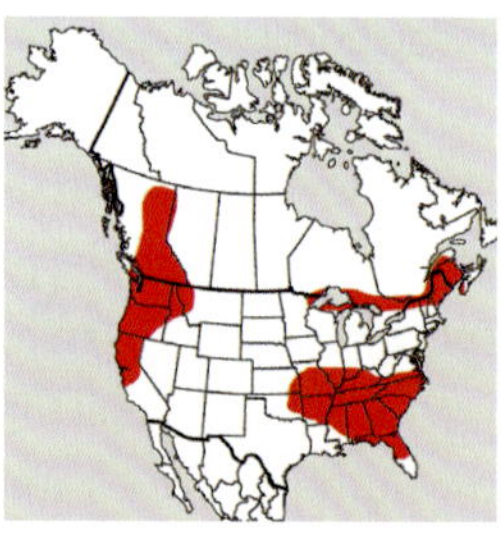

CHEMISTRY: No substances. Spot tests. K-, KC-, C-, P-, UV-.

NICHE: *Chaneothecopsis nana* is infrequent but widespread in eastern North America. It is typically found on the bark of pines in the southeastern United States, but this may be more of a perception than a truth. It is known from a small number of scattered occurrences in the Smokies but has likely been overlooked by most lichenologists, including us.

KEY FEATURES: Small, short, slender, black pins, persistent asci, simple, ellipsoid spores, black stalk, on wood and bark of trees.

Chaenothecopsis pusilla

Fairy Pins

McMullin 19052 (photo: Tripp)

NOTES: As the generic name implies, *Chaenothecopsis* "looks like" *Chaenotheca* but differs by having septate, 2-celled spores (for the most part) and in lacking a mazaedium. The genus also differs from the superficially similar *Calicium* by its lack of a mazaedium. From both genera, *Chaenothecopsis* species are further differentiated by their lack of a photobiont. Within *Chaenothecopsis*, *C. pusilla* is differentiable from other species by

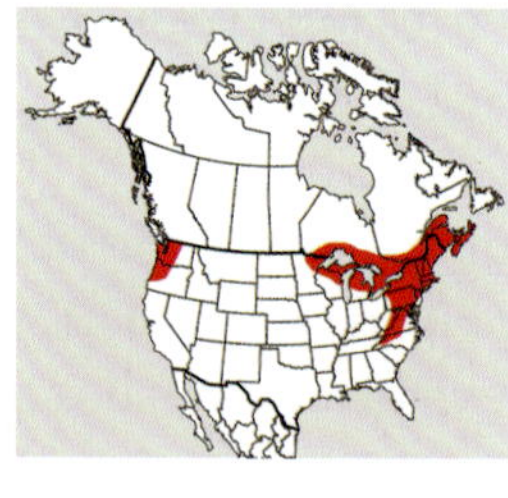

its spherical heads, its K+ capitulum reaction that does not bleed outside into a water mount (use a compound microscope!), and its occurrence on bark or wood. It is closely related to *C. nigra,* but the latter species has spores with septa that are darker than the cell wall.

CHEMISTRY: Unidentified pigments in capitulum. Spot tests. (capitulum) K- or K+ intensifying the unknown pigment, C-, KC-, P-, UV-.

NICHE: Do you know anything about *Chaenothecopsis* niches? We don't.

KEY FEATURES: Continuous white thallus (appearing as stain on substrate), brown, 2-celled spores, lacking a photobiont, corticolous on old, chunky-bark trees, difficult to photograph.

Chaenothecopsis savonica

Perplexion

McMullin 19002 (photo: McMullin)

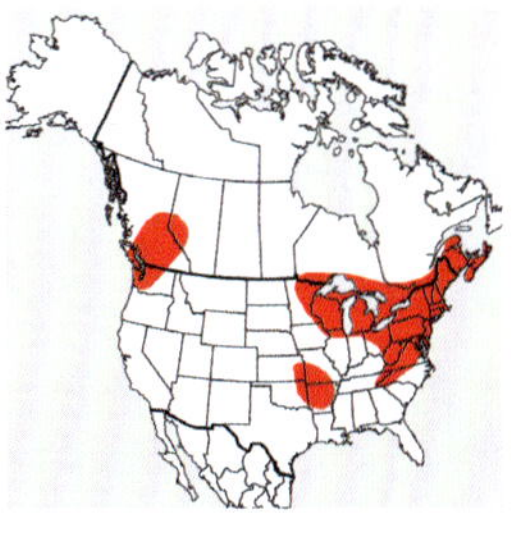

NOTES: Stubbles are fungi that generally everyone can love. An exception might be this species, which seems to cause endless confusion to all that try to learn it. Most species of *Chaenothecopsis* can be identified by their septate spores or by their colorful pruina. *Chaenothecopsis savonica* is an exception to this, and the black stalks with simple brown spores lead one to immediately think of *Mycocalicium*. Nearly every identification key distinguishes *C. savonica* from species like *M. subtile* or *M. fuscipes* by the presence of a narrow canal in the ascus tip of *Chaenothecopsis*. However, after years of searching, we can report that this character is almost impossible to observe no matter how hard you try. Instead, the best way to recognize *C. savonica* is by is slender, delicate stalks that tend to become somewhat branched, its globose rather than flattened capitulum, and the presence of football-shaped brown spores.

CHEMISTRY: No substances. Spot tests: K-, C-, KC-, P-, UV-.

NICHE: A widespread species of both conifer and hardwood snags, *C. savonica* is common and widespread throughout eastern North America, including in the Smokies.

KEY FEATURES: Non-lichenized thallus, slender, delicate black stalks, epruinose globose capitula, simple, brown, football-shaped spores, on snags at all elevations.

Chrysothrix chamaecyparicola

Moonglow

Lendemer 23648 (photo: Tripp)

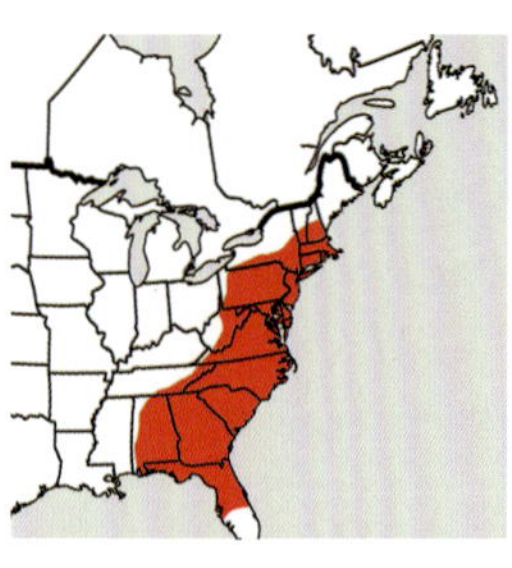

NOTES: *Chrysothrix chamaecyparicola* is one of those crustose lichens that is easy to overlook because the dull greenish-yellow color of the thallus, coupled with the small size of the thallus granules, would lead one to think it was an alga rather than a lichen. Indeed, despite being common and widespread, it was almost entirely overlooked until it was formally described in 2010. When in doubt as to whether this is really a lichen, you can check for the spot test reaction of the thallus under UV. The only species in the Smokies that could be confused with *C. chamaecyparicola* is *Psilolechia lucida*, another similarly pigmented lichen with a minutely granulose leprose thallus. However, that species occurs on sheltered rocks rather than conifer bark. Although one may be tempted to confuse *C. chamaecyparicola* with other members of *Chrysothrix*, it is readily distinguished from all of them by the extremely small size of the granules.

CHEMISTRY: Rhizocarpic acid. Spot tests: K-, KC-, C-, P-, UV+ dull orange.

NICHE: This species is almost entirely restricted to the bark of conifer trees in humid habitats such as along streams or in wet depressions. It is extremely common in the Coastal Plain, where it is often the dominant lichen species in swamps and covers the trunks of whole stands of trees. In the Smokies it occurs at middle-to-low elevations on pines and hemlocks, and it used to be common on the old hemlocks along streams near Cataloochee.

KEY FEATURES: Minutely granular leprose thallus, dull greenish-yellow in color, UV+ dull orange, on conifers at middle-to-low elevations.

Chrysothrix insulizans

Golden Isles

Collector 5096 (photo: Tripp)

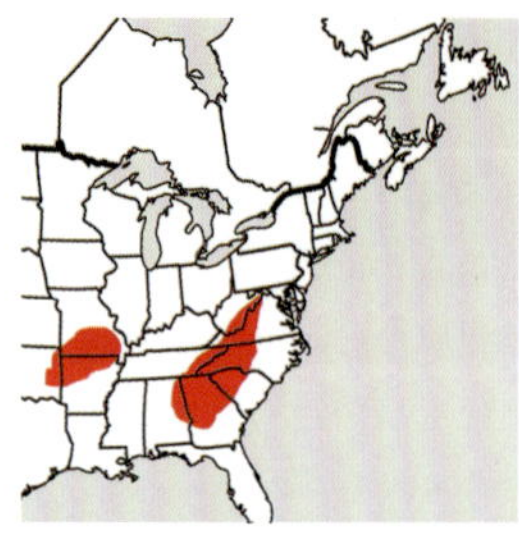

NOTES: From a distance, populations of *Chrysothrix insulizans* resemble a continuous layer of yellow paint on the rocks they inhabit. However, closer inspection will reveal the thallus to be comprised of many haphazardly arranged small piles ("Golden Isles") of bright yellow granules, some of which have grown together and merged. *Chrysothrix susquehannensis* is somewhat similar, but is very rare, duller green in color, and C+ pink. *Psilolechia lucida* could also be confused with *C. insulizans*, but the thallus in that species is continuous rather than island forming and is duller green in color.

CHEMISTRY: Calycin, leprapinic acid?, zeorin (trace). Spot tests: K+ fleeting reddish, C-, KC-, P-, UV+ dull orange.

NICHE: This species occurs throughout the Smokies on sheltered vertical faces and overhangs of non-calcareous rocks. If you are hiking along the Appalachian Trail and see a rock that looks like it is painted bright yellow, then it is probably this species. Looking down from the "The Jumpoff" near Charlies Bunion, one can easily see whole cliff faces covered with it. Outside of the Smokies, *Chrysothrix insulizans* occurs throughout the central and southern Appalachians, with a disjunct distribution in the Ozarks.

KEY FEATURES: Granular leprose thallus, forming island-like thalli, yellow in color, C-, on sheltered rocks and on sheer vertical non-calcareous rock faces at middle-to-high elevations.

Chrysothrix onokoensis

Limelight

Tripp 8673 (photo: Tripp)

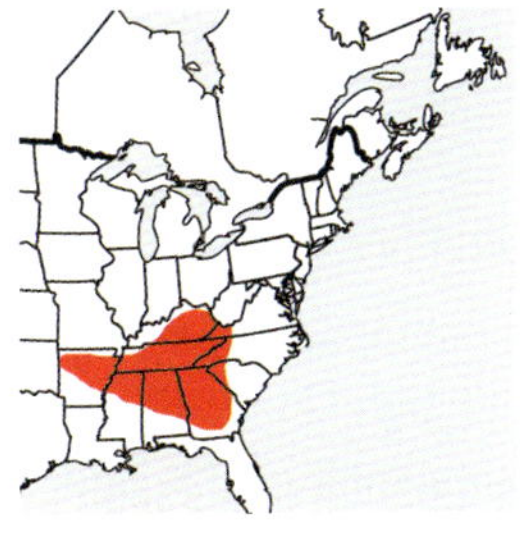

NOTES: Limelight is a wonderful secrete of the Smokies (and beyond!) but is neither common nor rare. It is one we urge you to find because it is both doable and beautiful! This species is characterized by having a lemon-lime colored thallus composed entirely of leprose granules, its occurrence on rock in nooks and other shaded crevices, and its C- and UV+ pink thallus. It also has a distinct, sordid white hypothallus (look closely!), and is relatively easily removed from its rock substrate owing to the thick, poofy colonies it forms. Among saxicolous members of the genus, you are most likely to confuse limelight for *C. susquehannensis*, but that species is C+ pink/red and also occurs on moss cushions or other organic matter (that, in turn, occurs on top of rock) rather than directly on rock. *Chrysothrix insulizans* can also be similar, but it has thinner continuous thalli that never become poofy and produces calycin rather than leprapinic acid. You might also confuse *C. onokoensis* for *Psilolechia lucida*, but that species is duller green in color, produces a different chemical (rhizocarpic acid), and never forms big, poofy thalli.

CHEMISTRY: Leprapinic acid. Spot tests: K-, C-, KC-, P-, UV+ pink.

NICHE: *Chrysothrix onokoensis* is a species primarily of the southeastern United States, where it occurs on shaded non-calcareous rocks in overhangs. Its epithet refers to its type locality, Glen Onoko, Pennsylvania, a gorge known to harbor several other taxa that are disjunct from more southerly locations.

KEY FEATURES: Bright lemon-lime-colored thallus of leprose granules, poofy colonies, C- thallus, on non-calcareous rocks at low and middle elevations.

Chrysothrix susquehannensis

Mountain Gold Dust

Collector Tripp 6065 (photo: Lendemer)

NOTES: You'll say, "I'm rich!" when you find this gold dust on one of your hikes in the Smokies. Of the *Chrysothrix* species that occur in the Park, this is the rarest and least conspicuous. As a result of its ecology, it is most likely to be confused with *C. insulizans*, which also grows on massive non-calcareous rock faces and has thalli comprised of dispersed piles of granules. However, *C. susquehannensis* differs from that species in having a greenish-yellow (vs. strong yellow) coloration and in reacting C+ pink due to the presence of gyrophoric acid.

CHEMISTRY: Rhizocarpic acid and gyrophoric acid. Spot tests: K-, KC+ pink, C+ pink, P-, UV+ dull orange.

NICHE: *Chrysothrix susquehannensis* is a North American endemic that is widely distributed in the Appalachian Mountains, with disjunct populations in mountainous areas the southwestern United States. It occurs on rocks and small pieces of organic matter in sheltered areas of vertical non-calcareous rock faces. In the Smokies, it is known from only a single location north of Fontana Dam.

KEY FEATURES: Granular leprose thallus, forming small island-like thalli, greenish-yellow in color, C+ pink, on sheltered rocks and organic matter on shear vertical non-calcareous rock faces at middle-to-low elevations.

Chrysothrix xanthina

Eastern Stardust

Tripp 1479 (photo: Deregibus)

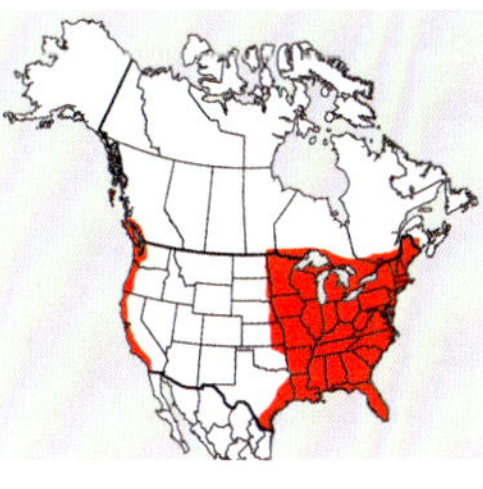

NOTES: *Chrysothrix xanthina* is one of the most common, widespread, and easy-to-recognize sterile crustose lichens in the Smokies and much of eastern North America. This species is very easy to identify by its bright yellow leprose thallus composed of relatively large granules. This species differs from two close relatives, *C. insulizans* and *C. onokoensis*, by its ecology (see Niche; the latter two are saxicolous). *Chrysothrix chamaecyparicola* is another relative but differs by having a thinner, more dispersed thallus of greenish-yellow granules that are much smaller in size (thus resembling an alga more so than a lichen) and its preference for conifers.

CHEMISTRY: Pinastric acid. Spot tests. K-, C-, KC-, P-, UV-.

NICHE: Eastern Stardust occurs on the bark of hardwoods as well as conifers, especially in humid forests, at all elevations. It is very common and very widespread in the Smokies as well as throughout the remainder of eastern North America. It is a weed, but a pretty one that you will soon have the pleasure of knowing, if you do not already.

KEY FEATURES: Bright yellow thallus of ecorticate (i.e., leprose) granules, occurrence on bark, very common and widespread.

Cladonia apodocarpa

Mouse Mats

Harris 56261 (photo: Lendemer)

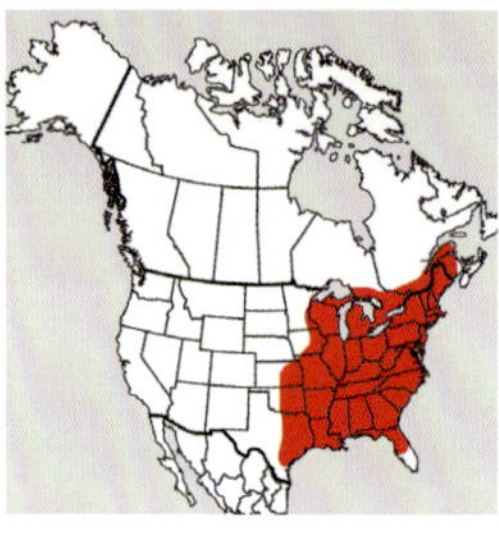

NOTES: Mouse Mats is one of several, persistently sterile terricolous *Cladonia* in the Smokies. Like some of the others, it tends to prefer disturbed soils such as trail sides. The species can be recognized by is its long, strappy lobes with bright white undersides. Nonetheless, don't trust your good looks or luck to confirm its identify based on these features alone. As with all of the frequently sterile *Cladonias* (including *C. stipitata*, *C. caespiticia*, and *C. sobolescens*), you need to run TLC to verify the identification.

CHEMISTRY: Atranorin, fumarprotocetraric acid. Spot Tests: K+ yellow, KC-, C-, P+ orange to red, UV-.

NICHE: *Cladonia apodocarpa* is a species primarily of the eastern United States. It is very common throughout the region and is certainly at home throughout the Appalachian Mountains. Look for this species on soil at low elevations, particularly on south-facing slopes. Mouse Mats *loves* the Lakeshore Trail, along with several other of its *Cladonia* friends . . .

KEY FEATURES: Persistently sterile *Cladonia* with strap-like lobes and bright white undersides, disturbed soils at low elevations, bearing atranorin and fumarprotocetraric acid.

Cladonia appalachensis

Anakeesta Cupless Cups

Lendemer 43817 (photo: Lendemer)

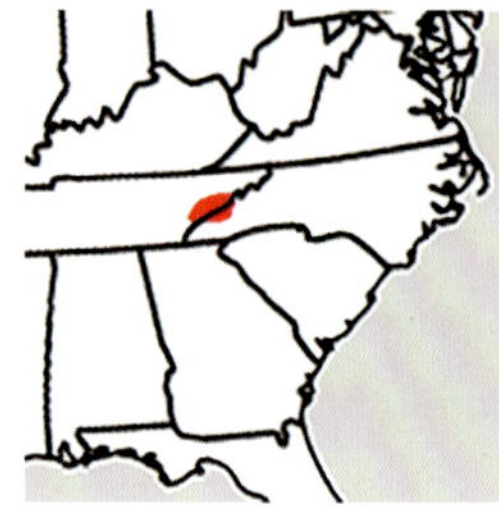

NOTES: This species has been living in the Smokies for hundreds (thousands?) of years but didn't have a name. The famous bryologist A.J. Sharp first collected it in the 1930s, but it remained nameless until 2013. *Cladonia appalachensis* is a highly distinctive member of the genus first and foremost by its characteristic greenish-brown color. Its slender, abundantly branched podetia, chemistry, and highly specialized ecology (see Niche) confirm the identity of this Smokies treasure. In the field, it looks superficially similar to *C. furcata*, but the podetia of the latter are far less delicate and slender (and are also P+ red and UV-).

CHEMISTRY: Squamatic acid. Spot tests. K-, C-, KC-, P-, UV+ blue-white.

NICHE: Narrowly restricted-endemic to Anakeesta Rock outcrops in the Smokies.

KEY FEATURES: Greenish-brown, slender, highly branched podetia, P- and UV+ blue-white thallus, restricted to Anakeesta Rock at high elevations.

Cladonia arbuscula

Mountain Bedazzlement

Tripp 5082 (photo: Lendemer)

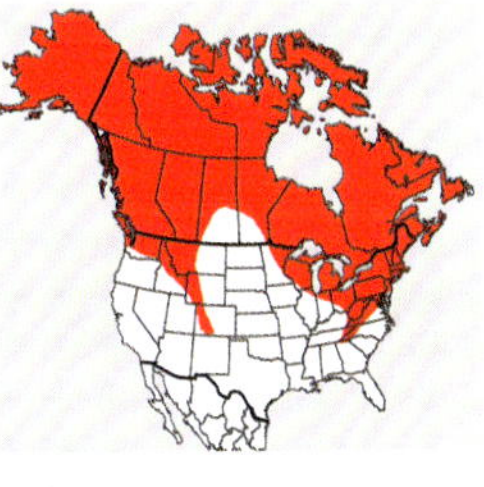

NOTES: *Cladonia arbuscula* is a highly charismatic member of the genus and is one of the "true" Reindeer Lichens (i.e., was formerly classified within *Cladina*). This species is instantly recognizable by its abundant, dichotomously branching, ecorticate podetia, its thalli that form cushions, and its blue-green or yellowish-green appearance (see Chemistry). In the Smokies, *C. arbuscula* is only likely to be confused with *C. subtenuis* (with a smaller thallus, more slender branching, and occurrence at low elevations), or *C. rangiferina* (with a whiter thallus containing atranorin instead of usnic acid).

CHEMISTRY: Usnic acid, fumarprotocetraric acid. Spot tests. K-, C-, KC+ yellow, P+ orange-red, UV-.

NICHE: *Cladonia arbuscula* is relatively common on organic matter at high-to-middle elevations in the Smokies. It is a classic boreal element of North America that makes its way down the Appalachians to its southernmost populations in the Smokies.

KEY FEATURES: Abundantly and dichotomously branching ecorticate podetia, KC+ gold and P+ red, high elevations on organic matter.

Cladonia caespiticia

Brown Town

Lendemer 32987 (photo: Tripp)

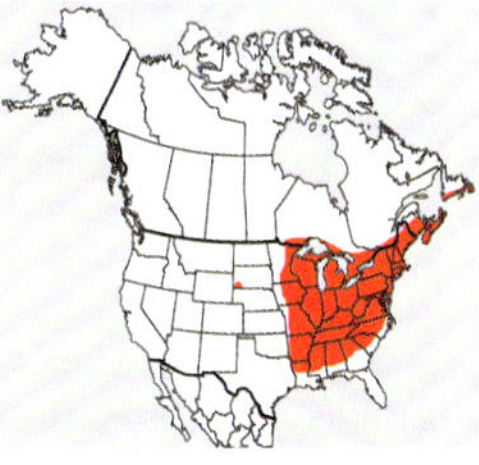

NOTES: *Cladonia caespiticia* is highly distinctive because of its relatively large but finely divided squamules and its apothecia that sit atop short translucent-to-white stipes instead of podetia. It can only be confused with *C. petrophila* or *C. stipitata*, both of which have stipes instead of podetia, and differ chemically, ecologically, and in having squamules that are never finely divided. The epithet "caespiticia" likely refers to its mounding apothecia.

CHEMISTRY: Fumarprotocetraric acid. Spot tests. K+/- dirty brown, C-, KC-, P+ orange-red.

NICHE: *Cladonia caespiticia* is, lucky for us, relatively common throughout the Smokies. It always occurs on rather large heaps of organic matter and occupies similar environments throughout its range in eastern North America. A great place to see this species is on the summit of Luftee Knob. It's true, we are fond of it. Just look at those stipes!

KEY FEATURES: Large, finely divided squamules, apothecia on stipes (not podetia), K-, P+ orange-red, common on heaps of organic matter.

Cladonia chlorophaea

Lichen Salutation

Lendemer 18942 (photo: Lendemer)

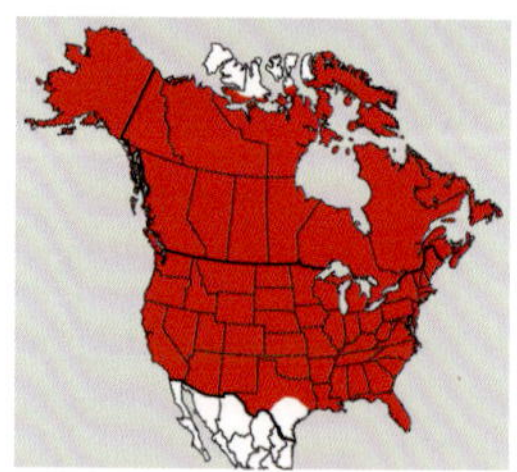

NOTES: *Cladonia chlorophaea* is a beautiful lichen that you will most likely find when in full salutation, as in this photograph. It is characterized by having short, stout podetia covered basally or entirely by coarse, bluish-gray soredia, which can be found lining the squamules, too. Lichen Salutation belongs to the *Cladonia chlorophaea* group, which consists of several closely related species that require Thin Layer Chromatography to identify with confidence. It is perhaps most likely to be confused with *C. grayi*, which is one of the most common lichens in the Smokies, but the latter is UV+ blue-white owing to the presence of grayanic acid. *Cladonia merochlorophaea* is similar but podetia in this species react KC+ pinkish red owing to merochlorophaeic acid (how convenient). Finally, don't confuse *C. chlorophaea* for *C. fimbriata*, which differs in having much taller and slenderer podetial, and in having much finer soredia.

CHEMISTRY: Fumarprotocetraric acid. Spot tests: K- to K+ brown, C-, KC-, P+ red, UV-.

NICHE: Lichen Salutation is relatively common in the high elevations of the Smokies and many places beyond, and thus is worth your while to learn! It grows on soil and on the bases of trees, typically at middle-to-high elevations. Look for it on the Boogerman Loop. Elsewhere in North America, it can be found in colder environments ranging through the Appalachians and Great Lakes and also occurring in the southern Rocky Mountains . . . all great places to travel to see lichens!

KEY FEATURES: Cup-forming *Cladonia*, short, stout podetial, bluish-gray, coarse soredia lining the squamules and covering all or portions of podetial, UV-, KC-, P+ red, on soil and tree bases, middle-to-high elevations.

Cladonia coccifera

Berry Tips

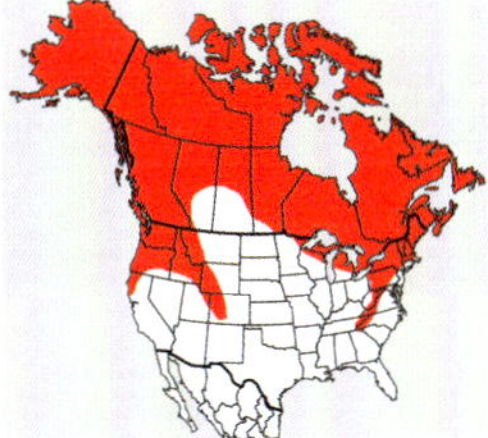

Tripp 5088 (photo: Lendemer)

NOTES: *Cladonia coccifera* can readily be identified by its usnic-acid containing thallus, which gives it a yellowish-green appearance, its cupped podetia that are covered by squamules (but otherwise ecorticate), and its red apothecia. This species could be confused with *C. cristatella*, which differs by having completely corticate podetia (see photo of that species), or *C. didyma*, which differs by lacking cups and having a different chemistry. *Cladonia coccifera* also has podetia that forms cups, unlike these other two species. The specific epithet likely refers to its red apothecia, which we suppose resembles a "berry." *Cladonia pleurota* is also similar; but in the Smokies, that species always has podetia that form cups and the cups are covered in soredia rather than squamules.

CHEMISTRY: Usnic acid, zeorin. Spot tests. K-, C-, KC+ gold, P-, UV-.

NICHE: *Cladonia coccifera* occurs on organic matter over non-calcareous rocks at high elevations, in both shaded and exposed habitats, but always in areas with cold air flows. This species is typical of arctic–boreal environments across northern portions of North America, making its way south into the Southern Appalachian as well as Southern Rocky Mountains . . . two of the greatest places on Earth to be a lichenologist.

KEY FEATURES: Ecorticate, cup-forming podetia covered by (corticate) squamules, red apothecia, K- and KC+ gold, high elevations.

Cladonia cristatella

British Soldiers

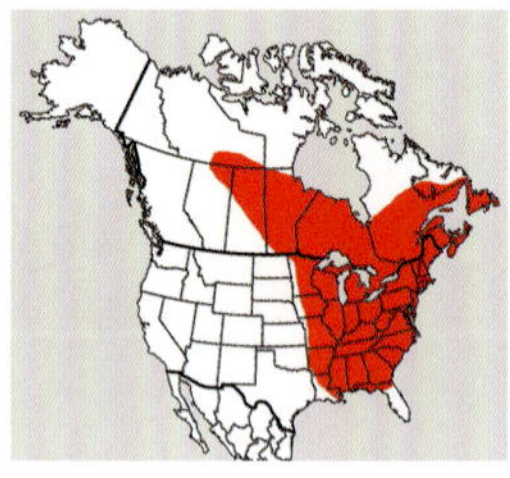

Tripp 3690 (photo: Lendemer)

NOTES: *Cladonia cristatella* has the unique distinction of being both beautiful and having one of the best-known common names of any lichen on the planet (we wish we could take credit but cannot). Beyond these facts that are fun to consider but don't aid in its identification, *C. cristatella* is very similar to and only potentially confusable with *C. coccifera*. Both have bright red apothecia but the latter has squamulose podetia that are otherwise ecorticate, whereas podetia of *C. cristatella* are covered by a continuous cortex.

CHEMISTRY: Usnic acid, barbatic acid, didymic acid. Spot tests. K-, C-, KC+ gold, P-, UV-.

NICHE: *Cladonia cristatella* is widespread and common throughout the Smokies, where it occurs on a variety of substrates, including soil, lignum, and bark. Elsewhere in eastern North America and into eastern Canada, it occurs on similarly diverse substrates and has a special knack for occupying polluted environments.

KEY FEATURES: Fully corticate podetia, red apothecia, usnic acid, barbatic acid, didymic acid, widespread and common on diverse substrates.

Cladonia didyma (w/ subspecies)

Anothery

Lendemer 33142 (photo: Tripp)

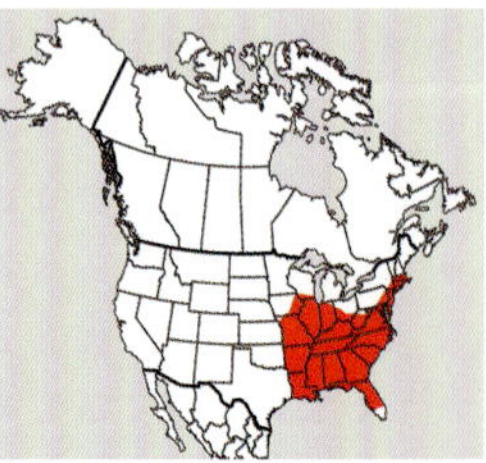

NOTES: Another red one. *Cladonia didyma* is most similar to *C. macilenta* but the latter species has sorediate instead of squamulose podetia that so typify *C. didyma*. *Cladonia didyma* is also superficially similar to *C. coccifera* in having red apothecia and squamulose podetia but podetia in the former are fully corticate versus ecorticate between the squamules in the latter. Furthermore, *C. didyma* is never cup forming whereas *C. coccifera* is. Their chemistries are also completely different and their distributions are non-overlapping. Finally, *C. didyma* is easily separated from *C. cristatella* by having podetia with squamules. See? The red ones aren't so bad afterall. Think of them more as an opportunity for celebration rather than a test of your patience.

CHEMISTRY: Thamnolic acid, barbatic acid, didymic acid, depending on variety. Spot tests. (podetia) K+ bright yellow or K-, C-, KC+ orange or KC-, P+ orange or P-, UV+ blue-white or UV-.

NICHE: This species is relatively common on organic matter, sandy soils, and rotting wood. It occurs in middle-to-low elevations of the Smokies and is also common throughout the southeastern United States.

KEY FEATURES: Squamulose, fully corticate, non-cup-forming podetia, red apothecia, diverse chemistries and spot tests depending on variety, common at low elevations on organic matter.

Cladonia digitata

Decorated and Distinguished

Lendemer 32932 (photo: Tripp)
Inset: Lendemer 32932 (photo: Tripp)

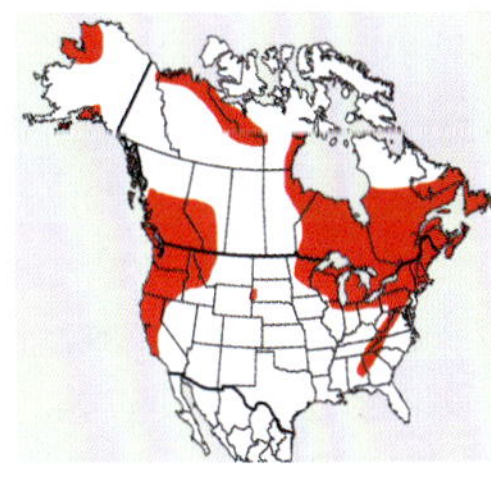

NOTES: Very soon we will have run out of red apotheciate *Cladonias* to talk about, we promise! But for now . . . *Cladonia digitata* is readily characterized by its large squamules, which are bordered by beautiful, marginal soralia; its cupped podetia, which are basally corticate and otherwise covered by large patches of soralia, especially toward the tips; and its red pycnidia (see the inset); and red apothecia. It is possible that some may confuse this species for the extremely common and widespread *C. ochrochlora*, but the latter doesn't have red reproductive structures and lacks soredia. *Cladonia macilenta* might also be mistaken for *C. digitata*, but that species does not have cupped podetia.

CHEMISTRY: Thamnolic acid. Spot tests. K+ bright yellow, C-, KC-, P+ orange, UV-.
NICHE: *Cladonia digitata* is restricted to the highest elevations in the Smokies, where it occupies organic matter and rotting wood. It is otherwise a boreal species, reaching its southern limit of distribution in the Smokies.
KEY FEATURES: Large squamules bordered by soralia, cupped podetia that are partially corticate and partially soraliate, red reproductive structures, thamnolic acid, high elevations.

Cladonia floerkeana

Flëchten Floerke

Tripp 3691 (photo: Lendemer)

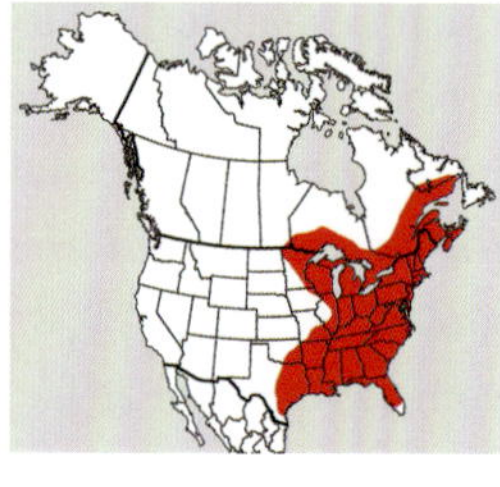

NOTES: *Cladonia floerkeana* is a good lichen to learn to identify by "gestalt," especially now that you may be reaching your upper limit of red *Cladonias*. Podetia of *C. floerkeana* resemble scraggly white fingers with lumpy surfaces. This appearance is due to its characteristic morphology, in which the podetia are covered by big, coarse, corticate mounds that peel off to reveal a fleshy stereome. This sounds disgusting, but it's true, and we hope the fun of saying its name out loud makes up for it. *Cladonia floerkeana* could be confused with other species with red reproductive structures, but *C. macilenta* and *C. didyma* are sorediate or squamulose rather than coarse and bumpy.
CHEMISTRY: Thamnolic acid. Spot tests. K-, C-, KC-, P+ orange, UV-.
NICHE: This species occurs only at the highest elevations in the Smokies, on humus over exposed rocks. It occupies similar habitats across the central and northern Appalachians and is much more common in northern portions of this range.
KEY FEATURES: Irregular, gnarled, sordid white, cupless podetia with bumpy corticate surfaces, red apothecia and pycnidia, thamnolic acid, exposed rocks, highest elevations.

Cladonia furcata

Woodland Feelgood

Lendemer 30282 (photo: Tripp)

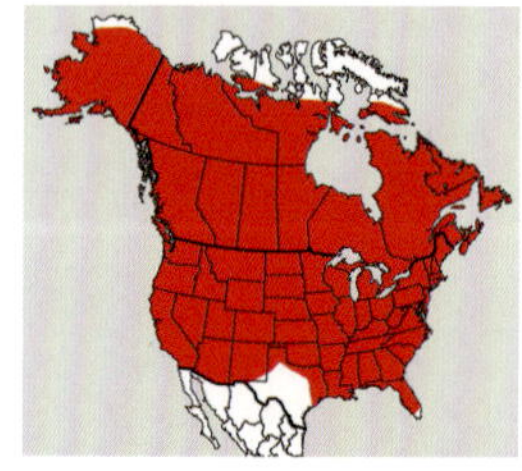

NOTES: Without a doubt, *Cladonia furcata* is one of the weediest fruticose lichens in the Smokies and is one that everyone can and will see on a typical day hike. This will make you feel good. It is easily recognized by its thalli, which form lax cushions of abundantly branching podetia that are corticate and often covered with small squamules, especially in the lower portions. One could confuse this species with the true reindeer lichens such as *C. rangiferina*, but those species all have ecorticate podetia giving the appearance of a fuzzy surface. *Cladonia appalachensis* is somewhat similar, but easily distinguished by its brownish color and UV+ blue-white thalli.

CHEMISTRY: Fumarprotocetraric acid. Spot tests. K+ dirty yellowish, C-, K-, P+ orange-red, UV-.

NICHE: This species is a good example of how the ecological principal of commonality versus rarity applies to lichens, too. *Cladonia furcata* is ubiquitous throughout the National Park, where it grows on rock and on the ground, especially along trails. It is equally common throughout the rest of its range in North America.

KEY FEATURES: Lax cushion-forming thalli, abundantly branching corticate podetia often covered by small squamules, bluish-gray in color contrasting with whiter portions of thallus, P+ red and UV-, absolutely everywhere.

Cladonia gracilis subsp. *turbinata*

Up Another Level Lichen

Tripp 5087 (photo: Lendemer)

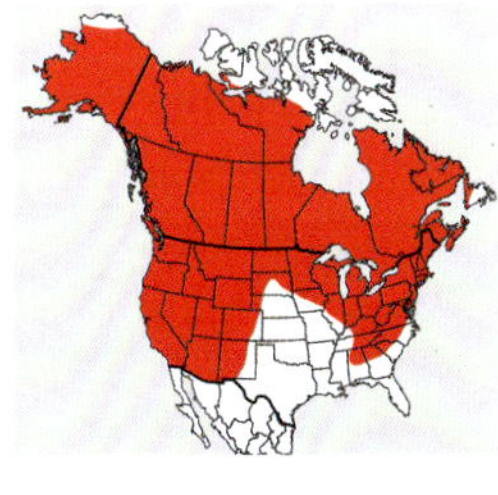

NOTES: *Cladonia gracilis* is a variable, brown-apotheciate species composed of several subspecies. Only *C. gracilis* subsp. *turbinata* is known from the Smokies, where it is easily recognized as one of the few cupped *Cladonia* with corticate podetia. It is most likely to be confused with *Cladonia verticillata*, but the latter differs in having cups that are consistently formed in tiers that arise from the centers of the cups versus margins of the cups, as in *C. gracilis* subsp. *turbinata*. The irregularly tiered podetia that proliferate from cup margins and the overall haphazard organization of the podetia serve to distinguish *C. gracilis* subsp. *turbinata* from all other *Cladonia* in the Smokies.

CHEMISTRY: Fumarprotocetraric acid. Spot tests. K+ dirty yellowish, C-, K-, P+ orange-red, UV-.

NICHE: In the Smokies, this species occurs only at the highest elevations, where it is found on shaded organic matter and rotten logs. It occurs sporadically throughout the southern and central Appalachians in similar habitats, becoming much more common in northern portions of its range.

KEY FEATURES: Corticate, cupped podetia with brown apothecia, irregular tiers forming from cup margins, highest elevations on organic matter.

Cladonia grayi

Appalachian Martini

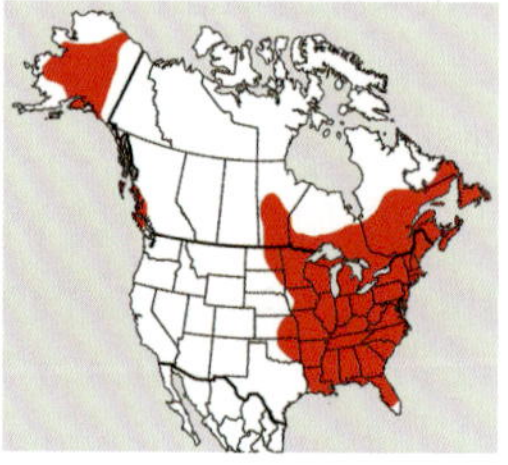

Lendemer 29547 (photo: Tripp)

NOTES: You might be thinking to yourself at this point, "Wow, all those cupped *Cladonia* species look the same." That might be a reasonable worldview, but of the cupped species, *Cladonia grayi* is by far the most common and likely to be seen. You will first recognize it in the field by its short, evenly sized cupped podetia, which occasionally are topped with large brown apothecia that give the cups a distorted appearance. This gestalt, coupled with the coarsely sorediate cups, irregular scattered squamules, and chemistry, distinguishes it from other similar species.

CHEMISTRY: Grayanic acid, +/- fumarprotocetraric acid. Spot tests. K-, C-, KC-, P+/- orange-red, UV+ blue-white.

NICHE: *Cladonia grayi* is a species that loves disturbance and is particularly abundant and widespread on trailside soil and organic matter at all elevations. It most likely keeps you company at your favorite backcountry campsite. The species is common and widespread throughout much of eastern North America.

KEY FEATURES: Sorediate, cupped podetia, often of even heights, conspicuous brown apothecia that distort the cups, irregularly scattered squamules, UV+ blue-white, very common.

Cladonia incrassata

Erin's Pocket Lichen

Tripp 5095 (photo: Lendemer)

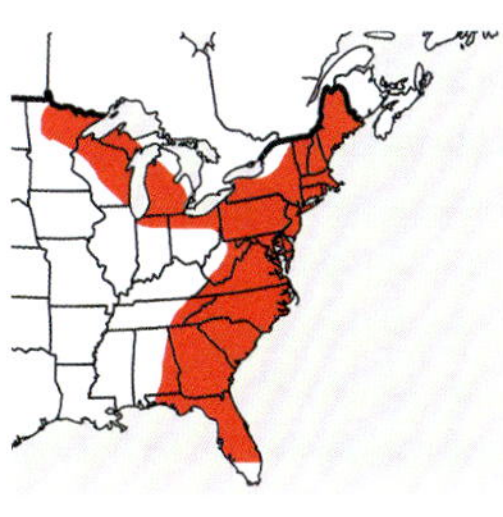

NOTES: We think this is one of the most charismatic and recognizable species of *Cladonia* in North America, and possibly the world. As the specific epithet implies, *Cladonia incrassata* can be recognized in the field by its fat podetia, which are short, fully corticate, and topped with large red apothecia. Even when podetia are absent, this species can be recognized by its large squamules, which are sorediate at the tips, often with red pycnidia on the surface, and by its chemistry. Once learned, nothing can be confused with this species!

CHEMISTRY: Usnic acid and squamatic acid. Spot tests. K-, C-, KC+ yellow, P-, UV+ blue-white.

NICHE: *Cladonia incrassata* occurs on rotting conifer logs and organic matter, especially in wet, humid habitats throughout the Smokies. It is also commonly found in Erin's pocket at the end of the day, but should not end up in yours unless you have a collection permit. The species is common and widespread throughout its range in eastern North America and justifies a trip to the Smokies if you are in need of any more excuses to get yourself to this most remarkable piece of Earth.

KEY FEATURES: Short, fat, corticate podetia, red apothecia and pycnidia, large sorediate squamules, UV+ bright blue-white, on rotting wood and organic matter throughout park.

Cladonia macilenta (w/ varieties)

Lichen Lipstick

Tripp 3494 (photo: Lendemer)

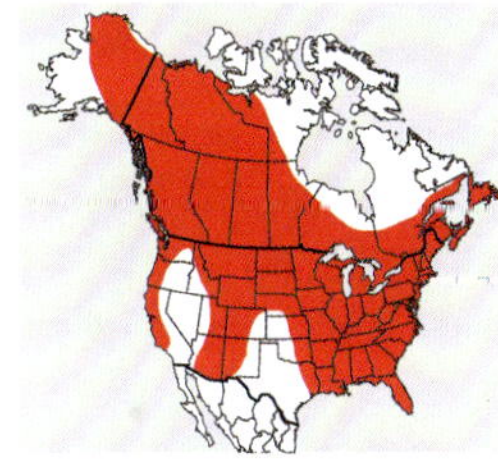

NOTES: *Cladonia macilenta* is one of several species whose chemical variants are recognized as subspecies or varieties. In this case, a chemotype with barbatic acid is recognized as variety *bacillaris* while one with thamnolic acid is recognized as variety *macilenta*. Both varieties can be distinguished from other red apotheciate species that occur in the Smokies by their conspicuously long and slender, cupless, sorediate podetia.

CHEMISTRY: Usnic acid, barbatic acid, or thamnolic acid. Spot tests. K- or + yellow, C-, KC- or + orange, P- or + orange, UV- or + blue-white.

NICHE: Both varieties of this species are common throughout the Smokies, where they can be found on rotting wood and on bark of both conifers and hardwoods. The species is similarly common throughout much of its range in North America.

KEY FEATURES: Long, slender podetia, these sorediate, cupless, and topped by pretty red apothecia and pycnidia, widespread and relatively common on rotting wood or bark of hardwoods and conifers.

Cladonia mateocyatha

Mixed Up Pixie Cup

Tripp 6055 (photo: Tripp)

Tripp 6055 (photo: Lendemer)

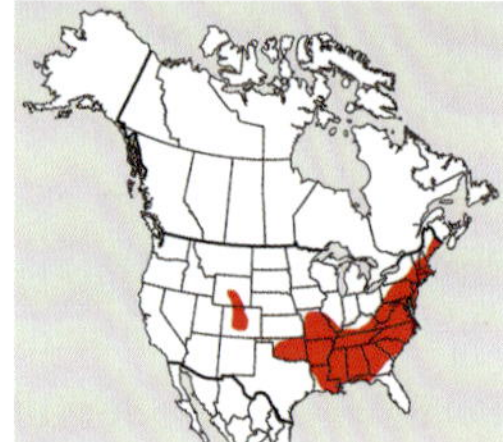

NOTES: There are populations of *Cladonia mateocyatha* in the Smokies that are *so big and lovely* that they barely resemble this species as we know it! As seen in the photograph, this species is characterized first by having a thallus composed of both podetia and squamules (vs. only podetia . . . the first couplet in any *Cladonia* key). Unfortunately, you are just as likely to find it with only squamules as you are with both squamules and podetia. If you find it with only squamules: look for its greenish-brown color, its squamules that are relatively large, cushion-forming, and not finely divided, and then by its P+ orange-red, K-, and UV- medulla. If you happen to find it with podetia in addition to squamules, the former should lack soredia and instead be corticate as well as have brown apothecia and/or pycnidia that sit atop irregularly shaped cups and sometimes proliferate from their margins and/or centers. This species is most likely to be confused with *C. sobolescens*, which is also primarily squamulose but differs in having much smaller squamules that do not form cushions.

CHEMISTRY: Fumarprotocetraric acid. Spot tests. K+ brown, C-, KC-, P+ red, UV-.

NICHE: Mixed Up Pixie Cup, which is a common name used by Brodo et al. (2001) and one we are quite fond of, occurs throughout the southeastern United States and into the northern Appalachians. It is quite uncommon in the Smokies, where we have collected it only once, on rock. Shuckstack Firetower and surrounding rock outcrops are a great place to see this species, which also has disjunct populations in the Southern Rocky Mountains of Colorado and Wyoming.

KEY FEATURES: Generally found only with large, undivided squamules, when podetiate with brown fruiting bodes, podetia often compound or proliferating, fumarprotocetraric acid, uncommon to rare in Park, on rock.

Cladonia ochrochlora

Nine Iron

Lendemer 32927 (photo: Tripp)

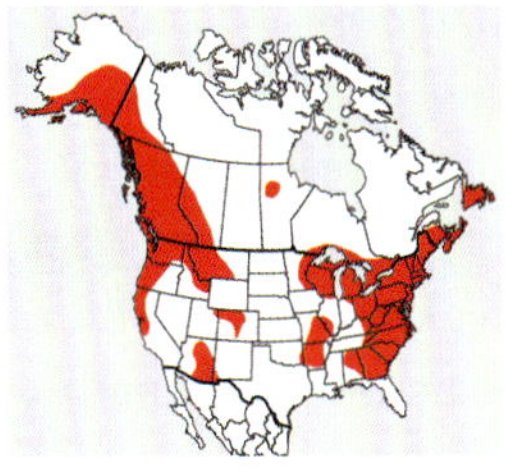

NOTES: Well, we have waited long enough to introduce you to *the* most common *Cladonia* in Great Smoky Mountains National Park and, indeed, much of eastern North America. Tee time! It's the loftiest iron in the bag and, just like number nine, it has quite the upward inclination. *Cladonia ochrochlora* is a very distinctive and common species of the Smokies that can be recognized by its relatively tall, slender podetia with a cortex that dissolves into soralia toward upper portions, its brown apothecia and pycnidia, and its large, esorediate squamules. In some ways, *C. ochrochlora* is most likely to be confused with *C. digitata*, which differs in having red apothecia and pycnidia, sorediate squamules, and a K+ yellow thallus. The name *C. coniocraea* was long used for *C. ochrochlora* and older guides use that name instead. There is debate as to what the correct name for the species is, however we use *C. ochrochlora* here.

CHEMISTRY: Fumarprotocetraric acid. Spot tests. K+ dirty yellow, C-, K-, P+ orange-red, UV-.

NICHE: This is a very common and widespread species that occurs throughout the Smokies on the bark of trees, on organic matter and lignum, and even on shaded rocks. It is common throughout much of eastern North America, and also widely distributed in the mountains of western North America. Make this the first species of *Cladonia* that you learn.

KEY FEATURES: Large, esorediate squamules, typically cupless, slender podetia that are basally corticate and dissolve into soredia in upper portions, brown apothecia and pycnidia, P+ red thallus, common.

Cladonia parasitica

A Shoulder To Lean On

Lendemer 32943 (photo: Tripp)

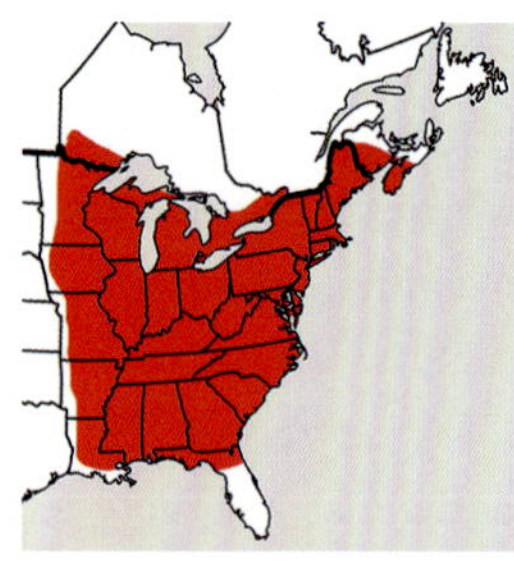

NOTES: *Cladonia parasitica* could be called one of the forgotten *Cladonias*, mainly because it is often inconspicuous and hidden among other physically, but not necessarily emotionally, larger species. Regardless, *C. parasitica* is a distinctive species characterized by its cupless podetia that are covered by tiny squamules, its primary thallus of squamules that fragment readily into an almost powdery mass of microsquamules, its brown apothecia and pycnidia, and its chemistry. *Cladonia parasitica* often grows with *C. ramulosa*, another relatively diminutive species that differs in having coarsely sorediate, irregularly shaped podetia and in producing fumarprotocetraric acid, which yields a K- and P+ red reaction.

CHEMISTRY: Thamnolic acid. Spot tests. K+ bright yellow, C-, KC-, P+ orange, UV-.

NICHE: *Cladonia parasitica* is common and widespread throughout the Smokies, where it is frequently found in mixed colonies with other species growing on old logs and on the bark of conifers.

KEY FEATURES: Minute squamulose thallus, cupless podetia covered with minute squamules, K+ yellow and P+ orange, on rotting wood and conifers.

Cladonia petrophila

Rock Adornment

Tripp 3643 (photo: Lendemer)

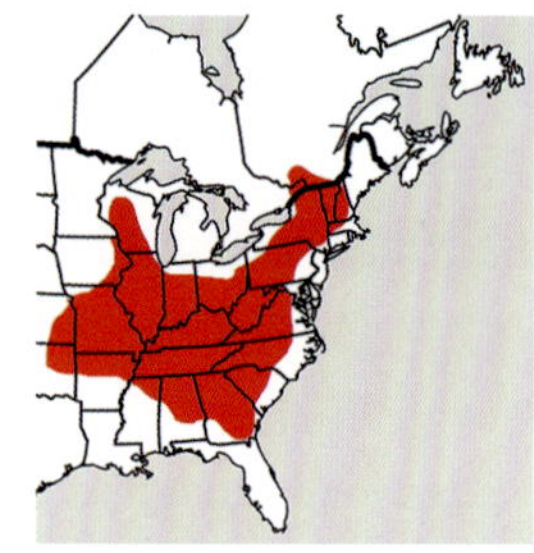

NOTES: *Cladonia petrophila* is a relatively easy-to-recognize species by its often-sterile occurrence, its finely divided gray squamules, and its niche (see below). It is one of four members of the genus in the Smokies that, when fertile, has short stipitate apothecia rather than apothecia atop podetia; these species are, furthermore, recognized by their unique chemistries and ecologies. It is possible to confuse *C. petrophila* with *C. caespiticia*, but that species has a UV-thallus and usually occurs on humus. *Cladonia stipitata* is similar but has shorter, less-divided squamules, a UV-thallus, and occurs on sun-exposed rocks. Finally, *C. apodocarpa* differs in occurring on soil, in having larger, strap-like squamules, and in having a UV-thallus.

CHEMISTRY: Sphaerophorin, +/- fumarprotocetraric acid, atranorin (traces). Spot tests. K-, C-, KC-, P- or + orange-red, UV+ blue-white.
NICHE: As the name implies, *Cladonia petrophila* loves and is typically found on rocks. It is rather common at middle-to-low elevations in the Smokies on shaded non-calcareous rocks in humid habitats. It is an eastern North American endemic species but is common throughout its range.
KEY FEATURES: Esorediate squamulose thallus with somewhat finely divided squamules, lacking podetia, UV+ blue-white, on shaded non-calcareous rocks, middle-to-low elevations.

Cladonia peziziformis

Round and Brown

Lendemer 33063 (photo: Tripp)

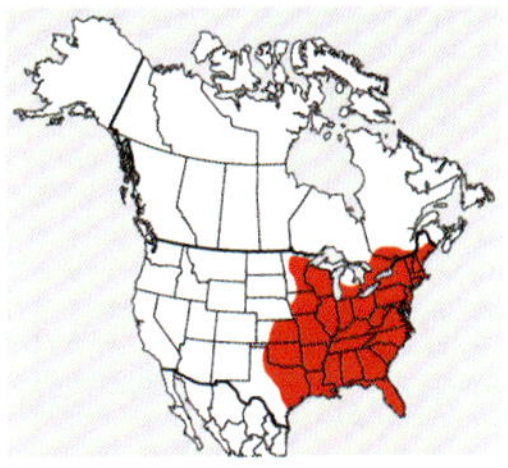

NOTES: *Cladonia peziziformis* is a distinctive species easily recognized by its short, thick podetia, which are covered with bumpy areoles and squamules and topped with large brown apothecia that are conspicuously wider than the stalks that support them. No other species in the Smokies shares this combination of features. Occasionally, one will encounter a variant of this species with pale-to-white rather than brown apothecia.
CHEMISTRY: Fumarprotocetraric acid. Spot tests. K+ dirty yellowish, C-, K-, P+ orange-red, UV-.
NICHE: This common and widespread species in the Smokies typically occurs on soil and organic matter in disturbed areas at middle-to-low elevations. It is frequently seen along trail sides, especially in areas such as the Lakeshore Trail north of Fontana Dam. *Cladonia peziziformis* is common in similar habitats throughout temperate North America.
KEY FEATURES: Short, thick podetia covered with bumpy areoles and squamules and topped with a single, unusually large brown apothecium, P+ red, on soil and organic matter at middle-to-low elevations.

Cladonia pleurota

Lichen Watchtower

Collector Tripp 2231 (photo: Lendemer)

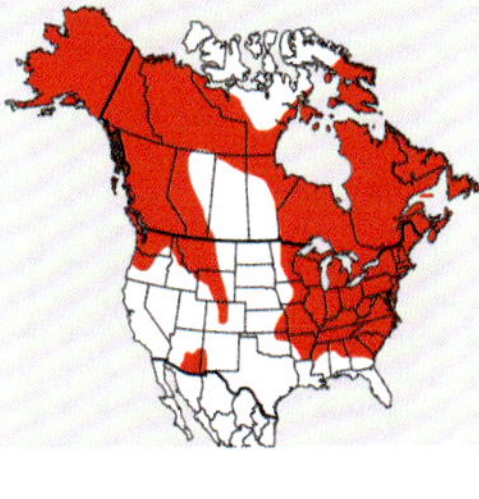

NOTES: It is always a pleasure to find *Cladonia pleurota* because its characteristic cupped, densely sorediate podetia with red apothecia and pycnidia have a way of brightening one's day. The most morphologically similar species is *C. digitata*, which differs chemically in having a K+ yellow thallus and sorediate squamules.

Cladonia coccifera is chemically identical to *C. pleurota* but differs in having cupped podetia covered with squamules and areoles rather than soredia.

CHEMISTRY: Usnic acid, zeorin. Spot tests. K-, C-, KC+ gold, P-, UV-.

NICHE: *Cladonia pleurota* is infrequent in the Smokies, where it is found at middle-to-high elevations on soil, organic matter, and sometimes rocks. It is similarly infrequent and widespread throughout large portions of North America. Our qualities of life would be improved if this species were more common.

KEY FEATURES: Cupped podetia covered densely by soredia, red apothecia and pycnidia, K- and UV-, middle-to-high elevations.

Cladonia polycarpoides

Brunette Toccata

Lendemer 26920 (photo: Lendemer)

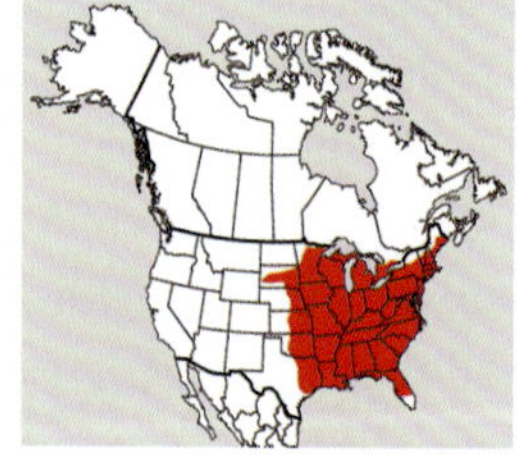

NOTES: So, you've found another *Cladonia*, and you don't know what on Earth to do next. Call your father? Call your senator? Call the *Cladonia* Hotline? Sit down on the side of the trail and eat another cookie? All of these sound like fun options, but none will help you solve this mystery unless your elected representative happens to be a lichenologist. Here is what we do: recite the various partitions of most *Cladonia* keys. You've got the species that basically lack squamules and are all podetial (nope); the ones that are all squamules and almost never bear any signs of reproductive structures (nope, although this species sometimes goes without); your cup-forming ones (nope); and your podetial ones that do not make cups and also have squamules (yep!). Within the latter, you think about fruiting bodies; red (nope) or brown (yep). See? Almost there. Within this group, *Cladonia polycarpoides* is identifiable in lacking any signs of soredia or granules on its podetial (this eliminates a large chunk of species!) and instead has fully corticate podetia that are UV-. Given these features, there are only a couple of species with which you might confuse it. *Cladonia peziziformis* and *C. sobolescens* both differ chemically and are P+ red. Some authors consider *C. polycarpoides* to be conspecific with *C. subcariosa*, but we continue to treat them as distinct based on their different chemistries and generally different geographic distributions in North America.

CHEMISTRY: Norstictic acid. Spot tests: K+ yellow turning red, C-, KC-, P+ yellow, UV-.

NICHE: Brunette Toccata occurs at low elevations in the Smokies, typically on south-facing slopes on acidic soils. Look for it near that tunnel of doom, or perhaps along the nearby Beauregard Ridge.

KEY FEATURES: Non cup-forming *Cladonia* with fully corticate podetia and brown pycnidia/apothecia, K+ red and P+ yellow podetia that are relatively tall and on the thin side, low elevations, on soil.

Cladonia pyxidata

Your Pixie-ness

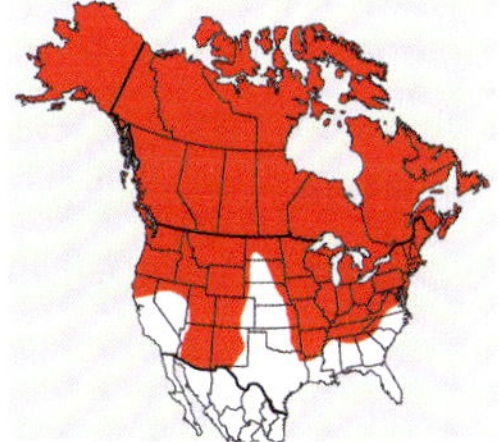

Tripp 6047 (photo: Lendemer)

NOTES: Well, you have arrived at the Crown Pixie of all Pixies. *Cladonia pyxidata* is an extremely wide-ranging species characterized by the following combination of characters: small squamules, frequent presence of relatively short, thick podetia that bear brown pycnidia and/or apothecia, and podetia that are cup-forming and very regular in shape and have a discontinuous, areolate cortex. This species also typically forms small, corticate squamules inside of the pixie cups. In the Smokies, *C. pyxidata* is most likely to be confused with *C. chlorophaea*, which differs primarily in having sorediate to granulose podetia (vs. corticate areoles in *C. pyxidata*).

CHEMISTRY: Fumarprotocetraric acid. Spot tests. K-, C-, KC-, P+ red, UV-.

NICHE: *Cladonia pyxidata* is uncommon in the Smokies, where it occurs on non-calcareous rock (or soil) at primarily middle elevations. This species is much more common in other portions of its native range, such as more northern parts of North America as well as the Rocky Mountains.

KEY FEATURES: Small squamules, short and stout podetia, brown pycnidia and/or apothecia, cup-forming, broken areolate cortex covering and inside of podetia.

Cladonia ramulosa

Sponge Cakes

Lendemer 33295 (photo: Tripp)

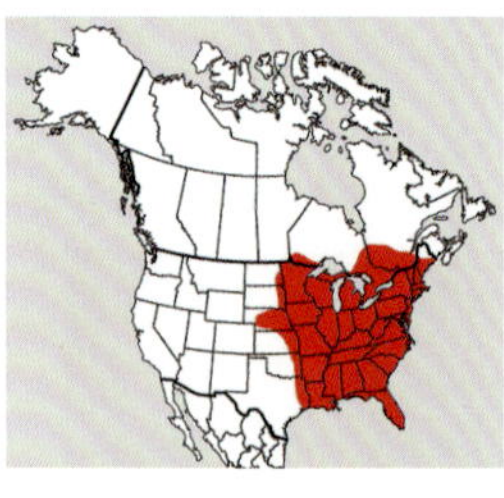

NOTES: *Cladonia ramulosa* is an endearing species easily recognized by its coarsely sorediate, short, fat, irregularly shaped and somewhat flaccid-in-appearance podetia. As in the photograph, the podetia often appear covered with small squamules, which are actually piles of coarse, ragged soredia that billow off the surface of podetia. The resemblance of *C. ramulosa* to *C. parasitica* sometimes causes confusion, especially because the two species often grow together in intermixed colonies. However, *C. parasitica* can easily be distinguished by its truly squamulose podetia and K+ yellow thallus.

CHEMISTRY: Fumarprotocetraric acid. Spot tests. K+ dirty yellowish, C-, K-, P+ orange-red, UV-.

NICHE: *Cladonia ramulosa* is infrequent but widespread throughout the Smokies at all elevations, usually growing on old rotting logs or humus. It occurs in similar habitats throughout its range in North America, where it is, at best, uncommon. This species sometimes competes with *C. incrassata* for space in Erin's pocket.

KEY FEATURES: Minute granulose thallus, cupless podetia covered with coarse, ragged soredia, K- and P+ red, on rotting wood and organic matter.

Cladonia rangiferina

Silent Raindrops
(alternate name: Reindeer Lichen)

Tripp 3950 (photo: Lendemer)

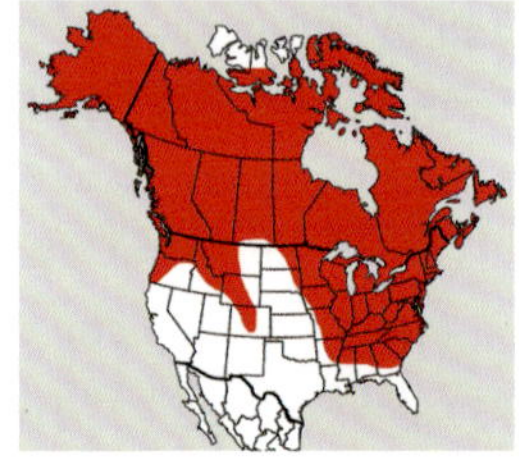

NOTES: *Cladonia rangiferina* is the iconic species commonly referred to as Reindeer Lichen in popular literature. It is one of several members of the genus typified by its cushion-forming thalli composed of ecorticate, abundantly branching podetia. *Cladonia rangiferina* can easily be recognized by these features in combination with its bone-white color (owing to its lack of a cortex) and its chemistry. *Cladonia rangiferina* could be confused for *C. stygia*, but that species is not known from the Smokies and, in any case, differs in having strongly blackened basal portions of the podetia (rather than pale to brownish in *C. rangiferina*).

CHEMISTRY: Atranorin and fumarprotocetraric acid. Spot tests. K+ yellow, C-, KC-, P+ orange-red, UV-.

NICHE: *Cladonia rangiferina* is infrequent at middle-to-high elevations in the Smokies, where it occurs on organic matter in open areas or on rocks. It is far more common in northern portions of its range, closer to the reindeer.

KEY FEATURES: Abundantly and dichotomously branching, ecorticate, bone-white podetia, K+ yellow and P+ red, middle-to-high elevations on organic matter.

Cladonia robbinsii

Sky Straps

Langdon s.n. (photo: Lendemer)

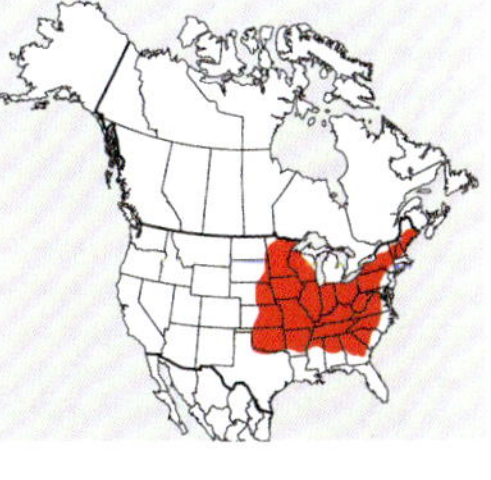

NOTES: Where would we be in life without species like *Cladonia robbinsii*? The yellowish-green strap-like squamules ascend to the sky, revealing their white, ecorticate bottoms. Morphologically, the most similar species to Sky Straps is *C. apodocarpa*, but the latter is blue-gray due to the presence of atranorin instead of usnic acid and the medulla is UV- due to the absence of barbatic acid. *Cladonia polycarpoides* and other members of the *C. subcariosa* group, likewise, can have large squamules, but they all produce other substances that make the medulla P+ yellow, orange, or red, and the squamules tend to curl rather than ascend strongly. The squamules of *C. robbinsii* often form small mounds or cushions, and then can also cause confusion with the much more common *C. strepsilis*. Thankfully, the squamules of *C. strepsilis* are usually plane and flat, shorter, and the medulla is always C+ green due to the presence of its namesake strepsilin.

CHEMISTRY: Usnic acid and barbatic acid. Spot tests: (cortex): K-, C-, KC+ yellow, P-, UV-; (medulla): K-, C-, KC-, P-, UV+ blue-white.

NICHE: Where, oh where, does *Cladonia robbinsii* occur? For such a pretty species, it seems to be fairly uncommon in the Smokies and eastern North America, generally, usually occurring on sun-exposed rock outcrops. To see it in true abundance and glory, one should travel to the glades and barrens of the Ozarks.

KEY FEATURES: Persistently squamulose *Cladonia*, cushion-like thalli, yellow-green upper surface, white lower surface, medulla K- and C- and UV+ blue-white, on exposed non-calcareous rocks at low elevations.

Cladonia squamosa

UV-Me

Tripp 3730 (photo: Lendemer)

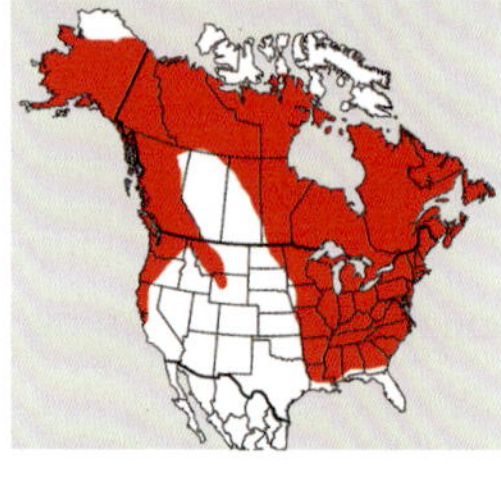

NOTES: Like Woodland Feelgood (aka: *Cladonia furcata*), *Cladonia squamosa* is one of the most common lichens in the Smokies. It is also *the* most common species of *Cladonia* in the Park, and for that reason, make it your #1 learning goal in this genus. *Cladonia squamosa* is easily recognized by its relatively tall, slender, squamulose podetia, which are UV+ blue-white and have poorly formed cup-like expansions at the tips. *Cladonia crispata* is chemically identical to *C. squamosa* and has similar podetia, but in that species podetia are fully corticate rather than covered with squamules.

CHEMISTRY: Squamatic acid. Spot tests. K-, C-, KC-, P-, UV+ blue-white.

NICHE: *Cladonia squamosa* is extraordinarily common and occurs throughout the Smokies on organic matter, rotting wood, and rocks. It is equally common throughout much of its range in North America.

KEY FEATURES: Tall, slender podetia covered with squamules, poorly formed cup-like expansions at podetial tips, brown apothecia and pycnidia, UV+ blue-white thallus, common as dirt, growing on dirt.

Cladonia stipitata

Silvering

Tripp 3731 (photo: Lendemer)

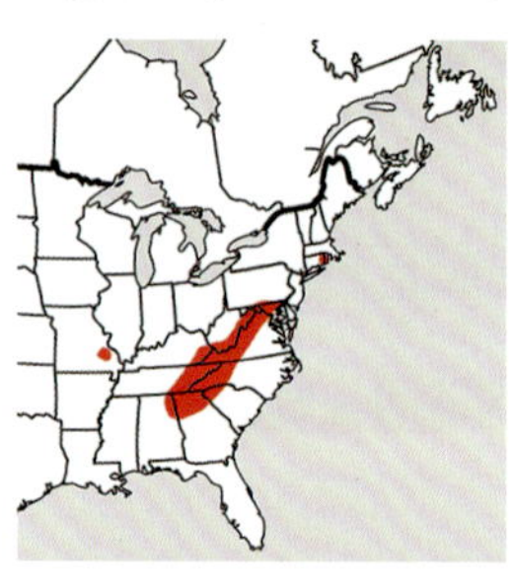

NOTES: *Cladonia stipitata* is one of several species that have conspicuous squamulose thalli and, where present, apothecia borne on short stipes rather than podetia. It is most readily recognized by its white, relatively undivided squamules that form small, very tight, very compact cushions, its chemistry, and its niche (see below). *Cladonia apodocarpa* is chemically identical to *C. stipitata* but has much larger, strap-like squamules and furthermore grows on soil or organic matter, especially in exposed habitats. Like *Cladonia stipitata*, *C. petrophila* occurs on rocks but has larger squamules and a thallus that is UV+ blue-white.

CHEMISTRY: Atranorin and fumarprotocetraric acid. Spot tests. K+ yellow, C-, KC-, P+ orange-red, UV-.

NICHE: *Cladonia stipitata* is uncommon in the Smokies, where it occurs only on exposed non-calcareous rocks in full sun at middle elevations. It occurs in similar habitats throughout the Appalachians and is especially common on granite domes in North Carolina.

KEY FEATURES: Esorediate thallus of densely packed, white, undivided squamules, forming small cushions, lacking podetia, UV-, on sunny non-calcareous rocks, middle elevations.

Cladonia strepsilis

Princess Shingles

Tripp 2643 (photo: Lendemer)

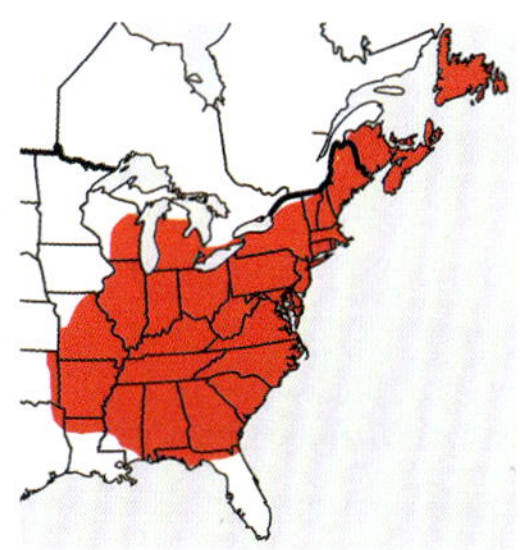

NOTES: *Cladonia strepsilis* is an attractive species that forms conspicuous cushions of large, strap-like squamules with occasional inflated and thorn-like podetia (where present, we think they sort of resemble phylogenetic trees). In addition to morphology, the chemistry of *C. strepsilis* is extremely distinctive (see below). This species could only be confused with *C. apodocarpa* or *C. robbinsii*, but neither of these species produces podetia. *Cladonia apodocarpa*, furthermore, differs in having a P+ red thallus, while *C. robbinsii* differs in having a KC+ gold thallus.

CHEMISTRY: Strepsilin, baeomycesic acid and squamatic acids. Spot tests. K-, C+ green, KC-, P+ yellow, UV+ blue-white.

NICHE: *Cladonia strepsilis* occurs on exposed rocks in sunny, open habitats at middle and low elevations in the Smokies. It is common and widespread throughout temperate eastern North America, where it occurs in similar habitats as well as other open areas such as sandhills and barrens.

KEY FEATURES: Esorediate squamulose thallus with strap-like squamules forming large cushions, podetia rare but inflated and sometimes bifurcating when present, C+ green, on sunny, non-calcareous rocks, middle-to-low elevations.

Cladonia subtenuis

Common Cushions

Lendemer 29584 (photo: Tripp)

NOTES: Another Reindeer Lichen, but don't worry, it's the last one. Compare the photo of *Cladonia subtenuis* here to that of *C. arbuscula* several pages back. The two species are similar and share many features including cushion-forming thalli, abundantly branching ecorticate podetia, and chemistry. In most scholarly works, *C. subtenuis* is distinguished from *C. arbuscula* by its slenderer podetia. In theory, the two have allopatric distributions, different colored pycnidial gel, and different branching patterns. We leave it to you to decide.

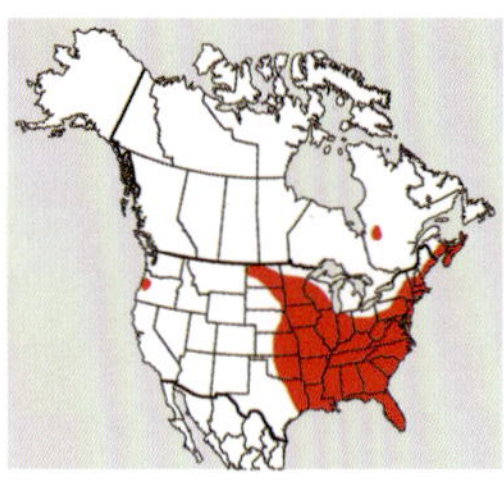

CHEMISTRY: Usnic acid, fumarprotocetraric acid. Spot tests. K-, C-, KC+ yellow, P+ orange-red, UV-.

NICHE: *Cladonia subtenuis* is infrequent but widespread at middle-to-low elevations in the Smokies, especially in western portions of the Park, where it occurs in open areas on soil and rocks. Outside the Smokies, it is common in the southeastern United States.

KEY FEATURES: Abundantly branching, ecorticate, slender podetia, KC+ yellow and P+ red, low-to-middle elevations on soil and rocks.

Cladonia uncialis

Love Me Lichen

Tripp 3946 (photo: Lendemer)

NOTES: What's not to love about *Cladonia uncialis*? It is a very attractive species with abundantly, dichotomously branching, corticate podetia that have white maculae toward upper portions and brown tips. The podetia are spiny and thorn-like, sending a friendly message "better not collect me" without a permit. In some ways, *C. uncialis* resembles *C. arbuscula* or *C. subtenuis*, but it can easily be distinguished by its corticate rather than ecorticate, spikey podetia.

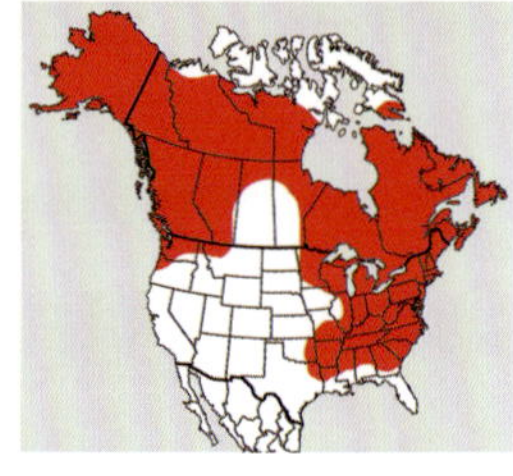

CHEMISTRY: Usnic acid, squamatic acid. Spot tests. K-, C-, KC+ yellow, P-, UV+ blue-white.
NICHE: *Cladonia uncialis* is infrequent in the Smokies and throughout the Appalachians, where it occurs on exposed rocks and organic matter at all elevations.
KEY FEATURES: Abundantly branching, corticate, spikey podetia with maculate tips, KC+ yellow and UV+ blue-white, on organic matter.

Cladonia verticillata

My Cup Runneth Over

Tripp 3689 (photo: Lendemer)

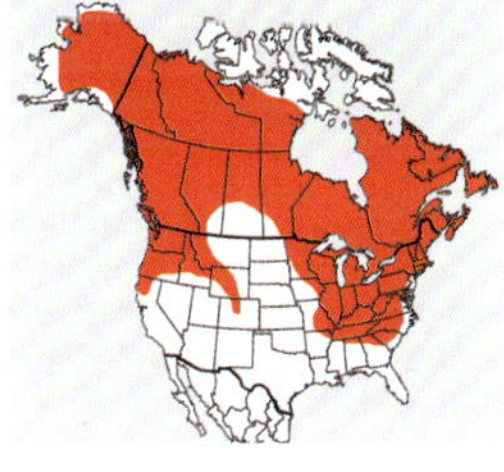

NOTES: *Cladonia verticillata* is one of the easiest members of the genus to learn, on account of its corticate podetia that form tiered cups, which proliferate from the centers of the cups. It could be confused with *C. gracilis* subsp. *turbinata*; but, in that taxon, the podetia are poorly organized and the cups haphazardly proliferate primarily from the margins of the cups. *Cladonia gracilis* subsp. *turbinata* is furthermore restricted to the highest elevations of the Smokies, whereas *C. verticillata* occurs elsewhere (see Niche).
CHEMISTRY: Fumarprotocetraric acid. Spot tests. K+ dirty yellowish, C-, K-, P+ orange-red, UV-.
NICHE: This species is infrequent but widespread at middle-to-low elevations in the Smokies, where it occurs on open soil, particularly in disturbed habitats such as along trails. It is relatively common in similar habitats throughout temperate eastern North America but its distribution is poorly understood elsewhere across the continent.
KEY FEATURES: Corticate, cupped podetia, regular tiers formed from the centers of the cups, brown apothecia and pycnidia, middle-to-low elevations, on organic matter and soil.

Clauzadea chondrodes

Murky Pits

Lendemer 44551 (photo: Tripp)

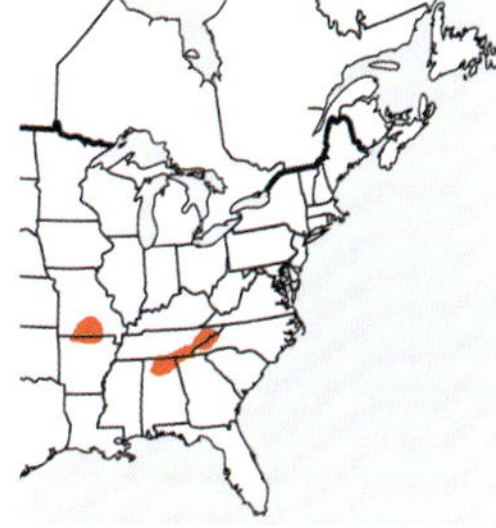

NOTES: *Clauzadea chondrodes* is not the most remarkable lichen on the planet but its niche (see below) readily helps to identify this species. It is characterized by its black apothecia against a poorly formed, dirty white thallus. It might be mistaken for *Kephartia crystalligera*, but that species has beautiful orange crystals in its hypothecium, which react K+ purple (the hypothecium is brown-to-colorless and K- in C. chondrodes). You might also be tempted to confuse *C. chondrodes* for *Baglietto a baldensis*, but the latter is perithecíate.

CHEMISTRY: No substances. Spot tests. K-, KC-, C-, P-, UV-.

NICHE: Restricted to limestone outcrops in the Southern Appalachians and disjunct on similar substrates (limestone, dolomite) in the Ozarks.

KEY FEATURES: Saxicolous and restricted to calcareous rock, black apothecia, simple, colorless spores, brown-to-colorless hypothecium.

Cliostomum griffithii

I Left My Heart In . . .

Lendemer 33287 (photo: Tripp)

NOTES: *Cliostomum griffithii* is an inconspicuous crustose lichen with colorless, 2-celled spores. It is here shown with a very weakly developed (and wet) thallus, but in other instances typically has a thin, gray-to-green thallus. It

could be confused with species of *Micarea*, but that genus has a characteristic micareoid photobiont that most people will never see. Our recommendation in this case: learn the species rather the genus!

CHEMISTRY: Atranorin. Spot tests. K+ weak yellow, C-, KC-, P+ weak yellow, UV-.

NICHE: *Cliostomum griffithii* occurs on lignum and bark of conifers in spruce-fir forests at high elevations in the Park. Elsewhere in North America, such as the Santa Monica Mountains in southern California, this species is downright weedy.

KEY FEATURES: Small, inconspicuous crustose lichen with pale, flesh-colored apothecia, tiny, 2-celled colorless spores, non-micareoid photobiont, high elevations on conifers and lignum.

Coccocarpia palmicola

Cascadas Aguas Azules

Tripp 3583 (photo: Lendemer)

NOTES: *Coccocarpia palmicola* is a large, conspicuous foliose cyanolichen that is unlikely to be confused with anything else in the Smokies. It is characterized by its slate-gray to blue-gray (when wet) thallus that always hosts

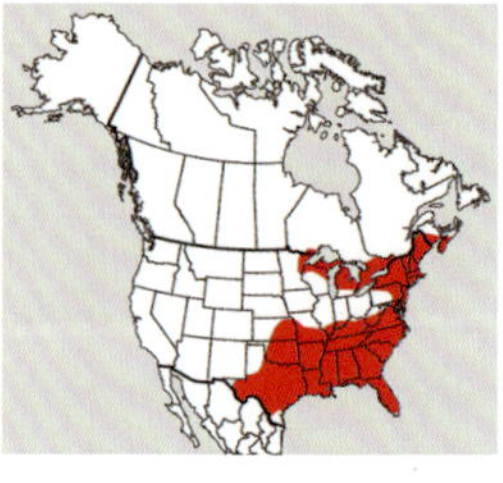

large, cylindrical-to-globose isidia and has a dense tomentum covering its lower surface. A casual observer might be tempted to confuse this species with *Parmeliella appalachensis* or *P. triptophylla*, but those species differ in having much smaller lobes, prominent black prothalli, and different cyanobacterial photobionts (vs. *Scytonema* in *Coccocarpia*). The color of *C. palmicola* reminds one of *Normandina pulchella*, but the latter has much smaller, nonisidiate thalli, and a green algal photobiont. The turquoise blue lobes of *C. palmicola* reminds us of one of the best swimming holes in Chiapas, Mexico–Cascadas Aguas Azules–go there, and you might find *Coccocarpia*, too!

CHEMISTRY: No substances. Spot tests. K-, C-, KC-, P-, UV-.

NICHE: *Coccocarpia palmicola* is common at low elevations in the Smokies, where it occurs on hardwoods and occasionally on conifers or rock. In the Neotropics, this species is common to even weedy.

KEY FEATURES: Large, slate-gray to blue-gray thallus with abundant isidia, dense tomentum on undersides, a *Scytonema* cyanobacterial photobiont, widespread at low elevations on bark.

Coenogonium luteum

Don't Ask, Don't Tell

Lendemer 30277 (photo: Tripp)

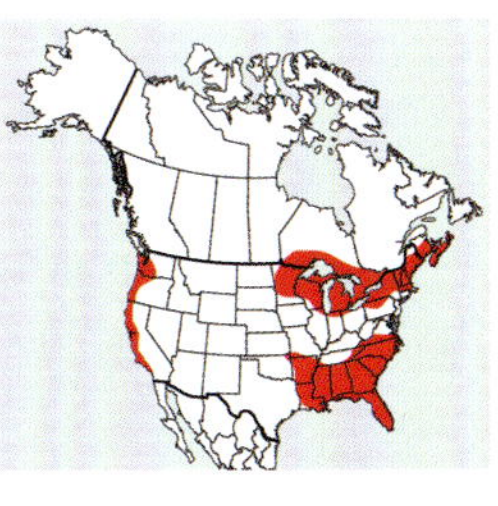

NOTES: *Coenogonium luteum* is one of two species in a genus that is otherwise highly emblematic of tropical environments. The genus is characterized by its fleshy orange apothecia, its *Trentepohlia* photobiont, and its 2-celled colorless spores. In the tropics, *Coenogonium* has a distinctively filamentous thallus composed primarily of algal cells netted together to form a shelf or mat of sorts; but our two Smokies species have a thin, continuous dirty-green crustose thallus, which led to their previous classification in *Dimerella*. The genus *Absconditella* is similar but differs in having a coccoid photobiont and an I- hymenium (vs. I+ blue in *Coenogonium*). *Coenogonium luteum* is easily differentiated from *C. pineti* by its larger, orange apothecia.

CHEMISTRY: No substances. Spot tests. K-, C-, KC-, P-, UV-.

NICHE: This species is relatively common in the Smokies, where it occupies the bark of hardwoods and conifers at middle-to-low elevations. Look for this species at The S.W.A.G. (circa Purchase Knob) or, if you are cold and need some hot sunshine, head to Cape Hatteras!

KEY FEATURES: Thin, dirty-green thallus, fleshy orange apothecia, 2-celled colorless pores, *Trentepohlia* photobiont, middle-to-low elevations.

Coenogonium pineti

Pale Pails

Tripp 2228 (photo: Lendemer)

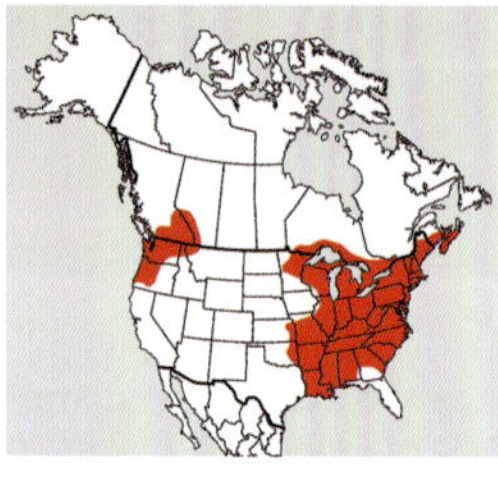

NOTES: *Coenogonium pineti* is a tiny, shiny, gem. This species is easily recognized by its continuous, dark green thallus over bryophytes or bark, its pale flesh-colored apothecia with conspicuous white margins, and its 2-celled colorless spores. It is not likely to be confused with *C. luteum* owing to the very differently colored and larger apothecia in the latter species. *Absconditella* is very similar to *C. pineti* but has a coccoid photobiont and an I- hymenium (vs. *Trentepohlia* photobiont and I+ blue in *C. pineti*).

CHEMISTRY: No substances. Spot tests. K-, C-, KC-, P-, UV-.

NICHE: Yes, it has one. This species "behaves" exactly like it does in the photo: it prefers bryophytes and organic matter over the trunks of hardwoods. You will almost certainly find it in wet forests throughout the Park. If, for some reason, you think that a vacation in Pennsylvania would be more desirable, well, you can find it there, too! In fact, it occurs just about everywhere in the East, in forests on the bases of trees and over mosses.

KEY FEATURES: Dark green, continuous thallus over bryophytes and organic matter, pale, small, flesh-colored apothecia with conspicuously white margins, 2-celled colorless spores, I+ blue hymenium.

Collema coccophorum

Frog Jelly

Lendemer 44539 (photo: Tripp)

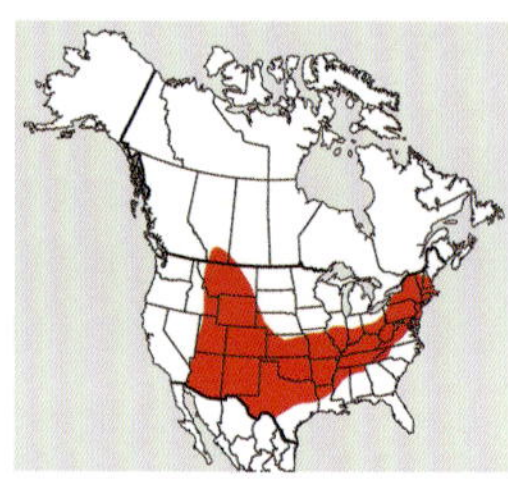

NOTES: Species of *Collema* can, in general, be recognized in the field by their foliose, dark-colored thalli. When wet, they can be difficult distinguished from *Leptogium*, so hang out long enough, wait for them to dry, and you will note how species of *Collema* always dry out to a blackish color (vs. dark, slate-gray typical of dry *Leptogium*). If there were ever a true Jelly Lichen, it would have to be *Collema coccophorum*. This species is distinctive because its lobes become large and swollen when wet; but, when dry, they shrivel up and give the appearance of a thick, snotty pile of algae. In addition to its ecology (see Niche) and swelling lobes, it differs from all other jelly lichens in the Smokies by the absence of isidia and usual presence of apothecia with 2-celled colorless spores.

CHEMISTRY: No substances. Spot tests. K-, C-, KC-, P-, UV-.

NICHE: Widely distributed on mosses over calcareous rocks throughout temperate eastern and central North America. In the Smokies, it occurs on the few existing dolomite outcrops and is readily seen as you hike through Ace Gap.

KEY FEATURES: Jelly lichen with large, swollen lobes (when wet), apotheciate, 2-celled colorless spores, muscicolous over calcareous rocks.

Collema conglomeratum

Spore Huddle

Lendemer 33106 (photo: Tripp)

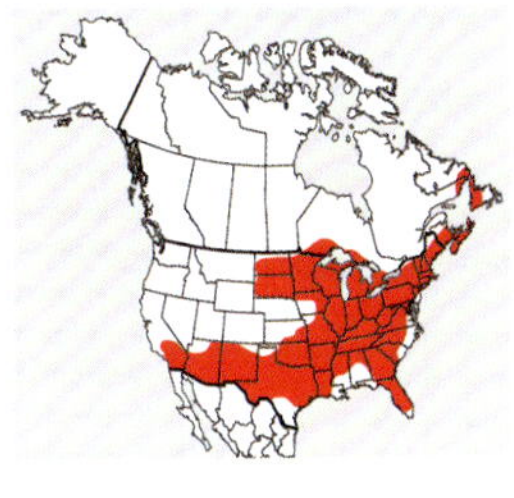

NOTES: *Collema conglomeratum* is a fitting epithet for a very fine jelly lichen characterized by its profusion of clustered apothecia, which more or less obscure the underlying thallus. Its cyanobacterial photobiont and general morphology make it distinct from all other jelly lichens in the Smokies, except for *C. coccophorum*, which occurs on calcareous rocks and soil (vs. trees in *C. conglomeratum*). Have you ever seen so much lichen sex in one place?

CHEMISTRY: No substances. Spot tests. K-, C-, KC-, P-, UV-.

NICHE: If you find this species, you should stop and ponder your luck at having arrived in one of the few places in eastern North America where this species still thrives. *Collema conglomeratum* is infrequent at middle-to-low elevations in the Smokies, where it occupies the bark of mature hardwoods. You can see this species near Horseshoe Bend on Eagle Creek Trail. The map pictured here shows a broad distribution where this species used to occur more commonly, in the past.

KEY FEATURES: Jelly lichen visible mostly as clustered apothecia that obscure a fringe of tiny, marginal lobes, 2-celled colorless spores, corticolous, middle-to-low elevations.

Collema subflaccidum

Wish I Was In Dixie

Tripp 3606 (photo: Lendemer)

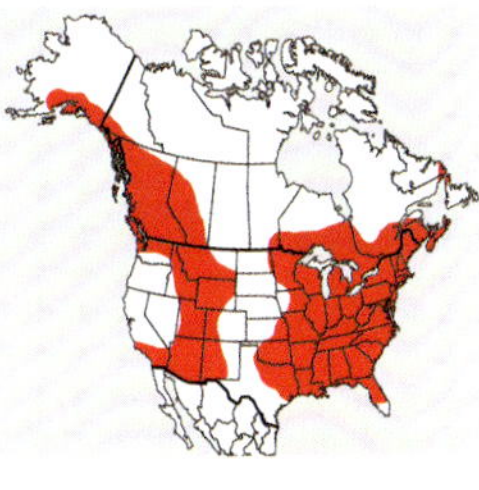

NOTES: *Collema subflaccidum* is the most common jelly lichen in the Smokies. It can be confused with other species of *Collema* that have isidia (e.g., *C. flaccidum*, *C. furfuraceum*) but is distinguished from these other species by its smooth thalli, which lack pustules and have short, globose isidia. We repeat: it has short, globose isidia. *Collema*

furfuraceum has conspicuous ridges and pustules on its thallus surface and cylindrical isidia. *Collema flaccidum* also has ridges and pustules on its upper surface but its isidia are flattened and resemble squamules. Spend a little time learning these three, and then whistle Dixie.

CHEMISTRY: No substances. Spot tests. K-, C-, KC-, P-, UV-.

NICHE: *Collema subflaccidum* enjoys the bark of hardwoods and non-calcareous rocks in humid forests at middle-to-low elevations in the Smokies. Just like us. It is one of the few jelly lichens that persists throughout much of its original range.

KEY FEATURES: Very common, large foliose jelly lichen, smooth upper surface without ridges, short globose isidia.

Collema pulcellum var. leucopepum

Sherman's March

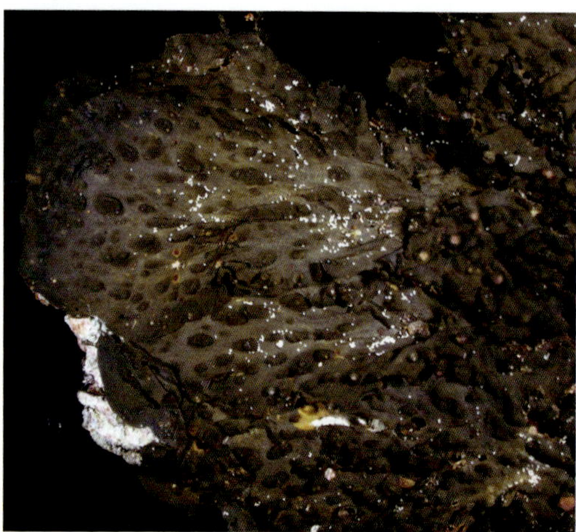

Dey 32692 (photo: Lendemer)

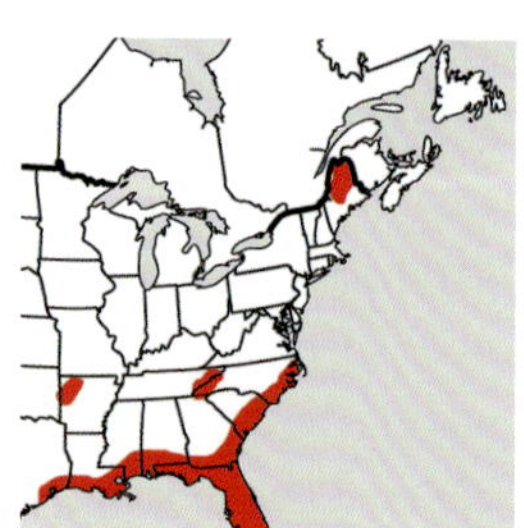

NOTES: *Collema* can be your enemy or your friend, depending on the day and your attitude. Whichever way you look at it, this species should be on your wish list rather than any other type of list, and you will be very lucky if you find it. Although very rare in the Park, it is extremely distinctive on account of its thallus, which lacks isidia, and its pruinose apothecia discs bordered by a proper thalline margin. It is also characterized by its brown-black warty thallus and its large lobes that, when dry, sit flat and closely appressed against the substrate. Please let somebody know if you find new Park records of this species, and then treat yourself to an ice cream (maybe a Dip Cone?) afterwards!

CHEMISTRY: No substances. Spot tests. K-, KC-, C-, P-, UV-.

NICHE: Sherman's March has been found only at very low elevations in the Smokies and is generally rare throughout the southeastern United States. It occurs elsewhere in eastern North America on the bark of hardwood trees, but again appears to be quite rare.

KEY FEATURES: Large, brown-black thallus that bears warts, broad, closely appressed lobes, no isidia, pruinose apothecia with thalline margins, very low elevations, rare.

Conotrema urceolatum

Sunday Communion

Lendemer 33056 (photo: Tripp)

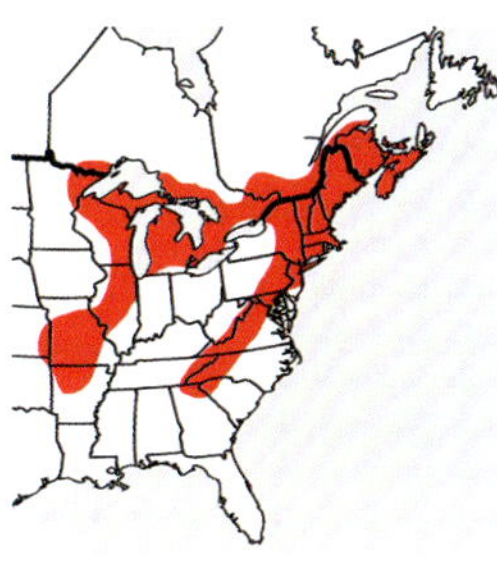

NOTES: This is a common crustose lichen on smooth bark trees, such as young Sugar Maple, and one that you can learn with little effort. Its uniquely urn-shaped black apothecia set against a continuous, smooth white thallus make *Conotrema urceolatum* immediately recognizable throughout the Smokies. Its spores are equally striking and distinctive: they are septate and worm-like! To see this species, take a slow, quiet stroll through Schoolhouse Gap on your next Sunday morning in the Smokies.

CHEMISTRY: No substances. Spot tests. K-, C-, KC-, P-, UV-.

NICHE: *Conotrema urceolatum* is a common crustose lichen that you find when you least expect it. It is usually right under your nose, even if you don't know it. This species can be found throughout the Park at middle-to-high elevations on the bark of hardwoods. Unfortunately, it has been extirpated from most of its native range. What a discovery it would be to find huge, healthy, living populations of this species elsewhere!

KEY FEATURES: White, continuous thallus, large urn-like black apothecia, worm-like spores, on smooth bark hardwoods, low-to-middle elevations.

Cresponea flava

Lemon Fissure Cookies

Tripp 3638 (photo: Lendemer)

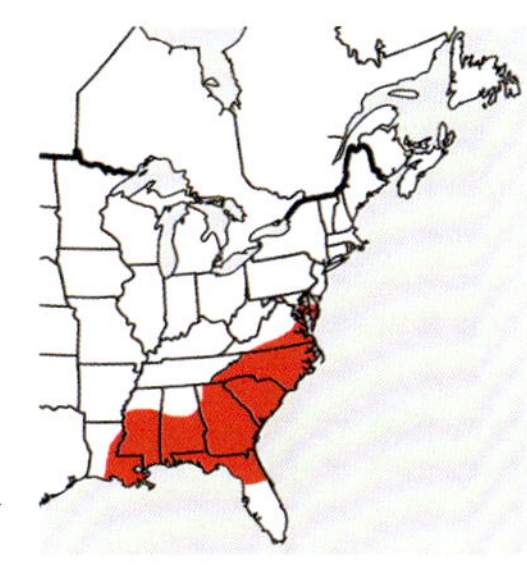

NOTES: *Cresponea flava* is a beautiful species that warrants two exclamation marks!! It is characterized by its thin, continuous, blue-gray thallus, dark black apothecia with cracked margins, and most notably but somewhat variably, yellow pruina on the discs. In the photo, the pruinose discs are visible on two apothecia toward the lower center as well as on the left. No other lichen in the Smokies is likely to be confused with this species.

CHEMISTRY: No substances. Spot tests. K-, C-, KC-, P-, UV-.

NICHE: This species occurs on large, old hardwoods; but wait, there's more: look for it specifically on dry, south-facing surfaces of the largest trees in the Smokies, especially Tulip Poplars, at low-to-middle elevations. *Cresponea flava* is rare throughout the southeastern United States, occurring only in remnant, mature hardwood forests.

KEY FEATURES: Thin, blue-gray thallus, black apothecia with cracked margins, +/- yellow pruinose discs (look closely), on old hardwoods, south-facing portions, at low elevations.

Cyphelium tigillare
Trail Blaze

Lendemer 33045 (photo: Tripp)

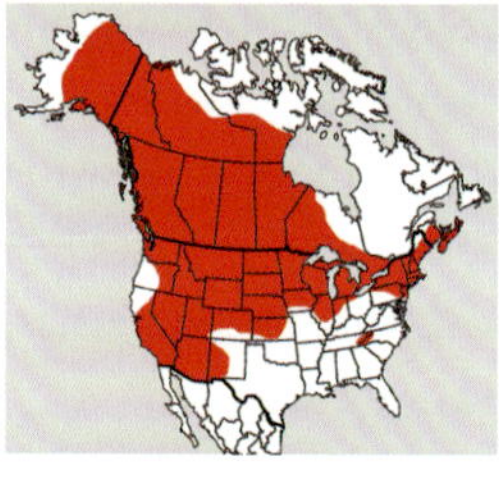

NOTES: Unassisted, this one is doable. Look for a bright yellow thallus, and then look closer for dark black apothecia with a mass of spores in a distinctive mazaedium. In the Smokies, only some of the pin lichens (*Calicium, Chaenotheca*) similarly produce spores in mazaedia, but these genera have stalked apothecia. Knowing this, *Cyphelium tigillare* is not confusable for any other lichen in the Smokies.

CHEMISTRY: Vulpinic acid. Spot tests. K-, C-, KC-, P-, UV-.

NICHE: *Cyphelium tigillare* is uncommon in the Smokies, restricted to suitable lignum substrates such as old sign posts and AT shelters at middle and upper elevations, particularly in western portions of the Park. There is a lovely population near the Laurel Gap Shelter on the way up to the summit of Luftee Knob (which, incidentally, is lichenologically one of the most interesting parts of the Smokies). This genus and species are far more common in the western than eastern United States. In any case, it is spectacular no matter which side of the Divide you find it on.

KEY FEATURES: Golden yellow crustose, areolate thallus with dark black, sessile apothecia, spores in a mazaedium, on lignum, middle-to-high elevations.

Cystobasidium hypogymniicola
Stuck on Hypogymnia

Lendemer 33009 (photo: Tripp)

NOTES: This is one of those times where we felt compelled to include a lichenicolous fungus in this *Field Guide*. This parasite is a great one to learn because it is readily identifiable and, to some, sort of pretty. *Cystobasidium hypogymniicola* is a heterobasidiomycete fungus characterized by the irregular galls that it induces on the thalli of its host. Those galls may or may not be accompanied by tiny, spore-filled basidia. If you want to learn more about this and related species, check out Paul Diederich's monograph on them, written in 1996.

CHEMISTRY: No substances. Spot tests. K-, C-, KC-, P-, UV-.

NICHE: *Cystobasidium hypogymniicola* seems to have a sporadic distribution that largely tracks its hosts across portions of the eastern and western United States. It parasitizes species of *Hypogymnia*, in particular *H. krogiae* and *H. incurvoides* in the Smokies. Look for it at middle-to-high elevations on the thalli of these easily identifiable macrolichens.

KEY FEATURES: Gall-like structures that occur on *H. krogiae* and *H. incurvoides* at middle-to-high elevations.

Cystocoleus ebeneus

Shady Rock Soot

Lendemer 53163 (photo: Tripp)

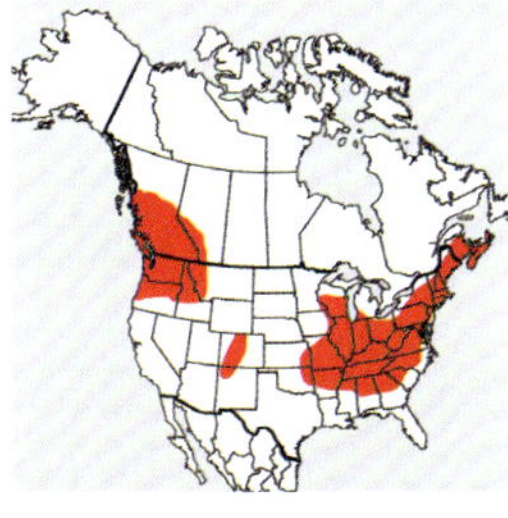

NOTES: There are truly very few lichens in the Smokies with which you could confuse this. *Cystocoleus ebeneus* is macroscopically characterized by its black, typically dime-sized, fruticose (downright bushy) thalli of fine, black filaments resembling hair. *Cystocoleus ebeneus* doesn't actually look very lichen-like, but it is. Shove a strand under a high-powered microscope to discover fungal hyphae encircling filaments of its photobiont partner, *Trentepohlia*. This species is most likely to be confused with species of *Ephebe* found in the Park, but those contain cyanobacteria as photobionts and have branches that are much thicker, and also a bit on the warty side (unlike the fine, thin, smooth branches of Shady Rock Soot). You might also confuse this species for free-living cyanobacteria. When in doubt, shove it under the scope.

CHEMISTRY: No substances. Spot tests. K-, KC-, C-, P-, UV-.

NICHE: Shady Rock Soot occurs throughout the Park on non-calcareous rocks. Look for it, in particular, on vertical faces in cool, humid areas. This species is relatively widespread across the United States and other areas worldwide.

KEY FEATURES: Fruticose thalli, black color, fine, thin filaments, *Trentepohlia* photobiont, on non-calcareous rocks, in cool humid habitats.

Dendriscocaulon intricatulum

50 Ways To Leave Your Lover

Tripp 5309 (photo: Lendemer)

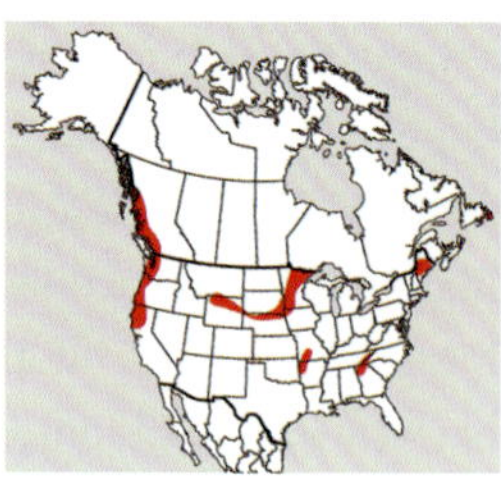

NOTES: Just when you thought lichenology couldn't get any weirder, enter *Dendriscocaulon*, a genus of "cyanomorphs" that are morphologically distinct versions of lichens that generally associate with green algal photobionts rather than cyanobacteria (in this case, *Nostoc*). *Yep: find a new partner, get a new look.* In some instances, you actually find chimeric individuals–half looking like one thing with one partner and half looking like something else, with another partner. *Dendriscocaulon intricatulum* is distinctive and easily recognized by its absolutely minute fruticose thallus. Some texts call this growth form "microfruticose." It was recently discovered to be the blue-green algal version of *Lobaria quercizans*, if you can fathom that!

CHEMISTRY: No substances. Spot tests. K-, C-, KC-, P-, UV-.

NICHE: This species is uncommon at middle elevations on the bark of hardwoods. Look for it in coves full of Sugar Maple dripping with other cyanolichens, such as species of *Pannaria*.

KEY FEATURES: Very small fruticose lichen with brownish-green branches, *Nostoc* photobiont, uncommon in high-quality, rich coves full of hardwoods and other cyanolichens.

Dermatocarpon luridum

Eastern Mountain Wave

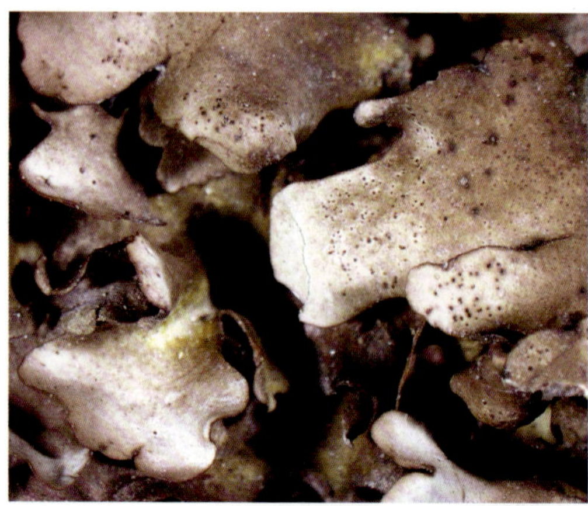

Tripp 1152 (photo: Lendemer)

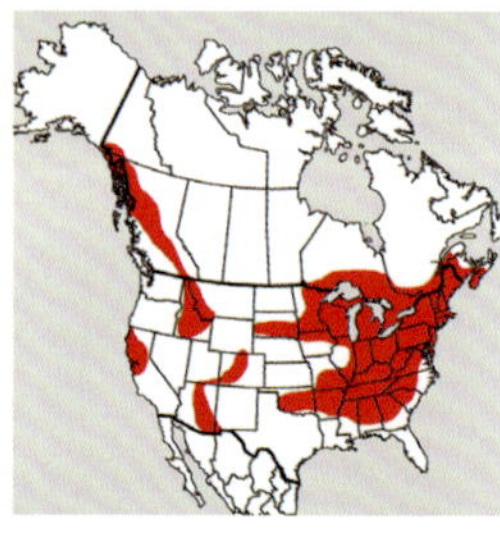

NOTES: *Dermatocarpon luridum* is one of two species in the genus presently known from the Smokies. We are okay with that, considering species identification in this lineage isn't the easiest task of your day. If you find a large foliose lichen with thick, gray, crunchy lobes (when dry) growing on a rock, it is likely to be either this species or its close relative, *D. muhlenbergii*. You can differentiate the two by niche: our species in question grows on wet, non-calcareous rocks, whereas the latter species occurs on dry, typically calcareous rock faces. *Dermatocarpon luridum* is also quite different from *D. muhlenbergii* in morphology; namely, it forms large, mat-like colonies of jumbled lobes with multiple attachment points to the substrate. In contrast, *D. muhlenbergii* usually has a thallus consisting of a single large lobe, with a single attachment point to the substrate. The difficultly comes when the one large lobe of *D. muhlenbergii* becomes frayed at the edges, folding in and around on itself, and causing the tangled appearance. The same can happen when multiple young thalli of *D. muhlenbergii* grow closely together.

CHEMISTRY: No substances. Spot tests. K-, KC-, C-, P-, UV-.

NICHE: Not surprisingly, Eastern Mountain Wave is a lover of mountainous portions of eastern North America, where it grows on non-calcareous or weakly calcareous rocks, where it is periodically submerged by water. It also occurs in scattered localities in western North America. In the Smokies, look for this species in or near streams, such as along the Rabbit Creek Trail.

KEY FEATURES: Relatively large foliose lichen, forming mats of thick, crunchy lobes, with perithecia, on wet, non-calcareous rocks often associated with streams.

Dermatocarpon muhlenbergii

Henry's Legacy

Tripp 2098 (photo: Lendemer)

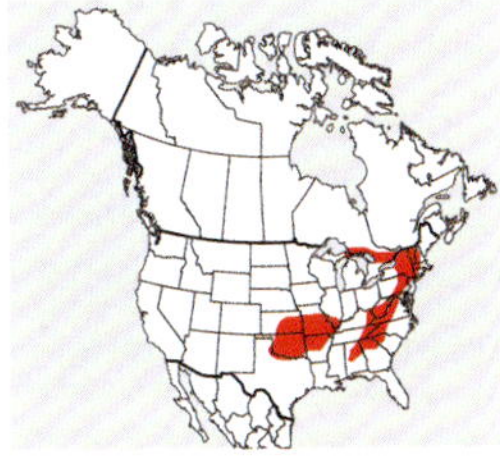

NOTES: Dreadful genus but honorable epithet that pays homage to Henry Muhlenberg (sans umlaut, people), who was an early American botanist and lichenologist of Lancaster, Pennsylvania and an inspiration to many. *Dermatocarpon muhlenbergii* is recognizable by its gray, thick, crunchy umbilicate thalli that are generally pruinose on the upper surfaces. From other members of the genus, this species can be identified by its ecology (see Niche). It differs from the far more common and superficially similar genus *Umbilicaria* in having perithecia instead of apothecia.

CHEMISTRY: No substances. Spot tests. K-, C-, KC-, P-, UV-.

NICHE: *Dermatocarpon muhlenbergii* occurs on both vertical faces and in cracks of non-calcareous rocks at middle-to-low elevations throughout the Smokies. Look for this species on a casual stroll from The S.W.A.G. down to Caldwell Fork. Don't miss *Rockefellera crossophylla* and *Caloplaca reptans* along the way. *Dermatocarpon muhlenbergii* is widely distributed in eastern North America.

KEY FEATURES: Crunchy, umbilicate thallus, perithecia, dry rock faces, occasional throughout the Smokies.

Dibaeis absoluta

Diamonds in the Dust

Tripp 5089 (photo: Lendemer)

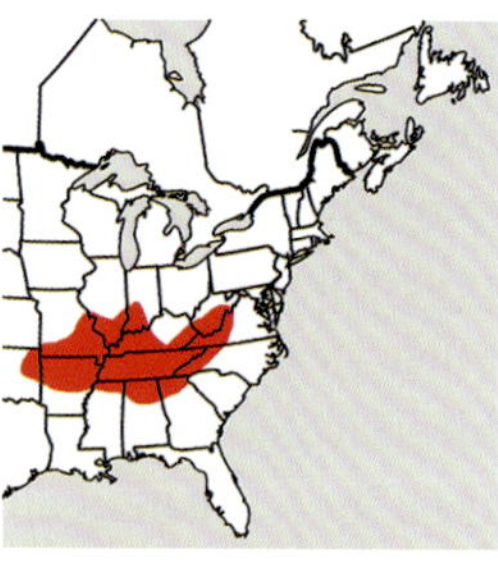

NOTES: *Dibaeis absoluta* is absolutely unmistakable for any other lichen in the Smokes. It is an unusual species that looks nothing like the other, much more common member of the genus, *D. baeomyces*, which has a thick, white thallus and spherical apothecia raised on tall, slender stalks. *Dibaeis absoluta* is more likely to be confused with *Icmadophila ericetorum*, from which it differs by its army-green thallus, which resembles a thick scum of green algae (scratch that: it *is* a thick scum of green algae) and its irregularly shaped, strongly convex apothecia, which are sessile or raised only on very short stipes, as seen in the photograph.

CHEMISTRY: Baeomycesic acid and squamatic acid. Spot tests. K-, C-, KC-, P+ yellow, UV+ blue- white.

NICHE: This species is rare throughout its range, including in the Smokies, where it is found on shaded, seepy, non-calcareous rocks, and in rock overhangs at middle-to-high elevations. Our recent fieldwork in northern Alabama has turned up several additional known populations of this species. In fact, this is a rare example of a lichen that seems happier outside of, rather than inside, the Smokies.

KEY FEATURES: Pink, sessile to short-stiped apothecia against a dull, army-green thallus of algal scum, non-calcareous rocks and dirt over rocks, especially in overhangs, at middle-to-high elevations.

Dibaeis baeomyces

Hayden's Surprise

Lendemer 26750 & Tripp 4964
(photos: Moroz & Lendemer)

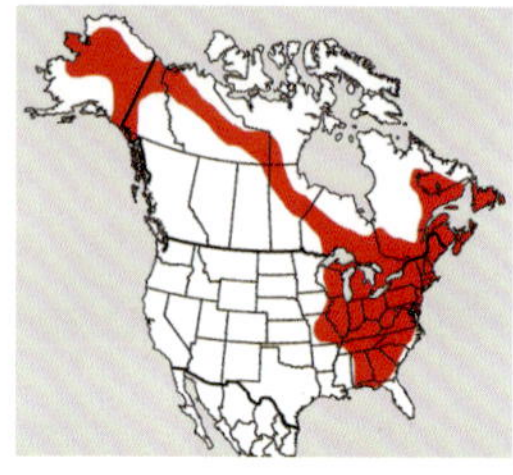

NOTES: *Dibaeis baeomyces* is an iconic lichen of eastern North America. It is also a species you probably already know because it is hard to ignore, when spotted. It is characterized by its generally thick, crusty, gray-to-green thalli (these forming hard pans of sorts, when dry), which bear numerous large, bright pink apothecia that sit atop tall, thick white stalks. If that wasn't enough to distinguish it from all other Smokies lichens, see further information regarding its growth and substrate preferences (under Niche).

CHEMISTRY: Baeomycesic acid and squamatic acid. Spot tests. K+ yellow, C-, KC-, P- yellow, UV+ blue-white.

NICHE: *Dibaeis baeomyces* prefers highly disturbed and eroded soils that border the sides of trails and other human interventions. This is probably why this otherwise extremely common species throughout eastern North America is relatively uncommon in the Smokies. Look for it in particular along the margins of hot, exposed trails at lower elevations. A good place to see this species is near the weather station just down the road from Purchase Knob.

KEY FEATURES: Thick, crusty gray-to-green thallus that bear huge pink apothecia atop thick white stalks, all elevations, on soils.

Dictyocatenulata alba

Mealy Bug Lichen

Lendemer 30283 (photo: Tripp)

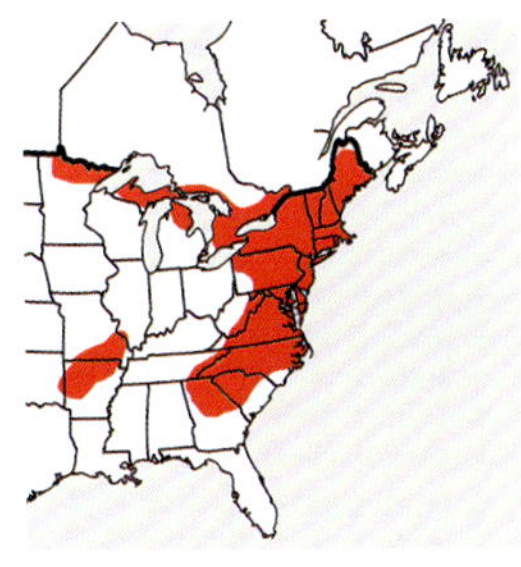

NOTES: *Dictyocatenulata alba* is a mysterious lichen that is instantly recognizable in the field, once learned. It resembles a tiny, white pin lichen with fuzzy white globose heads that sit atop flesh colored stalks. Rather than a mazaedium of spores, the fuzzy head is an asexual structure covered with loosely packed conidia. If you are a horticulturist and know mealy bugs, this is the mealy bug of the lichen world.

CHEMISTRY: No substances. Spot tests. K-, C-, KC-, P-, UV-.

NICHE: *Dictyocatenulata alba* is common throughout the Smokies and is frequently found on roots or shaded bases of Yellow Birch and occasionally on other hardwoods. Look for this species near The Jumpoff from the Boulevard Trail. Although often overlooked, the species is common throughout the Appalachians, Great Lakes region, and Ozarks.

KEY FEATURES: Tiny white pins with white, mealy heads and flesh-colored stalks, on roots and bases of hardwoods, especially Yellow Birch, at all elevations.

Dimelaena oreina

Close Encounter

Tripp 3655 (photo: Lendemer)

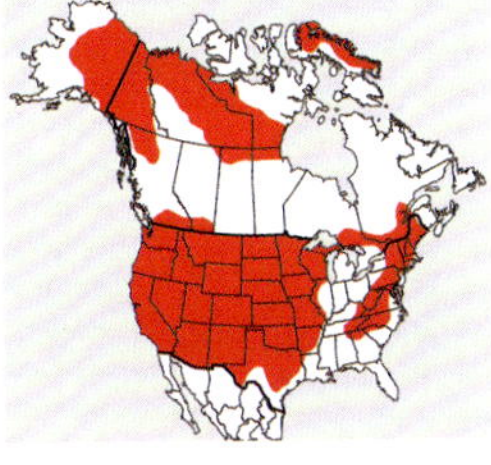

NOTES: This one is easy, as there are no other crustose lichens in the Smokies with yellow-green thalli that have radiating marginal lobes that are *very* tightly appressed to the substrate. These characters, together with the dark purple-brown discs that have lecanorine margins and C+ thallus make *Dimelaena oreina* instantly recognizable. While many lichen enthusiasts classify growth forms as fruticose, foliose, or crustose, this is an example of a species that blurs the boundaries of the latter two. This morphology sometimes gets called "placodioid," which is basically the name for when a crustose lichen develops structures that look like lobes. We apologize, but we didn't invent the term.

CHEMISTRY: Usnic acid, gyrophoric acid. Spot tests. K-, C+ pink, KC+ pink, P-, UV-.

NICHE: *Dimelaena oreina* is a common species that occurs on exposed, sunny, non-calcareous rocks throughout the Smokies. It is common in similar habitats throughout its range; however, to see the true colors of this species, go west where it and its numerous chemotypes are *much, much, much* more common. Within the Smokies, you can see this species on the popular Chimney Tops.

KEY FEATURES: Distinctive yellow-green placodioid thallus with radiating marginal lobes that are tightly appressed to the substrate, lecanorine apothecia, dark purple-brown discs, C+ pink, sunny, exposed non-calcareous rocks at all elevations.

Diploschistes actinostomus

Warm and Weathered Lichen

Tripp 3941 (photo: Lendemer)

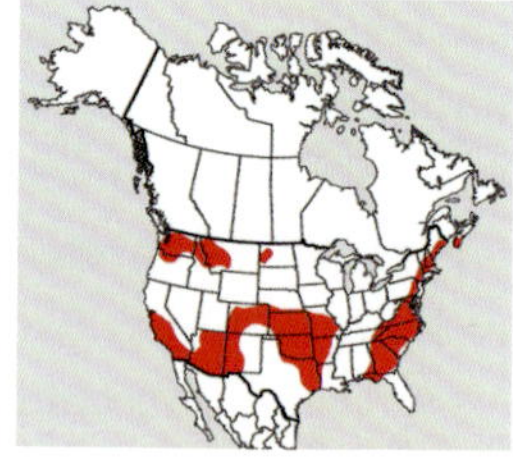

NOTES: We think you'll agree that *Diploschistes actinostomus* is a strange bird. (Well, a strange lichen.) Unlike most members of the genus, which have black apothecia with readily visible discs, this species has flask-like apothecia buried in a gray thallus and visible only as a pore, thus resembling perithecia. These characteristics, together with its brown, muriform spores and C+ pink thallus, make this species easy to recognize, even in the field. Note its crustose growth form and its white prothallus, too.

CHEMISTRY: Lecanoric acid. Spot tests. K-, C+ pink, KC+ pink, P-, UV-.

NICHE: *Diploschistes actinostomus* is uncommon in the Smokies, as is the case throughout much of its range, and is found on exposed, sunny, non-calcareous rocks at all elevations. The protologue from 1810 describes its original collection made by early mycologist Christiaan Persoon in "Gallia" as having been found in similar habitats.

KEY FEATURES: Thick gray areolate thallus, immersed apothecia visible as black pores and resembling perithecia, C+ pink, on sun-exposed non-calcareous rocks at all elevations.

Diploschistes muscorum

Lichen Sucker

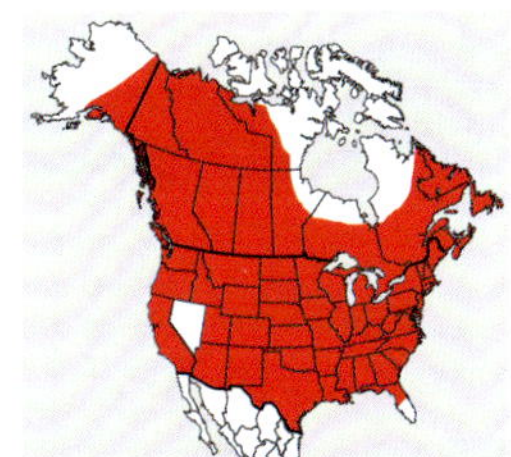

Tripp 5027A (photo: Lendemer)

NOTES: Unlike the photograph seen here, *Diploschistes muscorum* is rarely front and center. It is somewhat of an inconspicuous crustose lichen that begins life as a parasite on other lichens, most often on *Cladonia*, although here it is pictured growing on the closely related (to *Cladonia*) genus *Stereocaulon*. Eventually, *D. muscorum* overtakes the thallus of its host and forms its own independent, white colony. This parasitic lifestyle, combined with the black apothecia, brown, muriform spores, and C+ pink thallus readily serve to identify the species. This lifestyle serves to distinguish this species from all other *Diploschistes* in the Smokies.

CHEMISTRY: Lecanoric acid. Spot tests. K-, C+ pink, KC+ pink, P-, UV-.

NICHE: *Diploschistes muscorum* is a true winner, establishing itself on the thalli of other lichens and eventually becoming independent. Sounds like a good game strategy to us, anyway! It is infrequent but widespread in the Smokies, where it can be found on *Cladonia* and other saxicolous or terricolous lichens. Above, it can be seen overgrwoing basal portions of a thallus of *Stereocaulon*. The epithet derives from the assumption that the species grows on mosses initially rather than on lichens. We have seen much less evidence of the former compared to the latter.

KEY FEATURES: Juvenile parasite, thick gray thallus, immersed black apothecia, brown muriform spores, C+ pink, on other lichens, and eventually independent on soil and rocks at all elevations.

Diploschistes scruposus

Cowboy Float

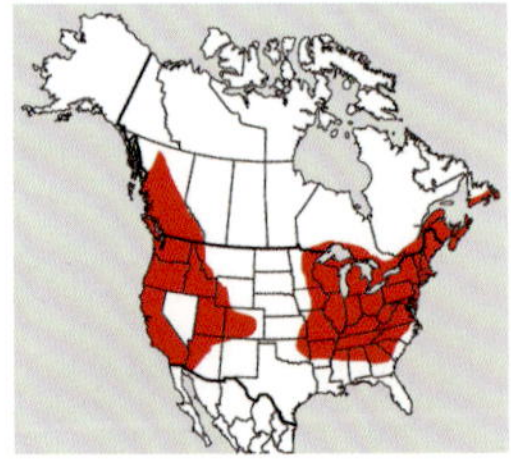

Tripp 3725 (photo: Lendemer)

NOTES: *Diploschistes scruposus* is a charismatic lichen not easily confused with any others in the Smokies. It can be recognized by its brownish areolate thallus with large, yawning black apothecia that have white pruinose discs and bear brown muriform spores. In addition to those characters, the thallus is C+ pink. When *Diploschistes muscorum* becomes independent of its host, it could be confused with *D. scruposus*. However, the thallus in *D. muscorum* is white rather than brown, and searching almost always reveals some remnants of its parasitic stage nearby the independent thalli. It is unlikely that *D. scruposus* could be confused for *D. actinostomus* because the latter has buried, flask-like apothecia rather than the huge, yawning, open discs of *D. scruposus*.
CHEMISTRY: Lecanoric acid. Spot tests. K-, C+ pink, KC+ pink, P-, UV-.
NICHE: This species is infrequent but widespread on sunny, non-calcareous rocks throughout the Smokies. It occurs in similar habitats throughout its range and is doubtless the most common member of the genus in North America. In general, western North America is a better place to see *Diploschistes* species than is the East—but hey, we'll take it.
KEY FEATURES: Thick, yellowish-brown thallus, immersed, yawning black apothecia, brown muriform spores, C+ pink, on sun-exposed, non-calcareous rocks at all elevations.

Dirina massiliensis f. *sorediata*

A Sheltered Life

Tripp 3663 (photo: Lendemer)

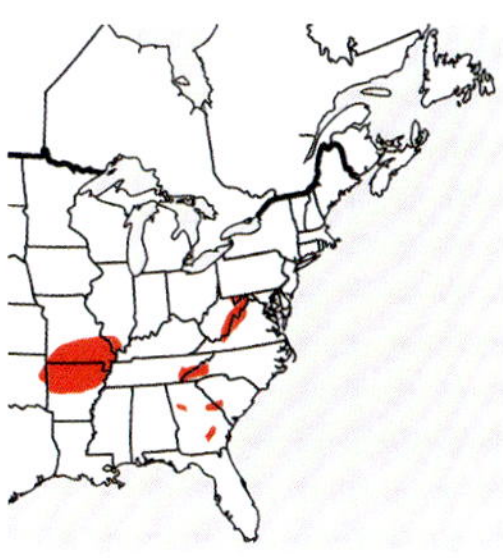

NOTES: This lichen is a bit on the silly side, but we think you will like it because it is easy to identify on account of its continuous blue-gray thallus, soralia that react C+ red, *Trentepohlia* photobiont, and ecology (see Niche). Indeed, *Dirina massiliensis* f. *sorediata* is one of many sorediate crusts in the Smokies that are accessible even to the beginner. Molecular studies on European material have shown that sorediate forms of this lichen should be recognized under the same taxonomic concept as esorediate forms. However, studies have yet to be conducted in North America, where esorediate populations have not yet been reported.

CHEMISTRY: Erythrin. Spot tests. K-, C+ red, KC+ red, P-, UV-.

NICHE: This taxon is characteristic of the few calcareous rock outcrops in the Smokies, where it grows in shaded, cool overhangs. Look for it among the calcareous cliffs above Lake Chilhowee.

KEY FEATURES: Thin, blue-gray, continuous thallus, C+ red soralia, *Trentepohlia* photobiont, on calcareous rocks in overhangs at scattered middle-to-low elevation locations.

Dirinaria frostii

Dapper Leggings

Lendemer 29514 (photo: Lendemer)

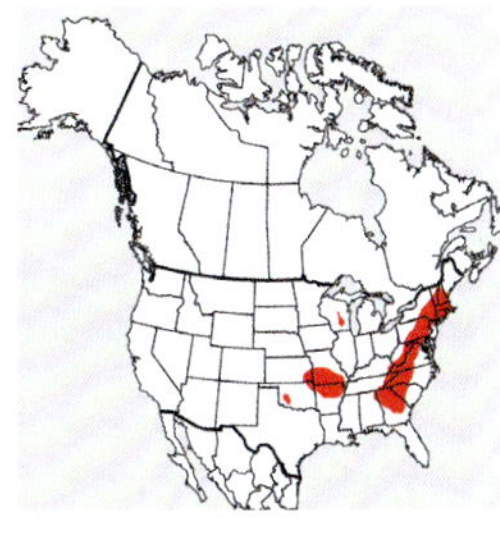

NOTES: *Dirinaria frostii* is truly a beautiful lichen, at least in our opinion. Maybe it is because one never expects to see a lovely foliose lichen hiding in rock overhangs. Or maybe it is the distinctive blue-gray coloration of the thallus and the laminal soralia. Regardless, this species is not to be missed. In many respects, it resembles species of *Physcia* such as *P. americana*, but can easily be recognized by its tightly adnate thallus and UV+ blue-white medulla. The overall appearance of the lobes is similar to species of *Pyxine* such as *P. sorediata*, particularly due to the tendency for the surface of the lobe tips to be white pruinose. Thankfully, species of *Pyxine* do not usually occur on sheltered rocks, and the medulla is both UV- and pigmented yellow or orange in all of the species that occur in the southern Appalachians. For those who may wonder, the species was named in honor of C.C. Frost, a botanist from the snowy climes of Vermont.

CHEMISTRY: Atranorin and divaricatic acid. Spot tests: (cortex) K+ yellow, C-, KC-, P-, UV-; (medulla) K-, C-, KC+ fleeting pinkish, P-, UV+ bright blue-white.

NICHE: This species occurs throughout eastern North America on non-calcareous rocks in cool, shaded overhangs or vertical cliff faces. Because most rocks in the Smokies are wet, and covered with vegetation or other lichens, it is rare in the Park. The best displays can be seen on the cliffs along Chillhowee Lake.

KEY FEATURES: Blue-gray, strongly adenate, foliose thallus, laminal soralia, pruinose lobe tips, UV+ blue-white medulla, on non-calcareous rocks in overhangs are on bluffs.

Distopyrenis americana

Abandoned Birch Spots

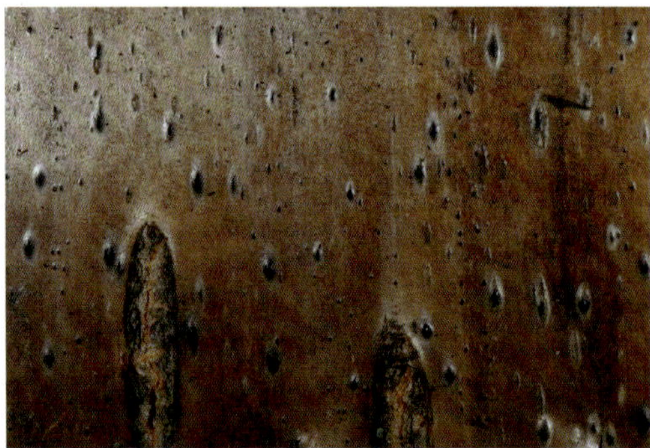

Tripp 2531 (photo: Lendemer)

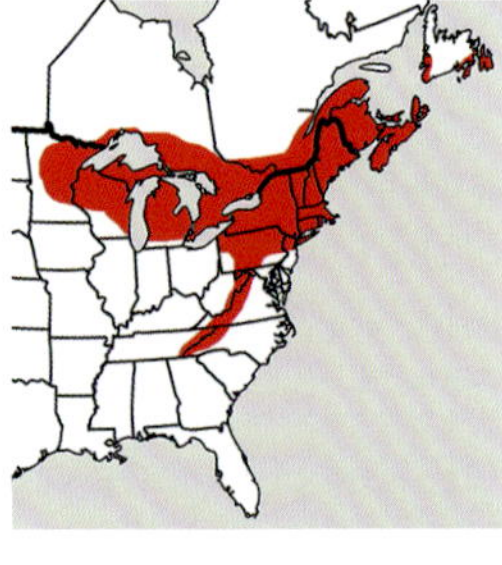

NOTES: *Distopyrenis americana* is an excellent example of an evolutionary transition from a lichenized to a non-lichenized lifestyle. The species belongs to the same family as *Pyrenula*, a genus with one too many species in the southeastern United States. You can recognize *D. americana* in the field by its restricted ecology (see Niche) and its small, black perithecia with 2-celled, polarilocular, brown spores. Note that this is one of the rare instances where polarilocular spores occur in a lineage unrelated to *Caloplaca* and friends. Otherwise, it looks more like lenticels on twigs.

CHEMISTRY: No substances. Spot tests. K-, C-, KC-, P-, UV-.

NICHE: *Distopyrenis americana* is narrowly restricted to the bark and branches of Yellow Birch, specifically portions that have not yet been colonized by lichens. It's distribution mirrors that of the host, which occurs at middle-to-high elevations of the Smokies. Outside of the Smokies, you can find this species anywhere its host also occurs.

KEY FEATURES: Non-lichenized thallus, small black perithecia often elongated with the grain of the bark, 2-celled, brown, polarilocular spores, on Yellow Birch at middle-to-high elevations.

Endocarpon pallidulum

Western Crepes (Alternative Name for Fast Typers: Western Creeps)

Lendemer 23611A (photo: Lendemer)

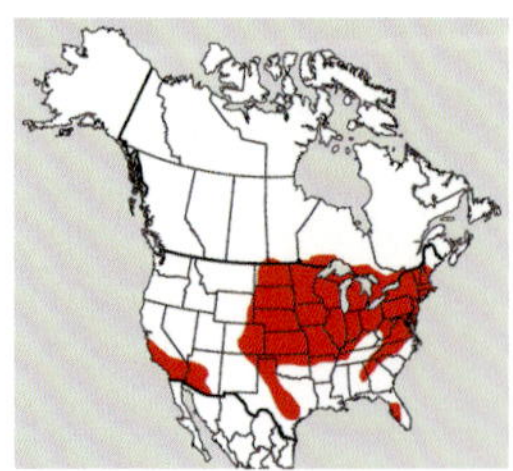

NOTES: Western Crepes is a highly distinctive species that you cannot confuse for anything else that occurs in our area. Here's why: first and foremost, it is one of only two species in the Park that has a hymenium inspersed not only with the usual spores (in this case, big, beautiful muriform spores), but also with green algae. Further, the thallus of *Endocarpon pallidulum* is definitively squamulose, albeit micro-squamulose, and grows primarily on calcareous rocks. The thalli range from chocolate-brown-to-bright-green, depending on whether dry or wet. Western Crepes is furthermore recognizable by its perithecia, which are slightly emergent above the thallus and sit more or less in the middles of the scales or squamules.

CHEMISTRY: No substances. Spot tests. K-, KC-, C-, P-, UV-.

NICHE: Western Crepes ranges from the desert southwestern United States through the Great Plains, and into the eastern United States, primarily in northern areas and south along the Appalachian Mountains. Look for it on calcareous rocks at low elevations. It is very uncommon in the Smokies.

KEY FEATURES: Green to brown microsquamulose thalli that bear slightly raised, dark brown perithecia, algae contained within the hymenium, muriform spores, on calcareous rocks at low elevations.

Enterographa hutchinsiae

Sneaky Cracks

Tripp 3670 (photo: Lendemer)

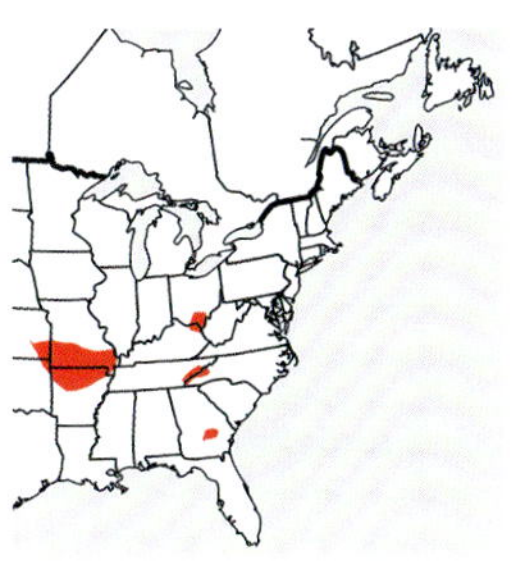

NOTES: *Enterographa hutchinsiae* superficially resembles species of *Arthonia* but differs from all members of that genus that occur in the Smokies by its ecology (see Niche), brown, cracked thallus, narrowly fusiform spores, and chemistry. *Enterographa hutchinsiae* could also be confused with species of *Opegrapha*, but the apothecia do not have black, carbonized margins typical of the latter genus.

CHEMISTRY: Confluentic acid. Spot tests. K-, C-, KC-, P-, UV+ dull blue-white.

NICHE: *Enterographa hutchinsiae* is a rare species in North America that is restricted to the Appalachians and Ozarks. It is rare in the Smokies and known only from non-calcareous rock overhangs at middle-to-low elevations. A great place to see this species is on the rock cliffs just above Ramsey Cascades Trail.

KEY FEATURES: Thin brown-gray thallus, immersed irregularly shaped brownish-black apothecia, narrowly fusiform colorless spores, on non-calcareous rocks in overhangs at low and middle elevations.

Ephebe americana

Halloween Hair

Tripp 3737 (photo: Lendemer)

NOTES: *Ephebe americana* is not the most attractive lichen you will encounter in the Smokies (you can't penalize it for its looks), but its specialness makes up for it. Count yourself among the few and the lucky if you find it. This species is easily differentiated from its two congeners in our area, *E. solida* and *E. lanata*, by its thalli that form rosettes, its frequent apothecia, and its occurrence on dry rocks (vs. pendulous, rare apothecia, and on riparian rocks). Besides the three *Ephebe*, there are only two other fruticose cyanolichens in the Park. *Thermutis velutina* looks like freshly cut, fallen black hair, and *Dendriscocaulon intricatulum* occurs on bark and looks nothing like *Ephebe*.

CHEMISTRY: No substances. Spot tests. K-, C-, KC-, P-, UV-.

NICHE: *Ephebe americana* always occurs on dry, exposed granitic rocks but is rare in the Smokies. Oh—and it is even rarer throughout the southern Appalachians, where it is endemic.

KEY FEATURES: Fruticose cyanolichen with finger-like lobes, rosette-forming thalli, frequent apothecia, occurrence on dry, exposed rocks, endemic to the Southern Appalachians.

Ephebe solida

After the Deluge

Tripp 4979 (photo: Tripp)

NOTES: *Ephebe solida* is one of three members of the genus in the Smokies, but of these, only *E. lanata* shares a distinctive preference for riparian habitats (see Niche) and a pendulous thallus. *Ephebe solida* can be differentiated from *E. lanata* by having dichotomous branching (vs. irregularly in *E. lanata*), we guess. In contrast, *E. americana* always occurs on dry rocks in exposed environments. Additionally, *E. americana* is typically found with apothecia, unlike *E. solida*. The genus as a whole is easily recognized by its fruticose growth form and cyanobacterial photobiont.

CHEMISTRY: No substances. Spot tests. K-, C-, KC-, P-, UV-.

NICHE: In the Smokies as well as other areas of the southern Appalachians, *Ephebe solida* occurs on wet rocks associated with riparian environments, at middle-to-high elevations. The full distribution of this species is poorly understood.

KEY FEATURES: Fruticose cyanolichen with finger-like lobes, pendant thalli, absence of apothecia, occurrence on rocks in riparian environments, endemic to southern Appalachians.

Erioderma mollissimum

Southern Refuge

McMullin 7971 (photo: McMullin)

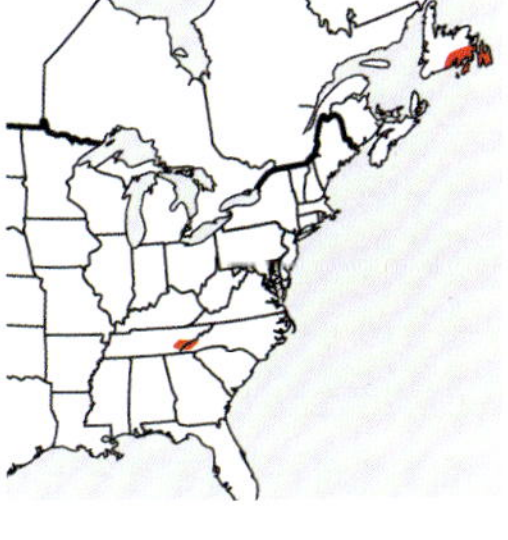

NOTES: This lichen is among the top ten unsolved lichenological mysteries in Great Smoky Mountains National Park. It is known from one—and only one—collection: a scrap of a specimen vouchered back in 1962. Fast forward 56 years later, and no one has seen it since. We consider it to be extirpated from the Park (and to see it, one must travel north to the Canadian Maritimes, where it is still a very rare species). It is highly improbable that, back in 1962, there was only one thallus growing on a tree. The population must have been larger, consisting of several to many individuals. But for unknown reasons, it has never been found again, including over the course of our extensive travel, both on trail and off trail. This species is highly distinctive and knowable on account of several features. First, it is a sorediate cyanolichen (but not a jelly lichen, as in species of *Leptogium* and *Collema*). Second, it lacks cyphellae or pseudocyphellae on the lower surface (typical of species of *Sticta* and *Pseudocyphellaria*, respectively). Third, its lower surface is marked by distinctive, raised veins, as in *Peltigera*, but *Erioderma mollissimum* differs in having minute trichomes or hairs that cover the upper surface. There is only one other lichen with this combination of characters, and that species (*Leioderma cherokeense*) is also extirpated based on current knowledge. *Erioderma mollissimum* differs from *L. cherokeense* in being sorediate. But if you find either in the Park or surrounding areas, mark it down as one of the best days of your life!

CHEMISTRY: No substances. Spot tests. K-, KC-, C-, P-, UV-.

NICHE: This species should be looked for in high-quality forests that are middle-to-high in elevation based on its known, extant distribution elsewhere (such as in Nova Scotia, Canada, where this photograph was taken; we apologize if you were hoping for a much less interesting photo of a dried herbarium specimen collected in the 1960s in the Smokies). It's only known from one occurrence in the Smokies, which was on the bark of a *Magnolia*.

KEY FEATURES: Exceptionally rare, presumably extirpated cyanolichen previously known from only one record at middle elevations, large gray foliose thallus, soredia, upper surface covered by minute hairs, lower surface with distinct pattern of raised veins, most likely on the bases of hardwood trees if ever rediscovered in the Park—in which case please do not collect it, but do take photos and record your coordinates!

Everniastrum catawbiense

Shelton's Bones

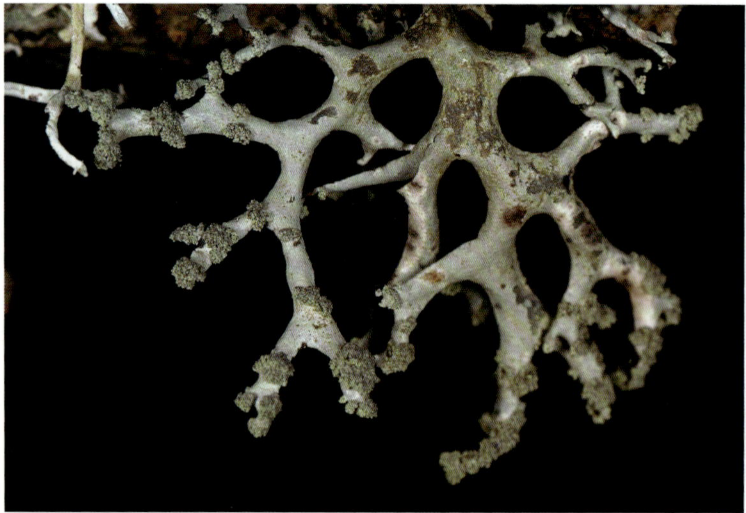

Tripp 2289 (photo: Lendemer)

NOTES: *Everniastrum catawbiense* is a charismatic and highly recognizable species in the Smokies and elsewhere. It is foliose (i.e., has a differentiable upper and lower cortex) but its growth pattern gives it an almost fruticose appearance. The green-gray thallus with very large, erupting soredia characterize this species, along with its ecology (see Niche). *Everniastrum catawbiense* is a wonderful lichen that cannot be confused with anything else in the Park, not even *Pseudevernia consocians*, whose isidia you should be jealous of.

CHEMISTRY: Atranorin, gyrophoric acid. Spot tests. K+ yellow, C+ pink, KC+ pale red or KC-, P-, UV-.

NICHE: *Everniastrum catawbiense* has a narrow niche in the Smokies but is very common where it occurs: at the highest elevations in the Park, on the branches of conifers and hardwoods. One often finds this species growing together with *Pseudevernia cladonia* and *Hypogymnia* spp. The Smokies is the center of distribution for *E. catawbiense*. It is uncommon-to-rare elsewhere in its range.

KEY FEATURES: Foliose but appearing fruticose, greenish-gray thallus, large, erupting soralia, on twigs at highest elevations in the Park.

Fellhanera bouteillei

Over Easy

Tripp 3895 (photo: Lendemer)

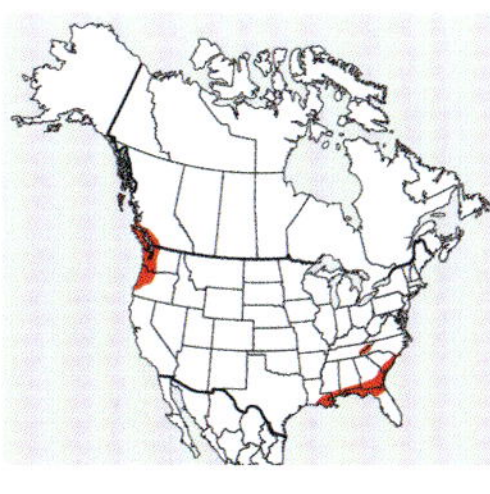

NOTES: *Fellhanera bouteillei* is a miniscule but very easy to identify species in the Smokies by its occurrence on leaves and its flesh-colored apothecia against a thin, continuous veneer of a gray-green thallus. That's a mouthful, but that's what it looks like. All other species of *Fellhanera* in the Park grow on other substrates. The only other relatively common foliicolous lichen in the Smokies is *Byssoloma subdiscordans*, but that species has a dark apothecial disc with distinctive byssoid margins, unlike those of *F. bouteillei*.

CHEMISTRY: No substances. Spot tests. K-, C-, KC-, P-, UV-.

NICHE: *Fellhanera bouteillei* is the most common foliicolous lichen in the Smokies. It can readily be found growing in high-humidity environments at all elevations, on evergreen leaves such as those of *Rhododendron*, *Kalmia*, and *Leucothoe*. This species also occurs in humid areas of the southeastern Coastal Plain and the Pacific Northwest.

KEY FEATURES: Foliicolous, flesh-colored apothecia, thin gray-green thallus, high-humidity environments throughout the Smokies.

Fellhanera eriniae

Erin's Mountain Treasure

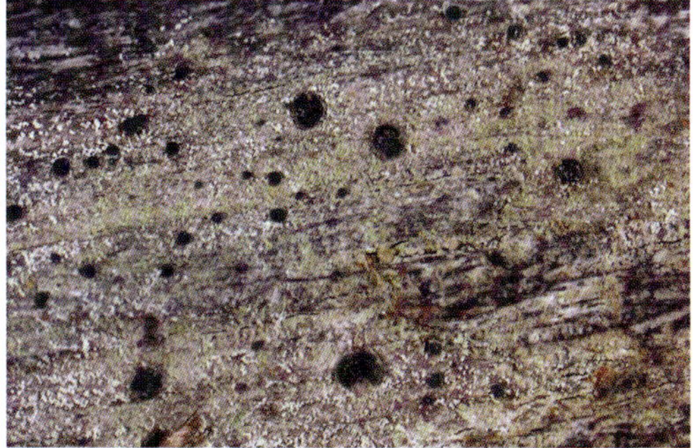

Lendemer 8164 (photo: Lendemer)

NOTES: On the lower slopes of Mount Sterling near Baxter Creek is a hectare of forest where we first started our work in the Smokies. One day of searching for lichens turned up many new species to science, including this one. *Fellhanera erinae* belongs to a group of species with small black apothecia and thin, greenish-gray thalli. It differs from the other, more common members of the group in growing on bark, having long, 6-8-celled spores, and having a blastidiate thallus. Other species, such as *F. granulosa* and *F. minisinkorum* have 4-celled spores. *Fellhanera montesfumosi* is another species with 6-8-celled spores, but it differs in lacking blastidia and in occurring on non-calcareous rocks.

CHEMISTRY: No substances. Spot tests. K-, KC-, C-, P-, UV-.

NICHE: One of the small number species that appears to be endemic to the Smokies, *Fellhanera erinae* is known only from a small area of Haywood County, North Carolina. It is known to grow on the branches of trees in high-humidity, mature forests.

KEY FEATURES: Greenish, crustose thallus, minute blastidia, small, dark brown-to-brown-black apothecia, 6-8-celled colorless spores, bacilliform conidia, on tree branches.

Fellhanera granulosa

Leprechaun Leavins

Tripp 3652 (photo: Lendemer)

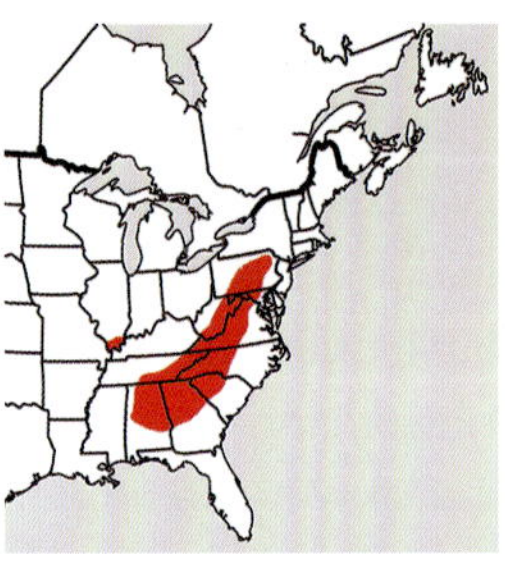

NOTES: *Fellhanera granulosa* is identifiable by its blastidiate thallus (as the epithet implies) and its black apothecia with 4-celled spores. Like all other *Fellhanera* in the Smokies, this species has a thin, dull green thallus with a conspicuous white prothallus and bowling-pin shaped conidia. Species of *Fellhanera* are often found sterile, and so one usually only has a thallus and pycnidia to work with. Look for the white prothallus to make an educated guess as to the genus. Getting to the species is more difficult in this group, it requires careful measurements and technical keys.

CHEMISTRY: No substances. Spot tests. K-, C-, KC-, P-, UV-.

NICHE: *Fellhanera granulosa* occurs on shaded, non-calcareous rocks at low-to-middle elevations in the Smokies. Its primary range is southern Appalachian (disjunct in southern Illinois), where it can be found in similar habitats.

KEY FEATURES: Dull-green blastidiate thallus, white prothallus, 4-celled colorless spores, bowling-pin-shaped conidia, shaded rocks, low-to-middle elevations.

Fissurina insidiosa

Insidious Intents

Tripp 5561 (photo: Lendemer)

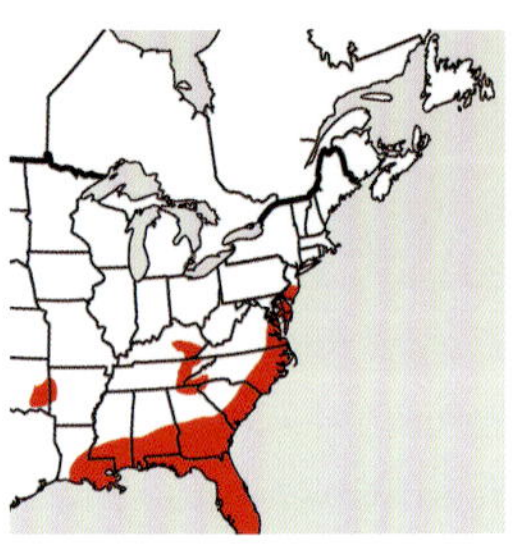

NOTES: *Fissurina insidiosa* is one of those *must-learn* lichens, no matter where you are in the southeastern United States. While there are a lot of lip lichens in our area, particularly toward the Coastal Plain and lower latitudes, this one stands out as highly identifiable based on its thick, smooth, continuous, green thallus, which bears numerous large, fissurine lirellae. You can immediately differentiate *F. insidiosa* from all lip lichens in *Graphis* and *Opegrapha* in our area because species in the latter genera have partially or fully carbonized (i.e., blackened) exciples. In *Fissurina*, exciples are, for the most part, non-carbonized and immersed in the substrate, thus giving them a class cracked or "fissured" look. Yep, it is that easy, assuming you are willing to make a cross section and study its anatomy. And then learn to recognize it by gestalt. Fun fact: many people erroneously believe that lichens inflict considerable harm on their host tree substrate. In fact, most lichens probably minimally impact their hosts in a negative way, if at all. This may be one exception. Notice that the bark sometimes buckles underneath thalli of *Fissurina insidiosa*. Way back in 1860, authors of the specific epithet "insidiosa" might have gotten this one correct.

CHEMISTRY: No substances. Spot tests. K-, KC-, C-, P-, UV-.

NICHE: This lichen is definitively representative of the coastal plain biogeographical element, but it makes its way into southern mountains, including the Smokies and to a much lesser extent, the Ozarks. In the Smokies, it can be found at low-to-middle elevations, especially in forests along streams, on smooth-barked trees, such as American Beech or Sweet Birch.

KEY FEATURES: Continuous, green, shiny thallus, lirellate apothecia with completely non-carbonized exciples, occurrence at low-to-middle elevations on smooth-barked trees.

Flakea papillata

A Snail's Feast

Tripp 5001 (photo: Lendemer)

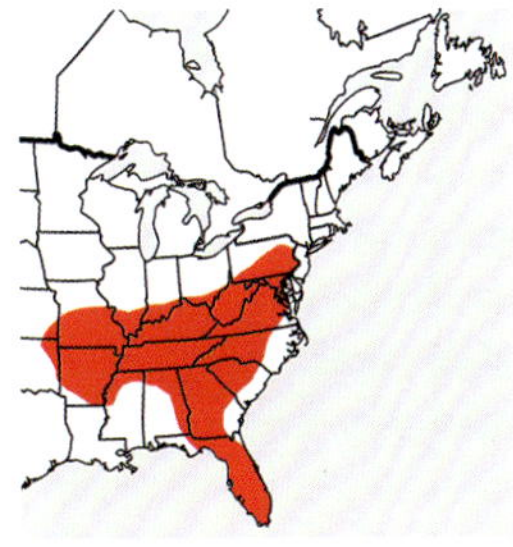

NOTES: *Flakea papillata* is an enigmatic lichen that was thought for the longest time to be a terrestrial alga. Wuuuuuuh? A dynamic duo of eminent bryologist and lichenologist (and Duke-ies) Bill Anderson and Bill Culberson knew better all along. *Flakea* is unmistakable for any other lichen, or alga for that matter. Just look at it! Oddly enough, and somewhat uncomfortably, it is related to *Verrucaria*, just like *Agonimia* and *Botryolepraria*, which are similar in color (bright green) when damp or wet. *Flakea* has papillose cells in the upper cortex, just like Pottiaceae, but we *promise* it is, in fact, a lichen!

CHEMISTRY: There is no chemistry in Verrucariaceae. Ever. Spot tests. K-, C-, KC-, P-, UV-.

NICHE: *Flakea papillata* is uncommon to rare in the Smokies, where it occurs in deep, shaded overhangs, usually together with *Botryolepraria lesdainii*. You'll only ever find *Flakea* off the beaten track. Look for it off trail near Greenbriar Pinnacle, or near the junction of Spring House Branch Trail and Forney Creek Trail.

KEY FEATURES: Bright green, small, strap-like thalli of dissected lobes, papillose upper surface, always sterile, in deep rock overhangs.

Flavoparmelia baltimorensis

Baltimore's Lichen

Tripp 3613 (photo: Lendemer)

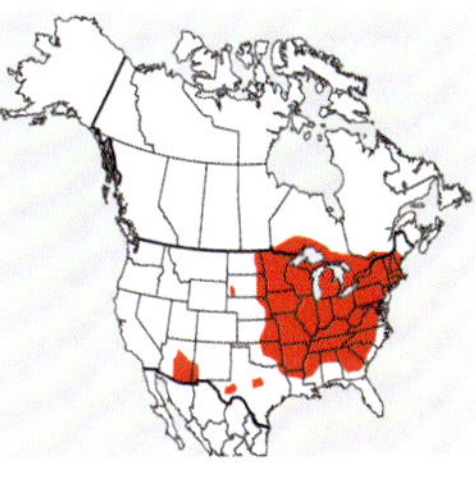

NOTES: *Flavoparmelia baltimorensis* is one of several common, greenish-yellow foliose lichens that occur on rocks in the Smokies. It differs from all others in having a pustulose upper surface versus upper surfaces with other kinds of structures, be they sexual or asexual. Like *F. caperata*, *F. baltimorensis* is maculate but differs in having coarse pustules rather than fine soredia. The abundance of usnic acid in thalli of *Flavoparmelia* makes them yellow. Yes, believe it or not, there is a species named after the City of Baltimore.

CHEMISTRY: Usnic acid, protocetraric acid. Spot tests. (medulla) K-, C-, KC-, P+ red, UV-.

NICHE: This species occurs almost exclusively on non-calcareous rocks, at middle-to-low elevations in the Smokies. It is very common elsewhere in eastern North America, where it occupies a similar niche.

KEY FEATURES: Large, green-yellow foliose thallus with pustules, on rock, relatively common.

Flavoparmelia caperata

Yellow Glory

Lendemer 32955 (photo: Tripp)

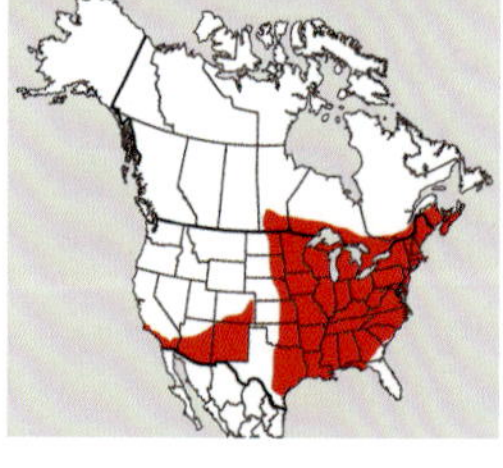

NOTES: This should be the first foliose lichen that you learn in the Smokies. We might have said that already, but this time we mean it. *Flavoparmelia caperata* is readily recognized by its greenish-yellow thallus with conspicuous soralia that contain fine, powdery soredia and cover substantial portions of the upper surface. It is differentiable from *F. baltimorensis* by these soredia (vs. pustules in the latter). *Flavoparmelia caperata* also grows on trees instead of rock, as in *F. baltimorensis* (typically). In the photograph, notice an unusual combination of simultaneous asexual (i.e., soredia) and sexual (i.e., apothecia) reproduction. Asexual species sometimes make sexual structures, but the converse is never true. This is one of the most common foliose lichens in eastern North America

CHEMISTRY: Usnic acid, protocetraric acid. Spot tests. (medulla) K-, C-, KC-, P+ red, UV-.

NICHE: Common. Widespread. Throughout the Smokies and beyond. You will find *Flavoparmelia caperata* on every available substrate. It may even latch onto your car as you drive through the Park. Don't stay too long, lest you become a substrate.

KEY FEATURES: Large, greenish-yellow foliose thallus with soredia, ubiquitous on bark throughout Park.

Flavopunctelia flaventior

Happy Ruffles

Lendemer 48570 (photo: Tripp)

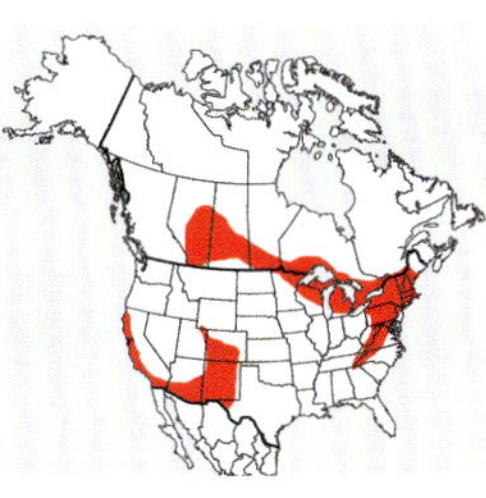

NOTES: When the only sorediate greenish-yellow foliose lichen you see day after day is *Flavoparmelia caperata*, you will be forgiven for missing this one. *Flavopunctelia flaventior* is not common in the Smokies, and where it grows it often hides among the *Flavoparmelia*. Nonetheless, you can recognize it even from several yards away by its dark green color, larger sizer, and its margins that are ruffled with soredia. Its most distinctive feature may, however, be its C+ red medulla, reminiscent of the strong C+ red reactions of the non-yellow *Punctelia*.
CHEMISTRY: Usnic and lecanoric acids. Spot tests. Cortex: K-, C-, KC+ gold, P-, UV-. Medulla: K-, C+ red, KC+ red, P-, UV-.
NICHE: It was only when we went to write this field guide that we discovered *Flavopunctelia flaventior* was almost entirely restricted to a small area of the Smokies on the long ridge between Paul's (Poles, Polls) Gap (it's true, three names over three iterations of Great Smoky Mountains National Park maps) and Purchase Knob. It is common there, in the northern hardwood forests, but, it seems, nowhere else. We have no further explanation.
KEY FEATURES: Big, greenish-yellow foliose thallus, ruffled sorediate margins, C+ red medulla, on hardwoods in northern hardwood forests near Purchase Knob and the S.W.A.G.

Fuscidea arcuatula

Kidney Bean Lichen

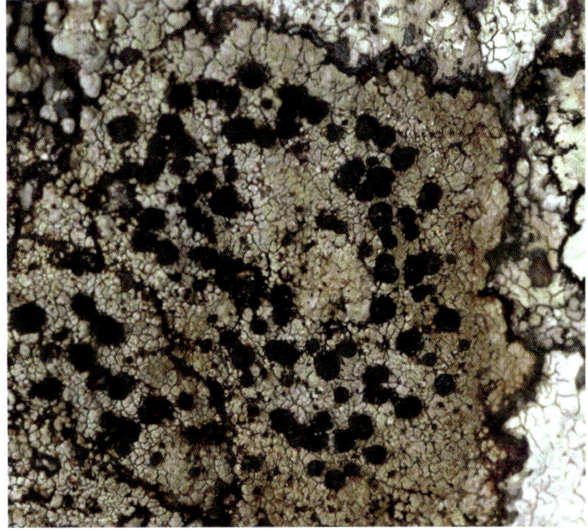

Tripp 2490 (photo: Deregibus)

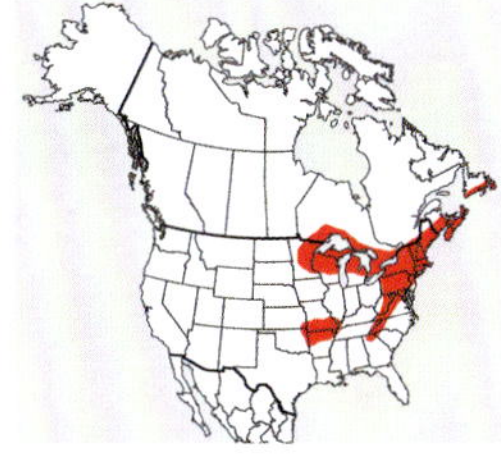

NOTES: *Fuscidea arcuatula* is one of the most common crustose lichens on rocks in the Smokies. It might not look like much from this photograph, but it does have adorable colorless spores that vary from simple to 2-celled and are bent to resemble kidney beans. Even in the field, you can recognize it by its gray-to-brown thallus bordered by a dark prothallus and its black apothecia. There are other species that look similar, like *Porpidia*, with its simple spores, or *Rhizocarpon*, with its larger 2-celled or muriform spores. But nothing can replace this once those kidney beans are seen. Also, check this one out under the UV lamp.
CHEMISTRY: Divaricatic acid. Spot tests. Medulla: K-, C-, KC+ pinkish, P-, UV+ blue-white.
NICHE: On exposed and shaded or sheltered non-calcareous rocks throughout the Smokies at nearly all elevations but especially the highest elevations. Outside of the Smokies, *Fuscidea*

arcuatula is common in the Appalachians, Great Lakes, and Ozarks.

KEY FEATURES: Gray-to-brown crustose thallus with black prothallus, black apothecia, colorless, bent spores that are simple or 2-celled, UV+ blue-white medulla, on non-calcareous rocks.

Fuscidea recensa

Zones of Incompatibility

Lendemer 53152 (photo: Tripp)

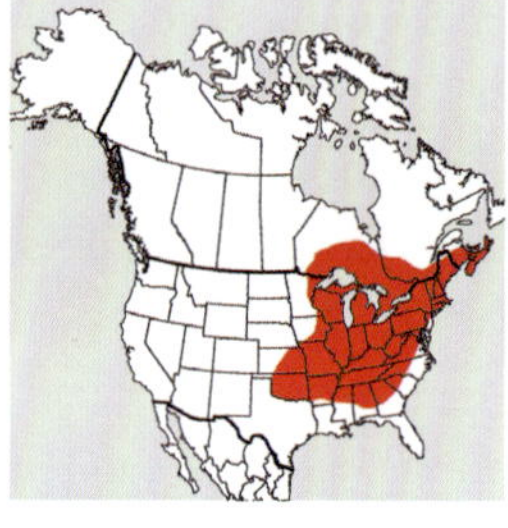

NOTES: *Fuscidea recensa* is a charismatic species that can be recognized by its thallus composed of gray-brown areoles that are thin and tend to be slightly dispersed across a thin black prothallus, small soralia with light greenish-blue to blue-gray soredia, and its niche. In the past, *F. arcuatula* was treated as an esorediate variety of *F. recensa*, but we believe they are distinct and can be distinguished even when growing side-by-side. *Fuscidea recensa* could be confused with *Opegrapha gyrocarpa*, *O. moroziana*, or *O. zonata*, but those species have a dark rusty-brown-colored thallus and soralia due to the presence of a *Trentepohlia* photobiont (vs. a green coccoid one in *F. recensa*). *Caloplaca reptans* can also resemble the species at a distance but differs in having larger areoles that are frequently lobed, and in producing no lichen substances. The common name of this species refers to the fact that it is most commonly found growing among several patches of itself, each with discrete lines of separation. In lichenology, these "zones of incompatibility" are largely the product of heterokaryon loci that help to specify "self" vs. "other."

CHEMISTRY: Divaricatic acid. Spot tests. Medulla: K-, C-, KC+ pinkish, P-, UV+ blue-white.

NICHE: This species is widely distributed in eastern North America, where it occurs throughout the Appalachian Mountains on non-calcareous rocks in cool, shaded overhangs or other sheltered microhabitats, such as cliff faces. In the Smokies it is uncommon and mostly found at high elevations where there are large rock outcrops, such as at Chimney Tops.

KEY FEATURES: Gray-brown, thin areolate thallus, black prothallus, minute soralia with light blue-gray soredia, C-, P- and UV+ bright blue-white, on rocks in overhangs at middle-to-high elevations.

Fuscopannaria frullaniae

Forever Nebulous

Tripp 4951 (photo: Tripp)

NOTES: *Fuscopannaria frullaniae* is one of the most enigmatic lichens in the Park. You win a gold star if you find it, and if you don't, we hope this entry and photograph satisfactorily share its story. This is a crustose cyanolichen composed entirely of tiny, bluish-gray granules that sometimes overlay a white prothallus. Beyond this and its ecology (see below), you have little else to go on. That being said, there is little that resembles it—whereas there are plenty of sterile leprose crusts in the Park, none of these is a cyanolichen.

CHEMISTRY: No substances. Spot tests. K-, KC-, C-, P-, UV-.

NICHE: Forever Nebulous is very rare in the Smokies, presently known only from a single population, at a high elevation, where thalli were found overgrowing mosses of the genera *Atrichum* and *Aulacomnium* (that in turn were growing on a damp, vertical rock). Given its very small size and general inconspicuousness, it is likely to occur elsewhere in the Park, but such discovery will require a very sharp set of eyes. This species was first described from Newfoundland and Nova Scotia overgrowing liverworts of the genus *Frullania*. It was later discovered in Portugal, also overgrowing *Frullania*. The present report of *F. frullaniae* from high elevations in the Smokies extends its known geographical range (to the mountains of the southeastern United States) as well as its known ecology (overgrowing mosses, too!).

KEY FEATURES: Sterile leprose cyanolichen composed entirely of bluish-gray granules, overgrowing bryophytes on rocks at high elevations, extremely rare.

Fuscopannaria leucosticta

White Eyes

Tripp 33140 (photo: Lendemer)

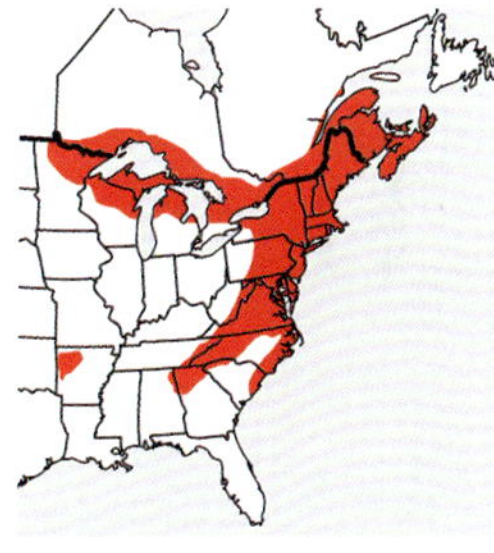

NOTES: The diversity and abundance of cyanolichens in the Smokies is truly remarkable, and this is one of the species that abounds here despite having fallen on hard times elsewhere across its range. You can recognize *Fuscopannaria leucosticta* by its thallus, which is composed of overlapping gray or brown squamules and by the frequent abundance of apothecia that have brown discs and conspicuously white-rimmed margins. All other similar species belong to *Pannaria* and differ in having truly foliose rather than squamulose thalli. Look closely, look frequently, and we think you will develop a similar opinion on squamulose versus foliose thalli.

CHEMISTRY: Terpenoids and fatty acids, if you believe the books. Spot tests. K-, C-, KC-, P-, UV-.

NICHE: Once widespread throughout temperate eastern North America but now in decline throughout much of its range, a lot like several other species of cyanolichens. This decline is the unfortunate result of habitat destruction and other forms of human-induced habitat alteration. The southern Appalachians are the last true stronghold where you can see *Fuscopannaria leucosticta* at middle-to-low elevations on the bark of mature hardwoods, more rarely on conifers or non-calcareous rocks along creeks and rivers.

KEY FEATURES: Cyanolichen with a squamulose thallus, lacking soredia or isidia, typically with abundant apothecia that have distinct white rims, P-, on mature hardwoods and conifers at middle-to-low elevations.

Fuscopannaria sorediata

A Pregnant Pause

Tripp 5002 (photo: Lendemer & Tripp)

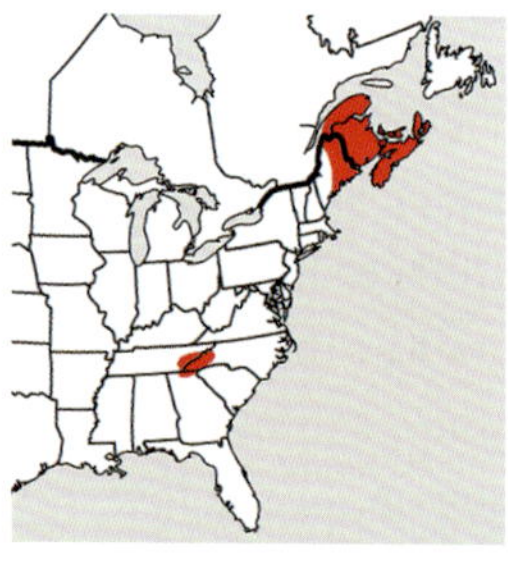

NOTES: This is a species that, despite being common in the southern Appalachians, was only recognized relatively recently. Like so many Pannariaceae, it defies clear understanding. Even though the species is currently treated in the squamulose genus *Fuscopannaria*, its thallus is more clearly foliose—a feature more typical of the genus *Pannaria*. The distinctive characters to help you identify it to species are, however, the presence of blue-gray soredia and its P- thallus. The distinction from *F. ahlneri* is unclear to us, and we are thankful that the latter species has not been reported from the Smokies. The only other similar species in the Smokies is *P. conoplea*, but the thallus in that species is larger and P+ orange-red.

CHEMISTRY: Terpenoids. Spot tests. K-, C-, KC-, P-, UV-.

NICHE: A species of the southern Appalachians and Canadian Maritimes. In the Smokies, it occurs on the trunks and bases of hardwood trees, especially at middle-to-low elevations.

KEY FEATURES: Squamulose-to-foliose cyanolichen with a brown thallus, blue-gray woolly soredia, P-, on hardwood trunks and bases mostly at middle-to-low elevations.

Glyphis cicatricosa

Pockmarks

Tripp 2641 (photo: Lendemer)

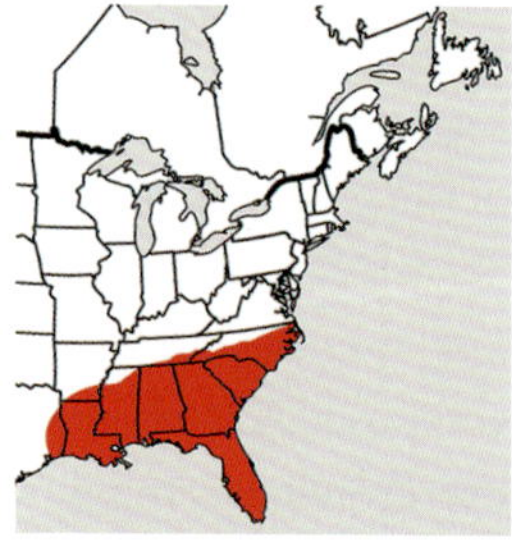

NOTES: It's hard to believe, but *Glyphis cicatricosa* is a script lichen just like *Graphis scripta*. But, instead of having perfectly defined and distinctly separated lirellae, the lirellae are aggregated together and raised into structures termed pseudostroma. That might sound strange, but the result is amazing, and we thank evolution for yielding endless forms most beautiful. No other crustose lichen in the Smokies is remotely similar in appearance, so no excuses! Learn this one the first time you see it! The specific epithet "-cicatr" is Latin for the word "scar."

CHEMISTRY: No substances. Spot tests. K-, C-, KC-, P-, UV-.

NICHE: A species more typical of warmer, subtropical and tropical latitudes. In the Smokies, you will find *Glyphis cicatricosa* only at the lowest elevations. It occurs on the bark of hardwood trees and shrubs, especially along creeks and rivers where there is ample humidity.

KEY FEATURES: Green to brown crustose thallus, lirellae aggregated and raised in pseudostroma and separated by patches of white tissue, on bark of trees and shrubs at low elevations.

Gomphillus americanus

'Merica! (alternate name: Cosmic Starbursts)

Lendemer 33158 (photo: Tripp)

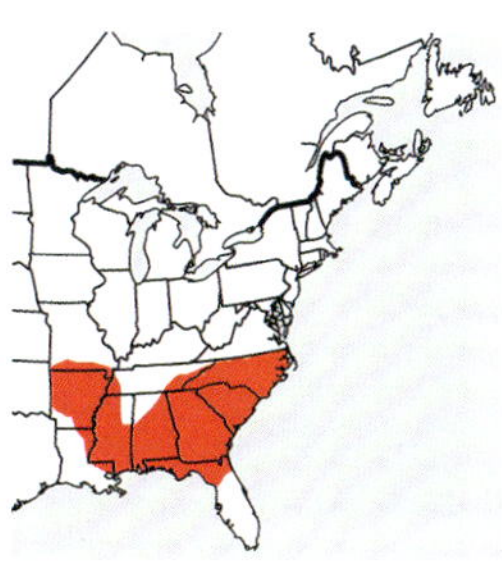

NOTES: What shouts "America" louder than a crustose lichen? Obviously, a crust that defies established order and grows stalks that stand tall and declare, "Here I am!" *Gomphillus americanus* is such a species, and one that deserves a salute if and when you find it. No other species in the Smokies has a shiny, film-like crustose thallus that produces tall, white stalks, the tops of which explode into cosmic starbursts.

CHEMISTRY: No substances. Spot tests. K-, C-, KC-, P-, UV-.

NICHE: Most relatives of *Gomphillus americanus* are tropical, which likely explains why the species occurs only in humid forests at middle-to-low elevations in the Smokies. 'Merica! is always found overgrowing mosses of the genus *Leucodon*, either in the canopies or overhanging branches of large hardwoods, especially oaks. *Gomphillus americanus* is endemic to southeastern North America, where it occurs from the coast to the mountains.

KEY FEATURES: Very shiny, film-like grayish-green thallus with star-like apothecia that sit proudly atop thin, white stalks, muscicolous, hard to find but if you manage it, one of your most memorable finds ever.

Gomphillus calycioides

Skinny Spindle Spores

Tripp 4999 (photo: Lendemer)

NOTES: This is a rare species in the Smokies that is almost as hard to find as it is to photograph. Like so many other members of the tropical family Gomphillaceae, *Gomphillus calycioides* forms a thin, green, shiny crustose thallus. In this species, the distinguishing character is its small, dark apothecia that sit on even smaller brown stalks. You might think that it is a miniature stubble lichen, but in fact it isn't even closely related to the calicioids. The spores are not produced in a mass at the top of the apothecia but instead inside of asci that are forced to grow to great lengths in order to accommodate the thread-like spores that are hundreds of microns in size. Wow!

CHEMISTRY: No substances. Spot tests. K-, C-, KC-, P-, UV-.

NICHE: A rare and elusive species found only on the bark of mature hardwood trees, especially oaks, in old growth forests at middle-to-high elevations throughout the Park. So far as known, this species is restricted to the southern Appalachians in North America, although it also occurs in Europe and southern South America.

KEY FEATURES: Thin, shiny green thallus, dark reddish-brown apothecia that are raised on short stalks resembling pins, very long, thread-like spores, on the bark of mature hardwood trees in old growth forests.

Graphis scripta

Scripts and Lips

Tripp 3805 (photo: Lendemer)

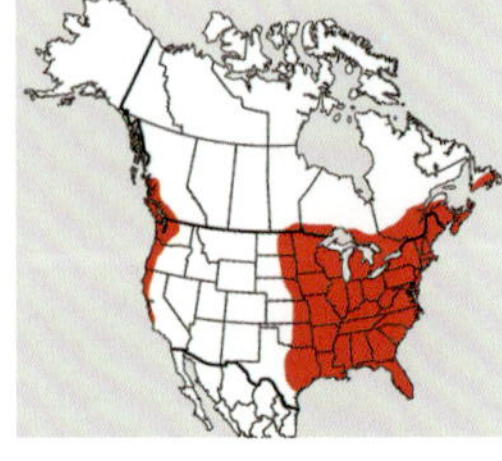

NOTES: Lovers of scripts and lips delight! This is the iconic crustose script lichen that you learned about in the *Audubon Field Guide*. It is the most common script lichen in the Smokies and the only one at middle-to-high elevations. If you see a script and call it *G. scripta* then you will be right 99% of the time. This species can easily be recognized by its black, script-like lirellae that have lightly to strongly white pruinose discs. No two individuals of this species look the same, as the size and shape of the lirellae, degree of immersion in the substrate, and amount of pruina all vary tremendously. There are only two species with which you are likely to confuse *G. scripta*: *G. endoxantha*, which has striate lips, and *G. lineola*, which has a hymenium inspersed with oil. *Graphis sterlingiana* is so entirely different that it really cannot be confused with *G. scripta*.

CHEMISTRY: No substances. Spot tests. K-, C-, KC-, P-, UV-.

NICHE: On bark and branches of just about every living tree and shrub in the Smokies. If you stand still long enough it will probably grow on you, too. You can find *Graphis scripta* from the lowest elevations to the highest peaks.

KEY FEATURES: White crustose thallus, black carbonized lirellae (aka: lips), white pruinose discs, transversely septate colorless spores, widespread on trees and shrubs throughout the Park.

Graphis sterlingiana

Mt. Sterling's Lips

Tripp 2379 (photo: Deregibus)

NOTES: *Graphis sterlingiana* is one of the most distinctive crustose lichens in the Smokies by its ecology (see Niche) as much as its morphology. This species is readily recognized by its striate labia, its large, colorless, densely muriform spores, its monosporous asci, inspersed hymenium, fully carbonized exciple, chemistry, and ecology. You would need to travel to the Neotropics before you could confuse this remarkable species with anything else. *Graphis sterlingiana* occasionally has strange, wart-like growths on its labia, but this is not the norm. In the Smokies, only *G. endoxantha* has similarly striate labia, but that species has transversely septate spores, naked lirellae that are not covered by thallus, and a very different ecology.

CHEMISTRY: No substances. Spot tests. K-, C-, KC-, P-, UV-.

NICHE. This species is currently known only from the Smokies, where it occurs exclusively on exposed roots or bases of very large, old growth Yellow Birch trees at high elevations. We consider this species to be narrowly endemic to a very specific type of habitat that is, on the whole, increasingly rare throughout the Appalachians, especially the southern mountains. This species deserves immediate protection under the US Endangered Species Program.

KEY FEATURES: Crustose script lichen with dirty white thallus, large muriform spores, fully carbonized exciple, no substances, exclusively on bases of large, old Yellow Birch trees at high elevations.

Gyalecta farlowii

Total Immersion

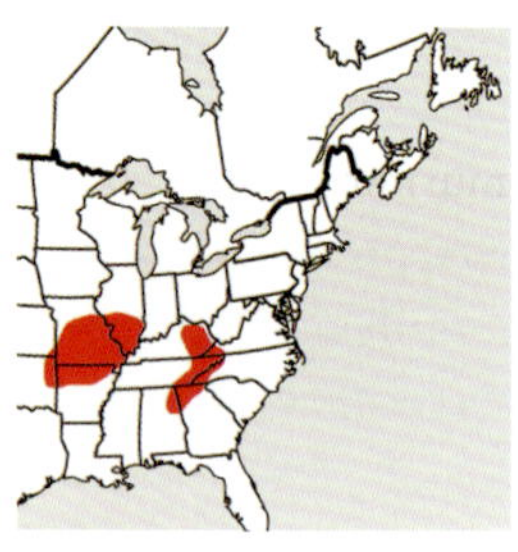

Lendemer 26857 (photo: Tripp)

NOTES: One of the most important lichen collections in North America is that of Harvard University, which like this species, was named in honor of W. G. Farlow, a well-known cryptogamic scientist of the 19th century. Although he did not publish very much on lichens, Farlow collected them far and wide, sending his specimens to experts for identification and description. *Gyalecta farlowii* is an easily recognized species on account of its distinctive ecology, colorless muriform spores, and apothecia, which are immersed in pits. The latter character means that confusion is much more likely with species such as *Bagliettoa baldensis*, *Clauzadea chondrodes,* and *Kephartia crystalligera*, all of which also occur on calcareous rocks and have fruiting bodies immersed in pits, than with other members of the genus *Gyalecta*. All the other species with immersed fruiting bodies can, however, easily be recognized by their simple spores.

CHEMISTRY: No substances. Spot tests. K-, KC-, C-, P-, UV-.

NICHE: A species of shaded calcareous rocks, *G. farlowii* is rare in the Smokies and in the southern Appalachians as a whole. It is much more common farther south and west in Alabama and Tennessee, where limestone or dolomite outcrops are common.

KEY FEATURES: Thin, immersed thallus, pale tan apothecia immersed in pits in the rock, *Trentepohlia* photobiont, muriform colorless spores, rare on calcareous rocks.

Gyalecta obesispora

Appalachian Broad Spore

Lendemer 33143 (photo: Lendemer)

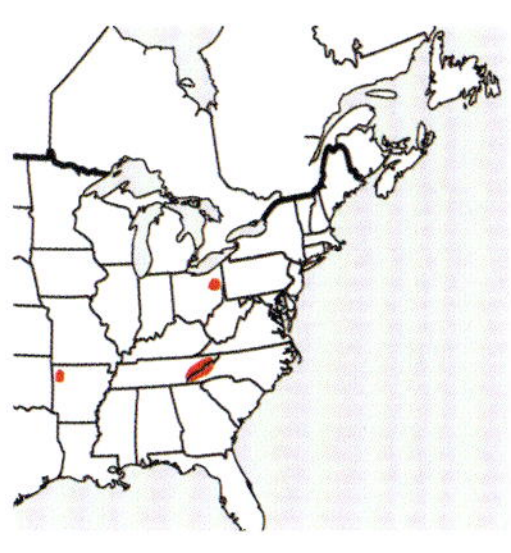

NOTES: *Gyalecta obesispora* is one of the tiny gems that makes studying lichens special. Like most members of the genus, it has pale tan-to-reddish-brown apothecia that are at least somewhat immersed in the substrate, and a *Trentepohlia* photobiont. However, it can be recognized by its occurrence on bark, coupled with its broad, rounded, muriform colorless spores. The most similar species is *Topelia aperiens*, which also occurs on bark, has muriform colorless spores, and a *Trentepohlia* photobiont. However, that species differs in having perithecioid apothecia that are not at all discoid. Some forms of *Coenogonium luteum* might be confused with *G. obesispora* because of their small pale apothecia, but that species differs noticeably in having 2-celled spores. *Thelopsis inordinata* is another species that does not occur in the Smokies, but could be confused with *G. obesispora* because of its photobiont and muriform spores. Like *T. aperiens*, it has perithecioid apothecia, and it differs from both *T. aperiens* and *G. obesispora* in having polysporous asci.

CHEMISTRY: No substances. Spot tests. K-, KC-, C-, P-, UV-.

NICHE: This species is rare in the southeastern United States, where it occurs on the bark of hardwoods and conifers in humid habitats, especially near streams or rivers. It appears to be rare in the Smokies, where it is known from low elevations along the Little Tennessee River.

KEY FEATURES: Thin, indistinct thallus, pale-tan-to-reddish-brown apothecia partially immersed in the substrate, globose, colorless, muriform spores, on bark at low elevations.

Gyalectidium appendiculatum

Flick-A-Tick

Lendemer 44582 (photo: Tripp)

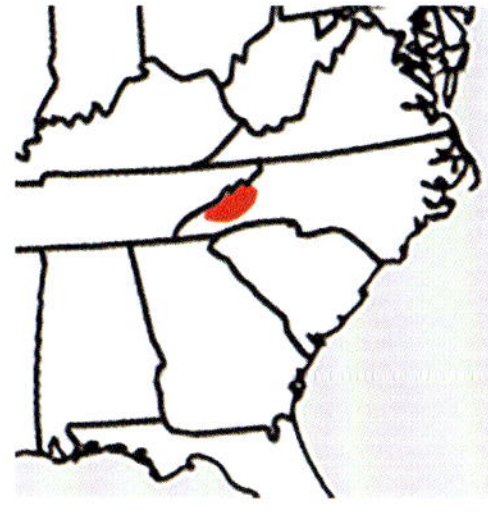

NOTES: Don't blink or you'll miss this species. *Gyalectidium appendiculatum* is one of those lichens that only reveals itself to those who bother to put a hand lens around their neck before heading out the door. It forms small, white, scale-like thalli on evergreen leaves and often grows with other members of the same tropical family (Gomphillaceae). If you are familiar with horticultural pests, look at evergreen leaves for something that resembles scale insects. What sets this species apart from all others are its scales, which are distinctly white, appear somewhat swollen, and always have a minute, fragile, glass-like hair extending upwards just ready to flick an offending tick.

CHEMISTRY: No substances. Spot tests. K-, C-, KC-, P-, UV-.

NICHE: This is an endemic of the southern Appalachians and is found only in nice, wet, secluded spots, in stream gorges and waterfalls off the beaten path. We first found it at the confluence of two tributaries to the Toxaway River in the Joccassee Gorges. It later turned up in the Smokies, growing on evergreen leaves. Look for it near Mingus Mill on *Rhododendrons* and other evergreen friends.
KEY FEATURES: Minute, white, scale-like thalli, foliicolous on hemlock needles and rhododendrons, extremely moist spray zones of waterfalls.

Gyalideopsis epicorticis
Dark Lashes

Lendemer 32926 (photo: Lendemer)

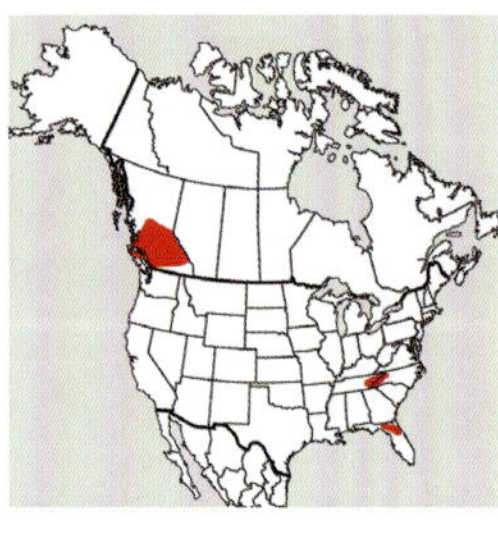

NOTES: *Gyalideopsis* species are easily recognized by their hyphophores and frequent occurrence on branches. This species is one of two known from the high elevation spruce-fir forests of the southern Appalachians, and can be recognized by its relatively large, lash-like, black hyphophores that are not bent downward or expanded at the tips. It often occurs with *G. piceicola*, which is much smaller in size and has hyphophores that are usually bent downwards as well as expanded at the apex. What does a hyphophore do? Why, it is a specialized structure that produces asexual hyphal propagules that resemble conidia and are released in droplet-like packets.
CHEMISTRY: No substances. Spot tests. K-, KC-, C-, P-, UV-.
NICHE: In the Smokies, this species seems to be restricted to the branches of spruces and firs at high elevations. It often grows near the tips of the branches, nestled between the needle bases at the junctions of branchlets from recent years of growth. Not surprisingly, it is restricted to spruce-fir forests in the southern Appalachians and has a disjunct distribution to oceanic areas of northern boreal forests.
KEY FEATURES: Thin, glossy thallus, *Trentepohlia* photobiont, large, black, lash-like hyphophores that are not strongly bent and not expanded at the apex, on spruce and fir branchlets at high elevations.

Gyalideopsis moodyae
Moira's Sails

Lendemer 29292 (photo: Lendemer)

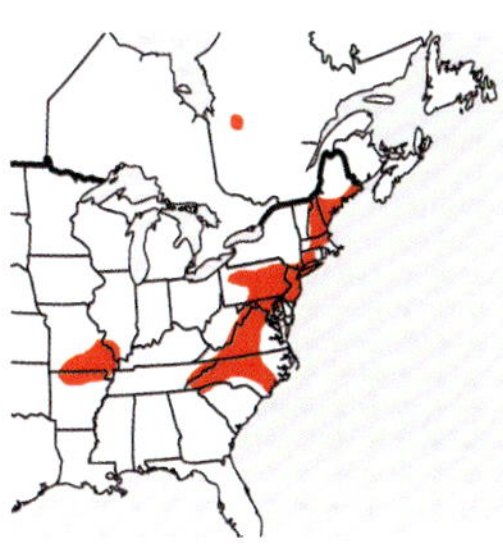

NOTES: In North America, the vast majority of *Gyalideopsis* species are corticolous and grow on the branches of trees or shrubs. However, this species stands out because it grows on organic matter and other similar substrates found on the ground. In addition to ecology, it can be recognized by its short, reddish-brown hyphophores that are typically bent and apically expanded. Like many members of the genus in the region, its apothecia are occasionally produced, but most thalli lack them and the hyphophores are almost always present. There are no species in the Smokies that can be confused with *G. moodyae*. Interestingly, it often occurs with *Diploschistes muscorum*, a parasitic lichen that grows on young thalli of *Cladonia* in similar habitats. When you find *Diploschistes*, you know that it is a good idea to search around for *G. moodyae*.

CHEMISTRY: No substances. Spot tests. K-, KC-, C-, P-, UV-.

NICHE: *Gyalideopsis moodyiae* grows on organic matter, bryophytes and dead grass tufts in disturbed or early successional habitats such as roadsides and powerline right-of-ways. It is common and widespread throughout eastern North America, but is rare in the Smokies, likely due to the overall lack of suitable habitat.

KEY FEATURES: Thin, glossy thallus, *Trentepohlia* photobiont, small, reddish-brown hyphophores that are typically bent and expanded at the apex, on organic matter and bryophytes in early successional habitats.

Gyalideopsis piceicola

Spruce Extensions

Tripp 3462 (photo: Lendemer)

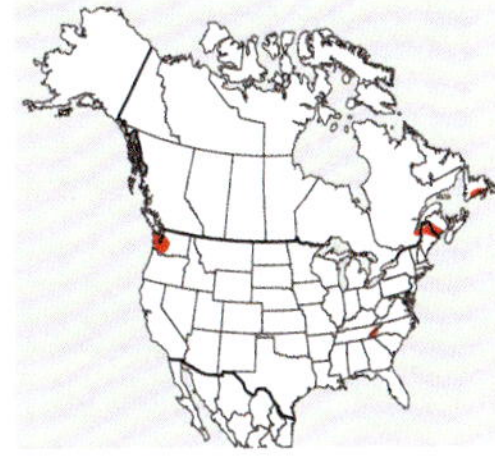

NOTES: If you want to find *Gyalideopsis piceicola* then you'll need to get out your hand lens, which you invariably packed before leaving the house for a lichenological foray this morning. It is one of several members of this tropical genus that occur in the Smokies. *Gyalideopsis piceicola* can be recognized by its niche, its shiny green film-like thallus, and short, slightly bent hyphophores that are initially pale but turn dark reddish-brown as they mature. This species often grows with *G. epicorticis,* but the latter is several times larger and has longer, eye-lash-like black hyphophores.

CHEMISTRY: No substances. Spot tests. K-, C-, KC-, P-, UV-.

NICHE: *Gyalideopsis piceicola* is widespread in limited portions of the Canadian Maritimes, Pacific Northwest, and southern Appalachians. It occurs in humid habitats where spruce and fir thrive. In that Smokies, this means that it is common in and around the high elevation peaks that are often immersed in smoky clouds. Despite the name, this species grows on both spruce and fir, always on the branches especially near the growing tips.

KEY FEATURES: Thin, shiny green thallus, short bent hyphophores that are initially dark at the tips and eventually become dark brown with age, on branches of spruce and fir at high elevations.

Hafellia subnexa

Needle-in-a-Haystack

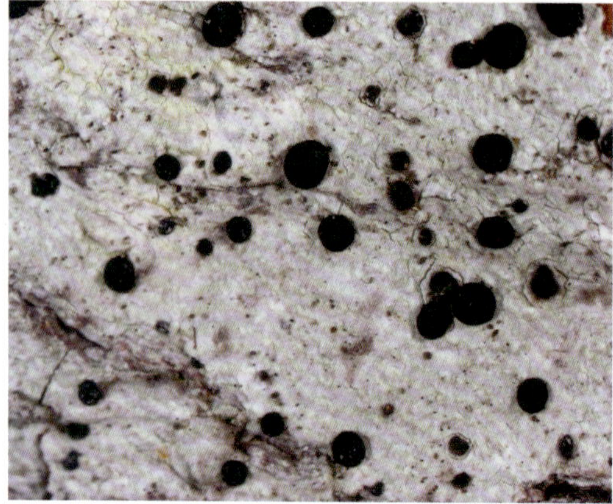

Tripp 1439 (photo: Lendemer)

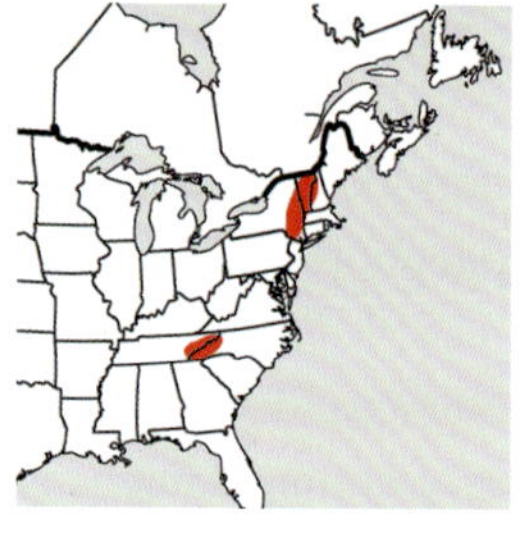

NOTES: There can be little doubt that the purpose of this species is to confuse the lichen lover who has finally become comfortable with crusts. It is externally indistinguishable from *Buellia stillingiana*, which is very common and widespread throughout the Smokies. Both species even have 2-celled brown spores and produce norstictic acid; however, the spores in *H. subnexa* have thickened walls with angular lumina while those of *B. stillingiana* do not have thickened walls. What makes *H. subnexa* stand out is that the asci are polysporous, and have 12–16 spores each (in contrast, *Buellia* species have eight-sored asci). *Amandinea polyspora* also has polysporous asci and 2-celled brown spores, but the spores of the latter are much smaller in size and the thallus is K-. *Buellia curtisii* is similar to *H. subnexa* in having spores with thickened walls and producing norstictic acid, but like *B. stillingiana*, it has eight-spored asci.

CHEMISTRY: Atranorin and norstictic acid. Spot tests: K+ yellow turning red, C-, KC-, P+ yellow, UV-.

NICHE: This species is rare in the Smokies, and generally elsewhere in the Appalachians where it occurs. It grows on the bark of hardwoods in a wide variety of habitats, although appears to be most common at middle elevations.

KEY FEATURES: White crustose thallus, black lecideine apothecia, 2-celled brown spores, polysporous asci, K+ yellow turning red, on hardwoods.

Halecania pepegospora

Rock Goo

Tripp 4943 (photo: Lendemer)

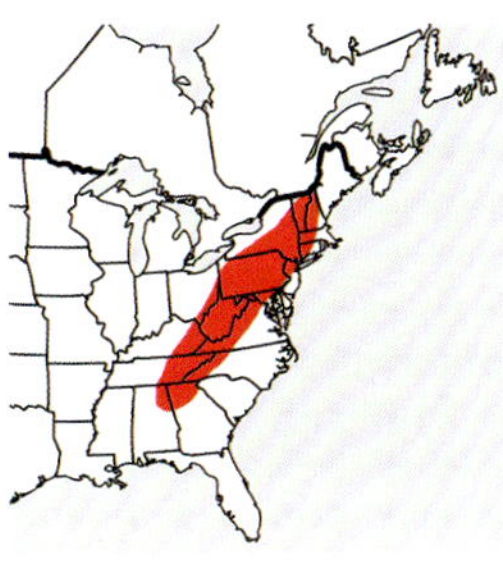

NOTES: *Halecania pepegospora* is something that most people would rightfully ignore. Even the best trained lichenologists avoid black goo on rocks, so we don't blame you. If you look closely, however, you will see that a subset of rock goo bears minute brown areoles covered with a black dusting. That black dust isn't a parasite but instead is the blastidia through which this species reproduces. The blastidia and P+ orange-red thallus identifies *H. pepegospora*, but finding small lecanorine apothecia might help, too. Once learned, we promise your life will be richer, and easier. Erin's certainly was.

CHEMISTRY: Argopsin. Spot tests. K-, C-, KC-, P+ orange-red, UV-.

NICHE: This species is overlooked but widespread in eastern temperate North America where it can be found on sun-exposed rocks, often in disturbed areas along roadsides. One finds *Halecania pepegospora* with some frequency in the Smokies, especially in the odd spot where sunlight bakes rocks along roads and trails.

KEY FEATURES: Greenish-black minute areoles that cover dark, dank rocks, often found with small apothecia, relatively common.

Herteliana schuyleriana

Who's on First

Tripp 2109 (photo: Lendemer)

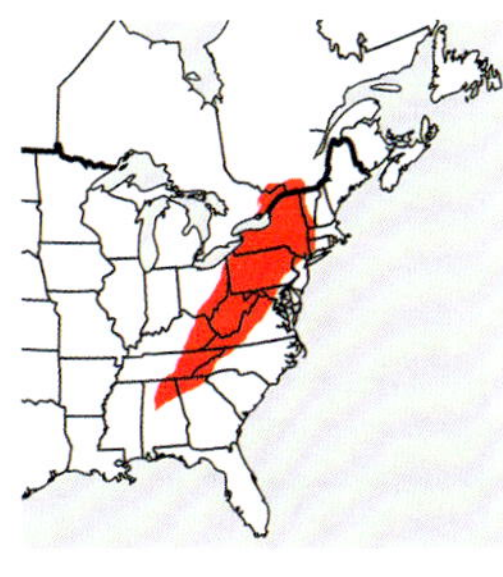

NOTES: Judging from the photograph, you can likely understand why it might have been overlooked until now. *Herteliana schuyleriana* is an unusual species in that it tends to blend in with its surroundings and is likely often assumed to be a sterile thallus of one of the other common species that it grows with, such as *Porpidia albocaerulescens*. In fact, it produces tiny propagules that bud from the surface of the thallus, and along the margins of the cracks that form in the surface. The propagules are called "blastidia," and this reproductive strategy, coupled with the pale blue-gray-to-greenish-gray color of the thallus is distinctive. This species was only recently described in 2016, and named in honor of Alfred Schuyler, botanist at the Academy of Natural Sciences of Philadelphia.

CHEMISTRY: Atranorin and roccellic acid. Spot tests. K+ yellow, C-, KC-, P-, UV-.

NICHE: *Herteliana schuyleriana* appears to be an endemic to the Appalachian Mountains, and although it is found in the Smokies, it is much more common further north in

Pennsylvania. In the Smokies, it has been found on shaded non-calcareous rocks mostly at middle elevations in the northern portions of the Park.

KEY FEATURES: Thick, pale blue-gray or green crustose thallus, abundant bud-like blastidia on the surface, K+ yellow, on shaded non-calcareous rocks.

Heterodermia albicans

Ashen Flaps

Lendemer 50551 (photo: Lendemer)

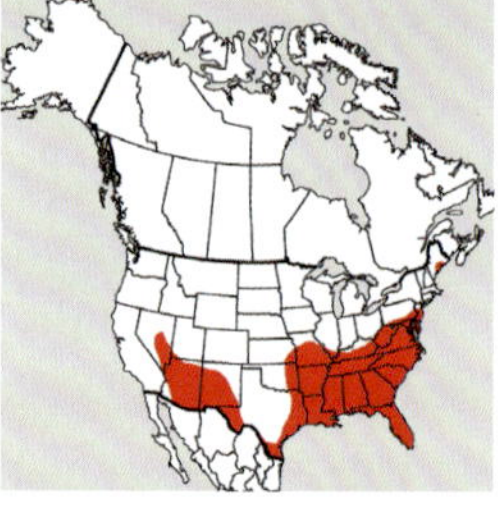

NOTES: First, you will learn to recognize this species as a *Heterodermia*, because *Heterodermias* just look like this . . . small foliose lichens with gray thalli where the lobes look like they just want to hug the substrate. Second, *Heterodermia albicans* is one of the few species in the genus present in our study area that has a fully corticate lower surface. In fact, with this in mind, you could confuse it only for *H. pseudospeciosa*, which has norstictic acid (confirmable via a K test under a mount of the thallus, or TLC) instead of salazinic acid in *H. albicans*. You might also confuse it for *H. speciosa*, which primarily has soralia at the lobe tips, versus soralia scattered about all margins of the thallus in *H. albicans*. *Heterodermia speciosa* also differs chemically in lacking salazinic acid. The latter two species also just look different, and in the end isn't that what matters?

CHEMISTRY: Atranorin, zeorin and salazinic Acid. Spot Test: (cortex): K+ yellow, C-, KC-, P-, UV-; (medulla): K+ red, KC-, C-, P+ yellow, UV-.

NICHE: The Smokies is a happy place for numerous species of *Heterodermia*, and Ashen Flaps is no exception. Look for it on the bark of hardwoods at the lowest elevations in Tennessee. Outside of that portion of the Park, it is common and widespread in the southeastern United States, especially the Coastal Plain.

KEY FEATURES: *Heterodermia* with a fully corticate lower surface, soralia that line the margins of the lobes, K+ red medulla, quite white, on hardwood bark at low elevations.

Heterodermia appalachensis

Appalachian Bootstraps

Tripp 5436 (photo: Lendemer)

NOTES: When you pull up your bootstraps before beginning a hike, think of *Heterodermia appalachensis*. You'll know it when you see it because of its long, strap-like lobes, the tips of which turn up to reveal a pad of soredia (gasp!) on the lower surface. This probably would have been illegal in the 1920s. If you look even closer, you'll see that the lower surface has a faint yellow pigment, reminiscent of butter (mmmmm). The only moderately similar species is *H. leucomela*, but that has longer, more delicate lobes that ascend away from the substrate and lack a yellow pigment underneath.
CHEMISTRY: Atranorin, zeorin, leucotylin. Spot tests. K+ yellow (cortex), C-, KC-, P-, UV-.
NICHE: This species is rare in the Smokies and throughout its range. It occurs on the bark of hardwoods in dry forests at middle-to-low elevations. It is endemic to the southern Appalachians, with disjunct populations in the mountainous regions of the Sonoran Desert.
KEY FEATURES: Strap-like lobes comprising the thallus, sorediate pads on undersides of lobe tips, faint yellow pigment on lower surface, rare and precious, better than gold . . .

Heterodermia crocea

Brier Rose

Lendemer 33101 (photo: Tripp)

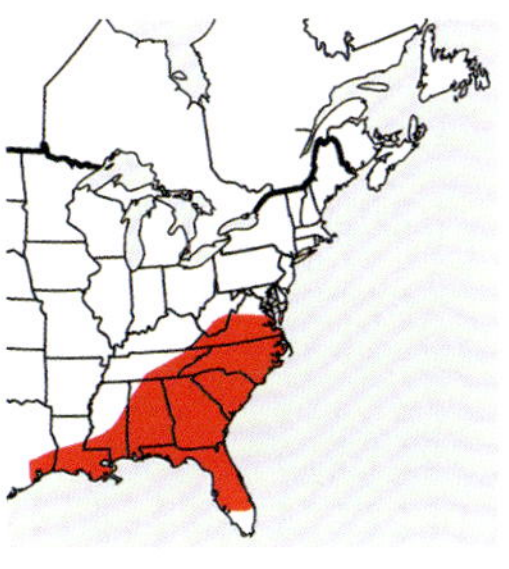

NOTES: *Heterodermia crocea* is a terrific entry-level macrolichen. It can be identified even in the field by its foliose thallus with narrow lobes, fragile isidia, and lower surface covered by an orange pigment that reacts K+ purple. *Heterodermia obscurata* also has an orange pigment, but instead of isidia, that species has large soralia on its lobe tips. *Heterodermia granulifera* is the only other isidiate member of the genus in the Smokies, but it lacks the orange pigment that makes *H. crocea* such a treat. A great way to remember this species is by its specific epithet, which refers to the orange-brown pigment present on the lower surface.
CHEMISTRY: Atranorin, zeorin, terpenoids, orange pigment. Spot tests. K+ yellow (cortex) and K+ purple (orange pigment on lower surface), C-, KC-, P-, UV-.
NICHE: This species is infrequent but widespread on the bark of hardwoods at low elevations in the Smokies. It occurs throughout the southeastern United States, where it is endemic.
KEY FEATURES: Gray thallus, narrow lobes, fragile isidia, K+ orange pigment on lower surface (which is ecorticate), occasional on the bark of hardwoods at low elevations.

Heterodermia echinata

Marginal Marvel

Lendemer 44627 (photo: Tripp)

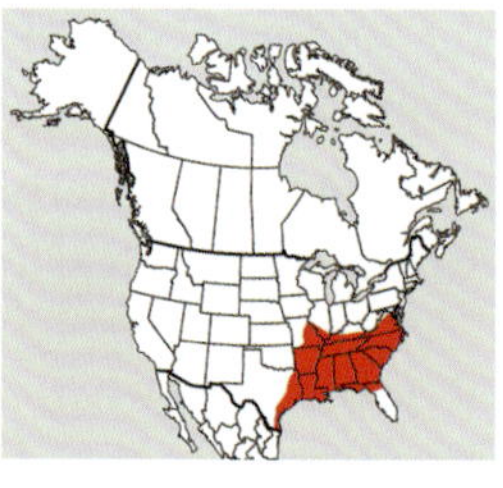

NOTES: Here is one of our favorite lichens in the Smokies! *Heterodermia echinata* is instantly identified in the field by its blue-gray thallus composed of flattened, hyper-erect lobes that have terminal apothecia with thick, pruinose discs. The margins of the apothecia are ragged and covered by the same, long cilia that also grow along lobe margins. No other species could be confused with *H. echinata*, even on a trail run!

CHEMISTRY: Atranorin, zeorin. Spot tests. K+ yellow (cortex), C-, KC-, P-, UV-.

NICHE: This species is frequent in low elevation forests in the Smokies, especially along Lakeshore Trail. It most often occurs on branches, especially in the canopies of conifers, which is why you often run into it on fallen material. Although widespread in southeastern North America, where it is endemic, *Heterodermia echinata* is relatively uncommon throughout this range. Check out Deborah Rabinowitz's seven forms of rarity for some potential ideas as to why.

KEY FEATURES: Foliose to fruticose in appearance, blue-gray lobes, terminal apothecia with heavily pruinose discs, ecorticate lower surface, dense cilia along lobe and apothecial margins.

Heterodermia granulifera

Humble Knobs

Lendemer 44566 (photo: Tripp)

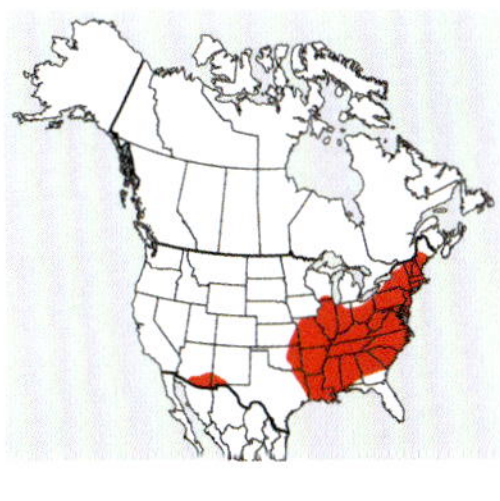

NOTES: *Heterodermia granulifera* is the humble cousin of the ostentatious *H. crocea*. It lacks the orange pigment that is present on the lower surface of *H. crocea*, is far less common, and overall has smaller thalli characterized by short, knobby, isidia. It also differs from *H. crocea* in having salazinic acid in the medulla, which reacts K+ red. We also think it has a better attitude. The specific epithet refers to its overall granular appearance, which works from a distance but is unfortunate upon closer inspection because the thalli aren't granular at all, but rather isidiate.

CHEMISTRY: Atranorin, zeorin, salazinic acid. Spot tests. K+ yellow (cortex), K+ red (medulla), C-, KC-, P+ yellow (medulla), UV-.

NICHE: This species is rare in the Smokies, where it is found at low elevations on the bark of hardwoods and over calcareous rocks. Outside of the Smokies it occurs throughout temperate eastern North America as well as mountainous regions of Central America. Like many other members of the genus, it is rare throughout its range.

KEY FEATURES: Gray thallus with short knobby isidia and a K+ red medulla. Unlike most other *Heterodermia*, this one has a corticate lower surface. Rare at low elevations on hardwoods.

Heterodermia hypoleuca

Tree Anemone

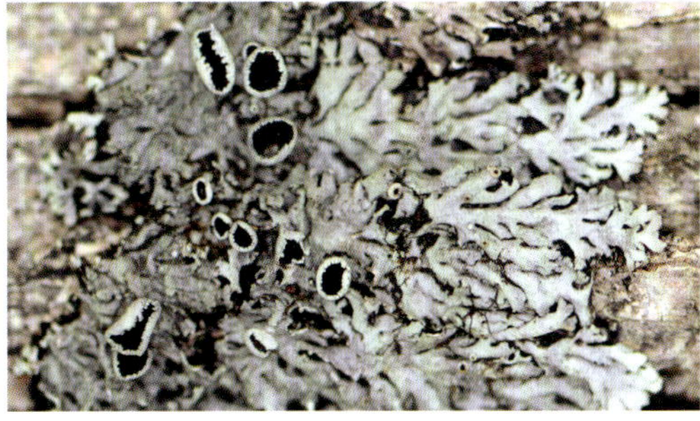

Lendemer 44538 (photo: Tripp)

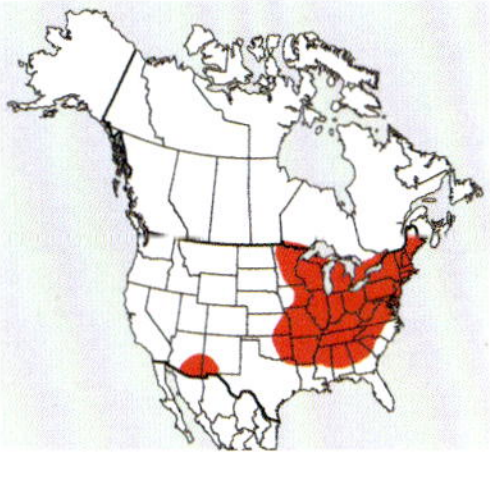

NOTES: *Heterodermia hypoleuca* can be recognized by its abundant apothecia, the margins of which are covered with luxuriant lobules. If apothecia are lacking, look for stiff, knobby lobules along lobes margins and the ecorticate, white lower surface (that likely inspired the specific epithet). This species must taste good, because a keen observer often finds evidence of invertebrate foraging, around which stiff regeneration lobules grow. These lobules might be confused with phyllidia of *H. squamulosa*, but the latter are fragile and finely divided. The regeneration lobules have also led to the misidentification of *H. hypoleuca* for the Japanese species *H. microphylla*.

CHEMISTRY: Atranorin, zeorin, leucotylin. Spot tests. K+ yellow (cortex), C-, KC-, P-, UV-.

NICHE: Uncommon but widespread at middle and low elevations of the Smokies, where it occurs on the bark of hardwoods. This species is also widespread but uncommon throughout eastern North America and in the mountains of Central America and eastern Asia.
KEY FEATURES: Abundant apothecia with conspicuous marginal lobules, white, ecorticate lower surface, thallus often damaged by invertebrates, which apparently induce production of regeneration lobules around damaged area.

Heterodermia japonica

A Difficult Lichen

Lendemer 26806 (photo: Moroz)

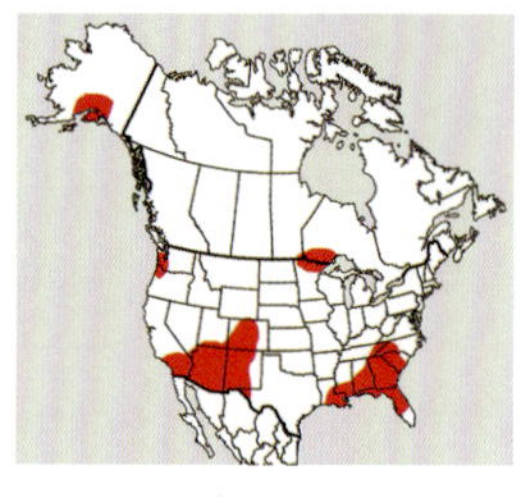

NOTES: Most *Heterodermias* are easy to differentiate from one another with some practice, but this one can be tricky. *Heterodermia japonica* is one of *several* sorediate species in the Smokies but differs from all the others in having an ecorticate lower surface that lacks any yellow or orange pigmentation. Among these, *H. neglecta* is probably the most similar but a quick flip of a lobe will reveal spots of orange pigment near the tip undersides. A much rarer species that also lacks lower surface pigments is *H. erecta*, but it occurs only in high elevation northern hardwood forests and has rosette-forming thalli with numerous short secondary lobes bearing terminal soralia.
CHEMISTRY: Atranorin, zeorin, +/- norstictic acid. Spot tests. K+ yellow (cortex), K- or K+ red (medulla), C-, KC-, P- or P+ yellow (medulla), UV-.
NICHE: Infrequent but widely distributed throughout the Smokies on the bark of hardwoods. Scattered populations occur in temperate and boreal North America, eastern Asia, and Central America.
KEY FEATURES: Gray-green thallus, sorediate, ecorticate lower surface lacking any orange pigmentation. Infrequent on bark of hardwoods.

Heterodermia langdoniana

Keith's Moustache Twirl

Tripp 3559 (photo: Lendemer)

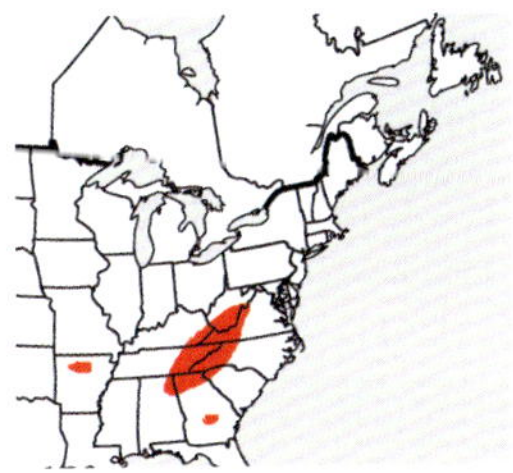

NOTES: This species was named in honor of the long time GSMNP natural resource manager Keith Langdon, well known for this distinctive moustache, which the slightly curled lobes with terminal soralia resemble. Although *Heterodermia* might be a difficult genus, especially in the Smokies where it seems like every species seems to occur, this one is fairly easy to recognize. The soralia set it apart from many esorediate or phyllidiate species with ecorticate lower surfaces, such as *H. hypoleuca* and *H. squamulosa*. Likewise, the multi-zoned pigments on the lower surface distinguish it from other sorediate species. Specifically, in this species, the lower surface is uniformly white near the margins, becomes orange-brown pigmented away from the margins, with that pigment being K-, and then purple-black pigmented in the central portions. *Heterodermia neglecta* and *H. obscurata* are most similar, but they have an orange pigment on the lower surface that reacts K+ purple. *Heterodermia japonica* and *H. erecta* have only the purple-black pigment and lack any sign of an orange or orange-brown pigment. In the past, *H. langdoniana* was confused with *H. casarettiana*, however that species has tiny spots of yellow pigment near the lobes tips and that pigment is K-.

CHEMISTRY: Atranorin, zeorin, +/- norstictic acid. Spot tests: (cortex) K+ yellow, C-, KC-, P-, UV-; (medulla) K- or K+ yellow turning red, C-, KC-, P- or P+ yellow, UV-.

NICHE: *Heterodermia langdoniana* is common and widespread in the southern Appalachian Mountains, where it is frequently found growing on rocks and tree bases at middle-to-low elevations in humid habitats. You can find it growing in all its glory along the north shore of Fontana Lake.

KEY FEATURES: Large, blue-gray foliose thallus, terminal soralia on secondary lobes, white medulla, ecorticate lower surface with a marginal unpigmented zone, inner orange-red pigmented zone that is K-, and central purple-black pigmented zone, on bark and rocks at middle-to-low elevations.

Heterodermia leucomela

Long Limbs

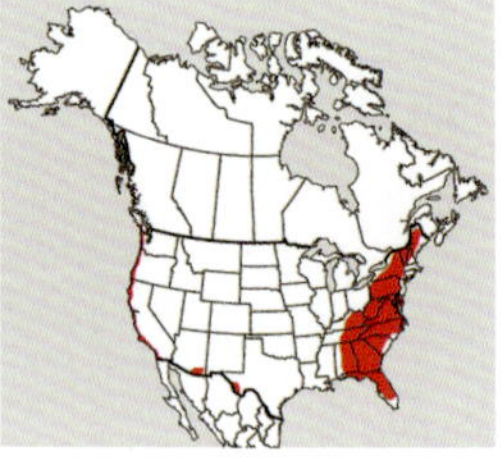

Lendemer 30271 (photo: Tripp)

NOTES: *Heterodermia* is a diverse, abundant, and wonderful genus in the Smokies. Species can be hard or easy to identify, in part depending on whether you are an optimist or pessimist and a perfectionist or not, but when it comes to *H. leucomela*, the pressure is off! This one is the easiest foliose species to name owing to its elongate, strap-shaped lobes that bear soredia and are decorated by profuse, marginal dark cilia. The only taxon that could be confused for this species is *H. echinata*, which is fertile, lacks soredia, has broader lobes, and is much rarer.

CHEMISTRY: Atranorin, salazinic acid. Spot tests. K+ yellow (cortex) K+ red (medulla), KC-, C-, P+ yellow (medulla), UV-.

NICHE: In the Smokies, *Heterodermia leucomela* can be found at all elevations, particularly mid elevations. It occurs on the bark of hardwoods, especially older trees. This species was recently found by Elisabeth Lay near sea level on the outer coastal plain of North Carolina, extending its known geographic range to a pretty cool part of the United States that will unfortunately likely be under water in our lifetimes owing to sea level rise.

KEY FEATURES: Foliose lichen appearing fruticose owing to highly elongate strap-like lobes, ecorticate lower surfaces that bear soredia, lack of apothecia.

Heterodermia neglecta

The Disregarded One

Lendemer 32936 (photo: Tripp)

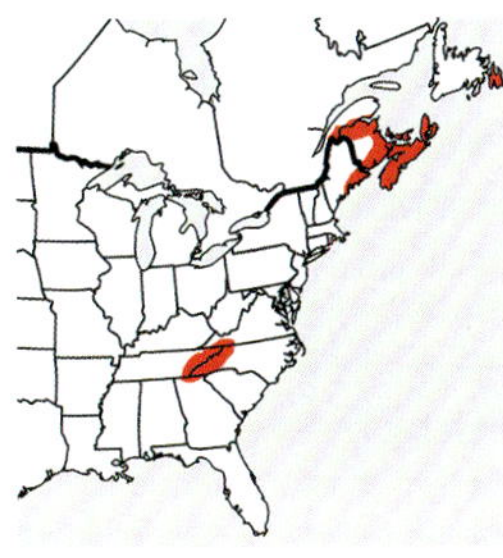

NOTES: *Heterodermia neglecta* looks just like *H. japonica* but has tiny orange spots (that react K+ purple) scattered near the lobe tips on the lower surface. *Heterodermia langdoniana* differs in having a continuous layer of orange (K-) pigment distributed across most of its lower surface. The other two *Heterodermia* in the Smokies with orange pigments that react K+ purple (*H. crocea* and *H. obscurata*) have this pigment distributed throughout their lower surfaces rather than confined to tiny spots near the lobe tips.

CHEMISTRY: Atranorin, zeorin, +/- norstictic acid. Spot tests. K+ yellow (cortex), K- or K+ red (medulla), C-, KC-, P- or P+ yellow (medulla), UV-.

NICHE: Widespread at middle-to-high elevations in the Smokies, occurring on the bark of hardwoods. *Heterodermia neglecta* also occurs throughout the southern Appalachians, with disjunct populations in the Canadian Maritimes. This species is also known from eastern Asia, temperate South America, and Great Britain.

KEY FEATURES: Gray-green thallus, sorediate, with an ecorticate lower surface that bears scattered, orange spots, these K+ purple, on bark of hardwoods at middle-to-high elevations.

Heterodermia obscurata

Rusty Bottoms

Lendemer 19179 (photo: Lendemer)

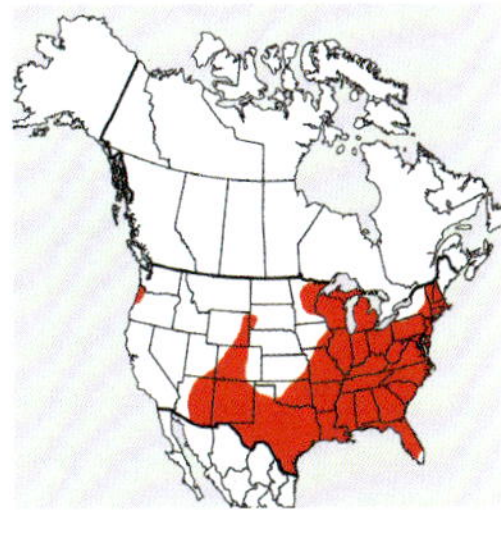

NOTES: If there is one species of *Heterodermia* in the Smokies that should not cause any problems in identification, it is this one. The sorediate thallus combined with a continuously orange-pigmented lower surface that is K+ purple are two very distinctive characters. The only similar species is *H. neglecta*, which occurs at higher elevations and differs in having the orange K+ purple pigment restricted to tiny spots near the lobe tips on the lower surface. *Heterodermia casarettiana* is also similar in having a pigment on the lower surface, but that pigment is yellow, K-, and restricted to tiny spots near the margins, as in *H. neglecta*. Confusion with *H. langdoniana* may also be possible, but the pigment in that species is confined to a middle zone of the lower surface, is darker brown, and K-.

CHEMISTRY: Atranorin, zeorin, flavobscurin. Spot tests: (cortex) K+ yellow, C-, KC-, P-, UV-; (lower surface): K+ purple, C-, KC-, P-, UV-.

NICHE: This very common and widespread species is found throughout most of the Smokies and most of temperate eastern North America. Although it usually grows on the bark and branches of trees in humid habitats, it is also occasionally found on rocks.

KEY FEATURES: Medium-to-small, blue-gray foliose thallus, terminal soralia on secondary lobes, white medulla, ecorticate lower surface with continuous orange pigment that is K+ purple, on bark and rocks throughout except at the highest elevations.

Heterodermia pseudospeciosa

Gone Fishin'

Tripp 3676 (photo: Tripp)

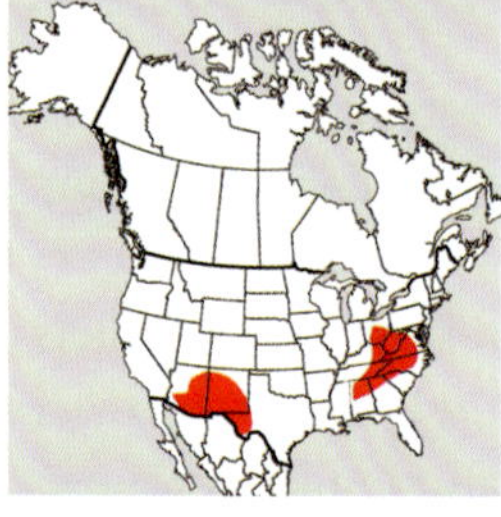

NOTES: There are too many "speciosa" lichens in the world. First you have *Heterodermia speciosa*, which is common in the Smokies and differs from the present species in having larger thalli, primarily terminal soralia, and a medulla that is K- due to the absence of norstictic acid, and in growing primarily on bark. Then you have *Physcia pseudospeciosa*, which vaguely resembles a *Heterodermia*, but differs in a number of characters, most notable from *H. pseudospeciosa* in having a K- medulla due to the lack of norstictic acid. *Heterodermia albicans* probably should have been called "pseudopseudospeciosa" instead just because it is also similar, differing from the present species in having salazinic acid in the medulla. Although both *H. albicans* and *H. pseudospeciosa* have a K+ medulla, the norstictic acid in *H. pseduospeciosa* can be detected by the red crystals it forms in a slide under the microscope (whereas salazinic acid turns red but does not make crystals).

CHEMISTRY: Atranorin, zeorin and norstictic acid. Spot tests: (cortex): K+ yellow, C-, KC-, P-, UV-; (medulla): K+ yellow turning red, C-, KC-, P+ orange, UV-.

NICHE: A lover of rivers and waterfalls, this species inhabits the best parts of the Smokies and southern Appalachians. Almost invariably it occurs on non-calcareous rocks near running water.

KEY FEATURES: Gray foliose thallus, narrow lobes, soralia along margins, K+ red medulla, on rocks along rivers and near waterfalls.

Heterodermia speciosa

Put A Hair Net On It

Tripp 3604 (photo: Lendemer)

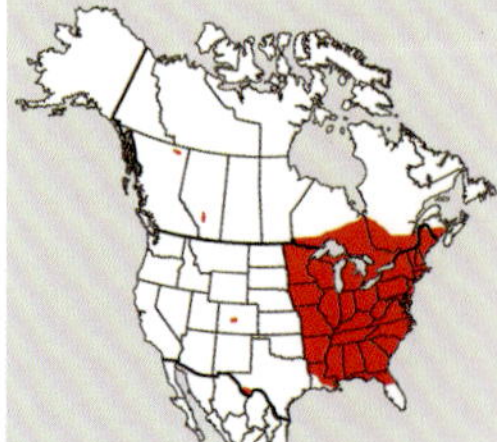

NOTES: *Heterodermia speciosa* is probably the most common member of the genus in the Smokies. We hate to admit it, but it's not a bad way to learn this one. It's also one of the rowdiest species. Just look at those pale, marginal cilia! Hence, it could use a hair net. In addition to these cilia, it is differentiated from most other *Heterodermia* by having a lower surface that is more or less corticate. And yes, it's sorediate, but then again so are most of the others. (You would be too, if you could . . .).

CHEMISTRY: Atranorin, zeorin. Spot tests. K+ yellow, C-, KC-, P-, UV-.

NICHE: Extremely common throughout middle-to-low elevations of the Smokies as well as most of the rest of eastern North America. Known from scattered other localities in western North America.

KEY FEATURES: Extremely common corticolous species throughout middle and low elevations, corticate lower surface, sorediate, charismatic, hyperactive marginal cilia.

Heterodermia squamulosa

I Could Play Tennis Wearing That

Tripp 3625 (photo: Lendemer)

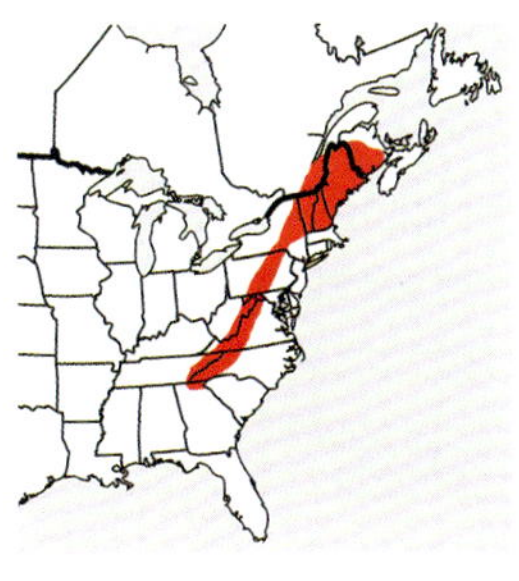

NOTES: There is absolutely no way to confuse this species for anything else in the genus, lucky for all of us. (Hey, at some point, even *we* hit *Heterodermia* saturation.) *Heterodermia squamulosa* is, as the specific epithet implies, a distinctive species by its fragile marginal squamules, or phyllidia, if you prefer. This photograph shows these squamules in a very advanced state. When they are on the immature side, sort of like us, they appear more lobule-like. In any case, you got this one! Sometimes, *H. squamulosa* also makes apothecia in addition to these asexual propagules.

CHEMISTRY: Atranorin, zeorin. Spot tests. K+ yellow, C-, KC-, P-, UV-.

NICHE: Common on the bark of hardwoods at middle-to-high elevations, also known from elsewhere in the Appalachians, the Canadian Maritimes region, and southeastern Arizona.

KEY FEATURES: The only *Heterodermia* with squamules (lobules), sometimes also with apothecia (the thought . . . can you imagine?), common at middle-to-low elevations.

Hyperphyscia syncolla

Neighborly Lobe

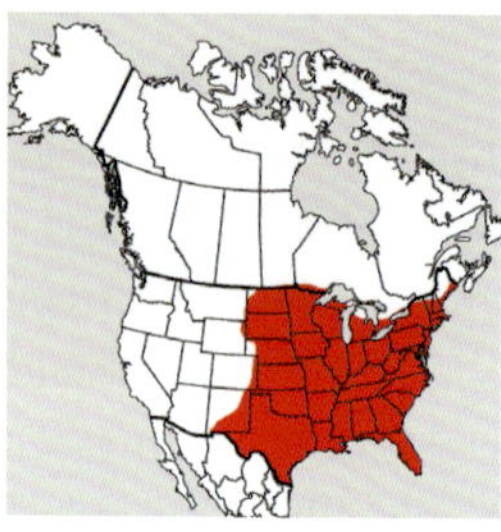

Sharp s.n. (photo: Lendemer)

NOTES: *Hyperphyscia syncolla* is one of those species that it pays to know before you have to identify it. We didn't make the rules, but we do have some tips that we hope help you cope with them. *Hyperphyscia syncolla* is a sexually reproducing species that is foliose but is so closely appressed to its substrate that it appears crustose. This closeness is perhaps the best means of identifying the species, after you have landed in the right "vicinity" via our dichotomous keys. Another foliose species with such fondness for its substrate that immediately comes to mind is *Dirinaria frostii*, but that lichen is sorediate. Confusion with other small esorediate foliose lichens such as *Phaeophyscia ciliata* or *P. hirtella* is possible, but they are all clearly foliose and not superficially crustose.

CHEMISTRY: No substances. K-, KC-, C-, P-, UV-.

NICHE: This is a species of forest edges and other habitats with ample sunlight. Not surprisingly, it is rare in the Smokies and generally throughout the southern Appalachians, where it usually occurs on the branches of hardwood trees, especially in old fields.

KEY FEATURES: Foliose thallus tightly appressed to substrate and appearing crustose, frequent apothecia, no soredia, on bark and branches of hardwoods, rare.

Hypocenomyce anthracophila
Pine Candle Wax

Lendemer 33078 (photo: Tripp)

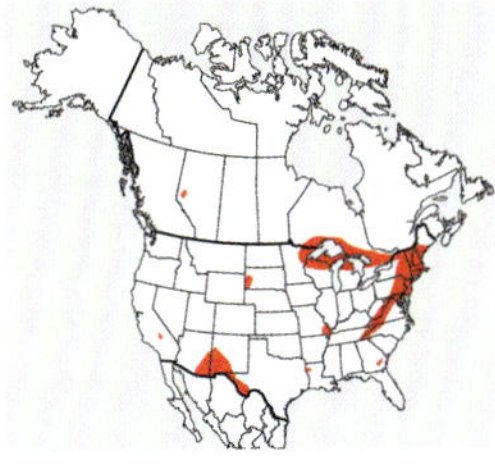

NOTES: This photo is not award winning (nor should you ever plan to get rich as a botanist or lichenologist) but the species is really rare, so you take what can you get. *Hypocenomyce anthracophila* is a strange species characterized by its dispersed small squamules, which are actually sorediate, but the latter are reliably inconspicuous. If found, it is easily distinguished from its close relative *H. scalaris* by its narrower and more dispersed squamules. The specific epithet refers to the propensity of this species (and other congeners in the Smokies) to grow on burnt ("charcoal") wood.

CHEMISTRY: Colensoic acid, fumarprotocetraric acid. Spot tests. K-, C-, KC-, P+ orange, UV+ blue-white.

NICHE: *Hypocenomyce anthracophila* is a curious species with a wide distribution but it seems to be rare, everywhere. In the Smokies, it occurs uncommonly on the bark of conifers at low elevations, or on old, sometimes charred wood. Let us know if you manage to learn more about the natural history of this species.

KEY FEATURES: Small squamules that are relatively narrow in size and dispersed across the thallus, sorediate, but the soredia inconspicuous, rare.

Hypocenomyce friesii
Shiny Shingles

Tripp 3588 (photo: Lendemer)

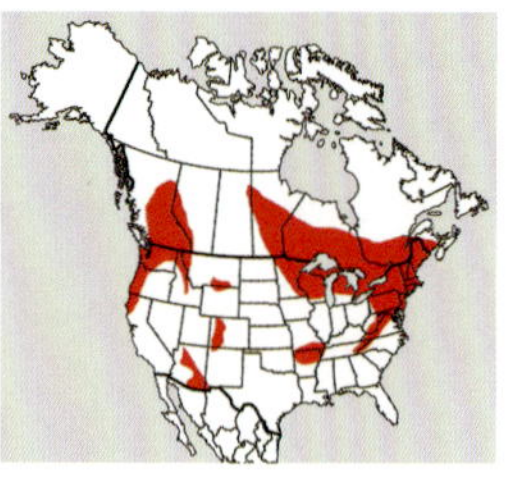

NOTES: Just like other species in this genus, *Hypocenomyce friesii* is characterized by having a thallus made up of squamules. Its one of those features best learned by gestalt (and thanks to the Germans for giving us that word). This species is easy to separate from its relatives because it is fertile, whereas *H. scalaris* and *H. anthracophila* are sorediate. Look for its black, fruiting bodies and the UV+ blue-white reaction.
CHEMISTRY: Friesiic acid (convenient). Spot tests. K-, C-, KC- (to occasionally KC+), UV+ blue-white.
NICHE: Lignicolous in middle-to-high elevation habitats throughout the Smokies and Appalachians. Look for this species elsewhere in North America, particularly in the Pacific Northwest and the upper Great Lakes. Always on lignum, often of the charred sort.
KEY FEATURES: Large, densely clustered squamules and black apothecia, UV+ blue-white, lignicolous at middle-to-high elevations.

Hypocenomyce scalaris (and *Clypeococcum hypocenomycis*)
Truffle Shuffle

Tripp 3584 (photo: Lendemer)

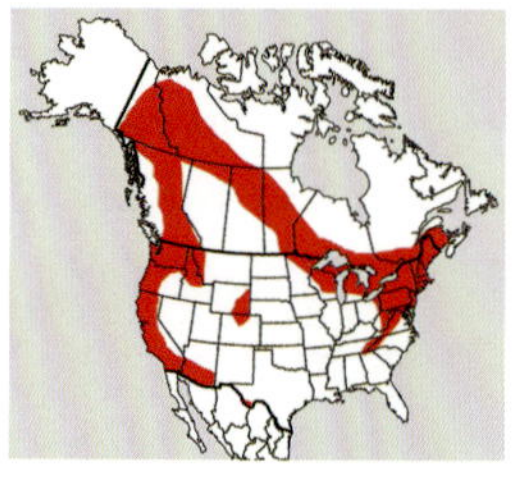

NOTES: This species is so commonplace (outside of the Smokies) that we decided to skip it and talk about its parasite instead. Well, ok: we will tell you that Hypocenomyce scalaris is squamulose, sorediate, and C+ red, and that it differs from *H. anthracophila* by its much larger squamules. If you must know more, read in *Lichens of North America*. So, its parasite is *Clypeococcum hypocenomycis*. In the photograph, you can see clear patches of black mycelia infecting its host. Buried in these mycelial patches are black fruiting bodies (perithecia). And that is just about where our knowledge of it starts and stops!
CHEMISTRY: Lecanoric acid. Spot tests. K-, C+ red, KC+ red, P-, UV-.
NICHE: Corticolous or lignicolous but you stand a better chance of finding it in crapped up habitats surrounding Philadelphia than you do in the Smokies, which is in general just too nice for *Hypocenomyce scalaris* . . .
KEY FEATURES: Broad, dispersed to clustered sorediate squamules. Always infected. The specific epithet is as catchy as the lichen itself.

Hypogymnia incurvoides
Crown Jewel of the Appalachians

Tripp 3513 (photo: Lendemer)

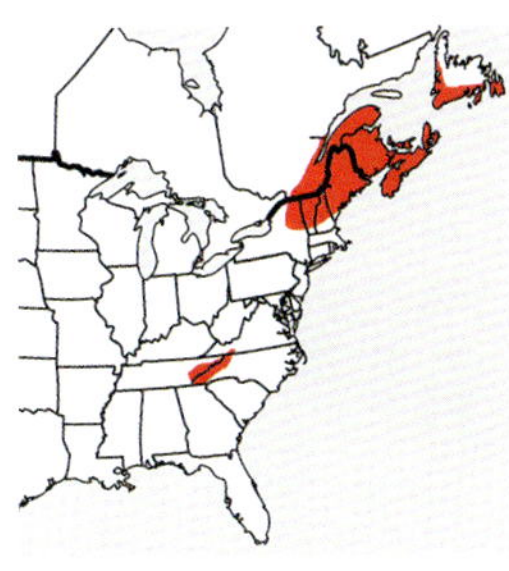

NOTES: We nearly faint each time we see this species. Just look at this beauty: a drop-dead knockout. Anyway (sigh . . . focus), enter *Hypogymnia*–the land of hollow, inflated lobes. Because of this feature, you cannot mistake this genus for anything else. *Hypogymnia incurvoides* is characterized by being soraliate on the lower surface of the lobe tips and having a K+ red medulla. Very rarely do you find it additionally producing apothecia, as in the above. You will be lucky to find this species, and if you do, you should organize a party in honor of the occasion. And invite us.

CHEMISTRY: Atranorin, physodalic acid. Spot tests. K+ yellow (cortex), K+ red-brown (medulla), C-, KC-, P+ red (medulla), UV-.

NICHE: *Hypogymnia incurvoides* is rare in the Smokies and occurs only in spruce-fir forests, which are highly threated ecosystems that may very well go extinct in our lifetimes. Elsewhere, this species occurs most commonly in oceanic regions of the Canadian Maritime and northeastern United States.

KEY FEATURES: Hollow, inflated lobes, soredia on lower lobe tips, K+ red medulla, lobe margins black, high-elevation spruce-fir forests. Corticolous, mostly on fir.

Hypogymnia krogiae

Fertile tubes

Tripp 3473 (photo: Lendemer)

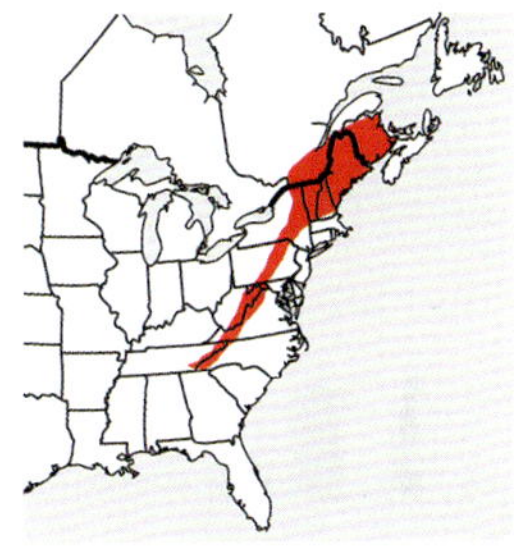

NOTES: *Hypogymnia krogiae* is easily recognizable because it is beautiful. In addition, it is the only fertile member of the genus in the Smokies (*H. incurvoides*, *H. physodes*, *H. tubulosa*, and *H. vittata* are all sorediate; very rarely, some of these also produce apothecia, but always in addition to soredia). See? Lichens of the Smokies—easy!

CHEMISTRY: Do you really need to know this? Atranorin, physodalic acid, physodic acid, protocetraric acid. Spot tests. K+ yellow (cortex), C-, KC+ pink, P+ red (medulla), UV-.

NICHE: This species is relatively common in, but restricted to, spruce-fir ecosystems in the Smokies. Look for it during your stay at LeConte Lodge, perhaps with a chilly glass of Stellenbosch Chardonnay in hand. *Hypogymnia krogiae* is also distributed in montane and oceanic habitats throughout the central and northern Appalachians.

KEY FEATURES: Hollow, inflated lobes, apotheciate, never sorediate, relatively common at high elevations in spruce-fir forests, most commonly corticolous on *Abies*.

Hypogymnia physodes

Trashy Tubes

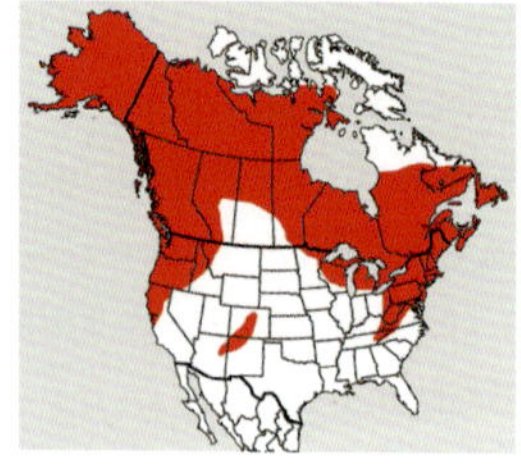

Tripp 2317 (photo: Deregibus)

NOTES: Hollow, inflated lobes just like the rest of its relatives characterize *Hypogymnia physodes*. This sorediate species is most likely to be confused with *H. incurvoides*, from which it differs in having a K- medulla (vs. K+ in *H. incurvoides*). *Hypogymnia physodes* might also be confused for *H. tubulosa*, but that species has soralia on its upper surface (vs. soralia on the lower surface in *H. physodes*).

CHEMISTRY: Atranorin, physodalic acid, physodic acid, protocetraric acid. Spot tests. K+ yellow (cortex), C-, KC+ pink (medulla), P+ red (medulla), UV-.

NICHE: Much trashier than *H. krogiae*. *Hypogymnia physodes* is very common throughout middle-to-high elevations in the Smokies and southern Appalachians, where it occurs on conifers as well as hardwoods. It also occurs broadly throughout north-temperate to boreal regions of North America.

KEY FEATURES: Hollow, inflated lobes, soredia on lower surface of lobe tips, K- medulla, very common at middle-to-high elevations, corticolous on conifers and hardwoods.

Hypogymnia tubulosa

A Corticolous Party

Tripp 3533 (photo: Lendemer)

NOTES: Squeezed between *Pseudevernia cladoniae*, *Flavoparmelia caperata*, and *Parmotrema arnoldii* is this very fine specimen of *Hypogymnia tubulosa*. This species is one of four sorediate *Hypogymnias* in the Smokies but is easily differentiated from the other three because its soralia sit atop the upper surface rather than on the undersides of the lobe tips. And it is more social. Look at all those lichens! A typical day in the life of a branch in the Smokies . . .

CHEMISTRY: Atranorin, physodic acid. Spot tests. K+ yellow (cortex), K- or K+ yellow turning brown (medulla), C-, KC+ red (medulla), P-, UV-.

NICHE: See Niche under *Hypogymnia physodes* . . . this one is the same. Elsewhere, *H. tubulosa* occurs in the northern Appalachians and Great Lakes region as well as the greater Pacific Northwest.

KEY FEATURES: Hollow, inflated lobes, soredia on the upper surface, corticolous on hardwoods and conifers at middle-to-high elevations in the Smokies.

Hypogymnia vittata

Little Bloops

Tripp 5038 (photo: Lendemer)

NOTES: *Hypogymnia vittata* is a lovely lichen and is the easiest to recognize species of *Hypogymnia*. It is characterized but its long, strappy lobes, a morphology that none of the other *Hypogymnias* comes close to approximating. *Hypogymnia vittata* also doesn't usually form large thalli. Rather, they are, as James says, more like "little bloops" (imagine hand and arm gestures here).

CHEMISTRY: Atranorin, physodic acid, 3-hydroxyphysodic acid. Spot tests. K+ yellow (cortex), C-, KC+ red (medulla), P-, UV-.

NICHE: *Hypogymnia vittata* occurs only in old growth spruce-fir forest. You won't find it in anything crappier, which is in part why we love it so much. It is corticolous on conifers, particularly on *Abies fraseri*. At the time of writing, there are very healthy populations of this species on Snake Den Trail, just before reaching Inadu Knob. Look for it while simultaneously keeping an open eye for *Graphis sterlingiana*.

KEY FEATURES: Hollow, inflated lobes, long strappy lobes, soredia on undersides of lobe tips, relatively small thalli, corticolous on conifers in old growth spruce-fir forest.

Hypotrachyna afrorevoluta

Bubbling Square Britches

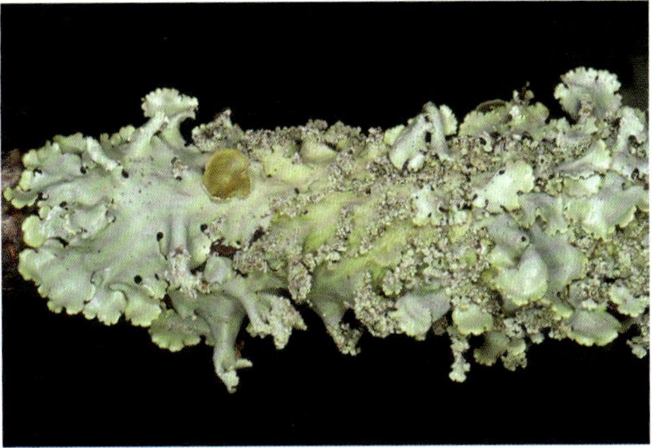

Tripp 3454 (photo: Lendemer)

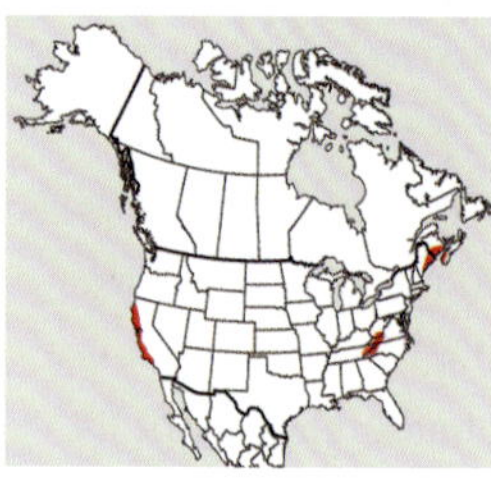

NOTES: Alright, so you bought this book and decided to try it out in the field. And then you realized there were 20 something different species of Square Britches in this one Park alone. Stay positive—with a little practice and good luck charm that the Smokies inevitably bring, these are doable. *Hypotrachyna afrorevoluta* is actually among the easier to identify species by its pustules that break apart to reveal a black lower cortex, seen from the upper surface (amazing . . . lichens do it all!). This species also has relatively broad lobes and a C+ red medulla. No other species of *Hypotrachyna* looks like this. Focus on the pustules.

CHEMISTRY: Atranorin, gyrophoric acid, hiascic acid aggregate. Spot tests. K+ yellow (cortex), C+ pink (medulla), KC+ pink (medulla), P-, UV-.

NICHE: *Hypotrachyna afrorevoluta* occurs primarily on the bark of hardwoods and conifers, but occasionally also on rocks, all of these at middle-to-high elevations. It is disjunct in the northern Appalachian Maritimes and also in coastal California.

KEY FEATURES: Broad, revolute lobes, pustules that break away to reveal the black lower cortex, C+ pink medulla, on bark at middle-to-high elevations.

Hypotrachyna croceopustulata

Stained Square Britches

Tripp 3952 (photo: Lendemer)

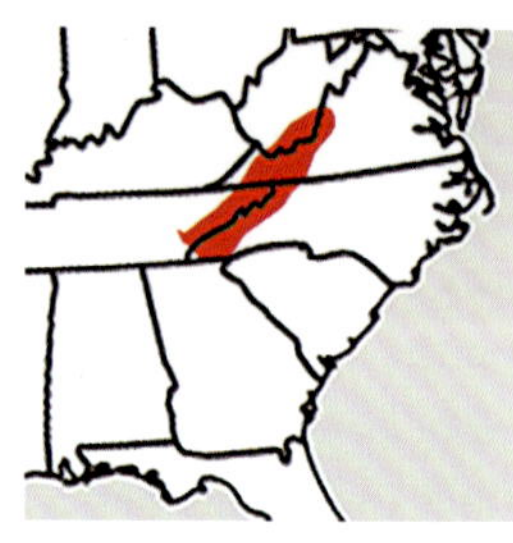

NOTES: Alright, so you are feeling good because you just successfully identified your first *Hypotrachyna*, *H. afrorevoluta*, in the Smokies. Luckily, you are in for more success. *Hypotrachyna croceopustulata* is the easiest to identify *Hypotrachyna* in the Smokies, and we mean it! Like *H. afrorevoluta*, *H. croceopustulata* is also pustulose but the latter is the only species that has an orange-pigmented medulla, visible under the pustules. Oh, that pigment by the way, it's called skyrin. Would make a nice name for a cat, no? *Hypotrachyna croceopustulata* is also C-, unlike *H. afrorevoluta*, which is C+ pink.

CHEMISTRY: Atranorin, protocetraric acid, skyrin. Spot tests. K+ yellow (cortex), K+ purple (medulla, where pigmented), C-, KC-, P+ red (medulla), UV-.

NICHE: Common to weedy (!) at high elevations throughout the Smokies. This species is corticolous on both hardwoods and conifers in spruce-fir forest. It is equally common at high elevations throughout the southern Appalachians and even in spruce forests of the central Appalachians.

KEY FEATURES: Pustules that break away to reveal an orange K+ purple medulla, otherwise a P+ red medulla where not pigmented orange, on bark at high elevations.

Hypotrachyna gondylophora

Surprise Square Britches

Tripp 3485 (photo: Lendemer)

NOTES: *Hypotrachyna gondylophora* is identifiable by its huge, capitate soralia that sometimes have an orange pigment underneath. It is easily differentiated from *H. croceopustulata* because the latter is pustulose, not sorediate. *Hypotrachyna gondylophora* is an attractive species guaranteed to make your already awesome hike in the Smokies that much better! You could mistake it for *H. oostingii* or *H. thysanota*, but those species have different chemistries and never hide an orange pigment underneath their soralia. *Hypotrachyna pseudosinuosa* is almost identical to *H. gondylophora* but occurs at low elevations, lacks orange pigment, and produces protocetraric acid in the medulla instead of fumarprotocetraric acid.

CHEMISTRY: Atranorin, fumarprotocetraric acid, skyrin. Spot tests. K+ yellow (cortex), C-, KC-, P+ red (medulla), UV-.

NICHE: This species occurs at high elevations throughout the Smokies and is corticolous on both hardwoods and conifers, but mostly occurs on *Abies*.

KEY FEATURES: Large thallus, wide lobes, capitate soralia on the lobe tips, orange, K+ purple pigment under the soralia, K+ red medulla where not pigmented, on bark in spruce-fir forests.

Hypotrachyna horrescens

Stubby Square Britches

Tripp 3562 (photo: Lendemer)

NOTES: *Hypotrachyna horrescens* is, like the name implies, somewhat on the grotesque side. The extremity of its isidia, these often with strange ciliate projections, serve to readily identify this species. As seen in this photograph, its lobes also tend to be rather short and stubby, as if they want to grow longer but just can't manage it. On the whole, this species produces rather small thalli and is frequently seen forming small rosettes, these often immature in appearance, on trees. This species is most likely to be confused with *H. minarum*, which also produces isidia, but the latter has a C+ pink medulla and lacks the small black cilia that typically adorn the isidia of *H. horrescens*. *Hypotrachyna horrescens* is actually an unforgettable and unmistakable species. For many among us, it was the first *Hypotrachyna* in eastern North America that we learned.

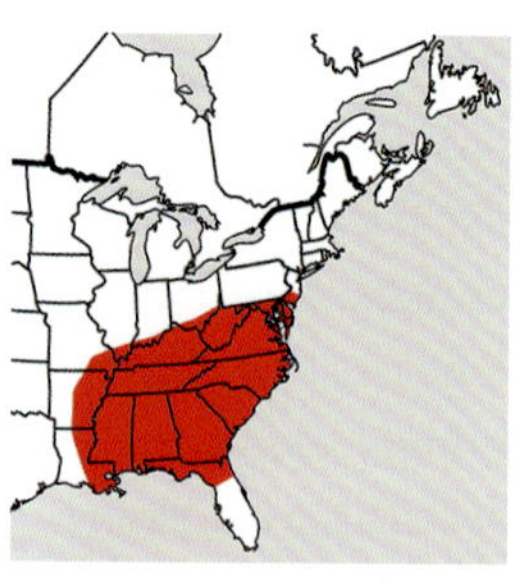

CHEMISTRY: Atranorin, 3-methoxy-2,4-di-O-methylgyrophoric acid, other substances but these are best ignored. Spot tests. K+ yellow (cortex), C-, KC+ pink (medulla), P-, UV-.

NICHE: Everywhere in the Smokies, low, middle, and high elevations, corticolous on both conifers and hardwoods and definitely not shy of seriously acidic substrates such as rhododendrons, pines, and hemlocks.

KEY FEATURES: Small thalli, short, stubby lobes, with isidia, some of which bear black, apical cilia, C- and KC+ pink medulla, on bark, especially conifers, at all elevations.

Hypotrachyna imbricatula

Great Horned Square Britches

Tripp 5391 (photo: Lendemer)

NOTES: In the Smokies, *Hypotrachyna imbricatula* is the only species in the genus that produces a big thallus, big lobes, big isidia, and fluoresces UV+ blue-white. The photo doesn't do quite do this species justice, but you can't win 'em all. This combination of features makes this species impossible to confuse for any other.

CHEMISTRY: Atranorin, barbatic acid aggregate. Spot tests. K+ yellow (cortex), C-, KC+ yellow (medulla), P-, UV+ blue-white (medulla).

NICHE: Corticolous on conifers at high elevations. You will only ever find *Hypotrachyna imbricatula* in mature spruce-fir forests. This species is a southern Appalachian endemic.

KEY FEATURES: Large thallus, wide lobes, big isidia, UV+ blue-white and KC+ yellow medulla, on conifers in spruce-fir forests.

Hypotrachyna kauffmaniana

Gary's Square Britches

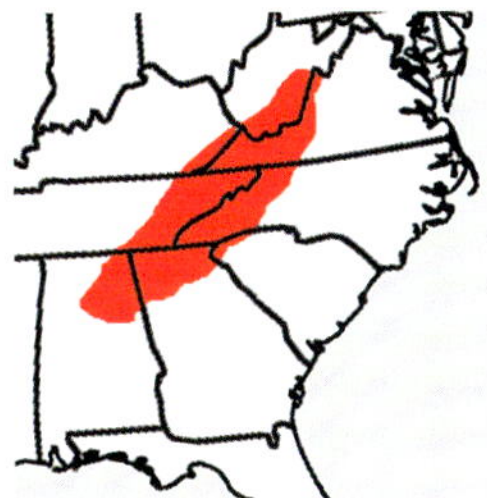

Lendemer 53161 (photo: Tripp)

NOTES: If one part of North America were blessed with *Hypotrachyna*, it would be the southern Appalachians, and the Smokies host a particularly high number of species. *Hypotrachyna kauffmaniana* belongs to a group of species that includes *H. afrorevoluta* and *H. revoluta*, all of which have revolute lobes and produce gyrophoric acid together with various accessory substances. The present species has most frequently been confused with *H. afrorevoluta* and *H. showmanii* because it is pustulose. However, it differs from both of those species in having ascending, secondary lobes that stick out perpendicular to the thallus and pustulose soralia that are more-or-less restricted to those projecting lobes. In contrast, both *H. afrorevoluta* and *H. showmanii* have flat, adnate lobes and pustules that are evenly distributed on the surface of the thallus. In many respects, *H. kauffmaniana* is most similar to *H. revoluta* because both species have strongly ascending secondary lobes and produce gyrophoric acid. However, *H. revoluta* differs markedly in having capitate soralia with fine soredia.

CHEMISTRY: Atranorin, 4-o-methylhiascic acid, gyrophoric acid. Spot tests: (cortex) K+ yellow, C-, KC-, P-, UV-; (medulla) K-, C+ pink, KC+ pink, P-, UV-.

NICHE: This species is common and widespread throughout Smokies, except at the highest elevations in spruce-fir forests. It grows on the branches and boles of trees, especially hardwoods, and can occasionally be found on shaded non-calcareous rocks. The species is essentially endemic to the southern Appalachian Mountains, with a small number occurrences slightly northward in the central Appalachians.

KEY FEATURES: Gray foliose thallus, revolute and ascending secondary lobes, pustulose soralia abundant on the subterminal portions of the secondary lobes, medulla C+ pink, on bark and branches throughout except the highest elevations.

Hypotrachyna livida

Clingy Square Britches

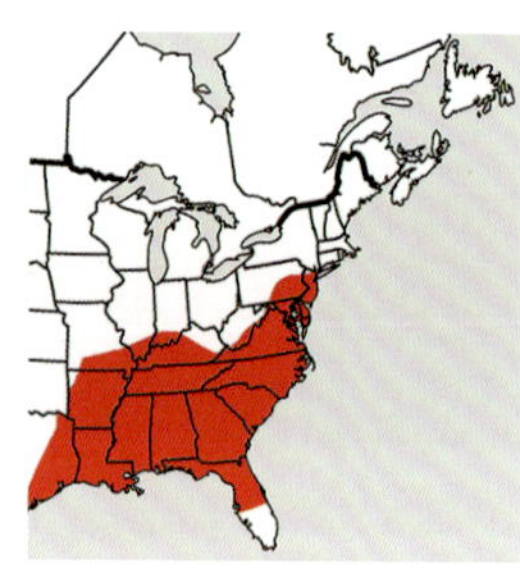

Tripp 3544 (photo: Lendemer)

NOTES: Okay, we have tried to be as honest as possible throughout our lifetimes but might have fibbed in saying that *Hypotrachyna croceopustulata* was the easiest member of the genus to identify in the Smokies. *Hypotrachyna livida* actually wins that award. Here is the only species that never produces asexual propagules, including soredia, isidia, or pustules. We can't believe it either. Other species can occasionally be found with apothecia, but always in addition to asexual propagules. The only macrolichen you could confuse this for is *Myelochroa galbina*, which differs in having smaller thalli with narrower lobes, and a P+ orange medulla that is also lightly yellow pigmented (especially near the margins of the apothecia). This species is often seen clinging tightly to tree branches and bark. It seems to nestle up closer to its substrates than other *Hypotrachynas*, which is sort of sweet.

CHEMISTRY: Atranorin, lividic acid. Spot tests. K+ yellow (cortex), K+ purple-brown (medulla), C-, KC-, P-, UV-.

NICHE: Quite common at middle-to-low elevations throughout the Smokies, nearly always corticolous on hardwoods, including twigs and branches.

KEY FEATURES: Absence of soredia and other lichenized diaspores, apothecia usually present, medulla P- and not yellow pigmented, on bark throughout except in spruce-fir habitats, tightly appressed to substrates.

Hypotrachyna lividescens

Lesser Square Britches

Lendemer 53264 (photo: Tripp)

NOTES: This isn't the most beautiful photo in the world, but . . . if you find another one we will include it in the 2nd edition. *Hypotrachyna lividescens* is quite rare and even if you think you find it, chemistry is needed to help confirm its identity. This species is characterized by small thalli and conspicuous capitate soralia near the lobe tips. It might be mistaken for *H. pseudosinuosa*, but the latter has a larger thallus and a P+ red medulla. In some ways, the most similar species to *H. lividescens* is *Myelochroa metarevoluta*, which differs in having ascending, revolute lobes and a P+ orange medulla that is also lightly yellow pigmented.

CHEMISTRY: Atranorin, olivetoric acid. Spot tests. K+ yellow (cortex), C+ red (medulla), KC+ red (medulla), P-, UV-.

NICHE: *Hypotrachyna lividescens* occurs on twigs of conifers and hardwoods, but perhaps more commonly makes its home in the canopy–when we find it, it is often on fallen branches. Additionally, look for it on shrubs in high elevation balds.

KEY FEATURES: Small thalli, narrow lobes, capitate soralia on lobe tips, C+ red medulla, on branches of trees and shrubs at middle-to-low elevations.

Hypotrachyna minarum

Simple Square Britches

Tripp 3451 (photo: Lendemer)

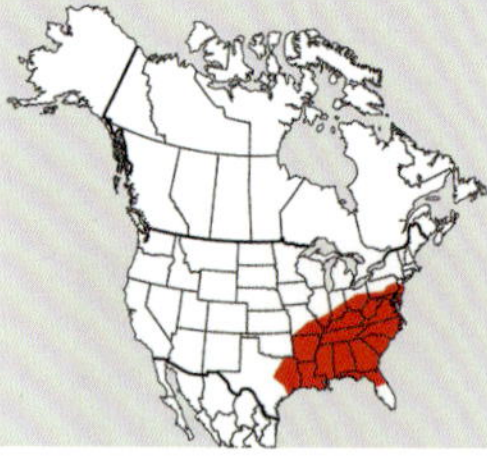

NOTES: *Hypotrachyna minarum* is most likely to be confused with *H. horrescens* but can be differentiated easily from that species because the latter has a C- medulla, in contrast to the C+ pink reaction in the medulla of *H. minarum*. The isidia in *H. horrescens* also tend to be adorned with apical cilia, however these structures are not always abundantly produced or conspicuous. *Hypotrachyna minarum* does not have isidia with apical cilia; however given the variability in *H. horrescens*, this character should not be used for identification in place of chemistry. If you choose to do that, we aren't responsible for the results.

CHEMISTRY: Atranorin, gyrophoric acid, hiascic acid aggregate. Spot tests. K+ yellow (cortex), C+ pink (medulla), KC+ pink (medulla), P-, UV-.

NICHE: Just like *Hypotrachyna horrescens*, *H. minarum* occurs just about everywhere in the Smokies: low, middle, and high elevations. It is corticolous on both conifers and hardwoods.

KEY FEATURES: Small thalli, narrow lobes, isidia without apical cilia, medulla C+ pink, on bark at all elevations.

Hypotrachyna oostingii

Oosting's Square Britches

Tripp 5386 (photo: Lendemer)

NOTES: We know what you are thinking at this point: "Dang, this *Hypotrachyna* looks like all the others." And, in this case, you would be absolutely correct. The trick to identifying *H. oostingii*, other than learning to speak lichen, is shoving this one under the UV lamp before puzzling over every other feature. Its medulla fluoresces blue-white, which might be illegal in some parts of the country. This species honors the late Dr. Henry Oosting, ecologist from Duke University who authored important works on Southern Appalachian spruce-fir forests.

CHEMISTRY: Atranorin, alectoronic acid, gyrophoric acid. Spot tests. K+ yellow (cortex), C+ pink (medulla), KC+ pink (medulla), P-, UV+ blue-white (medulla).

NICHE: At this point, you know where to head to get your *Hypotrachyna* fix: high-elevation spruce-fir forest! This habitat provides the opportunity to learn 10 species all at once! *Hypotrachyna oostingii*, like all the others, occurs on the bark of conifers and (sometimes) hardwoods.

KEY FEATURES: Large thallus, wide lobes, soralia usually with darkened soredia, medulla C+ pink and UV+ blue-white, on bark in spruce-fir forests.

Hypotrachyna osseoalba

Wake Turbulence

Lendemer 29579 (photo: Tripp)
Inset: Hollinger 20100411.7 (photo: Lendemer)

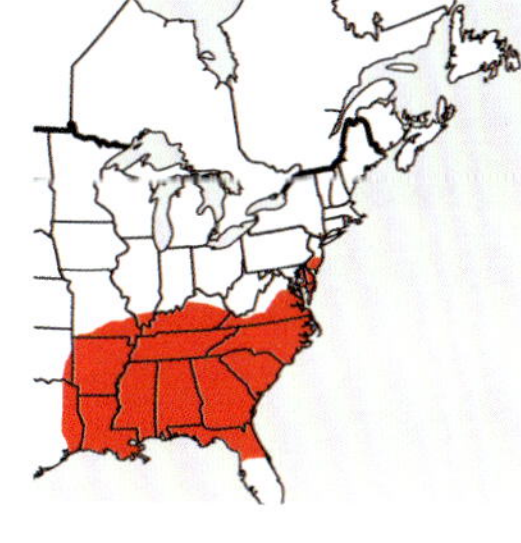

NOTES: There used to be a time when pustulose *Hypotrachynas* numbered one or two. Then three or four. We are now approaching a number that botanists tend to lose track of and just call it "infinity" (really, anything over seven . . .). But let's not overcomplicate this one: the world of pustulose *Hypotrachynas* is big, grand, and beautiful, but there is only one whose cortex screams BRIGHT YELLOW!!!!! at your face peering at it under the UV lamp, and that is *H. osseoalba*. The brightness of this reaction will likely leave you grounded, just like vortices coming off the wingtips of a giant Boeing 787. The specific epithet of this species refers to its thallus, which apparently resembles something bone white in color, although more often than not, it is bluish-gray, just like all the other *Hypotrachynas*. Don't confuse this for *Myelochroa aurulenta*, which is an extremely common foliose lichen throughout the Park. The latter differs by not having a bright yellow cortex under UV and in typically having a yellowish-orange pigmented medulla. *Hypotrachyna pustulifera* is similar, but differs readily in its UV-upper cortex.

CHEMISTRY: Atranorin and lividic acid aggregate. Spot tests: (cortex) K-, C-, KC-, P-, UV+ bright yellow; (medulla) K+ dark brownish, C-, KC-, P-, UV-.

NICHE: Wake Turbulence is widespread throughout the southeastern United States.

In the Smokies, look for it at low and middle elevations on various trees, but particularly on fallen hemlock and pine branches. It is neither common nor uncommon, but always a good day if you find it.

KEY FEATURES: Pustulose bluish-grayish-sometimes-whitish foliose thallus, UV+ bright yellow, on branches of acidic trees and also hardwoods, low and middle elevations.

Hypotrachyna prolongata

Fringed Square Britches

Tripp 5030 (photo: Lendemer)

NOTES: When it comes down to it, there are two kinds of *Hypotrachyna* in the world: green ones and gray ones. Look at this photograph and you will see a thallus transitioning from green to gray as it dries out. This is an excellent example of how lichens change in appearance, sometimes dramatically, when hydrated . . . just like people! *Hypotrachyna prolongata* is characterized by its *big* lobes with *fine* lobules that are usually developed to an excessive degree, making you gasp with anticipation at what other morphologies could possibly remain in the genus. Focus on the marginal fringe created by these fine lobules—they will help.

CHEMISTRY: Atranorin, anziaic acid. Spot tests. K+ yellow (cortex), C+ red (medulla), KC+ red (medulla), P-, UV-.

NICHE: Uncommon in the Smokies and only ever found in the nicest, highest-quality habitats. Corticolous on conifers, mostly *Abies*. You know the *Hypotrachyna* routine.

KEY FEATURES: Large thallus, wide lobes, finely divided marginal lobules, C+ red medulla, on conifer bark in mature sprue-fir forests.

Hypotrachyna revoluta

Flying Square Britches

Tripp 3572 (photo: Lendemer)

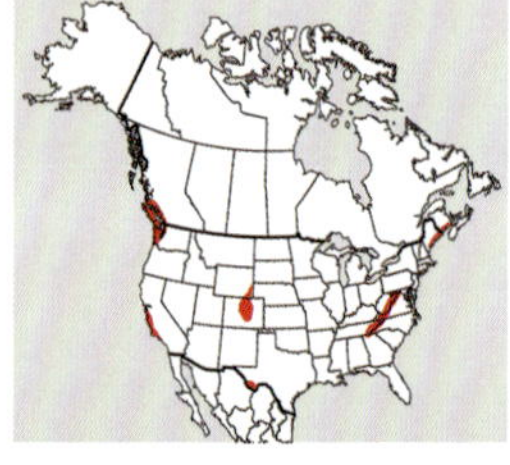

NOTES: *Hypotrachyna revoluta* is most likely to be confused not for another *Hypotrachyna* but for *Myelochroa metarevoluta*. Both species share squarrose lobes with capitate soralia, but the latter is C- with a yellow-pigmented medulla. *Hypotrachyna revoluta* might also be confused with *H. afrorevoluta*, but that species is pustulose. In fact, there has been a lot of confusion between *H. afrorevoluta* and *H. revoluta* in the past, but it's not our fault. Look for the revolute margins and ascending lobes while learning this species.

CHEMISTRY: Atranorin, gyrophoric acid. Spot tests. K+ yellow (cortex), C+ pink (medulla), KC+ pink, P-, UV-.

NICHE: *Hypotrachyna revoluta* occurs at all elevations throughout the Smokies. It is most commonly encountered on branches, flying from the canopy to the ground, but more rarely is found directly on the trunks of trees.

KEY FEATURES: Narrow revolute lobes that are often ascending away from substrate, soralia on upper surface of the lobe tips, medulla C+ pink, on bark, especially branches, at all elevations.

Hypotrachyna rockii

Ole Joe's Square Britches

Lendemer 30262 (photo: Tripp)

NOTES: *Hypotrachyna rockii* is yet another species with soralia, a big thallus, and big lobes. It is most similar to *H. tayloren-sis*, which differs in having pustules. And that's really all we have to offer. We leave you with a good luck wish in trying to tell *H. rockii* apart from all the other *Hypotrachynas* at high elevations.

CHEMISTRY: Atranorin, lecanoric acid, evernic acid. Spot tests. K+ yellow (cortex), C+ red (medulla), KC+ red (medulla), P-, UV-.

NICHE: *Hypotrachyna rockii* occurs at middle-to-high elevations throughout the Smokies and is usually corticolous on hardwoods. But don't make this the first *Hypotrachyna* that you learn–it is relatively rare, relatively unremarkable, and quite difficult . . . especially when presented with *H. prolongata*, which we should probably just keep talking about here, instead.

KEY FEATURES: Large thalli, wide lobes, presence of soralia, medulla C+ red, on bark and rocks at middle-to-high elevations.

Hypotrachyna showmanii

Ray's Square Britches

Tripp 3548 (photo: Lendemer)

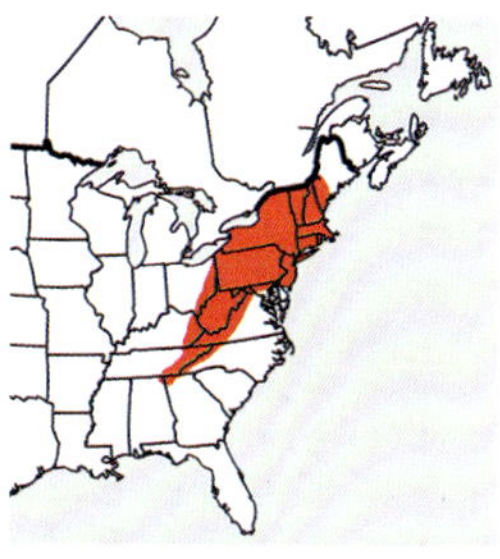

NOTES: *Hypotrachyna showmanii* is similar to *H. afrorevoluta* but has maculate lobe tips and pustules that do not fragment to reveal a lower black cortex. It also occurs at different elevations (see Niche). There has been ample confusion between these two species in the past, and probably will continue to be in the future. Another similar species is

H. spumosa, which differs in having smaller thalli with a distinctive yellowish hue and an unknown substance that is only detectable via TLC. *Hypotrachyna spumosa* also occurs only at lower elevations of the Smokies. The list of pustulose *Hypotrachyna* grows, but using the above combination of features, we think you can successfully identify *H. showmanii*. Make Ray proud.

CHEMISTRY: Atranorin, gyrophoric acid, hiascic acid aggregate. Spot tests. K+ yellow (cortex), C+ pink (medulla), KC+ pink (medulla), P-, UV-.

NICHE: This species occurs on the bark of hardwoods at low elevations, in contrast to *Hypotrachyna afrorevoluta* which is a high-elevation species. *Hypotrachyna showmanii* is endemic to the Appalachian Mountain chain.

KEY FEATURES: Maculate upper surface, large, coarse pustules that do not flake away to reveal a black lower cortex, C+ pink medulla, on bark at low elevations.

Hypotrachyna spumosa

Keep It In Your Square Britches

Lendemer 29621 (photo: Tripp)

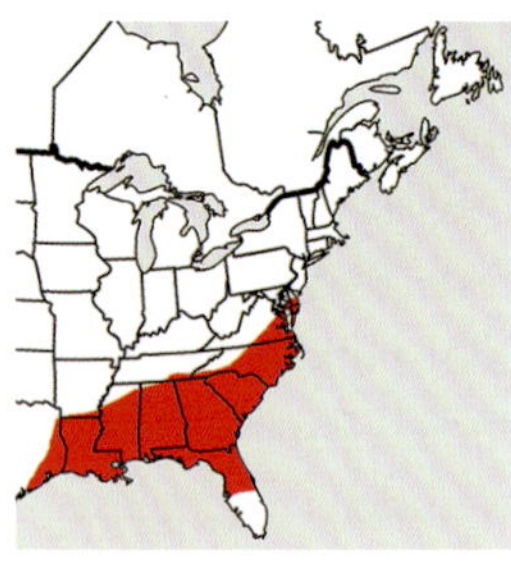

NOTES: *Hypotrachyna spumosa* is a pustulose species most likely to be confused with *H. showmanii*. *Hypotrachyna showmanii* is also a lover of low elevations but has larger thalli with wider lobes and coarse pustules that don't readily break apart. Advanced chemists can also use evidence from Thin Layer Chromatography, but the rest of us will just learn to differentiate the two by gestalt. Yep–size matters in *Hypotrachyna*.

CHEMISTRY: Atranorin, gyrophoric acid, hiascic acid aggregate, unknown compound. Spot tests. K+ yellow (cortex), C+ pink (medulla), KC+ pink (medulla), P-, UV-.

NICHE: *Hypotrachyna spumosa* is primarily a coastal plain species that, whenever it can, makes its way into the southern Appalachians, just like us. In the Smokies, it occurs throughout the Park at low elevations. Look for it around Fontana Lake while trying to avoid car sickness on "the dragon," especially if James is driving.

KEY FEATURES: Small thallus with narrow lobes, small fragile pustules, C+ pink medulla that also contains an unknown substance, on bark at low elevations.

Hypotrachyna taylorensis

Husky Square Britches

Lendemer 44652 (photo: Tripp)

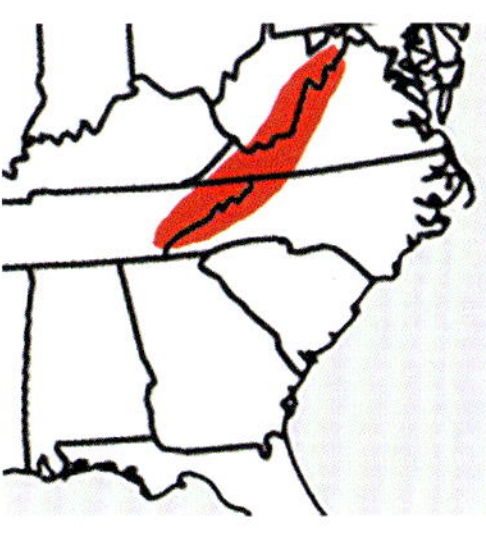

NOTES: *Hypotrachyna taylorensis* produces large thalli that bear conspicuous, flake-like pustules that do not fragment to reveal a black lower cortex, as in *H. afrorevoluta*. Spot tests will yield identical results, so use TLC to differentiate this one from the other C+ pink-to-red members of the genus. If it helps, *H. taylorensis* is probably the most common pustulose *Hypotrachyna* in northern hardwood forests. It was a special day when we photographed this one on Jenkins Ridge, which was far more pleasant an experience than was the giardia that Erin picked up from Blockhouse Mountain earlier that morning, which manifested itself two weeks later. Lesson (not) learned for a third time: never drink from the streams in the eastern United States, even when desperate.

CHEMISTRY: Atranorin, evernic acid, lecanoric acid. Spot tests. K+ yellow (cortex), C+ red (medulla), KC+ red (medulla), P-, UV-.

NICHE: Relatively common on bark of conifers and hardwoods at high elevations in northern hardwood forests; less common in spruce-fir forests.

KEY FEATURES: Large thallus, wide lobes, flaking pustules on the upper surface, C+ red medulla, on bark at high elevations.

Hypotrachyna thysanota

Mountain Fly Square Britches

Lendemer 5393 (photo: Tripp)

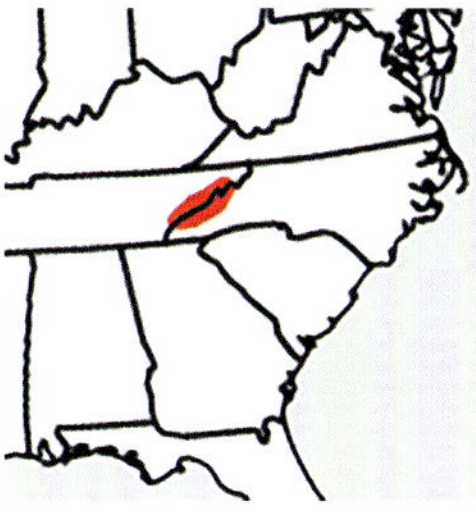

NOTES: *Hypotrachyna thysanota* is characterized by its big thalli, big soralia, and chemistry. It is most likely to be confused with *H. oostingii* or *H. gondylophora* but differs in its chemistry (see those species' entries and chemistry, below). In the photograph, disregard the small chunk of *Parmelia squarrosa* showing conspicuous maculae. Sometimes, you really can't help but to also photograph bycatch in the Smokies.

CHEMISTRY: Atranorin, echinocarpic acid, gyrophoric acid, microphyllinic acid. Spot tests. K+ yellow (cortex), C+ pink (medulla), KC+ pink (medulla), P+ orange, UV+ blue-white (medulla).

NICHE: *Hypotrachyna thysanota* occurs on the bark of conifers and is strictly a high-elevation species of spruce-fir forests in the southern Appalachians. Although it occurs at high elevations in other parts of the world, the Smokies and the Black Mountains are the only two areas where it makes its home in North America.

KEY FEATURES: Large thalli, wide lobes, soralia, C+ pink, P+ orange, and UV+ blue-white medulla, on bark in spruce-fir forests.

Hypotrachyna virginica

Virginia Square Britches

Tripp 5395 (photo: Lendemer)

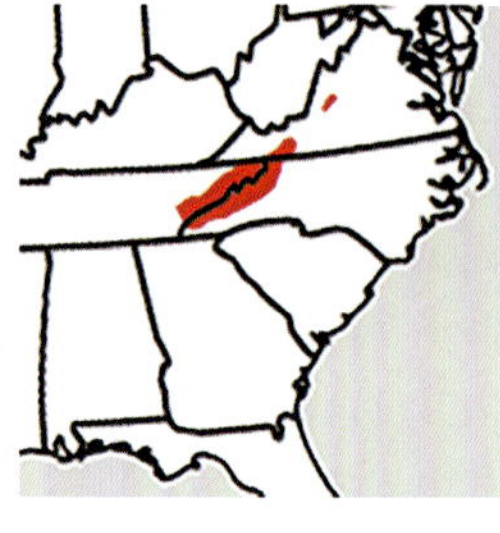

NOTES: You're probably thinking that *Hypotrachyna virginica* isn't anything special just because it has a common, unexciting specific epithet, but you would be wrong: it's one of our beloved, narrow endemics. *Hypotrachyna virginica* is easily recognized in the field by the white color of its thallus and its browned lobe tips, by the abundant small lobes produced in central portions of the thallus, and by the presence of large, flake-like pustules that peel away the upper surface of the thallus. Those characters, together with the presence of barbatic acid in the medulla, are distinctive for the species. And now, for patiently waiting, here is the full and final list of pustulose *Hypotrachyna* in the Smokies: *H. afrorevoluta*, *H. croceopustulata*, *H. osseoalba*, *H. pustulifera*, *H. showmanii*, *H. spumosa*, *H. taylorensis*, and *H. virginica*.

CHEMISTRY: Atranorin, barbatic acid aggregate. Spot tests. K+ yellow (cortex), C-, KC+ yellow (medulla), P-, UV+ blue-white (medulla).

NICHE: This species is a special treat to find in the forest. It is endemic to the southern Appalachians, where it occurs in high-elevation spruce-fir forests and shrub balds. Mason Hale described this species from Hawksbill Mountain in Virginia but the species no longer extinct in its type locality.

KEY FEATURES: Abundant, smaller secondary lobes in the center of the thallus, brown lobe tips, flakey pustules on upper surface, KC+ yellow and UV+ blue-white medulla, high elevations.

Icmadophila ericetorum

Soubrette

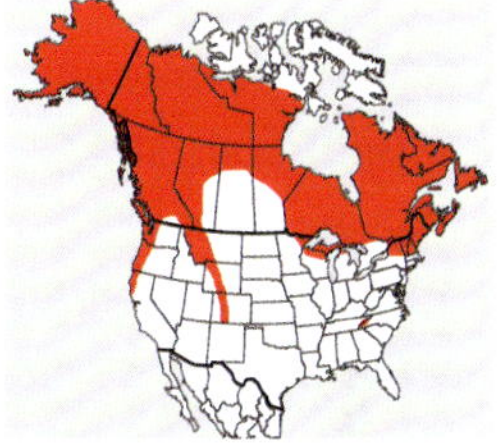

Tripp 5075 (photo: Lendemer)

NOTES: This is one of those lichens that you just have to love because its fun for the whole family. The large pink apothecia and thick blue gray-to-greenish thallus combined with its niche (see below) make *Icmadophila ericetorum* extremely easy to recognize in the field. There really isn't anything similar to it in the US of A. *Dibaeis absoluta* has similar apothecia but occurs on seepage faces of rocks and has a slimy film-like thallus dominated by algae.

CHEMISTRY: Perlatolic and thamnolic acids. Spot tests. K+ yellow, C-, KC-, P+ orange, UV+ blue-white.

NICHE: *Icmadophila ericetorum* can be found on rotting wood and humus at high elevations throughout the Smokies, especially in spruce-fir forests. As you can see from the map, the southern Appalachian populations are quite separated (disjunct) from the rest of the northern, boreal range of the species. This species also occurs in the southern, central, northern, and Canadian Rockies.

KEY FEATURES: Thick blue-gray thallus, large pink apothecia, on rotting wood and humus in spruce-fir forests.

Imshaugia aleurites

Henry's Heirloom

Tripp 3478 (photo: Lendemer)

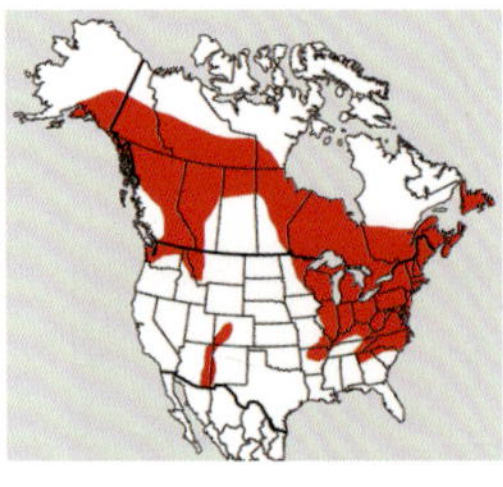

NOTES: Henry Imshaug was an important American lichenologist so it's fitting that this great genus of lichens honor him. Both *Imshaugia aleurites* and *I. placorodia* differ from other foliose lichens by having a pale lower surface and a K- cortex. In the field, you can always identify *I. aleurites* by these characteristics, combined with narrow lobes, presence of isidia, and distinctive white coloration of the thallus.

CHEMISTRY: Thamnolic acid. Spot tests. K+ yellow (medulla), C-, KC-, P+ orange (medulla), UV-.

NICHE: *Imshaugia aleurites* occurs throughout the Smokies on the bark of hardwoods, especially conifers. The southern Appalachians basically represent the southern edge of its range, and the species is much more common farther north. *Imshaugia aleurites* must like mountains (who doesn't, really) because it occurs in other, disjunct mountainous areas of the United States, such as the Sonoran Desert.

KEY FEATURES: White thallus, narrow lobes, isidia, pale lower surface, K+ yellow medulla.

Imshaugia placorodia

All Eyes On You

Lendemer 33163 (photo: Tripp)

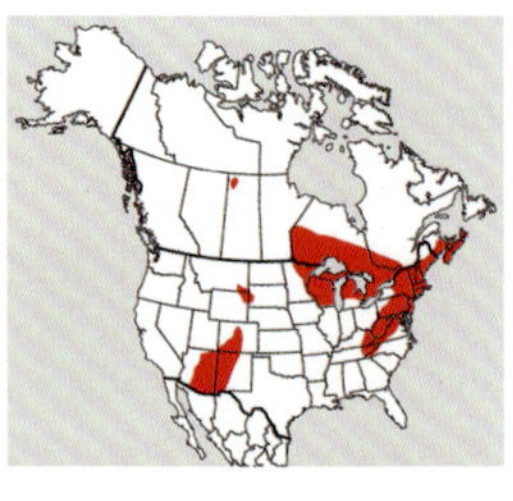

NOTES: *Imshaugia placorodia* is one of those species that you just have to see to believe. It's common elsewhere in the eastern United States, but in the Smokies you have to work hard to find it. This species is characterized by its pale brown lower surface, presence of apothecia, and K+ yellow medulla. You might be tempted to first confuse it with *Hypotrachyna livida* or *Myelochroa galbina*, but among many other differences, both of the latter species have black lower surfaces. *Imshaugia aleurites* is presumed to be the sister in this sexual–asexual species pair.

CHEMISTRY: Thamnolic acid. Spot tests. K+ yellow (medulla), C-, KC-, P+ orange (medulla), UV-.

NICHE: This species occurs on the branches of conifers throughout the Smokies and is actually quite rare. Despite being common in New England along the Atlantic Coast, it is found primarily at middle-to-high elevations in the southern Appalachians. Note the difference in the distributions of this species versus *I. aleurites*.

KEY FEATURES: White thallus, narrow lobes, absence of isidia, pale lower surface, K+ yellow medulla, on bark, especially conifers, at all elevations.

Ionaspis alba

Dry Eyes

Lendemer 44758 (photo: Tripp)

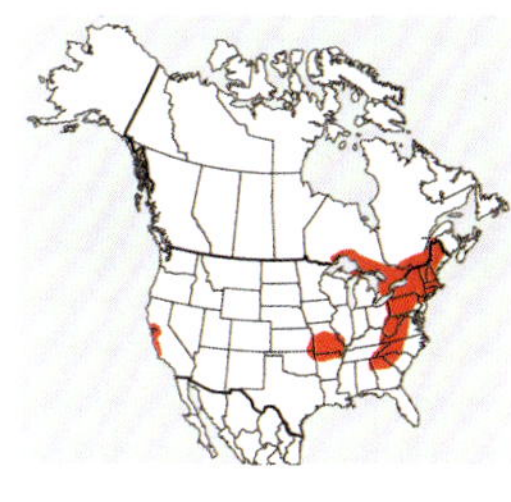

NOTES: Not many crustose lichens are defined by their absence of things, but this one is. *Ionaspis alba* has a pale white thallus, pale white immersed apothecia without any granules or pigments, and simple colorless spores. It doesn't even produce any flashy lichen substances. Despite these shortcomings, it's still a pretty thing to ponder, perhaps even more so than its bronzed cousin with which it cannot be confused owing to differences in thallus color, *I. lacustris.*

CHEMISTRY: No substances. Spot tests. K-, C-, KC-, P-, UV-.

NICHE: *Ionaspis alba* is an uncommon species in the Smokies and occurs on non-calcareous rocks at middle-to-high elevations. It is widely distributed in the Appalachians but also occurs in the Ozarks and coastal western North America. We suspect that it is more widespread and just overlooked because of its unassuming nature.

KEY FEATURES: White thallus, pale white, immersed apothecia, no granules in the epihymenium, simple spores, on rocks at middle-to-high elevations.

Ionaspis lacustris

The Lacustrine Lichen

Tripp 5094 (photo: Lendemer)

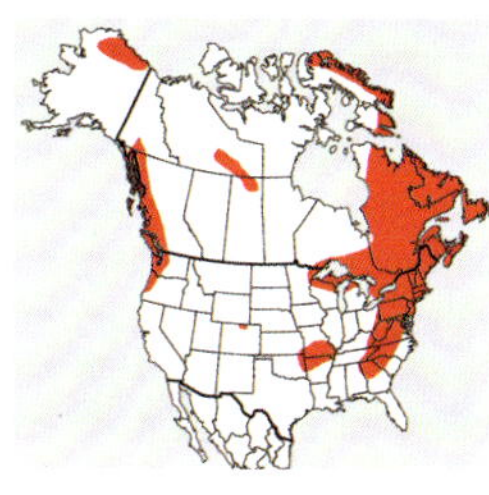

NOTES: *Ionaspis lacustris* is the bronze relative of *I. alba*, differing mostly in having a brown thallus and pigmented epihymenium inspersed with granules. In addition to these differences, which are readily seen in the field, the two species occupy separate niches (see below). We wish we had something more to say about this species and *I. alba*, but alas, it's Friday, and it's quittin' time.

CHEMISTRY: No substances. Spot tests. K-, C-, KC-, P-, UV-.

NICHE: Water is everywhere in the Smokies, and you can't blame anyone, let alone a crustose lichen, for loving these many rivers and streams. You'll find *Ionaspis lacustris* on non-calcareous rocks that are periodically inundated with water, or otherwise closely associated with nice humid spots. It is common throughout the Smokies, the Appalachians, and indeed much of temperate eastern North America.

KEY FEATURES: Brownish thallus, immersed, brown apothecia, granules in epihymenium, colorless simple spores, on non-calcareous rocks associated with water.

Jamesiella anastomosans

Little James

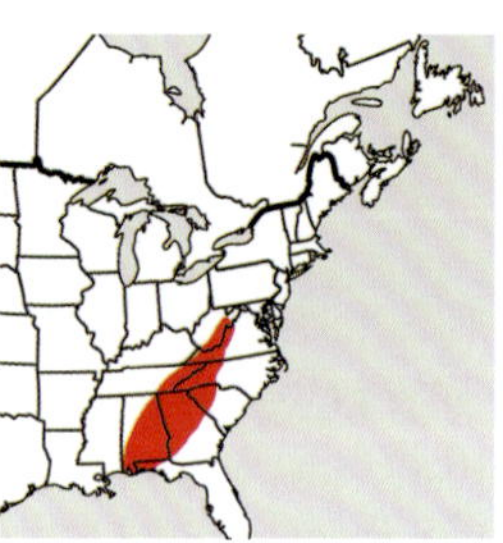

Lendemer 19228 (photo: Lendemer)

NOTES: When you see the small, circular thalli of *Jamesiella anastomosans* on your next bout of Rhodo-surfing, just remember that there is a Little James in every thicket. Although the species often looks like an unidentifiable, sterile crust from a distance, closer inspection will always reveal minute, short isidia that develop from the thallus surface (just look at how cute those isidia are!). Surprisingly, this species is easy to recognize in the field because it is one of the few lichens to grow in wet, humid, shaded forests laden with *Rhododendron*. It is easily detected because the thin, glossy thalli have a distinctly blue-green, almost emerald coloration, even from a distance.

CHEMISTRY: No substances. Spot tests: K-, C-, KC-, P-, UV-.

NICHE: A widespread species in oceanic areas of the Northern Hemisphere, *Jamesiella anastomosans* is common in the Smokies, where it occurs on bark and wood in humid habitats. It is particularly common on the trunks and branches of *Rhododendron* in thickets along streams or high mountain slopes.

KEY FEATURES: Thin, glossy, blue-green thallus, *Trentepohlia* photobiont, short isidia, on bark throughout in humid habitats.

Japewia tornoensis

Red Brown Spruce Town

Tripp 2428 (photo: Deregibus)

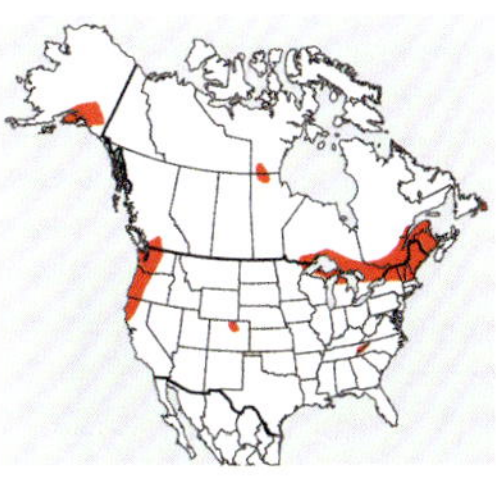

NOTES: This is one of those species you read about in a book and think to yourself, "How shall I know thee?" We often feel the same. Published descriptions regularly stress the details about asci and spores when discussing *Japewia tornoensis* and its relatives. Thankfully, none of its superficially similar relatives occur in the Smokies, and you can safely identify it by its niche (see below) and convex reddish-brown apothecia with thick-walled, colorless, simple spores.

CHEMISTRY: No substances. Spot tests. K-, C-, KC-, P-, UV-.

NICHE: *Japewia tornoensis* occurs on the bark of mature spruces in spruce-fir forests in the Smokies. It is not uncommon, but unique "stands" of *J. tonoensis* occur near Charlie's Bunion. Most populations of the species in North America are found in boreal New England and the Great Lakes, but as you can see from the map, it also occurs in scattered other boreal or montane areas.

KEY FEATURES: Small reddish-brown convex apothecia, simple colorless spores with thick walls, on spruce in spruce-fir forests.

Japewiella dollypartoniana

Dolly's Dots

Tripp 5521 (photo: Lendemer)

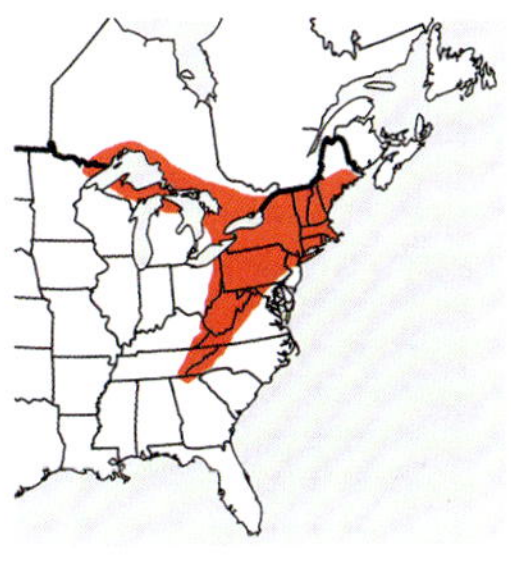

NOTES: What's the matter, you can't see a resemblance? Sometimes you have to look a little deeper. Just like its namesake, Dolly Parton, this species had its humble beginnings in Appalachia. You might pass it by on the trail, but that would be unfortunate. *Japewiella dollypartoniana* can be recognized by its coat of many colors, its soraliate thallus-bearing soredia that become variably pigmented orange or brown. This might not seem like much from the picture, but the real showstopper comes from a K+ red spot test. Few sorediate crusts in the Smokies react this way. Perhaps they are better behaved, or perhaps they are just boring, but in any case, none looks like *J. dollypartoniana*. If you're lucky enough to see this species in a fertile state, appreciate the small apothecia that are convex and red.

CHEMISTRY: Norstictic acid. Spot tests. K+ red, C-, KC-, P+ yellow, UV-.

NICHE: Not uncommon on hardwood bark in high-elevation northern hardwood forests, and particularly common on shrubs in *Rhododendron* balds in the Smokies and elsewhere in southern Appalachia.

KEY FEATURES: Sorediate thallus, variably colored soredia, K+ red reaction, on bark and shrubs at middle-to-high elevations.

Kephartia crystalligera

Horace's Ramble

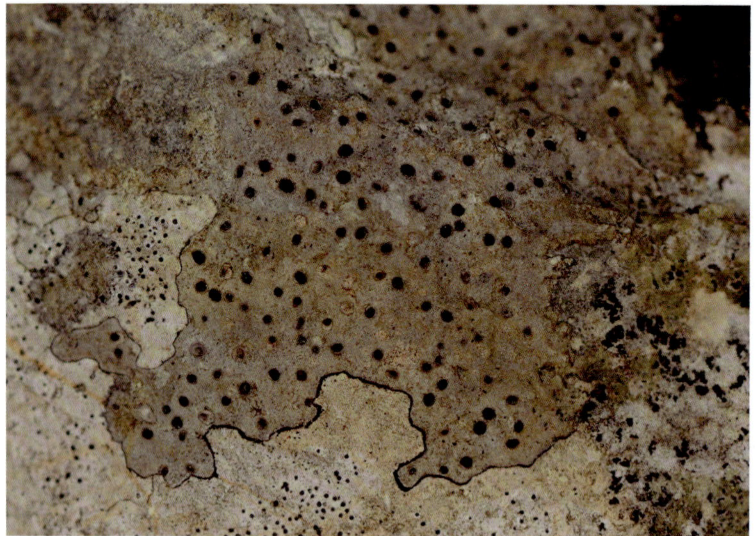

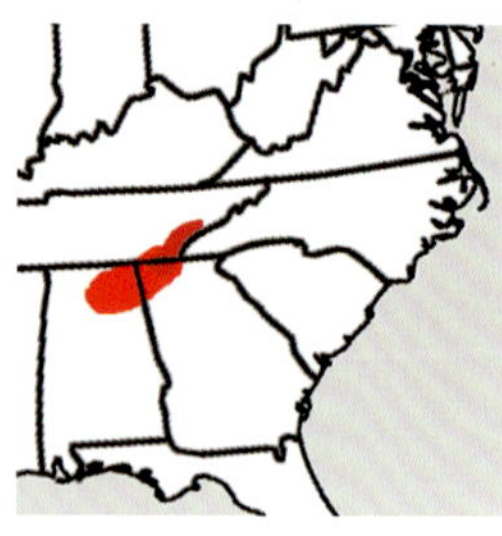

Lendemer 44552 (photo: Tripp)

NOTES: Here is a rare species that you have to go out of your way to see, which we think Kephart would approve of. *Kephartia crystalligera* is one of two species in the Smokies that occupies a similar niche (see below) and has flattened, dark purplish-brown apothecia. It can be separated from the other species, *Clauzadea chondrodes*, by the presence of orange, K+ purple crystals in the apothecia. In case you are wondering, this genus was named after Horace Kephart, a long-time proponent of establishing the Smokies as a National Park. He lived in a cabin at the end of the then-"road" on Hazel Creek, eventually moving to Bryson City.

CHEMISTRY: Orange crystals in the apothecia. Spot tests. K+ purple (crystals in apothecia), C-, KC-, P-, UV-.

NICHE: *Kephartia crystalligera* is restricted to limestone outcrops in the Smokies and is endemic to those habitats in the southern Appalachians.

KEY FEATURES: Immersed thallus, flattened, immersed, purplish-brown apothecia, orange, K+ purple pigments in apothecia, simple colorless spores, restricted to calcareous rocks.

Lasallia papulosa

Common Crinkles

Tripp 3581 (photo: Lendemer)

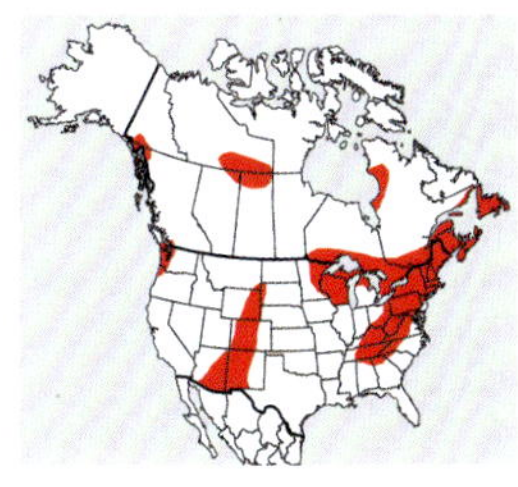

NOTES: Common crinkles is the most frequently encountered member of the informal group of lichens known as Rock Tripe. It can instantly be recognized by its foliose, umbilicate thallus with bumps or pustules that are visible on both the upper and lower surfaces, and by the pale brown coloration of the lower surface. *Lasallia pensylvanica* is similar in having bumps on both surfaces but is much less common and has a black lower surface. Species in the morphologically similar genus *Umbilicaria* are never bumpy like *Lasallia*, at least not in the Smokies.

CHEMISTRY: Gyrophoric acid. Spot tests. K-, C+ pink (medulla), KC+ pink (medulla), P-, UV-.

NICHE: *Lasallia papulosa* is common throughout the Smokies, where it occurs on non-calcareous rocks in both shaded and sunny environments. It is equally common throughout the Appalachians and Great Lakes. Disjunct populations occur in many other mountainous regions of North America, but its true home is in the East. This species was first discovered in Pennsylvania by the famous botanist and lichen lover Henry Muhlenberg.

KEY FEATURES: Foliose, umbilicate thallus, bumps on both the lower and upper surface, pale brown lower surface, on non-calcareous rocks at all elevations.

Lecanactis abietina

Blue White Spruce Light

Tripp 3489 (photo: Lendemer)

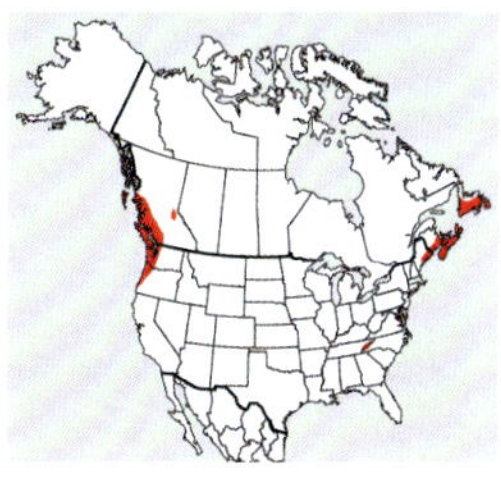

NOTES: *Lecanactis abietina* is a rather easy-to-recognize crustose lichen by its overall morphology and ecology. It produces a thin, continuous, white ecorticate thallus marked by conspicuous and pruinose white-tipped pycnidia that react C+ red (give it a try!). Sometimes, you also find this species with apothecia. You might be tempted to confuse it with *Micarea neostipitata* or *M. pycnidiophora*, but those species have pycnidia that are clearly raised on short stalks. Once learned, *L. abietina* won't be confusable for anything else. Learn this one by gestalt. We did.

CHEMISTRY: Lecanoric acid, schizopeltic acid. Spot tests. K-, C+ red (medulla), KC+ red, P-, UV+ blue-white.

NICHE: In the Smokies, *Lecanactis abietina* is restricted to the bark of spruce and fir at high elevations. Elsewhere, it occurs on the bark of hardwoods in dry, acidic forests, and occasionally on rocks.

KEY FEATURES: Thin white ecorticate thallus, white pruinose pycnidia that react C+ red, thallus UV+ blue-white, restricted to the bark of spruce and fir (mostly the former) at high elevations.

Lecania croatica

Can't Win 'Em All

Lendemer 446778 (photo: Tripp)

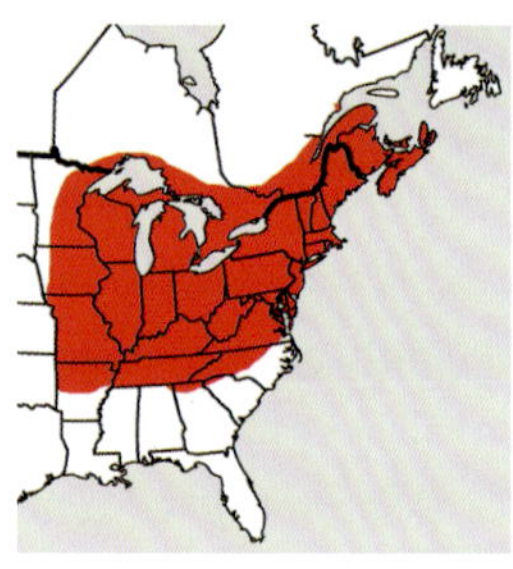

NOTES: *Lecania croatica* is a small crustose lichen characterized by its dispersed greenish areoles that bear very small yet conspicuous, discrete greenish-yellow soredia. It is most likely to be confused with sorediate species of *Biatora* such as *B. printzenii* or *B. pontica*, but those species have much larger soralia and also have secondary metabolites. Think of this as a small version of those asexual *Biatoras* that you will come to know and love if you hang around the Smokies long enough.
CHEMISTRY: No substances. Spot tests. K-, C-, KC-, P-, UV-.

NICHE: *Lecania croatica* is relatively common at middle-to-high elevations in the Smokies, where it occurs on the bark of hardwoods. It also occurs throughout eastern North America on similar substrates.
KEY FEATURES: Small greenish areolae that bear discrete yellow soredia, very small in size and lacking secondary metabolites, relatively common on the bark of hardwoods at and above middle elevations.

Lecanora albella (w/ varieties)

Foggy Eyes

Tripp 5067 (photo: Lendemer)

NOTES: So, you have decided to brave the *Lecanora* pages. That step, in and of itself, is commendable. Now it's time to get down and dirty. With over 30 species in the Park, it is important to stay positive at all times! *Lecanora albella* is characterized by its pale apothecial discs, which are covered in fine white pruina. Because of the latter, it is likely to be confusable only with *L. subpallens* or *L. caesiorubella*. *Lecanora albella* differs from *L. subpallens* in lacking a C+ yellow reaction in the epihymenium and differs from *L. caesiorubella* in other aspects of its chemistry (virensic acid in the latter compared to a whole pile of potential stuff present in *L. albella* (see Chemistry). The thallus of *L. albella* varies from white to yellow to bluish-gray in color.

CHEMISTRY: Atranorin, norstictic acid, +/- protocetraric acid. Spot tests. K+ yellow (thallus), K+ red (apothecia stipes, in section), C-, KC-, P+ yellow or red (apothecia), UV-.

NICHE: *Lecanora albella* is occasional in the Smokies at high elevations, where it occurs exclusively on the bark of northern hardwoods or on shrubs in balds.

KEY FEATURES: Thin thallus often with a white prothallus, pale apothecia with pruinose discs, corticolous on hardwoods at high elevations.

Lecanora anakeestiicola

Dangle Downs

Tripp 3517 (photo: Lendemer)

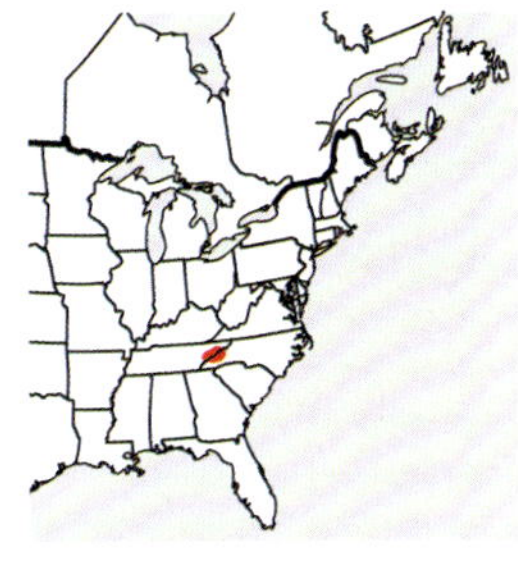

NOTES: This is a highly unusual species that is known only from the Smokies. While most species of *Lecanora* are crustose, this one makes shrub-like, fruticose branches that grow out of a yellow-green granular crust. Unlike most fruticose lichens, however, the branches of this lichen droop down under their own weight like a miniature, weeping tree. When we first found *Lecanora anakeestiicola*, we thought it was to represent the first record of *Leprocaulon microscopicum* in the Park, but that species has erect branches, and DNA analyses showed that the present species was instead a relative of *Lecanora*. Just goes to show–you can't always judge a lichen by its cover.

CHEMISTRY: Usnic acid and zeorin. Spot tests. K-, C-, KC+ gold, UV-.

NICHE: *Lecanora anakeestiicola* is, so far as known, endemic to Great Smoky Mountains

National Park in North Carolina and Tennessee where it is rare and found on shaded, overhanging rocks of the Anakeesta Formation at the highest elevations.

KEY FEATURES: Yellow-green granular thallus with flaccid, branch-like outgrowths that droop down, KC+ gold thallus, on Anakeesta rock at the highest elevations, very rare.

Lecanora appalachensis

Appalachian Fuzzy Do

Lendemer 44579 (photo: Tripp)

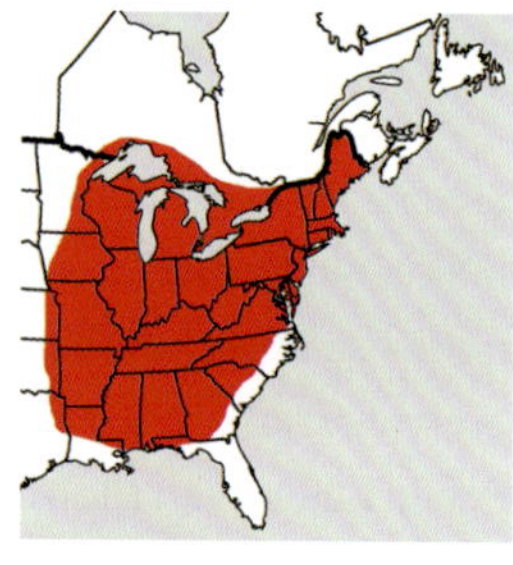

NOTES: Let us gather for a moment of silence to honor the most beautiful species of *Lecanora* in North America (and, with over 200 species in the genus known from this continent, that's no trivial recognition). Okay, it helps that the specimen photographed here happens to have both apothecia and soredia. Usually, however, you find it only with asexual reproductive structures (soredia), these rather coarse and bursting through a thick, bluish-green areolate thallus. In summary, *L. appalachensis* is yet another "ssc" (sterile sorediate crust) that will either make you love or hate the lichen diversity of Great Smoky Mountains National Park. For us, it's a turn on. *YOW!*

CHEMISTRY: Atranorin, zeorin. Spot tests. K+ yellow, C-, KC-, P- or P+ yellow, UV-.

NICHE: Relatively common on the bark of hardwoods as well as junipers at both middle and low elevations in the Smokies.

KEY FEATURES: Thick thallus that usually bears coarse, K+ yellow soredia, sometimes in addition to apothecia and if so, these with spectacular, sorediate margins, middle-to-low elevations on hardwoods and junipers.

Lecanora caesiorubella ssp. *caesiorubella*

Dust My Discs

Tripp 2222 (photo: Lendemer)

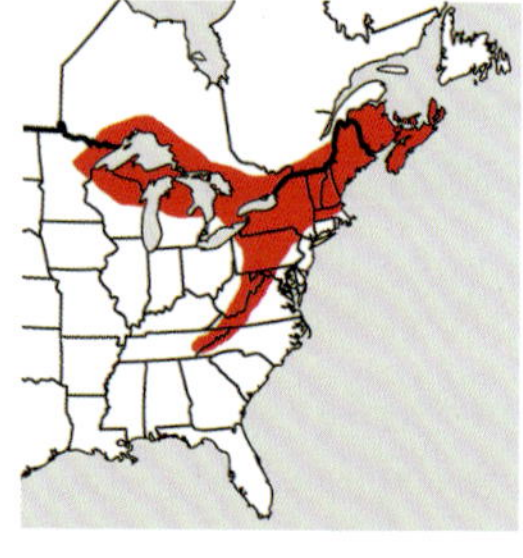

NOTES: *Lecanora caesiorubella* is one of three species in the Smokies with conspicuous white pruina covering the apothecial discs. This species in particular lacks a C+ yellow reaction in the epihymenium typical of *L. subpallens*, and differs from *L. albella* in other aspects of its chemistry (see entry for that species). In other news, *L. caesiorubella* has a continuous thallus that varies from white to bluish-gray in color.

CHEMISTRY: Atranorin and virensic acid. Spot tests. K+ yellow, C-, KC-, P+ red, UV-.

NICHE: This species is occasional in high-elevation northern hardwood forests and in shrub balds. Look for it corticolous on, and only on, hardwoods.

KEY FEATURES: Pale apothecia with pruinose discs, virensic acid, corticolous on hardwoods at high elevations.

Lecanora chlarotera

Lesser Dippy Dips

Tripp 5073 (photo: Lendemer)

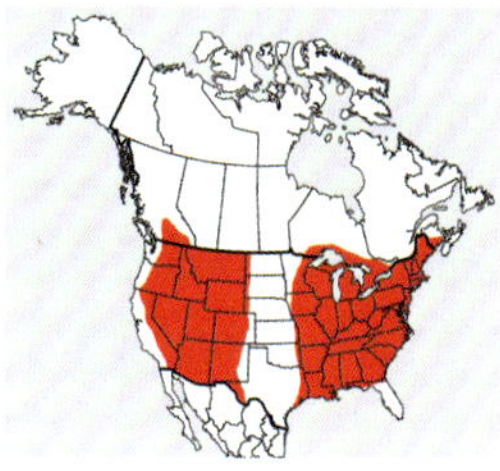

NOTES: *Lecanora chlarotera* is not a species that we have a lot of confidence in. It has a warty white thallus and is uncomfortably similar to *L. hybocarpa*, from which it differs only in having coarser crystals in its epihymenium. If this is disappointing news, you will at least find intrigue in the crystals–which many *Lecanora* have. To find these, do a cross section of the apothecium and apply polarizing light to your microscope. Cross your fingers, and maybe have a lichenologist queued up on your speed dial.

CHEMISTRY: Atranorin, gangaleoidin, roccellic. Spot tests. K+ yellow (thallus), C-, KC-, P-, UV-. Erin really wants this species to be fluorescent. But it isn't.

NICHE: *Lecanora chlarotera* occurs on trunks and branches of hardwoods at high elevations, particularly in open areas. The above photograph shows this species on *Sorbus* and was taken on Mt. LeConte. *Lecanora chlarotera* supposedly occurs throughout western North America as well, but again, we are pensive about its distinctiveness as a species.

KEY FEATURES: Warty white thallus, red discs with coarse crystals in the epihymenium, chemistry, corticolous on hardwoods at high elevations.

Lecanora cinereofusca

Dumpy Discs

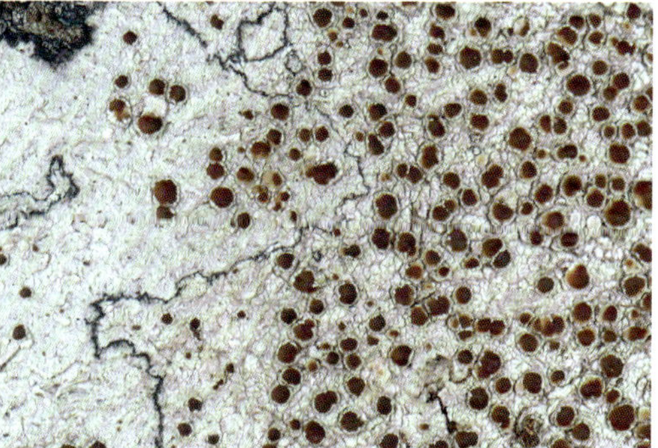

Tripp 2274 (photo: Lendemer)

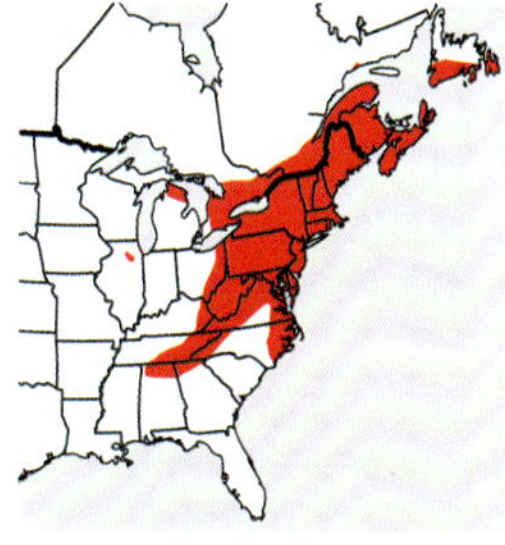

NOTES: *Lecanora cinereofusca* is one of the most common crustose lichens in Great Smoky Mountains National Park. Along with *L. thysanophora*, make this one of the first five crusts that you learn. *Lecanora cinereofusca* has a dull but relatively thick white thallus that is often irregularly lumpy, sort of like

its onlookers. Its apothecial margins are also bumpy (sometimes this gets called "beaded"), somewhat irregular in shape, and occasionally malformed. The discs vary from light to dark red and often appear damaged, as in several seen in this photograph. Talk to an invertebrate for answers. Also shown in this photograph are very neat black lines of sorts demarcating (presumably) different genetic individuals; these are called "zones of inhibition," and there is an entire set of genetic loci termed heterokaryon incompatibility loci that are said to govern processes related to self-recognition. Pretty heavy stuff for something seemingly so simple.
CHEMISTRY: Atranorin, pannarin, roccellic acid. Spot tests. K+ yellow (thallus), C-, KC-, P+ orange-red (epihymenium, in section), UV-.
NICHE: *Lecanora cinereofusca* is extremely common at all elevations in the Smokies, where it is corticolous on a wide variety of hardwoods, both on trunks and on branches. This species also occurs throughout the Appalachian Mountains and into the mid-Atlantic coastal plain and Canadian Maritimes, but in these locations, it is perhaps nowhere near as common as it is in the Smuh muh muh okies.
KEY FEATURES: Extremely common *Lecanora* on the bark of hardwoods at all elevations in the Smokies, irregularly lumpy thalli and discs, the latter sometimes malformed. Learn it by gestalt.

Lecanora darlingiae
Darling Dumplings

Lendemer 29662 (photo: Lendemer)

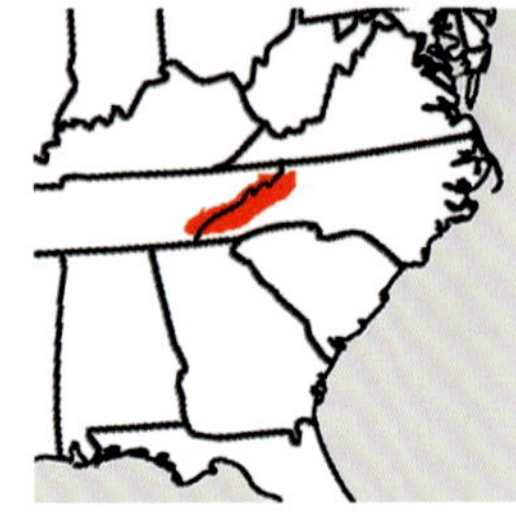

NOTES: *Lecanora darlingiae* is a wonderful species, just living its life in the cloud-covered peaks of the Smokies. We named it in honor of Emily Darling, an educator who worked for the Park and was among our first contacts at Purchase Knob. The species can be recognized, even in the field, by its blue-gray to whitish areolate thalli with conspicuous, black prothalluis and small soralia bearing lightly colored soredia. The most similar species is *Mycoblastus caesius*, which differs in producing only perlatolic acid and in being much more widespread. The thallus of *M. caesius* also tends to be darker gray in color, rather than the shades of blue or white. *Lecidea nylanderi* is another sorediate, areolate crustose lichen of high elevations that has a black prothallus. However, the latter can be distinguished by its more dispersed and well-developed areoles, and by the medulla that reacts UV+ bright blue-white due to the presence of divaricatic acid.

CHEMISTRY: Atranorin and 2'-o-methylperlatolic acid. Spot tests: K+ yellow, C-, KC-, P-, UV+ dull blue-white.
NICHE: *Lecanora darlingiae* is another southern Appalachian endemic that is particularly common in high elevation forests of the Smokies. It grows on the bark of hardwoods and conifers, but is particularly abundant on old Yellow Birches.
KEY FEATURES: Blue-gray to white-gray areolate thallus, black prothallus, small round soralia with light-colored soredia, K- and P-, on bark at high elevations.

Lecanora dispersa

Dead Reckoning

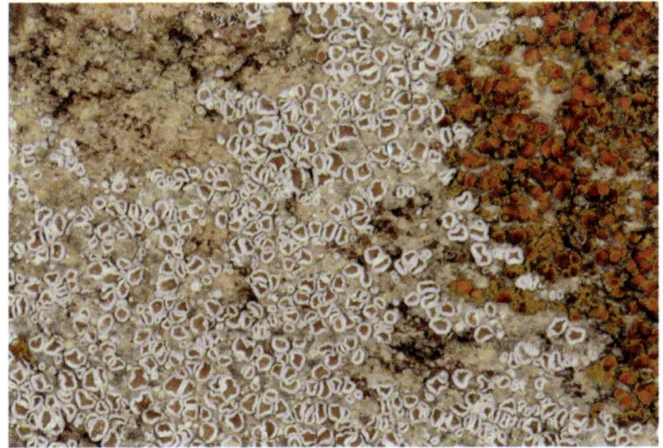

Lendemer 53169 (photo: Tripp)

NOTES: *Lecanora dispersa* is a remarkably easy-to-identify species in the Smokies. First, recognize it as a *Lecanora*. If you don't yet know the genus, it's time to learn it. *Lecanora* is the most species-rich genus of lichens in North America, and arguably one of the most so in the world. Most species share the classic "lecanorine margins" that stand out as very prominent features of the fruiting bodies, this species included. Much like dead reckoning is to the general aviation pilot, know this one by how you got there. It looks like a *Lecanora* based on those discs, right? Right. So, navigate your way to the genus you first thought about based on instinct, and know you have arrived at the correct location. Second, it is the only *Lecanora* in the Smokies that lacks all signs of a thallus. Sure, there are others that have minimal thallus material associated with the discs, such as *L. polytropa* or *L. sachsiana*, but none completely lacks a thallus on top of the substrate. Only *L. imshaugii* shares the stark white, ecorticate apothecia margins, but that species occurs on bark and has a blue-gray thallus. Also, this is the only *Lecanora* found in our area on concrete, although it often slides its way onto other nearby substrates (oops).
CHEMISTRY: 2,7-dichlorlichexanthone, ± pannarin. Spot tests: (apothecial margins) K-, KC-, C-, P- or P+ orange, UV+ dull and faint orange.
NICHE: This is an extremely common (and variable) species throughout North America. In the Smokies and elsewhere in the East, it is relatively uncommon in natural habitats and instead grows primarily on concrete and adjacent rock in places like parking lots. Look for it growing on concrete near High Rocks or on the Silers Bald Shelter, growing side by side with Orange Atoms.
KEY FEATURES: Our only *Lecanora* with a completely endolithic thallus, stark-white apothecia margins, pale reddish-brown discs, typically on artificial rock, such as concrete.

Lecanora glabrata

Lackluster Discs

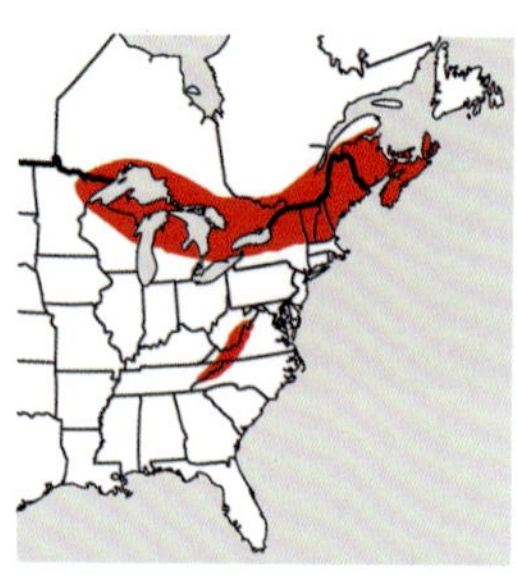

Tripp 2115 (photo: Deregibus)

NOTES: *Lecanora glabrata* may be one of those species that you just walk on by without noticing; that, in fact, you just missed an opportunity to add another number to your Lichen Life List. I mean, you wouldn't do that to a Cerulean Warbler, would you? Think about that for awhile. *Lecanora glabrata* looks a lot like other members of the red-brown discs of the *L. subfusca* group, especially *L. charlotera*, *L. hybocarpa*, and *L. pulicaris*. However, it differs in its niche (see below) and in having rather dark reddish-brown discs with an epihymenium that does not contain any POL+ crystals. That's right, *L. glabrata* simply lacks bling. The absence of crystals reminds one of *L. cinereofusca*, but that species is way more common, has apothecia with bumpy, beaded margins, and a has P+ orange epihymenium.

CHEMISTRY: Atranorin. Spot tests. K+ yellow (thallus), C-, KC-, P-, UV-.

NICHE: *Lecanora glabrata* occurs on the bark of hardwoods, especially American Beech and Yellow Birch, in northern hardwood forests at high elevations of the Smokies. It occurs in similar habitats elsewhere in the southern Appalachians.

KEY FEATURES: Quite dark, reddish-brown apothecia, lacking crystals in epihymenium (POL-), on bark of hardwoods in high-elevation northern hardwood forest.

Lecanora hybocarpa

Alabama Sunset

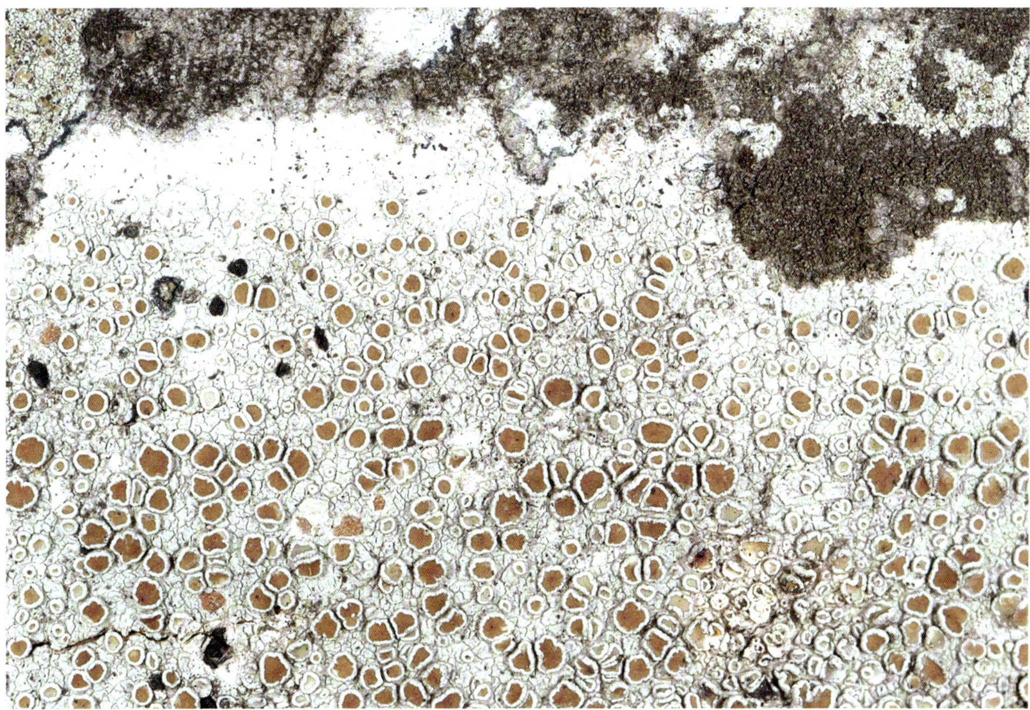

Lendemer 30281 (photo: Tripp)

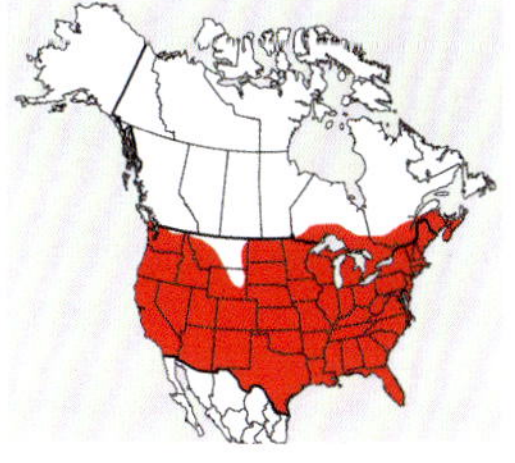

NOTES: There are just too many species of *Lecanora* in the Smokies, but we think there is enough time to learn them all! If you're out in the Smokies and you find a *Lecanora* with reddish-brown apothecial discs with bright white, smooth margins, it is probably *Lecanora hybocarpa*. You'll be forgiven (by . . . someone . . .) for not being able to separate this species from *L. charlotera*—we really can't either—which supposedly has coarse rather than fine POL+ crystals in the epihymenium.

CHEMISTRY: Atranorin, roccellic acid. Spot tests. K+ yellow, C-, KC-, P-, UV-.

NICHE: *Lecanora hybocarpa* is widespread on the bark of hardwoods throughout the Smokies. It is, however, much more common elsewhere in temperate eastern North America, such as northern Alabama. Watch for it on your drive home!

KEY FEATURES: Reddish-brown apothecia with small crystals embedded in the epihymenium (see these with polarizing light!), at all elevations in the Park on hardwoods.

Lecanora imshaugii

Exposed Henry

Lendemer 33118 (photo: Tripp)

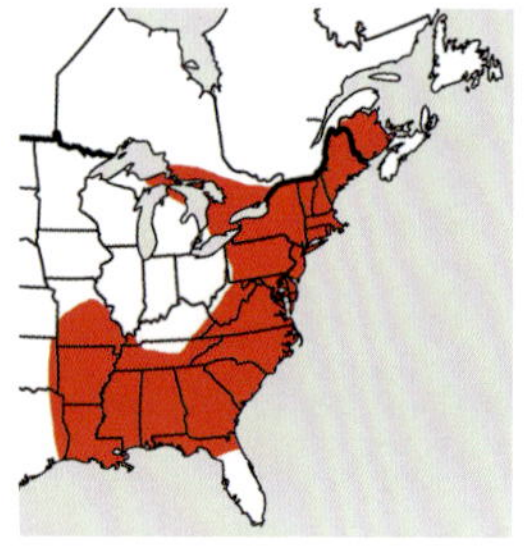

NOTES: *Lecanora imshaugii* is by far our favorite (yes, two opinions weighing-in, here) species of *Lecanora* in Great Smoky Mountains National Park. It is highly distinctive and absolutely unmistakable for any other *Lecanora* by its ecorticate, apothecial margins, which are stark white in appearance. More than other *Lecanoras*, these sexy margins will remind you of those of *Byssoloma subdiscordans*. This species was named in honor of the late Henry Imshaug—a leading figure in 20th-century lichenology.

CHEMISTRY: Atranorin, hypoprotocetraric acid, zeorin. Spot tests. K+ yellow (thallus), C-, KC-, P+ orange-red (thallus), UV-.

NICHE: *Lecanora imshaugii* occurs on the bark of hardwoods at low and middle elevations throughout the Smokies. This species is an example of a classic biogeographical disjunction between eastern North America and eastern Asia. See Hiromi Miyawaki's awesome table of ENA–EA lichen disjunctions in a 1994 publication appearing in *The Bryologist* (volume 97, pp. 409–411). It needs revising, but so do the rest of us.

KEY FEATURES: Absolutely distinctive *Lecanora* by its stark white, ecorticate apothecial margins. Corticolous on hardwoods except for the holotype, which was on pine, but that's Canada for you.

Lecanora intricata

The Intricate and Sophisticated

Tripp 5021 (photo: Lendemer)

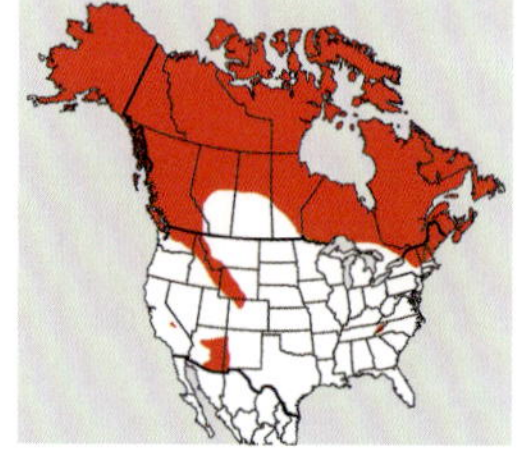

NOTES: *Lecanora intricata* is a pretty swell lichen. It is the only member of the genus that grows on rocks, has a thick, dirty yellow thallus, and dark brownish-yellow apothecia that usually have a persistent yellow margin. There isn't much else you could confuse it for, except maybe *L. polytropa*, which is far less spectacular. The two species are further differentiated

by the much better developed thallus in *L. intricata*, which gets a little lobe-y to almost placodioid in really well-developed material.
CHEMISTRY: Usnic acid, zeorin. Spot tests. K-, C-, KC+, P-, UV-.
NICHE: In the Smokies, *Lecanora intricata* is only found at the highest elevations on sun-exposed rocks in spruce-fir forests, frankly where such a sophisticated species should dwell. Look for it on LeConte's clifftops with many other very very special lichens. (Pop quiz: what is that crustose yellow species with black fruiting bodies, also seen in this photograph?) *Lecanora intricata* is more typical of arctic and alpine areas of North America, so thank your lucky stars you can see it this far south. At least for now.
KEY FEATURES: Well-developed, placodioid thallus, which basically makes this species distinctive from all others in the Smokies, brownish-yellow apothecia with persistent yellow margin, restricted to highest-elevation rock outcrops.

Lecanora layana

Elisabeth's Eyeshadow

Tripp 5310 (photo: Lendemer)

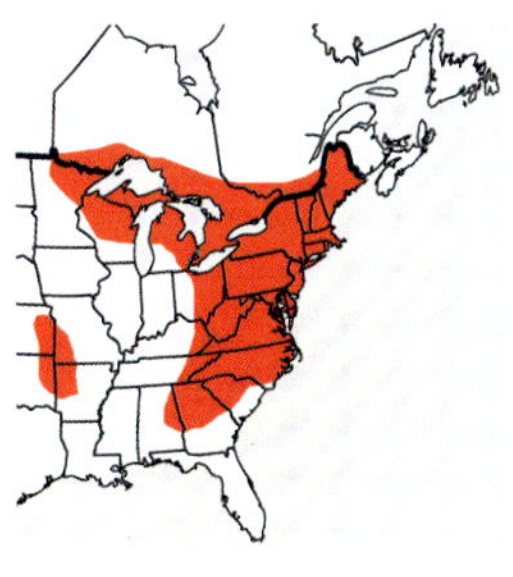

NOTES: *Lecanora layana* can be recognized by its crustose blue-gray sorediate thallus, which is often poorly developed and somewhat immersed in the substrate. It is almost identical to *L. nothocaesiella*, which differs in its P- thallus due to the absence of stictic acid. *Megalospora porphyritis* is also blue-gray and occurs on bark, however that species has a pustulose thallus and is K- due to the absence of atranorin. *Rinodina buckii* is also superficially similar but differs in having an areolate thallus with soralia that form centrally on the areoles. In addition *R. buckii* is also K- due to the absence of atranorin.
CHEMISTRY: Atranorin, zeorin and stictic acid. Spot tests: K+ yellow, C-, KC-, P+ orange, UV-.
NICHE: *Lecanora layana* is widely distributed in temperate eastern North America, although the majority of known occurrences are from the Appalachians and Great Lakes Region. It is common throughout the Smokies, and the southern Appalachians more generally, where it occurs on the bark hardwoods.
KEY FEATURES: Blue-gray crustose thallus, sorediate, K+ yellow and P+ orange, on hardwood bark throughout the Smokies.

Lecanora masana

Masa's Dots

Tripp 5064 (photo: Lendemer)

NOTES: *Lecanora masana* is one of numerous secrets of the Smokies that puzzles us as much as anyone. How are there so many species endemic to this small area? How is it possible that this species went unrecognized for so many decades? We "discovered it" and subsequently described it in our 2013 Smokies book, in honor of the famous Japanese-American photographer George Masa. Through his landscape photographs, Masa was instrumental in building support and enthusiasm for the establishment of the Smokies as a National Park. It is impossible to confuse this species for anything else because of its usnic acid-containing *big* apothecia, its thin thallus, and its niche (see below).

CHEMISTRY: Atranorin, usnic acid, 2'-O-methylperlatolic acid, xanthones. Spot tests. K+ weak yellow, C- or C+ yellow-orange (epihymenium, in section), KC+ yellow, P-, UV- or UV+ weak orange.

NICHE: At high elevations, this species is common on bark and twigs of *Abies* as well as hardwoods. You won't ever see it in any other ecosystem, that is, anything lower in altitude in the southern Appalachian Mountains.

KEY FEATURES: Large yellowish-orange apothecia that contain usnic acid, poorly developed thallus, quite common on bark and branches of firs and hardwoods but restricted to high elevations. So far as we know, a southern Appalachian endemic.

Lecanora minutella

No Place Like Cone

Lendemer 26974 (photo: Lendemer)

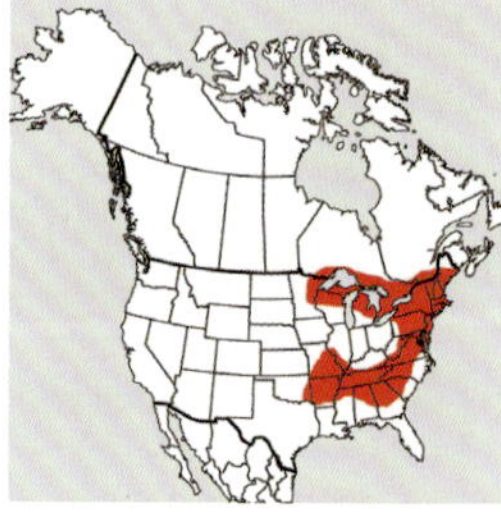

NOTES: You were looking for a miracle of a photograph and ended up with this? Well, if it helps, that's a single scale of a pine cone (a couple of millimeters wide) that you are looking at. It shouldn't come as a surprise that *Lecanora minutella* is one *minute* lichen, and we think this is quite a reasonable representation of it! This is a learnable species primarily owing to its very distinctive ecology: No Place Like Cone is one of the few lichens that preferentially occupies pine cones. In fact, if it has lecanorine apothecia and grows on cones, it is likely one of only two species: *L. minutella* or *L. strobilina*. *Lecanora minutella* is characterized by its tiny, reddish-brown apothecia that sit atop basically nothing other than the pine cone! This species more or less lacks all signs of a visible thallus. *Lecanora strobilina* differs in being yellow-green owing to the presence of usnic acid, usually having a well-developed thallus, and in having apothecia with ecorticate margins. Other than size and ecology, the two have little resemblance (note that *L. strobilina* grows like a weed on a wide variety of substrates while *L. minutella* sticks primarily with its cones. Finally, the color and size of the apothecia of *L. minutella* (reddish-brown, quite small) resemble the recently described *L. sachsiana*, however the latter differs in having a true thallus, pseudolecanorine apothecia margins, smaller spores, and a completely different ecology.

CHEMISTRY: Unidentified substances. Spot tests. K-, KC-, C-, P-, UV-.

NICHE: The type specimen of No Place Like Cone was collected on Lookout Mountain near Chattanooga, Tennessee, on a pine tree. In the Smokies, look for it at middle-to-low elevations, such as the Deep Creek area. Its known geographical range is that of the southern to northern Appalachians, Great Lakes, and Ozarks. A classic distribution, if we say so ourselves.

KEY FEATURES: Occurrence on pine cones, tiny reddish-brown apothecia, equally tiny colorless and simple spores, virtually no thallus, particularly common in the western and low elevations of the Park.

Lecanora nothocaesiella

False Dust

Lendemer 37924 (photo: Lendemer)

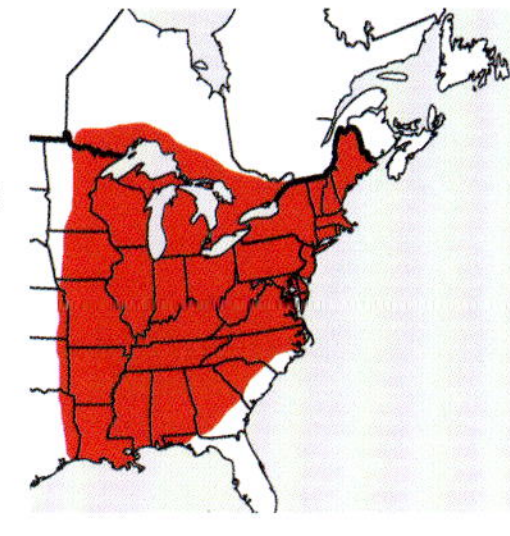

NOTES: There are a number of superficially similar sorediate and leprose crustose lichens with blue-gray crustose thalli in the Smokies, and this is one of them. *Lecanora nothocaesiella* is most likely to be confused with *L. layana*, which also has a mostly immersed thallus and irregularly shaped, eroding soralia, but *L. layana* differs in producing stictic acid. *Lecanora appalachensis* is chemically identical to *L. nothocaesiella*, but differs in having an areolate thallus with soralia that erupt from the tops of the areoles. The epithet "nothocaesiella" refers to the frequent confusion of this species with *Lepraria caesiella*, which differs in having a truly leprose thallus composed entirely of minute, ecorticate granules.

CHEMISTRY: Atranorin and zeorin. Spot tests: K+ yellow, C-, KC-, P- or P+ weak yellow, UV-.
NICHE: This species is widely distributed on the bark of hardwoods throughout the Smokies. It is common and widespread throughout much of temperate eastern North America.
KEY FEATURES: Blue-gray crustose thallus, mostly immersed in the substrate, erose soralia, frequent black prothallus, K+ yellow and P- or P+ weak yellow, on hardwood bark.

Lecanora oreinoides
Doll's Eyes

Tripp 3732 (photo: Lendemer)

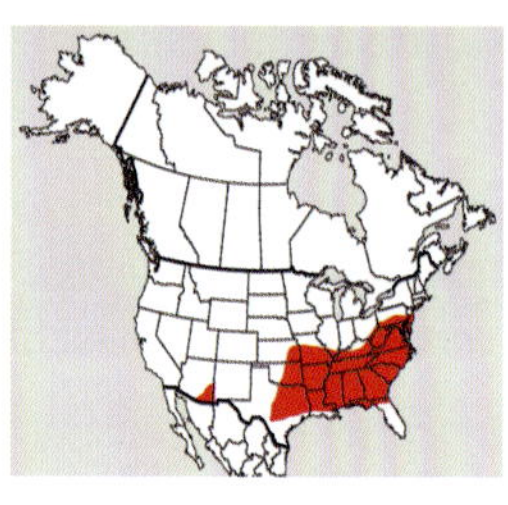

NOTES: *Lecanora oreinoides* is one of the most distinct members of the genus in the Smokies (and indeed, throughout the southeast). Its overall appearance, specifically its lecideine (rather than lecanorine) margins, suggests something other than *Lecanora* . . . such as *Lecidea*! In fact, it is most likely to be confused for *Lecidea tessellata* rather than another *Lecanora*. Neither species is common in the Smokies although both occur in similar habitats. They can be differentiated because *Lecanora oreinoides* has a shiny white cortex and lacks pruinose apothecia whereas *Lecidea tessellata* has a dull cortex with pruinose apothecia . . . at least the Appalachian version. Truth be told, *Lecidea tessellata* is a giant mess and needs an enterprising PhD student.
CHEMISTRY: Atranorin, confluentic acid. Spot tests. K+ yellow (cortex), C-, KC-, P-, UV+ blue-white (medulla).
NICHE: Uncommon but occasionally found on sun-exposed rock outcrops throughout the Park.
KEY FEATURES: Very thick, shiny white upper cortex with deep black apothecia that have lecideine rather than lecanorine margins, looks nothing like a *Lecanora*, occasionaln on sun-exposed rocks throughout the Smokies.

Lecanora polytropa
Yellow Acapella

Tripp 2464 (photo: Deregibus)

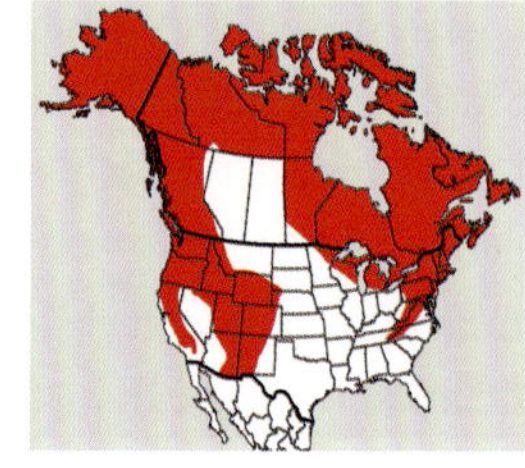

NOTES: This is an easy-to-identify, exceptionally common crustose lichen that you can impress your friends with next time you're up on Mt. LeConte. *Lecanora polytropa* is the only lichen in the Smokies that occurs on rocks, has an areolate, thin, yellowish-green thallus, and convex biatorine apothecia with waxy yellowish discs. You might confuse it with *L. intricata*, but that species is rarer, has apothecia that are darker in color, and a much more well-developed (and beautiful) thallus.

CHEMISTRY: Usnic acid, zeorin. Spot tests. K-, C-, KC+ yellow, P-, UV-.

NICHE: *Lecanora polytropa* can be found throughout the middle and high elevations of the Smokies on sun-exposed, non-calcareous rocks. It occurs in similar habitats throughout the Appalachians and elsewhere across huge parts of the rest of North America.

KEY FEATURES: Regular, convex, waxy yellow apothecia, on rocks throughout middle and high elevations of the Smokies.

Lecanora pseudistera

Cold Flow

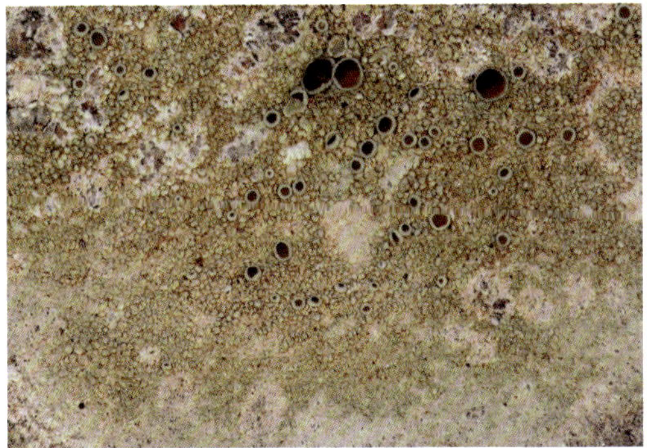

Lendemer 48566 (photo: Tripp)

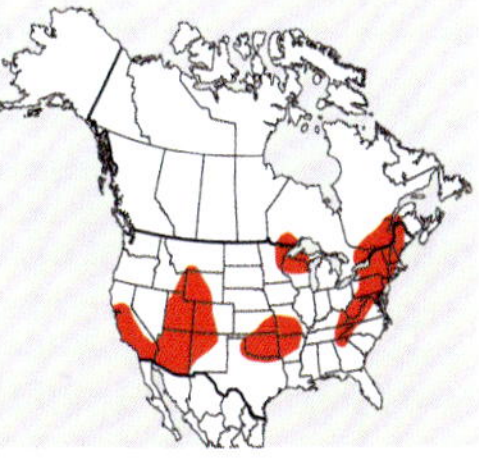

NOTES: There are a lot of *Lecanora* species in the Smokies, which you have now begun to appreciate. That being said, there are only a few that grow on rocks and have reddish-brown discs, clearly indicating their relationships to the *L. subfusca* group. *Lecanora pseudistera* is not uncommon on rock in the Smokies and can be recognized by its chemistry, reddish-brown discs, and non-immersed apothecia with smooth margins. *Lecanora saxigena* is similar, but that species has a whiter thallus, lumpy, beaded apothecial margins, and P+ red discs. *Lecanora subimmergens* is much less common, differs chemically, and typically has immersed apothecia.

CHEMISTRY: Atranorin and 2'-o-merthylperlatolic acid. Spot tests. K+ yellow, C-, KC-, P-, UV-.

NICHE: Widespread but infrequent in eastern North America and in the Smokies, where it grows on non-calcareous rocks, especially in protected or sheltered areas.

KEY FEATURES: Crustose thallus, sessile lecanorine apothecia with reddish-brown, P- discs with smooth margins, on protected faces of non-calcareous rocks at middle-to-low elevations.

Lecanora pulicaris

Log Sticker

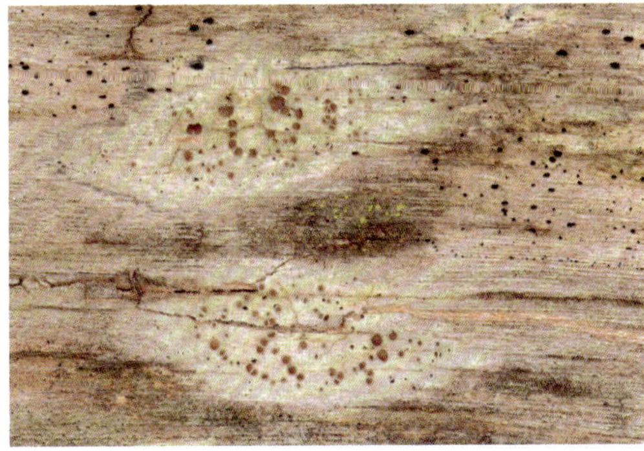

Lendemer 53140 (photo: Tripp)

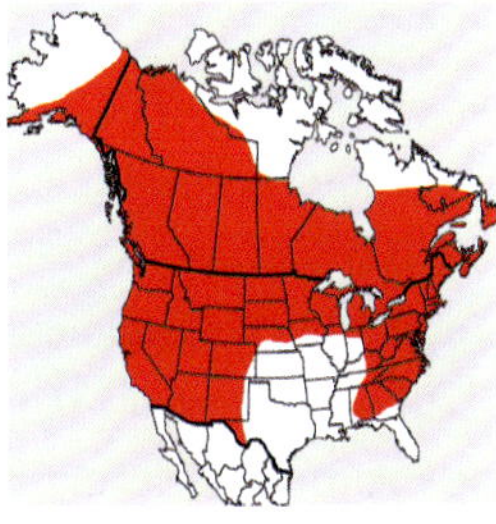

NOTES: *Lecanora pulicaris* is characterized by its thin, gray thallus and apothecia with reddish-brown discs and thin margins concolorous with the thallus that react P+ red. Although it is easy to miss, the apothecial margins of *L. pulicaris* also tend to have a distinct ring of tissue just adjacent to the discs, and this is not so clearly present in most other species. There are a lot of similar species in the Smokies and so you can be forgiven for not necessarily wanting to pick this one out of the pack. It is most likely to be confused with *L. hybocarpa*, which is common throughout the Park and differs most noticeably in having P- apothecial margins. *Lecanora chlarotera* and *L. circumborealis* are also similar, but both have P- apothecial margins and differ in various characters of the epihymenium and spores. Confusion with *L. argentata* may also be possible, but that species also has P- apothecial margins and the epihymenium is entirely lacking the numerous small crystals that are present in *L. pulicaris*. The only other P+ orange-red *Lecanora* species that occur on bark in the Smokies are *L. cinereofusca* and *L. insginis*, but in both of those species the reaction is due to the presence of pannarin in the epihymenium and not fumarprotocetraric acid in the margins. If you mount a section of the apothecium under the compound microscope and flood it with P, you can see the difference because the pannarin forms orange crystals while the fumarprotocetraric acid simply turns red within the margin of *L. pulicaris*.

CHEMISTRY: Atranorin, fumarprotocetraric acid and roccellic/angardianic acid. Spot tests: K+ yellow, KC-, C-, P+ red (apothecia margins), UV-.

NICHE: *Lecanora pulicaris* is restricted to high elevations in the Smokies, which is not surprising considering its preferred habitat is in more northerly areas (and extending into the desert southwestern mountains). Look for it growing on lignum and decorticate trunks (as seen in the photo) or on bark in spruce-fir forests.

KEY FEATURES: Gray crustose thallus, lecanorine apothecia with reddish-brown discs, P+ red margins, simple colorless spores, on bark or wood at high elevations.

Lecanora rugosella

Rhubarb Pie Lichen

Tripp 3488 (photo: Lendemer)

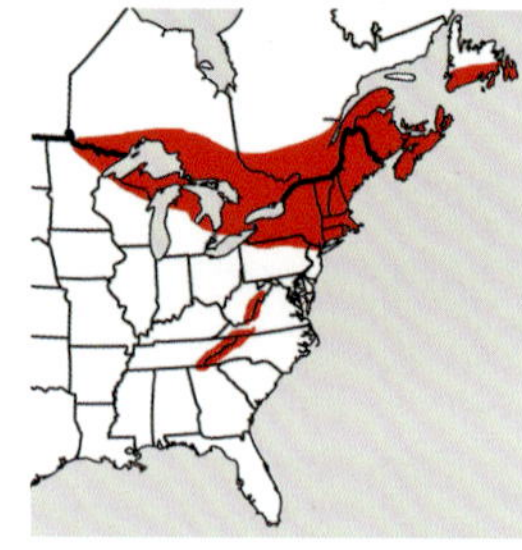

NOTES: That's right, my friend, another *Lecanora* species with reddish-brown discs! You should be excited about this one because it is easy to recognize by its niche and by its large apothecia with reddish-brown discs that are lightly white pruinose and basally constricted. You might confuse it for *L. masana*, which occupies a similar niche but has yellow-green apothecia, usnic acid, and *not much* in the way of a visible thallus. Actually, never mind. You really can't confuse these two species. The specific epithetic of this name means "folded" or "wrinkled."

CHEMISTRY: Atranorin, roccellic acid (again). Spot tests. K+ yellow (thallus), C-, KC-, P-, UV-.

NICHE: This is one of the most common lichens at high elevations in the Smokies. It grows on shaded branches of shrubs and conifers in spruce-fir forests and less frequently on the trunks of trees in the same habitat. It is common in similar environments throughout the Appalachians as well as Great Lakes and Canadian Maritimes.

KEY FEATURES: Large apothecia with reddish-brown, pruinose discs, these constricted at the base, on branches and bark in spruce-fir forest and not confusable with *L. masana*, which contains usnic acid.

Lecanora sachsiana

Susan's Sacs

Tripp 7769 (photo: Tripp)

NOTES: We enthusiastically present to you *Lecanora sachsiana* —aka Susan's Sacs (in references to the sacs in which meiospores form in ascomycete fungi)—one of the newest additions to the lichen biota of Great Smoky Mountains National Park. This species, which was recently described by us to honor long-term GSMNP staff member and Park educator Susan Sachs, is highly unlikely to be confused with any other crustose *Lecanora* in our area on account of its emergent, pseudolecanorine, reddish-brown apothecia and consistent production of a single, as yet unidentified compound. Its mature apothecia appear almost biatorine because thalline margins typical of the genus are observable only in early developmental stages. Don't confuse it for *L. symmicta*, which also lacks lecanorine margins but looks completely different owing primarily to its yellow-green thalli (the thallus of *L. sachsiana* is very thin and sordid white in color). Additionally, do not confuse the one and only Susan's Sacs for the common weed, *Pyrrhospora varians*, which differs in numerous features including having a well-developed, continuous thallus, apothecia that lack thalline margins at all stages of development, and in the production of xanthones.

CHEMISTRY: Unidentified compound. Spot tests (thallus): K- to K+ weak yellow, KC-, C-, P-, UV-.

NICHE: Susan's Sacs has a clear preference for cool, airy ridgetops at middle-to-high elevations where the lichen enjoys a great view of the Smokies. It has a particular affinity for growing on large, old wood, but also occurs on the bark of living Black Cherry and Red Oak. So far as known, this species is restricted to the southern Appalachian Mountains, primarily in North Carolina and Tennessee but with a slightly disjunct occurrence in Shenandoah National Park, Virginia.

KEY FEATURES: Thin inconspicuous gray-to-white crustose thallus, reddish-brown apothecia that mostly appear biatorine but have a pseudolecanorine margin in early development, chemistry, on large old wood or bark of *Prunus* or *Quercus rubra* in dry ridgetop forests at middle-to-high elevations.

Lecanora saxigena

Diamonds Are Forever

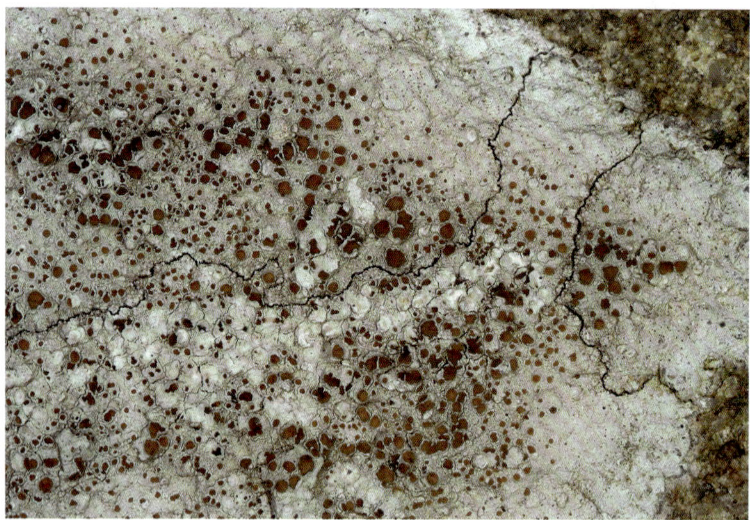

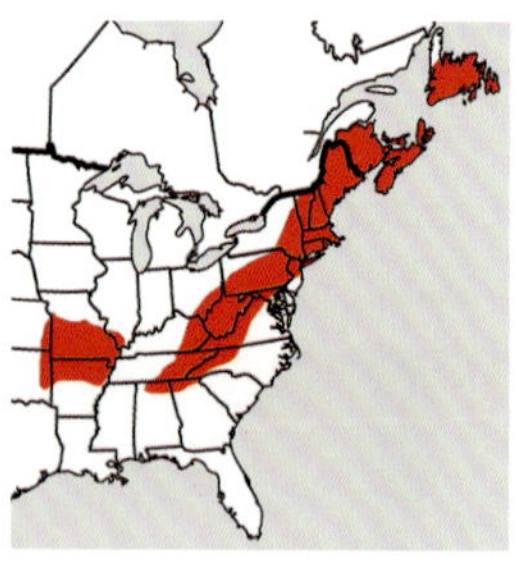

Lendemer 33186 (photo: Tripp)

NOTES: *Lecanora saxigena* has a long story involving a taxonomic name change, which—believe it or not, frustrates taxonomists as much as it does others. It is very similar in external appearance to *L. cinereofusca* in having reddish-brown apothecia with beady, white margins but differs in its chemistry and niche. When first given a name, the lichen here attributed to *L. saxigena* was recognized only as a variety of *L. cinereofusca*, which was unfortunate because it was named "appalachensis" even though it occurred commonly in other areas. It is only because we recently described a totally different, unrelated species (*L. appalachensis*) that we were forced to rename the taxon under discussion as *L. saxigena*. Diamonds might be forever, but lichen names can change. Something to not only get comfortable with if you stick around, but also learn to love.
CHEMISTRY: Atranorin, pannarin, roccellic acid, virensic acid. Spot tests. K+ yellow (thallus), C-, KC-, P+ orange (epihymenium), UV-.
NICHE: This species occurs on rocks at middle-to-low elevations in the Smokies and elsewhere in the Appalachian and Ozark Mountains. It is actually one of the most common rock-dwelling *Lecanoras* in the southern Appalachians as a whole.
KEY FEATURES: Reddish-brown apothecia with bumpy margins, epihymenium P+ orange, always saxicolous, which makes this one easy.

Lecanora strobilina

An Uncloudy Day

Lendemer 33167 (photo: Tripp)

NOTES: It's hard not to love a subset of our native weeds. Well, truth be told, *Lecanora strobilina* is not so weedy in the Smokies because the Park is just too nice! It's true. Everywhere else in eastern North America, *L. strobilina* is a corticolous weed, which is precisely why you should prioritize learning it. This species is identifiable by its always pale-yellow thallus with pale yellow, sun-like apothecia. If you look closely using the hand lens that you remembered to wear this morning, you will see that it also has ecorticate apothecial margins like those of *L. imshaugii*, but the latter are far more dramatic and sexy. So, we basically ignore this feature in *L. strobilina*, which is super easy enough to learn by gestalt in any case.
CHEMISTRY: Decarboxysquamatic acid, usnic acid, zeorin. Spot tests. K-, C-, KC+ yellow, P-, UV-.
NICHE: This species is relatively common at all elevations throughout the Smokies, on bark of hardwoods and on fallen branches. It is a weed elsewhere throughout its range, such as throughout eastern North America, where it has a preference for disturbed habitats.
KEY FEATURES: Light yellow thallus always decorated with sunny, yellow apothecia that have ecorticate margins. No other *Lecanora* in the Smokies looks like this.

Lecanora subpallens

Lesser Dust My Discs

Tripp 3543 (photo: Lendemer)

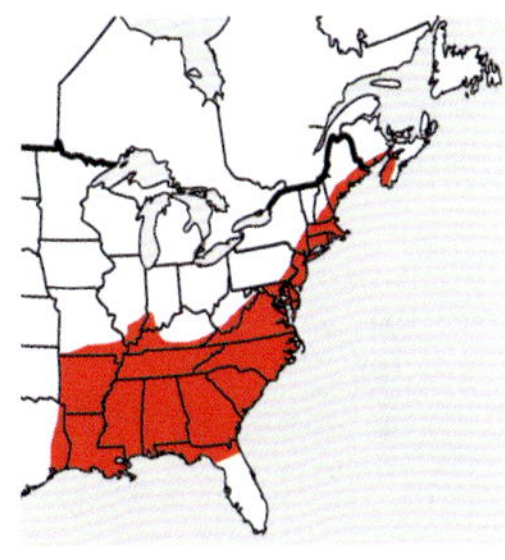

NOTES: Every now and then, we manage to take a photo that approximates the amazing forms most beautifully embodied by the lichen biota of the Smokies. Here is one we are proud of! *Lecanora subpallens* is quite the showstopper and also easy to learn. It has pale apothecial discs covered in white pruina (like *L. caesiorubella* and *L. albella*), but unlike those latter two, this species has a C+ yellow epihymenium. Do a cross section of the apothecium, put it under the compound scope, and do the C test. We recommend playing Nina Simone in the background. There are some additional differences in ecology (see Niche of each species).
CHEMISTRY: Atranorin, norstictic acid, sordidione. Spot tests. K+ red (apothecia, in section), C+ yellow (epihymenium, in section), KC-, P-, UV-.
NICHE: *Lecanora subpallens* is very common at low elevations throughout the Smokies. It is corticolous on a wide variety of hardwoods but is especially fond of Acer saccharum. Who isn't?
KEY FEATURES: Apothecia discs covered in white pruina, at low elevations on the bark

of hardwoods and particularly fond of sugar maple. (Do note that perfect sugar maple bark in the photograph . . . makes you want a waffle, right?)

Lecanora symmicta

Lost My Margins Lichen

Tripp 3945 (photo: Lendemer)

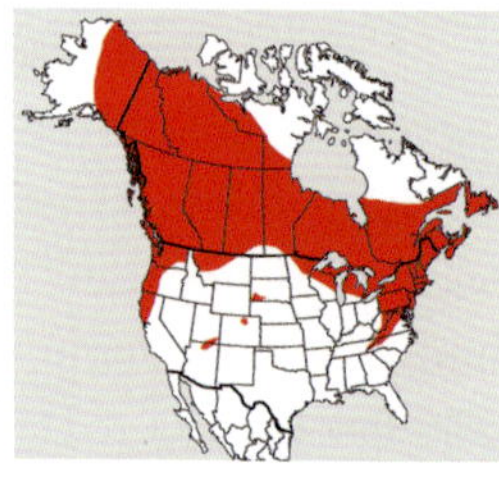

NOTES: *Lecanora symmicta* is ecologically and morphologically reminiscent of *L. strobilina*. It is, however, much less of a weed than is *L. strobilina*, which you have already read about and doubtless walked past in a parking lot in the Smokies. To identify *L. symmicta*, look for a yellow areolate thallus with convex, yellowish-brown, biatorine apothecia that do not have fuzzy, ecorticate margins. In contrast, *L. strobilina*, has a granular to leprose yellow thallus with lecanorine apothecia marked by their fuzzy ecorticate margins. Throughout North America, both are weedy. *L. symmicta* takes the high road (think: latitude and elevation), whereas *L. strobilina* takes the low road (think: boiled peanuts). We aren't entirely sure what the specific epithet ("sym" = together; "micta" = mixed) refers to, but maybe you can use these Greek roots to help tease apart *L. symmicta* from *L. strobilina*.

CHEMISTRY: Usnic acid, zeorin. Spot tests. K-, C-, KC+ yellow, P-, UV-.

NICHE: Uncommon at middle-to-high elevations in the Smokies, where it is lignicolous and corticolous on shrubs and twigs.

KEY FEATURES: Light yellow thallus with yellowish-red apothecia that are sometimes malformed (snail food?), apothecia with biatorine margins, uncommon to rare at high elevations, mostly lignicolous.

Lecanora thysanophora

Emerald Carpet

Lendemer 33195 (photo: Tripp)

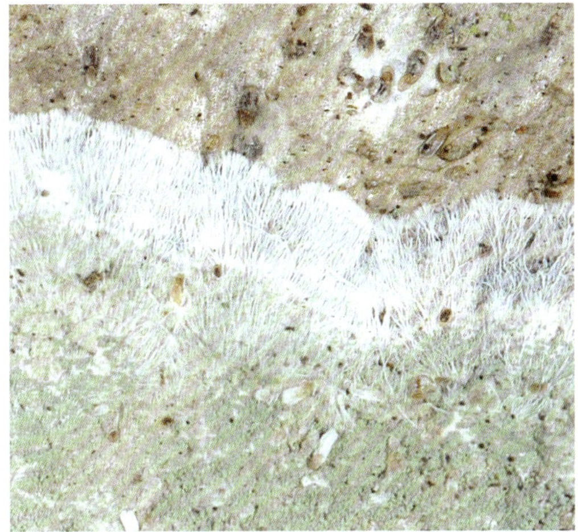

Tripp 2234 (photo: Lendemer)

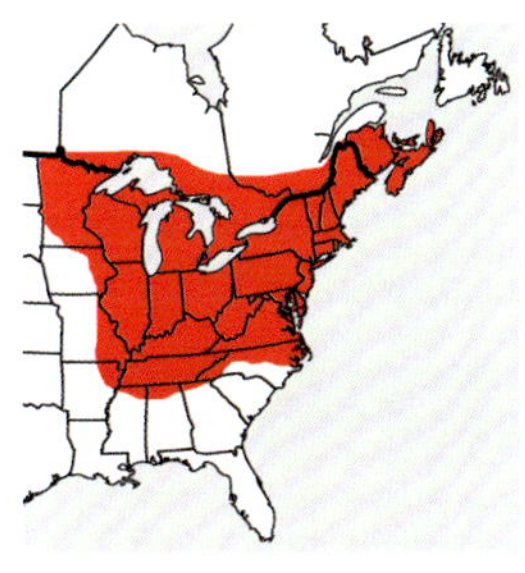

NOTES: *Lecanora thysanophora* is a *must* learn lichen of the Smokies—make it one of your first five, regardless of growth form. Seriously. Emerald Carpet is one of the most common and widespread species throughout the Park and is very easily recognized by its leprose thallus, which is composed almost entirely of densely packed granules that are seafoam green in color. Basically, the only other thing you are likely to notice about this species is its wickedly beautiful white prothallus (see both photographs above). There is a handful of other sterile, asexually reproducing *Lecanoras* in the Park such as *L. appalachensis* and *L. layana*, but *L. thysanophora* is not confusable with anything else, inside or outside of this genus

CHEMISTRY: Atranorin, usnic acid, zeorin. Spot tests. K+ yellow, C-, KC+ yellow, P-, UV-.

NICHE: *Lecanora thysanophora* occurs at all elevations throughout the Smokies but only on the bark of hardwoods. Thus, if you are up above 5,000 feet, look for it on *Betula alleghaniensis* or on *Amelanchier arborea*.

KEY FEATURES: Seafoam green leprose granules against thallus with thick and very conspicuous netted, white prothallus, one of the most common corticolous lichens on hardwoods at all elevations throughout the Smokies.

Lecidea ahlesii

Good For What Ails You Lichen

Collector 44651 (photo: Tripp)

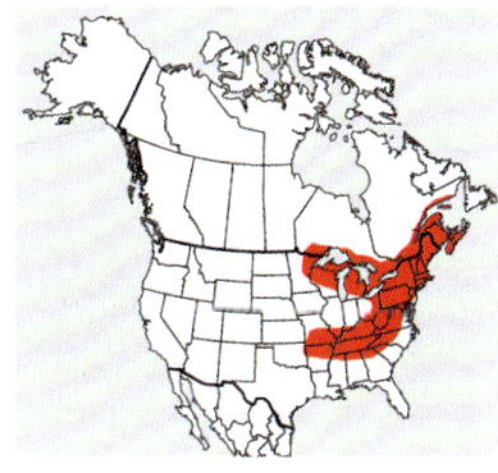

NOTES: *Lecidea ahlesii* is, as *Lecideas* go, a relatively common and easy-to-identify member of the genus. Sure, it looks like all of the other "dots on rocks" in the Park, upon first glance, but we encourage you to look a little deeper. This species is characterized by having a pale greenish-gray, continuous thallus upon which sits apothecia that are reddish-brown in color. As per most species of *Lecidea*, this one has apothecia bearing lecideine margins and simple, colorless spores. The reddish-brown color of the fruiting bodies is by far the best tip toward identifying this with confidence. *Lecidea ahlesii* could be confused with other rock-dwelling crustose lichens with lecideine apothecial margins and simple, colorless spores, such as members of

Porpidia, but none of the latter have reddish discs. This species is also potentially confusable for *Protoblastenia rupestris*, but the latter differs by its bright orange (vs. dull reddish-brown) discs. Members of the genus *Fellhanera* differ from *Lecidea ahlesii* by having primarily septate (vs. simple) spores.

CHEMISTRY: No substances. Spot tests. K-, C-, KC-, P-, UV-.

NICHE: This species occurs primarily at middle elevations among ridge forests dominated by oaks and maples. It occurs in other habitats, too, such as near waterfalls in buckeye-dominated forests. *Lecidea ahlesii* is common in similar habitats across the Appalachian-Great Lakes region. In the Smokies, look for it growing exclusively on non-calcareous rocks, perhaps on Beech Gap Trail between Straight Fork Road and Hyatt Ridge Trail. Didn't find it there? Try the Sunkota Ridge Trail. Didn't find it there? Try the herbarium at the New York Botanical Garden.

KEY FEATURES: Greenish-gray continuous thallus, reddish-brown apothecia, lecideine margins, spores > 10 µm long, 8/ascus, simple, colorless, non-calcareous rocks at middle elevations, relatively common.

Lecidea atrobrunnea

Chocolate Balds

Tripp 3735 (photo: Lendemer)

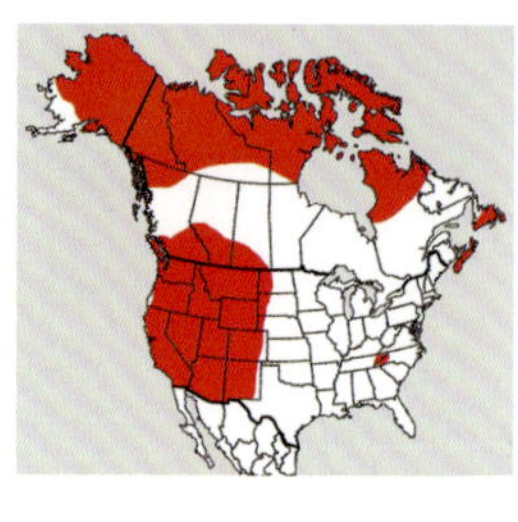

NOTES: We have a hard time introducing this species in a field guide to the lichens of Great Smoky Mountains National Park. *Lecidea atrobrunnea* is an absolute icon of the mountains of . . . not the Smokies, but . . . the Rockies. You cannot go anywhere in upland areas west of the Great Divide, or for that matter even east of and immediately proximal to the Great Divide, and avoid this species. Yet, somehow, it has a distant outpost in the southern Appalachians. Let us preface any further conversation by saying: *however, there is no genus in greater need of a doctoral dissertation than Lecidea (and Lecidella) in North America.* In the Smokies, this species is easily recognized by its chocolate brown, tile-like areoles that bear black apothecia that have lecideine margins. These areoles are sometimes pruinose, as seen in the photo. If you cut into the apothecia, you will find simple, colorless spores. This combination of features makes it unlikely to confuse *Lecidea atrobrunnea* for any other Smokies lichen. Some species of *Rhizocarpon* and *Buellia* look superficially similar, but these have brown spores that are almost always transversely septate.

CHEMISTRY: 2'-O-methylperlatolic acid. Spot tests. K-, C-, KC-, P-, UV-.

NICHE: *Lecidea atrobrunnea* is rare in eastern North America, including in the Smokies. Where found, it is usually at high elevations (or high latitudes) and occurs exclusively on non-calcareous rocks. If you are lucky, you might find this species near the Lonesome Pine Overlook in the Park.

KEY FEATURES: Chocolate brown, areolate thallus with black, lecideine apothecia, simple colorless spores, middle-to-high, dry elevations in the Park, very rare, on non-calcareous rock.

Lecidea berengeriana

Shrew Poo

Tripp 574 (photo: Tripp), including inset

NOTES: It feels strange to collect a species high in the alpine of Colorado's Rocky Mountains, and then find it again growing at middle elevations in the lush forests of the southern Appalachians. But, sometimes we just have to adjust. *Lecidea berengeriana* (think: *Beringia*) is a species that is quite at home in the tundra biomes of North America but makes it way down to southerly latitudes in both the Rocky and Appalachian Mountain chains. It is characterized by a lovely green (to greenish-gray), granulose thallus upon which jet-black apothecia sit and its substrate (see Niche). Like all *Lecidea*, the discs of the fruiting bodies have lecideine margins, and the spores are simple and colorless. If you look closely, you will also appreciate that it has capitate paraphyses. Due to its preference for overgrowing moss (or moss on rocks), *L. berengeriana* is perhaps most likely to be confused with *Bilimbia sabuletorum*, which differs in having septate, fusiform spores. *Lecidea berengeriana* might also be confused with *Lecidella elaeochroma*, but that species has a C+ yellow-orange thallus.

CHEMISTRY: No substances. Spot tests. K-, C-, KC-, P-, UV-.

NICHE: This species prefers to overgrow moss directly or grow on moss that is itself growing on rocks, sometimes weakly calcareous rocks. It is uncommon in the Smokies but becomes much more common in more northern, boreal portions of North America. In the Park, look for it in Bone Valley not far from the junction with the Hazel Creek Trail, or along Noland Divide Trail. That's right, we have collected at both low and high elevations in the Smokies.

KEY FEATURES: Crustose, green, granular thallus, jet black apothecia, lecideine margins, capitate paraphyses, colorless spores, uncommon, overgrowing mosses.

Lecidea fuscoatra

Polar Night

Lendemer 30419 (photo: Lendemer)

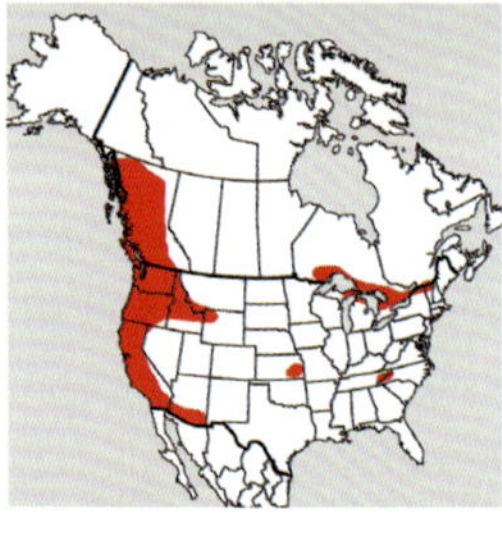

NOTES: The genus *Lecidea* is hard. We aren't going to lie. This is a species that might be best learned first by gestalt. It is relatively easy to distinguish from close relatives in the Smokies because of its thick, prominent, distinctive chocolate-brown, irregularly areolate thallus that bears jet black apothecia that are more or less immersed within, to very slight raised above, the thallus. The apothecia are flat to slightly convex and, if you look inside, you will find an olive greenish-brown epihymenium, an I+ blue hymenium, and simple colorless spores. Based on its outward morphology, you are likely to confuse *L. fuscoatra* only with *L. atrobrunnea*, which is a highly variable taxon that differs in lacking gyrophoric acid.

CHEMISTRY: Gyrophoric acid. Spot tests: (medulla) K-, C+ pink, KC+ pink, P-, UV-.

NICHE: Polar Night is a species of non-calcareous rocks that is most common in western North America. Our populations in the Smokies, if in fact truly conspecific with the western material, would be clearly disjunct from its western range. In the Smokies, this species is rather rare. Look for it at high elevation on Anakeesta Rock.

KEY FEATURES: Thick, areolate, chocolate-brown thallus, C+ pink, black apothecia +/- immersed, olive-green epihymenium, I+ hymenium, colorless simple, spores, rare on high elevation rock.

Lecidea plebeja

The Commoner Crust

Lendemer 26961 (photo: Lendemer)

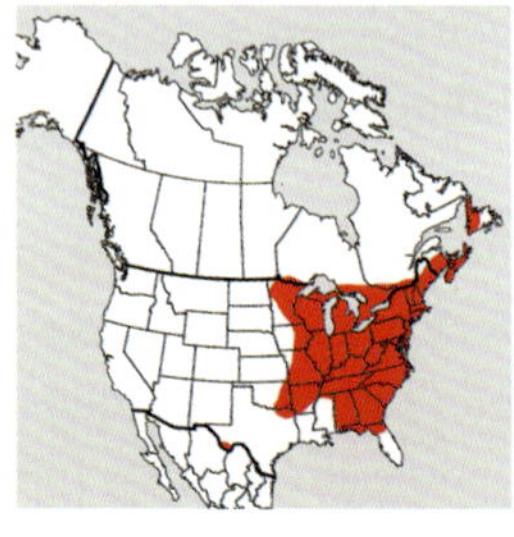

NOTES: A small, nondescript crust might be the best way to think of this species. *Lecidea plebeja* does not have much of a thallus to speak of—usually just small, scattered and poorly developed brownish areoles. What makes it distinctive are the dark brown-black lecideine apothecia that lack blue-green pigments in the epihymenium, have a brown hypothecium, and very small, simple, ellipsoid colorless spores. This is essentially a version of *L. cyrtidia* that grows on old conifer wood instead of non-calcareous rocks and pebbles. The most similar species is *Leimonis erratica*, which can readily be distinguished by its well-developed, gray thallus and blue-green epihymenium. Although *L. erratica* typically grows on non-calcareous rocks, it can occasionally be found on old fences or logs where *L. cyrtidia* might also occur.

CHEMISTRY: No substances. Spot tests: K-, C-, KC-, P-, UV-.

NICHE: *Lecicea plebeja* grows on the old wood of conifers at middle-to-low elevations throughout temperate eastern North America. It is rare in the Smokies, as is the case throughout much of its range.

KEY FEATURES: Poorly developed crustose thallus, dark brown-black lecidiene apothecia, small colorless simple spores, brown hypothecium, on old conifer wood at middle-to-low elevations.

Lecidea turgidula

Spruce Stardust

Lendemer 43338 (photo: Lendemer)

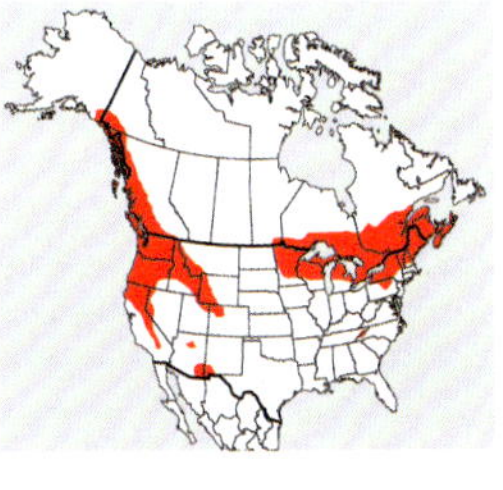

NOTES: We are rather fond of this species, probably because it is one of the many bark-dwelling lichens that was placed in the genus *Lecidea* long ago and doesn't really belong there. *Lecidea turgidula*, besides being fun to say out loud, is characterized by looking like little else beyond "little black dots" dusted with pruina. The thallus is immersed in the substrate and almost entirely invisible, which probably helps the blue-black apothecia that are lightly white pruinose stand out against the backdrop of wood. If there is something that this species would be confused with, it is probably that most people would leave it in the forest thinking it was a non-lichenized fungus. In addition, the overall appearance, the blue-green epihymenium and small, simple, colorless spores are distinctive. The most similar species is *L. leprarioides*, which is not known to occur in the Smokies, and differs in having a sorediate thallus.

CHEMISTRY: Placodiolic acid. Spot tests: K-, C-, KC-, P-, UV-.

NICHE: This is a boreal species that makes its way to southerly latitudes in both eastern and western North America via the tops of mountains. It grows on lignum or old wood, of conifers. In the Smokies, look for it in high elevation forests, almost always on wood of standing dead conifers such as spruce and fir.

KEY FEATURES: Indistinct crustose thallus, little blue-black apothecia with white pruina on lignum or old wood, dark blue-green epihymenium, small simple colorless spores, high elevations.

Lecidella carpathica

Lichen Moguls

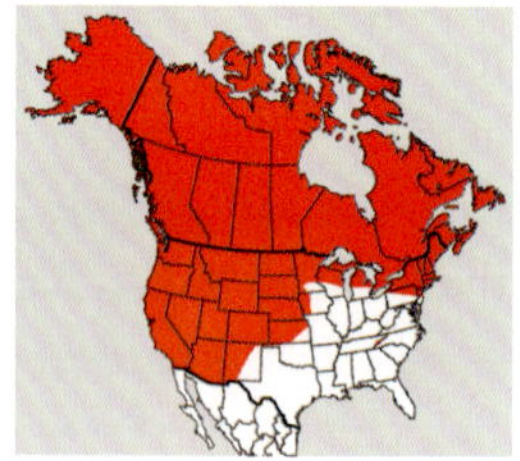

Lendemer 44852 (photo: Lendemer)

NOTES: If you are one of those overly enthusiastic natural historians who has already mastered the local *Solidago* diversity and is looking for the next path to self-destruction, *Lecidella* and *Lecidea* would be good places to start. However, for the real migraine, you must go West. Here in the East and in the Smokies, specifically, we have comparatively few species of these genera. As with nearly everything else in this guide, we suggest you learn the species themselves, rather than the genus. *Lecidella carpathica* can be readily identified by its saxicolous habit, areolate to bullate white thallus, black apothecia, blue-gray exciple, yellow-brown hypothecium, and chemistry. Species of *Lecidella* (as well as *Lecidea*) *always* have colorless spores.

CHEMISTRY: Does atranorin make you feel better? It has it. What about a xanthone? Spot tests. K+ yellow (cortex), C-, KC-, P+ yellow (cortex), UV-.

NICHE: *Lecidella carpathica* is rare in the Smokies, where it is restricted to non-calcareous rocks and occurs sporadically at all elevations. It occurs from the Smokies to the Great Lakes, across Canada, and throughout the Great American West.

KEY FEATURES: Bullate white thallus (that reminds one of moguls), deep black apothecia that bear colorless, simple spores, blue-gray exciple, K+ yellow thallus, non-calcareous rocks.

Lecidea cyrtidia

Pebble Pocks

Tripp 3926 (photo: Lendemer)

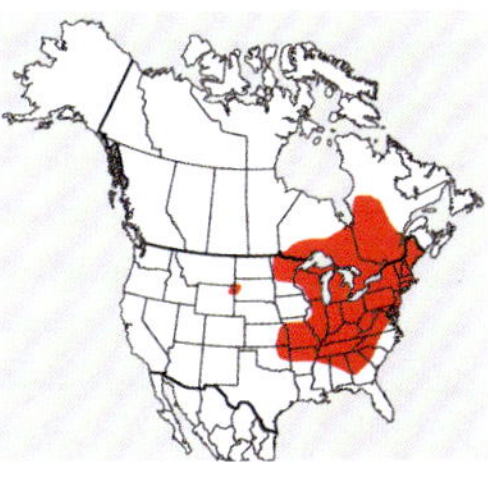

NOTES: *Lecidea cyrtidia* is characterized by, well, not much. It has a scuzzy brown, poorly developed thallus that is either continuous or irregularly areolate. Inside its black apothecia with lecideine margins, it has tiny, colorless spores. This species is most likely to be confused with *Leimonis erratica*, which differs in having a blue-green epihymenium as well as a shiny and well-developed gray thallus.

CHEMISTRY: No substances. Spot tests. K-, C-, KC-, P-, UV-.

NICHE: *Lecidea cyrtidia* is relatively common in the Smokies, where it occurs primarily at middle elevations in mesic forests dominated by Sugar Maple. It grows exclusively on non-calcareous rocks. Take a short hike to Spruce Mountain near Mt. Chiltoes to catch a glance.

KEY FEATURES: Poorly developed scuzzy brown thallus, black apothecia, tiny colorless spores, on non-calcareous rocks at middle elevations.

Lecidea nylanderi

William's Smoky Lichen

Tripp 3502 (photo: Lendemer)
Inset: Tripp 3610 (photo: Lendemer)

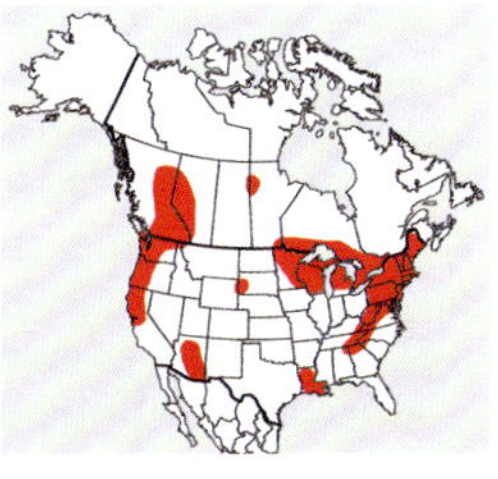

NOTES: How do you feel about polyphyly? Does it make you blue? *Lecidea nylanderi* is sorediate and grows on trees. Molecular studies have shown that it is that it is not closely related to all the rock dwelling species of *Lecidea* that lack soredia. This is the only species with this ecology that has a dark black prothallus with dispersed, sorediate blue-gray areoles. Every now and then, *L. nylanderi* makes sexual fruiting bodies (inset photograph), which have lecideine margins, are pale reddish-brown in color, and bare small, simple, colorless spores. You might want to save learning this one for a day when you have a lot of time to kill. It is perhaps a species that even William Nylander, famous Finnish lichenologist of the mid 1800s who invented the "spot test," might have accidentally overlooked.

CHEMISTRY: Divaricatic acid. Spot tests. K-, C-, KC-, UV+ blue-white.

NICHE: This species is common on bark of spruces in spruce-fir forests at high elevations in the Smokies, and a good place to look for it is on your next hike to Peck's Corner (plan on a *long* day trip). *Lecidea nylanderi* occurs on this same substrate throughout the high elevations of the southern Appalachians, where it is disjunct from the northern boreal forests.

KEY FEATURES: Tiny crustose lichen with black prothallus and bluish areoles covered in soredia, generally always makes small thalli, on bark of spruces, high elevations.

Lecidea roseotincta

Cookies n' Cream

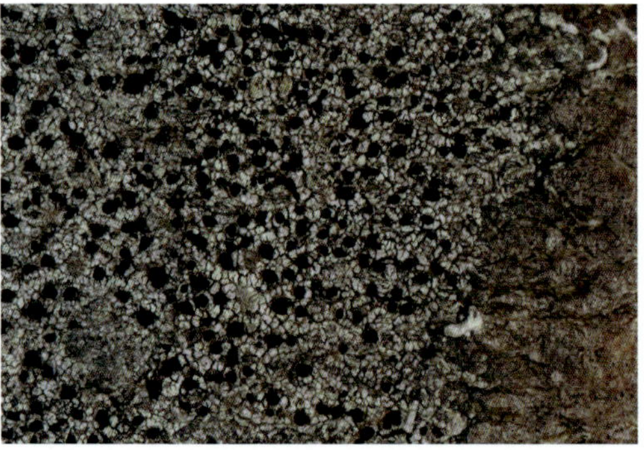

Collector Tripp 2469 (photo: Deregibus)

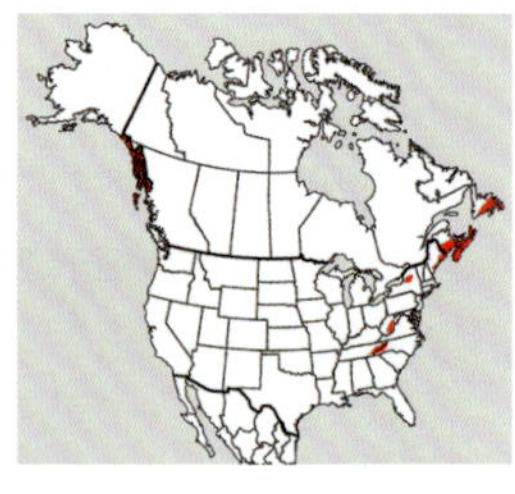

NOTES: This species is one of the many reasons why you should always check spores and chemistry if you want to be absolutely certain of an identification. In the field you might think you had found a form of *Buellia stillingiana* or *Hafellia subnexa* with small apothecia, but then you would be wrong. *Lecidea roseotincta* resembles those species, but differs in its niche, its chemistry (the others produce norstictic acid) and in having simple, colorless spores. Importantly, almost all of the populations of *L. roseotincta* from the Smokies are sterile, and so finding spores is often impossible. Thankfully that is not the case for *Buellia* and *Hafellia*, which almost always are fertile.

CHEMISTRY: Psoromic acid. Spot tests: K-, KC-, C-, P+ yellow, UV-.

NICHE: In the Smokies, this species is common in high elevation northern hardwood forests and shrub balds, where it occurs on the bark and branches of hardwood trees and shrubs. It was originally described from Scandinavia and later reported from oceanic areas of northeastern North America.

KEY FEATURES: White crustose thallus, small black apothecia, colorless simple spores when present, P+ yellow, on bark and branches of hardwoods at high elevations.

Lecidea tessellata

I Should Tell You That You Were My First Love

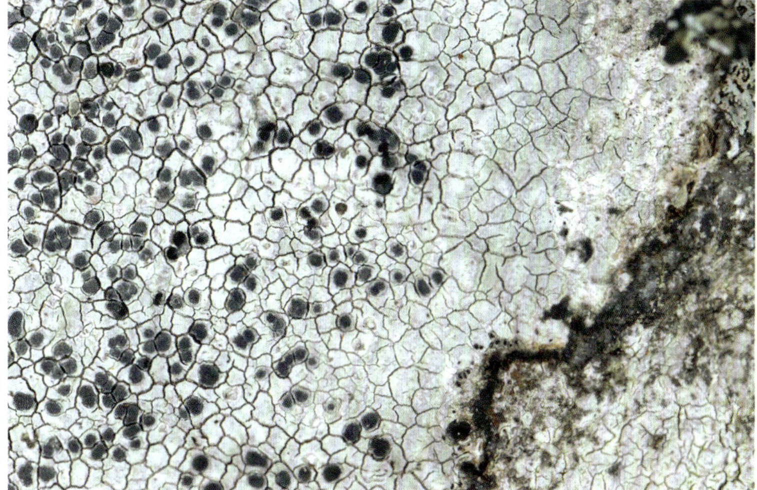

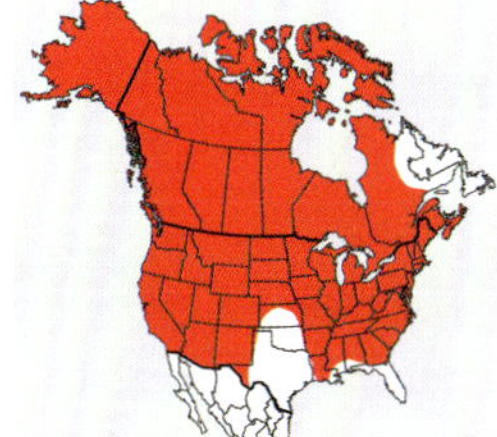

Tripp 3940 (photo: Lendemer)

NOTES: The only lichen that you are likely to confuse *Lecidea tessellata* with is *Lecanora oreinoides*. The two species have similar niches, often grow together, and have thick white thalli with immersed black apothecia. They also both have simple, colorless spores. But just when you thought it was looking bleak: *L. tessellata* can be distinguished from the very unrelated *L. oreinoides* by its white pruinose apothecia, its K- thallus, and its overall dull appearance compared to the super sheen of thalli of *L. oreinoides*.

CHEMISTRY: Confluentic acid. Spot tests. K-, C-, KC-, P-, UV+ blue-white (medulla).

NICHE: *Lecidea tessellata* occurs on sun-exposed non-calcareous rocks at middle and high elevations in the Smokies. It occurs in similar habitats elsewhere in the Appalachians, especially on talus slopes and other habitats frequented by rattlesnakes. *Lecidea tessellata* is "no big deal" east of the Mississippi River but if you head west, especially in the Rocky Mountains, be prepared to cope with an abundance of this species, which comes in several different chemical varieties. Lucky you!

KEY FEATURES: Continuous-to-slightly-cracked thick, white thalli, apothecia covered by white pruina, simple colorless spores, K- thallus, dull upper cortex, sunny talus slopes.

Lecidella elaeochroma

O-live Forever Lichen

Tripp 3587 (photo: Lendemer)

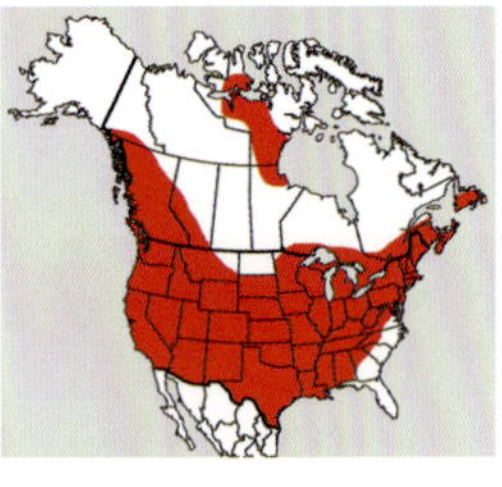

NOTES: *Lecidella elaeochroma* is a relatively inconspicuous crustose lichen characterized by its granular greenish-gray thallus, black apothecia with lecideine margins, 8 simple, colorless spores per ascus, and its green-to-black epihymenium that lacks crystals. A C+ spot test will potentially further help to identify this species. The specific epithet, which means "olive-colored," was aptly applied to this species.

CHEMISTRY: A lot of compounds, all of which are unpronounceable. Spot tests. K- yellow, C+ orange, KC+ yellow, P+ yellow, UV-.

NICHE: Look for *Lecidella elaeochroma* at middle-to-high elevations on the bark of northern hardwoods throughout the Smokies. Happen to be scoping out the old growth near Gabes Mountain Trail? Yep–it's there, too.

KEY FEATURES: Inconspicuous crustose lichen with thin, greenish gray thallus, black apothecia with lecideine margins, C+ orange thallus, corticolous on hardwoods throughout Smokies.

Leimonis erratica

Silver Sliver

Lendemer 33302 (photo: Lendemer)

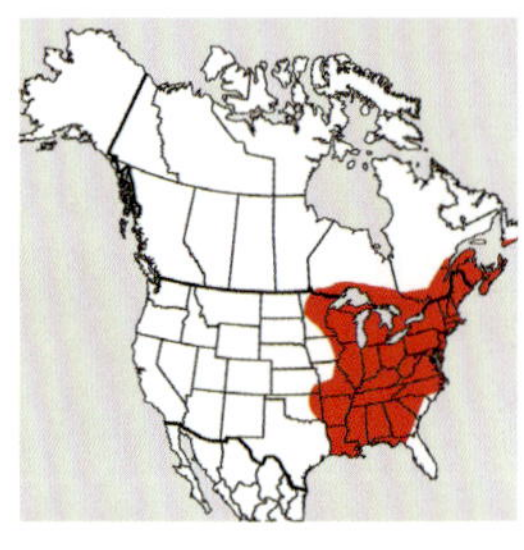

NOTES: *Leimonis erratica* is a distinctive little crust that can be easily recognized by its gray areolate thallus, black lecideine apothecia, and occurrence on non-calcareous rocks. The blue-green epihymenium, brown hypothecium, and small, simple, colorless, spores further confirm the identification. The most similar species is *Lecidea cyrtidia*, which has similarly sized spores but differs in having a poorly developed, brown thallus and dark brownish-black apothecia that lack blue-green pigments in the epihymenium. In the field, *L. erratica* might be confused with *Buellia badia*, which occurs in similar habitats but has brown, 2-celled spores. *Leimonis erratica* used to be treated in *Micarea*, and there are a lot of *Micarea* species in the Smokies. But none of them share the well-developed gray thallus, ecology, and characters of the apothecia and spores.

CHEMISTRY: No substances. Spot tests. K-, KC-, C-, P-, UV-.

NICHE: This species is common and widespread throughout temperate eastern North America, where it is found on sun-exposed, non-calcareous rocks, especially in early successional habitats. It is rare in the Smokies, but nonetheless occurs on both high elevation rock outcrops and pebbles at low elevations along the Foothills Parkway.

KEY FEATURES: Gray areolate thallus, black lecideine apothecia, blue-green epihymenium, brown hypothecium, small simple colorless spores, on non-calcareous rocks at low elevations.

Lepra amara

Bitter Bits

Tripp 2491 (photo: Deregibus)

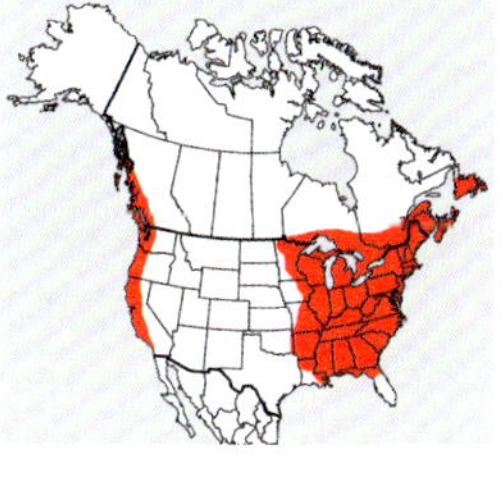

NOTES: Whenever someone asks us what this species is in the field, we give them a little piece and tell them to taste it. Even just a gentle lick of the white pruina on the discs often conveys an intense bitterness that conjures the taste of aspirin and is not easily forgotten. *Lepra amara* can be confused with several other members of the genus such as *L. multipunctoides* and *L. ophthalmiza*, but those differ in chemistry. It is also quite distinctive morphologically because the thallus is typically shiny and dark green-blue in color and is usually speckled with tiny white pseudocyphellae near the margins.

CHEMISTRY: Picrolichenic acid. Spot tests. Cortex: K-, C-, KC-, P-, UV-; medulla: K-, C-, KC+ fleeting red/purple, P-, UV-.

NICHE: This species is common and widespread in the Smokies, where it occurs on the bark and branches of hardwoods, and more rarely conifers, at all elevations. Outside of the Smokies, it is similarly widespread throughout temperate and boreal eastern North America, but in many areas, it appears to be uncommon and restricted to high-quality forests and mature trees. Thankfully, both of these are present in abundance in the Smokies.

KEY FEATURES: Greenish-blue crustose thallus with tiny white pseudocyphellae, densely pruinose apothecia that are frequently sterile and resemble soralia, K-, P-, KC+ fleeting red/purple, distinctly bitter taste, on bark at all elevations.

Lepra excludens

Rust in Winter

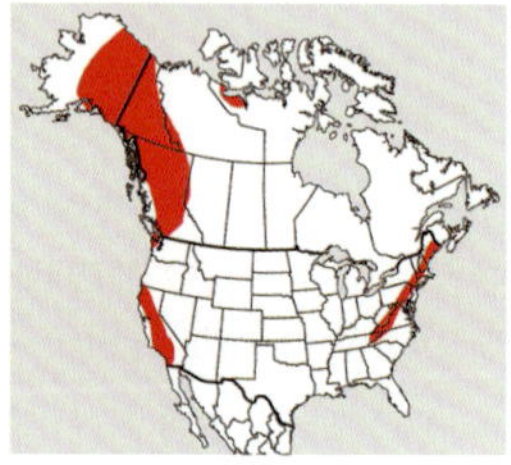

Lendemer 53145 (photo: Tripp)

NOTES: Due to the white thallus and the production of norstictic acid, there are relatively few species with which *Lepra excludens* is likely to be confused in the Smokies. Although it is chemically similar to *L. waghornei*, which also occurs at high elevations on bark, that species has a blue-gray thallus, the pruina on the apothecia tends to be dark gray in color, and the apothecia usually have well-developed hymenia with spores. In contrast, the apothecia of *L. excludens* are sterile, lack hymenia, and instead are much more likely to be mistaken for soralia. *Phlyctis speirea* is another corticolous species with a white thallus that produces norstictic acid, but it has apothecia with densely muriform spores. Although *P. petraea* is morphologically somewhat similar to *L. excludens*, and also produces norstictic acid, it is restricted to sheltered rocks and does not occur on bark.

CHEMISTRY: Norstictic acid. Spot tests: K+ yellow turning red, C-, KC-, P+ yellow, UV-.

NICHE: This species is infrequent at high elevations in the Smokies, where it grows on the bark and bases of trees such as Yellow Birch. It is a boreal-northern disjunct that is much more at home in Canada than in the southern Appalachians.

KEY FEATURES: Thick, white, continuous, crustose thallus; densely pruinose apothecia that resemble soralia, K+ yellow turning red, on bark at high elevations.

Lepra multipunctoides

Summer Snowballs

Tripp 2666 (photo: Lendemer)

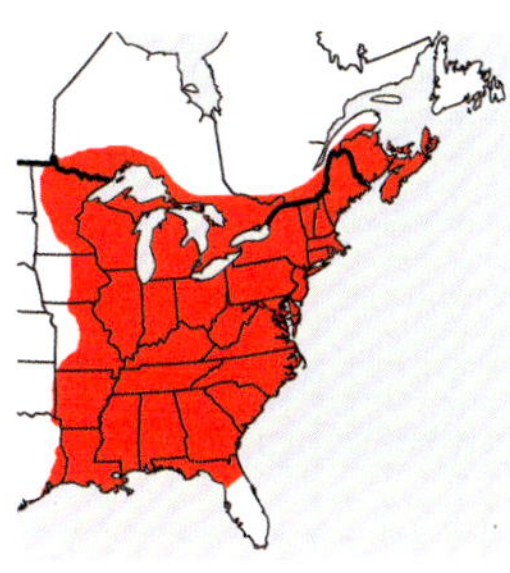

NOTES: If there were ever a lichen genus that conjures up images of disease it would be *Lepra*, which is unfortunate given how beautiful and diverse it is in the Smokies. *Lepra multipunctoides* is the most common species in our area and can readily be recognized by its continuous dark gray-to-blue-gray thallus that lacks pseudocyphellae, its chemistry, and the apothecia that often become so densely pruinose that the pruina overflows from the margins of the apothecia causing them to resemble soralia. *Lepra ophthalmiza* is similar but has a thallus that is lighter gray in color, and a medulla that is P-.

CHEMISTRY: Succinoprotocetraric acid. Spot tests. Cortex: K-, C-, KC-, P-, UV-; medulla: K+ dirty yellow-brown, C-, KC-, P+ red, UV-.

NICHE: *Lepra multipunctoides* is common and widespread in the Smokies, where it can be found on the bark of hardwoods and conifers at all elevations. Occasionally, it will also occur on large branches. Like other members of the genus, it tends to occur in high-quality forests and on mature trees outside of the Smokies. Interestingly, it is endemic to temperate and slightly subtropical eastern North America.

KEY FEATURES: Gray-to-blue-gray crustose thallus without tiny white pseudocyphellae, densely pruinose apothecia that are frequently sterile and resemble soralia, K-, P+ red, KC-, on bark at all elevations.

Lepra ophthalmiza

Tasteless Tops

Lendemer 32903 (photo: Tripp)

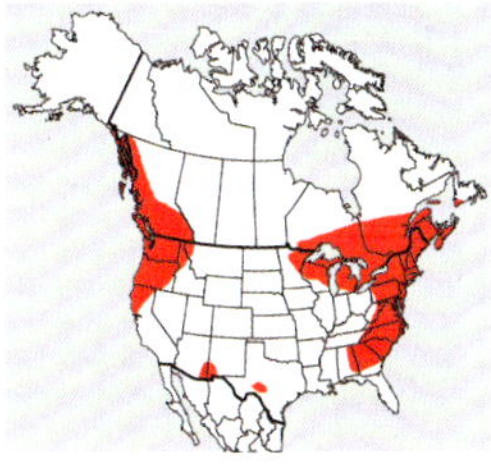

NOTES: In many respects, this species is likely to be confused with *Lepra amara* in cases where the bitter taste and KC+ reaction of *L. amara* are overlooked. Hence, the common name refers to the lack of bitterness in the medulla and pruina of the apothecia. *Lepra amara* also differs morphologically in having a darker, greenish, shiny thallus and tiny white pseudocyphellae. *Lepra multipunctoides* is also similar, and has mostly negative spot test reactions, but nonetheless is P+ red. *Lepra trachythallina* and *Lepra waghornei*, likewise, are also somewhat similar in overall appearance, but both have positive reactions with K, whereas *Lepra ophthalmiza* does not.

CHEMISTRY: Fatty acids. Spot tests. K-, C-, KC-, P-, UV-.

NICHE: *Lepra ophthalmiza* is infrequent but widespread on the bark and branches of trees throughout the Smokies. Although it often occurs on hardwoods, it is not atypical to find the species on conifers. Outside of the Smokies the species is similarly uncommon but widely distributed throughout temperate and boreal North America.

KEY FEATURES: Gray-to-blue-gray crustose thallus without tiny white pseudocyphellae, densely pruinose apothecia that are frequently sterile and resemble soralia, K-, P-, KC-, on bark at all elevations.

Lepra pustulata

Broken Dreams

Lendemer 29624 (photo: Tripp)

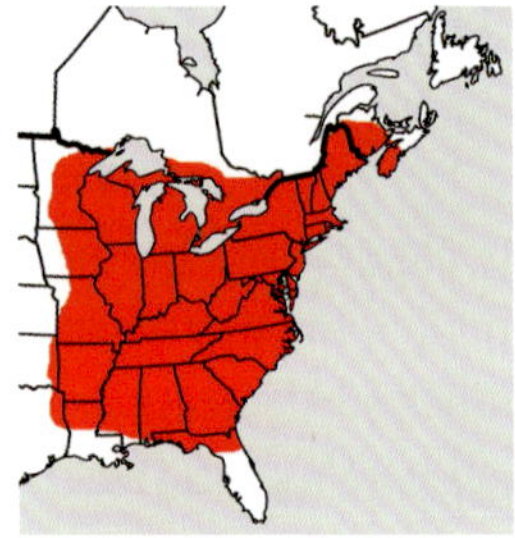

NOTES: You can be forgiven for not thinking this species belongs to *Lepra* because, unlike all the other members of the genus, it only rarely produces apothecia and their characteristic densely white pruinose discs. Instead, thalli of *L. pustulata* are covered with tiny pustules that break apart into fragments that disperse to form new individuals. The species is most likely to be confused with *Loxospora elatina*, which occurs at high elevations and is chemically similar in producing thamnolic acid. *Loxospora elatina*, however, has a dull rather than shiny thallus, and is sorediate instead of distinctly pustulose.

CHEMISTRY: Thamnolic acid. Spot tests. Cortex and medulla: K+ yellow, C-, KC-, P+ orange, UV-.

NICHE: This is one of the most common and widespread crustose lichens in both the Smokies and elsewhere in temperate eastern North America. It typically grows on the bark of hardwood trees but can also be found on conifers, branches of trees of all types, shrubs, and—rarely—even on shaded non-calcareous rocks. In the Smokies, it occurs in nearly every habitat, at all elevations, although it becomes very rare in the highest spruce-fir forests above 6,000 ft. elevation, where it is replaced by *Loxospora elatina*.

KEY FEATURES: Gray-to-blue-gray crustose thallus without tiny white pseudocyphellae, abundant pustules, K+ yellow, P+ orange, on bark at all elevations.

Lepra trachythallina

Sugar Pots

Tripp 2313 (photo: Deregibus)

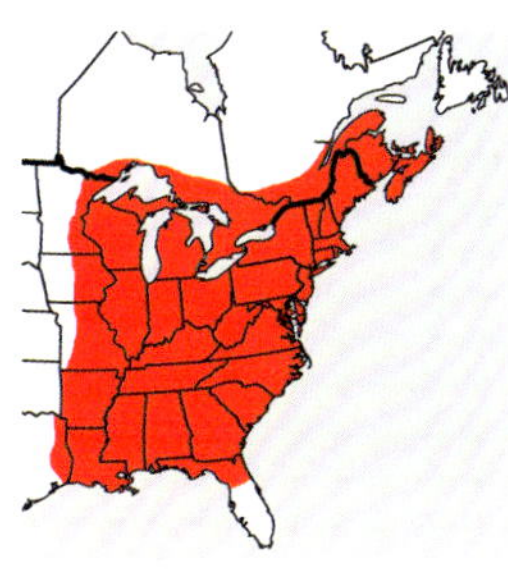

NOTES: This is one of the common crustose lichens of high elevations in the Smokies and can be recognized by its large, densely pruinose, discoid apothecia and by the intense K+ yellow reaction of the medulla and pruina. *Lepra waghornei* is superficially similar, but differs chemically in the production of norstictic acid, which reacts K+ yellow, turning blood red. Confusion between the two is possible, however, because the K+ yellow reaction of *L. trachythallina* eventually turns a dark brown color that can look somewhat reddish. That is why it helps to try the reaction in a squash mount under the microscope, where you can see that red crystals do not form after the application of K in *L. trachythallina*.

CHEMISTRY: Thamnolic acid with or without lichexanthone. Spot tests. Cortex: K-, C-, KC-, P-, UV- or UV+ bright yellow; medulla: K+ yellow turning brown, C-, KC-, P+ orange, UV-.

NICHE: *Lepra trachythallina* occurs throughout the Smokies, but only becomes common, even abundant, in the high elevations. Once you climb above 4,500 ft. elevation, it is likely to be found on the branches of nearly every tree and shrub. Outside of the Smokies, it is widely distributed in temperate and boreal eastern North America. The chemical variant without lichexanthone occurs in the northern portions of the range, while the variant with lichexanthone occurs in the southern portions of the range. It is only in the southern Appalachians where both variants can be found in a very small geographic area.

KEY FEATURES: Gray-to-blue-gray crustose thallus without tiny white pseudocyphellae, densely pruinose apothecia that are frequently sterile and resemble soralia, K+ yellow, P+ orange, KC-, on bark at all elevations.

Lepra waghornei

Waghorne's Watchin'

Lendemer 32977 (photo: Tripp)

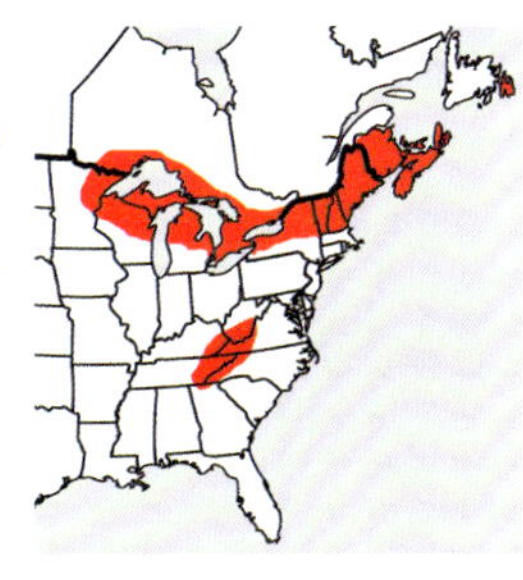

NOTES: *Lepra waghornei* gets its name from Arthur Waghorne, a minister who traveled from town to town in Newfoundland and Labrador, collecting lichens and bryophytes along the way. Unlike *L. trachythallina* and *L. multipunctoides*, the apothecial pruina tend to be only minimally developed in *L. waghornei*, and thus it is often possible to see the surface of the discs beneath their dusting. *Lepra waghornei* is easily differentiated from all the other members of the genus in the Smokies by the presence of norstictic acid, a substance that produces red crystals under the compound microscope if you apply K to a tiny piece mounted in water.

CHEMISTRY: Norstictic acid. Spot tests. Cortex: K-, C-, KC-, P-, UV-; medulla: K+ yellow turning red, C-, KC-, P+ yellow, UV-.

NICHE: This species is endemic to eastern North America and was originally described from Newfoundland, Canada. In the Smokies, like elsewhere in the southern Appalachians, it

occurs at middle-to-high elevations, primarily on the bark of hardwood trees and shrubs.
KEY FEATURES: Gray to blue-gray crustose thallus without tiny white pseudocyphellae, densely pruinose apothecia that are frequently sterile and resemble soralia, K+ yellow turning red, P+ yellow, KC-, on bark at middle-to-high elevations.

Lepraria arbuscula

Tree Embellishment

Tripp 5091 (photo: Lendemer)

NOTES: This is yet another secret of the Smokies. From a distance, it might look like all the other dust lichens, especially *Lepraria lanata*, with which it often commingles. However, *Lepraria arbuscula* can easily be distinguished by its short, cartilaginous strands that form stalks to which the dust-like granules adhere. Think of little, miniature trees, covered in dust or snow. This species was formerly classified in the genus *Leprocaulon*, which used to be the home for dust lichens that had fruticose branches. However, molecular studies have shown that this species is related to *Lepraria*, which makes sense if you look at the other photographs in this book. If you think the above picture resembles a diminutive *Stereocaulon*, you would be absolutely correct. The latter is the sister group to *L. arbuscula* and the rest of the dust lichens (i.e., *Lepraria*)!
CHEMISTRY: Roccellic acid, protocetraric acid. Spot tests. K-, C-, KC-, P+ red, UV-.
NICHE: If you find this species or something that you think might be this species, please take a picture and leave it be. If living on heavy-metal rich rocks on harsh cliffs in acid-fog-damaged spruce-fir forests isn't difficult enough, try being the only population of your species in all of North America. Then again (shout out to Robert Frost), one could be worse things than a clinger of rocks in the Smokies.
KEY FEATURES: Mini-fruticose lichen with short stalks covered in leprose granules, extremely rare, please do not disturb this one if found.

Lepraria caesiella

Blue Gray Bark Warbler

Tripp 5542 (photo: Lendemer)

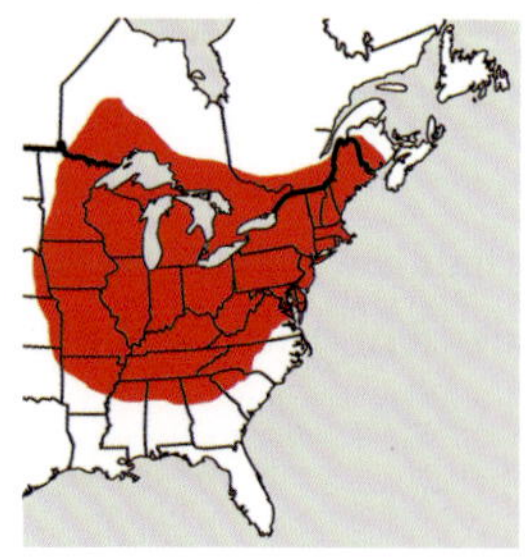

NOTES: *Lepraria caesiella* is your perfect introduction to the dust lichens, *Lepraria*—one of our favorite genera in the

Smokies. All members of this genus gave up sex long ago and now just form thalli of little dust balls that never bother with the eccentricities and burdens of apothecia and spores. *Lepraria caesiella* is one of the more common species in the Smokies and can be recognized by its thin, more-or-less dispersed thallus composed of blue-gray granules. It could be mistaken for *Lecidea nylanderi*, but that species has blue-gray areoles that erupt into soralia. The specific epithet of this species refers to its characteristic bluish-gray thallus. Don't you love it when scientific names make sense? We thank our colleague Dick Harris for that smart move.
CHEMISTRY: Atranorin, zeorin. Spot tests. K+ yellow, C-, KC-, P-, UV-.
NICHE: *Lepraria caesiella* is widely distributed throughout the Smokies on the bark of trees as well as on rocks in sheltered protected overhangs. It is endemic to eastern North America and is particularly common in the Appalachians, where it prefers the bark of hemlocks.
KEY FEATURES: Thin, characteristic blue-gray thallus, relatively common on bark and rock throughout the Smokies.

Lepraria cryophila

Dust Billows

Tripp 5024 (photo: Lendemer)

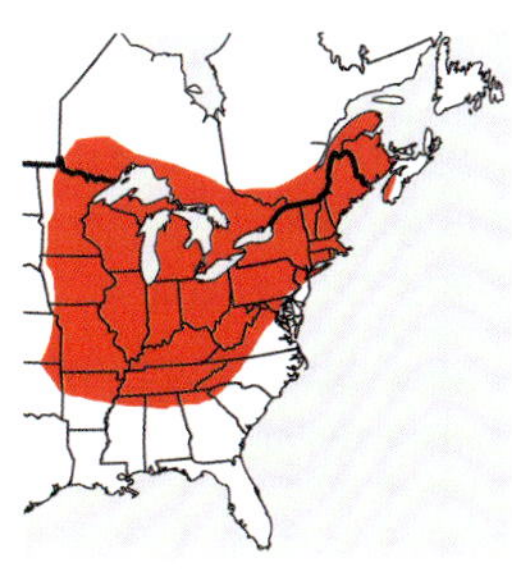

NOTES: You wouldn't be blamed if your initial impressions are that every dust lichen looks the same, but with a little work, we think you can learn to differentiate among them. *Lepraria cryophila* is one of those species you really have to hunt to find, mainly because of its inconspicuous niche (see below). But, it is the only member of the genus that forms shelf-like wefts under rock overhangs and has a UV+ blue-white thallus. Because of these features, it is only likely to be confused with *L. finkii*, but that species has P+ orange thallus.
CHEMISTRY: Divaricatic acid, nordivaricatic acid. Spot tests. K-, C+ pink, KC+ pink, P-, UV+ blue-white.
NICHE: *Lepraria cryophila* occurs on non-calcareous rocks in sheltered overhangs throughout the Smokies at all elevation as well as throughout the Appalachians and Ozarks. It is endemic to eastern North America.
KEY FEATURES: Rather poofy, thick, green thallus of leprose granules, under rock overhangs, UV+ blue-white.

Lepraria disjuncta

Fool's Gold Dust

Lendemer 53154 (photo: Tripp)

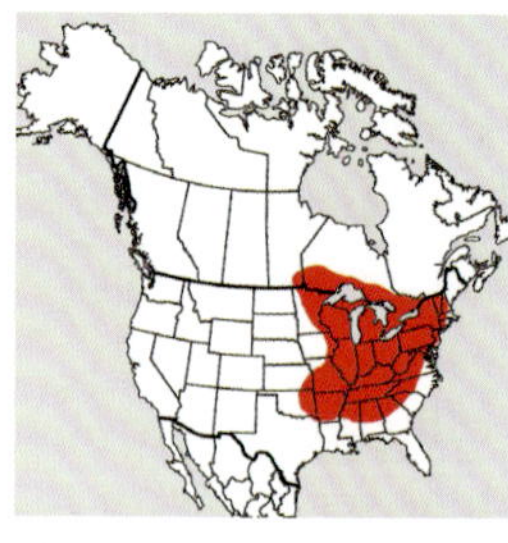

NOTES: The thick, greenish-yellow, leprose thallus of *Lepraria disjuncta*, combined with its distinctive substrate requirement, makes the species fairly easy to recognize, even in the field. It is most likely to be confused with either *L. finkii* or *L. vouauxii*, both of which have greenish leprose thalli, but differ markedly in producing substances that are P+ orange-red. *Lepraria xanthonica* and *Leprocaulon nicholsiae* are also similar in having greenish-yellow thalli due to the presence of usnic acid. However, both of those species have dispersed thalli composed of aggregated piles of granules, and they never form a thick, continuous crust. While *L. xanthonica* is chemically similar in producing a xanthone, *L. nicholsiae* differs further from *L. disjuncta* in not producing xanthones.

CHEMISTRY: Usnic acid, zeorin, two xanthones. Spot tests: K-, C-, KC+ yellow, P-, UV+ dull orange.

NICHE: *Lepraria disjuncta* is rare in the Smokes, largely because it is restricted to sheltered faces and overhangs of calcareous rock outcrops. Outside of the Park it is common and widespread throughout the Appalachian Mountains, where calcareous rocks are more ubiquitous.

KEY FEATURES: Continuous, thick, greenish-yellow, leprose thallus, P-, on sheltered calcareous or mineral enriched rocks.

Lepraria elobata

Glacier Spillover

Lendemer 33168 (photo: Tripp)

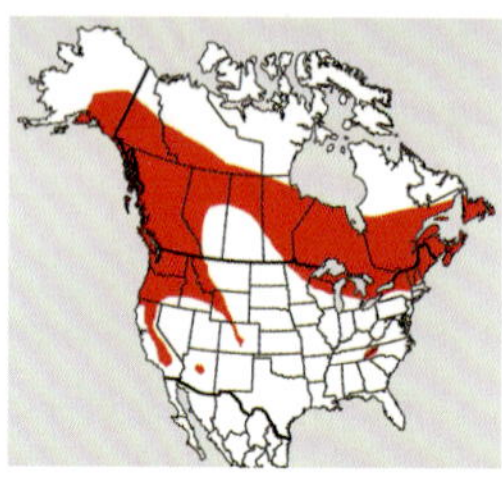

NOTES: *Lepraria elobata* is morphologically identical to *L. caesiella*, which is much more common and differs chemically in having a P- thallus, whereas the thallus of *L. elobata* is P+ yellow or orange depending on the substances present. *Lepraria finkii* is chemically identical to one of the chemotypes of *L. elobata* but differs in having a thick thallus with a well-developed white lower layer (we call it a hypothallus but others call it a medulla). The chemical variant of *L. elobata* that produces salazinic acid was originally recognized as a distinct species, *L. salazinica*, but molecular studies have shown that the two should be treated together. Both chemical variants occur in the Smokies and both are uncommon.

CHEMISTRY: **(1)** Atranorin, zeorin and stictic acid. Spot tests. K+ yellow, C-, KC-, P+ orange, UV-. **(2)** Atranorin, roccellic/angardianic acid and salazinic acid. Spot tests: Spot tests. K+ yellow turning red, C-, KC-, P+ yellow, UV-.

NICHE: This species is very rare in the Smokies and southern Appalachians in general, occurring on the bases of conifers in spruce-fir forests. On the converse, it is much more common in the northern boreal forests, where it is widespread. Again, thank the glaciers for pushing this one south for us southerners to enjoy. KEY FEATURES: Thin granular thallus, P+ orange, on the base of conifers at high elevations, truly a northern boreal species,

Lepraria finkii

Pookie

Tripp 4983 (photo: Lendemer)

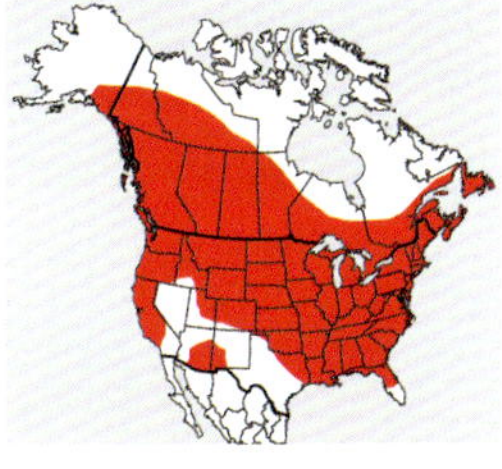

NOTES: Prepare to learn one of the top five most abundant (and arguably ecologically important) lichens in eastern North America. If you see a thick, gray-blue to green-blue dust lichen on a tree or rock, on a stone wall or a monument, there is a strong probability it is *Lepraria finkii*. The thallus pictured here is rather green in color, but that's what happens when lichens get wet (pull up a chair, wait for it to dry). *Lepraria finkii* is everywhere and is further recognizable by its thallus that is P+ orange. *Lepraria cryophila* differs in being P- while *L. vouauxii* is distinctly yellow in color, reacts P+ orange and K-, and is much rarer. CHEMISTRY: Atranorin, zeorin, stictic acid. Spot tests. K+ yellow, C-, KC-, P+ orange, UV-. NICHE: Ubiquitous is the only proper way to describe *Lepraria finkii*. You can find Pookie on bark, rock, and wood at every elevation in the Smokies, and more generally throughout eastern North America. It likes the West, too (don't we all?), but why isn't it in Nevada? KEY FEATURES: Thick granular leprose thallus that varies from blue to green (depending on hydration), P+ orange, one of the most common lichens on bark, rock, lignum, and human-made substrates across eastern North America.

Lepraria lanata

Appalachian Pillows

Tripp 3944B (photo: Lendemer)

NOTES: *Lepraria lanata* is a special treasure of the southern Appalachians and something to keep eyes wide open for. We actually contemplated making a separate field guide to specifically highlight these "Secrets of the Smokies," but haven't gotten

around to it yet. At first, you might be tempted into thinking that all dust lichens were more or less just variation on, well, dust, but *L. lanata* is set apart by its thallus that forms massive, grayish-white fluffy balls. It really looks nothing like any other lichen or organism for that matter except for, well, dust. *Lepraria normandinoides* and *L. oxybapha* are similar chemically, but their thalli are composed of smaller balls that form lips at the margins. The specific epithet means "wooly," in reference to its overall poofyness. This one is Erin's favorite *Lepraria* dust bunny!

CHEMISTRY: Roccellic acid, protocetraric acid. Spot tests. K+ yellow, C-, KC-, P-, UV-.

NICHE: In the Smokies, *Lepraria lanata* is found only at high elevations on non-calcareous rocks in sheltered overhangs. It is particularly common on Anakeesta rock outcrops and is endemic to the southern Appalachians, with only a few populations known outside of the Smokies. Did you really need any further evidence that the Smokies are *amazing*?

KEY FEATURES: Spectacular dust lichen characterized by its very thick, poofy thalli, nearly always on Anakeesta rock in dank places at high elevations, southern Appalachian endemic. Neo endemic? Paleo? To be determined . . .

Lepraria neglecta

The Neglected Dust Lichen

Tripp 3698 (photo: Lendemer)

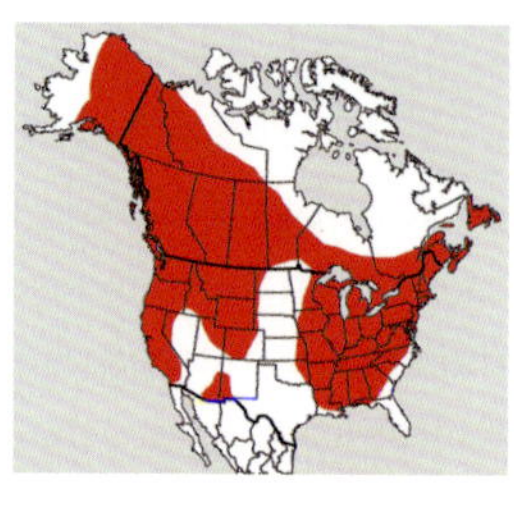

NOTES: *Lepraria neglecta* is a common species that can be easily recognized in the field by its rosette-forming thalli (somewhat apparent in the photograph) that often form concentric rings. It is further distinctive by its coarse, shiny granules. The reason for this disintctive appearance is that *L. neglecta* is the only dust lichen in the Smokies that has a superficially corticate thallus, and that thallus usually reacts KC+ red. One is most likely to confuse this species with *L. caesiella*, but the latter has a P- thallus and distinctly fuzzy granules that lack the pseudocortex typical of The Neglected Dust Lichen.

CHEMISTRY: Roccellic acid, alectorialic acid. Spot tests. K-, C-, KC+ red, P+ yellow, UV-.

NICHE: This species is common throughout the Smokies, where it occurs on the bark of trees and on sun-exposed rocks. It is equally common in the southern Appalachians and throughout eastern North America as well as in many other portions of the continent.

KEY FEATURES: Concentric, rosette rings of bluish granular thalli, superficial cortex, KC+ red (usually), common in the Smokies on both rocks and bark.

Lepraria normandinoides

Lichen Vulgaris

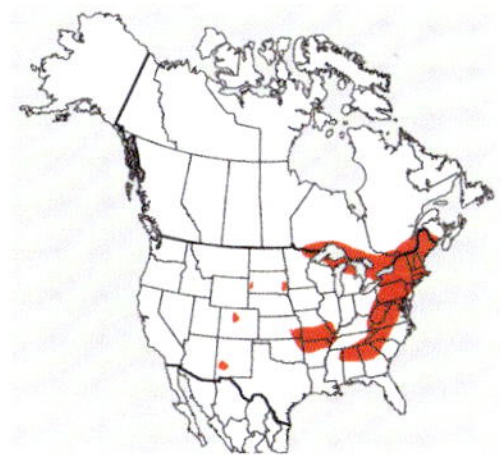

Tripp 2507 (photo: Tripp)

NOTES: *Lepraria normandinoides* is one of two members of the genus in the Smokies that can be instantly recognized by its thick, continuous, white or blue-white thalli that appear to have a rim or a crisped margin. The other species is *L. oxybapha*, which can only be distinguished using thin layer chromatography (it produces fumarprotocetraric acid instead of protocetraric acid). Those two substances differ only in a single chemical bond and yet data indicate they characterize two different species. We're amazed, too. As you might have guessed, the epithet "normandinoides" refers to the superficial resemblance of this lichen to *Normandina pulchella*.

CHEMISTRY: Atranorin, roccellic acid, protocetraric acid. Spot tests. K+ yellow, C-, KC-, P+ red, UV-.

NICHE: This species is widespread throughout the Smokies, where it occurs on the bark of trees and on non-calcareous rocks in overhangs. It is similarly very widespread throughout temperate eastern North America and has small populations scattered into the mountainous regions of the west. Look for this one on rock walls in urban areas, such as Wissahickon Schist, if you are spending time in Philadelphia.

KEY FEATURES: Blue-white granular thallus, crisped margins, on bark and rock at all elevations, undifferentiable from *L. oxybapha*, unless you have a home thin-layer chromatography setup (which doesn't take much beyond a pickle jar and some common chemicals).

Lepraria oxybapha

Lesser Lichen Vulgaris

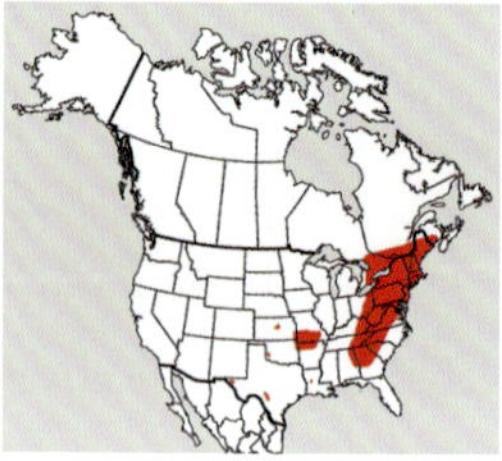

Tripp 3669 (photo: Lendemer)

NOTES: If you didn't know better, you might think that someone cruelly named this species "oxybapha" just to make sure that it would never be alphabetically separated from the morphologically nearly identical "normandinoides." Indeed, the two species are *together* easily recognized by their continuous, white or blue-white thalli that have distinctly "crisped" margins. However, you can't tell them apart in the field or even in the laboratory without chemical analysis. The epithet "oxybapha" refers to the saucer shaped "lobes" produced at the margins of the thallus.
CHEMISTRY: Atranorin, roccellic acid, protocetraric acid. Spot tests. K+ yellow, C-, KC-, P+ red, UV-.
NICHE: This species is widespread throughout the Smokies, where it occurs on the bark of trees and on non-calcareous rocks in overhangs. It is most common at high elevations, especially in spruce-fir forests. *Lepraria oxybapha* is similarly widespread throughout temperate eastern North America and has small populations scattered in the mountainous regions of the southwest.
KEY FEATURES: Blue-white granular thallus, crisped margins, on bark and rock of trees at all elevations, undifferentiable from *L. normandinoides* without TLC.

Lepraria vouauxii

Goblin Dust

Lendemer 33174 (photo: Tripp)

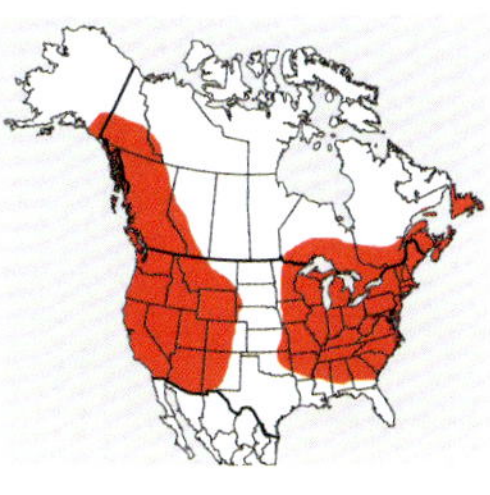

NOTES: We provided this species with the name of "Goblin Dust" because it seems to always be lurking but goes undetected. In the field, *Lepraria vouauxii* closely resembles *L. finkii*, which is so common you will be forgiven for overlooking the former. Even worse, the two species have similar spot test reactions, even though they produce very different chemical substances. However, if you look carefully, you can always identify *L. vouauxii* by its yellowish color, compared to *L. finkii*, but this enlightenment usually happens after you have already done thin-layer chromatography.

CHEMISTRY: Dibenzofurans. Spot tests. K-, C-, KC-, P+ orange, UV-.

NICHE: *Lepraria vouauxii* is rare but widespread in the Smokies, where it occurs on the bark of trees and on rocks in sheltered overhangs. You are most likely to find it at low elevations, and it is very common at lunch stops along Lakeshore Trail.

KEY FEATURES: Thin but distinctly colored, dirty yellow granular thallus, on sheltered rocks under overhangs and on bark at the bases of trees.

Lepraria xanthonica

Sometime Sunshine

Lendemer 53156 (photo: Tripp)

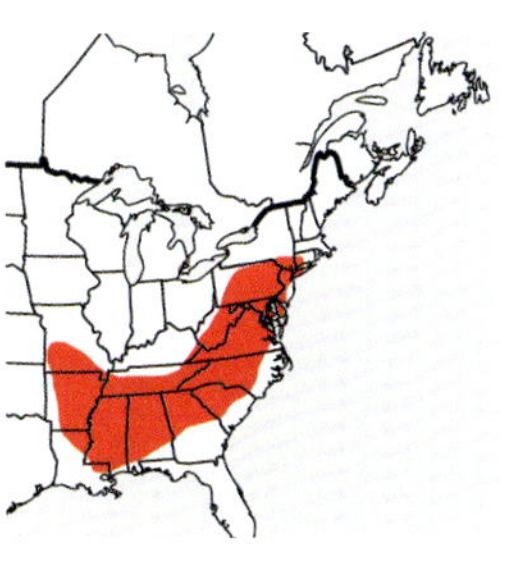

NOTES: Most lichens that produce xanthones reveal the presence of those substances by a bright UV reaction, or an intense color change with C. But that is not the case for *Lepraria xanthonica*. As a result of its lackluster spot test reactions, it is most likely to be confused with *Leprocaulon nicholsiae* because both species have thin, leprose thalli composed of dispersed aggregations of granules. The latter species differs chemically from *L. xanthonica* in lacking xanthones. Most of the time *L. xanthonica* grows on the bases of trees, especially over mosses, and so can be distinguished from *L. nicholsiae,* which grows only on rock. However, when *L. xanthonica* occurs on rock, it can be difficult to tell the two species apart without chromatography.

CHEMISTRY: Usnic acid, zeorin, one xanthone. Spot tests: K-, C-, KC+ yellow, P-, UV+ dull orange.

NICHE: This species is widespread at middle and low elevations throughout the Smokies and southern Appalachians, where it occurs on the bases of trees and on sheltered, non-calcareous rock outcrops.

KEY FEATURES: Dispersed, thin, greenish-yellow, leprose thallus, P-, on the bases of trees and sheltered non-calcareous rock outcrops at low and middle elevations.

Leprocaulon adhaerens

Stuck on Blue

Tripp 4963 (photo: Lendemer)

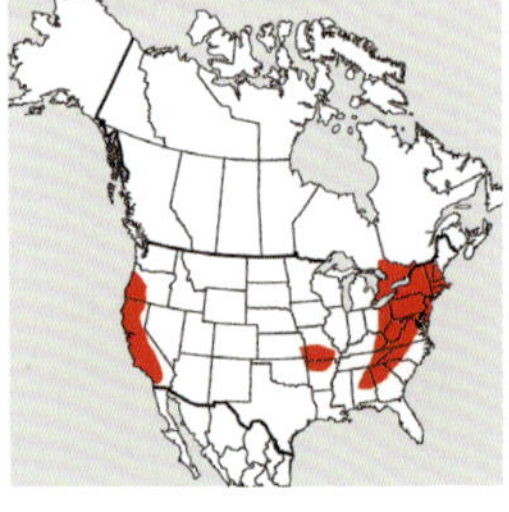

NOTES: At this point, you're probably thinking that there are more dust lichens than you know what to do with, right? Well, better too many than too few! *Leprocaulon adhaerens* is, in any case, both special and easy to identify. It is the only dust lichen in the Smokies that has a leprose thallus and has this distinctive shade of turquoise bluish-green in the field. The color actually resembles the soredia of a cyanolichen like *Fuscopannaria*, but dust lichens have a green algal photobiont and are unrelated to that genus. If you think you've found it on bark, then look for some areoles, which will confirm that you've probably found *Rinodina buckii*, instead.

CHEMISTRY: Pannarin, zeorin. Spot tests. K-, C-, KC-, P+ red, UV-.

NICHE: This species is infrequent but widespread at middle and low elevations in the Smokies, where it is found on non-calcareous rocks in overhangs. It occurs in similar habitats throughout temperate North America but is usually overlooked.

KEY FEATURES: TUR-QUOISE BLUE-GREEN DUST LICHEN. Need we say more? Nothing else looks like this. Thank your pannarin.

Leprocaulon nicholsiae

Becky's Lucky Dust

Lendemer 50060 (photo: Lendemer)

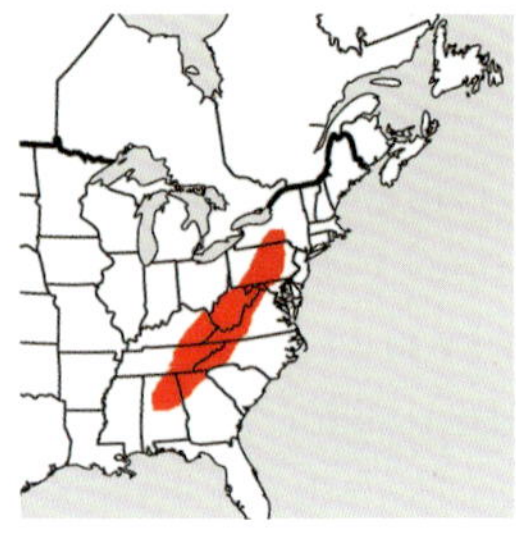

NOTES: Every day spent in the Smokies is special, and you can remember that with this lucky dust lichen. *Leprocaulon nicholsiae* is a common species at middle and low elevations that can easily be recognized by its thin, greenish-yellow, dust-like thalli that form extensive colonies on non-calcareous rock faces. It is most likely to be confused with *Lepraria xanthonica*, but that species tends to occur on mosses at the bases of trees and differs chemically in the production of a xanthone. Becky's Lucky Dust was named in honor of Rebecca Nichols, National Park Service entomologist who keeps track of all the species, lichens, and otherwise, reported to occur in the Smokies.

CHEMISTRY: Usnic acid and zeorin. Spot tests: K-, C-, KC+ yellow, P-, UV-.

NICHE: Becky's Lucky Dust is widespread and common on sheltered non-calcareous rock faces at middle and low elevations of the Smokies, and throughout the Appalachian Mountains. In fact, it forms extensive colonies on the rock outcrops along the Little River and can readily be seen on any drive into Cades Cove.

KEY FEATURES: Dispersed, thin, greenish-yellow, leprose thallus, P-, on sheltered non-calcareous rock outcrops at low and middle elevations.

Leptogium austroamericanum

Silk Cake Imagination

Tripp 1112 (photo: Tripp)

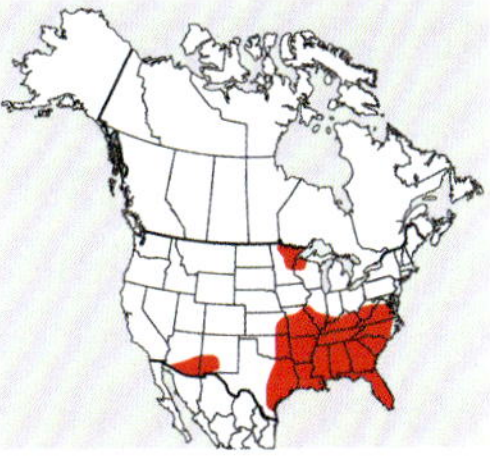

NOTES: First: start with jelly lichens (it's squishy looking, right?). You know immediately that this is either a *Leptogium* or *Collema*. Hopefully, you chose to hike in the sun rather than the rain, in which case you have most likely encountered a thallus that is dried out (rather than wet) to reveal a slate gray (rather than black) color. You have now arrived at *Leptogium* (getting closer). The first things to notice about *Leptogium austroamericanum* is that it produces isidia and lacks apothecia, lacks a dense tomentum covering its lower cortex, and has a conspicuously wrinkled upper cortex. This species is most often confused with *L. cyanescens*, which differs in having a smooth versus a conspicuously wrinkled upper surface. It is impossible to confuse this species with other isidiate, gray species owing to its broad lobes, which are never dissected. This species sort of reminds us of frosting on a cake; alternatively, imagine a piece of silk lying on a table, and envision bunching it up with the tips of your five fingers, and then flipping it over and looking at the side that used to lay against the table.

CHEMISTRY: No substances. Spot tests. K-, C-, KC-, P-, UV-.

NICHE: *Leptogium austroamericanum* occurs on the bark of hardwoods at low elevations in the Smokies, where it is uncommon. It seems to prefer portions of the Park that are slightly calcareous, such as White Oak Sinks. Alternatively, look for this species over on the Rabbit Creek Trail. Its center of distribution is the southeastern United States, with disjunct populations in the sky islands of Arizona.

KEY FEATURES: Gray (when dry) cyanolichen with broad, entire lobes, isidiate, no apothecia, no tomentum underneath, distinctly wrinkled upper surface, uncommon, on hardwoods at low elevations.

Leptogium chloromelum

With Olive Juice, And Make It A Double

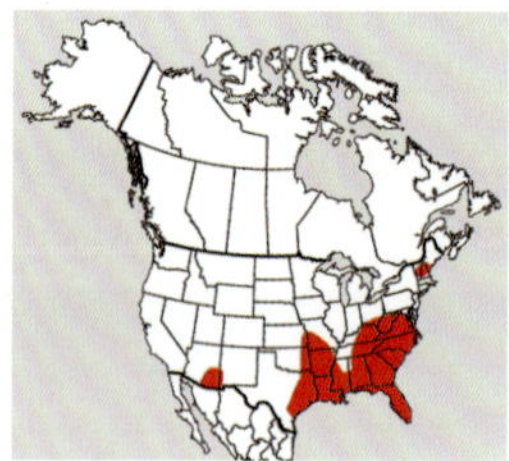

Tripp 3564 (photo: Lendemer)

NOTES: Species of *Leptogium* are, for the most part, slate gray, but sometimes a few species resemble something other in color, like olive juice. Primarily, we are referring to *Leptogium chloromelum* (that fortunately translates into something like "greenish brown" or "dark green"). This species varies across its range . . . sometimes looking slate gray and scuzzy (like the photograph), but sometimes looking more like olive juice. If you happen to encounter the latter in the field, this is probably the only member of the genus that you might mistake for a *Collema*. Regardless of its color, *L. chloromelum* can be easily recognized because it is the only fertile jelly lichen that has a thallus that is strongly wrinkled and becomes extremely swollen when wet. Other *Leptogium* species become swollen when wet, but all of these bear asexual propagules (namely: isidia). This species is furthermore characterized by its broad lobes that either appressed to or ascend away from the substrate (or both, on the same thallus!) and its big apothecia that are sessile atop the thallus and typically have thick, gray rims. When in doubt as to its identity, check for the clearly stratified thallus.

CHEMISTRY: No substances. Spot tests. K-, C-, KC-, P-, UV-.

NICHE: *Leptogium chloromelum* occurs primarily on the bark of hardwoods (especially oaks) and, to a lesser extent, on rocks. In the Smokies, it is found at low and middle elevations. With Olive Juice apparently likes people. Good places to see populations of it include: Sugarlands Visitor Center, The Park Headquarters at Twin Creeks, and Cades Code.

KEY FEATURES: Cyanolichen varying from slate gray to olive green in color, thallus differentiated into two layers, apothecia with thick gray rims, lower elevations in Park, on hardwoods.

Leptogium corticola

Grammy's Buttons

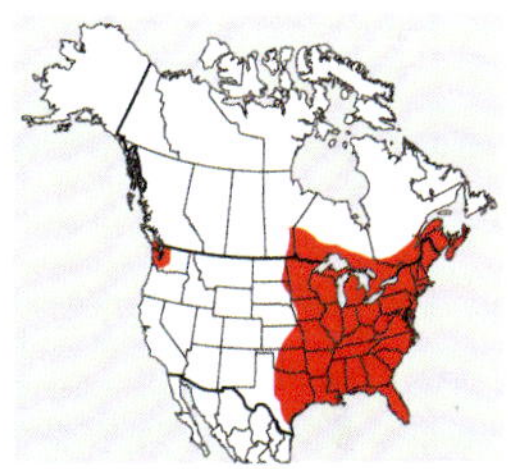

Tripp 3904 (photo: Lendemer)

NOTES: *Leptogium corticola* and the next species in this book, *L. cyanescens*, are arguably the two most common cyanolichens in Great Smoky Mountains National Park. Learn these—they are easy! Both are readily identifiable by their dark, greenish-brown thallus and jelly-like morphology when wet but beautiful slate-gray color when dry. *Leptogium corticola* is fertile (check out those super awesome apothecia!) and never with isidia, as in *L. cyanescens*. There are other fertile species of *Leptogium* in the Park, but 99 times out of 100, this is your top candidate. Check the key for confirmation.

CHEMISTRY: No substances. Spot tests. K-, C-, KC-, P-, UV-.

NICHE: *Leptogium corticola* is extremely common on the bark of hardwoods, especially hardwood bases, throughout all elevations in the Smokies. "Corticola" should tip you off as to its niche. This species is also common throughout eastern North American broadleaf forests and is disjunct in the Pacific Northwest.

KEY FEATURES: Apotheciate cyanolichen with a slate gray thallus (when dry), extremely common throughout the Park. Make this the first jelly lichen that you learn!

Leptogium cyanescens

An Isidiate Exhibit

Tripp 3922 (photo: Lendemer)

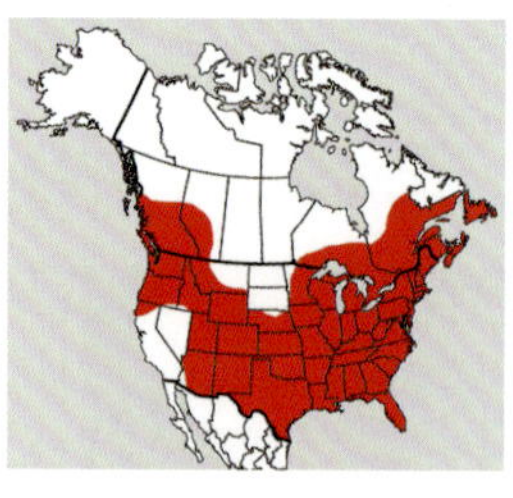

NOTES: Like *Leptogium corticola*, *L. cyanescens* is the other "most common cyanolichen in the Smokies." Unlike the former, *L. cyanescens* is isidiate, which makes it just a bit trickier to identify because of a few other isidiate *Leptogiums*. But it's doable. Note its entire margins (distinguishing it from *L. lichenoides*) and its smooth thallus surface (distinguishing it from *L. austroamericanum*, which is wrinkled). Finally, don't confuse *L. cyanescens* with species of *Collema*, the other "jelly lichen" genus in the Smokies, which is always much thicker and, when dry, blacker.

CHEMISTRY: No substances. Spot tests. K-, C-, KC-, P-, UV-.

NICHE: Just like *Leptogium corticola*, *L. cyanescens* is extremely common on the bark of hardwoods throughout the Smokies. But why stop there? This species is widespread throughout eastern North America and also occurs in forested habitats throughout North America.

KEY FEATURES: Isidiate cyanolichen with a slate gray thallus (when dry; "cyan" means dark blue in Greek for the record), extremely common. Make this the second jelly lichen that you learn, perhaps over lunch with a PB sandwich in hand!

Leptogium dactylinum

Grammy's Fingers

Tripp 5452 (photo: Tripp)

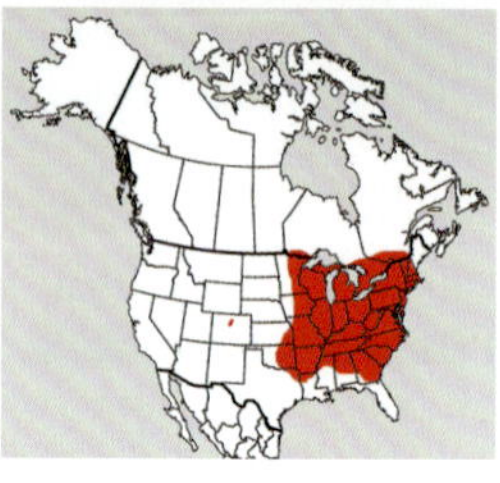

NOTES: *Leptogium dactylinum* is a relatively easy-to-identify member of the genus in the Smokies by its lobulate morphology and the olive-brown color of the thallus. The asexual reproductive propagules, which are situated on the margins rather than surfaces of the lobes, are actually often also accompanied by apothecia, which themselves have abundant lobules along the margins. Otherwise, look for a greenish-brown cyanolichen with a thallus differentiated into two layers. The specific epithet will help you remember its "fingers." Occasional forms of the normally isidiate *L. cyanescens* can develop lobules and thus become confused with *L. dactylinum*, but the former species is always slate gray in color rather than olive-brown.

CHEMISTRY: No substances. Spot tests. K-, C-, KC-, P-, UV-.

NICHE: *Leptogium dactylinum*'s heart lies in those beautiful broadleaf hardwood forests of eastern North America. In the Smokies, it is most often found among the mosses that coat the bases of hardwood trees at middle and high elevations.

KEY FEATURES: Sweet lobules arranged marginally on a two-layered brown thallus, sometimes accompanied by apothecia, occasional at middle-to-high elevations, most often corticolous on oaks.

Leptogium hirsutum

Grammy's Tomentum

Lendemer 44567 (photo: Tripp)

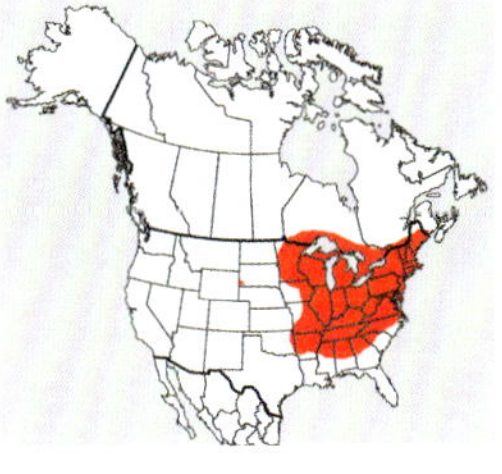

NOTES: *Leptogium hirsutum* is one of several isidiate species in the genus in the Smokies, but we think you will be able to readily put a name on this one owing to its lower surface, which is covered by an even tomentum of long hairs, and its similarly long, columnar isidia. It is most likely to be confused with *L. saturninum*, which has a cooler specific epithet but differs in having shorter, sometimes granular isidia. When in doubt, check the key and name it with confidence.

CHEMISTRY: No substances. Spot tests. K-, C-, KC-, P-, UV-.

NICHE: We have only seen this species a couple of times in the Smokies. Other collectors before us have found it with greater frequency. It seems to prefer central portions of the Park, from low to high elevations, usually always in very wet places (near streams or cloud-laden summits). Look for it on rocks, especially mineral-rich and calcareous ones.

KEY FEATURES: Long columnar isidia, dense tomentum on undersurface, saxicolous in wet areas from low to high elevations.

Leptogium laceroides

Woolly Stacks

Tripp 3577 (photo: Lendemer);
Inset: Tripp 4992 (photo: Lendemer)

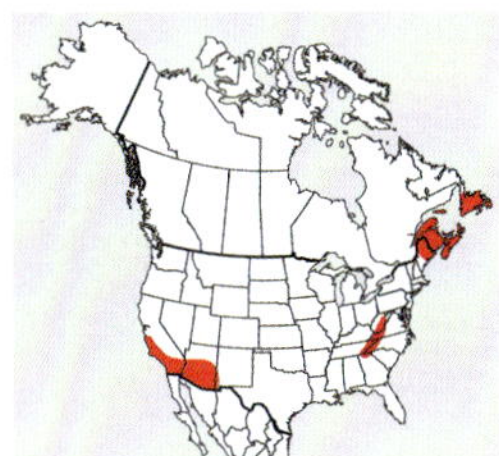

NOTES: *Leptogium laceroides* is relatively easy to identify, assuming you remembered your hand lens. Just flip over a lobe and look at its underside (apologies: we don't mean to be rude) and admire a dense, even tomentum of more or less short, white hairs (these < 1mm in length). Beyond this feature, Woolly Stacks is characterized by its gray to grayish-brown thallus, rather irregularly shaped lobes, lobules that cover primarily the margins of the lobes but sometimes creep onto lobe surfaces, and its ecology (see Niche). This combination of features makes it unlikely to confuse this lichen for any relatives, except perhaps *L. saturninum* and *L. hirsutum*, both of which have much longer tomentum hairs underneath.

CHEMISTRY: No substances. Spot tests. K-, C-, KC-, P-, UV-.

NICHE: Woolly Stacks likes to hang out at higher elevations in the Smokies, particularly on ridgetops that have ample exposure to the elements, including wind. After getting geared up for a long day hike, look for this species on the bark of hardwoods such as those that are densely covered by lichens in the vicinity of Eagle Rocks along the Appalachian Trail.

KEY FEATURES: Gray-to-brown thallus, irregularly shaped lobes, isidia along the margins and sometimes surfaces of lobes, dense tomentum of long white hairs underneath, high elevations, on hardwoods.

Leptogium lichenoides

Tattered Jellyskin

Lendemer 44533 (photo: Tripp)
Inset: Lendemer 44533 (photo: Tripp)

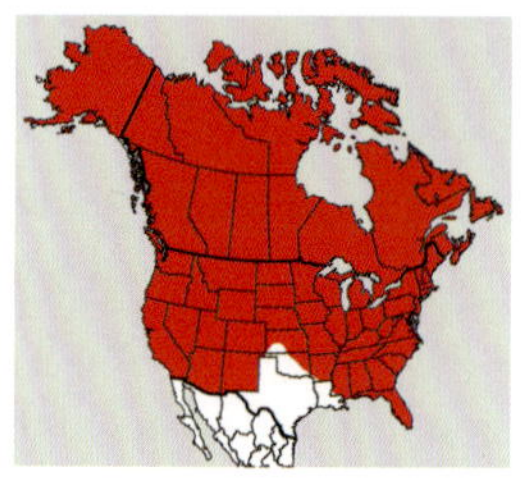

NOTES: *Leptogium lichenoides* is arguably the easiest-to-identify jelly lichen in the Smokies, and perhaps in eastern North America. It is quickly recognizable by its small, brownish-gray thallus with lobe tips that become finely dissected, thus resembling isidia. No other jelly lichen looks like this one. Other species such as *L. dactylinum* differ in having true lobules instead of very finely dissected lobe tips that look like isidia. We really like the common name employed in Brodo et al. (2001) and have retained its use here.

CHEMISTRY: No substances. Spot tests. K-, C-, KC-, P-, UV-.

NICHE: Tattered Jellyskin occurs primarily at low and middle (but sometimes high) elevations in the Smokies. It is most frequently found as small thalli overgrowing mosses on rocks, particularly weakly calcareous rocks. It is broadly distributed across all sides of the Park including places such as Big Creek, Heintooga Road, Abrams Creek, Deep Creek, and Rich Mountain Gap.

KEY FEATURES: Brownish-gray thalli with deeply dissected lobe tips that resemble isidia, no tomentum underneath, no other Smokies jelly lichen looks like this one, throughout Park, particularly low and middle elevations.

Leptogium millegranum

A Thousand Grains

Tripp 2632 (photo: Lendemer)

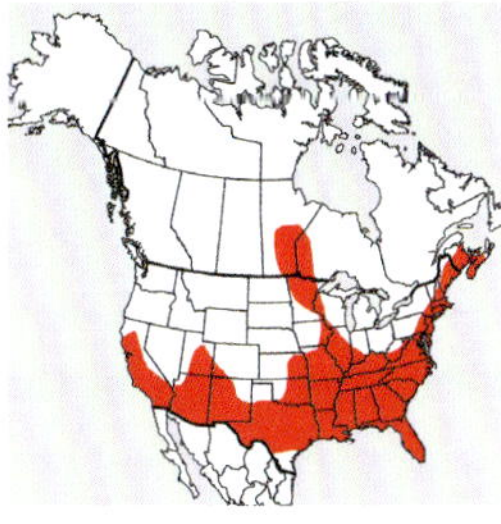

NOTES: If you spend enough time staring at them, and if you don't control for hypervariable morphology depending on whether the specimen is fully hydrated, dry, or somewhere in between, the jelly lichens will sooner than later vex you. But fear not: *Leptogium millegranum* is relatively easy to identify based on its olive-brown color (most *Leptogium* species, including the abundant ones in the Smokies, are slate gray), its lobe margins that are entire, and dense isidia that cover the entire surface, which is strongly wrinkled when dry. In fact, the specific epithet "millegranum" translates directly into "one thousand grains." Isn't that sweet? The most similar species is *L. chloromellum*, which also has a wrinkled surface but differs in lacking isidia.

CHEMISTRY: No substances. Spot tests. K-, KC-, C-, P-, UV-.

NICHE: A Thousand Grains can be found primarily at low elevations in the Smokies, where it occurs on the bark of hardwoods. It is widespread throughout southeastern North America, but never seems to be very abundant. Instead, it is just one of those pleasant species you encounter every now and then.

KEY FEATURES: Olive-brown thallus with entire lobes, strongly wrinkled surface covered by isidia, on bark of hardwoods, low elevations.

Leucodecton subcompunctum

Bourbon Sprawl

Lendemer 29363 (photo: Lendemer)

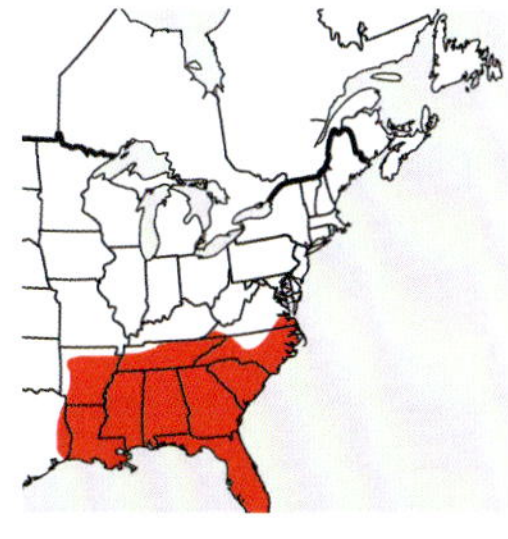

NOTES: *Leucodecton subcompunctum* is a crustose lichen that is quite easy to learn on account of its very thick, continuous thalli, which are a lovely brown in color (sometimes green when wet) and bear typically extensive, immersed apothecia resembling perithecia. These sunken apothecia

house beautiful, brown, submuriform (sub: think "sorta") spores, which is another path to knowing this species. Because of its conspicuously thick and continuous thallus, it stands out from many other lichens, but you might be tempted to confuse it for *Thelotrema subtile*. *Thelotrema subtile* is a close relative of *L. subcompunctum* and is similar in having apothecia immersed in pores, but it has colorless spores that are transversely septate. In some respects, the species is also similar to *Phlyctis speirea*, which occurs only at high elevations in the Smokies and differs in having large, densely muriform spores and producing norstictic acid.
CHEMISTRY: Stictic acid aggregate. Spot tests: K+ yellow turning dingy brown, KC-, C-, P+ orange, UV-.
NICHE: Bourbon Sprawl occurs widely across the southeastern United States and is primarily distributed in the Coastal Plain and Piedmont, only extending into the low and middle elevations of the southern Appalachians. It grows on the bark of hardwood trees, especially beeches, and tends to be restricted to mature, high-quality forests. Rumor has it that *L. subcompunctum* can grow on rocks, too, although it is left to be determined whether the latter entity is really the same species.
KEY FEATURES: Very thick, sprawling, continuous gray-to-green-brown thallus with immersed apothecia resembling perithecia, brown submuriform spores, K+ yellow and P+ orange, on smooth-barked trees in mature forest stands at low elevations.

Lobaria pulmonaria

Crown Jewel of America

Lendemer 29574 (photo: Tripp)

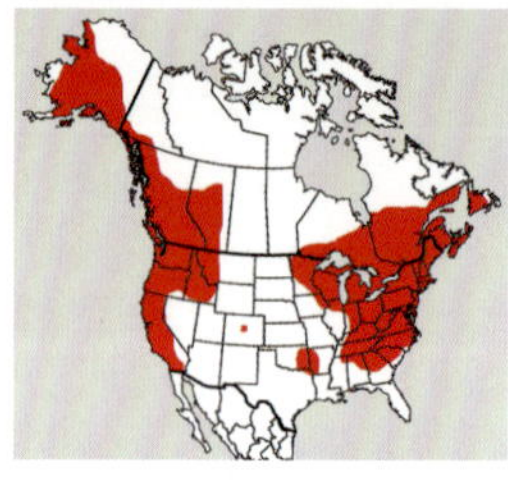

NOTES: So, you traveled all the way to GSMNP from International Falls, MN, just to see this showstopper. By the time you reached the Kentucky border, you finally thawed out. Welcome to truly hot, steamy, sexy, lichenology. The reproductive biology of *Lobaria pulmonaria* is as kinky as you might have guessed based on its thallus. This species is like no other. Learn it when it whispers to you. PS: We are sorry for failing to employ the much more common name of this lichen: Lungwort. This lichen has nothing to do with human anatomy and only minimal to do with old German botanical words (i.e., "wort").
CHEMISTRY: None of this matters. Spot tests. You will not need these to identify this incredible species. *Lobaria pulmonaria* for president!
NICHE: Relatively common on the bark of hardwoods at low and middle elevations, especially in older, mature forests. Of course, the Smokies has no shortage of these, so that clue won't help you much. Elsewhere, *L. pulmonaria* is widespread in portions of eastern and western

North America, but only in places that are ecologically nice enough for Crown Jewel of America. Fun fact: we recently discovered the first population of this species in Alabama, which was exciting and made us love the eastern temperate forests even more than we already do!

KEY FEATURES: Bright green, large "leafy" thalli, commonly embellished with both soredia and apothecia (I bet you can't do *that*, humans!), corticolous on hardwoods at low and middle elevations.

Lobaria quercizans

Our Nation's Queen

Tripp 1494 (photo: Moroz)

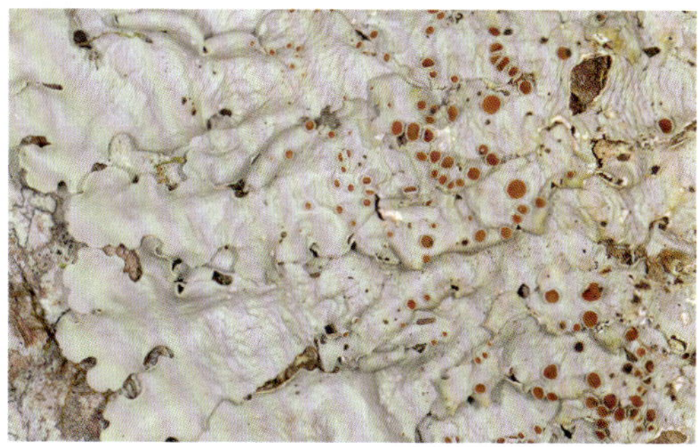

Lendemer 26684 (photo: Moroz)

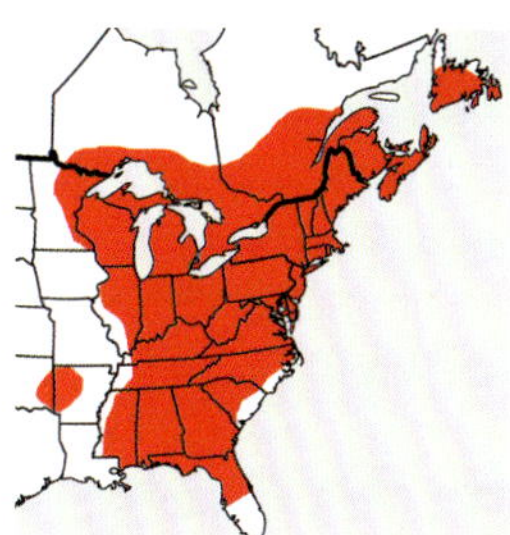

NOTES: Don't let the deep green color of this species fool you when it's wet–it is just as commonly light gray in color, that is, when the thallus is dry (see second photo). In any case, this is an easy species to put a name on owing to its very flat, smooth, thallus lobes, which readily distinguish the species from the only other green-algal *Lobaria* in the Park (*Lobaria pulmonaria*, i.e., Crown Jewel of America). *Lobaria quercizans* is also the "weediest" of our Smokies *Lobarias* . . . you are going to see it everywhere, almost always with the reddish-orange apothecia that it wears so well. Don't delay in learning it!

CHEMISTRY: Atranorin, gyrophoric, and 4-O-methyl-gyrophoric acids. Spot tests. Cortex K+ yellow; medulla K+ orange, C+ pink, KC+ red, UV-.

NICHE: Very common on hardwoods at low and middle elevations. This species was common in temperate woodlands throughout eastern North America, but now is common only in the southern Appalachians. Fun fact: on the same tree that we first discovered *Lobaria pulmonaria* in the State of Alabama on our recent collecting trip in Skyline Wildlife Management Area, we also found a population of the very rare (in that state) *Lobaria quercizans*. Coincidence? Or, something more to this such as photobiont genotype sharing?

KEY FEATURES: Large, flat, smooth lobes, these green to gray depending on hydration state, large rounded reddish-orange apothecia, very common on bark of hardwoods at low and middle elevations.

Lobaria scrobiculata

The Holy Land

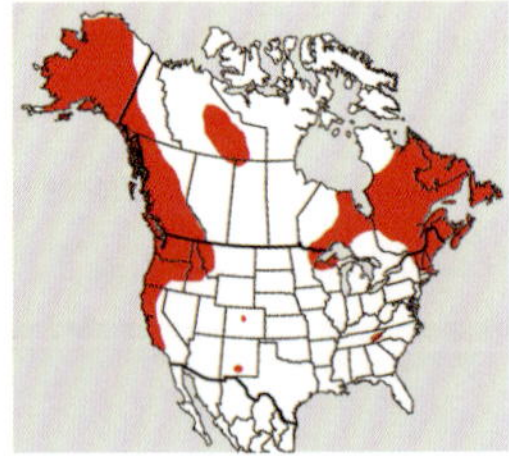

Lendemer 32985 (photo: Tripp)

NOTES: There are only three species of *Lobaria* in the Smokies, and they are *all* big and lovely. *Lobaria scrobiculata* is unmistakable among these three because it is the only one with a primary cyanobacterial photobiont (Crown Jewel of America and Our Nation's Queen both have a green algal photobiont); as such, it is the only one with a slate-gray thallus, which bears abundant soredia. *Lobaria quercizans* also has a gray thallus when dry, but bears only apothecia and never soredia.

CHEMISTRY: Stictic, constictic, norstictic, and usnic acids, scrobiculin. Spot tests. Medulla K+ yellow or orange, C-, KC- pink, P+ orange, UV-.

NICHE: This species is uncommon to rare in the Smokies, where it is disjunct from more northerly latitudes. Find it only in exceptionally high-quality forests, hanging out in the upper montane ecosystems with the spruce and fir. It is particularly fond of the latter as a substrate. A great place to see this species is near the summit of Luftee Knob.

KEY FEATURES: Large, gray thallus lobes covered by soralia, cyanobacterial photobiont, high-quality spruce-fir forests, corticolous primary on fir.

Lopadium disciforme

One-Spored Wonder

Collector Tripp 3500 (photo: Lendemer)

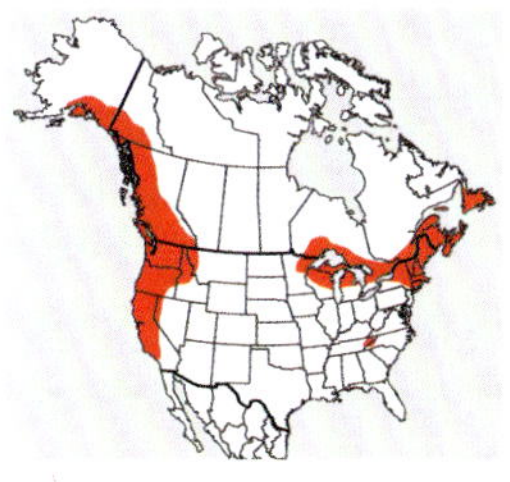

NOTES: *Lopadium disciforme* is an attractive crustose lichen with a distinctive ecology (see Niche). It is characterized morphologically by its green-to-olive, minutely areolate to granular thallus, its relatively large, black apothecia with lecideine margins, and its beautiful, colorless, muriform spores, which occupy an entire ascus (i.e., asci are 1-spored). It is most likely to be confused with bark-dwelling species in the genus *Lecidea*, but these never have muriform spores nor are they 1 spore to the ascus.
CHEMISTRY: No substances. Spot tests. K-, C-, KC-, P-, UV-.
NICHE: This species is common and widespread in high-elevation spruce-fir forests in the Smokies and elsewhere in the southern Appalachians, which represents a disjunction from the remaining range of this species (northern Appalachians plus portions of western North America). It grows primarily on large individuals of Red Spruce and can also be found on lignum of conifers at high elevations, such as Fraser Fir.
KEY FEATURES: Greenish, granular thallus, large black apothecia, lecideine margins, colorless muriform spores, spores 1 per ascus, high elevations on bark of spruce and on conifer lignum.

Loxospora cismonica

Lobster Spores

Lendemer 33251 (photo: Tripp)

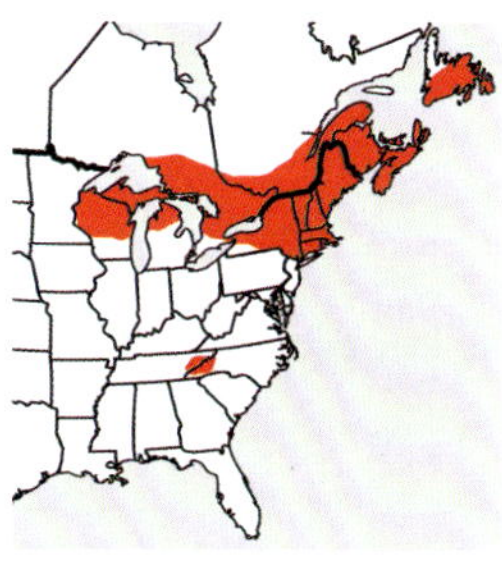

NOTES: The three *Loxosporas* currently known from the Smokies represent a lovely group of lichens, and one we think you will like. They are relatively common, easy to spot in a forest, and also easy to differentiate among one another. *Loxospora cismonica* is one of two fertile species and is easily distinguished from the other, *L. ochrophaea*, by its densely pruinose reddish-brown apothecia set against a gray-to-green continuous thallus. You will most likely confuse *Loxospora cismonica* with a *Lecanora*, but section the apothecia and see the magic! How could you not love a genus whose Latin origins mean "lobster spores," which is in reference to their marked, fusiform or spindle-shaped spores? Also, learn this one by its niche (see below).
CHEMISTRY: Thamnolic acid. Spot tests. K+ yellow, C-, KC-, P+ orange, UV-.
NICHE: On the bark of "She Balsam" at high elevations. If you aren't from the Southern Appalachians, this refers to *Abies fraseri*.

KEY FEATURES: Densely pruinose reddish-brown apothecia, continuous gray-to-green thallus, occasional on "She Balsam" at high elevations. Watch for the sparkle factor of those pruinose discs.

Loxospora elatina

Lobster Papillae

Tripp 5040 (photo: Lendemer)

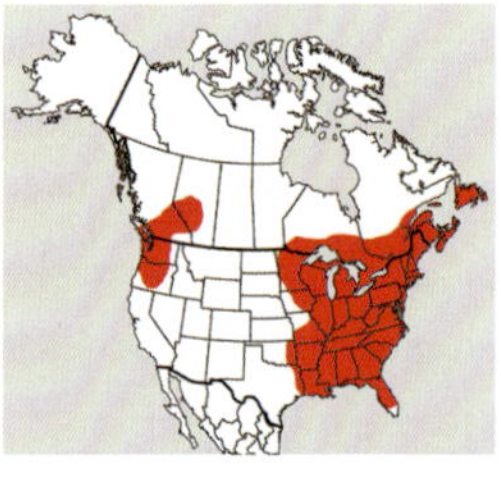

NOTES: The only of the three *Loxosporas* that isn't so loxosporous, *Loxospora elatina* is sterile. We can't help it. Sometimes you just have to accept the fact: there are a lot of sterile sorediate crusts in the Smokies. We think it is a key step in making America "Great Again," but in any case, our advice is to be optimistic. Look for this lichen at high elevations in the Park, where it is a bit on the rare side of the spectrum. This species is most likely to be confused with hundreds of other sterile sorediate crusts at high elevations, but don't despair. Look for its lemon-green, diffuse soralia, for starters.

CHEMISTRY: Thamnolic acid. Spot tests. K+ yellow, C-, KC-, P+ orange, UV-.

NICHE: Rare on bark at high elevations. Something to look up to.

KEY FEATURES: Send this one to an expert for identification, but only after you have received permission to collect it in the first place.

Loxospora ochrophaea

Superflesh

Tripp 3499 (photo: Lendemer)

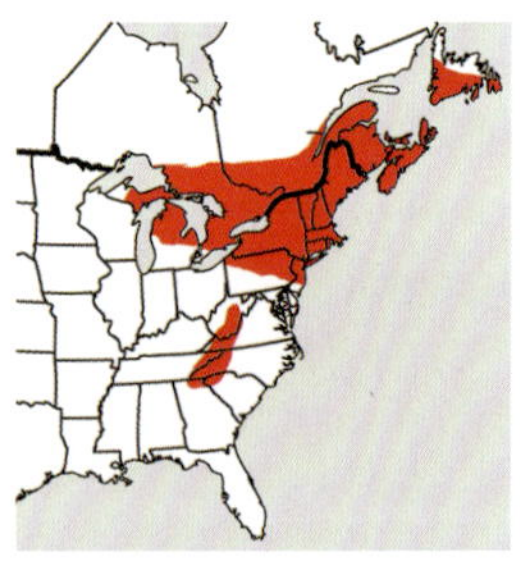

NOTES: *Loxospora ochrophaea* should be high on your bucket list while wandering through a spruce-fir forests of the Smokies. Its truly flesh-colored apothecia rimed by a white margin against a warty light gray-to-green, continuous thallus serve to identify this species. Its discs are never pruinose, thus readily separating this lobster spore from its close relative, *L. cismonica*. Much like the latter species, you are most likely to confuse *L. ochrophaea* with a high-elevation *Lecanora* but check those fusiform lobster spores and rest easy!

CHEMISTRY: Thamnolic acid, zeorin. Spot tests. K+ yellow, C-, KC+ yeller-brown, P+ orange, UV-.

NICHE: *Loxospora ochrophaea* is relatively common at middle-to-high elevations throughout the Smokies. Look for it in a happy place against the bark of *Abies fraseri*.

KEY FEATURES: Flesh-colored apothecia with white margins, gray-to-green continuous thallus, fusiform spores. On bark, middle-to-high elevations, particularly on fir, particularly pretty.

Maronea constans

Legume Lichen

Lendemer 53180 (photo: Tripp)

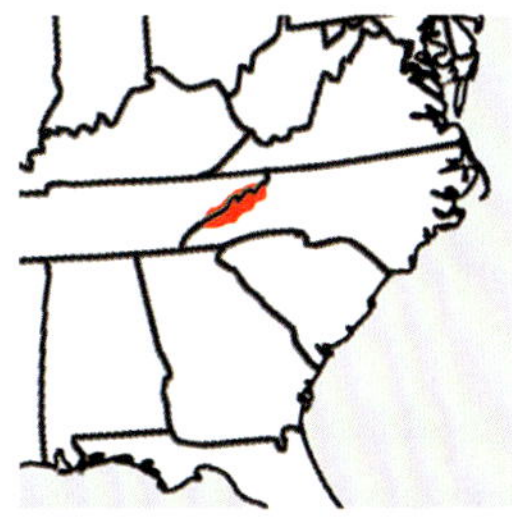

NOTES: The photo shown here is *precisely* what this species looks like. Very small, warty areolate thalli almost always on twigs. Always green (or occasionally graying green when in a desiccated state). Reminds one of green beans, to be sure. Also, look at how those apothecia margins are ever so slightly paler green than the thallus. And look at that brownish-black prothallus. Need more? Actually, you do need just a bit more. This species differs from its closest relative, *M. polyphaea*, in having an inspersed hymenium and a KC- thallus. The two also differ ecologically, too (see below). Beyond this congener, it is unlikely you will confuse Legume Lichen for anything else other than branch beautification. In addition to their distinctive external appearance, both species of *Maronea* have polysporous asci that contain hundreds of minute, ellipsoid, colorless, simple spores.

CHEMISTRY: Sekikaic acid. Spot tests: (medulla) K-, KC-, C-, P-, UV+ dull blue-white.

NICHE: Legume Lichen occurs at high elevations in the southern Appalachian Mountains, and so far as known, its primary center of distribution is in the Smokies. Recent fieldwork in Tennessee outside of the Smokies has, however, yielded discovery of several additional populations of this species, also at high elevations. In contrast, the much more common and widespread *M. polyphaea* occurs at middle and low elevations throughout the southeastern United States.

KEY FEATURES: Legume-green, warty to areolate thallus, lecanorine apothecia with dark discs, KC-, oil inspersed hymenium, polysporous asci containing numerous small simple spores, on twigs at high elevations.

Maronea polyphaea

Jolly Green

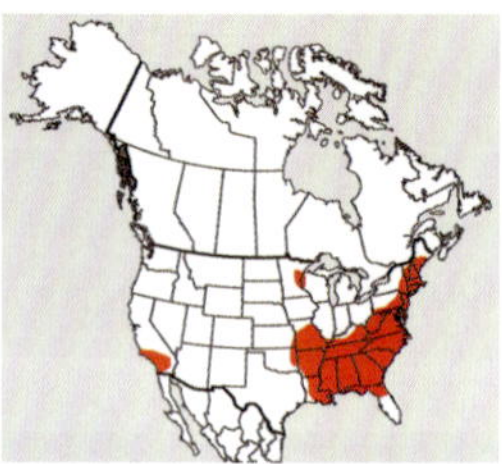

Lendemer 44629 (photo: Lendemer)

NOTES: There is little of the outward appearance of *Maronea polyphaea* that would lead to easy separation from its close relative, *M. constans* (the relatively darker, browner thallus seen here compared to *M. constans* is attributable to photography of a dried herbarium specimen). Both are lovely lichens characterized by having thick green thalli that bear large apothecia with dark discs and thick rims. In many respects the species resemble *Rinodina*, which differs internally in having brown, 2-celled spores with thickened walls. *Maronea* was treated within *Rinodina* historically, through one wonders why exactly that was given how different they are. *Maronea polyphaea* is only likely to be confused with *M. constans*, and it differs in having a clear hymenium that is not inspersed with oil droplets as well as in having a KC+ pink medulla. For now, don't sweat the differences, but rather focus on finding either Jolly Green or Legume Lichen on fallen branches and twigs. Your day will be brighter if you succeed.

CHEMISTRY: Submerochlorophaeic acid. Spot tests: (medulla) K-, KC+ pink, C-, P-, UV+ dull blue-white.

NICHE: Jolly Green is widespread at low and middle elevations in much of the southeastern United States, including the Smokies. This is in contrast to its relative, *M. constans*, which preferentially occupies much higher elevations. Look for both species on branches of hardwoods, such as on fallen twigs.

KEY FEATURES: Warty to areolate green thallus, lecanorine apothecia with dark discs, KC+ pink medulla, clear hymenium, polysporous asci containing numerous small simple spores, on twigs at middle and low elevations.

Megalaria beechingii

Sean's Delight

Tripp 4989 (photo: Tripp)

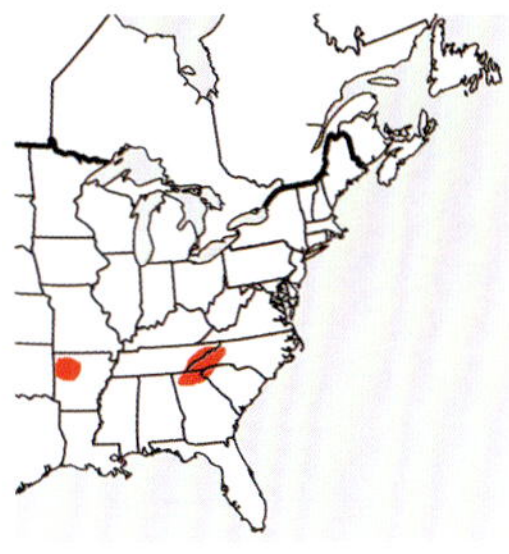

NOTES: *Megalaria beechingii* tells the story of many Smokies species that one has a hard time figuring out the answer to the following: why did it take centuries to name this thing? This species is an unmistakable crustose lichen that occurs only on humid rocks. You can recognize it by its white, fibrous prothallus, gray-to-greenish-gray sorediate thallus, and often but not always, the presence of black apothecia (not seen in collection photographed here). If you venture to look at the spores, they are colorless and 2-celled. This species was named after Sean Beeching from Atlanta, Georgia . . . a great carpenter, writer, recommender of books for the Tuckerman Workshop and, well, one our favorite people!

CHEMISTRY: Atranorin and zeorin. Spot tests. K+ yellow, C-, KC-, P-, UV-.

NICHE: This species occurs occasionally in the Smokies, always on rocks along lakeshores or in stream and river ravines. Keep an eye out for it on rocks if you go take a dip in Big Creek.

KEY FEATURES: Gray-to-greenish-gray crust with coarse soredia and a fibrous white prothallus, black apothecia, 2-celled colorless spores, on non-calcareous rocks at low elevations along rivers and creeks.

Megalospora porphyritis

Volcan-ito

Lendemer 32913 (photo: Tripp)

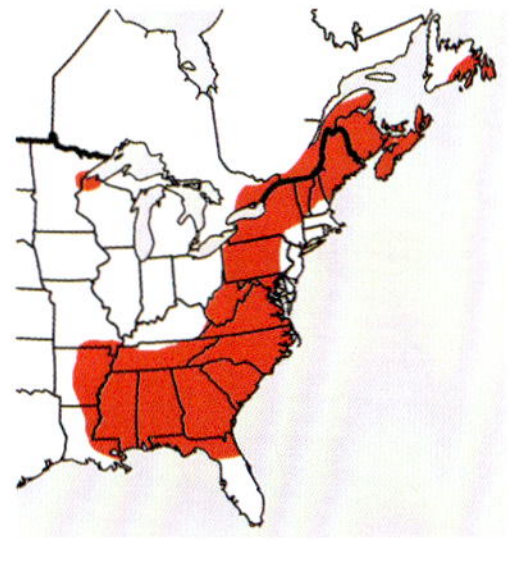

NOTES: This is one of those species that, once learned, you will wonder how you ever missed it in the past. You can recognize it by its continuous, shiny, bluish-gray thallus that produces abundant pustules that erupt into coarse, granular fragments—that, in addition to its chemistry and niche, make

Megalospora porphyritis unmistakable. This species is commonly also found with large black apothecia that sometimes have a white dusting on the discs.

CHEMISTRY: Pannarin and zeorin. Spot tests. K-, C-, KC-, P+ orange-red, UV-.

NICHE: *Megalospora porphyritis* occurs all around you—on hardwoods and conifers from the bases to the canopies from the lowest parts of Tennessee to the highest parts of North Carolina. In the Smokies, on any trail and in just about any forest type, you are probably always within a foot of this lichen.

KEY FEATURES: Shiny, continuous, bluish-gray crust, pustules erupting into coarse granular fragments, P+ orange-red and K-, on just about every tree in the Smokies.

Melanelia culbersonii

Petite Olivine

Tripp 3677 (photo: Lendemer)

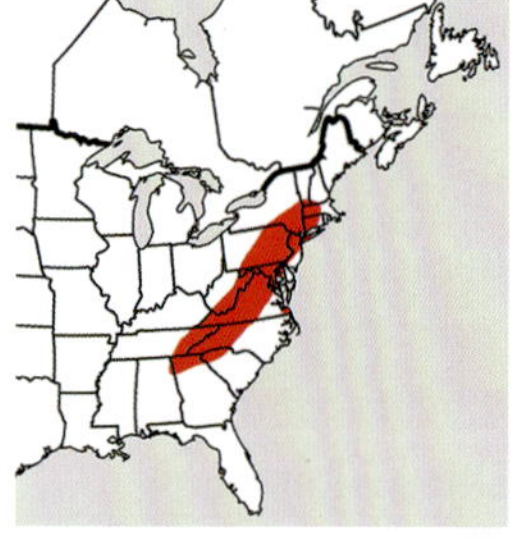

NOTES: Here is one truly beautiful lichen, understated but elegant. You can recognize it by its small brown thallus, which forms little rosettes, and the presence of interrupted sooty-colored soralia that line the lobe margins. It is fitting that when Mason Hale discovered that the old name for this species couldn't be used anymore, he renamed it in honor of the late Bill Culberson–eminent American lichenologist from Duke University.

CHEMISTRY: Stictic and norstictic acids. Spot tests. Cortex: K-, C-, KC-, P-, UV-. Medulla: K+ yellow turning red, C-, KC-, P+ orange, UV-.

NICHE: *Melanelia culbersonii* is endemic to the Appalachian Mountains, where it grows on exposed non-calcareous rocks on talus slopes, cliffs, and other rock outcrops. It is not common in the Smokies and is best left alone, if found.

KEY FEATURES: Small brown foliose thallus, sooty discontinuous soralia along the lobe margins, P+ orange medulla, on exposed non-calcareous rocks of talus slopes, cliffs, and other large rock outcrops.

Melanelia hepatizon

Uptown Olivine

Tripp 5090 (photo: Lendemer)

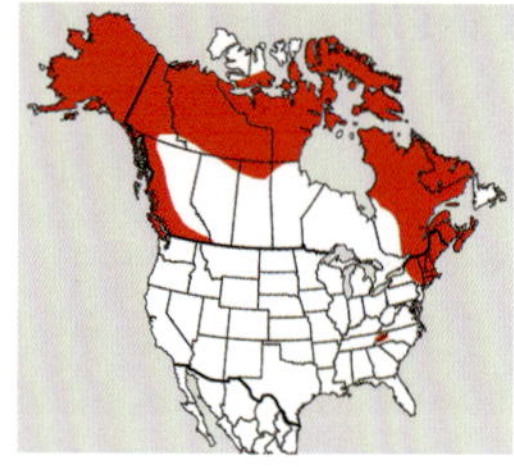

NOTES: Lichens don't get classier than this one, at least in our opinion. Just picture *Melanelia hepatizon* sunning itself on top of Mount LeConte, with its UV melanin pigment–protected thallus, narrow secondary lobes, and marginal white pseudocyphellae that resemble tiny dimples. The most morphologically similar species is *M. culbersonii*, which differs in having marginal soralia with sooty colored soredia. Elsewhere in its distribution, *M. hepatizon* produces abundant apothecia, but not so much in the Smokies. Why the difference?

CHEMISTRY: Stictic acid. Spot tests. Cortex: K-, C-, KC-, P-, UV-. Medulla: K+ yellow turning brown, C-, KC-, P+ orange, UV-.

NICHE: *Melanelia hepatizon* is common and widespread in the arctic and other northern portions of the world. Populations in the southern Appalachians are markedly disjunct from the rest of its range. In the Smokies, this species is found only on exposed, non-calcareous rocks at the tops of the highest peaks. Maybe it summers in the Smokies . . .

KEY FEATURES: Brown foliose thallus with smaller, narrow, secondary lobes, small white marginal pseudocyphellae, lack of soralia, P+ orange medulla, on exposed non-calcareous rocks on the highest summits in the Park.

Melanelixia glabratula

Rugged Traveler

Lendemer 53178 (photo: Tripp)

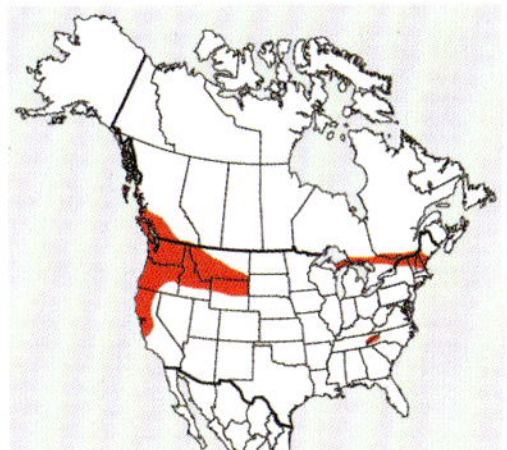

NOTES: The genus name says it all—well, sort of. This is one of the few foliose lichens with distinctly brown thalli in the Smokies. While it might be rare in the Park, it can easily be recognized by the white pseudocyphellae on the upper surface, cylindrical isidia, and C+ red medulla. The most similar species is *M. subaurifera*, which is also rare in the Smokies and differs primarily in having short, globose isidia that are easily broken and abraded over time, causing the thallus to appear sorediate. In contrast, the isidia of *M. fuliginosa* are not so easily broken and so tend to remain intact, giving the thallus a shaggy, rugged appearance. *Melanohalea halei* can be somewhat similar in appearance, particularly when it has abundant lobules, however the medulla in that species is C- and P+ red. If you know your brown foliose lichens and haven't seen this name before, it is because the species was previously included within the concept of *M. fuliginosa*. Recent studies, however, have shown that *M. fuliginosa* is restricted to rocks and does not appear to occur in the southern Appalachians.

CHEMISTRY: Lecanoric acid. Spot tests: (cortex) K-, C-, KC-, P-, UV-; (medulla) K-, C+ red, KC+ red, P-, UV-.

NICHE: Like most members of the genus, *M. glabratula* is primarily a species of northern temperate and boreal regions in North America. It is rare in the Smokies, and the southern Appalachians generally, where it occurs on the branches of trees and shrubs at high elevations.

KEY FEATURES: Brown foliose thallus, pseudocyphellae on upper surface, cylindrical isidia, C+ red medulla, on branches at high elevations.

Melanohalea halei

Beech Gap Olivine

Lendemer 48557 (photo: Tripp)

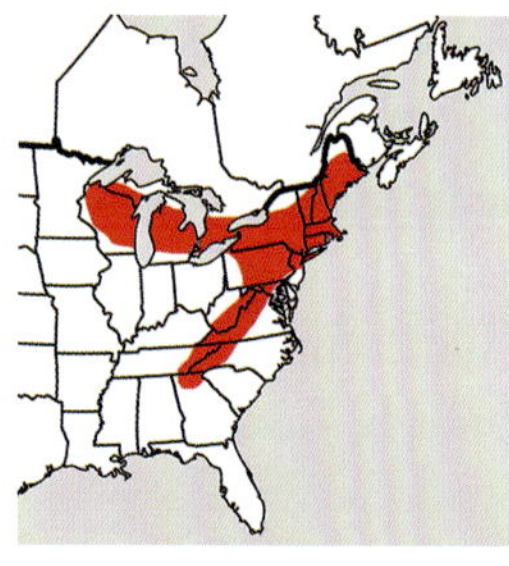

NOTES: *Melanohalea halei* is the most common brown foliose lichen in the Smokies and is easily identified in the field despite being exceptionally variable morphologically. The constants are that the lobes are always adnate to the substrate; the color is always a shade of brown or olive; and soredia and isidia are always absent. Other than these aspects, all bets are off: lobules and apothecia vary from abundant to absent. In the photograph, note the presence of tiny brown dots, which comprise pycnidia of this individual.

CHEMISTRY: Fumarprotocetraric acid. Spot tests. Cortex K-, C-, KC-, P-, UV-. Medulla: K+ yellow turning brown, C-, KC-, P+ red, UV-.

NICHE: This species is endemic to eastern North America, where it has an Appalachian-Great Lakes distribution. In the central and southern Appalachians, *Melanohalea halei* is found on the bark and branches of trees and shrubs at high elevations. When you see this species in the Smokies, you know you have passed into a biome typical of more northerly parts of our continent.

KEY FEATURES: Brown foliose thallus lacking isidia and soredia, either with abundant apothecia or lobules, medulla P+ red, on bark and branches at high elevations in the Smokies.

Menegazzia subsimilis

Perfectly Perforate

Tripp 3448 (photo: Lendemer)

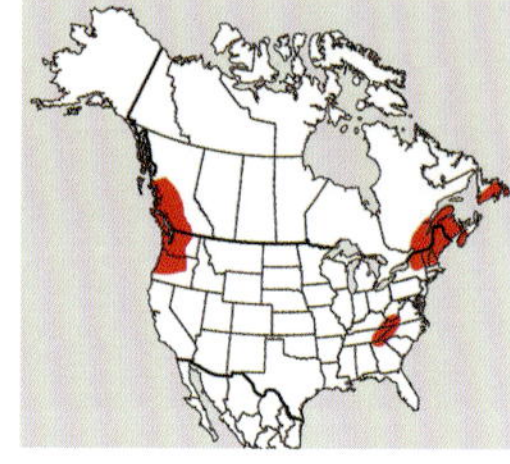

NOTES: Surprise! You might have thought this lichen with its inflated lobes was a *Hypogymnia*, but closer examination will reveal lobes with many circular perforations that provide a window into its dark heart (black inner cavity). In the past, lichenologists referred to many populations of this species as *Menegazzia terebrata*. *Menegazzia terebrata* can be separated from *M. subsimilis* by having soralia on the upper surface of the lobes while those of *M. subsimilis* are produced at the lobe tips and erupt with a hole in the center so that at least some resemble tiny trumpets.

CHEMISTRY: Atranorin, stictic acid, menegazziaic acid. Spot tests. Cortex: K+ yellow, C-, KC-, P-, UV-. Medulla: K+ yellow turning brown, C-, KC-, P+ orange, UV-.

NICHE: *Menegazzia subsimilis* occurs throughout the Smokies on the bark and branches of trees and shrubs, especially in humid habitats along creeks and rivers.

KEY FEATURES: Foliose lichen with inflated, perforated lobes, terminal erumpent soralia with a hole in the center, on bark and branches throughout the Smokies, especially in humid habitats.

Micarea bauschiana

Iron Distortion

Lendemer 44863 (photo: Tripp)

NOTES: *Micarea* is a difficult genus and is definitely not one of our favorites, which may or may not be related to its difficulty.

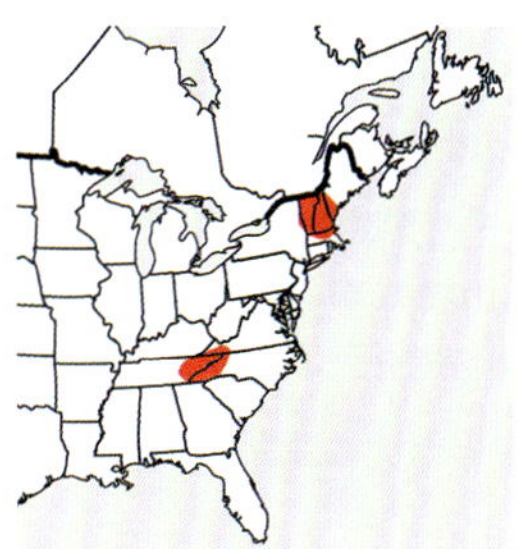

That being said, *Micarea bauschiana* is one of the easier-to-identify members of the genus, usually by its scurfy, greenish-to-brownish thallus and small, convex, variably pigmented apothecia, which bear very small, simple ellipsoid spores. There are not many other species that resemble *M. bauschiana* in the Smokies, and it is fairly common. Species of *Biatora* can be similar as they often have pale or tan apothecia, but the spores in all of those species are larger than those of *M. bauschiana*.

CHEMISTRY: No substances. Spot tests. K-, C-, KC-, P-, UV-.

NICHE: This species may be more widespread in eastern North America than our map indicates, but more collections are needed to confirm this suspicion. *Micarea bauschiana* is widespread in the Smokies and occurs in humid habitats along creeks and rivers as well as on cloud-immersed high-elevation mountain peaks. Although you may find it on a stray boulder, it thrives on metal-enriched rocks like the Anakeesta Formation and will even grow on old pieces of metal. Here, it is pictured growing on a rotting metal pipe in Bone Valley. Remember: lichens may grow on you, too, if you stand still long enough.

KEY FEATURES: Dirty green, crustose thallus, small, pale-to-dark convex apothecia, small colorless simple spores, on old metal or metal-enriched rocks.

Micarea chlorosticta

Lumpy Dumps

Lendemer 48564 (photo: Tripp)

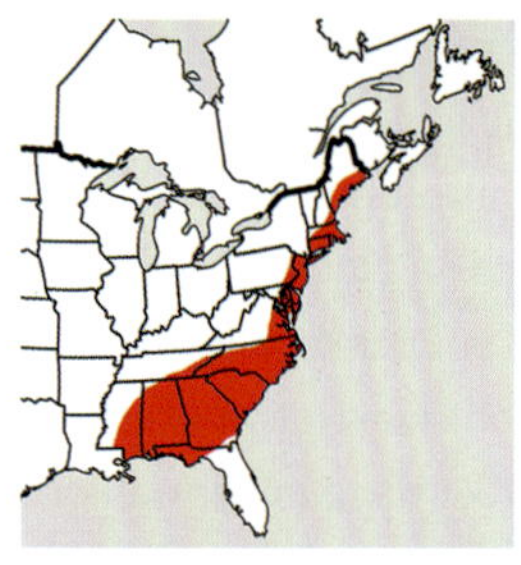

NOTES: *Micarea chlorosticta* is a widespread species in the Smokies that can easily be overlooked because of the small size of the apothecia and its brown thallus, which often blends in with the bark upon which it grows. Nonetheless, the species can be recognized by its minutely areolate greenish-brown thallus, small, colorless, needle-shaped spores, and apothecia that have a dark purplish-brown pigment in the hypothecium and epihymenium. *Micarea endocyanea* is superficially similar but differs in lacking secondary compounds and in having a smooth, thin thallus. The specific epithet of this species means something to the effect of "green" and "punctured"or "dappled," likely in reference to its minutely areolate thallus.

CHEMISTRY: Unidentified substance. Spot tests. K-, C-, KC-, P-, UV-.

NICHE: Though often overlooked, this species occurs on conifers throughout the Smokies. It typically grows on the trunks of trees in humid habitats such as along streams. Although most frequent in the low-elevation swamps of the Atlantic Coastal Plain, *Micarea chlorosticta* occurs up to approximately 5,000 ft. elevation in the Smokies.

KEY FEATURES: Crustose, minutely areolate greenish-brown thallus, small black apothecia with purplish-brown pigments, colorless, needle-shaped multi-celled spores, on conifers throughout the Smokies.

Micarea melaena

Malina

Tripp 3844 (photo: Lendemer)
Inset: Lendemer 32931 (photo: Tripp)

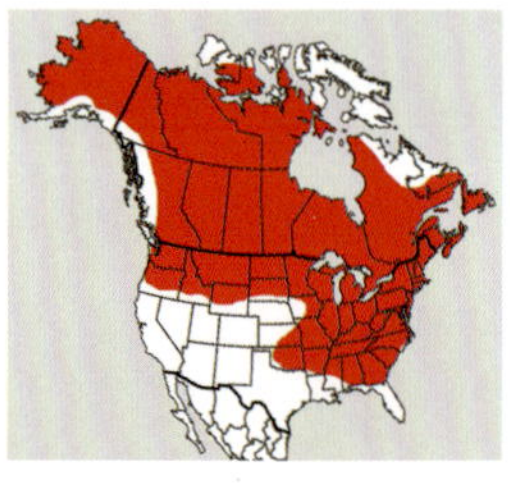

NOTES: You can be forgiven for thinking all *Micarea* species look more or less the same. We understand that it is just black dots and more black dots. However, if you take time to look inside, this one is a beauty. Externally, *M. melaena* resembles a dark form of *M. peliocarpa*, but the apothecia of *M. melaena* are all universally darker in color because they bear a purplish-brown pigment scattered throughout their fruiting bodies. *Micarea melaena* otherwise shares 4-celled spores and a C+ pink thallus with *M. peliocarpa*.

CHEMISTRY: Gyrophoric acid. Spot tests. K-, C+ pink, KC+ pink, P-, UV-.

NICHE: On rotting logs and old wood, especially conifers. In the Smokies, *Micarea melaena* is found primarily at middle-to-high elevations. Rarely, the species grows on shaded trunks of living conifers, especially spruce and hemlock, but also on fir and even more rarely on pine.

KEY FEATURES: Black convex apothecia with purplish-brown pigments throughout, 4-celled colorless spores, C+ pink (but almost impossible to see because this reaction is usually obscured by the pigments), on conifer lignum and trunks.

Micarea micrococca

Bitty Balls

Lendemer 33199 (photo: Tripp)

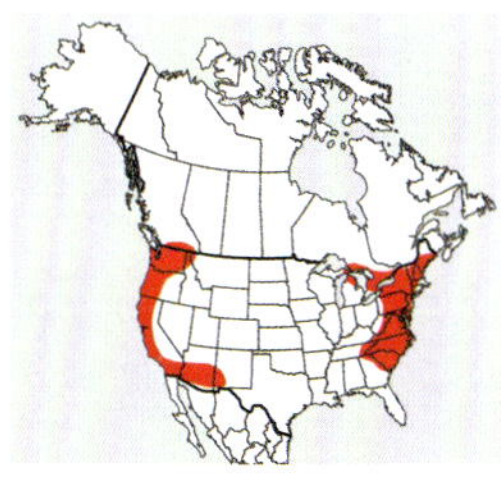

NOTES: Compare the photographs of this species, *Micarea micrococca*, with *M. prasina*. After doing so, we think you will agree that they are too similar to distinguish using only morphology, and will, too, turn to TLC for distinguishing features. In fact, the two species can grow together and—to make matters worse—often grow intermixed with C+ pink species of *Placynthiella* and *Trapeliopsis*. For more discussion, see entry under *M. prasina*, which is the name most specimens of this species would be filed under at your local museum. We could have in fact copied the entry for *M. prasina* here, but we didn't.

CHEMISTRY: Methoxymicareic acid. Spot tests. K-, C-, KC-, P-, UV-.

NICHE: Just like *Micarea prasina*, *M. micrococca* grows on shaded, decaying wood, humus, and bases of trees, especially conifers. If you see a green stain at the base of a pine tree, it is probably one of these species. It is common and widespread in the Smokies and throughout much of eastern North America.

KEY FEATURES: Scurfy, scuzzy green thallus composed of minute goniocysts (= thin, fungus enclosed plastic bags of algae), pale, convex apothecia, 2-celled colorless spores, methoxymicareic acid, on shaded wood, humus, and conifer bases.

Micarea neostipitata
White Stag

Lendemer 44527 (photo: Tripp)

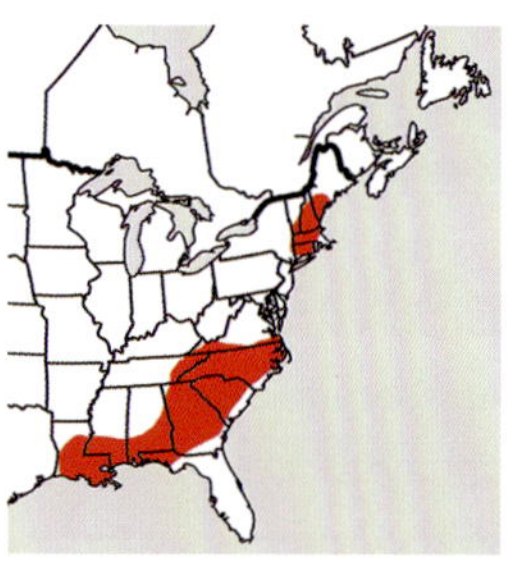

NOTES: This is probably one of the most common species of *Micarea* on conifers in the Smokies but it is usually overlooked because of its small size. Nonetheless, it can be easily recognized even in the field by its stipitate white pycnidia, which resemble small pale horns. The apothecia are black, and if you look inside, you will see that they contain purplish-brown pigments and narrow, needle-shaped colorless spores. The only similar lichen is *M. pycnidiophora*, but that species differs in having a P- thallus because it lacks fumarprotocetraric acid. The latter is also confined to the highest elevations of the Smokies.

CHEMISTRY: Fumarprotocetraric acid and lobaric acid. Spot tests. K-, C-, KC+ fleeting pink, P+ orange-red (especially the bases of the pycnidia), UV+ blue-white.

NICHE: On the bark of conifers, especially pine and hemlock throughout the Smokies. *Micarea neostipitata* is widely distributed and common in eastern North America but is generally overlooked because it is so humble.

KEY FEATURES: Black convex apothecia with purplish-brown pigments, needle-shaped spores, stalked white pycnidia, P+ orange-red, on conifers throughout the Smokies.

Micarea peliocarpa

Thare-n-Everywhare

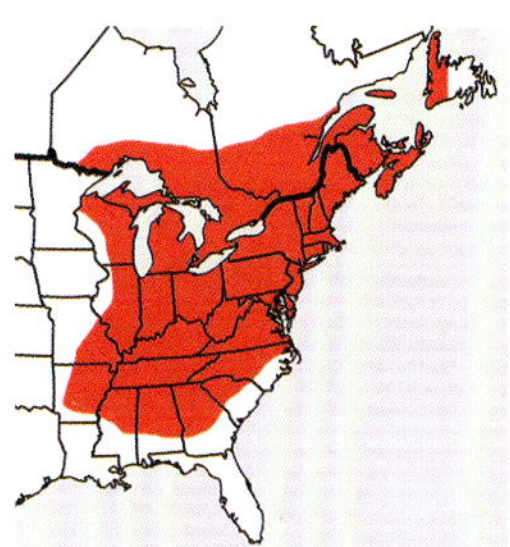

Tripp 3931 (photo: Lendemer)

NOTES: Another one of those pesky *Micarea* species! However, this one is common and easy to identify despite its variability. The apothecia of *Micarea peliocarpa* can vary from black to completely pale white, and the thallus can be indistinct or absent to thick and well-developed. The thallus photographed here represents the most common form you will see in the forest and shows the variability in apothecial morphology, even on a single thallus. Despite this variability, the species can be identified by its 4-celled colorless spores, C+ pink thallus and apothecia, which are internally pale pigmented except in the uppermost layers. The specific epithet "pelio" + "carpa" translates to "black and blue fruiting bodies."

CHEMISTRY: Gyrophoric acid. Spot tests. K-, C+ pink, KC+ pink, P-, UV-.

NICHE: All around you, everywhere close to the ground, in shaded places, on everything. Find a rotting log, it will be there. Find some mosses or humus along the trail, it will be there. Find a tree base or a stump, it will be there. Shaded non-calcareous rocks? Yep, there, too.

KEY FEATURES: Variably colored convex apothecia that are (allow us to propose a new contraction, suitable for the Smokies: "that are" = "thare") internally pale, 4-celled colorless spores, C+ pink, on all types of shaded substrates, close to the ground.

Micarea polycarpella

Carp Town

Tripp 3746 (photo: Lendemer)

NOTES: We have no idea what to write for this species. It is known from only one collection in the Smokies (and is a new report for North America), and there is no easy way to differentiate it from all the other *Micarea* species with black apothecia and simple spores. For more information, you will need to be able to translate Czech. Then, kindly inform us as to what is learned. Maybe you'll be the first to find more of it. Good luck!

CHEMISTRY: No substances. Spot tests. K-, C-, KC-, P-, UV-.

NICHE: The distribution of this species is poorly understood. It has been found on metal-rich rocks in disturbed and urban environments in Europe. In the Smokies, it was also found on iron-rich rocks near Lonesome Pine in North Carolina, but we would never go so far as to say the Smokies are either disturbed or urban!

WE THINK THE KEY FEATURES ARE: Indistinct thallus, black, convex apothecia, simple spores, C- and P-, on iron-rich rocks at middle elevations.

Micarea prasina

Log Leeks

Tripp 3844-A (photo: Tripp)

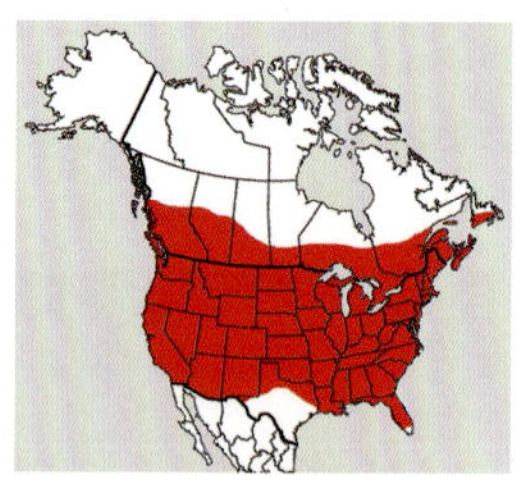

NOTES: Just when you thought we were confused . . . that's not an alga! All the thick, scurfy greenish scuz that you see on rotting logs and tree bases in the Smokies almost undoubtedly belongs to this species, at least in part. (Recall: different species of *Micarea* often grow together or in very very close proximity . . . sorry! On the bright side, it would make a *great* evolutionary system in which to study barriers to reproductive compatibility.) The thallus of *Micarea prasina* is composed of minute green structures termed goniocysts, which resemble thin plastic bags–like the kind you put your grocery store leeks in–filled with algae. If you search carefully, you will usually find at least some pale, convex apothecia associated with the crust, and if you continue to search carefully, you will find 2-celled, colorless spores inside these fruiting bodies. Despite all the talk, this species can only be differentiated from *M. micrococca* by TLC.

CHEMISTRY: Prasinic acid. Spot tests. K-, C-, KC-, P-, UV-.

NICHE: Widespread in the Smokies and beyond, on shaded, rotting logs, humus, and bases of conifers, especially pine. Be careful when trying to separate this species (and its niche) from other *Micarea* or even *Placynthiella* and *Trapeliopsis* species. It is all one big, confusing community where no one has any boundaries . . .

KEY FEATURES: Scurfy, scuzzy green thallus composed of minute goniocysts (algae bags), pale, convex apothecia, 2-celled colorless spores, prasinic acid, on shaded wood, humus, and conifer bases.

Micarea pycnidiophora

Rare Stag

Lendemer 53142 (photo: Tripp)

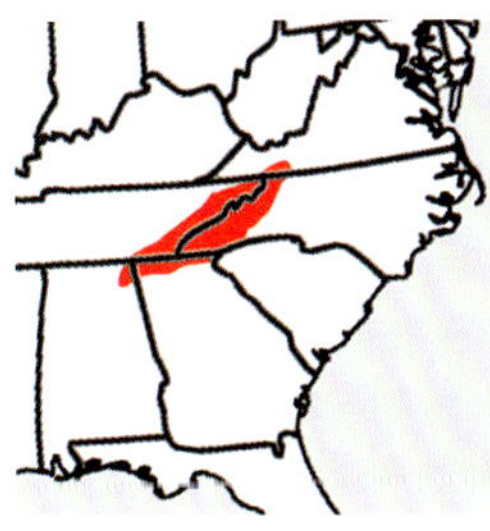

NOTES: *Micarea pycnidiophora* is rather like *M. neostipitata* but rarer and maybe ever so slightly more refined. It shares the pale, stipitate pycnidia and needle-shaped spores with the latter species but differs in having pale apothecia and lacking fumarprotocetraric acid (P-) in the thallus. To find Rare Stag, we suggest getting out your hand lens and start applying it to the bark of big conifers at high elevations. One has to be careful because the stipitate pycnidia are easily mistaken for isidia. If you mount one in water and examine it under the compound microscope, you will easily see a dollop of small, ellipsoid conidia that forms as they ooze out the tip of the pycnidia.

CHEMISTRY: No substances. Spot tests. K-, C-, KC-, P-, UV-.

NICHE: This species is rare and easily overlooked but apparently restricted to old growth forests at high elevations in the Smokies, where it grows on the bark of mature conifers and more rarely on Yellow Birch. Keep an eye out for it the next time you make the long, day trip to Pecks Corner.

KEY FEATURES: Gray-to-greenish areolate thallus, pale convex apothecia, needle-shaped spores, stalked white pycnidia, P-, on mature conifers and hardwoods at the highest elevations in the nicest forests.

Micareopsis irriguata

The Irrigated Micarea Look Alike Lichen

Tripp 3645 (photo: Lendemer)
Inset: Tripp 4980 (photo: Lendemer)

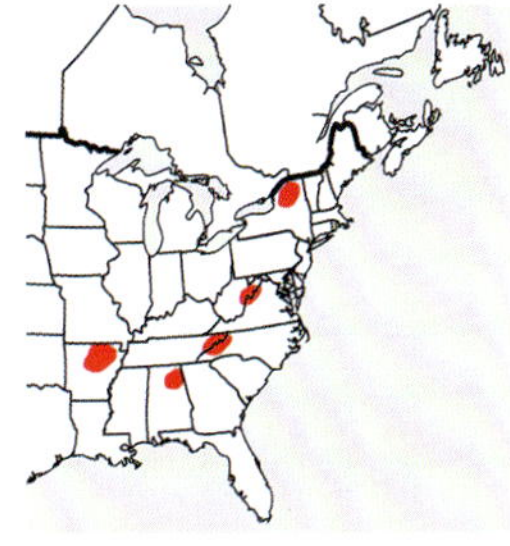

NOTES: Sometimes the best things come where you least expect them. Round a corner twenty miles in on the Lakeshore Trail, and you might find this species on a shaded vertical rock face near a roaring river or tumbling fork. The most striking feature of *Micareopsis irriguata* is its bright green soredia, which contrast the dull, dark green areoles and black prothallus. When you are done admiring it in the field, shine a UV lamp on it and win a free light show on par with the Northern Lights!

CHEMISTRY: Sphaerophorin. Spot tests. K-, C-, KC-, P-, UV+ blue-white.

NICHE: Rare to infrequent but nonetheless widespread throughout the Appalachians and Ozarks. *Micareopsis irriguata* is more common in the Smokies than it is in other parts of its range, but you will still have to hunt to find it. This species is always on shaded, vertical rock faces or sheltered rock overhangs, most often in riparian corridors and on seepage faces.

KEY FEATURES: Black prothallus, dull areolate thallus, bright green soredia erupting from areoles, UV+ blue-white, black, convex apothecia, on non-calcareous rocks in humid habitats at middle and low elevations.

Microcalicium ahlneri

Blurps

McMullin 19051 (photo: McMullin)

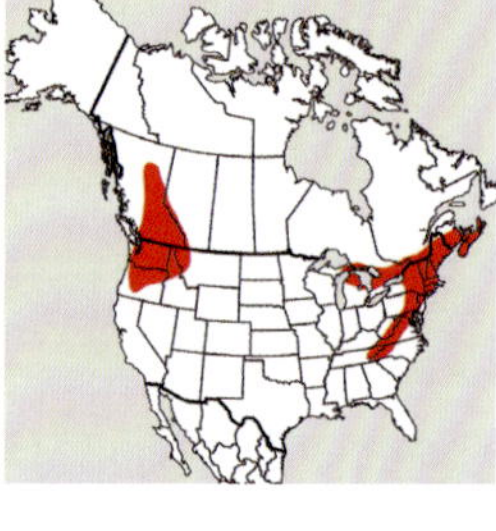

NOTES: This species is silly from the start; was there really a need to give it a genus name recognizing its short stature? Aren't just about all the stubbles "micro" in some regard? If you can get past the name, then you might be able to appreciate the unusual blue-green color of the spore mass that tends to make the genus recognizable in the field. That color, combined with the septate blue-brown colored, transversely septate spores makes the genus even more recognizable. This species differs from the others in having short stalked pins and not occurring on rocks or lichens growing on rocks.

CHEMISTRY: No substances. Spot tests. K-, KC-, C-, P-, UV-.

NICHE: If the morphology weren't strange enough, this ecology is even weirder. *Microcalicium ahlneri* grows on the soft, decaying wood of standing dead trees. In contrast, most other caliciodis grow on bark or hard, dry wood. The species is most often found on the wood of oaks. As it is easily overlooked, the exact distribution is not fully understood. Nonetheless, it appears to be widespread in temperate and boreal North America.

KEY FEATURES: Minute, short-stalked apothecia, blue-green spore mass, transversely septate blue-brown spores, on soft decaying wood of snags.

Miriquidica leucophaea

Tender Tiles

Lendemer 29717-A (photo: Tripp)

Tripp 5018 (photo: Tripp)

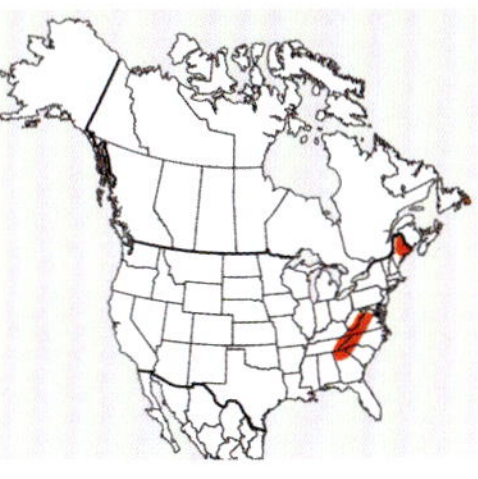

NOTES: Tender Tiles refers to the precious, interlocking weathered clay-colored areoles that surround the distinctive reddish-brown convex apothecia of this species. Upon first glance, *Miriquidica leucophaea* might resemble one or more species of *Trapelia* in the field, but members of that genus are C+ pink. Even though you can identify Tender Tiles by gestalt based on this photograph, you would be wise to check that it has simple, colorless spores and, if you have access to TLC, contains miriquidic acid. The other species of *Miriquidica* that occurs in the Smokies has an indistinct white areolate thallus and black apothecia.

CHEMISTRY: Miriquidic acid. Spot tests. K-, C-, KC-, P-, UV-.

NICHE: This species is an infrequent member of rock lichen communities at the highest elevations of the southern Appalachians. In the Smokies, it is most frequent around Clingmans Dome but can be found elsewhere on non-calcareous rocks. Like so many other species that occur in this park, its primary center of distribution is somewhere farther north.

KEY FEATURES: Weathered-clay-colored contiguous areoles, reddish-brown convex apothecia, simple colorless spores, miriquidic acid, on non-calcareous rocks at high elevations.

Montanelia disjuncta

Dark Stretch

Hollinger 2677 (photo: Lendemer)

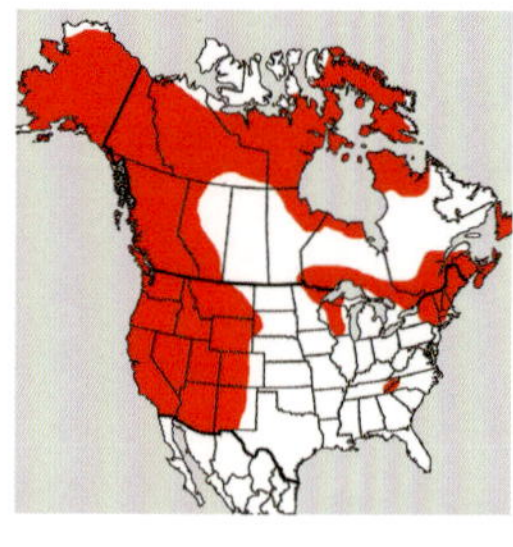

NOTES: There are not very many sorediate foliose lichens with brown thalli in the Smokies and, as such, *Montanelia disjuncta* is not likely to be confused with anything else in the region. *Melanelia culbersonii* is the most similar species but differs in having marginal soralia and a medulla that is P+ orange due to the presence of norstictic and stictic acids. As the genus name suggests, these are mostly species of montane habitats in the Appalachian Mountains. *Montanelia panniformis* and *Melanelia stygia* are two other rock-dwelling species that could be confused with *M. disjuncta*, but both of those are esorediate and instead have thalli with numerous small secondary lobes. *Melanelia stygia* also differs in having a P+ red medulla, while the chemistry of *M. panniformis* is that same as that of *M. disjuncta*.

CHEMISTRY: Perlatolic acid. Spot tests: (cortex) K-, C-, KC-, P-, UV-; (medulla) K-, C-, KC-, P-, UV+ dull blue-white.

NICHE: This is a species of exposed, non-calcareous rocks in boreal and arctic regions of North America. Its distribution extends down the Appalachian Mountains to the Smokies, where it is known from a single high-elevation location in North Carolina.

KEY FEATURES: Small brown foliose thallius, laminal soralia with coarse soredia, UV+ blue-white medulla, on exposed non-calcareous rock at high elevations.

Montanelia panniformis

Fudgy Fingers

Lendemer 37924 (photo: Lendemer)

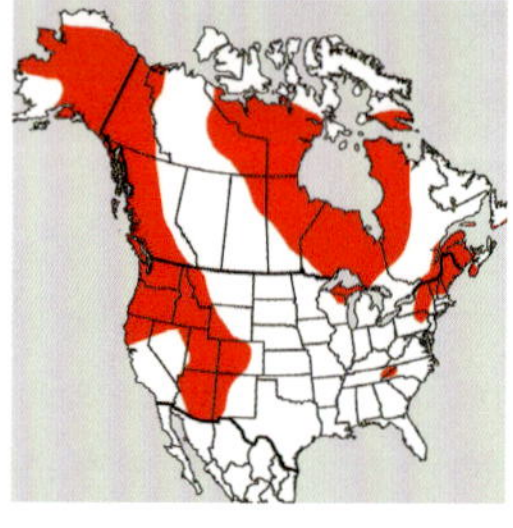

NOTES: *Montanelia panniformis* has a very distinctive brown thallus with small secondary lobes that seem to fan out over the surface, giving the appearance of a mat. In some respects, it is most likely to be confused with *Melanelia stygia*, but that species has a darker, almost black thallus and the medulla is P+ red due to the presence of fumarprotocetraric acid. *Montanelia disjuncta* is also similar in having small, narrow lobes and a brown thallus, but is sorediate and thus not likely to be confused with *M. panniformis*. In some respects, confusion with the normally corticolous *Melanohalea halei* is possible, but that species has a P+ medulla due to the presence of fumarprotocetraric acid, and the secondary lobes are shorter, never forming an elegant mat.

CHEMISTRY: Perlatolic acid. Spot tests: (cortex) K-, C-, KC-, P-, UV-; (medulla) K-, C-, KC-, P-, UV+ dull blue-white.

NICHE: This is a northern species of boreal and arctic regions that hops down the Appalachian Mountains, occurring rarely on exposed rocky summits at high elevations. It is extremely rare in the Smokies.

KEY FEATURES: Dark brown foliose thallus, numerous small secondary lobes, no isidia or soredia, medulla P- and UV+ blue-white, on exposed rocky summits.

Multiclavula mucida

Frail Bones

Lendemer 33279 (photo: Tripp)
Inset: Tripp 3900 (photo: Lendemer)

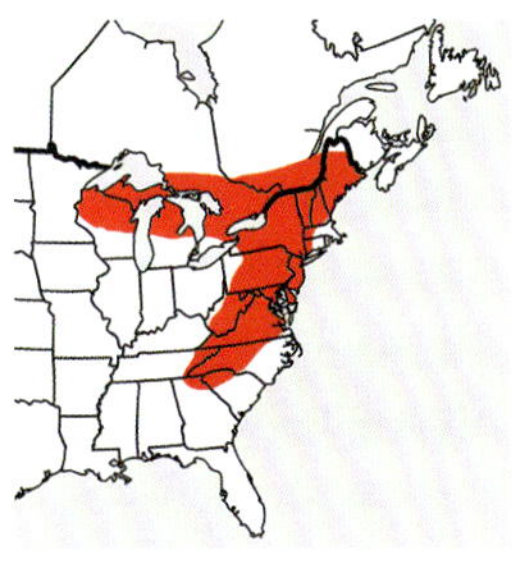

NOTES: Basidiomycete lichens are few and far between in this part of the world, and we may be better off for it. This is one of the lucky few that grows in our area, and it is relatively common throughout the Smokies. You can easily recognize *Multiclavula mucida* by its tall, pale white basidiomes, which resemble clubs or fingers growing out of a rather unattractive green algal crust. As seen in this photograph, the tips of these fruiting bodies are often slightly yellow or brown. However, like mushrooms, the fruiting bodies of this species don't persist: the basidiomes wither and discolor when they dry. So, enjoy it in the field and take only photographs!

CHEMISTRY: No substances. Spot tests. K-, C-, KC-, P-, UV-.

NICHE: Look at that map for a classic Appalachian-Great Lakes distribution in North America. *Multiclavula mucida* is widespread and common in eastern North America, where it grows on moist, shaded rotting wood.

KEY FEATURES: Unattractive green crust (the specific epithet means "moldy" or "mucus-y"), pale white club-like fruiting bodies (the generic name means "many clubs") with yellow or brown tips, on moist rotting wood and logs at all elevations.

Mycoblastus caesius

Blue Luster

Tripp 2466 (photo: Lendemer)

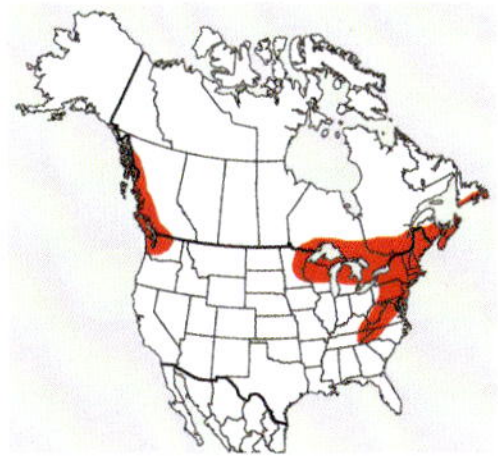

NOTES: At this point you may be wondering whether we are simply snapping random photos of wet, rotting wood, and calling it a lichen, but you would be wrong.

Mycoblastus caesius is one of those species that sometimes blends in with its surroundings, but can be picked out of the background noise with close inspection. It is easily recognized by its dark, blue-black prothallus (hence: "caesius") and white-to-gray areolate thallus bearing fine, powdery, greenish soredia. These characters combined with its chemistry are distinctive. You might confuse *M. caesius* with *Ochrolechia arborea*, but the latter species is UV+ yellow and lacks a dark prothallus. Several sorediate *Lecanora* species are also similar to *M. caesius*, but they are UV- and K+ yellow, the latter owing to their production of atranorin. *Mycoblastus caesius* is also restricted to the highlands.

CHEMISTRY: Perlatolic acid. Spot tests. K-, C-, KC-, P-, UV+ blue-white.

NICHE: This species is common and widespread at high elevations throughout the southern Appalachians, where it occurs on the bark of hardwoods such as cherry and birch as well as on bark of conifers such as spruce and fir. It also frequently occurs on shrubs in shrub balds.

KEY FEATURES: Blue-black prothallus, white-to-gray areolate thallus, fine, greenish soredia, UV+ blue-white, on bark and branches of trees and shrubs at high elevations.

Mycoblastus sanguinarioides

A Good Day

Tripp 5068 (photo: Lendemer)
Inset: Tripp 3535 (photo: Lendemer)

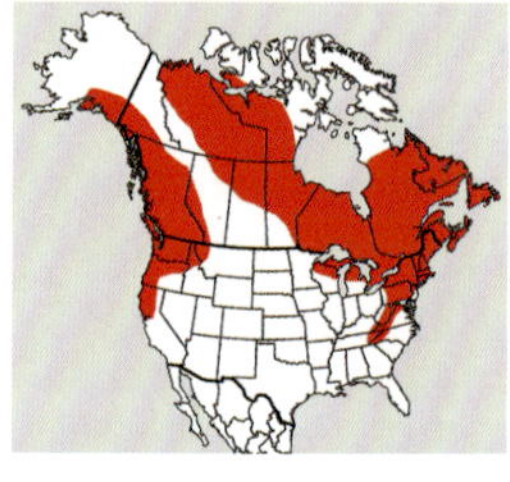

NOTES: *Mycoblastus sanguinarioides* is a good example of a species that was sitting right under our noses until someone finally pointed it out to us. The species was originally described from Tasmania but has since been shown to be widespread in many parts of the world. It was overlooked in the past because when lichenologists saw a crustose lichen with large black apothecia, simple colorless spores, and a blood-red pigment at the base of the apothecia, they simply called it *M. sanguinarius*. Both species are actually present in the Smokies, but *M. sanguinarioides* can be distinguished from *M. sanguinarius* by the presence of POL+ granules in the hymenium of the apothecia.

CHEMISTRY: Atranorin and fatty acids. Spot tests. K+ yellow, C-, KC-, P-, UV-; red pigments K+ purple.

NICHE: *Mycoblastus sanguinarioides* occurs on bark and wood at high elevations in the Smokies and elsewhere in the southern Appalachians. The species is much more common in the boreal forests than it is here in the southland.

KEY FEATURES: White crustose thallus, large black apothecia with blood-red pigment at the bases, simple colorless spores, K+ yellow thallus, POL+ granules in hymenium, on bark and wood at high elevations.

Mycoblastus sanguinarius

Lichen Blood and Guts

Tripp 5411A (photo: Lendemer)

NOTES: Life used to be easy. There was a time when any *Mycoblastus* with red pigments at the bases of the apothecia would have been called *M. sanguinarius*. It was a simpler time back then, before you had to use polarizing light filters to discern whether the hymenium was inspersed with granules or not. Now of course, life is so complicated. Both this species and *M. sanguinarioides* occur in the Smokies and the only way to distinguish them is to confirm the absence of POL+ granules in the hymenium of *M. sanguinarius*. However, if it helps, our fieldwork throughout the Smokies seems to suggest that this species is confined to the highest peaks in the central portions of the Park.

CHEMISTRY: Atranorin and fatty acids. Spot tests. K+ yellow, C-, KC-, P-, UV-; red pigments K+ purple.

NICHE: This species occurs on bark and wood at the highest elevations of spruce-fir covered peaks in the Smokies. It is otherwise widespread throughout arctic boreal forests.

KEY FEATURES: White crustose thallus, large black apothecia with blood-red pigment at the bases, simple colorless spores, K+ yellow thallus, no granules in hymenium, on bark and wood at high elevations.

Mycocalicium subtile

Weevil Parasols

Tripp 5286A (photo: Lendemer); Inset: McMullin 19122 (photo: McMullin)

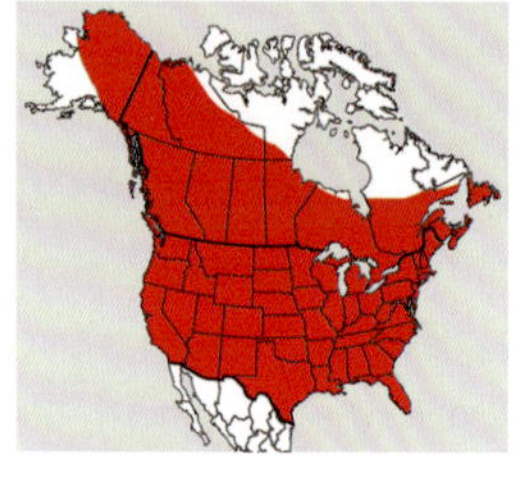

NOTES: *Mycocalicium subtile* is one of the fungal allies that lichenologists befriended a long time ago, even though they lack photobionts. This species can be recognized by its white, stain-like thallus, its black, pin-like ascomata, and its simple brown spores. Several other species of *Mycocalicium* have been reported from the Smokies, but they are difficult to tell apart without careful anatomical sectioning (when in doubt, check the dichotomous key). *Chaenothecopsis savonica* is morphologically also similar to Weevil Parasols but differs in characters of the ascus tips that are almost impossible to observe. If you call every non-lichenized stubble fungus with simple brown spores *M. subtile*, we doubt anyone will file a formal complaint.

CHEMISTRY: No substances. Spot tests. K-, C-, KC-, P-, UV-.

NICHE: *Mycocalicium subtile* occurs on dry wood of standing, decorticate trees and logs, especially at middle-to-low elevations in the Smokies. Outside of the Smokies, the species is widespread and common, exactly as the distribution map indicates.

KEY FEATURES: White stain-like thallus, no photobiont, pin-like stalked apothecia, brown, simple spores, on dry wood.

Mycoporum acervatum

Maple's Sugar

Lendemer 53183 (photo: Tripp)

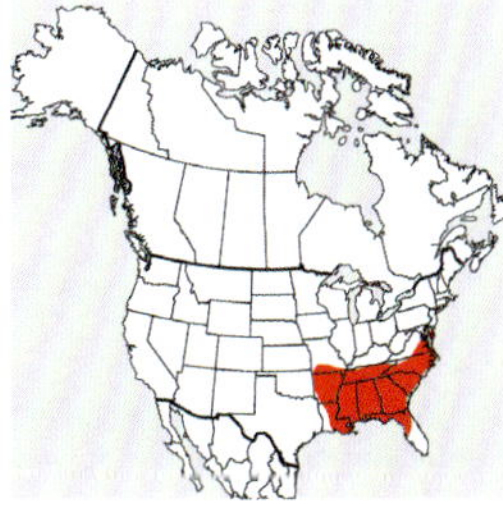

NOTES: You might be tempted to think that all species of *Mycoporum* look the same: some tiny crustose lichen (sometimes, not a lichen but a saprobe) with tiny black perithecia. Sometimes, we feel that way, too. But not in the case of Maple's Sugar! Whereas the thallus in most material of this species is on the white side, material in the Smokies is distinctively thick and yellowish in color. Know Maple's Sugar by this feature, together with its compound perithecia that bear muriform spores. On account of its muriform spores, you are most likely to confuse this species with *M. compositum* or *M. pycnocarpoides*, but both of those species have larger spores and lack the well-developed thallus typical of Smokies material of *M. acervatum*. You might also be tempted to confuse this species with *M. eschweileri*, which similarly has compound perithecia, but the latter has 4-celled instead of muriform spores.

CHEMISTRY: No substances. Spot Tests: K-, KC-, C-, P-, UV-.

NICHE: *Mycoporum acervatum* is a species of the southeastern United States, particularly in Coastal Plain habitats. It has been collected on a variety of hardwoods including oaks, black cherries, and maples, but we think it looks most at home on big giant Sugar Maples in the Smokies. The specific epithet actually refers to the fact that the perithecia are "heaping" (i.e., in small clustered) rather than the fact that, in the Smokies, this species preferentially grows on old, large Sugar Maples. Look for it at medium-to-high elevations in the Park, such as the lovely population photographed not far from the Flat Creek Trailhead on Balsam Mountain Road.

KEY FEATURES: Black, compound perithecia against a well-developed yellowish leprose thallus, relatively small muriform spores, occurrence on hardwoods at medium to high elevations.

Mycoporum biseptatum

Forney Flecks

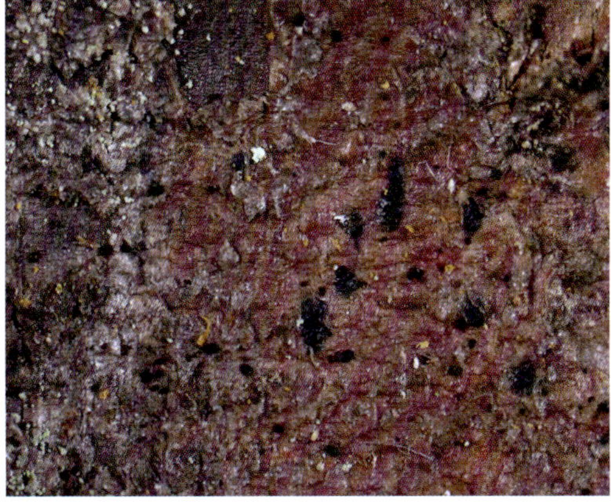

Lendemer 32776 (photo: Lendemer)

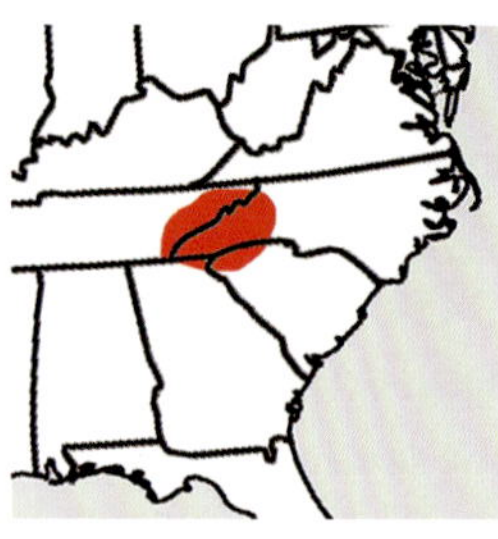

NOTES: *Mycoporum biseptatum* is a non-lichenized species that is nonetheless identifiable based on its lack of a photobiont, its compound, black perithecia, which are somewhat flattened, its relatively large, colorless, three-celled spores, and its ecology (see below). If found, the species will most likely be confused with members of *Arthonia*, however the spores of *M. biseptatum* are considerably larger than those of any *Arthonia* currently known from the southern Appalachians. *Mycoporum biseptatum* might additionally be confused for other members of this genus, but most of the species that occur in the southern Appalachians have muriform spores and occur on the bark of trees.

CHEMISTRY: No substances. Spot tests: K-, C-, KC-, P-, UV-.

NICHE: The type specimen of *Mycoporum biseptatum* was collected by Gunner Degelius at relatively high elevations along Forney Ridge, in 1941. Since then, it has been found only very rarely, in open habitats such as shrub balds at high elevations. Look for Forney Flecks on smooth bark and twigs of fast-growing shrubs such as *Viburnum*, or perhaps on young *Prunus* branches.

KEY FEATURES: Non-lichenized crustose thallus, compound, black, flattened perithecia, relatively large, colorless, 3-celled spores, high elevations, bark and branches of fast-growing shrubs.

Mycoporum compositum

High Lonesome

Lendemer 32899 (photo: Tripp)

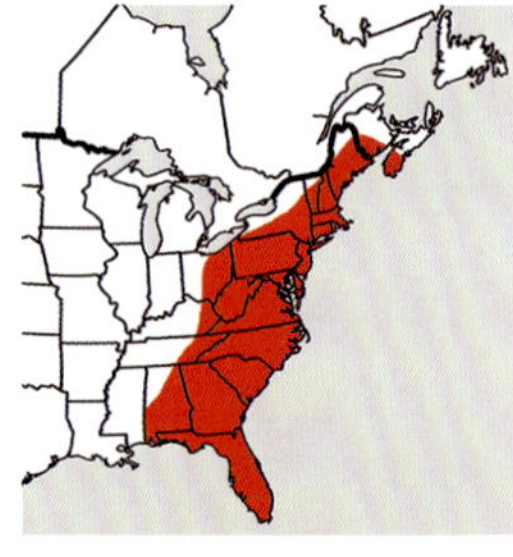

NOTES: At first glance in the field, you might be tempted to think that this is an *Arthonia* but if you look closer, you will see that the fruiting bodies are really aggregated clusters of perithecia, each bearing a tiny ostiole at the apex. Although species of *Mycoporum* are not truly lichenized, *M. compositum* and *M. pycnocarpoides* both occasionally form loose associations with algae (see the greenish patches in the photograph). In addition to the perithecia, this species can be recognized by its muriform, colorless spores that are smaller and have fewer cells than those of *M. pycnocarpoides*.

CHEMISTRY: No substances. Spot tests. K-, C-, KC-, P-, UV-.

NICHE: Widespread on the bark of hardwoods, especially maple. In the Smokies, *Mycoporum compositum* is most frequent in high-elevation northern hardwood forests but occasionally also occurs at middle and lower elevations.

KEY FEATURES: White crustose thallus, no photobiont, compound black perithecia, colorless muriform spores (these ~20-38 x 12-17 µm), on hardwoods, mostly at high elevations.

Myelochroa aurulenta

Blue Creeper

Tripp 3653 (photo: Lendemer)

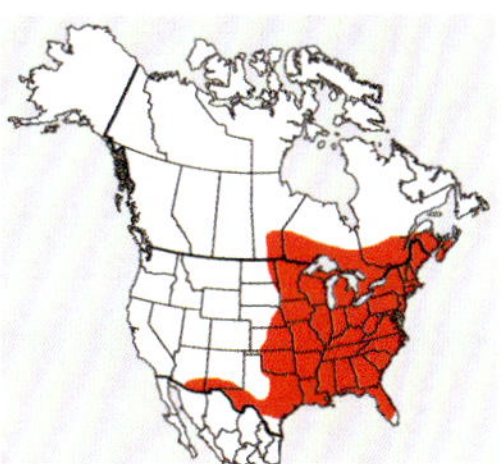

NOTES: Every time you see this species, you will remark to yourself how blue it is in color–in fact, one of those true blues that you see so rarely in lichens. This, together with squarrose lobe tips, pustular soralia on the thallus surface, and light-yellow-pigmented medulla (this visible among the pustules in the photograph) make *Myelochroa aurulenta* easily recognized in the field. When growing on rocks, the lobes appear to undulate and creep across the substrate, as if looking for a moss or another lichen to overtake.

CHEMISTRY: Atranorin, zeorin, leucotylin, secalonic acid. Spot tests. Cortex: K+ yellow, C-, KC-, P-, UV-. Medulla: K+ weak yellow, C+ weak yellow, KC+ weak yellow, P-, UV-.

NICHE: Blue Creeper is extremely common throughout the Smokies on bark and branches of trees and shrubs as well as on non-calcareous rocks. The most striking individuals, like the one photographed here, occur on shaded rocks in forests. Make this one of the top 10 foliose lichens that you learn in the Smokies. It is elsewhere common in similar habitats in eastern North America.

KEY FEATURES: Blue-gray foliose thallus, squarrose, browned lobe tips, laminal pustular soralia, light yellow pigmented medulla, on rocks and bark throughout the Smokies, exceptionally common.

Myelochroa galbina

After The Storm

Tripp 3863 (photo: Lendemer)

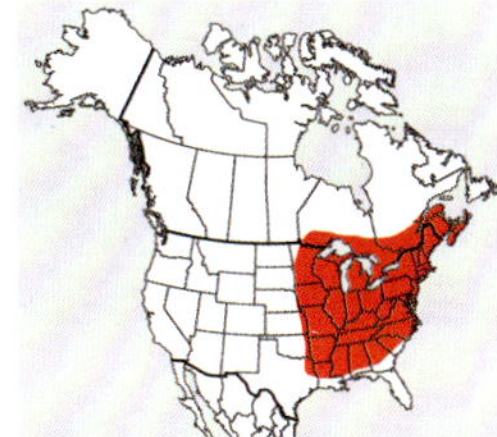

NOTES: Once common and widespread in temperate eastern North America, *Myelochroa galbina* has become rare in this region over the last century. In fact, the only place it is common is in the southern Appalachians. This species quite resembles *Hypotrachyna livida* but is smaller, has a light-yellow-pigmented medulla, and produces galbinic acid in the medulla, which reacts P+ orange. The specific epithet "galbina" means yellow, which might seem helpful at first, but the epithet "aurulenta" (as in *Myelochroa aurulenta*) also means yellow. Moreover, this yellow pigment in the medulla *of M. galbina* can be hard to see, but is most easily detected near the margins of the apothecia.

CHEMISTRY: Atranorin, zeorin, leucotylin, galbinic acid, secalonic acid. Spot tests. Cortex: K+ yellow, C-, KC-, P-, UV-. Medulla: K+ yellow turning dark brown, C-, KC-, P+ orange, UV-.

NICHE: *Myelochroa galbina* occurs on bark and branches of hardwoods and conifers, especially at low and middle elevations in the Smokies. It is most frequently found covering fallen branches that litter the forest floor after a storm.

KEY FEATURES: Blue-gray foliose thallus, absence of isidia and soredia, light-yellow-pigmented medulla, medulla P+ orange, on bark, frequently on canopy branches throughout the Smokies.

Myelochroa metarevoluta

Highball

Lendemer 18952 (photo: Tripp)

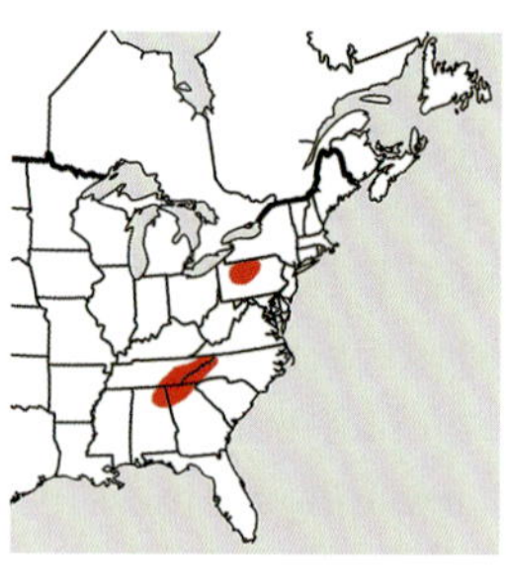

NOTES: *Myelochroa metarevoluta* is a foliose lichen that loves to hide in plain sight. It blends in with other foliose lichens in habitats where it grows and could be mistaken for *Hypotrachyna revoluta* if chemistry is overlooked. The most distinctive features of *M. metarevoluta* are its lightly yellow-pigmented medulla, conspicuous capitate soralia on the lobe tips, and ascending strap-like lobes.

CHEMISTRY: Atranorin, zeorin, leucotylin, galbinic acid, secalonic acid. Spot tests. Cortex: K+ yellow, C-, KC-, P-, UV-. Medulla: K+ yellow turning dark brown, C-, KC-, P+ orange, UV-.

NICHE: *Myelochroa metarevoluta* is widespread but overlooked on branches of trees and shrubs overhanging creeks and rivers in the Smokies. It also occurs throughout the Appalachians at middle-to-low elevations. Keep an eye out for this species next time you visit your favorite swimming hole. It is a typical example of an Appalachian-east Asian disjunction, as it also occurs in Japan.

KEY FEATURES: Gray to blue-gray foliose thallus, ascending strap-like lobes, conspicuous capitate soralia, light yellow-pigmented medulla that is P+ orange, on shrubs and branches along riparian corridors.

Myelochroa obsessa

Impossibly Adpressed

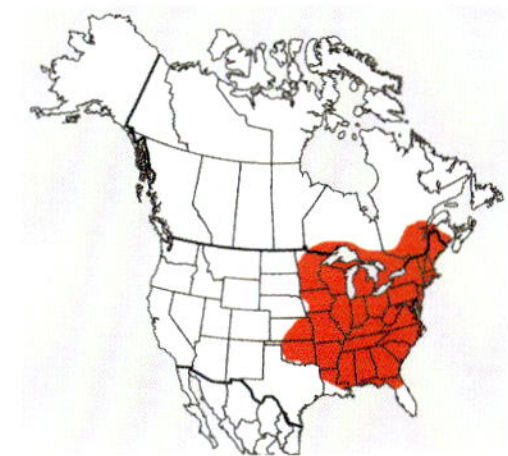

Lendemer 33111 (photo: Tripp)

NOTES: This is one of our favorite foliose lichens in the Smokies–you might even say we are obsessed with it. The tightly attached, flat lobes with brown tips, short isidia, and weakly yellow-pigmented medulla are distinctive traits of this species. The only species with which *Myelochroa obsessa* could be confused are members of the genus *Xanthoparmelia* (all yellow-green instead of blue-gray) or *Imshaugia aleurites* (medulla K+ strong yellow, pale lower surface, rarely on rock). You might also be tempted to call this lichen a species of *Bulbothrix*, but those have characteristic marginal cilia that are inflated at their bases ("bulbils").

CHEMISTRY: Atranorin, zeorin, leucotylin, secalonic acid. Spot tests. Cortex: K+ yellow, C-, KC-, P-, UV-. Medulla: K+ weak yellow, C+ weak yellow, KC+ weak yellow, P-, UV-.

NICHE: *Myelochroa obsessa* is widespread at middle-to-low elevations on non-calcareous rocks throughout the Smokies. It is also common throughout the central and southern Appalachians, but becomes rarer farther north. Enjoy this one on your next hike along Fontana Lake.

KEY FEATURES: Blue-gray thallus, tightly attached lobes, short isidia, pale-yellow-pigmented medulla, on non-calcareous rocks throughout middle and low elevations.

Nadvornikia sorediata

Puddin' Pots

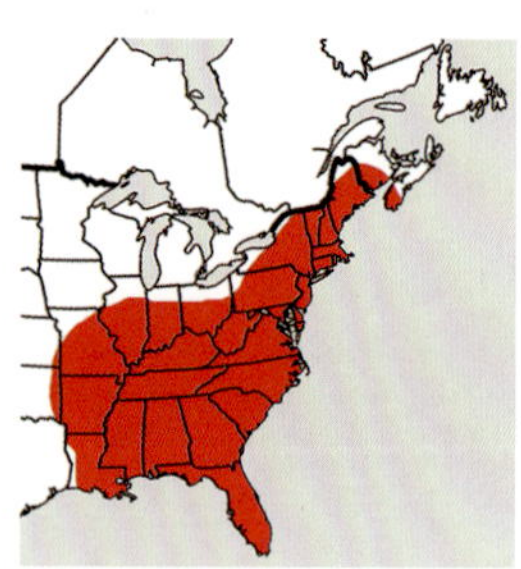

Tripp 3607 (photo: Lendemer)

NOTES: *Nadvornikia sorediata* is easy to take for granted. With its shiny gray, continuous thallus and white to gray soredia, Puddin' Pots is fairly non-descript as far as lichens go. But what if we told you it was one of the most common crustose lichens in the Park? We now invite you to take a closer look at a species that holds several surprises. First, it has a *Trentepohlia* photobiont, which gives the thallus a bronze hue whenever slightly abraded. Second, the soralia are punctiform and reminiscent of what pudding looks like after you stick your finger in the bowl. Third, the chemistry of this species is both simple and highly distinctive. The most similar sterile sorediate crust is *Lecanora layana*, but the latter associates with a coccoid green alga and produces atranorin and zeorin in addition to stictic acid.

CHEMISTRY: Stictic acid. Spot tests. K+ yellow turning dirty brown, C-, KC-, P+ orange, UV-.

NICHE: On the bark and branches of conifers and hardwoods throughout the Smokies and across temperate eastern North America.

KEY FEATURES: Gray, shiny continuous crust, punctiform soralia, white to gray soredia, *Trentepohlia* photobiont, P+ orange, on bark and branches throughout the Smokies at all elevations.

Nectriopsis rubefaciens

Orange Horror

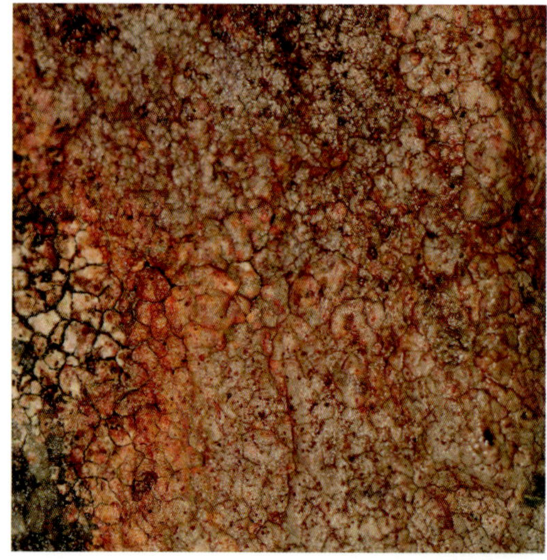

Tripp 4957 (photo: Lendemer)

NOTES: *Nectriopsis rubefaciens* is one of the lichenicolous fungi that we decided to include in this book because it is both conspicuous and common and thus can be readily spotted. In the field, one can find this species by the relatively large, discolored patches that form on the thalli of its host, *Aspicilia* (typically *A. cinerea*). Closer inspection will reveal a peppering of minute, orange perithecia scattered across the host thallus. In addition to the host, as well as size and color of the perithecia, a distinguishing character of this species are the small 2-celled colorless spores.

CHEMISTRY: No substances. Spot tests. K-, C-, KC-, P-, UV-.

NICHE: *Nectriopsis rubefaciens* forms conspicuous infections on thalli of *Aspicilia*, on shaded, non-calcareous rocks throughout eastern North America. In the Smokies, it is scattered but widespread. Its geographical range still eludes us.

KEY FEATURES: Orange-to-red, discolored patches on *Aspicilia* on non-calcareous, small, orange perithecia, 2-celled colorless spores, throughout the Smokies.

Nephroma helveticum

Bottoms Up

Lendemer 33033 (photo: Tripp)
Inset: Tripp 3719 (photo: Lendemer)

NOTES: *Nephroma helveticum* is a relatively common cyanolichen in the Smokies. All members of the genus produce apothecia on the lower surfaces of the lobe tips, making it impossible to confuse for any other genus of cyanolichen. This feature is seen in the photograph of *N. helveticum* in this guide. However, this species is quite variable and sometimes lacks apothecia entirely, instead producing abundant lobules among the lobe margins as in the inset photograph. These two extremes might look very different, but intermediates exist, which is likely why the lobulate version doesn't have a different name.

CHEMISTRY: Terpenoids. Spot tests. K-, C-, KC-, P-, UV-.

NICHE: This species is found throughout the Smokies, from the highest to lowest elevations. It grows on the bark and branches of trees and shrubs as well as on shaded non-calcareous rocks, especially in humid habitats. Despite being common in the Smokies, it is rare in many other parts of its range across North America.

KEY FEATURES: Foliose cyanolichen, brown thallus, apothecia produced on lower surface of lobe tips, often with abundant lobules along lobe margins, on bark and branches as well as non-calcareous rocks throughout the Park.

Nephroma parile

Whiskey River

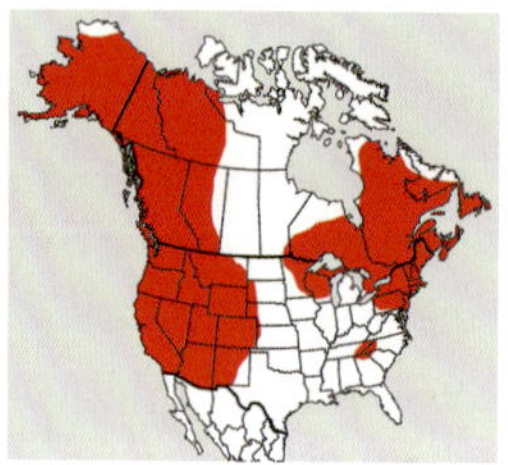

Tripp 3493 (photo: Lendemer)

NOTES: A visitor from the north decided that it liked the Smokies so much that it stayed here after the glaciers retreated. *Nephroma parile* is an easily recognized foliose cyanolichen that has a large brown thallus and abundant soralia. There are no other lichens in the Smokies with which it could be confused, except perhaps for wet thalli of *Lobaria scrobiculata*, which are instantly distinguished when they dry by the yellowish-green color of the thalli.

CHEMISTRY: Zeorin. Spot tests. K-, C-, KC-, P-, UV-.

NICHE: This is an example of a species that is relatively frequent in northern boreal forests as well as in coniferous forests of the southern Rocky Mountains, but rare in the southern Appalachians, where it only occurs at the highest elevations in mature forests. Whiskey River is rare in the Smokies–we have ourselves seen it only a couple of times. Look for it on the bark of high-elevation hardwoods, such as *Acer spicatum* or *Sorbus americana*, perhaps around Myrtle Point, or Chiltoes Mountain.

KEY FEATURES: Large brown thalli with abundant laminal and marginal, gray soralia, corticolous on high elevation hardwoods, rare in the Smokies.

Normandina pulchella

Blue Heart

Tripp 4974 (photo: Lendemer)

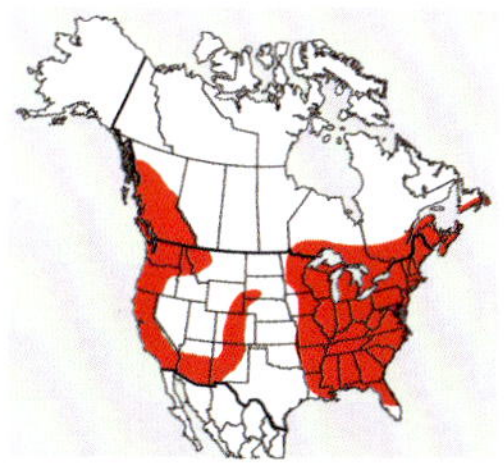

NOTES: *Normandina pulchella* is one of the most distinctive and unusual lichens in the Smokies. It is closely related to pyrenolichens of the Verrucariaceae but doesn't look much at all like other members of that family (see *Endocarpon*, *Placidium*, *Verrucaria*, *Willeya*). You can instantly identify Blue Heart by its small, blue-to-jade-colored, scallop-shaped squamules that sometimes have marginal soredia. The generic name was the inspiration for the specific epithet of *Lepraria normandinoides*, whose margins resemble this species.

CHEMISTRY: No substances. Spot tests. K-, C-, KC-, P-, UV-.

NICHE: Blue Heart is widespread in various parts of North America but is often overlooked unless you happen upon an abundant patch where the color catches the eye. This species is similarly common and widespread throughout the Smokies, from the highest to lowest elevations and on bark as well as bryophytes growing over bark. When feeling blue, *N. pulchella* grows on top of, or in the folds of, the lobes of foliose lichens.

KEY FEATURES: Blue-to-jade-colored, scallop-shaped squamules, occasional, marginal soredia, P-, on bark and bryophytes throughout the Smokies, always an excellent day in the field if fournd.

Ochrolechia arborea

Ultraviolet Glow

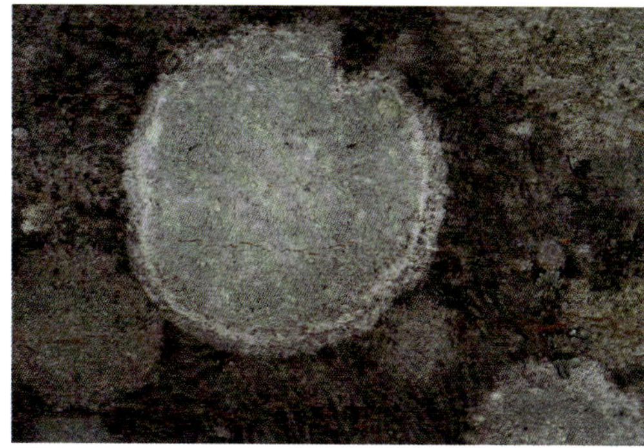

Tripp 3589 (photo: Lendemer)

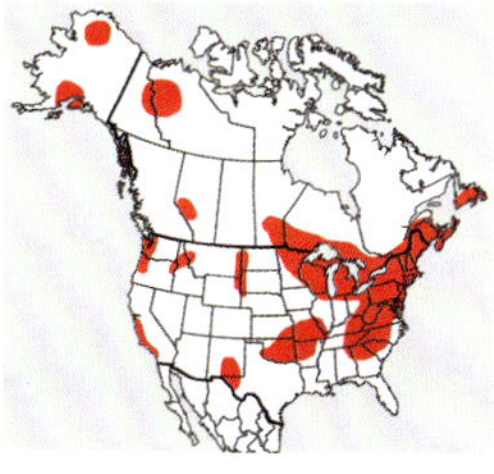

NOTES: For those who like to go out into the forest at night with a UV light, this is a species to watch out for. There may be other sorediate crustose lichens with blue-gray thalli in the Smokies, but this is the only one that will put on a major light show and also react C+ pink (back in the lab). Some forms of *Varicellaria velata* can be confused with *Ochrolechia arborea*, but that species has pruinose apothecia, not soralia. The only truly similar species is *O. mahluensis*, which is very rare and differs in being UV- as sadly, it lacks lichexanthone.

CHEMISTRY: Lichexanthone and gyrophoric acid. Spot tests. K-, C+ pink, KC+ pink, P-, UV+ bright yellow.

NICHE: *Ochrolechia arborea* is frequent and widespread at middle-to-high elevations in the Smokies and elsewhere in the southern Appalachians. It occurs on the bark and branches of trees, but is also very common on shrubs, such as *Rhododendron*.

KEY FEATURES: Blue-gray sorediate thallus, UV+ yellow, C+ red, on bark and branches of trees and shrubs at high elevations.

Ochrolechia mexicana

Mexican Sunshine Lichen

Lendemer 53181 (photo: Tripp)

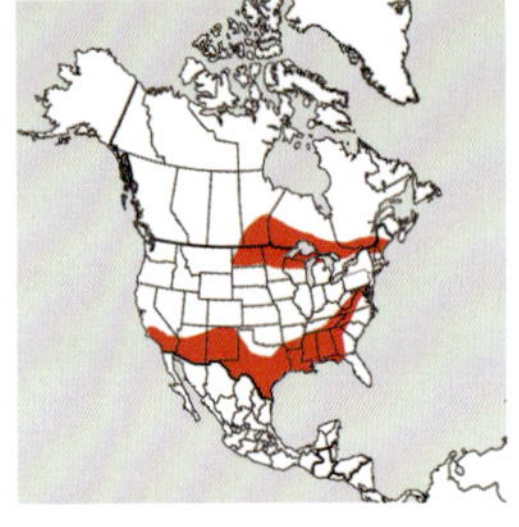

NOTES: You should probably check your vitals if you find this species and don't fall immediately and passionately in love. *Ochrolechia mexicana* is extremely easy to identify . . . that is, if you find it! It isn't the most common species in the Park, but hey–you can't win them all. First, learn *Ochrolechia*, which is easy! To do so, try "gestalt" (and yes, those broad discs surrounded by very thick margins help, as does the typical wagon-wheel patterning across the disc). Second, do a C test; if the cortex of the apothecial margin is C+ pink, then your lichen is either *O. mexicana* or the closely related and much more common and widespread *O. trochophora*. Third, if the medulla of the apothecial margin is C+ pink then you have *O. mexicana*. Isn't it strange that *Ochrolechia* species have partitioned their secondary compounds in the cortex, the medulla, or both? This species also often fluoresces UV+ bright yellow sort of like warm, Mexican sunshine.

CHEMISTRY: Lichexanthone and gyrophoric acid. Spot tests: (cortex) K-, C+ pink, KC+ pink, P-, UV+ bright yellow or UV-; (medulla) K-, C+ pink, KC+ pink, P-, UV-.

NICHE: Mexican Sunshine Lichen has an unusual distribution in North America, where despite the name, it is particularly common and abundant in northern temperate areas surrounding the Great Lakes. It occurs in mountainous areas in the Sonoran Desert, and from there ranges down into Central America. The species is rare in the Smokies, where it almost always occurs on the canopy branches of trees at high elevations.

KEY FEATURES: White-gray crustose thallus, large apothecia with thick rims, C+ red apothecial cortex and medulla, often UV+ bright yellow, on canopy branches at high elevations.

Ochrolechia pseudopallescens

Pale Fire

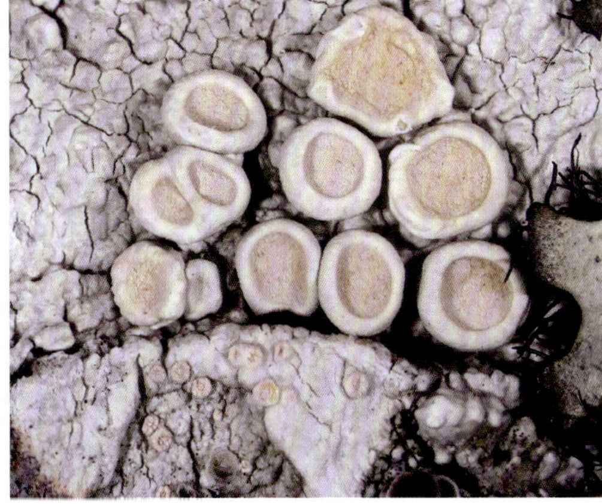

Lendemer 33338 (photo: Lendemer)

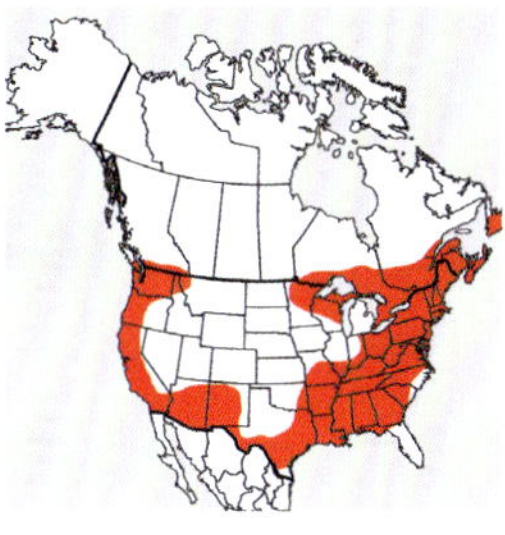

NOTES: We have good news and bad news. The bad news is that there aren't that many species of *Ochrolechia* in the Park. The good news is that this makes your life easier! *Ochrolechia pseudopallescens* is one of four species present in the Smokies that look superficially similar to one another, and differences can be best discerned through spot tests of the various parts of the apothecia. *Ochrolechia pseudopallescens* lacks any sort of reaction in either the cortex or medulla of the apothecial margins. But, it does yield a positive, fiery C+ pink reaction on the disc of the apothecium, so don't confuse this with the apothecial margins. All of the other similar species in the Smokies differ in having positive C+ pink reactions in either the apothecial cortex and/or the medulla.

CHEMISTRY: Gyrophoric acid, lecanoric acid, variolaric acid, lichesterinic acids agg. Spot tests: (cortex and medulla): K-, C-, KC-, P-, UV-; (apothecial discs): K-, C+ pink, KC+ pink, P-, UV-.

NICHE: Pale Fire grows on a variety of substrates in a variety of elevations but tends to prefer twigs and bark of acidic species such as *Rhododendron*, *Tsuga*, or *Pinus*. Look for it along Pretty Hollow Gap, not far from the beautiful old growth forests that flank the trail as you approach Sterling Ridge. It is endemic to eastern North America, but most common in the Mid-Atlantic and New England.

KEY FEATURES: Gray crustose thallus, C- apothecial cortex and medulla, C+ red apothecial disk, broadly distributed but with some preference for acidic bark.

Ochrolechia trochophora

An Unrequited Love

Lendemer 32937 (photo: Tripp)
Inset: Lendemer 33050 (photo: Tripp)

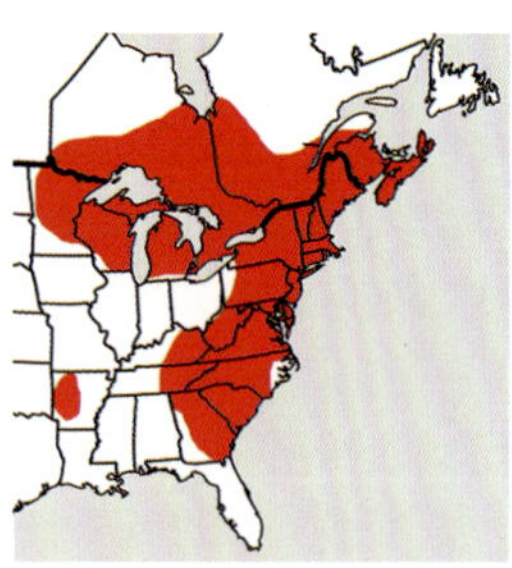

NOTES: *Ochrolechia trochophora* is a variable crustose lichen that you will invariably fall in love with, although we are unsure whether this love will be returned. The apothecia can be large or small, the disc can be simple or divided like a pinwheel, and the thallus can be relatively smooth to extremely lumpy. Nonetheless you can always recognize this species by its gray-to-greenish-gray thallus and apothecia with a margin cortex (but not its medulla!) that is C+ pink. The specific epithet derives from "the thing that bears" (=*phora*) the "wheel" (=*trocho*), in reference to its often wheel-like, divided apothecial discs.
CHEMISTRY: Gyrophoric acid. Spot tests. Cortex: K-, C+ pink, KC+ pink, P-, UV-. Medulla: K-, C-, KC-, P-, UV-.
NICHE: *Ochrolechia trochophora* is relatively common in the Smokies and more generally in eastern North America but is usually restricted to the bark of mature hardwood trees in older forest stands. This species prefers middle-to-high elevation habitats.
KEY FEATURES: Gray-to-blue-gray crustose thallus, lecanorine apothecia with C+ pink cortex and C- medulla, apothecia discs often wheel-like in appearance, simple spores, on bark of mature hardwoods at middle-to-high elevations.

Ochrolechia yasudae

Isidiate Thicket

Tripp 3683 (photo: Lendemer)

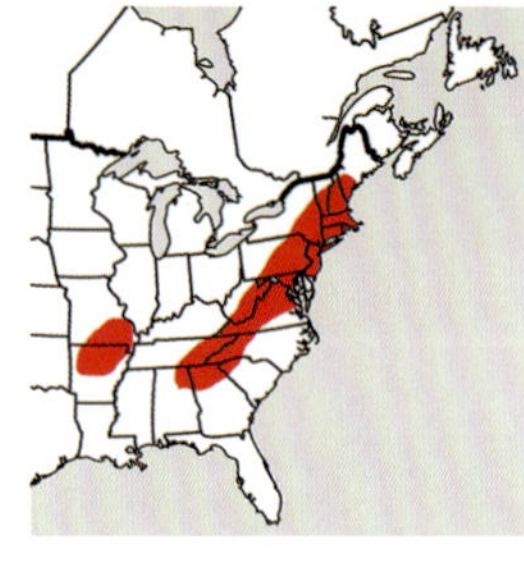

NOTES: Have you ever thought about how rare crustose lichens with isidia are? Of all the species treated in this book, very few of them (< 10) are isidiate. *Ochrolechia yasudae* is impossible to confuse with any other because, in addition to having a thick, beautiful gray thallus bearing nothing short of a fleet of isidia, its cortex reacts C+ pink. There is absolutely no other species in the Smokies that is remotely similar. *Pertusaria globularis* is also isidiate and grows in similar habitats as *O. yasudae*, but that species is C-, has short, stumpy isidia, is much smaller in size, and typically grows on mosses. *Ochrolechia yasudae* sometimes produces sexual spores (see apothecia in the photograph) in addition to the hyperabundant asexual propagules (i.e., isidia). Two different ways of getting your genome out into the world, simultaneously nonetheless. Sounds like a winning strategy to us!
CHEMISTRY: Gyrophoric acid. Spot tests. Cortex: K-, C+ pink, KC+ pink, P-, UV-. Medulla: K-, C-, KC-, P-, UV-.

NICHE: *Ochrolechia yasudae* is common on shaded, non-calcareous rocks in nearly all habitats below the spruce-fir zone in the Smokies. Its affinity for rock is rather striking, and if you find a thick, gray crust with isidia on such a substrate, this is almost certain to be *O. yasudae*. This species is also common in the central and southern Appalachians but becomes less frequent further north.

KEY FEATURES: Gray-to-blue-gray crustose thallus, abundant isidia, C+ pink cortex but C- medulla, on shaded, non-calcareous rocks mostly below the spruce-fir zone.

Opegrapha corticola

Bronze Medal

Tripp 3632 (photo: Lendemer)

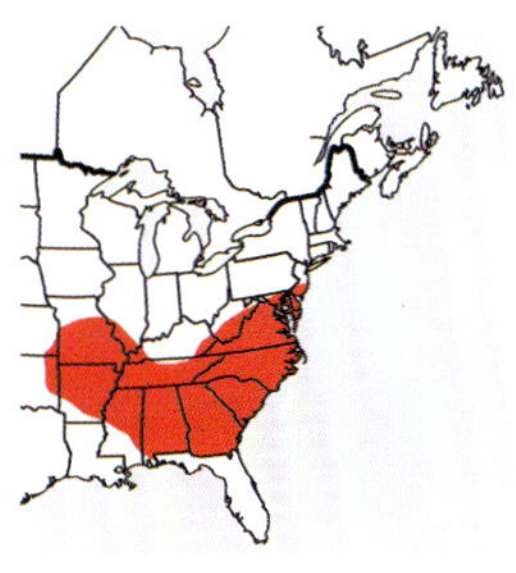

NOTES: Yep, folks, that's right: another inconspicuous sorediate crust that is easily overlooked. *Opegrapha corticola* has a thin, shiny thallus that is slightly bronze or brown and appears as a modest discoloration of the bark on trees. It can be recognized by its erose soralia that contain yellowish-to-bronze-colored soredia. The coloration of the thallus and soredia is due to the presence of a *Trentepohlia* photobiont that produces abundant carotenoids. Only *Caloplaca chrysophthalma* might be confused for *O. corticola*, but it has a K+ thallus, and you would be extremely lucky to find that species in the first place!

CHEMISTRY: No substances. Spot tests. K-, C-, KC-, P-, UV-.

NICHE: *Opegrapha corticola* occurs on the trunks and bases of hardwoods and more rarely on conifers, especially in humid habitats. It is common throughout much of temperate eastern North America but has mostly been overlooked. In the Smokies, it occurs across all elevations but is most frequent at lower altitudes in the Park.

KEY FEATURES: Inconspicuous brown-to-bronze thallus, erose, yellow-to-bronze-colored soredia, K-, on bark of hardwoods and rarely conifers, throughout the Smokies.

Opegrapha gyrocarpa

Stony Sleep Lichen

Lendemer 29649 (photo: Lendemer)

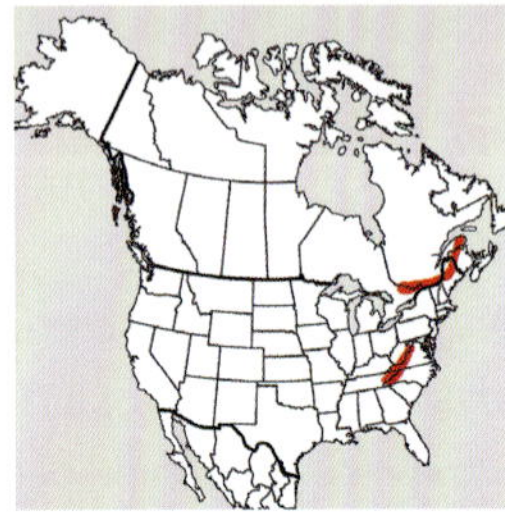

NOTES: *Opegrapha gyrocarpa* is, at present, one of three known sorediate saxicolous members of the genus in eastern North America. Prior to our work in the Smokies, it was known only from northern, boreal habitats. As a sterile sorediate crust, there isn't an easy way to know this as an *Opegrapha* other than learning the species. That's just the way it goes with asexual lichens sometimes. *Opegrapha gyrocarpa* is characterized by having a thin, orangish-brown thallus that bears pinkish-brown soredia. The thallus is often bordered by a black prothallus. Apothecia are rare but, where present, are black and bear 4-celled colorless spores. This species differs from the far more common *O. zonata*, which also occurs at high elevations, primarily in chemistry (presence of gyrophoric acid vs. confluentic acid in *O. zonata*). When in doubt, do a C test; if it's C-, you found the more common of the two. *Opegrapha moroziana* is the third known saxicolous, sorediate member of the genus, but that species manufactures psoromic acid and is P+ yellow. Finally, don't confuse *O. gyrocarpa* for *Dirina masilliensis* f. *sorediata*, which bears white soredia instead of pinkish-brown soredia in *O. gyrocarpa*. Although the two species have similar spot tests, *Dirina* mass f. *sored* produces erythrin instead of gyrophoric acid.

CHEMISTRY: Gyrophoric acid. Spot tests: K-, KC+ pink, C+ pink, P-, UV-.

NICHE: *Opegrapha zonata* is very rare in the Smokies, where it is known only from high elevations in cool rock overhangs. Look for these overhangs when you peek over the ridges of the Tennessee Divide.

KEY FEATURES: Small crustose lichen with thin, orangish-brown thallus, black prothallus, pinkish-brown soredia, rare presence of black fruiting bodies and 4-celled colorless spores, C+ pink (gyrophoric acid), very rare in rock in cool overhangs at high elevations.

Opegrapha varia

Mouthy Lips

Tripp 3601 (photo: Lendemer)
Inset: Tripp 3919 (photo: Lendemer)

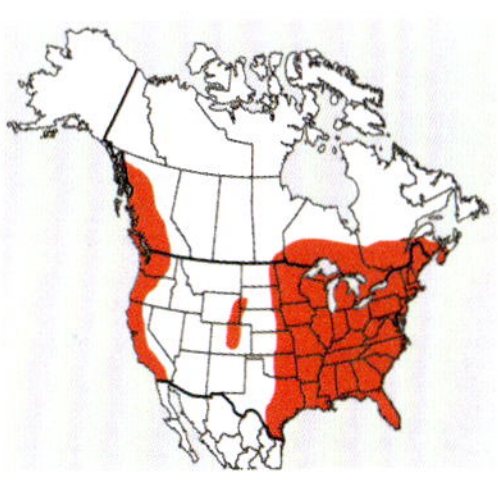

NOTES: The name says it all: this one comes in many forms. The most typical form of *Opegrapha varia* is reflected in the main photograph shown here. This species can be recognized by its short, simple lirellae, and discs that are blue-green pruinose. A less frequent form is pictured in the inset and comprises more elongate, branched lirellae that are epruinose. The species can readily be separated from other members of the genus that occur in the Smokies because the lips of the lirellae are parted to expose the disc. In fact, consider this one typical of the genus, at least based on its Greek roots: "*ope*-" means open and "*graph*-" means writing, as in the lirellae of the script lichens. Literally: open lips.

CHEMISTRY: No substances. Spot tests. K-, C-, KC-, P-, UV-.

NICHE: This species is widespread throughout eastern North America, with disjunct populations in western North America. It is infrequent in the Smokies but can be found on the bark of mature hardwoods, especially oak and hickory, in older forests at middle and low elevations.

KEY FEATURES: Indistinct crustose thallus, short to elongate lirellae with exposed discs, these usually blue-green pruinose, 5- to 7-celled colorless spores, on bark of mature hardwoods.

Opegrapha viridis

Bark Commas

Lendemer 46005 (photo: Lendemer)

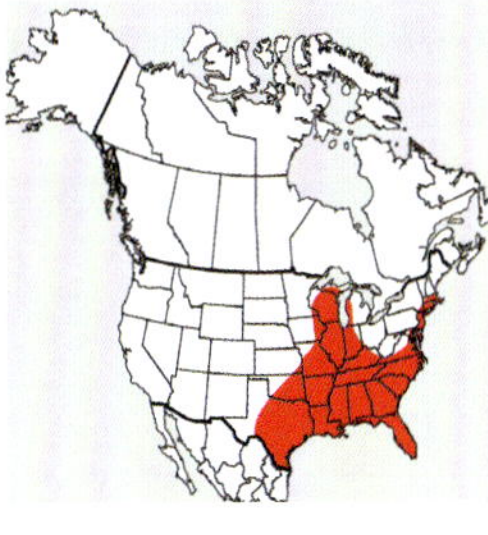

There are only three relatively common species of *Opegrapha* in the Smokies that grow on bark and are esorediate. *Opegrapha viridis* can easily be recognized by its short, relatively unbranched lirellae with concealed discs that resemble commas, and relatively slender spores with 12–14-celled spores. You don't even need to count the number of cells because all the other members of the genus that occur in the Smokies max out at six cells. *Opegrapha varia* is similar in that it also has short lirellae, but the discs are exposed, frequently blue-green pruinose, and the spores are much shorter. *Opegrapha vulgata* tends to be smaller and more runty in appearance, and its spores are very narrow as well as shorter. Due to the number of cells in the spores, *O. viridis* could be confused with members of the genus *Graphis*. However, in the Smokies, the only *Graphis* with a fully carbonized exciple is *G. sterlingiana*, which has large, muriform spores and a completely different overall appearance.

CHEMISTRY: No substances. Spot tests: K-, C-, KC-, P-, UV-.

NICHE: This species is common and widespread throughout southeastern North America, where it grows on the bark of hardwoods and conifers, often in humid habitats such as swamps. In the southern Appalachians, it tends to occur at lower elevations, and this is certainly the case in the Smokies.

KEY FEATURES: Brown-to-gray crustose thallus, black lirellae, fully carbonized exciple, 12–14-celled colorless spores, on bark at low and middle elevations.

Opegrapha vulgata

Common Lips

Tripp 3639 (photo: Lendemer)

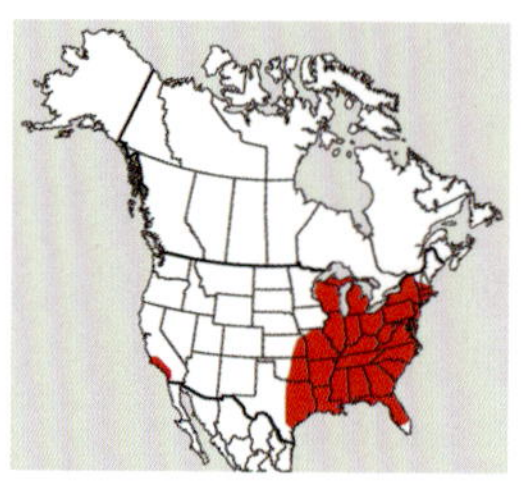

NOTES: *Opegrapha vulgata* resembles some species belonging to the genus *Graphis* but it is easily distinguished by its very narrow spores, poorly developed thallus, and additional anatomical characters. *Opegrapha vulgata* is also somewhat similar to *O. viridis*, but that species has spores with 10 to 12 cells, while those of *O. vulgata* have fewer than six cells. Other than those two species, *O. vulgata* is not easily confused with other lichens in the Smokies.

CHEMISTRY: No substances. Spot tests. K-, C-, KC-, P-, UV-.

NICHE: *Opegrapha vulgata* is widespread in temperate eastern North America, as the map indicates. It is also widespread in the Smokies and is relatively common at low and middle elevations, where it grows on the bark of hardwoods.

KEY FEATURES: Poorly developed crustose thallus, relatively unbranched lirellae with closed discs, narrow, 4- to 6-celled colorless spores, on the bark of hardwoods at middle and low elevations.

Opegrapha zonata

Slow Crossing, Opegrapha Zone

Lendemer 48568 (photo: Tripp)

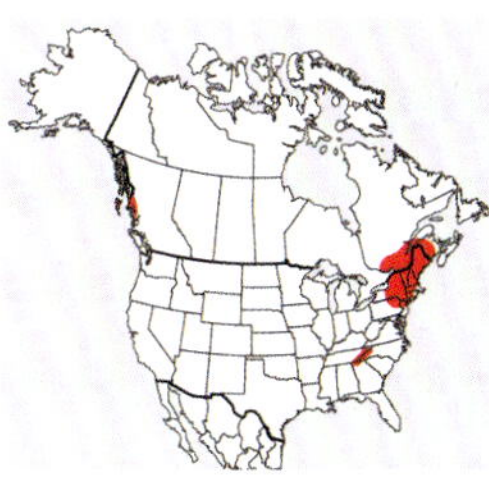

NOTES: If you are brown, sorediate, C-, P-, and growing on sheltered rocks in overhangs, then I'm your gal. *Opegrapha zonata* can only be confused with *Opegrapha gyrocarpa*, which looks very similar but is C+ pink due to the presence of gyrophoric acid. The dark, often bronze-brown color of the thallus and soredia of *O. zonata* is due to its *Trentepohlia* photobiont. *Opegrapha moroziana* is also somewhat similar and, in the southeastern United States, could also be confused with this species, but it is P+ yellow due to the presence of psoromic acid. We aren't sure that it helps, but the specific epithet "zonata" refers to a girdle, belt, or a zone. Use your imagination?

CHEMISTRY: Confluentic acid. Spot tests. K-, C-, KC-, P-, UV+ dull blue-white.

NICHE: *Opegrapha zonata* grows on non-calcareous rocks in overhangs and other sheltered microhabitats. It is common in northeastern North America and occasional in the Pacific Northwest. In the southern Appalachians, it is restricted to, and fairly uncommon in, high elevation habitats.

KEY FEATURES: Dark brown sorediate thallus, C-, P-, UV+ dull blue-white, on non-calcareous rocks in overhangs and other protected, shaded microhabitats, high elevations.

Pannaria conoplea

Shadowland Luxury

Lendemer 32820 (photo: Tripp)

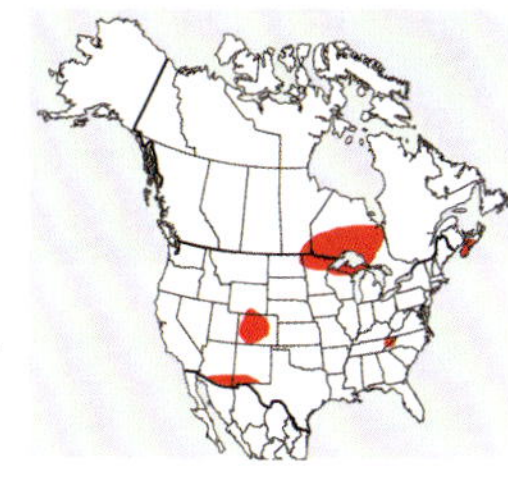

NOTES: *Pannaria conoplea* is a true mountain treasure. There are not many lead blue-gray colored foliose cyanolichens in the Smokies—and even fewer that are sorediate. This one can be easily recognized by its marginal soralia and P+ orange-red chemistry. The most morphologically similar species is *Fuscopannaria sorediata*, but that species differs in having a more diminutive thallus that approaches a squamulose growth form, and in lacking a P+ reaction. The name of the genus translates to "cloth" or "rag" in Greek, likely in reference to its relatively large, gray, foliose morphology.

CHEMISTRY: Pannarin. Spot tests. K-, C-, KC-, P+ orange-red, UV-.

NICHE: This is a very rare species in most of its range in North America. In the southern Appalachians, *Pannaria conoplea* is restricted to the bark of mature hardwoods in humid, high elevation ridge-top (think: windy and well-ventilated) forests. How's that for specific? See, lichens have ecology, too.

KEY FEATURES: Blue-gray foliose cyanolichen, marginal soralia, P+ orange-red, rare on hardwoods at high elevations.

Pannaria rubiginosa

Tea Cups

Tripp 2407 (photo: Lendemer)

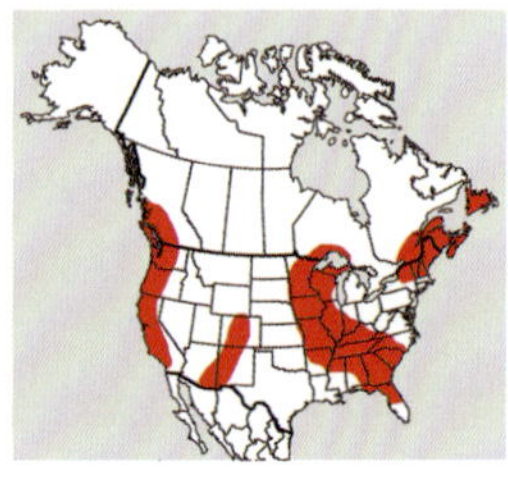

NOTES: All species of *Pannaria* are relatively rare in the Smokies, and this one is no exception. *Pannaria rubiginosa* can be identified by its small, rosette-forming thalli, minute knob-like sublobes near the central portions of the thallus, absence of soredia and isidia, and P+ chemical reaction that is especially conspicuous near the lobe tips. *Pannaria subfusca* is similar in appearance but can easily be distinguished by a P- reaction. *Fuscopannaria leucosticta* is also superficially similar but differs in having a squamulose thallus that is P-. The specific epithet of this species likely refers to its apothecia that, when present, are reddish in color.

CHEMISTRY: Pannarin. Spot tests. K-, C-, KC-, P+ orange-red, UV-.

NICHE: This species is rare but widespread in various portions of North America. In the southern Appalachians, it is known from a small number of scattered occurrences at middle and low elevations, where it has been found growing on the bark of hardwoods.

KEY FEATURES: Brown-gray foliose cyanolichen, lacking soredia and isidia, apothecia usually present, discs dirty red, knob-like secondary lobes, P+ orange-red cortex, on hardwoods at middle and low elevations.

Pannaria subfusca

Tiny Tea Cups

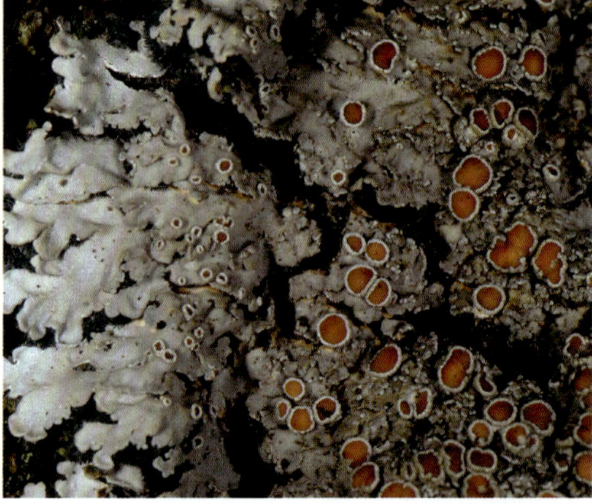

Tripp 3923 (photo: Lendemer)

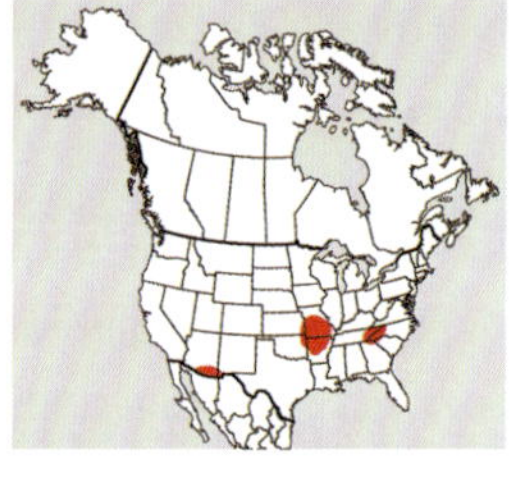

NOTES: Here is yet another rare cyanolichen from the Smokies. *Pannaria subfusca* closely resembles *P. rubiginosa* but has somewhat of a smaller thallus that always reacts P- (vs. P+ orange-red). These two species have been confused historically, and many older records will be found under the name *P. rubiginosa.*

CHEMISTRY: No substances. Spot tests. K-, C-, KC-, P-, UV-.

NICHE: *Pannaria subfusca* is distributed primarily in the Ozarks and southern Appalachians, with disjunct populations in southern Arizona and Mexico. It is rare in the southern

Appalachians and in the Smokies and only occurs at low elevations on the bark of hardwoods. Look for this one the next time you are hiking in the vicinity of Ace Gap.

KEY FEATURES: Brown-gray foliose cyanolichen lacking soredia and isidia, apothecia usually present, knob-like-secondary lobes, P- cortex, on hardwoods at middle and low elevations.

Pannaria tavaresii

Spanish Tea Cups

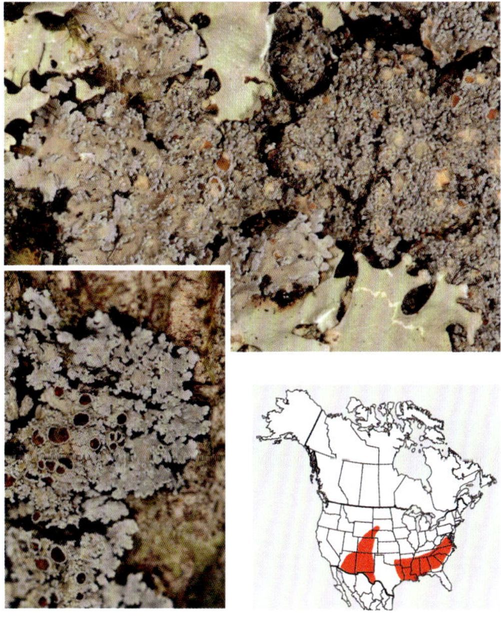

Lendemer 26981 (photo: Tripp)
Inset: Lendemer 44635 (Tripp)

NOTES: *Pannaria tavaresii* is an infrequent cyanolichen (we're on a roll!) in the Smokies that can be confused with several others on account of its morphological variability. The species is effectively defined by having a foliose, P+ thallus that bears marginal isidia. Many individuals are clearly isidiate, with cylindrical isidia produced along the margins of the thallus. Nonetheless, we have encountered others where the marginal isidia become flattened and resemble lobules. Still, more have very short isidia that approach the knobby secondary lobes of *P. rubiginosa*. Taking full morphological variability into consideration, *P. tavaresii* is most likely to be confused with *P. rubiginosa*, but that species lacks isidia and has larger spores.

CHEMISTRY: Pannarin. Spot tests. K-, C-, KC-, P+ orange-red, UV-.

NICHE: *Pannaria tavaresii* is widespread in the southern United States but is found only infrequently. In the Smokies, it occurs in middle- and low-elevation habitats, where it grows primarily on the bark of hardwoods.

KEY FEATURES: Brownish-to-gray foliose cyanolichen, variably shaped marginal isidia, P+ orange-red, on the bark of hardwoods at middle and low elevations.

Parmelia saxatilis

Forking Splendor

Tripp 5017 (photo: Lendemer)

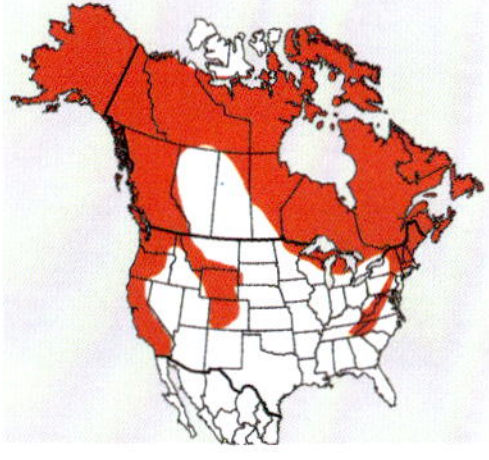

NOTES: *Parmelia* was once a very species-rich genus, but over the decades it was whittled away to a shadow of its formal self through removal and reclassification of many other foliose lichens treated in this book. The small group that remains is composed of blue-gray species with pseudocyphellae on the upper surfaces of the lobes, most of which produce salazinic acid. This is one of two isidiate members of the genus in the Smokies and can be separated from the much more common *P. squarrosa* by its furcating rhizines rather than bottle-brush-shaped squarrose rhizines.

CHEMISTRY: Atranorin and salazinic acid. Spot tests. Cortex: K+ yellow, C-, KC-, P-, UV-. Medulla: K+ yellow turning red, C-, KC-, P+ orange, UV-.

NICHE: *Parmelia saxatilis* is a northern species that is disjunct in the high elevations of the southern Appalachians. In the Smokies, it occurs in on non-calcareous rocks and on the bark of Fraser Fir. This species is also very common in the West, particularly the southern Rocky Mountains.

KEY FEATURES: Gray foliose lichen, pseudocyphellae on upper surface, isidiate, furcate rhizines, medulla K+ yellow turning red, on rocks and bark at high elevations.

Parmelia squarrosa

Familiar Square Lobes

Tripp 2339 (photo: Deregibus)

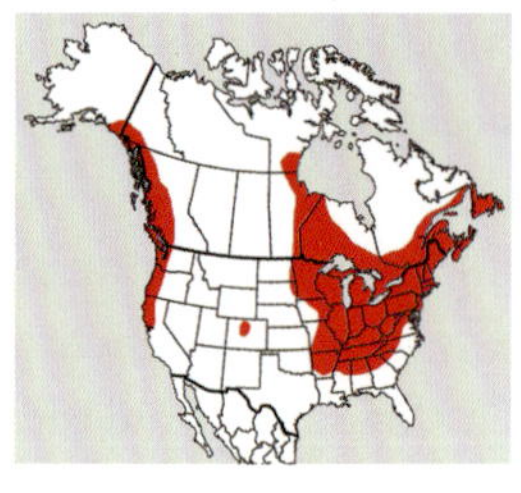

NOTES: This species was confused historically with *Parmelia saxatilis* but was recognized as distinct by the late great American Parmeliologist Mason Hale, who described the species formally based on material from the Shenandoahs. *Parmelia squarrosa* and *P. saxatilis* are outwardly similar, but *P. squarrosa* differs in having squarrose rhizines that resemble bottle-brushes with many short side branchlets. Care should be taken when examining these because the short branchlets may only be present on some of the rhizines, or only present on central or older portions of the thallus.

CHEMISTRY: Atranorin and salazinic acid. Spot tests. Cortex: K+ yellow, C-, KC-, P-, UV-. Medulla: K+ yellow turning red, C-, KC-, P+ orange, UV-.

NICHE: *Parmelia squarrosa* is common and widespread throughout temperate eastern North America. It is one of the most common corticolous foliose lichens in the southern Appalachians, including the Smokies.

KEY FEATURES: Gray foliose lichen, pseudocyphellae on upper surface, isidiate, squarrosely branched bottle-brush rhizines, medulla K+ yellow turning red, on rocks and bark throughout Park.

Parmelia sulcata

Bottom Brushes

Lendemer 32947 (photo: Tripp)

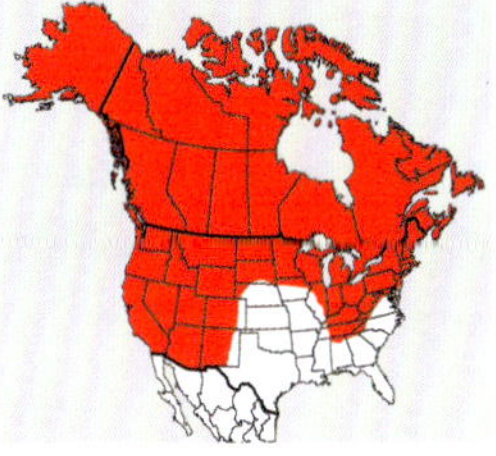

NOTES: This is the only sorediate member of the genus *Parmelia* in the Smokies. It is also the only foliose lichen with pseudocyphellae on the upper surface, soralia, and a K+ yellow turning red medulla. Even though this character combination distinguishes *Parmelia sulcata* from other species in the region, it is important to always check the rhizines to make sure they are squarrose (i.e., bottlebrush-like).

CHEMISTRY: Atranorin and salazinic acid. Spot tests. Cortex: K+ yellow, C-, KC-, P-, UV-. Medulla: K+ yellow turning red, C-, KC-, P+ orange, UV-.

NICHE: *Parmelia sulcata* is an extremely common species throughout much of North America (*look* at that distribution map) and even grows in polluted urban centers. Given its frequency elsewhere, it might be initially surprising that it quite uncommon in the Smokies, where it is generally restricted to the branches of trees and shrubs at high elevations. But then you stop for a PB & J, reflect on your day of lichenizing, and finally realize that, once again, the Smokies are just too nice for trash species.

KEY FEATURES: Gray foliose lichen, pseudocyphellae on upper surface, sorediate, squarrosely branched rhizines, medulla K+ yellow turning red, uncommon on rock and bark at middle and high elevations.

Parmeliella appalachensis

Appalachian Bedroll

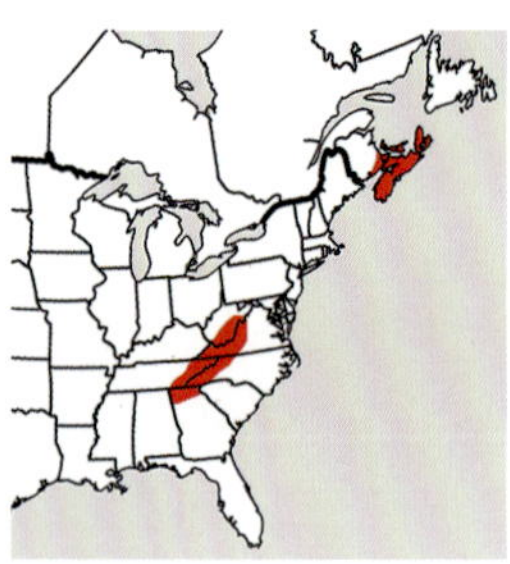

Tripp 3911 (photo: Lendemer)

NOTES: When you see this species in the Smokies, you might wonder to yourself: *Why did so many beautiful lichens end up with "appalachian" epithets?* We do not have an answer that other than to say: there are a lot of lichens in the Appalachian Mountains, and most of them are beautiful! This lovely foliose cyanolichen was initially overlooked and confused with other species until relatively recently, despite the fact that it is common and abundant in its happy places. You can recognize *Parmeliella appalachensis* by its foliose, brown-to-gray-brown thallus, its striking, carpet-like fibrous prothallus, and its abundant marginal lobules. The only similar lichen is *P. triptophylla*, but that species differs in having a smaller thallus composed of squamules, a film-like prothallus, and isidia instead of lobules.

CHEMISTRY: No substances. Spot tests. K-, C-, KC-, P-, UV-.

NICHE: *Parmeliella appalachensis* has a southern Appalachian, northern hardwood forest center of distribution. The species is especially fond of mature oaks. It was originally described from the central Appalachians, where it is very rare, but has outpost populations in the Canadian Maritimes. Did it evolve there and migrate southward with the glaciers? Did it evolve in the Smokies and migrate northward after glacier retreat? We await population genetic data to shed light on the answer.

KEY FEATURES: Foliose brown-to-gray-brown cyanolichen, felt-like fibrous prothallus, marginal lobules, P-, on the bark of mature hardwoods in high elevation northern hardwood forests.

Parmeliella triptophylla

Unfinished Puzzle Lichen

Tripp 4990 (photo: Lendemer)

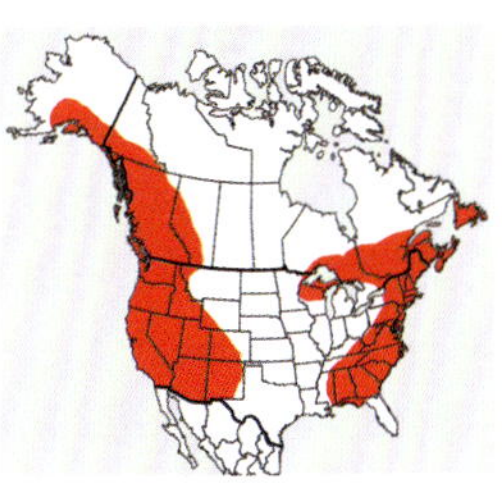

NOTES: *Parmeliella triptophylla* is an understated and diminutive cyanolichen that can be overlooked in the field. From a distance, this species resembles a black mark on the bark of a tree and only on closer inspection do its minute squamules become evident. It is most likely to be confused with *P. appalachensis*, which differs in having larger, distinctly foliose thalli, lobules instead of isidia, and a felt-like prothallus instead of one that is shiny and film-like in *P. triptophylla*. Confusion between these two species is especially possible when the isidia of *P. triptophylla* are short and/or poorly developed, or when the lobules of *P. appalachensis* are small and only somewhat flattened. In such cases, the thallus and prothallus characters can be used to distinguish the two.

CHEMISTRY: No substances. Spot tests. K-, C-, KC-, P-, UV-.

NICHE: *Parmeliella triptophylla* occurs on hardwood trunks and bases, especially at middle and low elevations in humid habitats. Rarely, it can be found growing on shaded, non-calcareous rocks. This species is neither common nor uncommon in the Smokies, but it is certainly more frequent in the Park than many of its close relatives.

KEY FEATURES: Small, squamulose brown-to-gray-brown cyanolichen, film-like prothallus, isidiate, P-, on the bark of hardwoods throughout but especially at middle and low elevations in humid habitats.

Parmotrema arnoldii

Light Bright

Tripp 3459 (photo: Lendemer)

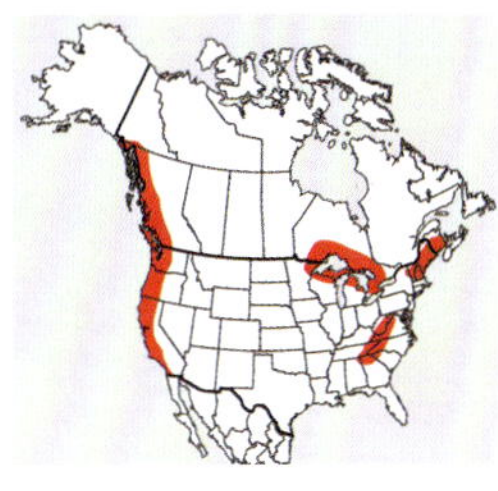

NOTES: And so begins your tour of the genus *Parmotrema*. This is a very important (and easy) genus to learn in the Smokies because the species are *everywhere*. You will feel good at least being able to get these to genus with relative ease! Naming the species is also doable with a bit of practice. All members of the genus can be recognized by their broad, rounded, ruffled lobes, and the vast majority of Smokies *Parmotrema* also have conspicuous, marginal cilia (just like in the photograph of *P. arnoldii*, here). What sets this species apart from the rest is its conspicuous, capitate, profuse soralia on the tips of the secondary lobes and its UV+ bright blue-white medulla. *Parmotrema reticulatum* looks similar but it has maculate upper surfaces, rhizines that extend to the margins (vs. absent from a broad zone near the lobe margins), and a K+ yellow to red medulla. *Parmotrema perlatum* is also similar but has a UV- and P+ orange medulla. When learning *Parmotremas*, be sure to have the full complement of spot test reagents ready to use.

CHEMISTRY: Atranorin and alectoronic acid. Spot tests. Cortex: K+ yellow, C-, KC-, P-, UV-. Medulla: K-, C-, KC-, P-, UV+ blue-white.

NICHE: Light Bright occurs on the bark and branches of trees and shrubs at high elevations in the Smokies. It is particularly abundant in spruce-fir forests.

KEY FEATURES: Blue-gray foliose thallus, ruffled lobes with conspicuous marginal cilia, capitate marginal soralia, UV+ blue-white medulla, on conifers and hardwoods at high elevations.

Parmotrema crinitum

Highland Wallpaper

Tripp 3740 (photo: Lendemer)

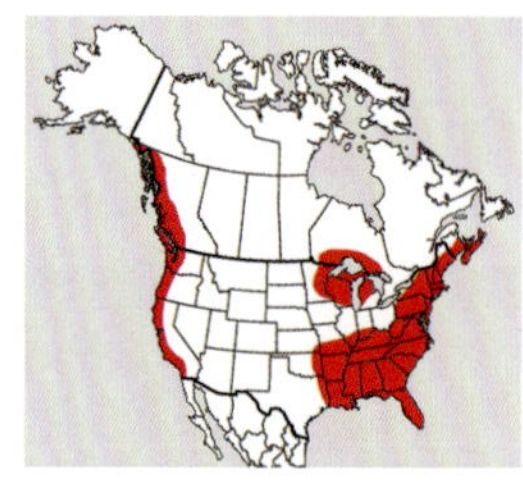

NOTES: *Parmotrema crinitum* may very well be the most common species of *Parmotrema* in the Smokies. It can easily be recognized by its marginal cilia, abundant isidia that occur on the surfaces ("lamina') of lobes and often bear apical cilia, P+ orange medulla, and rhizines that are lacking from a broad zone near the margins of the lobes. *Parmotrema ultralucens* and *P. subisidiosum* are similar species but both have a medulla that is K+ yellow turning red. *Parmotrema mellissii* is also similar, but the medulla in that species is UV+ bright blue-white.

CHEMISTRY: Atranorin, stictic acid and associates. Spot tests. Cortex: K+ yellow, C-, KC-, P-, UV-. Medulla: K+ yellow turning brown, C-, KC-, P+ orange, UV-.

NICHE: *Parmotrema crinitum* is very common and widely distributed on the bark and branches of conifers, hardwoods and shrubs throughout the Smokies. Lichens belonging to this species are particularly frequent at middle and high elevations.

KEY FEATURES: Blue-gray foliose thallus, ruffled lobes with conspicuous marginal cilia, laminal isidia often with apical cilia, medulla P+ orange, on conifers and hardwoods throughout the Smokies.

Parmotrema hypotropum

Luscious Lobes

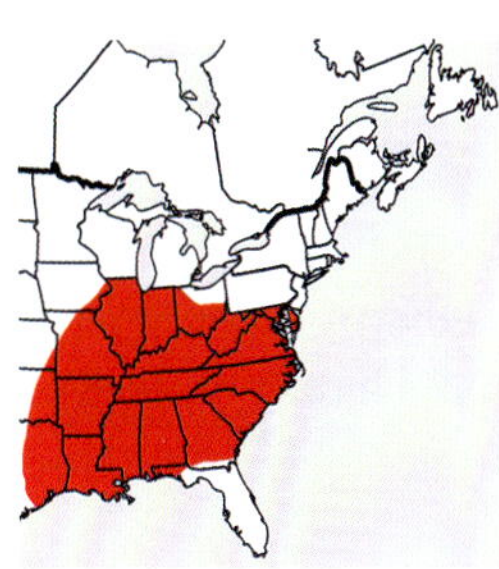

Tripp 1500 (photo: Lendemer)

NOTES: *Parmotrema hypotropum* is a species that is worthwhile to learn in the field. It belongs to a group of *Parmotrema* species that have ascending lobes with white blotches on the lower surface where it grows free of the substrate. This growth form, combined with the presence of marginal cilia and soralia as well as K+ yellow turning red medulla, makes it easy to identify this member of the group to species. Very rarely, the thallus of *P. hypotropum* will grow slightly adnate to the substrate and in such cases, the white blotches on the lower surface will be less conspicuous. *Parmotrema hypotropum* is perhaps most closely related to *P. perforatum*, which differs in lacking soredia.

CHEMISTRY: Atranorin and norstictic acid. Spot tests. Cortex: K+ yellow, C-, KC-, P-, UV-. Medulla: K+ yellow turning red, C-, KC-, P+ yellow, UV-.

NICHE: *Parmotrema hypotropum* is common on bark and branches of trees and shrubs, particularly at middle and low elevations in the Smokies. This species is widespread in the northeastern United States and commonly occurs down the Appalachians, but as soon as you hit the lowlands, it is replaced by *P. hypoleucinum*.

KEY FEATURES: Blue gray foliose thallus, ruffled lobes with conspicuous marginal cilia, lower surface with white blotches, marginal soralia, medulla K+ yellow turning red, on bark especially at low elevations.

Parmotrema margaritatum

Black and Tan

Lendemer 44568 (photo: Tripp)

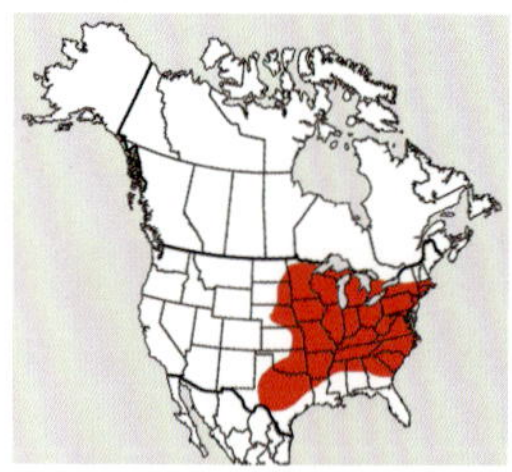

NOTES: *Parmotrema margaritatum* is a very confusing species that defies classification despite numerous attempts. It is most similar to *Parmotrema reticulatum* in having capitate soralia on the tips of the secondary lobes, a K+ yellow turning red medulla, and a maculate upper surface. The difference is essentially that the rhizines are absent from a broad zone near the margins of the lobes while they are present in the area in *P. reticulatum*. Keeping this in mind it then becomes difficult to distinguish *P. stuppeum* from *P. margaritatum*. The primary difference between *P. stuppeum* and *P. margaritatum* seems to be the absence of marginal cilia and emaculate upper surface in the former.

CHEMISTRY: Atranorin and salazinic acid. Spot tests. Cortex: K+ yellow, C-, KC-, P-, UV-. Medulla: K+ yellow turning red, C-, KC-, P+ orange, UV-.

NICHE: On the bark of hardwoods throughout temperate eastern North America, but the distribution is poorly understood. Further study is needed here!

KEY FEATURES: Blue-gray foliose thallus, ruffled lobes usually with conspicuous marginal cilia, capitate soralia, upper-surface maculate, medulla K+ yellow turning red, rhizines absent near margins, rare.

Parmotrema mellissii

Crumbling Towers

Tripp 3909 (photo: Lendemer)

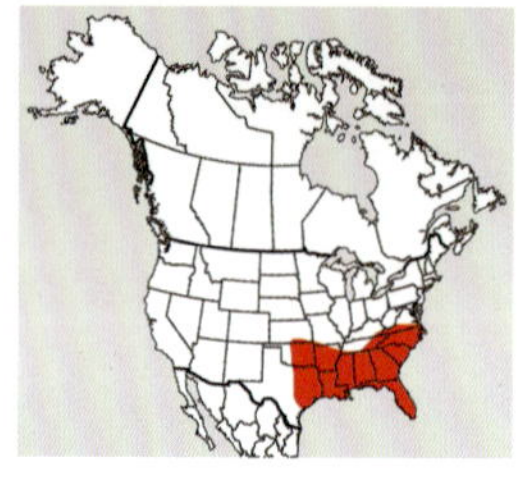

NOTES: We are quite fond of isidiate macrolichens, especially when they are as bold and beautiful as *Parmotrema mellissii*. The species resembles several others in the genus, such as *P. crinitum*, *P. subisidiosum*, and *P. ultralucens*, but differs from all in having a UV+ bright blue-white medulla. The isidia of *P. mellissii* have a habit of breaking apart and crumbling, causing them to resemble coarse soredia. In such cases, the species can be confused with *P. arnoldii* even though the latter is restricted to high elevations and always has conspicuous capitate soralia with fine rather than course soredia.

CHEMISTRY: Atranorin and alectoronic acid. Spot tests. Cortex: K+ yellow, C-, KC-, P-, UV-. Medulla: K-, C-, KC-, P-, UV+ blue-white.

NICHE: *Parmotrema mellissii* occurs on the bark and branches of trees and shrubs throughout the Smokies (and elsewhere in eastern North America) but almost always at middle and low elevations in the Park. Occurrences at high elevations are very rare.

KEY FEATURES: Blue-gray foliose thallus, ruffled lobes with conspicuous marginal cilia, laminal isidia often with apical cilia, medulla UV+ blue-white, on conifers and hardwoods throughout the Smokies.

Parmotrema perforatum

Rain Drain

Tripp 1502 (photo: Moroz)

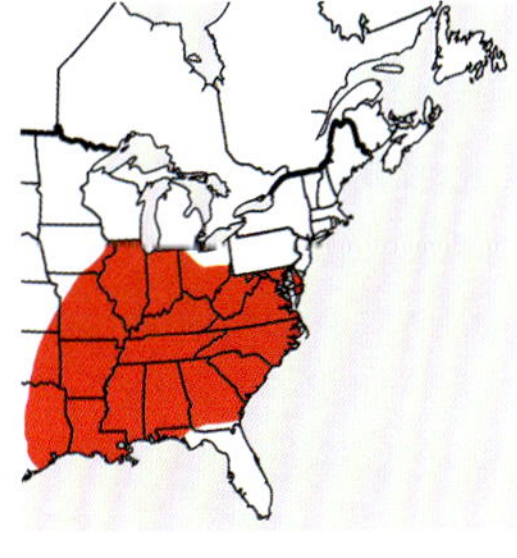

NOTES: Here is a species in which the name says it all. *Parmotrema perforatum* is one of the few members of the genus that has apothecia that are perforated via a hole right through their centers. This species can furthermore be recognized by its ascending lobes (like *P. hypotropum*) that have white blotches on the lower surface, but clearly differs from *P. hypotropum* by its absence of soralia. *Parmotrema perforatum* is instead most likely to be confused with other species that have perforate apothecia, such as *P. cetratum*. *Parmotrema cetratum* differs by having lobes that are adnate to the substrate and lack white blotches as well as in having an upper surface that is maculate and different chemistry. Finally, *P. submarginale* is superficially similar to *P. perforatum* but can be distinguished by its hole-less apothecia. Another species with perforate apothecia that occurs primarily along the coast is *P. subrigidum*, but that species differs chemically in having a UV+ bright blue-white medulla due to the presence of alectoronic acid. *Parmotrema* means "shield" + "hole," referring to the holes in the apothecia. Guess what the type species of the genus is? Surprise—it's *P. perforatum*!

CHEMISTRY: Atranorin and norstictic acid. Spot tests. Cortex: K+ yellow, C-, KC-, P-, UV-. Medulla: K+ yellow turning red, C-, KC-, P+ yellow, UV-.

NICHE: *Parmotrema perforatum* occurs extensively on bark and branches at middle and low elevations throughout the Smokies. It is similarly common elsewhere in the southeastern United States.

KEY FEATURES: Blue-gray foliose thallus, ruffled lobes with conspicuous marginal cilia, lower surface with white blotches, esorediate, perforate apothecia, medulla K+ yellow turning red, on bark throughout the Park.

Parmotrema perlatum

Emaculate Conception

Lendemer 44631 (photo: Tripp)

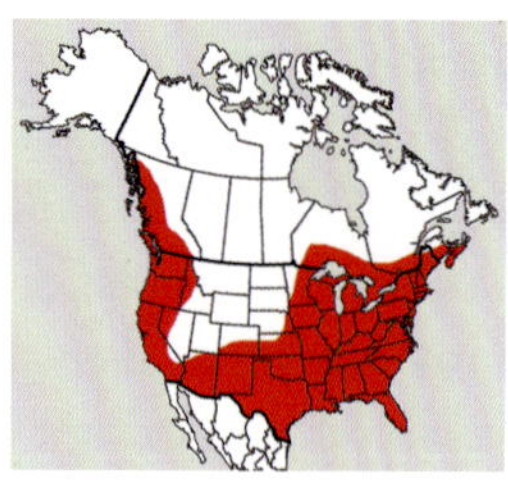

NOTES: *Parmotrema perlatum* flies under the radar because it closely resembles *Parmotrema reticulatum* from a distance. Indeed, both species have capitate soralia on the tips of the secondary lobes. However, if you flip over the thallus and look at the lobes, you will see that *P. perlatum* has a broad zone near the margins of the lobes that is free of rhizines, while the same area of *P. reticulatum* bears rhizines. The upper surface of *P. reticulatum* is also maculate while that of *P. perlatum* is not. The easiest way to tell these two species apart, however, is that the medulla of *P. perlatum* is K+ yellow turning brownish rather than blood-red.

CHEMISTRY: Atranorin, stictic acid and associates. Spot tests. Cortex: K+ yellow, C-, KC-, P-, UV-. Medulla: K+ yellow turning brown, C-, KC-, P+ orange, UV-.

NICHE: *Parmotrema perlatum* occurs throughout the Smokies on hardwoods, conifers, and shrubs. It is particularly abundant on branches in humid habitats at high elevations.

KEY FEATURES: Blue-gray foliose thallus, ruffled lobes usually with conspicuous marginal cilia, capitate soralia, maculate upper surface, medulla P+ orange, rhizines absent near margins, high elevations.

Parmotrema reticulatum

Camper's Crackling

Tripp 3596 (photo: Lendemer)

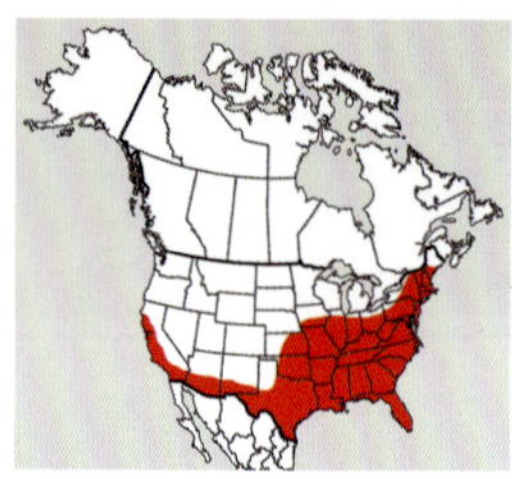

NOTES: If *Parmotrema crinitum* is the most common isidiate *Parmotrema* in the Smokies, then *Parmotrema reticulatum* is the most common sorediate species. It can be immediately recognized by its maculate upper surface, capitate soralia on the tips of secondary lobes, K+ yellow turning red medulla, and rhizines that grow all the way to the margins of the lobes on the lower surface. *Parmotrema margaritatum* and *P. stuppeum* differ in having a broad zone near the lower surface of the lobes entirely devoid of rhizines. Confusion with *P. subisidiosum* will likely occur once or twice, but careful searching will always reveal at least some isidia in the latter species.

CHEMISTRY: Atranorin and salazinic acid. Spot tests. Cortex: K+ yellow, C-, KC-, P-, UV-. Medulla: K+ yellow turning red, C-, KC-, P+ orange, UV-.

NICHE: *Parmotrema reticulatum* occurs on the bark and branches of conifers, hardwoods, and shrubs throughout the Smokies, especially at middle and low elevations.

KEY FEATURES: Blue-gray foliose thallus, ruffled lobes usually without conspicuous marginal cilia, capitate soralia, maculate upper surface, K+ yellow turning red medulla, *very* common throughout the Park.

Parmotrema simulans

False Camper's Cracklings

Lendemer 39577 (photo: Tripp)

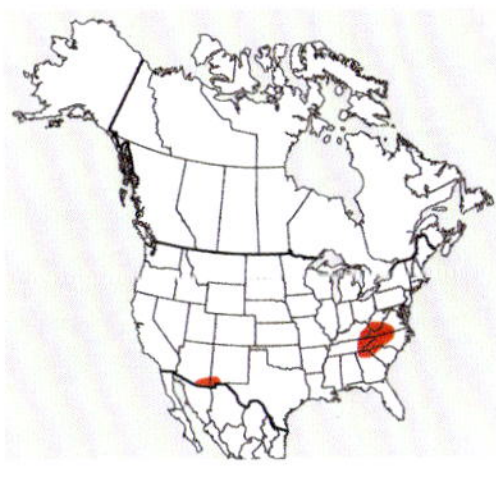

NOTES: Externally, *Parmotrema simulans* is identical to *P. reticulatum*, and there is no way to easily distinguish the two without studying chemistry to confirm that the medulla of *P. simulans* is K- (vs. K+ yellow turning red in *P. reticulatum*). *Parmotrema diffractaicum* is also superficially identical to these two species, and like *P. simulans* has a K- medulla. However, instead of having a fatty acid in the medulla, *P. diffractaicum* produces diffractaic acid, which reacts CK+ yellow and UV+ dull blue-white.

CHEMISTRY: Atranorin and caperatic acid. Spot tests. Cortex: K+ yellow, C-, KC-, P-, UV-. Medulla: K-, C-, KC-, P-, UV-.

NICHE: *Parmotrema simulans* is a very rare species that is restricted to the southern Appalachians in eastern North America, where it occurs only at the low elevations. It is most often found in acid forests in humid habitats, such as along creeks and rivers. Look for it along Fontana Lake in the Smokies.

KEY FEATURES: Blue-gray foliose thallus, ruffled lobes usually without conspicuous marginal cilia, capitate soralia, maculate upper surface, K- and P- medulla, rare on hardwoods at low elevations.

Parmotrema stuppeum

Morning Bird

Lendemer 26707 (photo: Moroz)

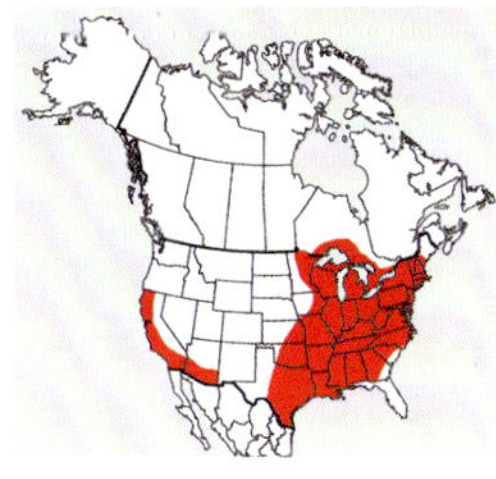

NOTES: *Parmotrema stuppeum* is a relatively common species at high elevations in the Smokies. It is instantly recognizable in the field because of its large thalli, its lobe tips that turn up like a morning bird to reveal brown margins of the lower surface, a lower surface that lacks marginal rhizines, and discontinuous marginal soralia that form along the secondary lobes. The most similar species is *P. margaritatum*, which differs primarily in having a distinctly maculate upper surface. Two other species are also similar to *P. stuppeum* but differ in key features. *Parmotrema reticulatum* has a maculate upper surface and rhizines all the way to the margins of the lower surface. *Parmotrema perlatum* has a K+ yellow turning brown medulla.

CHEMISTRY: Atranorin and salazinic acid. Spot tests. Cortex: K+ yellow, C-, KC-, P-, UV-. Medulla: K+ yellow turning red, C-, KC-, P+ orange, UV-.

NICHE: *Parmotrema stuppeum* grows the bark of hardwoods at high elevations in the Smokies. This is a typical species of northern hardwood forests and high elevation oak ridges.

KEY FEATURES: Blue-gray foliose thallus, ruffled, ascending lobes with conspicuous marginal cilia, marginal soralia, emaculate upper surface, medulla K+ yellow turning red, high elevations in the Park.

Parmotrema subisidiosum

Daily Deception

Tripp 3957 (photo: Lendemer)

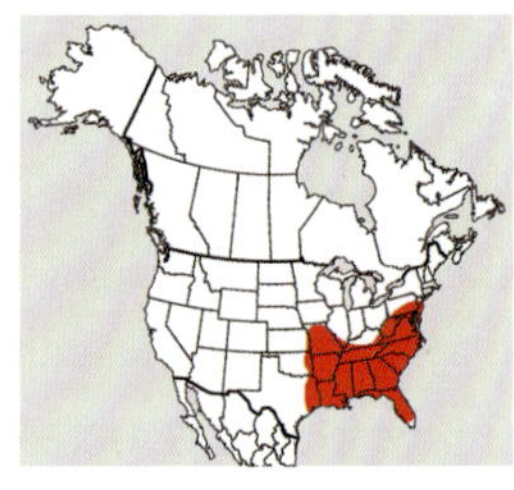

NOTES: We are sad to report that *Parmotrema subisidiosum* can easily be mistaken for many other *Parmotremas* in the Smokies because of its abundant isidia, which are especially concentrated along the margins of the lobes, and its conspicuous white maculate upper surface. This being said, however, isidia of this species often break down and crumble to resemble coarse soredia, making it likely to be confused with *P. reticulatum*, which has capitate soralia on the tips of the secondary lobes. Therefore, care must be taken to make a diligent search for at least some remnant isidia. *Parmotrema ultralucens* is also very similar to *P. subisidiosum* but the thallus in the former species always has a slight greenish or yellowish hue and the medulla is a remarkable UV+ bright yellow.

CHEMISTRY: Atranorin and salazinic acid. Spot tests. Cortex: K+ yellow, C-, KC-, P-, UV-. Medulla: K+ yellow turning red, C-, KC-, P+ orange, UV-.

NICHE: Daily Deception occurs on the bark and branches of trees and shrubs throughout the Smokies. More rarely, it is found on shaded, non-calcareous rocks in humid habitats.

KEY FEATURES: Blue-gray foliose thallus, ruffled lobes usually without conspicuous marginal cilia, isidiate, maculate upper surface, medulla K+ yellow turning red, rare on hardwoods at low elevations.

Parmotrema submarginale

Fixin' To Lichen

Lendemer 33170 (photo: Tripp)

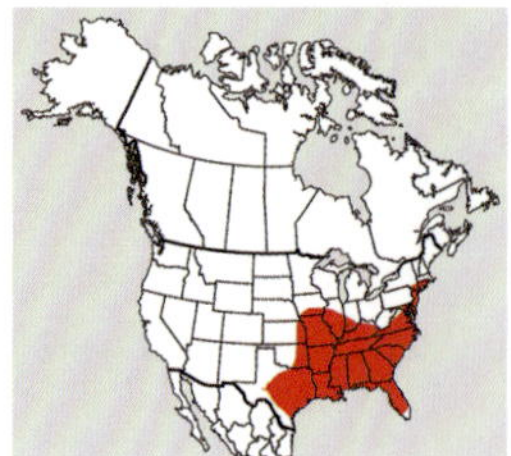

NOTES: Here is a lovely foliose lichen that is just fixin' to play tricks on you. *Parmotrema submar-*

ginale can be recognized by its K+ yellow turning brownish medulla, apothecia that are not perforate, and adnate lobes. Nonetheless, if you glance too quickly at this species, you might think it is *P. perforatum*, which differs in having perforate apothecia, ascending lobes, and a K+ yellow turning red medulla. Although the lobes of Fixin' To Lichen are adnate and often flat against the substrate, their morphology can become distorted when this species grows on thin branches, in which case its lobes may appear to be somewhat ascending. Always check chemistry and apothecium perforation to confirm your *Parmotrema* identifications!

CHEMISTRY: Atranorin and protocetraric acid. Spot tests. Cortex: K+ yellow, C-, KC-, P-, UV-. Medulla: K+ yellow turning brown, C-, KC-, P+ red, UV-.

NICHE: *Parmotrema submarginale* occurs on bark and branches of trees and shrubs at low elevations throughout the Smokies and, more generally, across the southeastern United States.

KEY FEATURES: Blue gray foliose thallus, ruffled adnate lobes without conspicuous marginal cilia, esorediate, apothecia not perforate, medulla P+ red, on bark at low elevations.

Parmotrema subsumptum

Land of Cotton

Tripp 3637 (photo: Lendemer)

NOTES: *Parmotrema subsumptum* belongs to a group of *Parmotrema* species that have completely brown lower surfaces and maculate upper surfaces. It is the only sorediate species with this combination of characters in the Smokies. Confusion with *P. hypotropum* is possible, however, that species has white blotches on the lower surface, ascending lobes, a different chemistry, and an emaculate upper surface.

CHEMISTRY: Atranorin and salazinic acid. Spot tests. Cortex: K+ yellow, C-, KC-, P-, UV-. Medulla: K+ yellow turning red, C-, KC-, P+ orange, UV-.

NICHE: This species is confined to the southern Appalachians in eastern North America, where it is found only at the lowest elevations. Sounds like it wants to get out of the mountains, right? So why doesn't it? Not enough evolutionary time to expand its range, perhaps? In the Smokies, *P. subsumptum* is fairly common on the low elevation slopes above Fontana Lake. Look for it on the bark of hardwoods in acid cove forests.

KEY FEATURES: Blue-gray foliose thallus, ruffled lobes with conspicuous marginal cilia, marginal soralia, brown lower surface, medulla K+ yellow turning red, without norlobaridone, on bark at low elevations.

Parmotrema subtinctorium

Great Horned Lichen

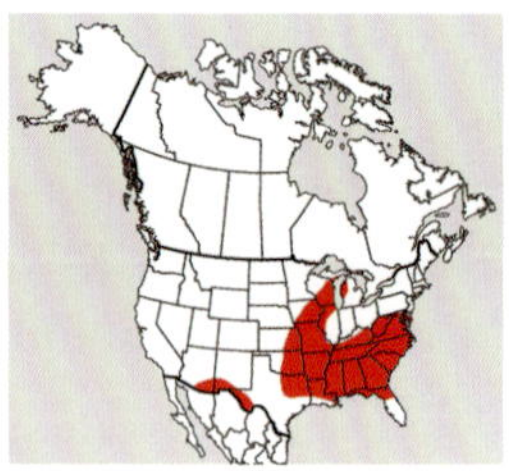

Tripp 3599 (photo: Lendemer)

NOTES: *Parmotrema subtinctorium* belongs to the same group of species as *P. subsumptum*, all of which have completely brown lower surfaces and maculate upper surfaces (it also belongs to the somewhat irksome group of *Parmotremas* that start with the letter "s," this makes number six). *Parmotrema subtinctorium* differs from *P. subsumptum* in having isidia instead of soralia and in producing norlobaridone in addition to salazinic acid. It is morphologically identical to *P. neotropicum*, which differs only in the absence of norlobaridone; these two species can only reliably be distinguished with TLC. *Parmotrema ultralucens* and *P. subisidiosum* might additionally be confused with *P. subtinctorium*, but both have a lower surface that is black toward the center of their thalli.

CHEMISTRY: Atranorin, norlobaridone and salazinic acid. Spot tests. Cortex: K+ yellow, C-, KC-, P-, UV-. Medulla: K+ yellow turning red, C-, KC+ pinkish, P+ orange, UV-.

NICHE: *Parmotrema subtinctorium* occurs on the bark of hardwood trees at middle and low elevations in the Smokies. It is common and widespread throughout southeastern North America.

KEY FEATURES: Blue-gray foliose thallus, ruffled lobes with conspicuous marginal cilia, laminal isidia, brown lower surface, medulla K+ yellow turning red, with norlobaridone, on bark at low elevations.

Parmotrema ultralucens

Southern Lights

Tripp 3568 (photo: Lendemer)

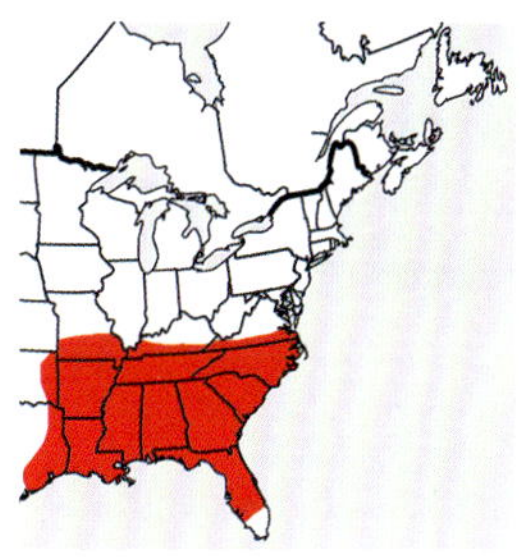

NOTES: The penultimate *Parmotrema* of the Smokies, and one of our finest members of the genus–period. *Parmotrema ultralucens* is a show stopper easily recognizable by its big ostentatious thallus, isidia, marginal cilia, maculae, and K+ red medulla. Oh, it might also have the most remarkable UV+ bright yellow reaction of any macrolichen, owing to an abundance of lichexanthone in the medulla. Southern Lights is most likely to be confused with *P. subisidiosum*, which can easily be separated by its UV- medulla. The story goes that Mason Hale initially overlooked the lichexanthone in the medulla; Hildur Krog pointed it out to him, then described the species. Not one to be outdone, Hale combined it into *Parmotrema*. Let us now gather for a moment of silence in honor of the Greek translation of "ultralucens": "beyond shining."

CHEMISTRY: Atranorin, lichexanthone and salazinic acid. Spot tests. Cortex: K+ yellow, C-, KC-, P-, UV-. Medulla: K+ yellow turning red, C-, KC-, P+ orange, UV+ bright yellow.

NICHE: Southern Lights is common on the bark of hardwoods and conifers at low elevations, especially in humid habitats. It is similarly common throughout southeastern North America, and reminds us of the tropics.

KEY FEATURES: Blue-gray foliose thallus, ruffled lobes with conspicuous marginal cilia, laminal isidia, UV+ bright yellow medulla, K+ yellow turning red medulla, on conifers and hardwoods at low and middle elevations.

Parmotrema xanthinum

Lime Light

Tripp 2692 (photo: Lendemer)

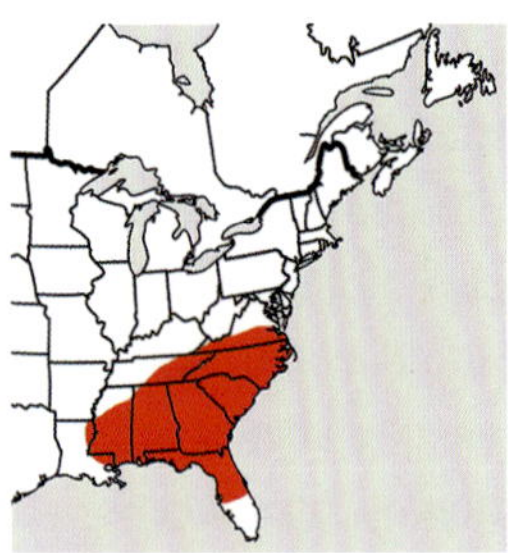

NOTES: This species allows us to end *Parmotrema* on a high note. *Parmotrema xanthinum* is one of the easiest members of the genus to recognize because it has a yellow-green upper cortex owing to the presence of usnic acid in addition to abundant isidia. In the past, populations ascribable to this name were treated as two different species based on differences in chemistry, but recent molecular study has failed to yield support for a distinction.

CHEMISTRY: Usnic acid, fatty acids, +/- gyrophoric acid. Spot tests. Cortex: K-, C-, KC+ gold, P-, UV-. Medulla: K-, C- or C+ pink, KC- or KC+ pink, P-, UV-.

NICHE: Lime Light is widespread throughout southeastern North America. In the Smokies, this species occurs at middle and low elevations on the bark and branches of trees and shrubs. Two different chemotypes occur in the Smokies, and they have very similar distributions, sometimes even occurring at the same site.

KEY FEATURES: Yellow-green foliose thallus, ruffled lobes with conspicuous marginal cilia, laminal isidia often with apical cilia, C- or C+ pink medulla, on conifers and hardwoods at middle-to-low elevations.

Peltigera didactyla

Tawny and Brawny

Lendemer 23274 (photo: Lendemer)

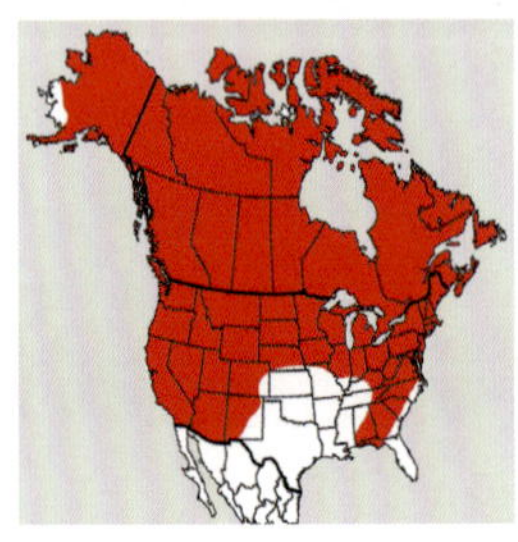

NOTES: *Peltigera.* Life could be worse, but life could also be better. Fortunately, *P. didactyla* is among the easiest members of the genus to identify in the Smokies on account of its small, scallop-like or curled lobes and laminal soralia. The only other sorediate species of *Peltigera* in the region is *P. extenuata*, which is basically a much larger version of *P. didactyla* that has abundant rhizines and occurs only at high elevations. While the soredia of *P. didactyla* are distinctive, they are also not always present. Sometimes *P. didactyla* is fertile and, in that case, it loses its soredia and develops large, reddish-brown, saddle-shaped apothecia on the lobe tips. When that happens, you can still recognize Tawny and Brawny based on the very short lobes, sparse rhizines and tomentose upper surface.

CHEMISTRY: No substances. Spot tests: K-, KC-, C-, P-, UV-.

NICHE: This species is common in northeastern North America, where it occurs in both disturbed and non-disturbed habitats, growing on rocks and soil. It is likely widely distributed at middle and high elevations of the Appalachians but has been relatively little collected because of its small size. In the Smokies, you can look for it on your next hike up to Purchase Knob.

KEY FEATURES: Foliose lichen with small, scallop-like or cupped gray lobes, tomentose upper surface, laminal soralia or reddish-brown saddle shaped apothecia, on soil and on rock at middle-to-high elevations.

Peltigera evansiana

Tawny Titivation

Lendemer 44536 (photo: Tripp)

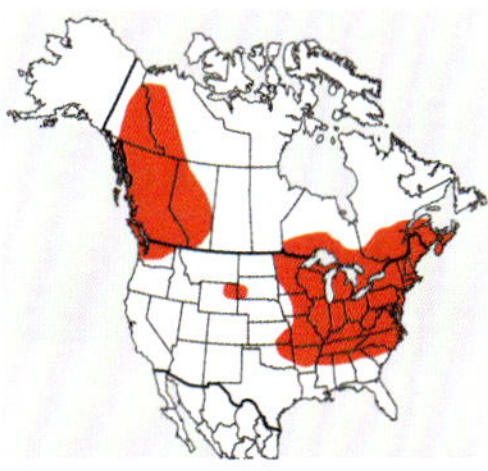

NOTES: Although species of *Peltigera* are generally easy to recognize to genus, especially after confirming the presence of a cyanobacterial photobiont (applicable to almost all our Smokies species and the vast majority worldwide), many of them are difficult to identify to species. Thankfully, that does not apply to *P. evansiana* because it is the only *Peltigera* in the Smokies that produces isidia throughout the upper surface of the thallus. Sometimes, *P. didactyla* and *P. extenuata* appear superficially similar to *P. evansiana* if their soredia become darkened and coarse, however the thalli in those species tend to be composed of solitary, large, bowl-like lobes and they are, as mentioned, sorediate (with discrete soralia) rather than isidiate. Ragged thalli of *P. praetextata* can also be confused for *P. evansiana*, but that species produces lobules rather true isidia.

CHEMISTRY: No substances. Spot tests. K-, C-, KC-, P-, UV-.

NICHE: This relatively rare species is nonetheless widespread throughout the Smokies, where it grows on mosses over tree bases and rocks.

KEY FEATURES: Large brown foliose thallus with broad lobes and laminal isidia, cyanobacterial photobiont, infrequent on mosses over tree bases and rocks throughout.

Peltigera hydrothyria

Clear and Cold

Tripp 4979 (photo: Lendemer)

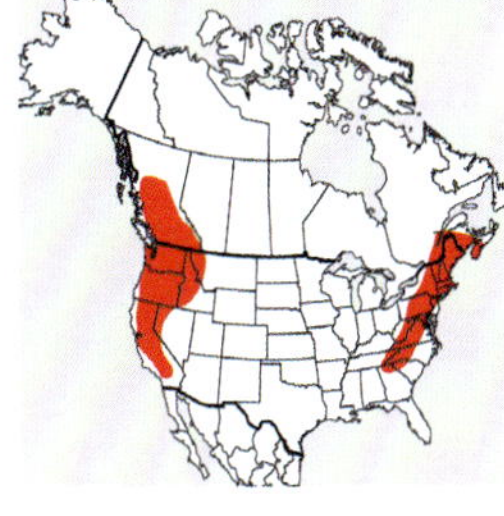

NOTES: This is not a pretty lichen, at least when out of the water. And you won't find it unless you go looking for it. Remember the days of flipping stones in the creek in search of salamanders? The good news is that you can do that in search of lichens, too. *Peltigera hydrothyria* is one of very few truly aquatic lichens. In North America, it is one of three such species. The other two are the closest sister taxa to *P. hydrothyria*, but do not occur in eastern North America. *Peltigera hydrothyria* can be found in clear-flowing, typically cold streams throughout the Appalachian Mountains to the Gaspe Peninsula. It is characterized by its black thallus that, when wet, resembles seaweed, and when dry, resembles something that you wouldn't want to pick up with your hands. To confirm it is in fact a lichen, you'll need to put it under your compound microscope to look for the fungal-cyanobacterial association. Specifically, look for *Nostoc* as a photobiont. Molecular phylogenetic studies have shown that Clear and Cold, together with its two aquatic relatives, form a clade that is early diverging with respect to all other *Peltigera*. Could it be that this group originated in or near the water and then later moved onto land?

CHEMISTRY: No substances (but variable in some populations). Spot Test: K-, KC-, C-, P-, UV-.

NICHE: Clear and Cold just about sums it up. Look for this species in suitable habitat anywhere in/among the Appalachian Mountains.

KEY FEATURES: Black foliose lichen that is fully aquatic, found in submerged waters of creeks that are clear flowing and typically cold.

Peltigera leucophlebia

Green Acres

Tripp 8666 (photo: Tripp)

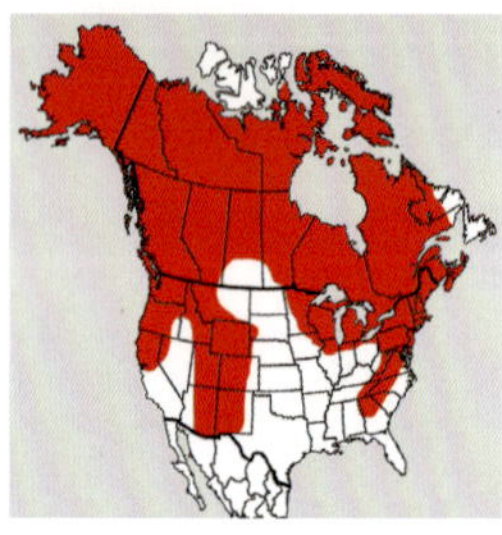

NOTES: Okay, it is true that *Peltigera* isn't always a kind friend in the taxonomic sense. This is an easy exception (and with nearly 1,000 species of lichens in the Smokies, we like exceptions!). *Peltigera leucophlebia* is extremely easy to identify because it is the only member of the genus in the southern Appalachians that has a green alga as its primary photobiont. See? Green Acres (the acres refers to its very large, expansive lobes). Additionally, this lichen harbors cyanobacteria in the form of conspicuous bluish-black cephalodia scattered across the upper surface (symbiosis: what a great way live). Need help getting to genus? Just flip a lobe over and look for the distinctive veins. This species is not at all common in the Smokies. We have seen it only a few times. If you encounter Green Acres, consider photo-documenting it and sending us the geographical coordinates!

CHEMISTRY: Methylgyrophate, tenuiorin, terpenoids. Spot tests: K-, KC-, C-, P-, UV-.

NICHE: Green Acres has a broad distribution that includes much of northeastern North America, extending south along the Appalachian Mountain chain, and into much of western North America, primarily mountainous areas. In the southern Appalachians, look for it at low-to-middle elevations on rocks in wet, riparian corridors.

KEY FEATURES: Bright green *Peltigera* with very broad lobes that bear conspicuous and persistent bluish-black cephalodia, on rocks in wet places, low-to-middle elevations.

Peltigera neckeri

Staccato

Buck 56282 (photo: Lendemer)

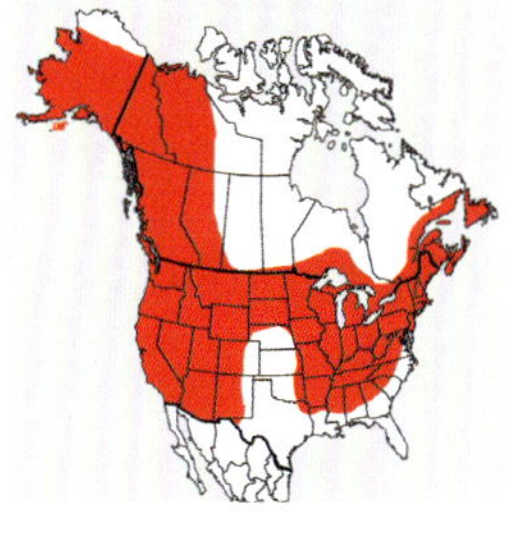

NOTES: As *Peltigeras* go, *P. neckeri* is among the easy pickings. Its black apothecia separate it from most others, and this feature—together with its characteristic lobe tips dusted by pruina and darkened veins on the lower surfaces—seal the deal. These black apothecia, which rather remind one of short, dumpy, black keys on a piano, will almost be what draws your eye to this species in the field. Don't confuse this for *P. phyllidiosa*, which also has black apothecia; the latter typically is decorated by minute phyllidia lining the margins. However, *P. phyllidiosa* is sometimes lacking such structures and in these cases, study the lobe tips: those of *P. phyllidiosa* lack any sort of pruina or tomentum (in contrast to *P. neckeri*).

CHEMISTRY: There seems to be some disagreement as to the chemistry of this species based on the literature. Spot tests: K-, KC-, C-, P-, UV-.

NICHE: *Peltigera neckeri* is not a common species, but it can be found in a wide variety of habitats. It is also widely distributed across the United States. Go figure. Look for it low, look for it high, search on soil, search on rocks. Sounds like a strategy that yields commonness rather than uncommonness. Hmmmpf!

KEY FEATURES: Large foliose lichen, black apothecia, pruinose lobe tips, dark veins on the lower surface, on soil or rock at all elevations.

Peltigera neopolydactyla

Runner Up

Tripp 3838 (photo: Lendemer)

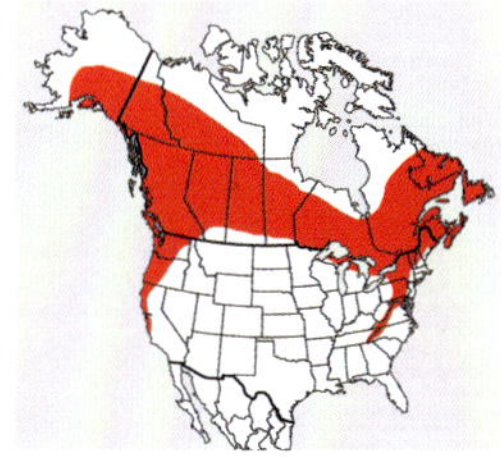

NOTES: If there is one *Peltigera* that you can't help but notice in the Smokies (and trust us, we try our best to do so), it will be this one. Of all the members of the genus that occur here, *Peltigera* "new many fingers" (neopolydactyla) has the broadest lobes, which have a distinctly shiny surface when both wet or dry, no trace of tomentum near the lobe tips, and bear distinctive, upturned, saddle-shaped apothecia that are light-to-dark reddish-brown in color (see inset). *Peltigera seneca* is smaller in size but otherwise comparable in most aspects.

CHEMISTRY: Tenuiorin, methyl gyrophorate, gyrophoric acid, triterpenes. Spot tests. Cortex: K-, C-, KC-, P-, UV-. Medulla: K-, C-, KC-, P-, UV-.

NICHE: Although largely restricted to the Appalachian Mountains in temperate eastern North America, this species is very abundant and common, especially in high elevation habitats. In the Smokies, *Peltigera neopolydactyla* occurs throughout the Park, where it often forms extensive colonies on logs, on soil along trails, and on old, wet, rock walls.

KEY FEATURES: Large brown-to-gray-brown thallus, exceedingly broad lobes, shiny upper surface lacking tomentum, reddish-brown, saddle-shaped apothecia, common in high, wet areas throughout the Park.

Peltigera phyllidiosa

Considerably Pleated

Tripp 4939 (photo: Lendemer)

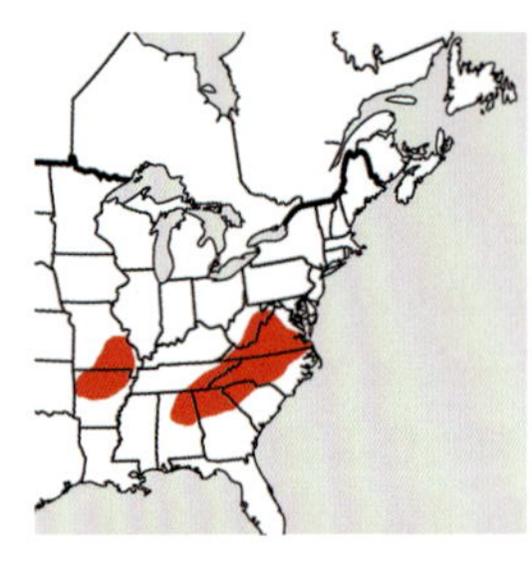

NOTES: There can be little doubt that *Peltigera phyllidiosa* is the most beautiful *Peltigera* in the Smokies. It can be recognized by its distinctive, dark brown coloration, abundant tiny lobules produced along the lobe margins and on the thallus surface, a lower surface that is dark in color except for a regular pattern of ellipsoid white patches and, when present, black, saddle-shaped apothecia. It is most likely to be confused with *P. praetextata*, but that species has a morphologically different lower surface and reddish-brown apothecia. *Peltigera neckeri* has similarly colored apothecia, but lacks the phyllidia that give *P. phyllidiosa* its name.

CHEMISTRY: Tenuiorin, methyl gyrophorate, gyrophoric acid, zeorin. Spot tests. Cortex: K-, C-, KC-, P-, UV-. Medulla: K-, C-, KC-, P-, UV-.

NICHE: This infrequent species is endemic to the Appalachian Mountains and Ozark Highlands. It is widespread in the Smokies, particularly at middle-to-low elevations. Unlike other members of the genus, *Peltigera phyllidiosa* tends to occur directly on tree bark or calcareous rocks, rather than growing on a carpet of mosses.

KEY FEATURES: Large brown thallus, with broad lobes and phyllidia on both the surface and margins, lower surface with white, ellipsoid patches between the dark veins, infrequent on tree bases and trunks.

Peltigera praetextata

Chill Toes

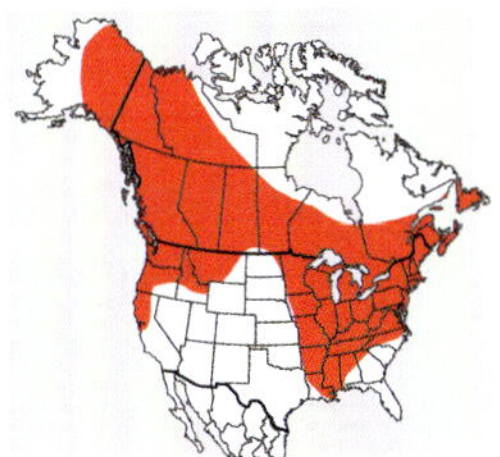

Tripp 3683 (photo: Lendemer)

NOTES: The most distinctive features of *Peltigera praetextata* are its broad lobes with ruffled margins that appear to be lobulate and its tips of lobes that appear frosted due to the presence of a light tomentum. Indeed, the frosted look of the lobes gives the impression that this species always has cold feet. *Peltigera neopolydactyla* and *P. seneca* are morphologically similar to *P. praetextata*, but both of those species have shiny upper surfaces, even at the lobe tips. *Peltigera phyllidiosa* is also similar but has a black-and-white-patterned lower surface (vs. pale to dirty brown and distinctly veined, as in *P. praetextata*). *Peltigera phyllidiosa* also has blackish apothecia (vs. reddish-brown, as in *P. praetextata*).

CHEMISTRY: No substances. Spot tests. K-, C-, KC-, P-, UV-.

NICHE: As is the case throughout temperate eastern North America, this species is common and widespread throughout the Smokies. Chill Toes occurs on mosses growing over logs, tree bases, and soil along roadsides and trails.

KEY FEATURES: Large brown-to-gray-brown thallus, broad lobes with ruffled margins that are usually lobulate, upper surface with tomentum near tips, reddish-brown, saddle-shaped apothecia, common throughout the Smokies.

Peltigera seneca

Lesser Runner Up

Lendemer 33030 (photo: Tripp)

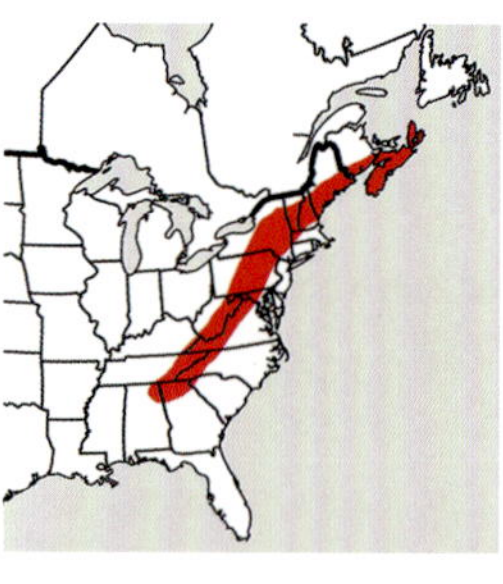

NOTES: Just when you thought that *Peltigera* couldn't get more confusing, they go and change the names on you. For many years, a *Peltigera* resembling *P. neopolydactyla* but with smaller, more narrow lobes would have been called *P. polydactyla*. However, recent studies have shown that *P. polydactyla* doesn't actually occur in North America and, instead, specimens from eastern North America previously attributed to this name should be treated under the concept *P. seneca*. While many questions remain—such as "Will taxonomists ever get it right?"—at least we can be confident *P. seneca* occurs in the Smokies.

CHEMISTRY: Tenuiorin, methyl gyrophorate, peltidactylin, dolichorhizin, zeorin. Spot tests. Cortex: K-, C-, KC-, P-, UV-. Medulla: K-, C-, KC-, P-, UV-.

NICHE: Due to its previous confusion with *Peltigera polydactyla*, the distribution of *Peltigera seneca* is not clear. It appears to be infrequent but widespread throughout the Smokies, where it grows in wet areas on mosses over soil, tree bases, logs, and organic matter.

KEY FEATURES: Large brown-to-gray-brown thallus, relatively broad lobes, shiny upper surface lacking tomentum, reddish-brown, saddle shaped apothecia, infrequent on logs and organic matter.

Pertusaria appalachensis

Sunshine Sneak

Tripp 2671 (photo: Lendemer)

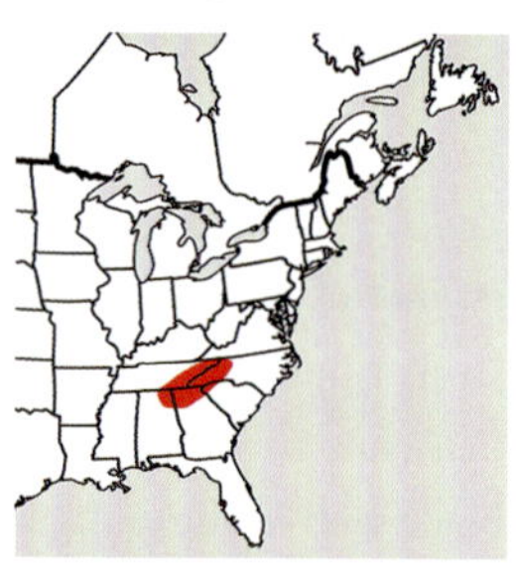

NOTES: *Pertusaria appalachensis* is a rare species that tends to blend in with a plethora of other *Pertusaria* species in the Smokies. (Not to worry: unlike *Peltigera*, species of *Pertusaria* are actually fun to identify! We blame it on growth form . . . macrolichens are *difficult*.) *Pertusaria appalachensis* can be recognized, first, to genus by gestalt (learn those beautiful warts) and, second, to species by yellowish areas that reliably characterize the tops of the warts. Chemistry of the medulla further confirms the identity of Sneak Sunshine. The species mostly likely to be confused with *P. appalachensis* are *P. epixantha* and *P. texana*, both of which have UV+ bright orange cortices and tend to occur at lower elevations. *Pertusaria neolecania* and *P. pustulata* can also be somewhat similar in appearance, but these two have smaller, more conically shaped warts and differ in their chemistries.

CHEMISTRY: Arthothelin and other accessory xanthones. Spot tests. Cortex: K-, C-, KC-, P-, UV-; medulla K-, C+ yellow, KC+ orange, P-, UV+ bright orange.

NICHE: Sunshine Sneak is a rare southern Appalachian endemic confined to hardwood trees at middle-to-high elevations.

KEY FEATURES: Crustose thallus, large, flat-topped warts with distinctive yellow area around the ostiole, UV- cortex, P- / KC+ orange / UV+ bright orange medulla, 8-spored asci, on bark at middle-to-high elevations.

Pertusaria epixantha

Simple Dimples

Lendemer 33133 (photo: Tripp)

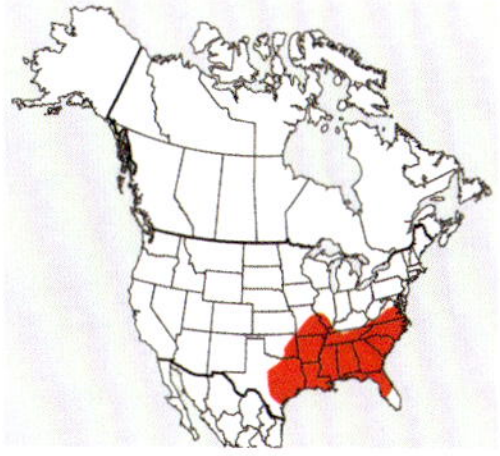

NOTES: Although there are a lot of species of *Pertusaria* in the Smokies, we hope you will (1) begin to learn them using our keys, (2) make use of our photographs in providing further identification assistance, and (3) learn to love this beautiful genus. *Pertusaria epixantha* can be recognized by its large, flat-topped warts that have a depressed yellowish area around the ostiole. It is only likely to be confused with *P. appalachensis*, which has a UV- cortex, or with *P. texana*, which has a P+ orange medulla. Appreciate that the specific epithet of this species translates to "yellow on top."

CHEMISTRY: Thiophanic acid and a substance resembling variolaric acid (hey, who's perfect?). Spot tests. Cortex: K-, C-, KC-, P-, UV+ bright orange; medulla: K-, C-, KC-, P-, UV-.

NICHE: *Pertusaria epixantha* is common and widespread throughout southeastern North America where it is common from the low elevations of the Appalachians and Ozarks southward and westward to the coastal plain. In the Smokies, Simple Dimples is encountered primarily at low elevations, especially along Fontana Lake and the valleys south of Cades Cove in Tennessee.

KEY FEATURES: Crustose greenish-yellow thallus, large flat-topped warts with pale yellow area around the (single, simple) ostiole, UV+ bright orange cortex, P- medulla, 8-spored asci, on bark at middle-to-low elevations.

Pertusaria globularis

Fragile Fingers

Tripp 3616 (photo: Lendemer)

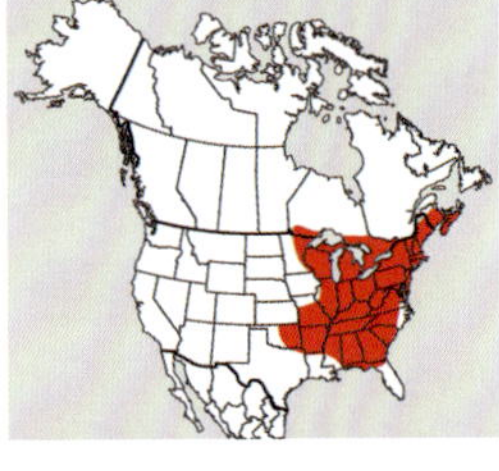

NOTES: *Pertusaria globularis* is an inconspicuous crustose species with short isidia that grow on raised warts and are fragile, tending to crumble into bits over time. The mealy appearance leads to the resemblance of this species with *Megalospora porphyritis*, but the latter is pustular-sorediate and P+ orange-red. In many respects, *Ochrolechia yasudae* is morphologically the most similar species to Fragile Fingers, and it often grows with it on rocks. However, the former differs in having a thick, cream colored thallus with more robust isidia and a C+ pink cortex. It also forms much, much larger thalli than those of Fragile Fingers. Once learned, you will have no problems calling this one out in the field during subsequent trials.

CHEMISTRY: 2,7-dichlorolichexanthone, 2'-o-methylperlatolic acid. Spot tests. Cortex: K-, C-, KC-, P-, UV+ dull pink; medulla: K-, C-, KC-, P-, UV+ pale blue-white.

NICHE: This species is widespread throughout temperate eastern North America, where it occurs on the bark of hardwoods, especially on the bases of mature trees. It also occurs overgrowing bryophytes and on non-calcareous rocks. In the Smokies, it can be found at all elevations.

KEY FEATURES: Crustose thallus, small fragile isidia, fertile warts rare, rounded, 4-spored asci, cortex UV+ dull pink, medulla UV+ pale blue-white, throughout the Smokies on the bark of trees, on bryophytes, and rarely rocks.

Pertusaria macounii

Canadian Warts

Tripp 2692 (photo: Lendemer)

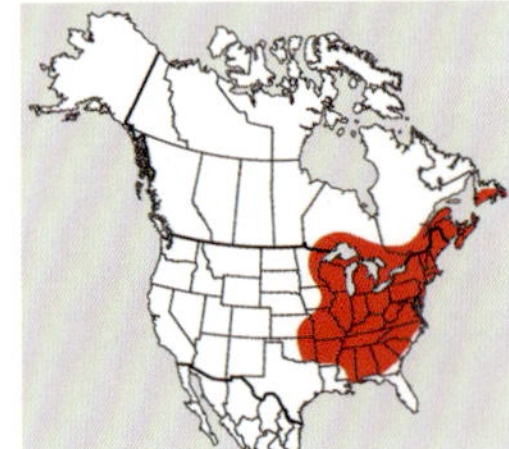

NOTES: *Pertusaria macounii* was named by the late Ivan Mackenzie Lamb in honor of the naturalist John

Macoun, who collected many lichens throughout Canada. The species can be recognized by its blue-gray thallus, relatively large, flat-topped warts bearing multiple ostioles, 2-spored asci with spores that often turn brownish with age, and by its chemistry. Superficially similar species include *P. appalachensis*, which has 8-spored asci, yellow areas around the ostioles, and a K- medulla, and *P. consocians*, which also has 2-spored asci, but with spores that remain pale, and a dark brown epihymenium that reacts K+ purple (vs. K- in *P. macounii*).

CHEMISTRY: 2,7-dichlorolichexanthone, stictic acid aggregate. Spot tests. Cortex: K-, C-, KC-, P-, UV+ dull pink; medulla: K+ yellow, C-, KC-, P+ orange, UV-.

NICHE: Widespread and common in northern temperate eastern North America, but restricted to middle and high elevations in the Smokies. Canadian Warts typically occurs on hardwood bark and branches. If you were a corticolous lichen, which tree would you inhabit?

KEY FEATURES: Crustose gray thallus, large, flat-topped warts with pale area surrounding multiple, clustered ostioles, UV+ dull pink cortex, P+ orange medulla, 2-spored asci, on bark at middle-to-high elevations.

Pertusaria neolecania

Northern Crawler

Lendemer 29338 (photo: Lendemer)

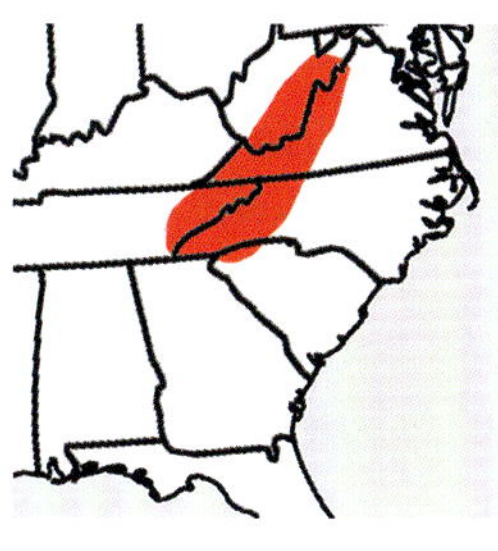

NOTES: *Pertusaria neolecania* is among the seven or so members of the genus currently known in the Smokies that have asci containing two to four spores each. Among these, there are only two species that produce thiophanic acid in the cortex and thus yield a very strong UV+ orange reaction: *P. neolecania* and *P. pustulata* (do not confuse strong orange with dull pink to orange, which is typical of several of the other species). *Pertusaria neolecania*, like *P. pustulata*, has solitary fruiting warts with dark ostioles, but can be differentiated by producing norstictic acid (instead of stictic acid in *P. pustulata*).

CHEMISTRY: Thiophanic acid and norstictic acid. Spot tests: (cortex) K-, C+ yellow/orange, KC+ yellow/orange, P-, UV+ bright orange; (medulla) K+ yellow turning red, C-, KC-, P+ yellow, UV-.

NICHE: This species was first described from mountainous forests of western Mexico. Its presence in North America north of Mexico was discovered a half decade ago, from material collected in the Smokies. In North America, it remains known from scattered locations in northern hardwood forests throughout the southern Appalachian Mountains. Look for it on branches and bark of hardwoods at middle-to-high elevations.

KEY FEATURES: Yellowish-gray continuous thallus, solitary warts with dark ostioles, 2(-4) spores per ascus, K+ yellow to red medulla, UV+ bright orange cortex, on branches and bark of hardwoods, middle-to-high elevations.

Pertusaria neoscotica

Bloody Blisters

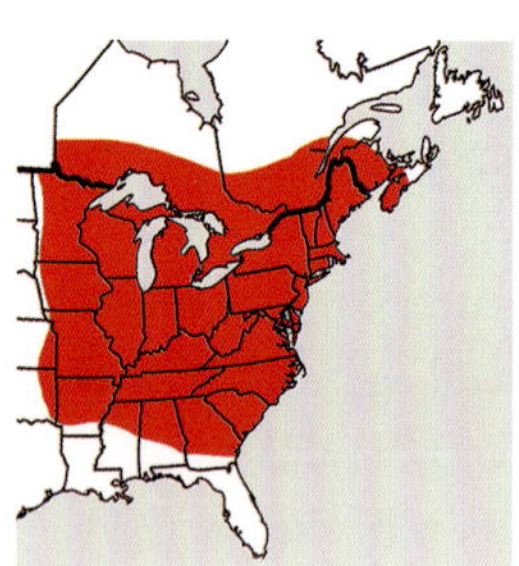

Tripp 2232 (photo: Lendemer)

NOTES: Think of this one on your next long hike in wet shoes (that, unlike your wet shoes in the Rocky Mountains, never dry out in the East). *Pertusaria neoscotica* has a distinctive, pale, white coloration to it that looks as though it has been bleached. However, when you cut it open and apply some K, it turns blood red. The color and chemistry, together with its large, flat warts, are distinctive. *Pertusaria plittiana* has a similar chemistry and color but occurs only on rocks.

CHEMISTRY: Norstictic acid aggregate. Spot tests. Cortex: K-, C-, KC-, P-, UV-; medulla: K+ yellow turning red, C-, KC-, P+ yellow, UV-.

NICHE: Although the name implies that this species tends to occur in more northern areas such as Nova Scotia, it has actually been found to be widespread throughout much of temperate eastern North America. In the Smokies, Bloody Blisters occurs at all elevations, where it is found on bark and on canopy branches, primarily of hardwood trees.

KEY FEATURES: Crustose, sometimes bone-white thallus, large, flat-topped warts with pale area surrounding multiple ostioles, UV- cortex, K+ blood red medulla, 2-spored asci, on bark throughout the Park.

Pertusaria obruta

Buggin' Out

Lendemer 33061 (photo: Tripp)

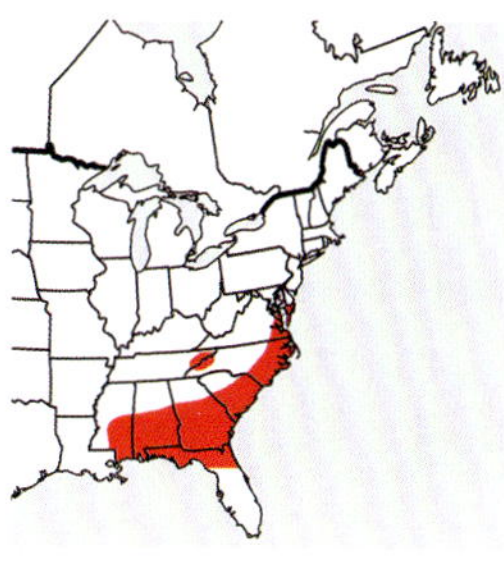

NOTES: Don't you hate inherent biases? You fear bugs, and so you find this bugged out lichen to be a turn off? We are the first to admit that this may not be the sexiest lichen in the book. But stop for a moment anyway and appreciate its characteristic dark green thallus, which is covered with minute, white cracks, along with warts that are only slightly raised and resemble bug eyes. No other species of *Pertusaria* in the Smokies quite looks like *P. obruta*. Some can come close, but all differ in chemistry and in the number of spores per ascus. This species has 2-per ascus, and after studying the key, you will know why we continue to emphasize this character in learning *Pertusaria* of the Smokies.

CHEMISTRY: Variolaric acid. Spot tests. K-, C-, KC-, P-, UV-.

NICHE: Buggin' Out is yet another friend that hails from the southeastern Coastal Plain. This species occurs at low elevations in the southern Appalachians, including the Smokies. It is most often found on the bark of hardwoods, and you are likely to see it during a hike along Lakeshore Drive.

KEY FEATURES: Crustose, dark-green thallus with white cracks, small, semi-immersed warts with pale areas around large, dark, minimally raised ostioles, UV- cortex and medulla, 2-spored asci, on bark at low elevations.

Pertusaria paratuberculifera

Elite Eight

Lendemer 33117 (photo: Tripp)

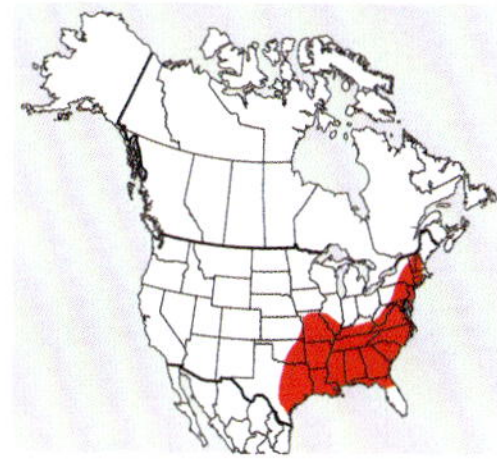

NOTES: *Pertusaria paratuberculifera* is a treat for all people and all seasons and, in general, just makes you feel good. It is one of the few members of the genus in the Smokies that yields a *bright* UV+ yellow reaction across its thallus. This feature, combined with the large, more-or-less conical warts and 8-spored asci, makes the species extremely easy to identify. The only *Pertusaria* likely to be confused with Elite Eight is *P. valliculata*, which is very rare and differs in having 4-spored asci. If one missed the UV+ reaction (how *could* you?), then this species could be confused with *P. tetrathalamia*, which is UV+ dull orange and has 4-spored asci.

CHEMISTRY: Lichexanthone and 2'-o-methylperlatolic acid. Spot tests. Cortex: K-, C-, KC-, P-, UV+ bright yellow; medulla: K-, C-, KC-, P-, UV+ dull blue-white.

NICHE: As is the case throughout its range, *Pertusaria paratuberculifera* is common and widespread in the Smokies, with the exception of the highest spruce-fir peaks. It typically occurs on the bark hardwoods, especially larger, more mature trees.

KEY FEATURES: Crustose gray-to-greenish-gray thallus, large +/- conical warts with pale area surrounding multiple ostioles, UV+ *bright yellow* cortex, K- medulla, 8-spored asci, on bark throughout the Park.

Pertusaria plittiana

Plitt's Platters

Lendemer 33190 (photo: Tripp)

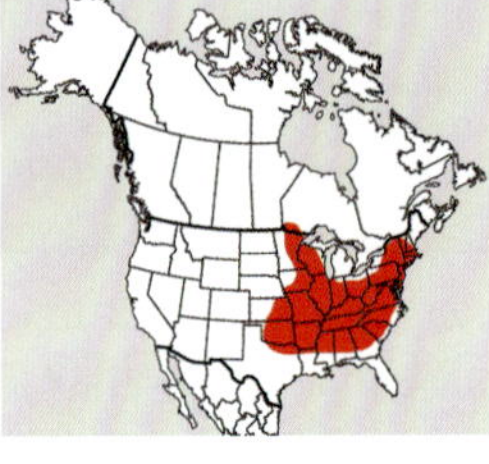

NOTES: This species is named in honor of the botanist and public school teacher Charles Plitt from Baltimore, Maryland. It can easily be recognized by its ecology (see Niche) combined with the pale-gray to blue-gray thallus, flat-topped warts, 2-spored asci, and distinctive chemistry. *Pertusaria neoscotica* is chemically similar to Plitt's Platters but occurs on bark and has a bone-white thallus. No other species of *Pertusaria* is more or less restricted to rock in the Smokies.

CHEMISTRY: Norstictic acid aggregate and perlatolic acid. Spot tests. Cortex: K-, C-, KC-, P-, UV-; medulla: K+ yellow turning red, C-, KC-, P+ yellow, UV+ blue-white.

NICHE: *Pertusaria plittiana* is a species of temperate eastern North America that occurs on non-calcareous rocks, especially on sandstone. In the Smokies, it is found throughout the Park, but is particularly common at middle and low elevations.

KEY FEATURES: Crustose gray to blue-gray thallus, small, flat-topped ostioles with pale area surrounding each cluster of ostioles, UV- cortex, K+ red medulla, 2-spored asci, on non-calcareous rock throughout the Park.

Pertusaria propinqua

False Flasks

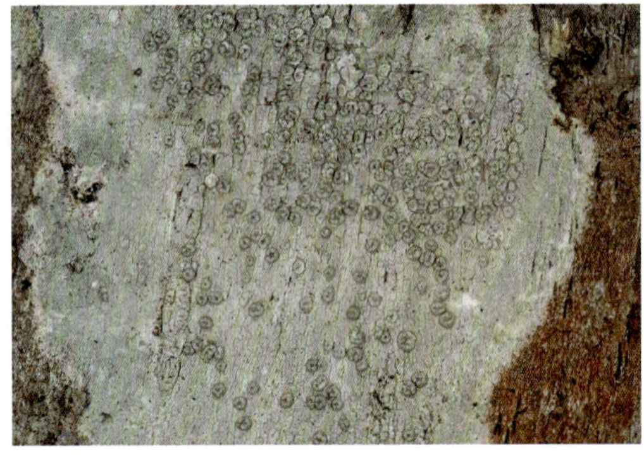

Tripp 2210 (photo: Lendemer)

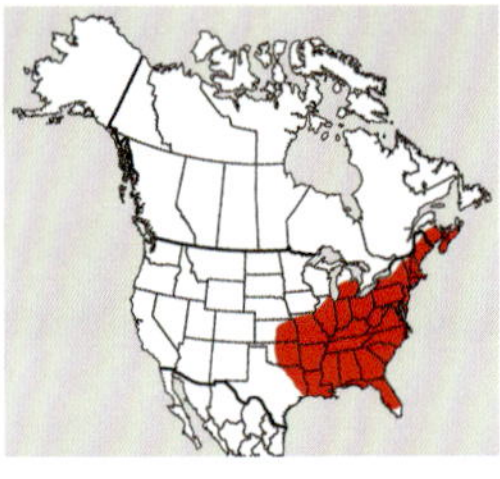

NOTES: *Pertusaria propinqua* nicely illustrates the complexity of fruiting warts of *Pertusaria*: their distinctive shape reminds beginners of perithecia, but these warts are distinctly apotheciate. In False Flasks, the ostioles become broadly expanded and thus give the appearance of a lecanorine apothecium with an overgrown margin. In addition to this feature, *P. propinqua* can be identified by its UV-cortex, 8-spored asci, and spot test reactions in the medulla (see Chemistry). *Pertusaria rubefacta* has 8-spored asci and the same chemistry in the medulla but differs in having a UV+ bright yellow cortex. Confusion with *P. neoscotica* is also possible, but that species has 2-spored asci, a white thallus, and flattened warts.

CHEMISTRY: Norstictic acid aggregate. Spot tests. Cortex: K-, C-, KC-, P-, UV-; medulla: K+ yellow turning red, C-, KC-, P+ yellow, UV-.

NICHE: This species is widespread throughout temperate eastern North America and occurs at all elevations in the Smokies. It is typically found on the bark of hardwood trees.

KEY FEATURES: Crustose gray-to-blue-gray thallus, large +/- flat-topped warts with broad ostioles, UV-, K+ red medulla, 8-spored asci, on bark throughout the Park.

Pertusaria pustulata

Branch Bumps

Lendemer 53159 (photo: Tripp)

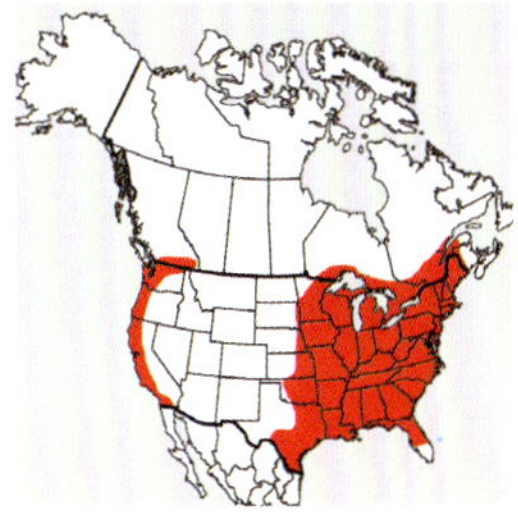

NOTES: *Pertusaria pustulata* is a bit of a mystery in the Smokies, and the southern Appalachians, more broadly. It is common in some areas of eastern North America and yet, when it comes to the Smokies, it is exceedingly rare. There isn't much to confuse it with because the thin, yellowish-gray continuous thallus and small, solitary thalline warts with ostioles are usually dark in color. The fruiting bodies of this species, like all *Pertusaria*, make the group as a whole very easy to identify. First do a UV test, and you will see that *P. pustulata* is characterized by having a bright UV+ orange thallus. In section, the asci contain two spores, rarely up to four, per ascus. The only other species with similar spores and chemistry is *P. neolecania*, which differs in the production of norstictic acid and thus has a K+ yellow turning red medulla. In contrast, *P. pustulata* produces stictic acid, and thus has a K+ yellow medulla. By the way, under *Fuscidea rescensa*, we mentioned zones of incompatibility among lichen thalli. This branch is a great example of several of them. It's sort of like putting fences around your yard to separate yourself from your neighbor, but in this case, the fences are made of hyphae.

CHEMISTRY: Thiophanic acid and stictic acid. Spot Tests: (cortex) K-, C+ yellow/orange, KC+ yellow/orange, P-, UV+ bright orange; (medulla) K+ yellow turning dingy brown, KC-, C-, P+ orange, UV-.

NICHE: *Pertusaria pustulata* is an extremely wide-ranging species in eastern North America with disjunct occurrences on the west coast. It almost always occurs on branches of trees or shrubs and seems to occur sporadically at all elevations.

KEY FEATURES: Small, yellowish-gray continuous thalli, wart-like apothecia usually with dark ostioles, UV+ bright orange, 2(-4) spores per ascus, on branches throughout.

Pertusaria rubefacta

Yellow Dumplings

Tripp 5482 (photo: Lendemer)

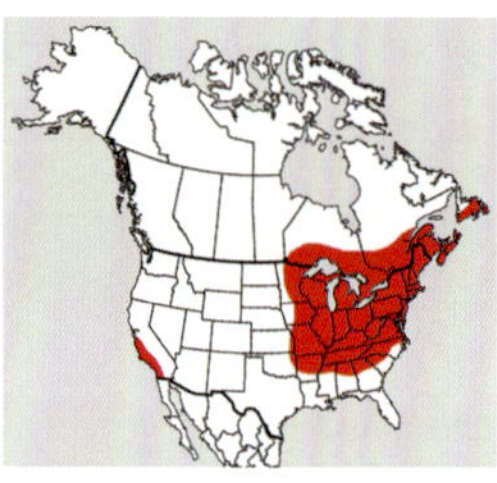

NOTES: *Pertusaria rubefacta* is essentially a more beautiful version of *P. propinqua*. Rather than being dull in color and lackluster under UV, *P. rubefacta* has a yellowish colored thallus, which is due to the presence of xanthones, which yield the bright UV+ orange reaction in this species. Yellow Dumplings could be confused with *P. appalachensis*, but that species lacks norstictic acid and has a UV+ orange reaction in the medulla, but not the cortex.

CHEMISTRY: Thiophanic acid and norstictic acid aggregate. Spot tests. Cortex: K-, C+ yellow, KC+ orange, P-, UV+ bright orange; medulla: K+ yellow turning red, C-, KC-, P+ yellow, UV-.

NICHE: In the Smokies, *Pertusaria rubefacta* is most often found on the bark of mature hardwoods, especially oaks, in middle-to-high elevation ridge top forests. Like several other species included in this book, Yellow Dumplings has an unusual disjunct distribution between temperate eastern North America and coastal southern California.

KEY FEATURES: Crustose, yellowish thallus, large +/- flat-topped warts with broad ostioles, UV+ bright orange, K+ red medulla, 8-spored asci, on bark throughout Park but especially at middle elevations.

Pertusaria subpertusa

So Happy Together

Lendemer 26719 (photo: Tripp)

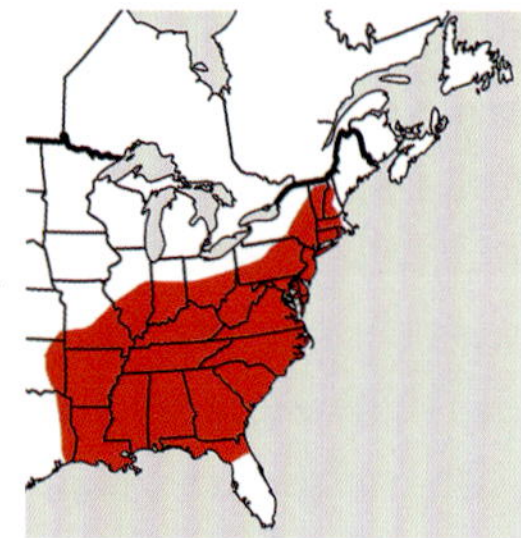

NOTES: Eminent lichenologist Irwin Brodo described this species as new to science after collecting it for the first time during his graduate work in the New

York City area. It may be somewhat surprising to many to learn that it is also one of the most common *Pertusaria* species in the Smokies. As is illustrated in the photograph, it can easily be recognized by its gray thallus and flat-topped warts that each bear multiple ostioles. The identification of *Pertusaria subpertusa* can be confirmed by finding 2-spored asci as well as a K+ reaction in the medulla. *Pertusaria neoscotica* is similar in appearance and also has 2-spored asci, but differs chemically.

CHEMISTRY: Xanthone and succinoprotocetraric acid. Spot tests. Cortex: K-, C-, KC-, P-, UV+ dull pink; medulla: K+ yellow turning brown, C-, KC-, P+ red, UV-.

NICHE: This species is widespread and common throughout the Smokies at high elevations. It typically occurs on the bark and branches of hardwoods.

KEY FEATURES: Crustose gray to blue-gray thallus, large, flat-topped pale area surrounding multiple ostioles per wart, UV+ dull pink cortex, P+ red medulla, 2-spored asci, on hardwoods throughout Park.

Pertusaria superiana

Paul's Super Lichen!

Tripp 6033 (photo: Lendemer)

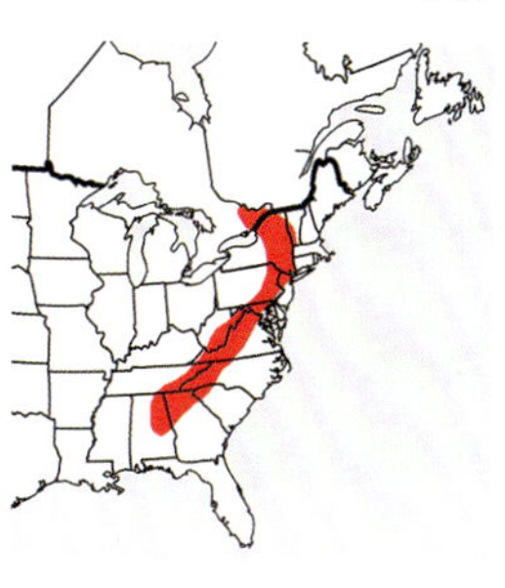

NOTES: *Pertusaria* is one of the most entertaining (and, for that matter, gratifying) lichen genera in the Smokies, and this species is no exception. *Pertusaria superiana*, named in honor of long-term Park biologist Paul Super, is a recent addition to the relatively long list of *Pertusaria* diversity present in the Park. Lucky for all of this, Paul's Super Lichen! is easy to identify based on its relatively thick bluish-gray (when dry; green when wet, like everything else) continuous thallus that is typically bumpy across the surface, its spores that number four per ascus, and its chemistry (a weak and dull UV+ orange reaction in the cortex). *Pertusaria neoscotica* is similar but produces norstictic acid (and reacts K+ yellow to red) in the medulla. *Pertusaria subpertusa* tends to have clustered ostioles and is P+ red. *Pertusaria ostiolata* is somewhat similar in appearance but has eight spores per ascus. *Pertusaria obruta* is another species with similar spot tests, but it has low thalline warts that are mostly immersed in the bumpy thallus, primarily two-spored asci, and differs chemically in lacking 2'-*O*-methylperlatolic acid. Although yet to be tested phylogenetically, *P. superiana* seems to be the sexual counterpart to the common and widespread asexual, isidiate species *P. globularis*. The two are friends and often grow together.

CHEMISTRY: 2,7-dichlorolichexanthone and 2'-*O*-methylperlatolic acid. Spot tests: (cortex) K-, C-, KC-, P-, UV+ weak dull orange; (medulla) K-, C-, KC-, P-, UV+ dull blue-white.

NICHE: This species has a classic Appalachian-Great Lakes distribution, but is clearly most at home in the southern Appalachians, just like the man after whom it was named. It occurs on the bark of hardwoods (occasionally also on conifers), especially near their bases. Look for it across a broad range of elevations, but almost always in mesic habitats.

KEY FEATURES: Relatively thick, continuous, often warty thallus, four spores per ascus, UV+ dull orange cortex, P- all over, broadly distributed in humid environments, bark of hardwoods.

Pertusaria tetrathalamia

Four-Spore, and Seven Xanthones Ago

Tripp 2622 (photo: Lendemer)

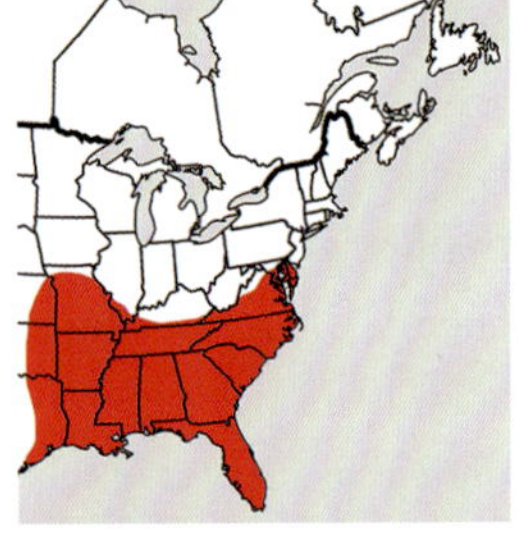

NOTES: Externally, *Pertusaria tetrathalamia* is most similar to *P. paratuberculifera*, but at this point in the *Field Guide*, do you really want to start judging a lichen by its cover? *Pertusaria paratuberculifera* differs from *P. tetrathalamia* in having 8-spored asci and a UV+ bright yellow cortex. This species could also be confused with *P. macounii*, but the latter has spores that tend to turn brown with age and react K+ purple. Although *P. tetrathalamia* generally has 4-spored asci, the number of spores per ascus is sometimes two or three.

CHEMISTRY: 2,7-dichlorolichexanthone, stictic acid aggregate. Spot tests. Cortex: K-, C-, KC-, P-, UV+ dull pink; medulla: K+ yellow, C-, KC-, P+ orange, UV-.

NICHE: This species is widely distributed in the southeastern United States; in the Smokies, it tends to be restricted to low and middle elevations where it occurs on the bark of hardwoods.

KEY FEATURES: Crustose yellowish thallus, large flat-topped warts bearing multiple ostioles each, UV+ dull pink cortex, P+ orange medulla, (2-)4-spored asci, on bark at middle-to-high elevations.

Pertusaria texana

Texas Tops

Tripp 3571 (photo: Lendemer)

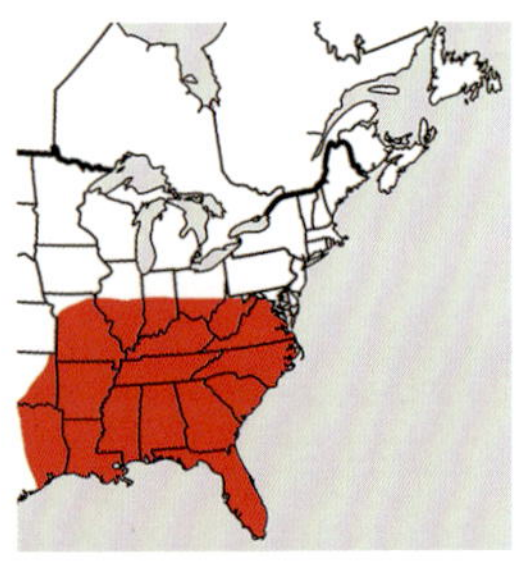

NOTES: *Pertusaria texana* is an extremely variable species, and the photograph presented here may or may not be typical of what you will find while hiking through America's favorite National Park. Texas Tops can *always* be recognized by its chemistry, coupled with its 8-spored asci. Often, the ostioles of this species have a distinctive, raised, yellow protrusion that resembles a nipple although in some cases this appears to be absent (as in the photograph). *Pertusaria pustulata* is probably the most similar species but differs in having 2-spored asci. *Pertusaria epixantha* shares a UV+ bright orange cortex and 8-spored asci, but differs in having a P- medulla.

CHEMISTRY: Thiophanic acid and stictic acid aggregate. Spot tests. Cortex: K-, C+ yellow, KC+ orange, P-, UV+ bright orange; medulla: K+ yellow, C-, KC-, P+ orange, UV-.

NICHE: This species is common and widespread in southeastern North America and is largely restricted to low and middle elevations in the Smokies. It typically occurs on hardwood bark and branches. Texas Tops becomes even more common as one travels south and west from the Park.

KEY FEATURES: Crustose yellowish thallus, small +/- conical warts with a raised yellow nipple at the top of each ostiole, UV+ bright orange cortex, P+ orange medulla, 8-spored asci, on bark at middle-to-low elevations.

Phaeocalicium polyporaeum

Bracket Lashes

Lendemer 33208 (photo: Tripp)

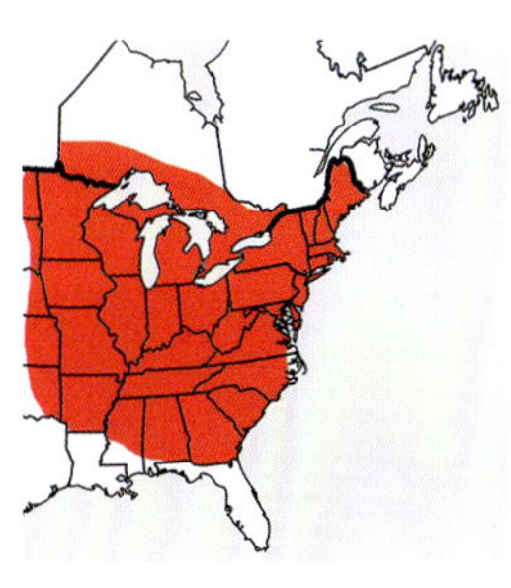

NOTES: *Phaeocalicium polyporaeum* is an honorary lichen; although it lacks a photobiont, this deficiency is made up by its unusual ecology. The species always grows on the upper surface of shelves of the bracket fungus *Trichaptum biforme*, but specifically just behind the growing edge, where it is often protected from the elements by another shelf just above. Species of *Phaeocalicium* are generally host specific, and all of the others that occur in the Smokies are very rare and found on non-fungal substrates. The only other lichen that occurs on bracket fungi is *Chaenotheca balsamconensis*, which differs in its larger pins, simple, globose, brown spores (vs. 2-celled, ellipsoid in *P. polyporaeum*) and occurrence on *T. abietinum*, which grows on conifers rather than hardwoods.

CHEMISTRY: No substances. Spot tests. K-, C-, KC-, P-, UV-.

NICHE: This species is always found on the outer edges of individuals of the bracket fungus *Trichaptum biforme*, which grows on dead or diseased hardwood trees. Both species are common and widespread throughout the Smokies and much of temperate eastern North America.

KEY FEATURES: Small black pins, 2-celled brown spores, on *Trichaptum biforme* on dead hardwoods, throughout the Smokies (elsewhere, also on *T. abietinum*).

Phaeographis inusta

Burnt Lips

Lendemer 33126 (photo: Tripp)

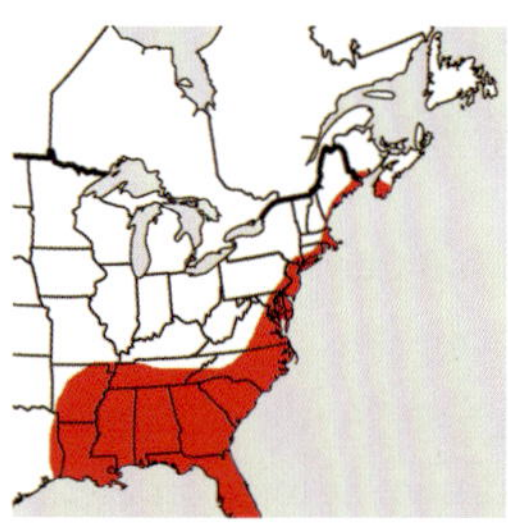

NOTES: *Phaeographis inusta* is a very easy lichen to learn by sight recognition. It belongs to the group of crustose species that we informally refer to as "Scripts and Lips." This species is characterized by its thick, dull-white-to-greenish-white thallus, which bears conspicuous lirellae often arranged in little rosettes (and remind one of something imprinted in a fossil, perhaps). You might be tempted to confuse this species for other script/lip lichens such as several species of *Graphis*, but the latter typically have closed (or comparatively closed lips that do not broadly expose the discs and have colorless spores (vs. open lips that expose the disc and brown spores in *Phaeographis*). The only species with which *P. inusta* can be confused in the Smokies is its close relative, *P. brasiliensis*, which differs in having a non-inspersed hymenium (vs. a hymenium inspersed with oil droplets in *P. inusta*). Also, *P. inusta* has margins of its discs that are apically blackened ("carbonized"), unlike *P. brasiliensis*. The specific epithet "inusta" means "burned."

CHEMISTRY: No substances. Spot tests. K-, C-, KC-, P-, UV-.

NICHE: Burnt Lips occurs at low elevations in the Park, where it grows on the bark of hardwoods, particularly plate-like flakes of bark such as on *Acer saccharum*. Most populations of this species derive from southern portions of the Smokies, such as along Lakeshore Trail, particularly near the end of the "Road to Nowhere."

KEY FEATURES: Thick, dirty white thallus, rosette-like lirellae that have open, black discs, oil droplets inside hymenium, and brown spores, low elevations on bark of hardwoods.

Phaeophyscia adiastola

Lost in the Keys

Lendemer 44561 (photo: Tripp)

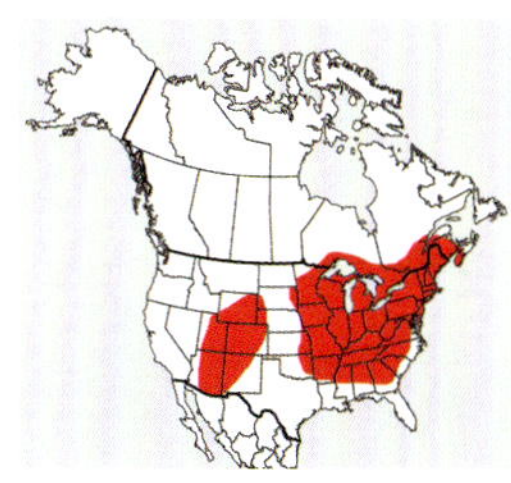

NOTES: *Phaeophyscia* is one of those genera that can be difficult to key in a book, but is relatively easy to recognize by appearance. Members of the genus *Physcia* can be superficially similar, but they always have a K+ yellow upper cortex due to the presence of atranorin. Generally speaking, *Phaeophyscia* tends to be darker gray in color than *Physcia*, and often the thalli are even brownish in color. *Phaeophyscia adiastola* is quite variable in overall appearance (size and color of the thalli), but can be recognized by the sorediate thallus, K- upper cortex, black lower surface, and white medulla. *Phaeophyscia rubropulchra* is externally similar, but differs in having an orange-pigmented medulla. *Phaeophyscia pusilloides* has a white medulla, but differs in having distinctive capitate soralia that form on the tips of the secondary lobes in the central portions of the thallus.

CHEMISTRY: No substances. Spot tests. K-, C-, KC-, P-, UV-.

NICHE: Lost in the Keys is common and widespread throughout the Smokies and can be found on the bark of both hardwoods and conifers as well as on shaded calcareous and non-calcareous rocks. It is likewise widely distributed in inland areas of temperate eastern North America, with disjunct populations in the Rocky Mountains and other mountainous areas of the southwestern United States.

KEY FEATURES: Brown-gray to blue-gray foliose thallus, black lower surface, marginal and laminal soralia, white medulla, on bark and rocks throughout.

Phaeophyscia endococcinoides

Greater Red Heart

Tripp 1532 (photo: Lendemer)

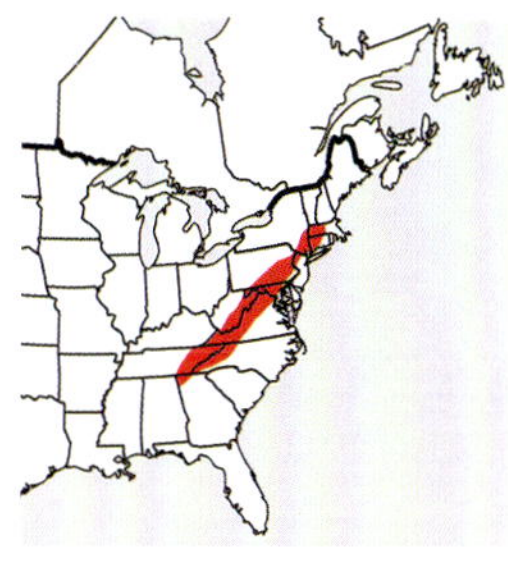

NOTES: This is one of those special foliose lichens that you hope to see on every hike and are thrilled to find when you finally encounter it. Like other species of *Phaeophyscia*, it has a dark gray-to-brown-gray upper surface, a black lower surface, and a K- cortex. But what makes Greater Red Heart special is the orange-pigmented medulla and the lack of soralia. It could be confused with *P. rubropulchra*, but that species always has soralia and rarely produces apothecia. *Phaeophyscia ciliata* is also similar, but that species has a white medulla, tiny stiff hairs on the upper surface near the lobe tips, and typically occurs on bark rather than rock.

CHEMISTRY: Skyrin. Spot tests. Cortex: K-, C-, KC-, P-, UV-; medulla: K+ purple, C-, KC-, P-, UV-.

NICHE: *Phaeophyscia endococcinoides* is a rare species in the Smokies and throughout its range in North America. It occurs on shaded rocks at middle-to-low elevations and is often found in humid habitats along streams.

KEY FEATURES: Brown-gray to blue-gray foliose thallus, black lower surface, no soralia or phyllidia, orange medulla, on rocks.

Phaeophyscia hispidula

Cat's Meow

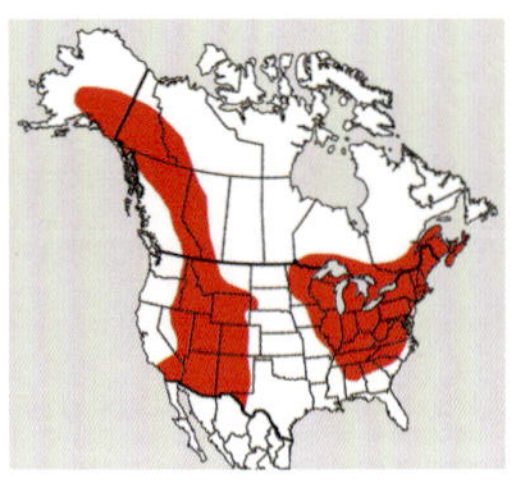

Lendemer 48558 (photo: Tripp)

NOTES: *Phaeophyscia hispidula* resembles a thallus of *P. adiastola* on steroids. It is easily twice, if not three times, larger than *P. adiastola* and always has a distinctive carpet of black rhizines that extends outward from the edge of the thallus, completely surrounding every inch of the thallus. The large size and the carpet of rhizines make it difficult to confuse *P. hispidula* with any other species once learned in the field. Once you know the species is out there, you will keep looking at *P. adiastola* wondering if the lobes are big enough to qualify as *P. hispidula*. However like true love, you'll know it when you see it.

CHEMISTRY: No substances. Spot tests. K-, C-, KC-, P-, UV-.

NICHE: This species has a scattered distribution in North America, where it is known from a number of different mountainous areas. In the southern Appalachians it is very rare, but seemingly occurs at both middle and high elevations on the bark of hardwoods. In the Smokies it is known from only two locations, and we have only seen it once near Hemphill Bald.

KEY FEATURES: Brown-gray to blue-gray foliose thallus, black lower surface, marginal soralia, white medulla, large lobes with a complete fringe of black rhizines, on bark.

Phaeophyscia pusilloides

Top of the Rock

Lendemer 32907 (photo: Tripp)

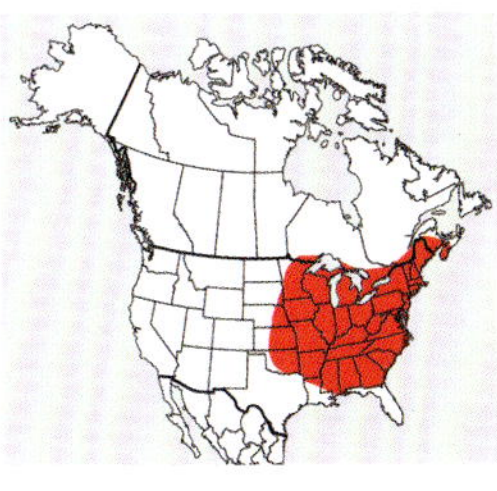

NOTES: *Phaeophyscia pusilloides* is an easily recognized foliose lichen on account of its dark gray thallus, black lower surface, and capitate soralia that form on the tips of the secondary lobes and contain light green soredia whose color contrasts strongly with the other portions of the thallus. *Phaeophyscia adiastola* is most likely to cause confusion, but the soralia in that species are never capitate and instead form on the margins, and occasionally the upper surface, of the lobes. Some forms of *Physciella chloantha* with particularly hemispherical soralia could also be confused with *P. pusilloides*, however the lower surface in *P. chloantha* is pale brown rather than black.

CHEMISTRY: No substances. Spot tests. K-, C-, KC-, P-, UV-.

NICHE: This species is common and widespread on the bark of hardwood trees and shrubs throughout the Smokies, although it is particularly abundant at middle and high elevations. It is similarly widely distributed throughout much of temperate and boreal eastern North America, although it is much more common in the northern portions of its range.

KEY FEATURES: Brown-gray to blue-gray foliose thallus, black lower surface, capitate soralia on tips of secondary lobes, white medulla, on bark throughout.

Phaeophyscia rubropulchra

Red Heart

Tripp 2506 (photo: Tripp)

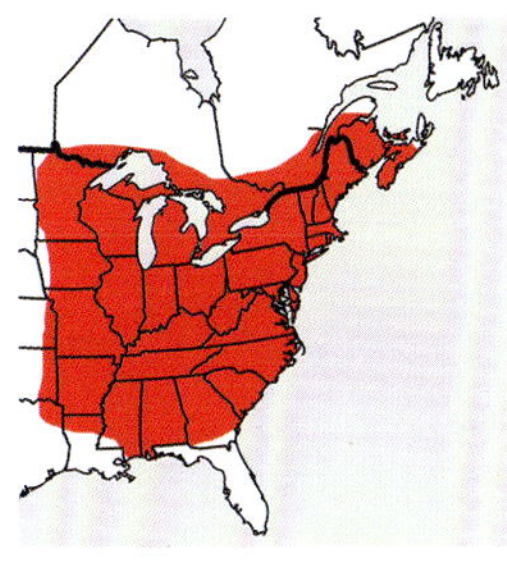

NOTES: This is one of the easiest (and weediest) lichens to identify in the Park. It is a very small, gray, foliose species immediately recognizable by its bright red medulla. This feature can almost always been seen directly in the field (assuming you brought your hand lens) by looking for areas where the cortex has become cracked or otherwise broken to reveal the bright red heart of *Phaeophyscia rubropulchra*. There are *lots* of other Physciaceae in the Smokies, but none of them has a medulla that bleeds for you. This species reproduces primarily through soredia, but can also be found with apothecia (always in addition to soredia) on occasion. The specific epithet translates to "beautiful red."

CHEMISTRY: Skyrin. Spot tests. Cortex: K-, C-, KC-, P-, UV-; medulla: K+ purple, C-, KC-, P-, UV-.

NICHE: Red Heart occurs throughout the Smokies, particularly at low and middle elevations. It is exceptionally common here and elsewhere throughout its range, where it can be found growing on the bark of numerous different species of hardwoods as well as sometimes on their branches.

KEY FEATURES: Very small, gray, foliose lichen with bright red medulla, sorediate (sometimes also with apothecia, depending on whether it is happy or unhappy, and whether you are an optimist or a pessimist), very common on bark of hardwoods at low elevations.

Phaeophyscia sciastra

Lichen Shadowlands

Lendemer 53148 (photo: Tripp)

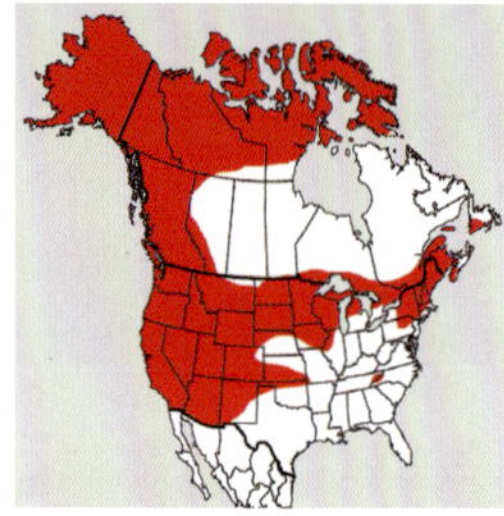

NOTES: *Phaeophyscia sciastra* is one of several species in the genus that occur in the Smokies, but it is much rarer than most of the others and also easy to identify (if found) owing to the fact that it is the only known isidiate member of the genus in the Park. It is further characterized by its orbicular thalli, which are green (when wet) or gray (when dry) that bear very skinny lobes and are covered by extremely dense isidia that are somewhat darker in color than the thallus. These isidia are rather variable in shape and size, ranging from nearly granular to columnar. You might be tempted to confuse this species for the more common *P. rubropulchra*, *P. squarrosa*, *P. pusilloides*, or *P. adiastola*, but all four of these are sorediate, not isidiate, and differ in additional characters such as in having striking, red medullae (*P. rubropulchra*) or in having soredia that are raised on small doo-dads that stick up from the thallus (*P. pusilloides*). The specific epithet of this species means "shadow," in reference to its overall dark appearance. We think "sciastra" has a nice ring to it.

CHEMISTRY: No substances. Spot tests: K-, KC-, C-, P-, UV-.

NICHE: *Phaeophyscia sciastra* is a species primarily of western North America extending into northeastern North America. Populations in the Smokies are disjunct, presumably from the north, but as usual, we need more data to test this. Lichen Shadowlands occurs on rocks of various kinds, typically in shaded environments. In the Smokies, it is known only from high elevations.

KEY FEATURES: Grayish green orbicular thalli bearing skinny, strap-like lobes and variably shaped isidia, rare on sheltered rocks at high elevations.

Phaeophyscia squarrosa

Lobble Bobble

Lendemer 44562 (photo: Tripp)

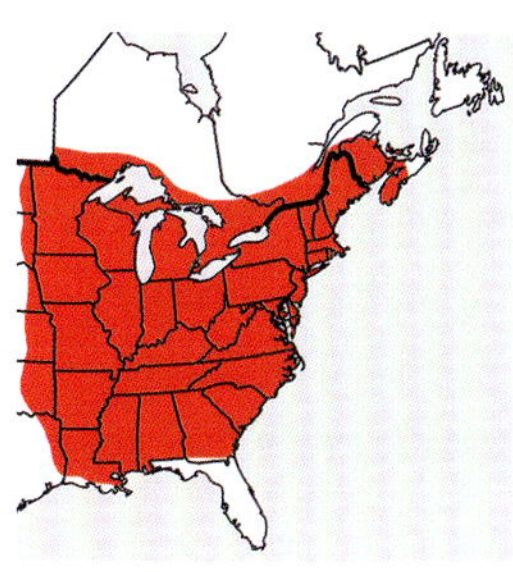

NOTES: *Phaeophyscia squarrosa* is bound to cause you trouble. Although the lower surface is centrally black, it is often very pale near the margins. The abundant phyllidia (tiny lobules) that form all along the margins of the lobes and sublobes are reminiscent of *Heterodermia squamulosa*, which has a completely ecorticate lower surface that is white toward the margins of the thallus and pigmented purple-black toward the center. Nonetheless, the upper cortex of *H. squamulosa* is K+ yellow because it contains atranorin, and although the lower surface is centrally dark in color, it is ecorticate while that of *P. squarrosa* is corticate. Confusion with *Anaptychia palmulata* is also possible; however, in that species the lower surface is uniformly pale, and the lobules are typically elongate and robust, whereas those of *P. squarrosa* are short and fragile.

CHEMISTRY: No substances. Spot tests: K-, C-, KC-, P-, UV-.

NICHE: Although this species is widely distributed throughout temperate eastern North America, it is never common and only rarely locally abundant. The same is true in the Smokies, where Lobble Bobble tends to be found at low-to-middle elevations and is usually associated with the mossy bases of mature trees, or with moss-covered surfaces of calcareous rocks.

KEY FEATURES: Brown-gray to blue-gray foliose thallus, black lower surface, fine marginal phyllidia, white medulla, on bark and rocks throughout.

Phlyctis boliviensis

A.T. Blaze

Lendemer 33200 (photo: Tripp)

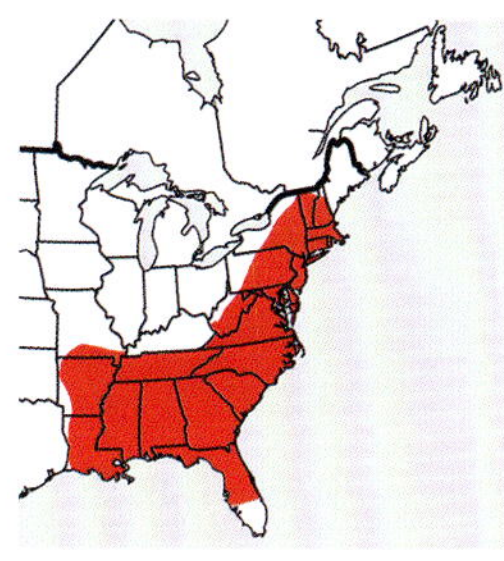

NOTES: Move all other lichens that were previously at the top of your *Must Learn* list down. Do it! This one can now move up, especially if you are trying to learn the most common crustose lichen in the Smokies. There are a million technical things that we could tell you about this species, and if we can muster it, we might. But for now, all you need to know is that this lichen looks like paint–period! This feature is attributable to its very, very thick thallus. If you see bluish-white paint on a tree, it is almost certainly *Phlyctis boliviensis* (unless it is a trail marker, which the Smokies tries to avoid). If you see bluish-white paint on a rock, then it is likely the close relative of *P. boliviensis* (i.e., *P. petraea*). In addition to being everywhere and paint-like, *P. boliviensis* often has a continuous, white fibrous prothallus upon which bluish-green granules are scattered across. The thallus is also K- and P+ yellow, a combination not repeated in our other two species of *Phlyctis* in the Smokies (see those entries).

CHEMISTRY: Psoromic acid. Spot tests. K-, C-, KC-, P+ yellow, UV-.
NICHE: *Extremely* common on all of the most common hardwoods in the Smokies, such as Sugar Maple, Tulip Poplar, and various species of oak.
KEY FEATURES: Paint-like, on bark!

Phlyctis petraea
Eggshell Rock Blaze

Tripp 2103 (photo: Lendemer)

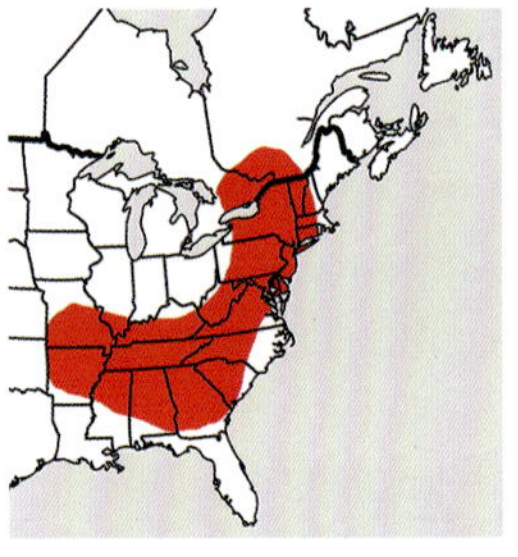

NOTES: If you thought the paint-like qualities of *Phlyctis boliviensis* were marked, this one is extreme. Just like its close relative on the preceding page, *Phlyctis petraea* cannot be confused for any other lichen because of its paint-like qualities. This species is characterized by its thick, blue to bluish-white thallus that bears regular or irregular (i.e., diffuse) soralia. Unlike *P. boliviensis*, you will only ever find Rock Blaze on . . . you guessed it: rock (now, can you take a guess as to what "petraea" means?). Rather than *P. boliviensis*, *P. petraea* is most likely to be confused for another rock-dwelling species of *Phlyctis* in the Smokies that is also K+ yellow turning red: *P. speirea*. The two are however easily differentiable because *P. petraea* actually lacks a proper cortex and is *way* more common. You might additionally be tempted to misidentify *P. petraea* as *Variolaria waghornei*, but that species has very discrete, regular, disciforme soralia, unlike species of *Phlyctis*. PS: Does anyone know what that cute, woolly creature crawling across the thallus is?
CHEMISTRY: Norstictic acid with or without stictic acid. Spot tests. K+ yellow turning red, C-, KC-, P+ yellow or orange, UV-.
NICHE: Widespread on non-calcareous rocks, especially in overhangs or shaded places, throughout the Park. Rock Blaze is similarly common throughout its range, which is the greater Appalachian Mountain area.
KEY FEATURES: Paint-like, on rocks!

Phlyctis speirea
Lesser Rock Blaze

Lendemer 48563 (photo: Tripp)

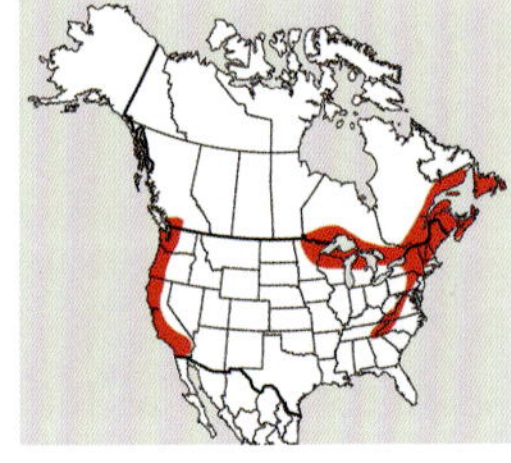

NOTES: We don't have extensive things to say about Lesser Rock Blaze other than that it greatly resembles Greater Rock Blaze morphologically, ecologically, and chemically, except that it has a corticate surface that is usually distinctly sordid white (vs. bluish-white) in color. On second thought, those little structures that look like soralia are actually apothecia (surprise!), but we key this species out in the asexual lichens section because most people will mistake these doo-dads for soralia.

CHEMISTRY: Norstictic acid. Spot tests. K+ yellow turning red, C-, KC-, P+ yellow, UV-.

NICHE: *Phlyctis speirea* occurs on both bark and non-calcareous rocks (the latter, particularly in overhangs at high elevations in the Smokies). Look for this species in northern hardwood forests such as near the summits of Chimney Tops or Mt. Sterling.

KEY FEATURES: Crustose lichen with thick, continuous, sordid white, corticate thallus, soraliate-like apothecia, K+ yellow turning red, on bark of hardwoods (or in rock overhangs) at high elevations.

Phyllopsora confusa

Emerald Remembrance

Lendemer 33091 (photo: Tripp)

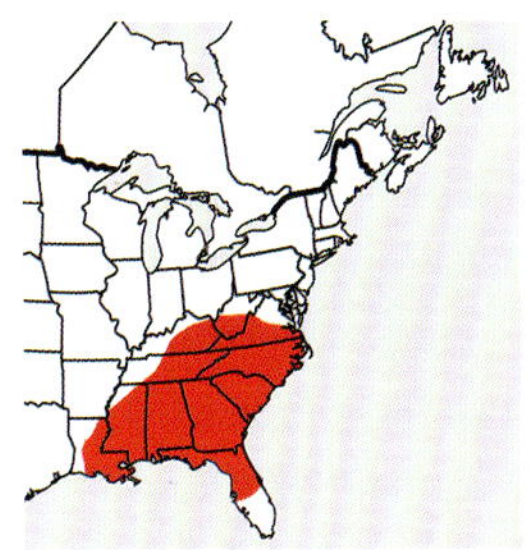

NOTES: As the name suggests, the species can be confusing at first. Forms of *Phyllopsora corallina* wherein the isidia are poorly developed and shortened could be misinterpreted as lobulate. However the lobules of *P. confusa* are quite distinctive and give the appearance of the areoles throwing off tiny shingles into the forest. *Phyllopsora confusa* also often has a yellowish or brownish hue, while *P. corallina* is always dull green. In some respects, the most similar species to *P. confusa* are members of the genus *Hypocenomyce*, but these differ in having at least one positive spot test. Likewise some species of *Rinodina* could be similar, such as *R. excrescens* and *R. bullata*, but those species produce rounded blastidia instead of scale-like lobules.

CHEMISTRY: No substances. Spot tests. K-, C-, KC-, P-, UV-.

NICHE: *Phyllopsora confusa* is widespread in southeastern North America, where it is commonly found on the bark of hardwoods in humid habitats. In the Smokies, it is infrequent, but nonetheless is found throughout the Park at middle and low elevations.

KEY FEATURES: Areolate greenish thallus, abundant lobules or phyllidia, white fibrous prothallus, all reactions negative, on bark of hardwoods and conifers at middle-to-low elevations.

Phyllospora corallina

Grassy Knees

Tripp 3650 (photo: Lendemer)

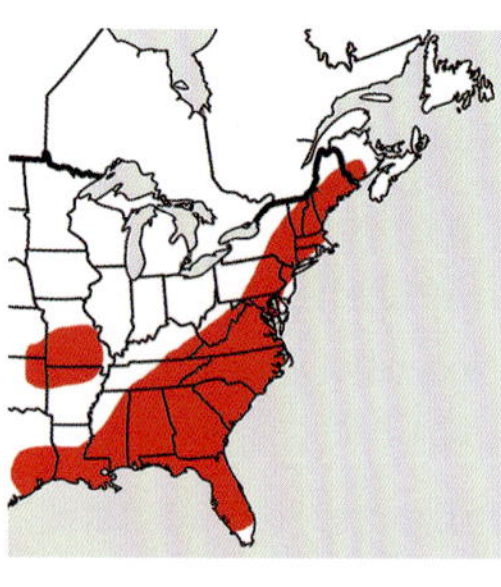

NOTES: *Phyllospora corallina* can be recognized by its green, areolate thallus that always has a distinct fibrous prothallus extending beyond the edges of the thallus, and by the presence of tiny fragile isidia that are usually brighter green in color when compared to the areoles out of which they grow. *Phyllopsora kalbii* is also isidiate, but differs in having a dull, whitish-green thallus with short, squat, globose isidia. Both species of *Phyllopsora* are most likely to be confused with sterile thalli of *Bacidia schweinitzii* wherein apothecia are lacking. Such thalli of *B. schweinitzii* can however, be distinguished from *Phyllospora* by the frequent presence of black pycnidia and by the production of atranorin, although the latter is usually detectable only with thin layer chromatography.
CHEMISTRY: No substances. Spot tests. K-, C-, KC-, P-, UV-.
NICHE: *Phyllopsora corallina* is common throughout the southern temperate regions of North America where it occurs on the bark of both hardwoods and conifers in humid habitats. It is infrequent in the Smokies, but nonetheless widespread at middle-to-low elevations.
KEY FEATURES: Areolate greenish thallus, abundant fragile light-green isidia, white fibrous prothallus, all reactions negative, on bark of hardwoods and conifers at middle-to-low elevations.

Physcia americana

Liberty Lichen

Tripp 6037 (photo: Lendemer)

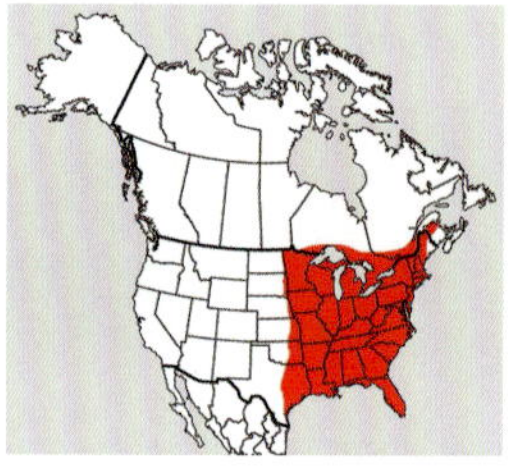

NOTES: If there were ever an American lichen, it would be *Physcia americana*. The species can easily be recognized by its gray foliose thallus, laminal soralia, which burst from the pruinose upper surface of the lobes, K+ yellow cortex, and pale lower surface. Ask yourself if there is anything more patriotic? *Physciella chloantha* is somewhat similar, but that species has marginal soralia and a K- upper cortex that does not contain atranorin. *Physcia caesia* is another species with laminal soralia, however that species grows on rocks at only the highest elevations in the Smokies and is very rare.
CHEMISTRY: Atranorin, unidentified terpene. Spot tests. Cortex: K+ yellow, C-, KC-, P-, UV-; medulla: K+ yellow, C-, KC-, P-, UV-.
NICHE: This species is widespread and common throughout much of temperate eastern

North America. In the Smokies, like elsewhere in the southern Appalachians, it is common at all elevations and usually found on the bark of branches of hardwoods.

KEY FEATURES: Medium gray foliose thallus, pruinose upper surface, pale corticate lower surface, laminal soralia, K+ yellow cortex, K+ yellow medulla, common on hardwoods throughout.

Physcia millegrana

Pucker Rubble

Tripp 1393 (photo: Lendemer)

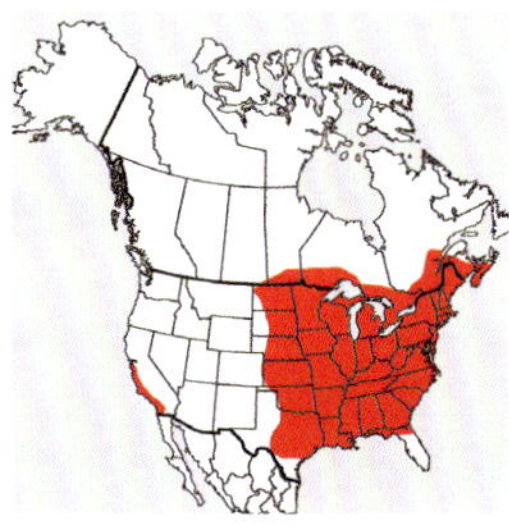

NOTES: There are a lot of *Physcia* species on the planet—we won't lie. We view this as something to celebrate and hope you will agree that learning *Physcia millegrana* will be a breeze with such a good attitude. Although the vast majority of species are (1) small, (2) gray, and (3) sorediate, and although *P. millegrana* falls into this majority, it is very easily differentiated from all others owing to its very coarse soredia, which are better thought of as blastidia that bud from the margins of the lobes. The density of these blastidia often obscures the overall shapeliness of the lobes, giving them a rather ruffled appearance. If you find a gray lichen growing on a tree with a pale, corticate lower surface, a K- and C- medulla, and these marginal, blastidiate sorediate, you've probably landed on *P. millegrana*. In addition to seeing blastidia, you might encounter this species fertile, too. If you find it growing on a rock, and with seemingly narrower lobes, it's probably the externally similar *P. thomsoniana*.

CHEMISTRY: Atranorin. Spot tests: (cortex): K+ yellow, C-, P-, UV-; (medulla) K-, C-, KC-, P-, UV-.

NICHE: *Physcia millegrana* is an extremely common foliose lichen across much of eastern North America but is conspicuously uncommon within the borders of Great Smoky Mountains National Park. Why, you ask? The Smokies are just too nice! It's true. This is about as trashy of a lichen species as you can get in the eastern United States, and we don't have much of that habitat in the Park. Look for populations along Wolf Ridge Trail not far from Parsons Bald, or in other perturbed areas, whether natural or human-mediated. It grows primarily on hardwoods and, while occasionally can be found on rocks, that habitat is dominated by the close relatives, *P. subtilis* and *P. thomsoniana*.

KEY FEATURES: Small gray foliose lichen, pale underside, K+ yellow cortex, K- medulla, marginal, blastidiate soredia, sometimes fertile, uncommon in Park but exceptionally common elsewhere in eastern North America.

Physcia pseudospeciosa

Celebration Day

Unvouchered (photo: Lendemer)

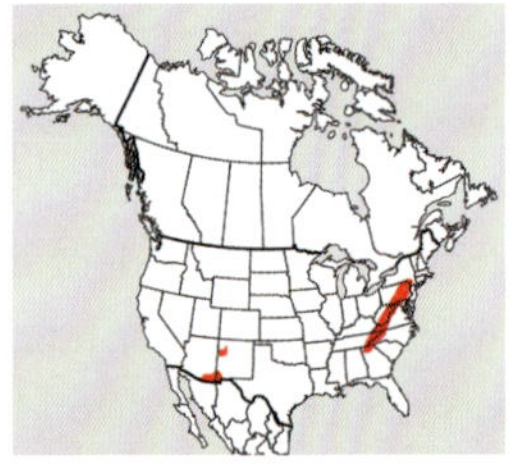

NOTES: As the name implies, this species is most likely to be confused with *Heterodermia speciosa* because both species have relatively large gray thalli, marginal soralia, a K+ yellow cortex and medulla, and a pale lower surface. Nonetheless when you see *Physcia pseudospeciosa* in the field there is something about it that gives you pause before considering *Heterodermia*. Unless you want to examine the anatomy of the thallus in detail under the compound microscope, *P. pseudospeciosa* can be distinguished by the frequently maculate upper surface and the coarsely granular soredia. We suggest that you compare the photographs of the two species produced here in this book.

CHEMISTRY: Atranorin and zeorin. Spot tests. Cortex: K+ yellow, C-, KC-, P-, UV-; medulla: K+ yellow, C-, KC-, P-, UV-.

NICHE: *Physcia pseudospeciosa* is a rare species that occurs in scattered locations throughout the central and southern Appalachians, with disjunct populations in the high mountains of the Sonoran Desert. It is always found on shaded non-calcareous rocks in humid habitats, often associated with streams, rivers, or waterfalls. In the Smokies, it is rare and mainly known primarily from northern portions of the Park in North Carolina and Tennessee.

KEY FEATURES: Large gray foliose thallus, pale corticate lower surface, marginal soralia, K+ yellow cortex, K+ yellow medulla, rare on shaded rocks at middle elevations.

Physcia pumilior

Lesser Gray Legs

Lendemer 33120 (photo: Tripp)

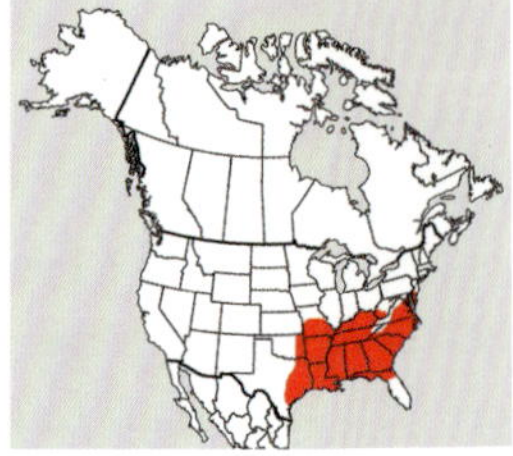

NOTES: While *Physcia stellaris* aims high, this species goes low. *Physcia pumilior* is superficially similar to *P. stellaris*, and differs

primarily in having a K- medulla. The spot test reaction in the medulla must be very carefully applied, however, because a stray drop on the cortex can cause the atranorin present there to react K+ yellow and bleed onto the adjacent medulla, giving a false positive reaction. The best way to separate the two is to perform thin layer chromatography to detect the presence of zeorin in *P. pumilior*, or the absence in *P. stellaris*. However, this is above and beyond the capacity of the typical field guide user, and so we suggest going with your elevation and your gut.
CHEMISTRY: Atranorin and zeorin. Spot tests. Cortex: K+ yellow, C-, KC-, P-, UV-; medulla: K+ yellow, C-, KC-, P-, UV-.
NICHE: *Physcia pumilior* is a species of southern temperate eastern North America whose distribution extends inland into the middle and low elevations of the southern Appalachians. In the Smokies, it is widespread at those elevations and is typically found on the branches of hardwood trees and shrubs.
KEY FEATURES: Small gray foliose thallus, esorediate, frequent apothecia K+ yellow cortex, K+ yellow medulla, on branches of hardwood trees and shrubs at middle-to-low elevations.

Physcia stellaris

Greater Gray Legs

Tripp 3898 (photo: Lendemer)

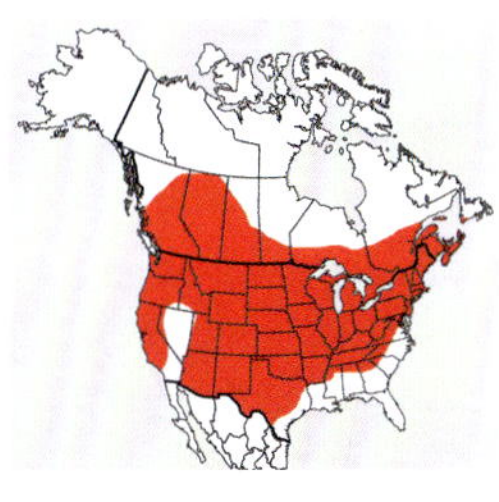

NOTES: Have you noticed that there are only a small number of typically fertile, foliose lichens that lack soredia, isidia, or lobules? *Physcia stellaris* is one of them, and it can be recognized by its small, gray foliose thallus, pale underside, K- medulla, and frequent occurrence on branches at high elevations. *Physcia pumilior* is a superficially similar species that lurks at middle and low elevations. It differs from *P. stellaris* in having a K+ yellow medulla and a more southern distribution. While confusion with species of Parmeliaceae such as *Hypotrachyna livida*, *Imshaugia placorodia*, and *Myelochroa galbina* may be possible, all of those species have colorless simple spores. Likewise, while other foliose species with brown 2-celled spores such as *Phaeophyscia ciliata* and *P. erythrocardia* have similar spores, they all have a K- cortex because they do not produce atranorin.
CHEMISTRY: Atranorin. Spot tests. Cortex: K+ yellow, C-, KC-, P-, UV-; medulla: K-, C-, KC-, P-, UV-.
NICHE: *Physcia stellaris* is common and widespread in many regions of North America, however in the central and southern Appalachians it is restricted to high elevations. In the Smokies, it generally occurs above 4,500 feet elevation and is commonly found on the branches of hardwood trees and shrubs.
KEY FEATURES: Small gray foliose thallus, esorediate, frequent apothecia K+ yellow cortex, K- medulla, on branches of hardwood trees and shrubs at high elevations.

Physcia subtilis

Subtley Doo

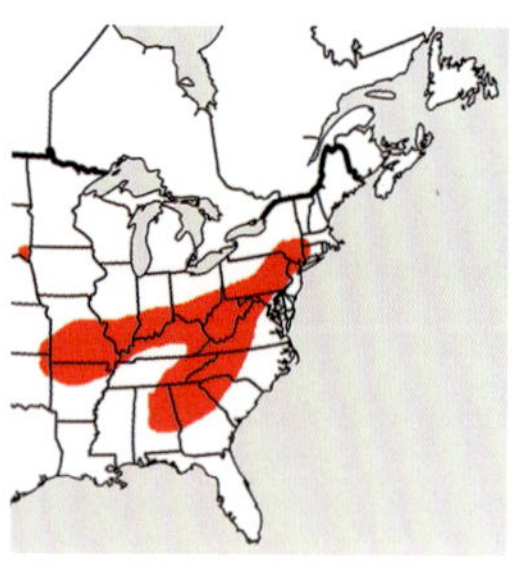

Lendemer 33112 (photo: Tripp)

NOTES: When Gunnar Degelius first found this species in the Smokies, he didn't have to travel far to discover new species. The holotype, or the museum specimen on which the name is based, was collected just above the Park office in 1939. Indeed, the species is common in the Smokies and often grows with *P. thomsoniana*, the species that was long confused with *P. subtilis*. Although as described in the entry for *P. thomsoniana*, the two species differ fundamentally in the anatomy of the thallus, the result is that they typically look completely different even from a distance in the field. *Physcia subtilis* has a thallus composed entirely of a single layer of fungal tissue and, as such, the entire thallus appears to be extremely appressed to the surface of the substrate and the lobes cannot easily be removed or separated. In contrast, *P. thomsoniana* is more loosely attached to the substrate and can frequently be removed.

CHEMISTRY: Atranorin. Spot tests. Cortex: K+ yellow, C-, KC-, P-, UV-; medulla: K-, C-, KC-, P-, UV-.

NICHE: This species is widespread in the central and southern Appalachian Mountains as well as the Ozarks. It is likely also widespread in the intervening area between the two regions and has simply been overlooked. It is particularly common in the southern Appalachians and occurs throughout the Smokies on exposed non-calcareous rock outcrops.

KEY FEATURES: Small gray foliose thallus, lobes completely and strongly appressed to the substrate, marginal blastidia, K+ yellow cortex, K- medulla, on exposed rocks at all elevations.

Physcia thomsoniana

Thomson's Diddley Doo

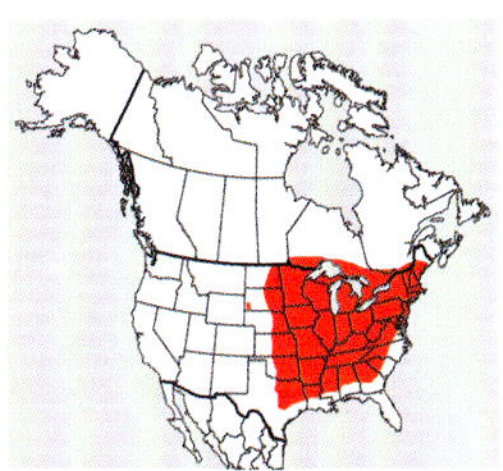

Tripp 3736 (photo: Lendemer)

NOTES: *Physcia thomsoniana* is a recently described species that was previously confused with *P. subtilis*, the latter having been described from the Smokies by Gunnar Degelius. The two species are quite similar in having small gray foliose thalli with marginal granules or blastidia, a pale lower surface, and a K- medulla. *Physcia thomsoniana* differs from *P. subtilis* in having a thallus with a distinct medulla rather than one that is composed of a single layer of fungal hyphae that is not differentiated into layers. Although minute anatomical difference may be difficult to observe without a compound microscope, it results in *P. thomsoniana* having a clearly foliose appearance wherein the thallus can often be removed from the substrate and the lobes lift away from the substrate, at least at their margins.
CHEMISTRY: Atranorin. Spot tests. Cortex: K+ yellow, C-, KC-, P-, UV-; medulla: K-, C-, KC-, P-, UV-.

NICHE: This species is common and widely distributed throughout temperate eastern North America where it often forms extensive colonies on exposed non-calcareous rocks, even in urban areas such as Manhattan in New York City. It is common in the Smokies and can be found throughout the Park at all elevations.
KEY FEATURES: Small gray foliose thallus, lobes not completely and strongly appressed to the substrate, marginal blastidia, K+ yellow cortex, K- medulla, on exposed rocks at all elevations.

Physciella chloantha

St. Patty's Snowballs

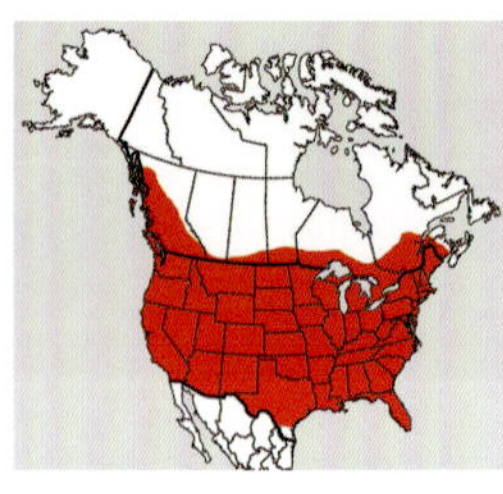

Tripp 2567 (photo: Lendemer)

NOTES: *Physciella chloantha* is a small, gray foliose lichen that has a pale, corticate lower surface and an upper surface characterized by soralia formed in discrete regions. These soralia, which bear very small, fine soredia, are often terminal on the lobe tips and, by some researchers, get described as labriform ("chlo-anth" specifically refers to the budding nature of these soredia). Across the United States, *P. chloantha* is most likely to be confused with other members of the genus that are not yet reported in the Smokies, such as *P. melanchra*. Within the Smokies, you might be tempted to confuse *P. chloantha* for *Physcia americana*, but the latter has laminal (vs. marginal) soralia. *Physcia pseudospeciosa* is somewhat similar but has a K+ upper cortex and, let's face it, is way prettier. Species of *Phaeophyscia* are also common throughout the Park and might be confused with *P. chloantha*, but most sorediate members of that genus, including *Phaeophyscia pusilloides*, *P. adiastola* have black (instead of pale) lower surfaces.

CHEMISTRY: No substances. Spot tests: K-, C-, KC-, P-, UV-.

NICHE: *Physciella chloantha* is uncommon in the Smokies and, if you find it, it will most likely be in areas of above average alkalinity, such as near Chillhowee Lake. The species is much more common elsewhere in North America, such as the Great Plains and southern Rocky Mountains.

KEY FEATURES: Small, gray foliose lichen, pale lower surface, K- cortex and medulla, marginal (typically terminal), fine soredia in discrete, labriform soralia, rare in Park at low elevations.

Physconia leucoleiptes

Frosty Pants

Lendemer 44557 (photo: Tripp)

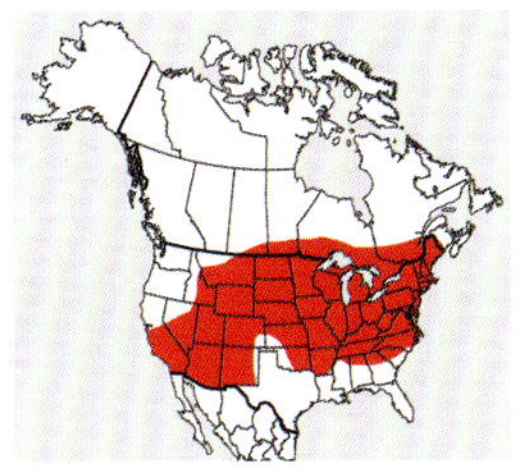

NOTES: *Physconia leucoleiptes* is a small, gray, foliose lichen that you will first recognize by its frosted appearance. That is because most of the genus is characterized by its lobe tips, which are densely white pruinose. There are very few other macrolichens in the Smokies with this "look," among them *Pyxine sorediata*, which differs from *Physconia leucoleiptes* by having a yellowish-orange medulla. *Physconia leucoleiptes*, too, has a yellowish-orange, pigment, but in this case it is restricted to the soredia (and the species otherwise has a white medulla). Please accept our apology that this sorediał pigment is not apparent from the photograph. There is only one other species of *Physconia* reported from Great Smoky Mountains National Park. That species is *Physconia subpallida*, and differs in having densely pruinose apothecia (and in lacking asexual propagules). We have seen the latter only once in the Smokies, on a day when we did not have our camera. *Physconia subpallida* is exceptionally rare and endangered and should not be collected if found.

CHEMISTRY: Secalonic acid A. Spot tests. Cortex: K-, C-, KC-, P-, UV-; medulla: K-, C-, KC-, P-, UV-; soralia: K+ yellowish, C-, KC+ yellow to yellow-orange, P-, UV-.

NICHE: This species occurs primarily at low elevations in the Park. It grows on the bark of hardwoods such as on oaks, elms, dogwoods, in rich cove to upland ridgetop forests. It seems especially fond of portions of the Park that are slightly calcareous, such as Wear Cove Gap and Cades Cove. *Physconia leucoleiptes* is relatively widespread across the continental United States in similar habitats, but avoids the deep south.

KEY FEATURES: Small, gray foliose lichen, narrow lobes frosted with dense white pruina at their tips, white medulla, yellowish-orange pigmented soredia in crescent slits, bark of hardwoods at low elevations, especially in calcareous areas.

Physconia subpallida

Fog Line

Tripp 1430 (photo: Lendemer)

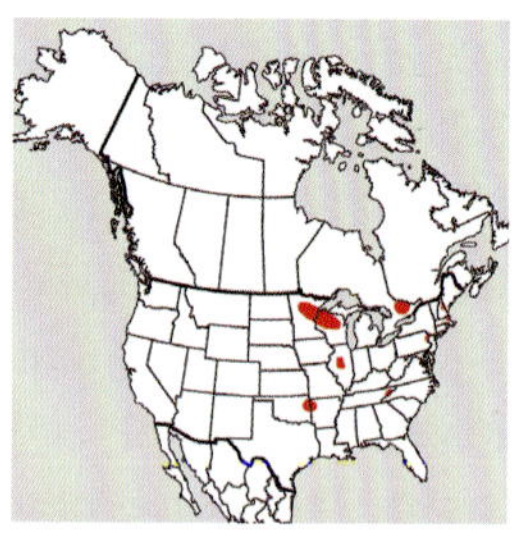

NOTES: *Physconia subpallida* is a very rare lichen that we have seen only once in Great Smoky Mountains National Park. If you find it, do not collect it, but kindly photograph and record geographical coordinates. This is a species characterized by its narrow, gray, often strappy lobes completely covered by a dense, white pruina. If found fertile, its apothecia are typically densely pruinose, too, and often bear small lobules. These features more or less set *P. subpallida* apart from all other species in eastern North America. *Physcia aipolia*, *P. pumilior*, and *P. stellaris* are morphologically similar and can (but do not always) have densely pruinose discs, but none of these has thallus lobes that are also densely pruinose, and they also lack lobules surrounding the apothecia. *Anaptychia palmulata* is the most similar species and has caused much confusion with *P. subpallida*, for some. In the Smokies, *A. palmulata* is extremely common and can usually be easily distinguished by the absence of pruina on the thallus and apothecia.

CHEMISTRY: No substances. K-, C-, KC-, P-, UV-.

NICHE: There are very few extant populations of *Physconia subpallida* in the United States, which includes locations in the Ozarks, the Appalachians, and the Great Lakes Region. It almost certainly deserves conservation protection throughout this its range. To our north, *P. subpallida* is member to the Canadian federally endangered species list (COSEWIC), where it occurs primarily in Ontario. We know from a very detailed ecological study in Canada that the species prefers mature hardwood forests with low stem density, and sites characterized by high local humidity (i.e., not far from bodies of water). We also know from this work that the species prefers growing on trees with flaky bark, such as on Black Gum. What makes a lichen very rare versus very common? More research would help the world understand . . .

KEY FEATURES: Gray foliose lichen, strappy lobes, entire thallus (plus apothecia, where present) often covered by dense white pruina, apothecia (where present) typically with lobules, extremely rare, on flaky bark of hardwoods in old/mature forests.

Pilophorus fibula

Appalachian Matchsticks

Tripp 4981 (photo: Tripp) inset: Tripp 2236 (photo: Tripp)

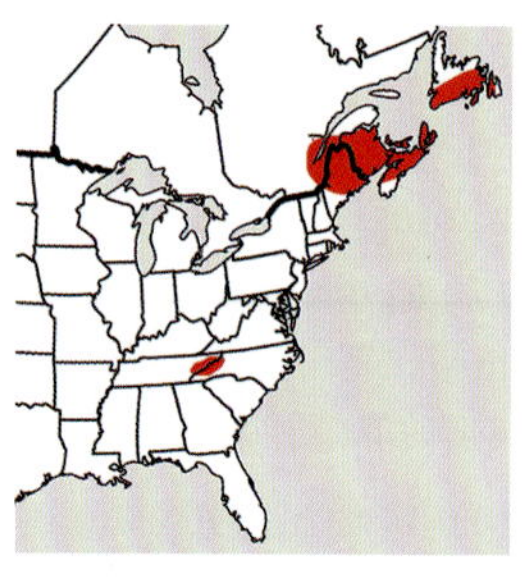

NOTES: *Pilophorus fibula* is a very rare lichen in the Smokies, characterized first and foremost by its fruticose growth form. Once learned, *Pilophorus* is impossible to confuse for any other genus. Its thallus is composed of small, corticate areoles that are army green in color and continue to march their way up to cover the stalks that bear shiny, black apothecia. The presence of corticate areoles on these stalks is distinctive for this species and readily separates it from its close relative, *Pilophorus cereolus*, which also occurs in the Smokies.

CHEMISTRY: Atranorin. Spot tests. K+ yellow, C-, KC-, P-, UV-.

NICHE: Appalachian Matchsticks is very rare in the Smokies as well as elsewhere throughout its range. It is restricted to rocks at high elevations. We have seen it on only a handful of occasions between Andrew's Bald and Mt. LeConte. If you find this lichen, take a GPS waypoint and let it be.

KEY FEATURES: Fruticose growth form, army green areoles on the thallus as well as apothecial stalks, shiny black apothecia, very rare on rocks at the highest elevations.

Placidium arboreum

Autumn Squamules

Lendemer 44559 (photo: Tripp)

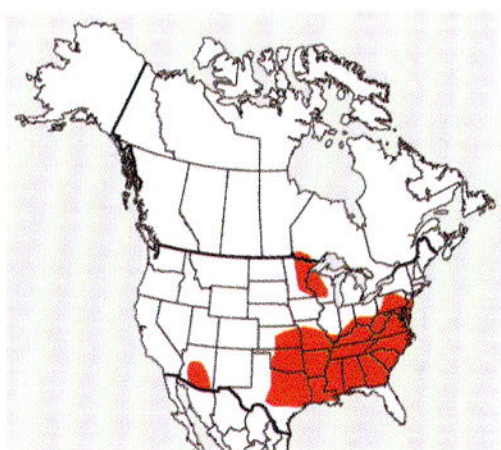

NOTES: *Placidium arboreum* is a very easy to identify lichen in the Smokies because there are very few other species that are characteristically squamulose and have squamules as large as this species. Owing to this, it is unlikely to be confused for anything else in this part of the world (out west: there are squamulose lichens everywhere . . .). If you happen to find this species fertile, look for brown-to-blackish perithecia that dot the surface of the squamules. The color of *Placidium arboreum* nicely captures the changing of the seasons in the Smokies. They are green when fresh and brown when dry, just like the beginning and the end of October in the Park. We are always pleased to find this species, no matter the time of year, and hope you will be, too.

CHEMISTRY: No substances. Spot tests. K-, C-, KC-, P-, UV-.

NICHE: Autumn Squamules occurs on the bark of hardwoods throughout the Smokies but is most abundant at middle elevations. The species is especially fond of crevices in bark, such as is typical of older Tulip Poplars or oaks. Look for it especially in areas of the Park where limestone lurks.

KEY FEATURES: Large-green-to-brown squamules that bear perithecia, in crevices of bark of hardwoods throughout the Park.

Placynthiella icmalea

Brittle Brown Thorns

Tripp 3504 (photo: Lendemer)

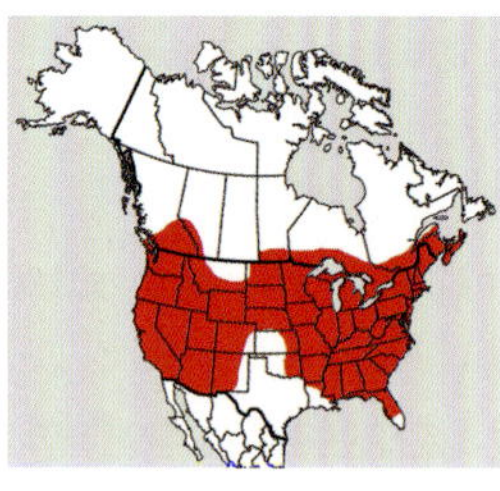

NOTES: There are relatively few isidiate crustose lichens, and certainly almost none in the Smokies that is brown in color and reacts C+ pink due to the presence of gyrophoric acid. Unfortunately, the C+ pink reaction is always obscured by the dark color of the thallus, and thus has to be performed on a squash of the thallus under the compound microscope. The only similar species is *P. dasaea*, which differs in having a sorediate rather than isidiate thallus. Both species often occur together, and even grow intermixed on the same log or patch of organic matter. The soredia in *P. dasaea* are typically light green or yellowish-brown in color, and so typically stand in contrast to the uniform coloration of the isidia in *P. icmalea*.

CHEMISTRY: Gyrophoric acid. Spot tests. K-, C+ pink (in mount!), KC+ pink, P-, UV-.

NICHE: *Placynthiella icmalea* is a common species in many regions of North America that is often found on organic matter, rotting logs, old dry wood, and even rarely on rocks. It can be found throughout the Smokies on those substrates.

KEY FEATURES: Brown isidiate crustose thallus, C+ pink, on organic matter or logs at all elevations.

Placynthiella uliginosa

Tar Heel

Tripp 3693 (photo: Lendemer)

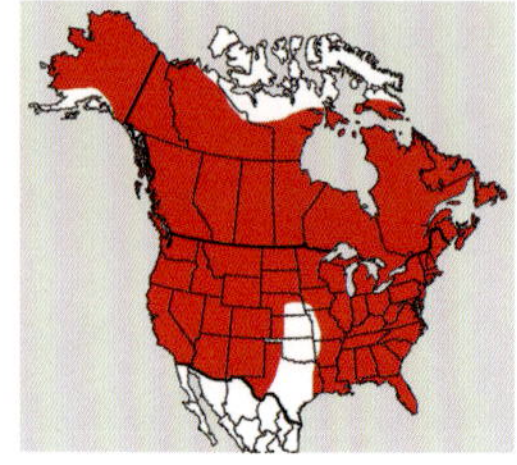

NOTES: *Placynthiella uliginosa* is often referred to as the tar spot lichen because of the black, tar-like stains that it makes on sand in the Coastal Plain. However, we thought that Tar Heel was more appropriate, particularly given that the dark patches it forms on organic matter and soil often resemble a trail of heel-like footprints. The dark color of the thallus, which is composed of minute areoles, is distinctive when combined with the dark reddish-brown apothecia and small colorless simple spores. *Placynthiella oligotropha* is a similar species that does not occur in the Smokies, but differs in having a thallus composed of larger areoles that are olive brown in color. *Placynthiella hy-*

porhoda is most similar in appearance, but also does not occur in the Smokies and differs in having a pigment in the apothecia that bleeds a cloud of purple when it reacts with K.
CHEMISTRY: No substances. Spot tests. K-, C-, KC-, P-, UV-.
NICHE: This species is widely distributed throughout North America, where it occurs on rotting logs, organic matter, and even nutrient-poor soils, including sand. In the Smokies, it occurs throughout the Park, although it is most often found in sun-exposed areas.
KEY FEATURES: Brown to black, inconspicuous crustose thallus, dark brownish-black apothecia, C-, on organic matter or logs at all elevations.

Placynthium nigrum

Cabinet Curiosity

Lendemer 44560 (photo: Tripp)

NOTES: *Placynthium nigrum* is a small cyanolichen characterized by a squamulose thallus that is dark-brown-to-black in color and an unusual

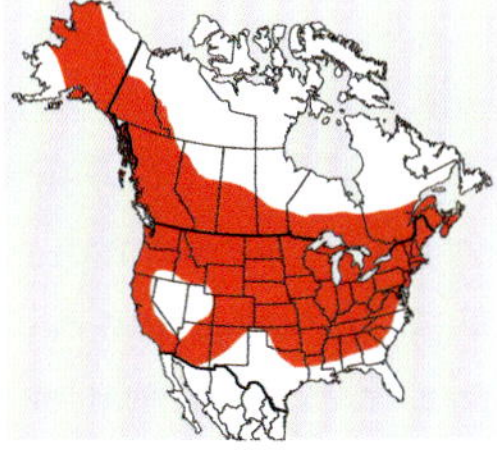

propensity to produce both isidia and sexual fruiting bodies at the same time. The isidia are generally long and cylindrical and can be erect or flattened against the thallus. Where found, apothecia of this species are distinctly convex, dark-brown-to-black in color, have lecideine margins, and are typically shiny. If you do a cross-section of these fruiting bodies, you will find colorless, transversely septate spores. There are few other micro-squamulose species in the Smokies that also have a cyanobacterial photobiont. When it doubt, look for a rather distinctive bluish-black prothallus that further characterizes *P. nigrum*.
CHEMISTRY: No substances. Spot tests. K-, C-, KC-, P-, UV-.
NICHE: This species occurs on rocks that are strongly to weakly calcareous. It is widespread throughout North America, but we have seen it only a few times in the Smokies, both times at relatively low elevations in sunny areas in the Rich Mountain region.
KEY FEATURES: Micro-squamulose lichen with brown-to-black thallus, present photobiont a cyanobacterium, isidia and apothecia sometimes born together on the same thallus (otherwise: isidiate), shiny, convex apothecia with colorless, septate spores, on sunny, calcareous rocks, uncommon.

Placynthium petersii

Ragged Rosettes

Lendemer 44544 (photo: Tripp)

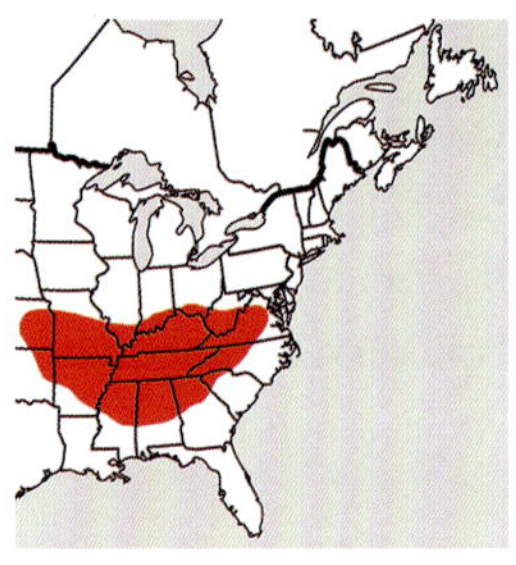

NOTES: Like *Placynthium nigrum, P. petersii* is characterized by its small thalli that bear a cyanobacterial photobiont. However, thalli of *P. petersii* have distinct, radiating marginal lobes, and whether to call this lichen foliose or crustose would depend on whether it has a true lower cortex (the answer is yes and no, but we aren't sure that it really helps you identify it). *Placynthium petersii* also lacks the isidia typical of *P. nigrum*. If you find a cyanobacterial lichen with tiny, radiating lobes . . . one that forms small rosettes with center portions that often appear to be eaten out (usually they are, by snails) . . . one that lacks isidia but has black apothecia . . . and one that occurs on calcareous rocks, it is almost certain to be *P. petersii*. Unlike *P. nigrum*, thalli of *P. petersii* generally lack the bluish-black prothallus.

CHEMISTRY: No substances. Spot tests. K-, C-, KC-, P-, UV-.

NICHE: This species is distributed in the central and southern Appalachians, with additional populations in the northern Appalachians and Ozark Mountains. It is not common in the Smokies, likely because of the paucity of calcareous rocks. Like other limestone lovers, look for this species near Ace Gap.

KEY FEATURES: Small rosette-like thalli with radiating lobes and center portions often eaten by invertebrates, no isidia, black apothecia, cyanolichen on sunny, calcareous rocks, uncommon.

Platismatia glauca

Summit Jumper

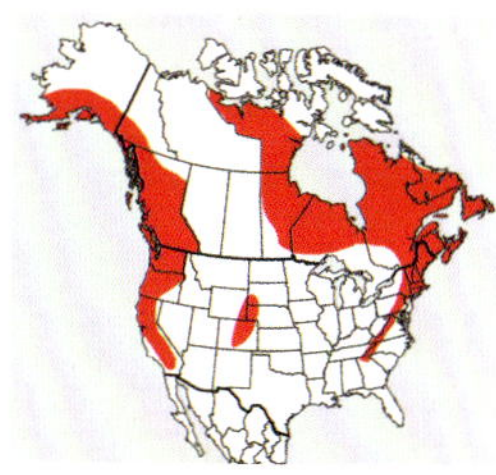

Tripp 3479 (photo: Lendemer)

NOTES: Species of *Platismatia* should be really easy to identify and, once learned, they are! Beginning lichenologists will tend to confuse them for other big macrolichens (we aren't sure why other than that it happened to us, too and, well, we think macrolichens are harder than microlichens). *Platismatia glauca* is readily spotted by its large, yellowish-brown foliose thalli with irregularly shaped, broad lobes that are lined by beautiful isidia and pale brown underneath. These isidia also make their way onto the surfaces of the lobes, as seen in this photograph. *Platismatia glauca* is further characterized by its rather narrow niche (see below).

CHEMISTRY: Atranorin and caperatic acid. Spot tests. Cortex: K+ yellow, C-, KC-, P-, UV-; medulla: K-, C-, KC-, P-.

NICHE: Summit Jumper is common at the highest elevations in the park. The places where we have collected it should tell you something: Mt. Amber, Mt. Sterling, Mt. LeConte, Pecks Corner, Mt. Buckley, Old Black, Tricorner Knob. Look for it on the branches of spruce and fir.

KEY FEATURES: Large, yellowish-brown thalli with pale undersides, irregular lobes with abundant, marginal and laminal isidia, on branches and bark of spruce and fir at high elevations.

Platismatia tuckermanii

Emperor of the Canopy

Tripp 3469 (photo: Lendemer)

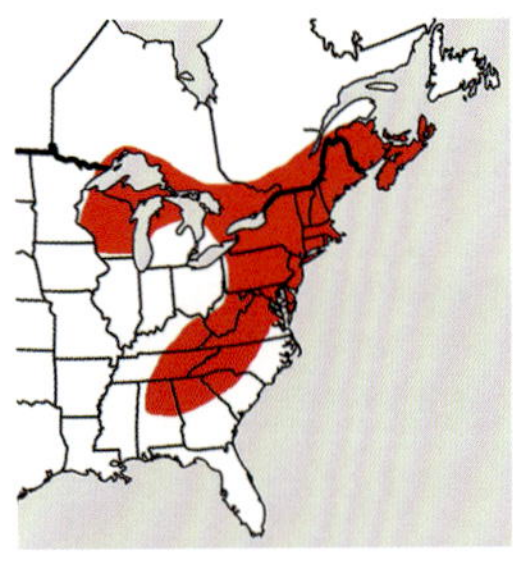

NOTES: *Platismatia tuckermanii* is definitely in the top ten list of *must learn* macrolichens in the Smokies. It is huge, and it is everywhere. This species is easily identified by its large, greenish-gray thalli that become increasing less thallus-like and more lobe-like toward the growing tips (look at the ruffled-like appearance of these lobes in the photograph; Smokies specimens look just like this), its densely white-maculate upper surface that lacks isidia but bears abundant apothecia, and its pale undersides. Additionally, margins of the lobes are sometimes blackened in appearance, as in the photograph. The apothecia are generally shiny brown in color.

CHEMISTRY: Atranorin and caperatic acid. Spot tests. Cortex: K+ yellow, C-, KC-, P-, UV-; medulla: K-, C-, KC-, P-.

NICHE: *Platismatia tuckermanii* is very common throughout the Smokies, but particularly common at high elevations. One most frequently finds this species fallen from the canopy and lying on the ground, just waiting for someone to pick it up and study it.

KEY FEATURES: Very large, greenish-gray thalli, white maculae on top, pale brown below, no isidia, with abundant apothecia, throughout the Park.

Polymeridium proponens

Blow-Me-Downs

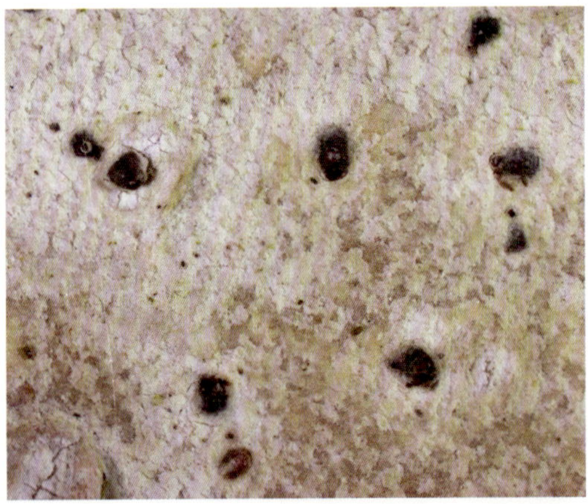

Tripp 2092 (photo: Lendemer)

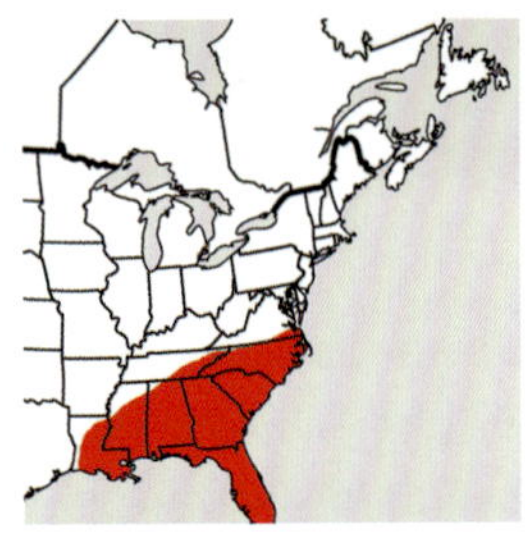

NOTES: To date, just about every pyrenolichen lichen you have encountered in this book has had the same boring old perithecia, solitary, and with the ostiole sticking right up out of the top. Well, that is not always the case, and *Polymeridium proponens* is an example in which the perithecia are oriented on their sides with a lateral (rather than apical) ostiole. This makes the perithecia look more like scale insects than the fruiting bodies of a lichen. From a distance, they look like a tiny forest of perithecia they blew down in a windy storm. There are no similar species in the Smokies, especially not when you take into account the chemistry and the presence of colorless, muriform spores that stain intense purple in iodine.

CHEMISTRY: Lichexanthone. Spot tests: K-, C-, KC-, P-, UV+ bright yellow.

NICHE: *Polymeridium proponens* is a species of subtropical regions of southeastern North

America, primarily in the Coastal Plain. It is rare in the Smokies, and the southern Appalachians in general, where it grows on the bark of hardwood trees at low elevations.

KEY FEATURES: White, ecorticate, crustose thallus, horizontal perithecia, muriform, colorless spores, UV+ yellow, on bark at low elevations.

Polysporina simplex

The Simple, Easy, Many-Spored Lichen

Tripp 3704 (photo: Lendemer)

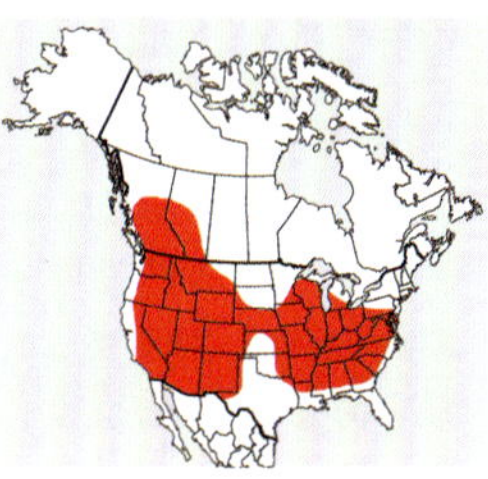

NOTES: "Nothing but little black dots on a rock" . . . should be really difficult to identify, right? Wrong! And this is precisely why we *love* lichens and *love love love* Great Smoky Mountains National Park, which should probably be renamed Great Lichen Mountains National Park. In any case, *Polysporina simplex*, indeed, looks like little more than black dots on a rock, but if you peer closely with the hand lens you remembered to put around your neck, you will see deeply carbonized discs that are almost always cracked to gyrose (yep, spinning around in circles as if it had a tail to chase). Cut into one of these apothecia with a sharp razor blade and you are in for further surprises: asci chock full of hundreds (or at least 100+) tiny, colorless spores that barely exceed 4 microns in length. "*Poly*" + "*sporina*" + "*simplex*" . . . get it? The simple, easy, many-spored lichen.

CHEMISTRY: No substances. Spot tests. K-, C-, KC-, P-, UV-.

NICHE: *Polysporina simplex* occurs sporadically throughout the Park in middle-to-high elevation habitats. It is not common here, especially compared to elsewhere across its range in North America (see map). The Simple, Easy, Many-Spored Lichen *always* occurs on rocks, usually mineral-rich to slightly calcareous rocks.

KEY FEATURES: Completely endolithic thallus to an occasional, dull, thin white thallus, essentially black dots on a rock, apothecia strongly carbonized throughout, cracked, gyrose, or otherwise "special," polysporous asci, uncommon on mineral-rich rocks in middle-to-high elevations.

Porina heterospora

Orange-Glaze

Lendemer 33183 (photo: Tripp)

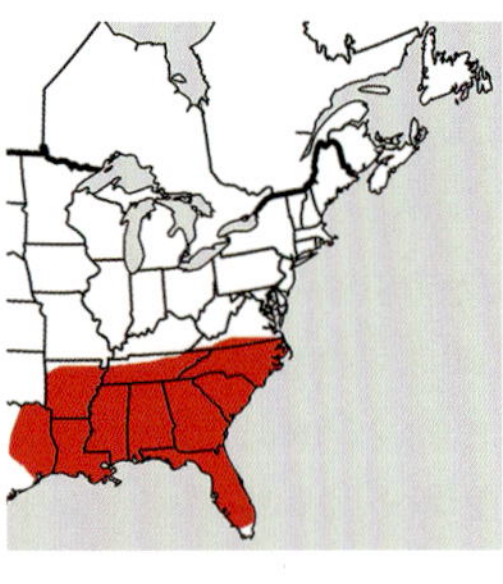

NOTES: *Porina heterospora* and its asexual counterpart *P. scabrida* (to have an asexual counterpart: not too shabby!) are actually very easy to identify crustose lichens that you should prioritize learning because they are both very common and often grow together. First: these things are orange, owing to the presence of a *Trentepohlia* photobiont. *Orange orange orange.* Beyond that, *P. heterospora* has wart-like—scratch that—doughnut-like perithecia that are very regular in shape and conspicuously sessile atop a thin thallus. If you care for more data, this species makes fusiform to clavate, transversely septate spores. It doesn't look like any other pyrenolichen in the Smokies, and beyond.

CHEMISTRY: No substances. Spot tests. K-, C-, KC-, P-, UV-.

NICHE: Orange-Glaze is both common and widespread throughout the Smokies, where it occupies the bark of hardwoods and is especially fond of hickories and low elevations. This species often co-occurs with *P. scabrida*, which we think is pretty nifty.

KEY FEATURES: Relative large, orange perithecia that sit atop an orange thallus, common on the bark of hardwoods, especially low elevations.

Porina scabrida

Orange Orchards

Tripp 2516 (photo: Deregibus)

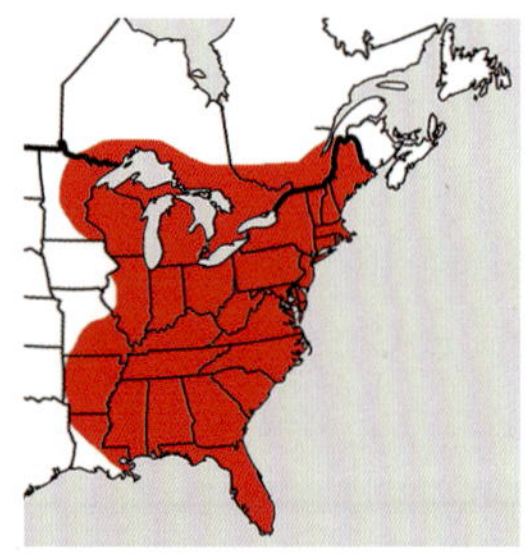

NOTES: *Porina scabrida*, together with its sexual counterpart, *P. heterospora*, represent a very easy "species pair" to learn because they are bright orange in color, thanks to having *Trentepohlia* for a photobiont. *Porina scabrida* is very easy to recognize because it forms large, dense orchards of orange isidia that sit atop a thallus that typically cannot be seen because its isidia are too dense. There are no comparable, densely isidiate orange lichens in the Park. If this thing had a green alga for photobiont, such as *Trebouxia*, it might start to resemble species of *Phyllopsora*, such as *P. confusa*. But it doesn't.

CHEMISTRY: No substances. Spot tests. K-, C-, KC-, P-, UV-.

NICHE: Little Orchards is very common and widespread throughout the Smokies, especially at low elevations. It inhabits the bark of hardwoods and in particular likes to make a home in the darker, humid crevices of cracks in the bark.

KEY FEATURES: Dense orchard of isidia yielding a golden orange sheen, common at low elevations on the bark of hardwoods.

Porpidia albocaerulescens

Granite Bejewelment

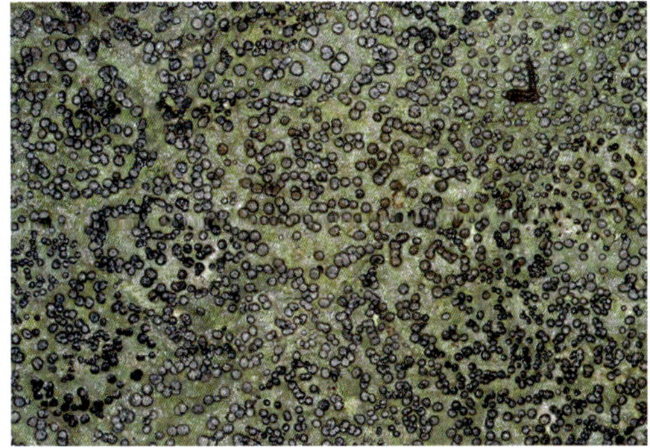

Lendemer 29553 (photo: Tripp)

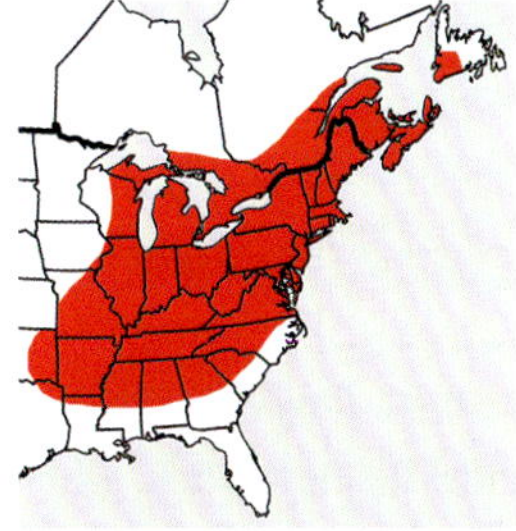

NOTES: So, we know that it feels like we have been telling you to "put this on your list of top 10 lichens" throughout this *Field Guide*, but if there is ever a time when we have been serious, it is now. *Porpidia albocaerulescens* is the most common crustose lichen on rocks in eastern North America. This species is so easy to identify owing to thick, greenish-white thalli that bear black apothecia with densely pruinose discs, which are invariably present on every thallus except the absolute youngest thalli that you probably won't see anyway. The pruina sometimes appears bluish, in which case you are having a very good day. We decided to use a photograph taken of this species on a wet day to illustrate that this dense pruina is apparent no matter what the forecast is. Sometimes rainy, sometimes sunny, sometimes partially cloudy, but always bejeweled.

CHEMISTRY: Stictic acid. Spot tests. K+ yellow, C-, KC-, P+ orange, UV-.

NICHE: We would not be surprised to learn that Granite Bejewelment occurs on >90% of all boulders in Great Smoky Mountains National Park. Sometimes, it feels like the rocks just can't get a break. Look for this on similar, non-calcareous rocks throughout eastern North America, where you will almost inevitably be successful.

KEY FEATURES: Thick greenish-white thallus with black apothecia covered by a dense, white pruina. The most common lichen on non-calcareous rocks throughout the Smokies and elsewhere in eastern North America.

Porpidia contraponenda

Winter's Remembrance

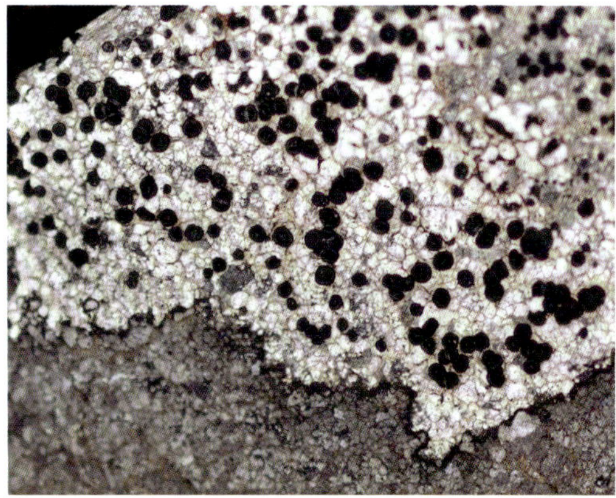

Tripp 5025 (photo: Lendemer)

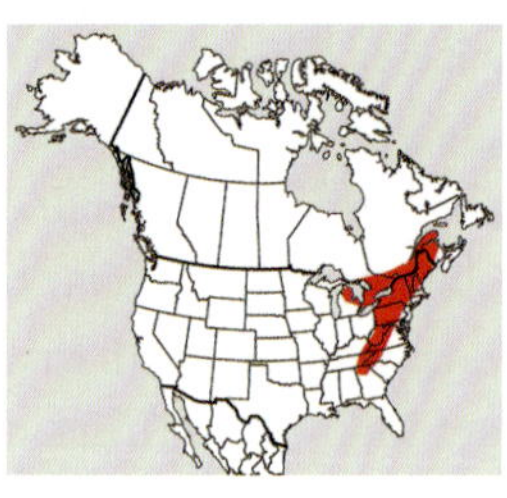

NOTES: In addition to *Porpidia albocaerulescens*, which really doesn't need any further introduction beyond the above, are five other species of *Porpidia* that are rock-loving and relatively common in the Smokies (*P. contraponenda, P. crustulata, P. degelii, P. macrocarpa, P. subsimplex*) plus one more than is very rare (*P. tuberculosa*). This group of six is relatively easy to differentiate among one another. Four of these (*P. contraponenda, P. crustulata, P. macrocarpa, P. subsimplex*) are apotheciate whereas the other two (*P. degelii, P. tuberculosa*) are sorediate. Of the four fertile species, *P. contraponenda* can be distinguished by having a thick, white, well-developed thallus that is continuous to areolate and bears large black apothecia with thick margins. This species is also distinguished by its production of confluentic acid. *Porpidia contraponenda* might be confused for species of *Rhizocarpon*, but the latter differ by having brown, 2-celled spores. Additionally, spores of species of *Porpidia* are usually relatively large and are distinctly halonate.

CHEMISTRY: Confluentic acid and 2'-*O*-methylmicrophyllinate. Spot tests. K-, C-, KC-, P-, UV+ dull blue-white.

NICHE: *Porpidia contraponenda* is a relatively rare species that, in the Smokies, is restricted to high-elevation forests dominated by Red Spruce, Fraser Fir, Yellow Birch, and Mountain Maple. This species is disjunct from more northern portions of the Appalachians, and also occurs in the Pacific Northwest.

KEY FEATURES: Well-developed thick white thallus, black apothecia with prominent margins, simple, colorless, halonate spores, UV+ blue-white, on rock, rare at high elevations.

Porpidia crustulata

Not My Problem

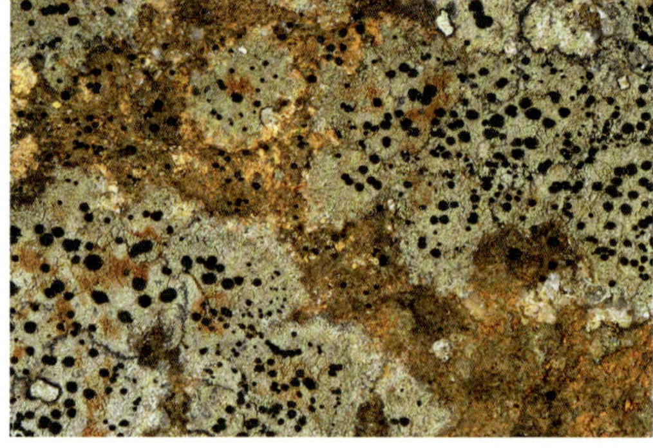

Tripp 3617 (photo: Lendemer)

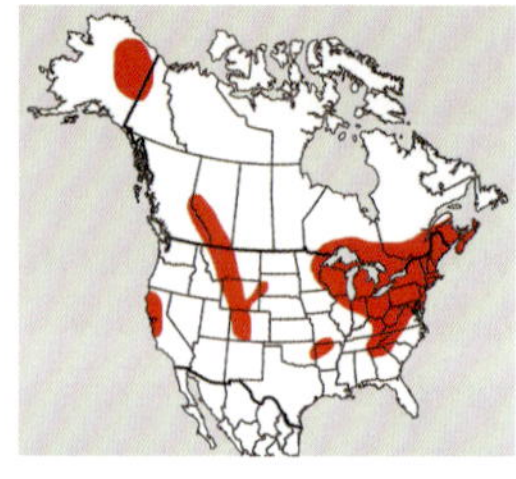

NOTES: *Porpidia crustulata* is one of five sexually reproducing (i.e., apotheciate) species in the Smokies. From *P. albocaerulescens*, it differs in not having conspicuously pruinose apothecia discs. From *P. contraponenda*, it differs in not having a thick, well-developed white thallus and in being UV-. From *P. macrocarpa*, it differs in having rather small apothecia (always < 1.5 mm in diameter) and smaller spores (these ca. 10-17 x 5-9 um). From *P. subsimplex*, it differs in having a weakly carbonized exciple that is smooth and not cracked. When in doubt, use the key. Also, see the short summary of *Porpidia* species present in the Smokies under the *P. contraponenda* entry.

CHEMISTRY: Stictic acid. Spot tests. K+ yellow, C-, KC-, P+ orange, UV-.

NICHE: *Porpidia crustulata* occurs primarily at middle elevations, where it is relatively common on non-calcareous rocks, particularly in wet, riparian environments dominated by an understory of Doghobble. Look for this species along western portions of the Boogerman Loop, especially near Caldwell Fork. This

species also occurs in acidic, southern hardwood forest and reaches up to spruce-fir forest on occasion. Elsewhere in North America, this species occurs erratically in suitable habitats.
KEY FEATURES: Weakly developed thallus, small apothecia with smooth margins, weakly carbonized exciple, small spores, broach niche consisting of low to high elevations, particularly riparian environments at middle elevations, always on rock.

Porpidia degelii

Upper Crust

Tripp 5093 (photo: Lendemer)

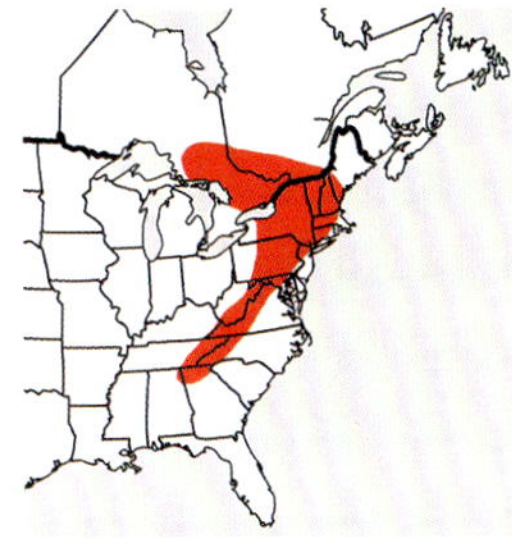

NOTES: *Porpidia degelii* is one of two sorediate species in the genus present in the Smokies. This species is characterized by having a thin, dirty, grayish-white thallus with similarly colored soredia. Although the thallus is often partially rust colored when the species grows on metal-rich rocks such as those of the Anakeesta Formation, you would not know it as a member of *Porpidia* except for when it simultaneously produces apothecia, as is shown in the photograph (but do read the short summary of *Porpidia* species present in the Smokies under the *P. contraponenda* entry). See those densely pruinose discs near 11 o'clock and 6 o'clock? What does that remind you of? If you guessed *P. albocaerulescens*–the most common lichen growing on rocks in eastern North America–you would be correct! Recent molecular phylogenetic studies have shown *P. degelii* and *P. albocaerulescens* to be each other's closest relatives, and the two species are "reciprocally monophyletic," supporting the hypothesis that species pairs consisting of one sexual and one asexual member exist in *Porpidia*. When present only with soredia, it is not likely to be confused with the other asexual species in the genus (i.e. *P. tuberculosa*, because the latter has a different chemistry).
CHEMISTRY: Stictic acid. Spot tests. K+ yellow, C-, KC-, P+ orange, UV-.
NICHE: *Porpidia degelii* is an uncommon (but not rare) species that grows exclusively at high elevations in spruce-fir forest on Anakeesta rock. Look for it on Clingman's Dome. This is an Appalachian species whose distribution barely extends north beyond the White Mountains of New Hampshire.
KEY FEATURES: Crustose grayish-white to rusty brown thallus, sorediate, P+ orange, high elevations, on Anakeesta rock.

Porpidia macrocarpa

Big Gal

Tripp 4984 (photo: Lendemer)

NOTES: *Porpidia macrocarpa* is one of five apotheciate/sexual members of *Porpidia* in the Smokies. *Porpidia albocaerulescens* differs in having densely pruinose apothecia. *Porpidia contraponenda* differs in having a very thick white thallus (and in chemistry). *Porpidia crustulata* has a thin, poorly developed thallus and a weakly carbonized exciple with smooth margins like *P. macrocarpa* but differs in having much smaller apothecia and smaller spores (see key). Finally, *P. subsimplex* also has a thin, poorly developed thallus but differs in having a strongly carbonized exciple with radially cracked margins. As the specific epithet implies, *P. macrocarpa* often has very large apothecia. Be sure to pay attention to the chemistry of various fertile species of *Porpidia*, and see the short summary of *Porpidia* species present in the Smokies under the *P. contraponenda* entry for additional insight.

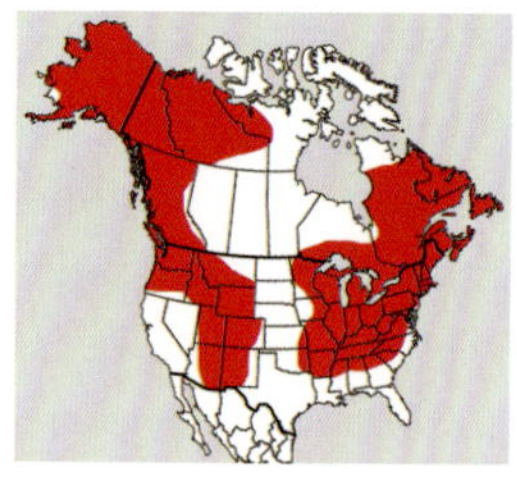

CHEMISTRY: Stictic acid. Spot tests. K+ yellow, C-, KC-, P+ orange, UV-.

NICHE: Big Gal is a relatively common species that occurs primarily in middle elevations (but does range from high to low altitudes) in humid, streamside habitats. It grows primarily on non-calcareous rocks but also occurs on mineral-rich rock such as Anakeesta formation. Look for a nice population of this species on Forney Creek Trail between its junction with White Oak Branch Trail and its crossing over Bear Creek. This species is rather widespread in suitable habitats throughout North America.

KEY FEATURES: Thin, poorly developed gray-to-greenish-white thallus, large black apothecia, weakly carbonized exciple with smooth margins, relatively large apothecia with relatively large spores, riparian environments at all elevations, always on rock.

Porpidia subsimplex

Cracked Pots

Tripp 3580 (photo: Lendemer)

NOTES: Porpidia *subsimplex* is our final stop on the tour of fertile species of *Porpidia* in Great Smoky Mountains National Park. Fortunately (because both you and I are losing steam), this one is easily differentiated from all the others in having a strongly carbonized exciple with margins that are distinctly cracked radially. Given the frequency at which this species lacks all signs of a visible thallus, it is easy to confuse *P. subsimplex* with species of *Sarcogyne* or *Lecidea*. *Sarcogyne* differs in having polysporous asci, and *Lecidea* differs in having smaller, non-halonate spores. See the short summary of *Porpidia* species present in the Smokies under the *P. contraponenda* entry.

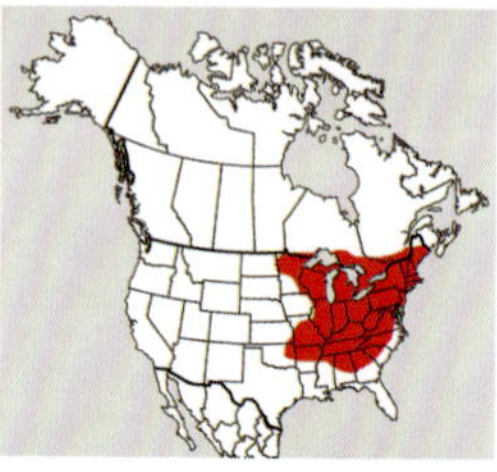

CHEMISTRY: No substances. Spot tests. K-, C-, KC-, P-, UV-.

NICHE: This species occurs at all elevations but especially middle altitudes in the Smokies. It can be found in a broad variety of environ-

ments, ranging from mixed hardwood forests of Buckeye, Sugar Maple, Yellow Birch, American Beech, and Silverbell to acid cove forests of Sourwood, Hickory, Ironwood, and an assortment of pines and oaks. Head to the summit of Brushy Mountain for a look at this species.
KEY FEATURES: Crustose, poorly developed (to completely endolithic) thallus, black apothecia, strongly carbonized exciple, cracked margins, relatively large, halonate spores, on non-calcareous rocks, widespread and common in Park.

Porpidia tuberculosa
North Wind

Tripp 3512 (photo: Lendemer)

NOTES: *Porpidia tuberculosa* is a very rare species that you are unlikely to encounter in the Smokies. It is sorediate and easily differentiated from *P. degelli* by a P- thallus that lacks stictic acid. Rather than other species in the genus, *P. tuberculosa* might be confused with *Fuscidea recensa*, which differs chemically in the production of divaricatic acid and morphologically in having a thallus composed of somewhat dispersed, flattened areoles that rest atop a dark prothallus. Also, see the short summary of *Porpidia* species present in the Smokies under the *P. contraponenda* entry.
CHEMISTRY: Confluentic acid. Spot tests. K-, C-, KC-, P-, UV+ dull blue-white.
NICHE: This species is restricted to high-elevation spruce-fir forests, where it occurs on Anakeesta rock outcrops. We have collected it precisely once in the Smokies. Look for it more commonly in northern portions of the Appalachians.
KEY FEATURES: Crustose gray thallus, sorediate, P- and UV+ blue-white, on Anakeesta rock, high elevations, very rare.

Protoblastenia rupestris
Rock Buddy

Harris 56345 (photo: Lendemer)

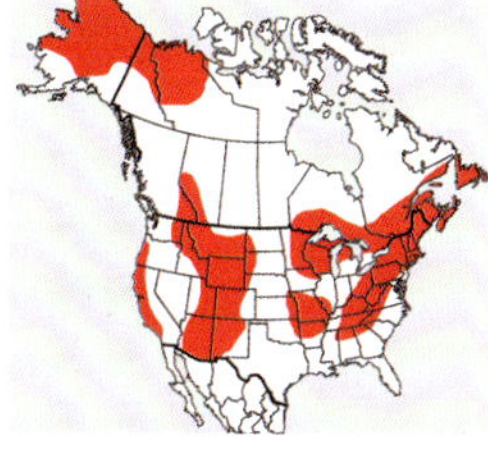

NOTES: Would you believe us if we told you that the thallus of this species always looks like scum? It's true, and that is precisely how you can identify Rock Buddy. We actually had in hand a different photograph of a "cleaner" thallus but chose to include this scummy one as your best means of recognizing *Protoblastenia rupestris* in the field. This is one of the most common lichens on limestone (and other calcareous substrates) in our area. It is characterized by a continuous to areolate, greenish-gray thallus that makes up for its appearance in the form of its beautiful, orange apothecia, which sit sessile atop the thallus. The only lichens with which you could confuse this species are members of the genus *Caloplaca*, nearly all sexual members of which have similarly bright orange apothecia. It's true that *P. rupestris* has anthraquinones like *Caloplaca*, and thus reacts strong purple in K; however, spores in the former are simple and reside inside fruiting bodies that are biatorine (vs. polarilocular spores and fruiting bodies with lecanorine or lecideine margins in *Caloplaca*). See how easy lichen identification can be? It's better than cleaning the scum from your bathroom, that's for certain.

CHEMISTRY: Anthraquinones (apothecia). Spot tests: (thallus) K-, C-, KC-, P-, UV-; (apothecia) K+ purple, C-, KC-, P-, UV-.

NICHE: Since Rock Buddy is restricted to calcareous rocks, there are a limited number of locations in which the species can be seen in the Smokies. Think: Ace Gap! And while there, look for *Sarcopyrenia calcarea*, a pretty sweet lichenicolous fungus that grows on top of it.

KEY FEATURES: Continuous to areolate, cruddy thallus, bright orange biatorine apothecia that are K+ purple, simple, colorless spores, on calcareous rocks.

Pseudevernia cladonia

Thallus Cladogenesis

Tripp 2263 (photo: Lendemer)

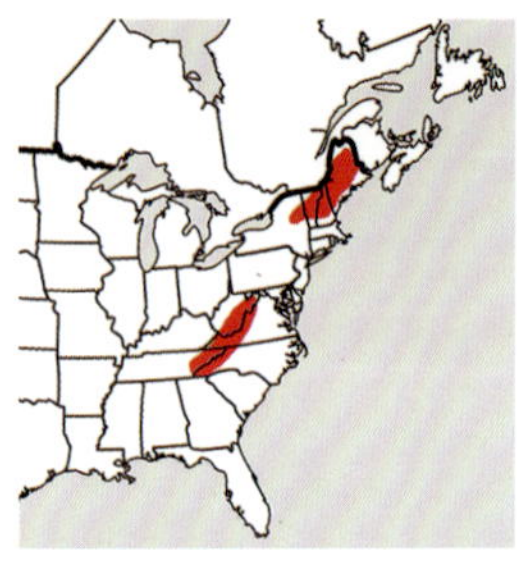

NOTES: *Pseudevernia cladonia* and its close relative *P. consocians* are very easy-to-identify macrolichens because they look like gray shrubs growing directly out of tree bark. In reality, both species are technically foliose because they have distinguishable upper versus lower cortices, but let's be serious: they otherwise look completely fruticose so let's call them fruticose. *Pseudevernia cladonia* cannot be confused with *P. consocians* because it lacks isidia (it also often lacks fruiting bodies, too, but that is out of our control). These two species differ ecologically, too (see Niches). *Pseudevernia cladonia* is, furthermore, characterized by its regular, dichotomously branching lobes that resemble phylogenetic trees, and thus make us love this species even more.

CHEMISTRY: Atranorin and lecanoric acid. Spot tests. Cortex: K+ yellow, C-, KC-, P-, UV-; medulla: K-, C+ red, KC+ red, P-, UV-.

NICHE: Thallus Cladogenesis is restricted to high-elevation habitats, where it is rather common and inhabits twigs of trees, especially spruces and firs. It occurs on bark and in simi-

lar environments throughout its range: high altitudes of the Appalachian Mountains.
KEY FEATURES: Gray fruticose lichen lacking both sexual and asexual reproductive propagules, regular, dichotomous branching, on twigs at high elevations, common.

Pseudevernia consocians

Bifurcation Nation

Lendemer 26780 (photo: Moroz)

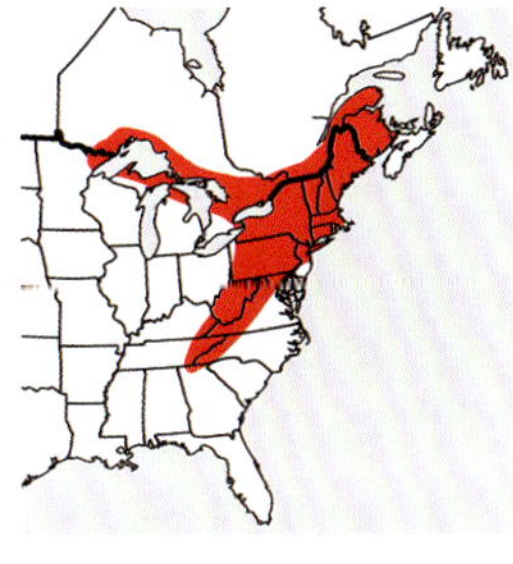

NOTES: Superficially, *Pseudevernia consocians* looks a lot like *P. cladonia*, especially when it comes to having a gray, fruticose thallus, dichotomous branching (a little less regular in *P. consocians*), and its twig affinity. However, the former is easily separated from the latter by dense, spike-like isidia that cover the surfaces of the thallus. Rather than confusing this species for *P. cladonia*, it is perhaps instead most similar to *Everniastrum catawbiense*, which differs in producing soralia instead of isidia. If you bring this book with you on your next hike in the Smokies, we guarantee you will learn this species! It is everywhere!
CHEMISTRY: Atranorin and lecanoric acid. Spot tests. Cortex: K+ yellow, C-, KC-, P-, UV-; medulla: K-, C+ red, KC+ red, P-, UV-.
NICHE: *Pseudevernia consocians* occurs throughout the Park but is most abundant in middle-to-high elevation habitats. It is particularly fond of acidic, dry environments such as Rhododendron slicks or Mountain Laurel gaps, where it grows on branches and between plates of bark, such as on pines. This species occurs in similar niches throughout the Appalachian Mountains
KEY FEATURES: Large, gray fruticose lichen covered in abundant isidia, at all elevations, on bark and branches, especially in acid-cove forests.

Pseudocyphellaria aurata

Succulent Sunshine

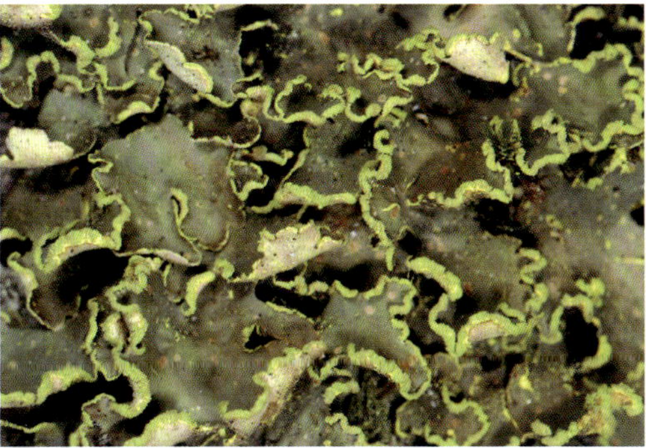

Lendemer 33173 (photo: Tripp)

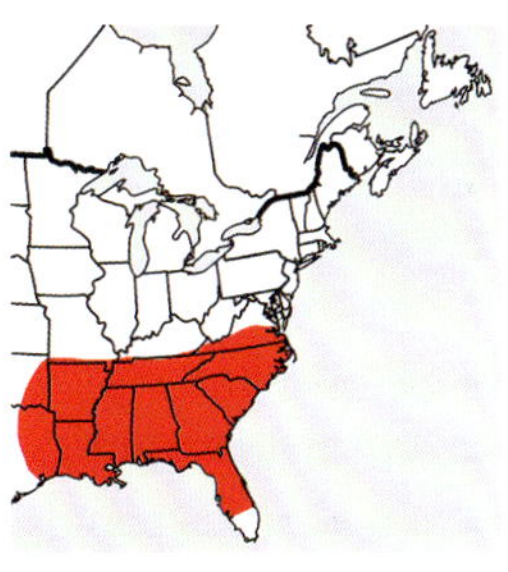

NOTES: You are going to *love* this species. *Pseudocyphellaria aurata* is one of the most beautiful macrolichens in eastern North America. Here are the features: large, bright green foliose thallus whose margins are paved with bright yellow soredia; underneath this layer, a bright yellow medulla; and underneath this layer, a lower cortex punctuated by large, conspicuous pseudocyphellae (hence the generic name). There is nothing on Earth like it, not even its close relatives *P. crocata* and *P. perpetua*, both of which have a cyanobacterial photobiont unlike the green algal photobiont that characterizes *P. aurata*.

CHEMISTRY: Pulvinic acid and related pigments. Spot tests. Cortex: K-, C-, KC-, P-, UV-; medulla: K-, C-, KC-, P-, UV-.

NICHE: *Pseudocyphellaria aurata* occurs only at low elevations in high-quality, mature, humid forests such as those near rivers and streams. It is not common in the Smokies or anywhere else in its range for that matter, but it is also not exceptionally rare. Uncommon enough to remain an exciting find every time encountered.

KEY FEATURES: Large, green foliose thallus with marginal soredia, bright yellow medulla, and pseudocyphellate lower cortex. Big, beautiful, and succulent.

Pseudocyphellaria perpetua

Nothing Is Forever Lichen

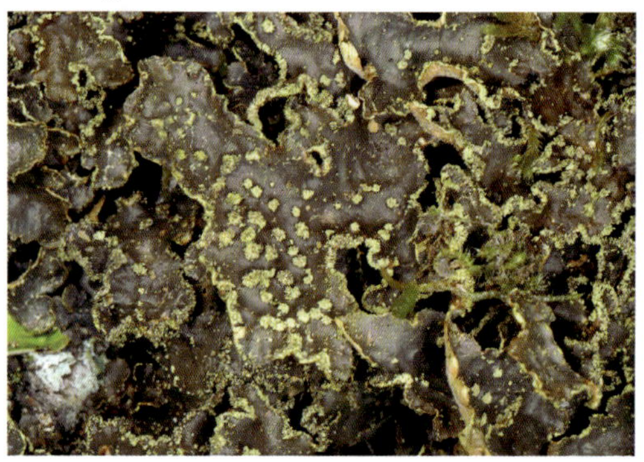

Tripp 3292 (photo: Tripp)

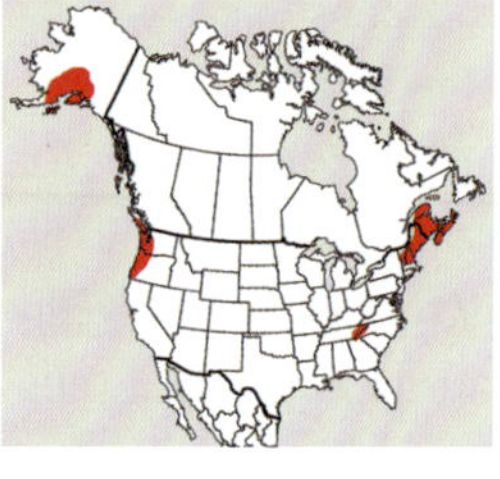

NOTES: *Pseudocyphellaria* is a spectacular, show-stopping foliose lichen, just like its relatives. This species, like *P. crocata*, is easily distinguished from *P. aurata* in having a cyanobacterial photobiont (vs. green algal in *P. aurata*). This photobiont difference contributes to the conspicuous brown color of thalli of *P. perpetua* and *P. crocata*, in contrast to the vibrant green color of thalli of *P. aurata*. Superficially, *P. perpetua* is similar to its closest relative, *P. crocata*, but differs in having mostly marginal soralia in contrast to marginal + laminal soralia in *P. crocata*. The two species form reciprocally monophyletic groups, and populations of *P. perpetua* from the Smokies form a clade with populations from the Pacific Northwest, but we note that only PNW specimens of *P. crocata* were sampled in the phylogenetic study of Miadlikowska et al. (2001). All three species in the Smokies share a beautiful yellow medulla and a lower surface punctuated by pseudocyphellae.

CHEMISTRY: Pulvinic acid and related pigments. Spot tests. Cortex: K-, C-, KC-, P-, UV-; medulla: K-, C-, KC-, P-, UV-.

NICHE: This species occurs on the bark of hardwoods and conifers primarily at high elevations in the Park. It prefers habitats that are mature, high in quality, and humid. It is neither common nor uncommon in the Smokies. Elsewhere, this species has a patchy distribution in oceanic areas of New England, the Pacific Northwest, and coastal Alaska.

KEY FEATURES: Large brown foliose thallus with marginal-only soralia, bright yellow medulla, pseudocyphellate lower surface, occasional on bark of hardwoods and conifers at upper elevations.

Pseudosagedia cestrensis

C'est Joli

Lendemer 29464 (photo: Lendemer)

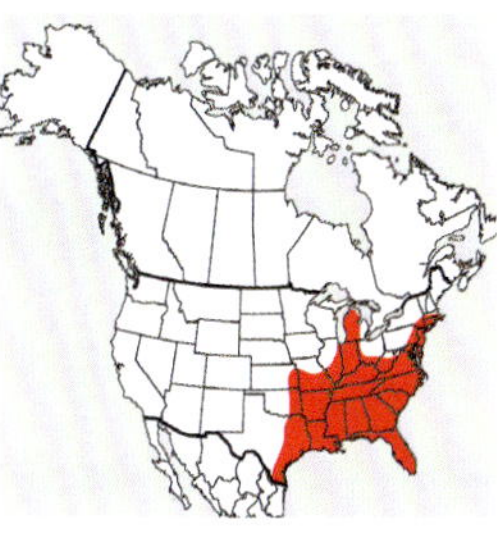

NOTES: As we have attempted to convey elsewhere with regard to *Pseudosagedia*, once learned, your life will be so much richer. Plus, it's really fun to say the name of the genus out loud, and impress your friends (or at least learn who your real friends are). *Pseudosagedia cestrensis* is one of two known sexual members of the genus in the Smokies that occur on bark. You cannot confuse it for the other, *P. rhaphidosperma*, because the latter has filiform spores . . . that is, very, very long, and very, very skinny spores (as opposed to clavate-fusiform spores of *P. cestrensis*). There are numerous peritheciate lichens that grow on bark, only some of which have colorless spores like *Pseudosagedia*. These include members of *Anisomeridium*, *Arthopyrenia*, *Strigula*, and *Mycoporum*. Species of *Anisomeridium* and *Strigula* differ in having 2-celled spores. Species of *Arthopyrenia* differ in lacking a photobiont and having 2-celled spores. Finally, some species of *Mycoporum* have muriform spores and thus are easily differentiated from *Pseudosagedia*. Others including *M. eschweileri* and *M. antecellens* have transversely septate spores like *P. cestrensis*, however these are 2–4-celled in contrast to spores that are always > 4-celled in *P. cestrensis*. And anyway, *Mycoporum* species just aren't very pretty. True.

CHEMISTRY: No substances. Spot tests. K-, C-, KC-, P-, UV-.

NICHE: *Pseudosagedia cestrensis* occurs throughout the southeastern United States, extending northward into the Mid-Atlantic Coastal Plain. It occurs on a variety of hardwood substrates, primarily on the trunks and roots. Look for it at low elevations throughout the Park.

KEY FEATURES: Thin brown crustose thallus, small black perithecia with fully carbonized exciples, multi-septate, clavate-fusiform, colorless spores, on bark of hardwoods, typically at low elevations.

Pseudosagedia guentheri

On the Rocks

Lendemer 8233 (photo: Lendemer)

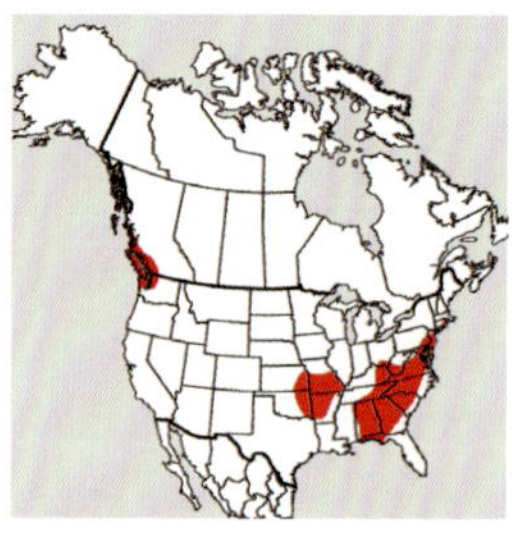

NOTES: There's nothing like a good old *Pseudosagedia* to start your day off on the right foot. It, a cup of coffee, the newspaper—you know. *Pseudosagedia guentheri* is an excellent species to put on your top 20 "I can do this peritheciate lichen thing" checklist. Why? Because it's beautiful. Why else? Because it's easy. Look for continuous, brownish gray thalli on shaded rocks that bear relatively small, black perithecia. A look at the inside of these perithecia will reveal magnificent, multi-septate (usually 6 to 9) colorless spores. These are long, skinny, and tapering to one end ("clavate-fusiform") and, once learned, you will know a *Pseudosagedia* of almost any sort by its spore morphology (think about your own family: in mine, the three daughters all have big foreheads; in others, all the siblings have large floppy ears; in *Pseudosagedia*, most species have fusiform spores . . . got it?). Among species of *Pseudosagedia* present in the Smokies, the only other member that similarly grows on rock is *P. chlorotica*, which differs in having 4-celled (vs. multi-septate) spores. Two other common species in the Park, *P. cestrensis* and *P. rhaphidosperma*, both grow on bark.
CHEMISTRY: No substances. Spot tests. K-, C-, KC-, P-, UV-.
NICHE: On the Rocks grows on . . . Guess what? Rocks! This species can be found throughout the eastern United States on shaded non-calcareous rocks, typically at low-to-middle elevations. Once learned, it is pretty easy to understand its generalist ecology, and be able to spot it in future lichen outings.
KEY FEATURES: Continuous, thin brown thallus with small black perithecia, long fusiform, and multi-septate spores, on shaded non-calcareous rocks, typically low-to-middle elevations.

Pseudosagedia isidiata

Animal Rub

Tripp 3629 (photo: Lendemer)

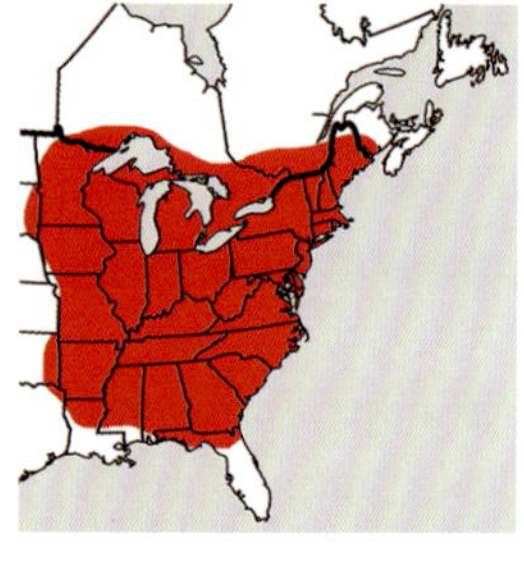

NOTES: *Pseudosagedia isidiata* is a beautiful, albeit very frequently overlooked, asexual species that with a little patience and curiosity, we think you might be able to spot. It is characterized by tiny, brownish-gray isidia that essentially comprise the entire (tiny) thallus. There are two other lichens in the Park that have a similar size, ecology, and gestalt: *Porina scabrida* is essentially a vivid orange version of *Pseudosagedia isidiata*, whereas *Phyllopsora confusa* is the bright green version, and both of the latter species are generally much larger in size than *P. isidiata*. *Pseudosagedia isidiata*

might also be confused for *Clathroporina isidiifera*, but that species is rusty-orange in color and grows on rocks in the Smokies. The dull brownish-gray color of *Pseudosagedia isidiata* is distinctive and can be attributed to its *Trentepohlia* photobiont. Keep eyes peeled for small flakes of bark that look like some furry animal (touché!). If found, as we like to say, "put a hand lens on it."

CHEMISTRY: No substances. Spot tests. K-, C-, KC-, P-, UV-.

NICHE: In the Smokies, Animal Rub occurs at low and middle elevations in both rich cove hardwood forest and acid cove forest. Look for it growing in its tiny manner on the bark of hardwoods, primarily oaks, hickories, and maples. This species is widespread in similar habitats throughout eastern North America but is often overlooked and thus under-collected.

KEY FEATURES: Tiny thallus composed essentially entirely of brownish-gray isidia, on flakes of bark at low and middle elevations, more common than collections indicate, but overlooked because of its very small size.

Psilolechia lucida

Clear, Light, & Shining

Lendemer 33172 (photo: Tripp)

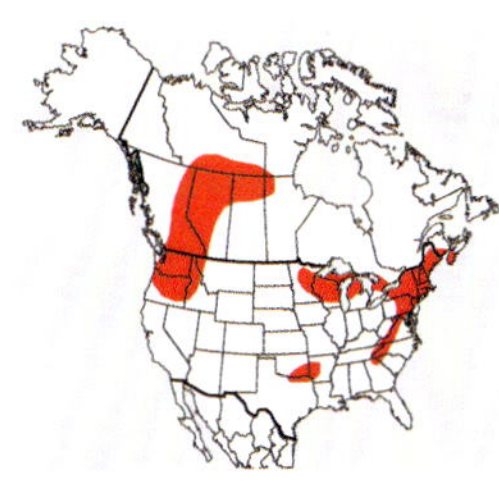

NOTES: *Psilolechia lucida* is an attractive, leprose lichen with a highly distinctive morphology ecology that, when learned in tandem, make for easy identification of this species. The thallus is composed of dull, sulfur yellow leprose granules that form a powder across the substrate. This species sometimes bears orange, biatorine apothecia, although more often than not is found sterile. You are most likely to confuse this species with members of the genus *Chrysothrix*, especially the yellowish-green tinted members *C. chamaecyparicola* or *C. onokoensis* (vs. *C. xanthina*, which is always bright yellow), but *C. chamaecyparicola* differs by being restricted to the bark of conifers, and *C. onokoensis*, although occurring on rocks, always has a thick, distinctive hypothallus. The common name we gave to this species refers to its Latin translation of "lucida."

CHEMISTRY: Rhizocarpic acid. Spot tests. K-, C-, KC-, P-, UV+ dull to bright orange.

NICHE: Clear, Light, & Shining has a very distinctive niche preference: it grows almost exclusively in deeply shaded, humid overhangs, where it can be found covering the surfaces of rock, sometimes large swaths of rock. This species often co-occurs with other lichens that share this niche, such as *Flakea papillata*. It is relatively common in this type of environment in the Smokies as well as elsewhere across its range, which is Appalachian–Great Lakes.

KEY FEATURES: Sulphur yellow leprose thallus, rarely also with yellow apothecia, always on rocks in deep overhangs, relatively common.

Punctelia appalachensis

The Incredible Hulk

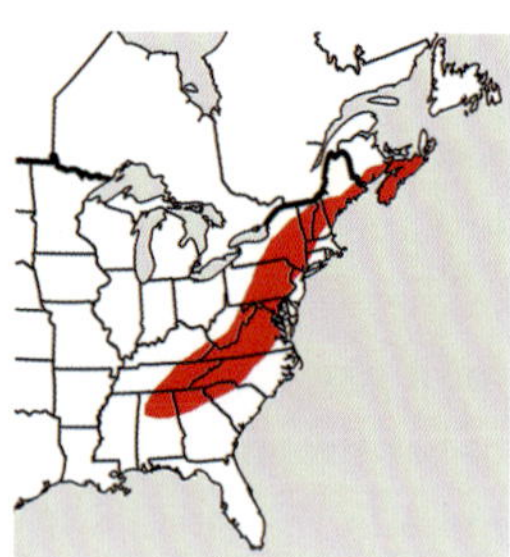

Tripp 3579 (photo: Lendemer)

NOTES: *Punctelia appalachensis* is a charismatic macrolichen that is easy to distinguish from other macrolichens by its bluish-gray upper surface (when dry, otherwise green when wet), dark lower surface, dense phyllidia that cover the thallus, especially central portions of the thallus, and pseudocyphellate upper surface. The dark lower cortex readily separates this species from two other common species of *Punctelia* in the Smokies: *P. caseana* and *P. rudecta*, both of which have pale undersides. Moreover, the latter two species are sorediate/isidiate, not phyllidiate. *Punctelia bolliana* is phyllidiate like *P. appalachensis*, but differs by its pale undersides. Rather than other *Punctelia*, *P. appalachensis* is perhaps most likely to be confused with species of *Cetrelia* that similarly have dark undersurfaces and white pseudocyphellae covering their upper cortices. However, they all differ in chemistry and have either a C+ red or UV+ blue-white medulla. They are also sorediate.

CHEMISTRY: Atranorin, protolichesterinic acid and related substances. Spot tests. Cortex: K+ yellow, C-, KC-, P- UV-; medulla: K-, C-, KC-, P-, UV-.

NICHE: *Punctelia appalachensis* is widespread at middle and high elevations in the park. It is found primarily on tree trunks and bases. This species has an Appalachian Mountain chain distribution. Learn it in the Smokies then take your new skills to Acadia National Park!

KEY FEATURES: Large, bluish-gray (or green if wet) foliose lichen, lower surfaces dark, upper surfaces with phyllidia and pseudocyphellae, widespread across all elevations, on bark, common.

Punctelia caseana

Wine Time

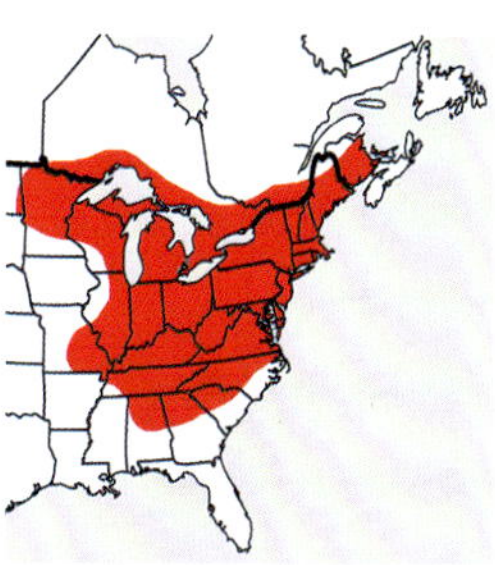

Tripp 3449 (photo: Lendemer)

NOTES: Many years ago, Erin first learned the genus *Punctelia* via the following trick: "P" is for pale, and "P"unctelia species have pale undersurfaces. This is very useful trick in separating most *Punctelia* from *Cetrelia*, which are similar in that species in both genera have upper surfaces with white pseudocyphellae; however, species of *Cetrelia* have dark undersurfaces in contrast to pale-to-white undersurfaces of species of *Punctelia* (note that this does not apply to *P. appalachensis*, which is dark underneath—never a dull moment in lichenology!). *Punctelia caseana* is a beautiful and easy-to-identify macrolichen by the following combination of features: blue-gray upper surface (when dry) with white pseudocyphellae, pale undersurface, sorediate, and C+ red medulla. It differs from *P. rudecta* because the latter has isidia, not soredia.

CHEMISTRY: Atranorin and lecanoric acid. Spot tests. Cortex: K+ yellow, C-, KC-, P- UV-; medulla: K-, C+ red, KC+ red, P , UV .

NICHE: This species is known from numerous collections in the Smokies, ranging from low to high elevations. It can be found on a broad diversity of hardwoods as well as conifers and also on rotting logs, at times. It is common in the Park. Look for a nice population of it on Lakeshore Trail, very close to and just southwest of the junction with Lost Cove Trail.

KEY FEATURES: Large macrolichen, blue-gray upper cortex (to green when wet) with white pseudocyphellae and soredia, pale lower surface, C+ red medulla, on hardwoods, common across all elevations.

Punctelia reddenda

Mountain Magnificence
(Alternative Name: To Dend Again)

Lendemer 53167 (photo: Tripp)

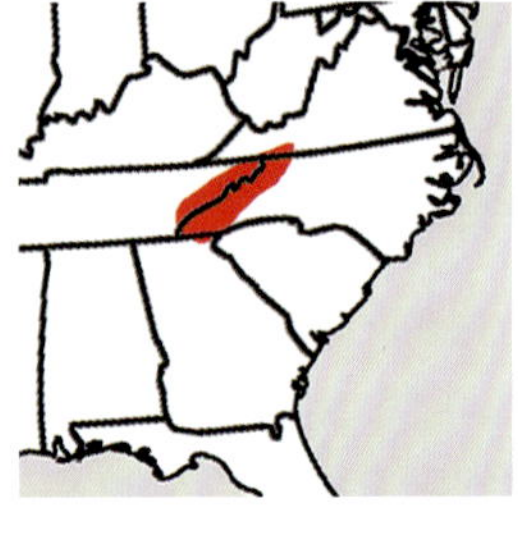

NOTES: *Punctelia* is an wonderful genus of macrolichens whose species are quite at home in the Smokies, like many of the rest of us. Each has a unique combination of characters, making identification and separation of the seven species currently known from the Park rather simple. *Punctelia reddenda* is beautiful lichen whose center of distribution in North America is the Smokies and surrounding southern Appalachian Mountains. It is characterized by its gray thallus bearing pseudocyphellae, its bright, almost glistening soredia, its C- medulla, and its undersurfaces that are black. It is the only C- and sorediate member of the genus in our area (all others being C+ pink or red owing to gyrophoric or lecanoric acid).

CHEMISTRY: Atranorin, protolichesterinic acid and related substances. Spot tests. Cortex: K+ yellow, C-, KC-, P- UV-; medulla: K-, C-, KC-, P-, UV-.

NICHE: *Punctelia reddenda* occurs only at high elevations in the Park. Look for it in montane northern hardwood forest or spruce forest, above 5,000 ft. Perhaps while hiking or trail running Jenkins Ridge, not far from the summit of Blockhouse Mountain

KEY FEATURES: Grey foliose lichen, upper surface covered by pseudocyphellae and strikingly beautiful soredia, black undersides, and a C- medulla, restricted to the bark of hardwoods in high-elevation habitats.

Punctelia rudecta

Backyard Buddy

Tripp 3450 (photo: Lendemer)

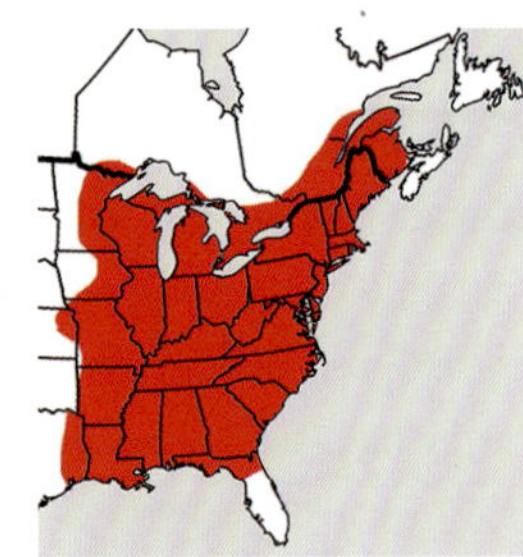

NOTES: *Punctelia rudecta*: arguably the weediest macrolichen east of the Mississippi River. Think we are kidding? Look for it outside your dentist's

office, in your local dog Park, near the gas pump, or perhaps in your back yard. Learn this species as quickly as possible via the following features: large macrolichen (albeit small and scraggly in highly polluted, disturbed areas), bluish-gray thallus (when dry, otherwise green when wet), upper surface with white pseudocyphellae and isidia, lower surface that is pale, and a bright C+ red medulla. This combination should lead you directly to *P. rudecta*. *Punctelia caseana* differs in being sorediate, and *P. appalachensis* differs in being phyllidiate and in having a dark undersurface. Do not be tempted to confuse this species with species of *Cetrelia*, which have dark undersides, and never grow in such crappy places as does Backyard Buddy.

CHEMISTRY: Atranorin and lecanoric acid. Spot tests. Cortex: K+ yellow, C-, KC-, P- UV-; medulla: K-, C+ red, KC+ red, P-, UV-.

NICHE: *Punctelia rudecta* grows on everything, in all kinds of forests, at all kinds of elevations. It is probably one of the top five most commonly collected macrolichens in eastern North America.

KEY FEATURES: Large, bluish-gray upper cortex (when dry), upper surface with white pseudocyphellae and isidia, pale lower surface, medulla C+ bright red, extremely abundant everywhere (at all elevations, in all habitats) in the Smokies.

Pycnora praestabilis

1-718-817-8700

Tripp 3456 (photo: Lendemer)

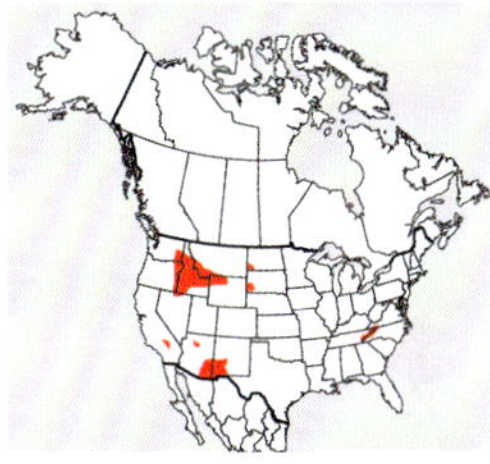

NOTES: *Pycnora praestabilis* is an inconspicuous lichen because in the southern Appalachians it tends not to make apothecia, but rather just pycnidia. The esorediate waxy yellow-gray thallus that reacts C- and KC+ red as well as P+ yellow is distinctive, particularly coupled with the unusual niche. Thalli of *Trapeliopsis* that have not yet formed soralia might be confused with *P. praestabilis* if they were found on old wood; however they produce gyrophoric acid and are P-. A second species of *Pycnora*, *P. sorophora*, is known from only the highest elevations of Mount Mitchell in the Black Mountains. It is chemically similar to *P. praestabilis*, but sorediate and with a very poorly developed thallus.

CHEMISTRY: Alectorialic acid. Spot tests. K-, C-KC+ red, P+ yellow, UV-.

NICHE: This species is known from scattered mountainous regions in North America, mostly in the western portions of the continent. In eastern North America, it is known only from the southern Appalachians and is particularly common in a small area of the

Smokies between Polls Gap and Purchase Knob.

KEY FEATURES: Light yellow-gray areolate thallus, esorediate, abundant black pycnidia, C+ red, KC+ red, P+ yellow, on old dry logs and fences at high elevations.

Pycnothelia papillaria

Clown Caps

Tripp 3715 (photo: Lendemer)

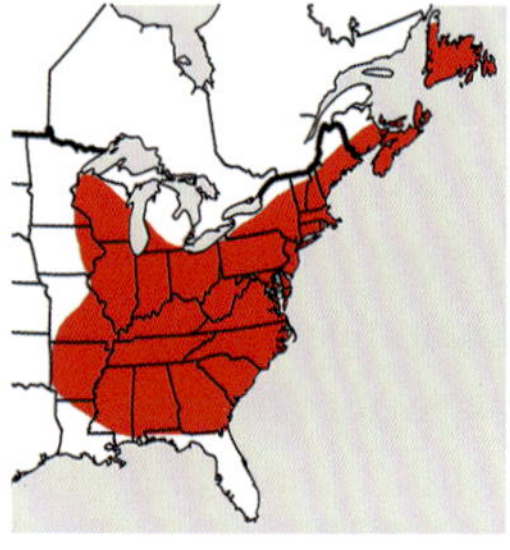

NOTES: As silly as it may be, we think that the common name captures the spirit of this lichen. It looks like two parts that don't belong together: a crustose thallus with inflated, thorn-like podetia that have dark brown caps, and growing on soil or organic matter no less! Even more surprising is the fact that it is closely related to *Cladonia*—although this makes sense given that some *Cladonia* species have short, thorny podetia, and others have a crustose primary thallus that quickly disappears after the podetia begin to form. But, at the same time, there is something implausible here, as though the lichen didn't know what to do and so decided to try everything at once. Thalli of *Trapeliopsis granulosa* without soredia might be confused with *P. papillaria*, but that species is C+ pink. Likewise, thalli of *Dibaeis baeomyces* that have not developed their pink stalked apothecia might also be confused, but that species is P+ yellow.

CHEMISTRY: Atranorin, protolichesterinic acid, and related substances. Spot tests. Cortex: K+ yellow, C-, KC-, P- UV-; medulla: K-, C-, KC-, P-, UV-.

NICHE: Clown Caps is widely distributed in temperate and boreal eastern North America, where it occurs on organic matter and soil in sun exposed habitats. It is rare in the Smokies, but that is largely because so much of the Park is shaded by forest.

KEY FEATURES: Gray crustose thallus, inflated thorn-like gray podetia with brown tips, K+ yellow, on soil and organic matter.

Pyrenidium aggregatum

Blue Nile

Collector: Lendemer 33105 (photo: Tripp)

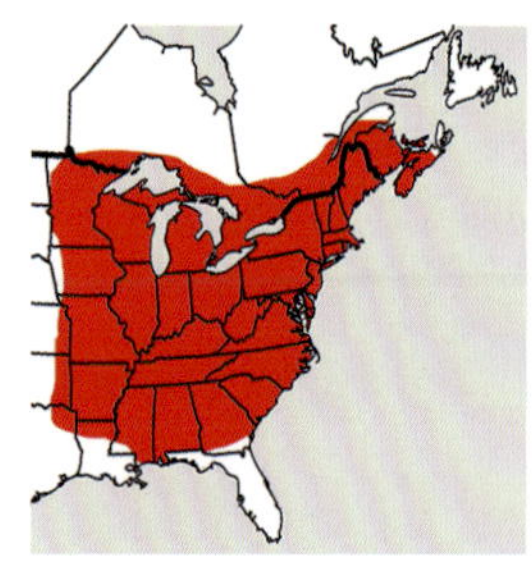

NOTES: As a general rule, we did not include lichenicolous fungi in this field guide, however some of the most unusual or common species did make the cut. *Pyrenidium* is one such species that induces the formation of large, raised, hemispherical galls on thalli of *Phaeophyscia*. Each gall contains one to several perithecia, which in turn contain asci with eight brown submuriform spores. The photograph included here illustrates the aspect of the fungus perfectly. The only similar species are members of the genus *Lichenochora* that can form similar galls on *Physcia* and *Phaeophyscia* species, including *Phaeophyscia rubropulchra*; however, the spores of that genus are colorless and 2-celled.

CHEMISTRY: No substances. Spot tests. K-, C-, KC-, P-, UV-.

NICHE: *Pyrenidium aggregatum* is a parasitic fungus that typically grows on thalli of *Phaeophyscia* and is widely distributed throughout eastern North America. It is widespread in the Smokies and most frequent on thalli of *P. rubropulchra*.

KEY FEATURES: Large, rounded galls formed on *Phaeophyscia* thalli, perithecia, submuriform brown spores, throughout the Smokies.

Pyrenula citriformis

Coastal Plain Cones

Tripp 2630 (photo: Lendemer)

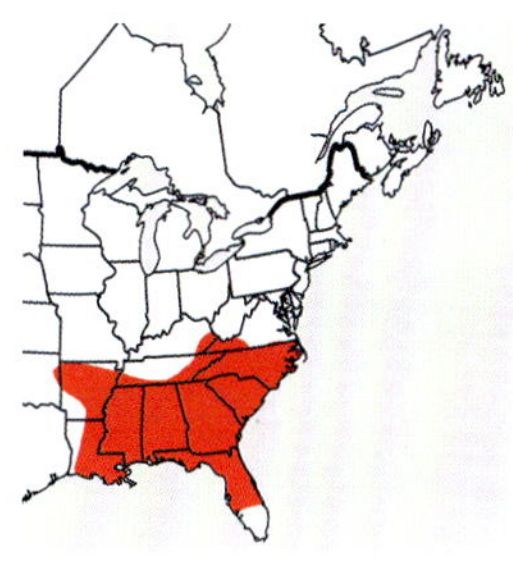

NOTES: *Pyrenula*, *Pyrenula*, oh to be a *Pyrenula* and wile away the hours in the warm lowlands of the Smoky Mountains. Once learned, this genus will never be forgotten. As a whole, this group is characterized by its smooth, continuous thalli, dark-colored (brown to black) but conspicuous and sometimes downright *big* perithecia, and brown, transversely septate (to more rarely muriform) spores. We have numerous species in the Smokies, and these can be identified by various features pertaining to the arrangement of perithecial ostioles, spore septation, spore size, hymenium, and chemistry. *Pyrenula citriformis* is easily distinguished from all other relatives in the Park by its 4-celled, relatively small spores (see the Key) and its hymenium that lacks oil inspersion.

CHEMISTRY: No substances. Spot tests. K-, C-, KC-, P-, UV-.

NICHE: Coastal Plain Cones is endemic to the southeastern United States. Like all good *Pyrenula* species, it has an affinity for coastal plain habitats, but like the best *Pyrenula* species, it also makes its way to the Smokies. Here, it is rare, known only from a few collections at low elevations. Look for it on the bark of hardwoods such as Ironwood and Sugar Maple.

KEY FEATURES: Thin, continuous, olive green to gray to inconspicuous UV- thallus, black perithecia, brown, small, 4-celled spores, non-inspersed hymenium.

Pyrenula leucostoma

White-Eyed Angler

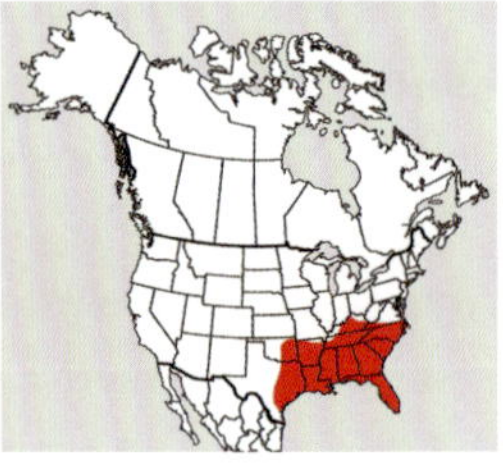

Lendemer 33182 (photo: Tripp)

NOTES: *Pyrenula leucostoma* is probably the easiest member of genus to learn in the Smokies on account of its brown, muriform spores inside of solitary perithecia (i.e., these scattered across the thallus). This species is most likely to be confused with *P. ravenelii*, which also has brown muriform spores but differs in having distinctly clustered perithecia. You have to cut into *Pyrenula* perithecia in order to identify them with confidence. That's just the way that it goes. As the specific epithet implies, this lichen sometimes has a "white mouth" surrounding the bases of the black perithecia.
CHEMISTRY: No substances. Spot tests. K-, C-, KC-, P-, UV-.
NICHE: *Pyrenula leucostoma* is a species that can be found commonly at low elevations in humid, streamside forests throughout the Smokies. It occurs in similar habitats throughout the southeastern United States. To see the species, take a hike on Little Bottoms Trail between Copper Road Trail and Hatcher Mountain Trail, but don't be late in returning to meet your group for lunch.

KEY FEATURES: Thin, continuous, olive-green thallus, black, solitary perithecia sometimes with faint white ring surrounding their bases, brown muriform spores, low elevations, on hardwoods, common in riparian forests.

Pyrenula pseudobufonia

Sort of Frog-Like

Tripp 4991 (photo: Lendemer)

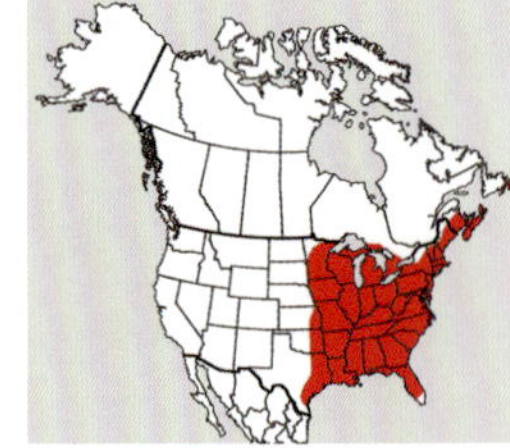

NOTES: What's not to love about the "sort of frog-like" *Pyrenula*? We aren't entirely sure what Rehm was thinking when we assigned the epithet to this species, but it sure makes this one memorable. *Pyrenula pseudobufonia* is readily distinguished from all other species in the Park by its thallus that is bright UV+ yellow owing to the presence of lichexanthone. Not that anything else matters, but it also has 4-celled spores nested within a hymenium that is inspersed with oil.

CHEMISTRY: Lichexanthone. Spot tests. K-, C-, KC-, P-, UV+ bright yellow.

NICHE: *Pyrenula pseudobufonia* is common in both low and middle elevations in the Smokies. Like all other members of the genus, it occurs on the bark of hardwoods. This one is particularly fond of American Beech. How's about a field trip to Madcap Branch Creek to see it? This species can be found commonly throughout the temperate forests of eastern North America.

KEY FEATURES: Thin, continuous, olive green thallus, UV+ bright yellow, black perithecia, 4-celled spores, inspersed hymenium, on bark throughout.

Pyrenula ravenelii

Scatter Shot

Lendemer 26873 (photo: Lendemer)

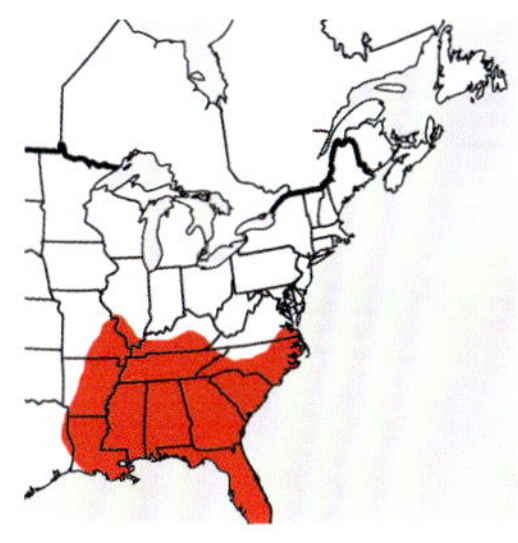

NOTES: *Pyrenula* is one of those genera where you just never know what you are going to get. The lichens can look externally different and be internally the same, or the same inside and very different outside. *Pyrenula ravenelii* is an example of one that looks different from the outside. Internally it is similar to *P. leucostoma* in having large, muriform, brown spores. But, while *P. leucostoma* has perithecia that tend to be solitary and with apical ostioles, those of *P. ravenelii* tend to be aggregated together and oriented horizontally on the substrate with the ostioles aiming to fire out the side rather than top. It seems sort of silly if you ask us. A mutation gone bad . . . perhaps one that should have been selected out, but one that wasn't.

CHEMISTRY: No substances. Spot tests. K-, C-, KC-, P-, UV-.

NICHE: Like so many other species, Scatter Shot is confined to the bark of hardwoods at low elevations in the Smokies. It is found in similar habitats elsewhere in the southern Appalachians, and not surprisingly is more common elsewhere further south in the Coastal Plain.

KEY FEATURES: Corticate, crustose thallus, perithecia with lateral ostioles and often joined together, muriform, brown spores, UV-, on bark at low elevations.

Pyrenula subelliptica

Fertile Foothills

Tripp 4986 (photo: Lendemer)

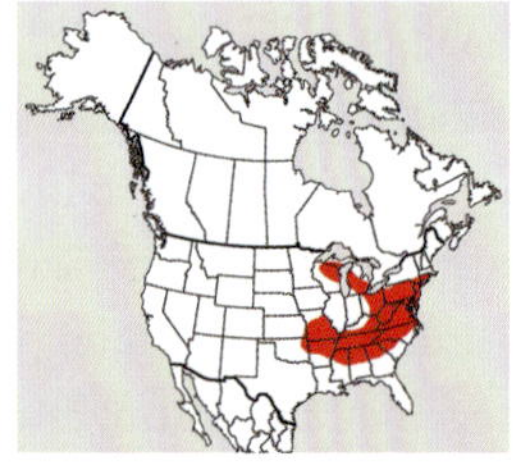

NOTES: *Pyrenula subelliptica* is identifiable owing to one particularly distinctive anatomical feature: spore lumina that are elongate compared to spores of other species. If you fail to check this feature, you might confuse this species for *P. punctella* or *P. macounii*, both of which have spores with angular lumina. Note the relatively small size of the perithecia in this species, too.

CHEMISTRY: No substances. Spot tests. K-, C-, KC-, P-, UV-.

NICHE: *Pyrenula subelliptica* is a common species at middle elevations throughout the Park but seems to prefer habitats that are slightly calcareous. Not surprisingly, this species is slightly more tolerant of higher elevations in the Park compared to other species: study the distribution maps and you will see that its primary center of distribution is farther north than most. In the Smokies and elsewhere, *P. subelliptica* can be found growing on a diversity of hardwoods, in particular Buckeye and Sugar Maple.

KEY FEATURES: Thin, continuous, olive-green thallus, small black perithecia, spores with elongate lumina, middle elevations, all habitats but in particular calcareous areas, on bark of hardwoods.

Pyrrhospora varians

Coat of Many Colors

Lendemer 32965 (photo: Tripp)

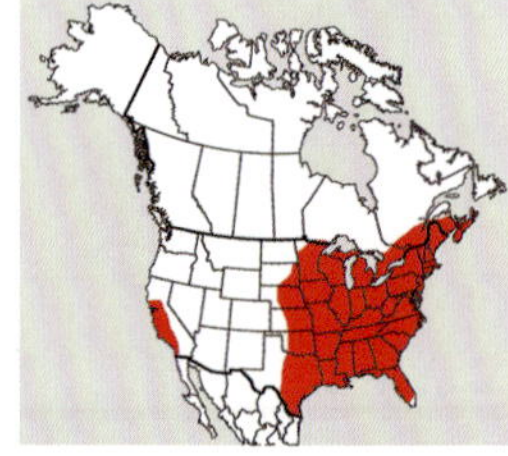

NOTES: Here is the moment you have all been waiting for: we are happy to announce this decade's award for weediest lichen east of the Mississippi River: *Pyrrhospora varians*! (Crowd goes wild). This lichen is characterized by its extreme variability (we hate to start with that, but it's true), especially with respect to the colors of its apothecial discs, which are sometimes red, sometimes yellow, sometimes brown, sometimes peach, and all shades in between, are often epruinose but sometimes pruinose, and are sometimes plane or sometimes convex. The thallus of *P. varians* is equally variable but revolves around irregular areoles atop a more or less continuous, dirty white or green or gray or yellow thallus. The spores of this species aren't particularly helpful during identification: they are simple and colorless. You are most likely to learn *Pyrrhospora varians* if you (a) have an open mind and (b) know where to look–its ecology is rather distinctive (see below). You might also try a C, KC, and UV test, which help to confirm its identity.

CHEMISTRY: Atranorin and xanthones. Spot tests. K+ yellow, C+ yellow, KC+ orange, P-, UV+ dull orange.

NICHE: Miss Everywhere grows on nearly every twig you will ever see or pick up off the ground, in nearly every habitat and especially in disturbed places, in the Smokies and beyond. It also grows on trunks, bases, and roots. On conifers? Check. On hardwoods? Check.

KEY FEATURES: Extremely variable crustose lichen with a +/- continuous, dirty white, gray, or green thallus, ample apothecia with variable colored discs, C+ yellow, KC+ orange, UV+ orange, on nearly every hardwood twig in the Smokies.

Pyxine sorediata

Hoar Frost

Tripp 3568 (photo: Lendemer)

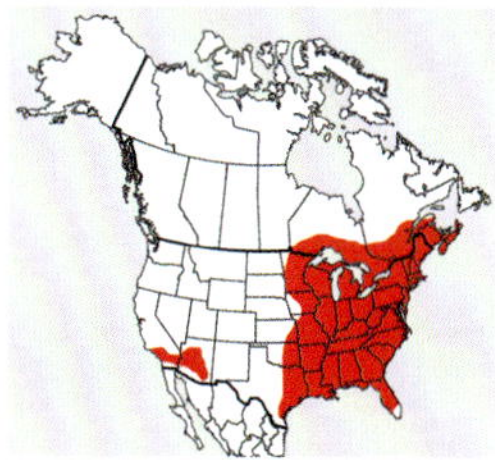

NOTES: This was the first species of lichen that many of us learned in temperate, broadleaf forests of eastern North America. For that reason alone, it will always be special. But it is special for so many other reasons. *Pyxine sorediata* is quite easy to identify by small, gray, foliose thalli that are very tightly appressed to their (bark) substrates, have abundant, discrete patches of soredia lining the margins of the lobes, and have dense pruina covering the tips (and only the tips!) of the lobes. If you look a little closer, you will find a medulla that is mustard-yellow in color. The only lichen in the Smokies with which this could be confused is *P. subcinerea*, but that species has a cortex that is bright UV+ yellow owing to the presence of lichexanthone.

CHEMISTRY: Atranorin, yellow pigments, unidentified terpenes. Spot tests. Cortex: K-, C-, KC-, P-, UV-; medulla: K-, C-, KC-, P-, UV-.

NICHE: *Pyxine sorediata* is widespread throughout low and middle elevations in the Smokies, where it occurs on the bark of a broad diversity of hardwood trees. This species is similarly common across eastern temperate forests.

KEY FEATURES: Small, gray, sorediate thalli, dense pruina covering lobe tips, closely appressed to substrate, very distinctive once learned, widespread throughout the Park, especially at middle and low elevations, on bark of hardwoods.

Pyxine subcinerea

Crop Dusters

Lendemer 25373 (photo: Lendemer)

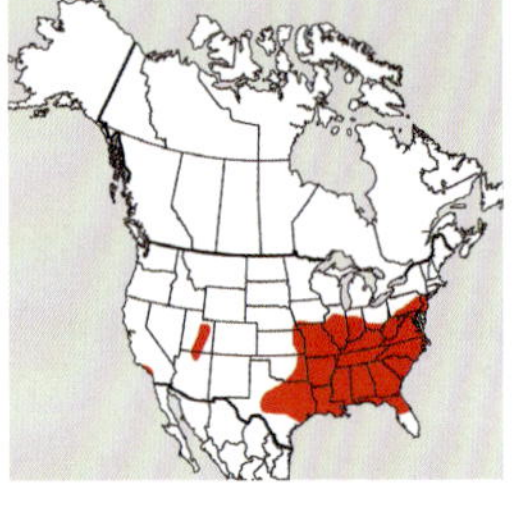

NOTES: *Pyxine subcinerea* is characterized by its flattened, gray lobes that are always closely appressed to the substrate and pruinose toward the tips. The soredia of this species are marginal and, if you look closely, often reveal a yellowish tint to the medulla. The most similar species is *P. sorediata*, which is much more common and tends to have larger, more robust thalli. Although both species have an intensely yellow pigmented medulla, *P. subcinerea* is "knock your socks off" UV+ loud yellow, in contrast to *P. sorediata*, which got the short end of the UV stick and doesn't fluoresce at all. One could confuse *P. subcinerea* with sorediate species of *Physcia* such as *P. americana*, but members of that genus are UV- and lack pigment in the medulla.

CHEMISTRY: Lichexanthone, terpenes and an unidentified pigment. Spot tests: (cortex) K-, C-, KC, P-, UV+ bright yellow; (medulla) K-, C-, KC-, P-, UV-.

NICHE: Are you questioning whether chemistry alone should be used to distinguish different species, that is, whether separating *P. subcinerea* and *P. sorediata* based on lichexanthone is really the wisest of decisions? We would be willing to entertain that conversation, but the problem is: these two entities are clearly differentiated ecologically, too. Crop Dusters has very high affinity to the southeastern United States, where it grows on hardwoods at low elevations. In contrast, *P. sorediata* occurs throughout eastern North America but is as happy at far more northerly latitudes (and at higher elevations) as it is in the south (and at lower elevations).

KEY FEATURES: Gray foliose lichen, flattened, closely appressed lobes, pruinose lobe tips, marginal soralia, yellow medulla, UV+ bright yellow, uncommon at low elevations on hardwoods.

Ramalina americana

Made In America

Lendemer 33025 (photo: Tripp)

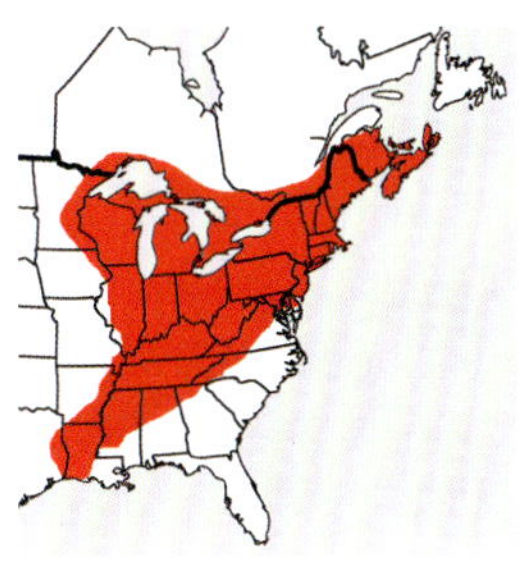

NOTES: *Ramalina americana* is a very common macrolichen in the Smokies characterized by its fruticose thallus with flattened branches (vs. round in cross section, as in *Alectoria*), its branches that are often ridged and marked by white pseudocyphellae, and its lack of soredia. Instead, thalli of *R. americana* are frequently adorned by copious fruiting bodies situated near the tips of the lobes. This species is likely to only be confused for *R. culbersoniorum*, but the latter differs chemically. Conducting TLC to identify secondary metabolites is a *must* in *Ramalina* taxonomy.

CHEMISTRY: Usnic acid. Spot tests. Cortex: K-, C-, KC+ yellow, P-, UV-; medulla: K-, C-, KC-, P-, UV.

NICHE: *Ramalina americana* is one of the most common fruticose lichens in the Smokies. It can be found at high elevations, and grows primarily on branches and on the bark of hardwoods. It is similarly widespread throughout eastern North America and into northwestern portions of North America.

KEY FEATURES: Yellowish-green fruticose thallus, no soredia, usually with abundant apothecia near tips of branches, chemistry, extremely common on branches throughout Park at high elevations.

Ramalina culbersoniorum

Bill's Branches

Lendemer 33049 (photo: Tripp)

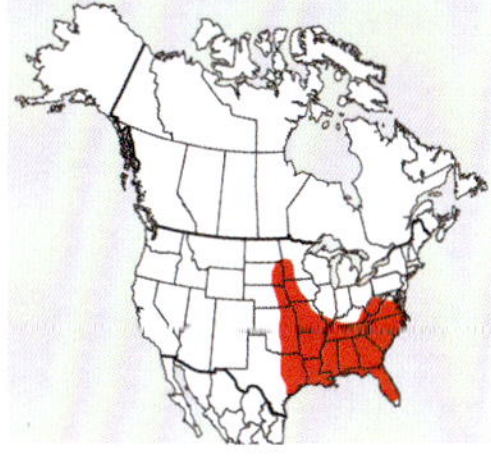

NOTES: *Ramalina culbersoniorum* looks essentially identical to *R. americana*. It is differentiated almost entirely by chemistry. The "additional substances" noted below refer to a combination of divaricatic, norbarbatic, 4-*O*-demethylhypoprotocetraric, stenosporic, and/or glomelliferic acids. Sorry you asked? Actually, questions are free. TLC is not. Sorry you collected it? We've been there . . .

CHEMISTRY: Usnic acid with various combinations of additional substances in the medulla. Spot tests. Cortex: K-, C-, KC+ yellow, P-, UV-; medulla: spot tests variable.

NICHE: Like *Ramalina americana*, *R. culbersoniorum* is an extremely common fruticose lichen in the Smokies and elsewhere in eastern North America, particularly the Southeast into the Mississippi River Valley. In the Park, it can be found at all elevations, particularly low and middle elevations, and grows primarily on branches and on the bark of hardwoods.

KEY FEATURES: Yellowish-green fruticose thallus, no soredia, usually with abundant apothecia near tips of branches, chemistry, extremely common on branches throughout Park.

Ramalina petrina

Rock Squats

Tripp 2104 (photo: Lendemer)

NOTES: As a genus, *Ramalina* is very easy to identify because it is a group of yellowish-green fruticose macrolichens that could only be confused with *Usnea* (but these have central cords) or *Alectoria* (but these are round in cross section). Within *Ramalina*, identification of species depends on whether they are asexual or sexual and secondarily on chemistry. *Ramalina petrina* is the only sorediate species in the Smokies that produces protocetraric acid. Give it a P test and look for the red reaction!

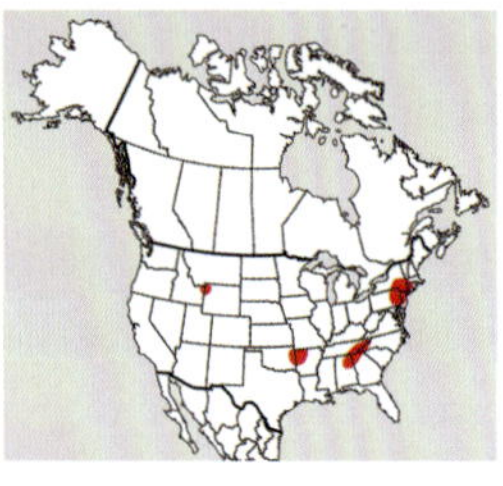

CHEMISTRY: Usnic acid and protocetraric acid. Spot tests. Cortex: K-, C-, KC+ yellow, P-, UV-; medulla: K+ yellow, C-, KC-, P+ red, UV-.

NICHE: This species occurs primarily at middle elevations in the Park. Most of its known populations are in southern portions of the Smokies, between Straight Fork Road and Purchase Knob, where it can be found growing on rock. It has disjunct populations elsewhere in North America including the Black Hills, the Ozarks, and the mid- to northern Atlantic Coastal Plain.

KEY FEATURES: Yellowish-green fruticose macrolichen that is not round in cross section and lacks a central cord, with soredia, P+ red, on rock.

Ramalina labiosorediata

Mealy Lips

Tripp 2107 (photo: Lendemer)

NOTES: *Ramalina labiosorediata* is characterized primarily by its sorediate thallus, in which the soredia occur in discrete, expanded soralia that are often terminal on the undersides of the branch tips. It is most likely to be confused with *R. intermedia*, a species that is also sorediate but lacks discrete soralia and also has soredia scattered across the thallus branches rather than pushed toward the tips. *Ramalina petrina* is also sorediate but has a P+ red medulla.

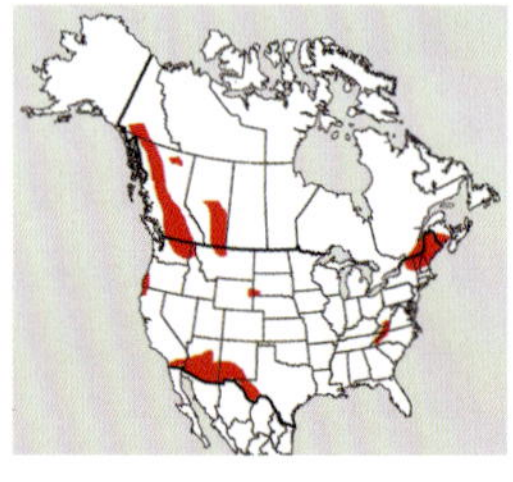

CHEMISTRY: Usnic acid and evernic acid. Spot tests. Cortex: K-, C-, KC+ yellow, P-, UV-; medulla: K-, C-, KC-, P-, UV-.
NICHE: *Ramalina labiosorediata* is an uncommon to rare species in the Smokies. It is known from fewer than five collections, which is not likely attributable to overlooking this species given it is a relatively large and distinctive macrolichen. Data indicate that this species occurs on a wide diversity at hardwoods, primarily at middle elevations. Look for it in rich cove forests of Hickory, Buckeye, Maple, Black Cherry, and Sweet Birch, but also in slightly more acidic forests of Black Gum, Chestnut Oak, Sourwood, and *Rhododendron*.
KEY FEATURES: Yellowish-green fruticose thallus, soredia in discrete soralia pushed toward branch tips, on bark of hardwoods, middle elevations, rare.

Ramboldia russula

Red Heads

Lendemer 29537 (photo: Tripp)

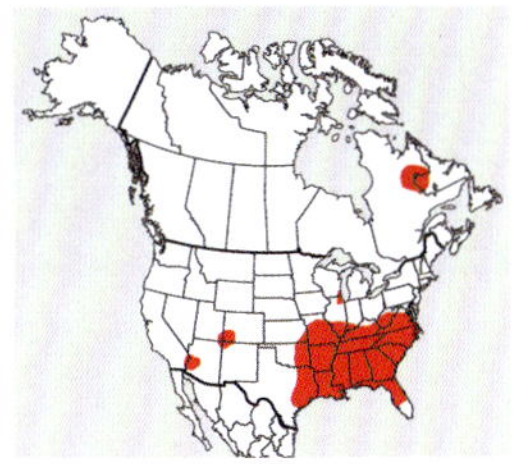

NOTES: This photograph is proof that, no matter how hard you try, you will eventually get caught in a rainstorm in the Smokies. Even though sopping wet, we hope you will have no trouble appreciating that *Ramboldia russula* is one of the most morphologically distinctive crusts in the National Park. Despite its showy (and sometimes, downright gaudy) appearance, Red Heads was surprisingly not reported from the Park boundaries until our 2013 book. We aren't sure what to make of that other than a lack of sufficient lichen inventory work prior to recent years, despite the Smokies being our nation's most popular (that is, most visited) National Park. In any case, *R. russula* is characterized by its crustose, continuous, bluish-gray thallus that bears bright red (to bright orange when wet) apothecia. If you cut into these fruiting bodies that have lecideine margins, you will find simple, colorless spores. This species is impossible to confuse for anything else.
CHEMISTRY: Lichexanthone and fumarprotocetraric acid. Spot tests. Thallus: K+ yellow, C-, KC-, P+ orange, UV+ bright orange; apothecia: K+ purple, C-, KC, P-, UV-.
NICHE: This species occurs infrequently in the Smokies on the bark of hardwoods (especially twigs and branches) at low elevations. Its center of distribution in the Park seems to be the Lakeshore/Tunnel Bypass Trail area, similar to so many other species that only occur in the Park's lowest elevations on south-facing slopes. One cannot help but wonder just how many more species would be known from the region had the Fontana Dam not been built.
KEY FEATURES: Bright-red-to-orangish-red apothecia against a crustose thallus, usually forming small patches on twigs or branches fallen from the canopy, always on hardwoods, always at low elevations, uncommon but spectacular if found.

Ramonia microspora

Bark Stars

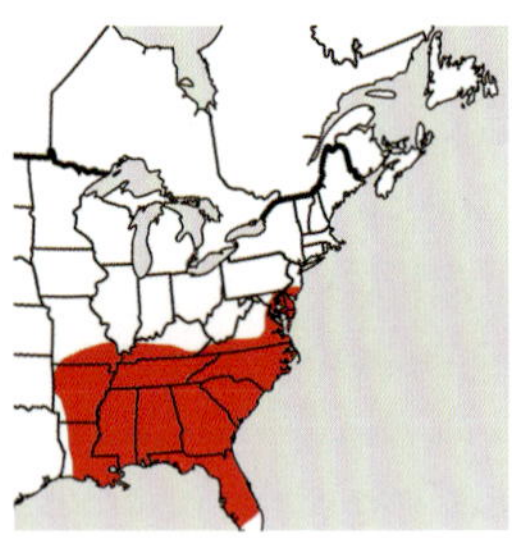

Lendemer 29493 (photo: Lender)

NOTES: There is something absolutely captivating about this tiny crust. It is a species that you wouldn't likely find unless you went looking for it. In fact we had to mount a search party to locate it and other low elevation species along the Foothills Parkway. Even though it is small, it can easily be recognized by its pale, crater-like apothecia, which look as though they have just burst through the surface of the bark. The apothecia, coupled with the presence of a *Trentepohlia* photobiont and asci containing hundreds of small, colorless, simple spores further distinguish it. The photobiont should give you the impression that it is more at home in the subtropical climates closer to the coast.

CHEMISTRY: No substances. Spot tests. K-, C-, KC-, P-, UV-.

NICHE: A species of hardwood bark throughout the southeastern United States, particularly in swamps or other humid habitats. In the Smokies, *Ramonia microspora* appears to be rare, found only at low elevations, such as along the Little Tennessee River near the Foothills Highway.

KEY FEATURES: Immersed crustose thallus, immersed, crater-like apothecia, many tiny, simple spores per ascus, on hardwood bark at the lowest elevations.

Rhizocarpon cinereovirens

Tax Man

Tripp 2420 (photo: Deregibus)

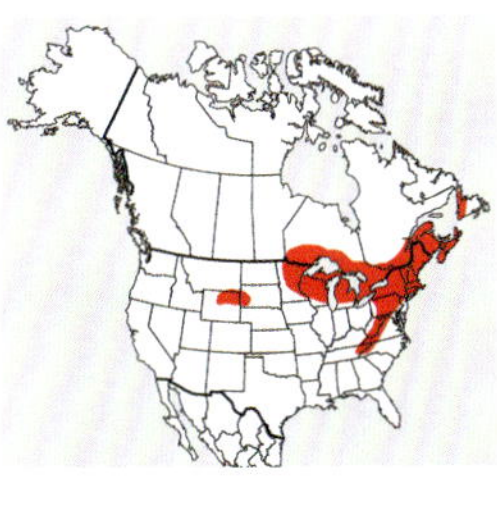

NOTES: *Rhizocarpon cinereovirens* is a poorly known species characterized by its gray-brown, areolate thallus (often with a black prothallus), black apothecia, its colorless, 2-celled spores, and chemistry. It is most likely to be confused with *R. infernulum f. sylvaticum*, but the latter is K- and occurs on deeply shaded rocks in contrast to this species (see Niche). Don't confuse this species for the far more common *Fuscidea arcuatula*, which has spores shaped like kidney beans.

CHEMISTRY: Norstictic acid or stictic acid. Spot tests. K+ yellow or K+ yellow turning red, C-, KC-, P+ yellow or orange, UV-.

NICHE: Tax Man is a very rare species known only from a few collections in the Smokies. It occurs at middle-to-high elevations on sun-exposed, non-calcareous rocks. Let us know if you find additional populations of this species!

KEY FEATURES: Brown, areolate crustose thallus, black prothallus, black apothecia, colorless, 2-celled spores, K+ yellow or yellow turning red, on sunny rocks, middle-to-high elevations, very rare.

Rhizocarpon geographicum

Peakbagger

Tripp 5097 (photo: Lendemer)

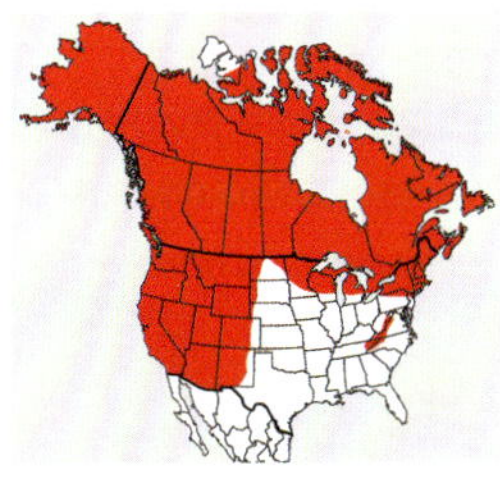

NOTES: We *love* to tell the story of this lichen. *Rhizocarpon geographicum* is not an eastern species. It definitely calls western North America (and into the Canadian arctic) home. But it makes its way to the Smokies (and very few other southern Appalachian localities) only via the highest peaks. This species is unmistakable for any other lichen in our area. It is characterized by its *bright* yellow, areolate crustose thallus with black apothecia interspersed among the areoles. If you section the fruiting bodies, you will find brown muriform spores. No other species comes remotely close to this character combination, except for perhaps *Rinodina chrysomelaena*, which has 2-celled spores and is one of the rarest species in North America, and *Buellia vernicoma*, which has 4-celled transversely septate spores and isn't nearly as flamboyant as *R. geographicum*.

CHEMISTRY: Rhizocarpic acid and psoromic acid. Spot tests. Cortex: K-, C-, KC-, P-, UV+ orange; medulla: K-, C-, KC-, P+ yellow, UV-.

NICHE: *Rhizocarpon geographicum* is a cold-loving arctic-boreal species that makes its way into the Smokies via fewer than five of the Park's high summits, including Mt. LeConte. It grows exclusively on rock, specifically exposed outcrops of the Anakeesta formation. Elsewhere across North America, this species is abundant on granitic rocks both of the Appalachian and Rocky Mountains. It is one of the most abundant saxicolous species of montane landscapes of the West.

KEY FEATURES: Bright yellow areolate thallus, black apothecia, brown muriform spores, exclusively on high-elevation Anakeesta Rock, very rare (if you find this species, please let it be).

Rhizocarpon grande

Peanut Butter Goes Hiking

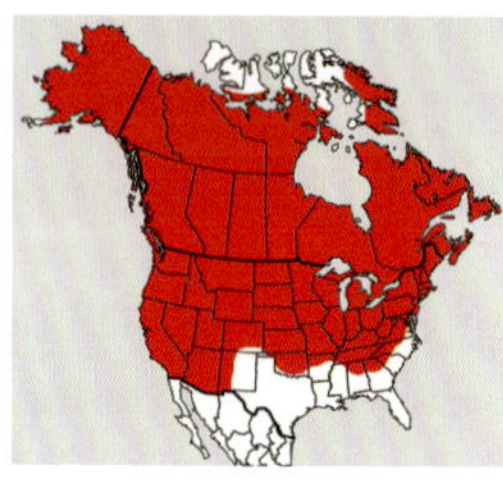

Tripp 3749 (photo: Lendemer)

NOTES: *Rhizocarpon grande* is an uncommon but rather striking species that you will be both lucky and happy if you find. Like *R. geographicum*, it is absolutely unmistakable by a unique character combination, in this case: a chocolate brown, distinctly areolate thallus (usually with a black prothallus), muriform spores, and a C+ pink and K+ yellow thallus reaction owing to its chemistry.

CHEMISTRY: Gyrophoric acid and stictic acid. Spot tests. K+ yellow, C+ pink, KC+ pink, P+ orange, UV-.

NICHE: This species is *not* common in the Smokies—we have seen it on fewer than seven occasions. If found, you will most likely be in a middle-to-high-elevation acid cove or ridgetop forest (think: Black Gum, Scarlet Oak, Southern Red Oak, White Pine, Mountain Laurel). Look for this species on sheltered rock faces.

KEY FEATURES: Chocolate brown areolate thallus, black prothallus, brown, muriform spores, K+ yellow, C+ pink.

Rhizocarpon infernulum f. *sylvaticum*

Forest Form

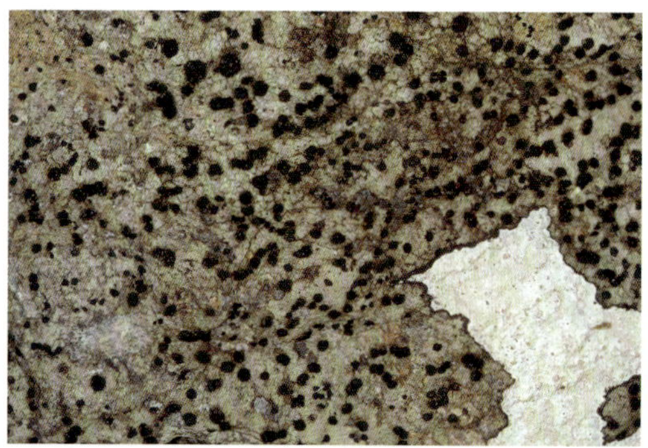

Tripp 2523 (photo: Tripp)

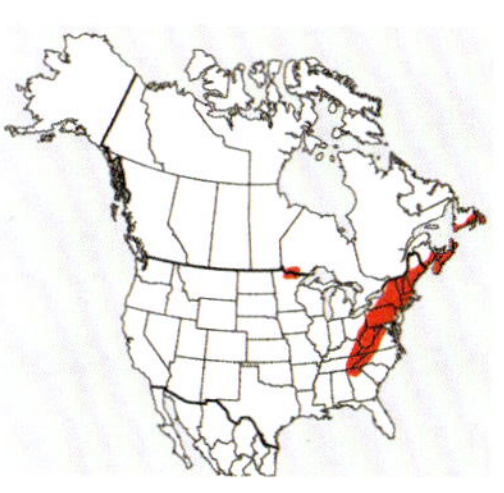

NOTES: *Rhizocarpon infernulum.* The forest form. There isn't a lot here to go on, and gestalt really helps. But before you have this one down by sight recognition, try these features: crustose lichen on rocks with a continuous, grayish-brown-to-sordid-white thallus with black apothecia, relatively distinctive ecology (see Niche), and colorless, 2-celled spores. Note this last clue: whereas many species of *Rhizocarpon* have brown spores, some—including several in the Smokies—instead have colorless spores, as in this species. It is perhaps most likely to be confused with *Fuscidea arcuatula*, which grows in similar environments, but that species has kidney-bean-shaped spores. It might also be confused for *Rhizocarpon cinereovirens*, which also has colorless spores and occurs on rock, but that species has a K+ yellow or K+ red thallus.

CHEMISTRY: No substances. Spot tests. K-, C-, KC-, P-, UV-.

NICHE: *Rhizocarpon infernulum* f. *sylvaticum* is a relatively common species that occurs exclusively on rocks in middle-to-high elevation habitats, always in deep shade. Look for this species particularly in upland forest with Chestnut Oak, Tulip Poplar, and Eastern Hemlock (at least what is left of them).

KEY FEATURES: Dirty gray-brown-white continuous, crustose thallus, colorless, 2-celled spores, middle-to-high elevations, non-calcareous, always shaded rocks, relatively common.

Rhizocarpon oederi

Rusty Riches

Tripp 3511 (photo: Lendemer)

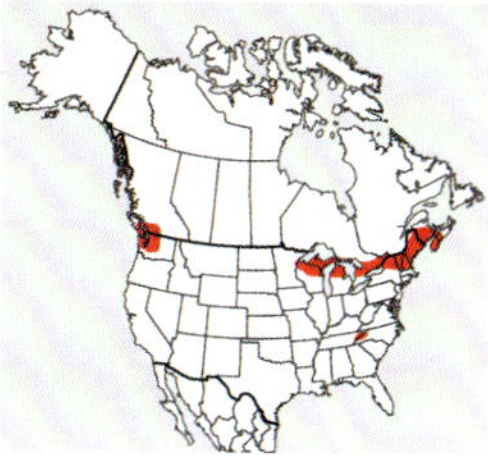

NOTES: The rusty-orange-to-reddish-brown thallus of *Rhizocarpon oederi* is quite distinctive, and when combined with the black apothecia and colorless septate spores, makes the species difficult to confuse with any others. Many crustose lichens that grow on metal-rich rocks also have rusty colored thalli; however, no other such species are known from the Smokies except for *Acarospora sinopica*, which is very different in appearance. The other crustose lichens that grow on Anakeesta rock seem to be able to deal with its unusual chemistry.

CHEMISTRY: No substances. Spot tests. K-, C-, KC-, P-, UV-.

NICHE: Rusty Riches is restricted to non-calcareous rocks that are rich in metal, particularly iron or copper. In North America, it occurs in a relatively narrow belt of humid boreal forests from the Atlantic Coast to the Great Lakes, and then is disjunct in the Pacific Northwest. The populations in the southern Appalachians are disjunct from the boreal region, and all occur in the spruce-fir zone of the Smokies, especially on Anakeesta Rocks.

KEY FEATURES: Rusty-orange-brown crustose thallus, black lecideine apothecia, colorless 4-celled to submuriform spores, on metal-rich rocks at high elevations.

Rhizocarpon reductum

The Abyss

Lendemer 29631 (photo: Tripp)

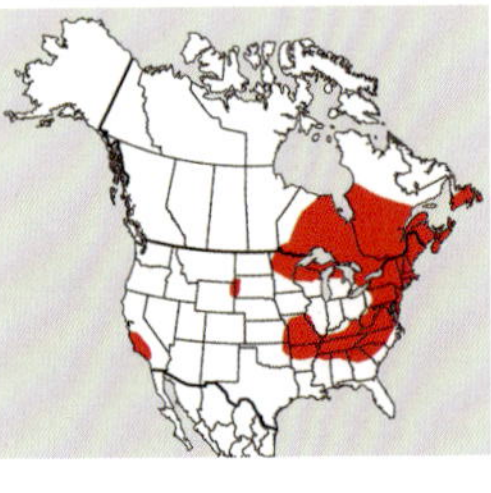

NOTES: *Rhizocarpon reductum* is a very uncommon species that probably won't be the first you learn in the Smokies. It is similar to *R. rubescens* in being crustose, growing on rock, and having dark-brown-to-black apothecia that contain colorless, muriform spores, but differs by its brownish-green thallus that is either K- or K+ yellow (vs. whitish and K+ yellow turning red). Note in the photograph how the apothecia of this species tend to "take a bit of thallus with them" (thus appearing a bit ragged) as they emerge and become upward bound.

CHEMISTRY: Stictic acid or no substances. Spot tests. K+ yellow or K-, C-, KC-, P+ orange or P-, UV.

NICHE: This is a rare species in the Smokies where it is known from fewer than five, rather disparate localities ranging from Hannah Mountain Trail to the Mt. LeConte area. It has been collected at middle and high elevation forests ranging from spruce-fir to oak-pine habitats. It occurs primarily on Anakeesta Rock. The primary center of distribution of this species is in colder temperate areas of eastern North America.

KEY FEATURES: Brownish-green thallus, K- or K+ yellow, black apothecia, colorless, muriform spores, on rock, rare.

Rhizocarpon rubescens

Red Stud

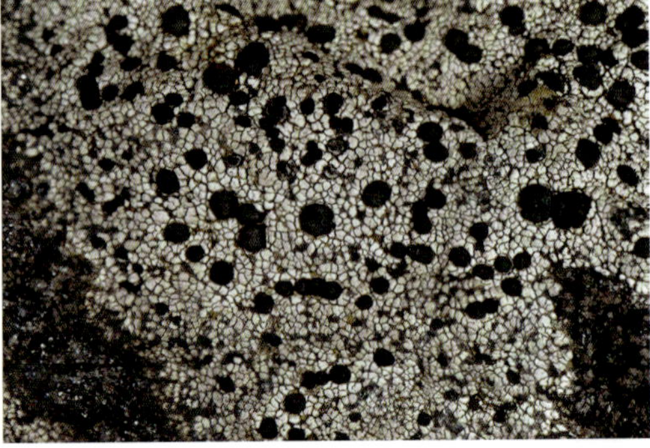

Tripp 3561 (photo: Lendemer)

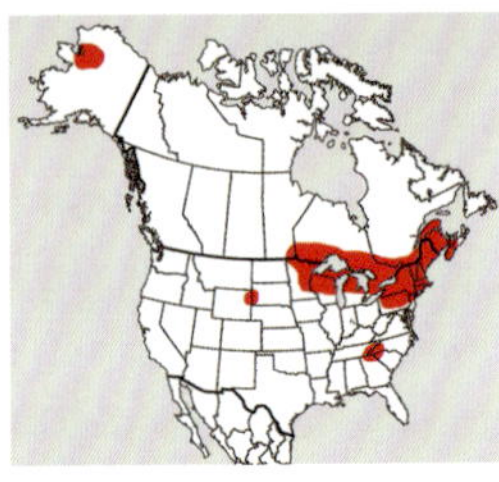

NOTES: *Rhizocarpon rubescens* is yet another rock-dwelling member of the genus characters by an areolate, crustose thallus that bears ample black apothecia. Really, almost all of them have this in common. This species, unlike its relative *R. reductum*, has a distinctly white thallus that is K+ yellow turning red owing to the presence of norstictic acid. With *R. reductum*, *R. rubescens* shares colorless, muriform spores.

CHEMISTRY: Norstictic acid. Spot tests. K+ yellow turning red, C-, KC-, P+ yellow, UV-.

NICHE: *Rhizocarpon rubescens* is uncommon in the Smokies, where it occurs in middle-to-high elevation habitats, primarily spruce-fir forests. It grows exclusively on rocks and is particularly fond of the Anakeesta formation. Look for this species on top of Clingman's Dome or Chimney Tops, along with a large assortment of other amazing, very special lichens that love Anakeesta. Populations in the Smokies are disjunct from the primary center of distribution of this species, which is the northern Appalachians and northern Great Lakes region.

KEY FEATURES: Crustose, areolate white thallus, K+ yellow turning red, black apothecia, colorless, muriform spores, on Anakeesta rock at middle-to-high elevations, uncommon.

Rhizoplaca subdiscrepans

Thalline Valentine

Tripp 3710 (photo: Lendemer)

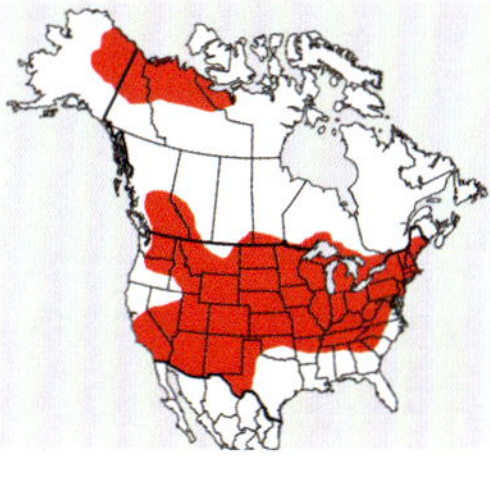

NOTES: *Rhizoplaca subdiscrepans* is a rare but readily identifiable species by its very thick, yellowish-green bullate areoles, its large apothecial discs (these often pruinose) with lecanorine margins, and its simple, colorless spores. It doesn't look like anything else morphologically (out West, one might confuse *Rhizoplaca* with placodioid species of *Lecanora*, but our Smokies lichen biota largely lacks the latter growth forms). This is a species to learn by gestalt, for sure.

CHEMISTRY: Usnic acid and pseudoplacodiolic acid. Spot tests. Cortex: K-, C-, KC+ yellow, P-, UV-; medulla: K-, C-, KC-, P-, UV-.

NICHE: *Rhizoplaca subdiscrepans* is relatively common on exposed, non-calcareous rocks in the central and northern Appalachians through western North America, but is quite rare in the Smokies. Please let us know if you find a population! As a whole, the genus *Rhizoplaca* is really a western phenomenon. Head to the Rockies to practice sight recognition of this lineage.

KEY FEATURES: Thick, yellow-green bullate thallus, on non-calcareous rocks, middle elevations, very rare in Park.

Rimularia badioatra

Chocolate Tiles

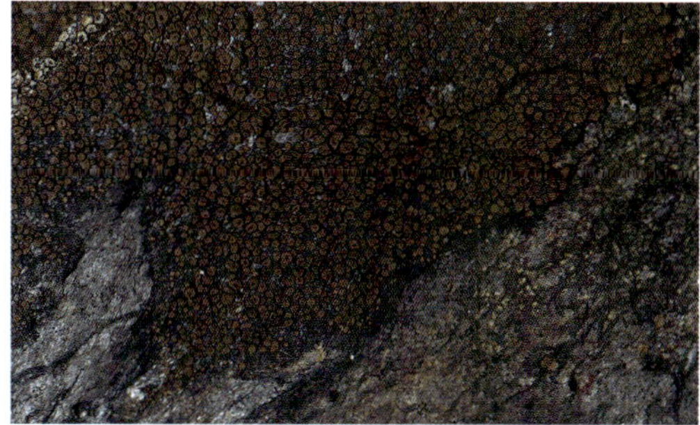

Tripp 3747 (photo: Lendemer)

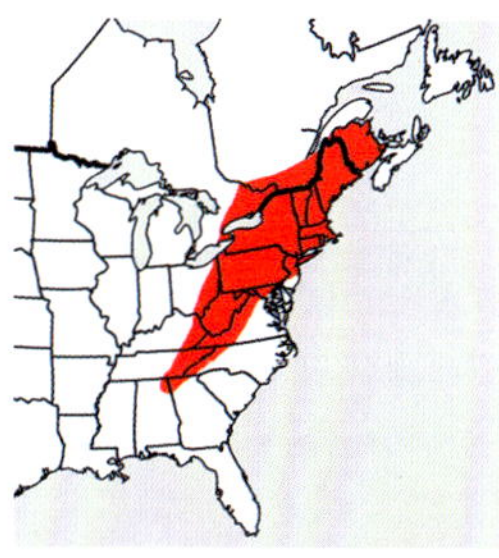

NOTES: *Rimularia badioatra* is likely to be overlooked by many because the apothecia immersed in the areoles are often small and inconspicuous, thus causing one to mistake the species for a poorly developed *Rhizocarpon*. Indeed, *R. grande* has a brown areolate thallus that is C+ pink and frequently pruinose. However, *R. grande* also produces stictic acid and is P+ orange. It also has muriform brown spores rather than colorless simple spores. If the apothecia and spores of *R. badioatra* are found, then confusion with species of *Trapelia* is possible. However, none of the species of that genus that occurs in the Smokies have thalli that are composed of such flat, regularly spaced areoles and are as dark brown in color.

CHEMISTRY: Gyrophoric acid and zeorin. Spot tests. K-, C+ pink, KC+ pink, P-, UV-.

NICHE: This species is widespread in the Appalachian Mountains, where it occurs on non-calcareous rocks, often in shade or on sheltered, shaded outcrop faces. It is rare in the Smokies and known from only a small number of middle-to-high elevation rock outcrops. You can view it from the chair left at High Rocks in Swain County.

KEY FEATURES: Brown areolate thallus that is often lightly pruinose, minute brown immersed apothecia, colorless simple spores, C+ pink, on non-calcareous rocks at middle-to-high elevations.

Rinodina adironadackii

Big-Spored Pepperpot

Tripp 5313 (photo: Lendemer)

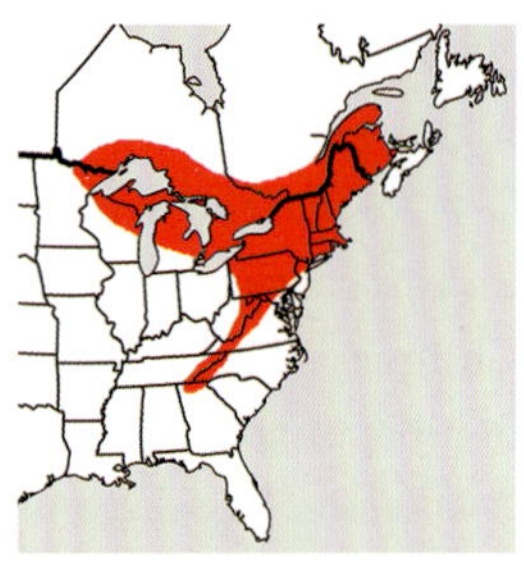

NOTES: *Rinodina* species are referred to as the pepperpot lichens because the brown spores are often ejected onto the surface of the apothecial discs and thus resemble pepper cracked over a pot of soup. When preparing sections of the apothecia for study, more of the spores are almost always present, and characters of the spores are often integral to identification. This species can be distinguished from others in the Smokies by its thin, continuous, P+ orange-red thallus and by its large, brown, 2-celled spores with thickened walls. *Rinodina subminuta* is superficially similar, but differs in having a P- thallus, smaller spores, and apothecia are erupt out of the thallus. Another species with large spores is *R. ascociscana*, but that species also differs in having a P- thallus.

CHEMISTRY: Pannarin. Spot tests. K-, C-, KC-, P+ orange-red, UV-.

NICHE: This species is endemic to eastern North America, where it is distributed primarily in the Appalachian Mountains and Great Lakes Region. It is rare throughout its range, including in the Smokies, where it has been found on bark at the bases of hardwood trees, primarily at lower elevations.

KEY FEATURES: Thin, blue-gray crustose thallus, esorediate, P+ orange-red, large brown 2-celled spores with thickened walls, on bark of hardwoods at all elevations.

Rinodina ascociscana

Mountain Pepperpot

Lendemer 44675 (photo: Tripp)

NOTES: This species was originally described by Edward Tuckerman, the father of American lichenology, based on collections he made in New England during the early 1800s. It can be distinguished from all other pepperpot lichens in the Smokies by its thick, glossy, brown-gray thallus that is P- and large 2-celled brown spores. The apothecia of *R. ascociscana* also often have distinct white striations on margins, which causes them to appear ashen or gray from a distance. *Rinodina dolichospora* is a rarer species that is somewhat similar, but has smaller spores. Similarly, *R. adirondackii* also has large spores, but the thallus in that species if thin, lighter gray, and P+ orange-red.

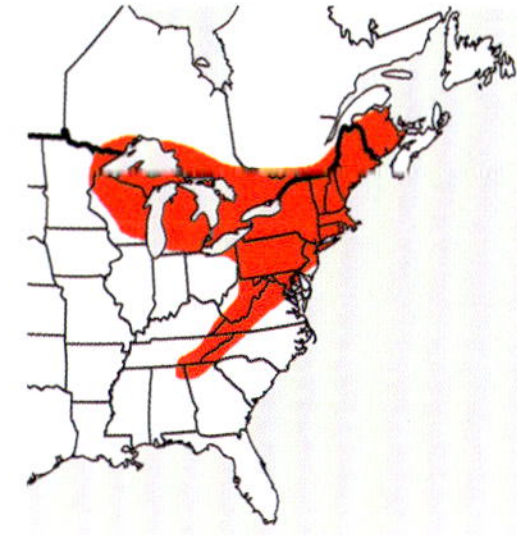

CHEMISTRY: No substances. Spot tests. K-, C-, KC-, P-, UV-.

NICHE: Moutain Pepperpot has a classic Appalachian-Great Lakes distribution and is another eastern North American endemic. It uncommon but widely distributed in the central and southern Appalachians, including the Smokies, and is typically found on the bark of mature hardwood trees. Despite being primarily distributed in northern temperate regions, the species is not restricted to high elevations in the Smokies, and instead can be found throughout the Park.

KEY FEATURES: Thick, brown-gray crustose thallus, esorediate, P-, large brown 2-celled spores with thickened walls, on bark of hardwoods at all elevations.

Rinodina brodoana

Brodo's Pepperpot

Tripp 5312 (photo: Lendemer)

NOTES: Brodo's pepperpot was named in honor of Irwin Brodo, a talented lichenologist who wrote *Lichens of North America*, a book that inspired both of us to produce the present field guide. This species is one of many that we discovered for the first

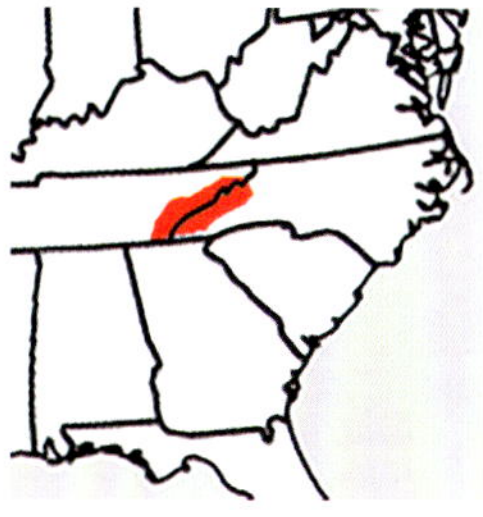

time as part of our work in the Smokies, and it is quite rare. Even in the field, *R. brodoana* can be recognized by its blue-gray areolate sorediate thallus, which most closely resembles *Trapeliopsis granulosa*. However *R. brodoana* is C- while *T. granulosa* is C+ pink, and *R. brodoana* grows on the bark at the bases of trees whereas *T. granulosa* grows on rotting wood and organic matter, often at high elevations in the Smokies. As strange as it may sound, there is no other crustose sorediate lichen in the Smokies that closely resembles *R. brodoana*.

CHEMISTRY: Atranorin. Spot tests. K+ yellow, C-, KC-, P-, UV-.

NICHE: Brodo's Pepperpot is narrowly endemic to the southern Appalachian Mountains, where it is known from a small number of middle elevation locations and grows on the bases of mature oak trees. In the Smokies it is rare, but you may be able to catch a glimpse of it on a hike up Jenkins Ridge Trail toward Thunderhead Mountain.

KEY FEATURES: Light blue-gray areolate crustose thallus, marginal soredia, K+ yellow and P-, on the bases of mature oaks at middle elevations.

Rinodina buckii

Buck's Pepperpot

Tripp 4944 (photo: Lendemer)

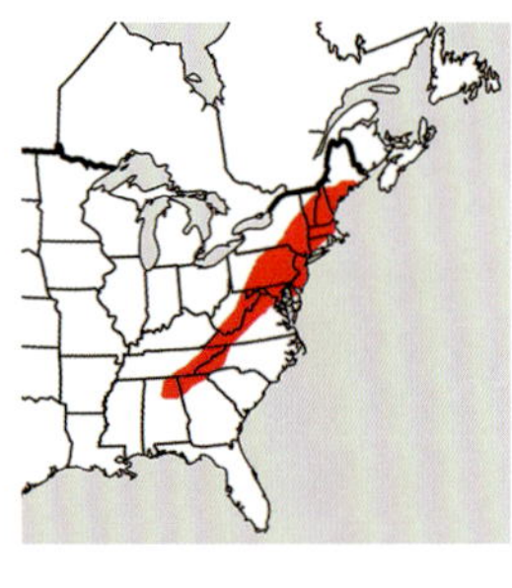

NOTES: Buck's pepperpot was named in honor of William R. Buck, a bryologist with a keen eye for small and unusual lichens. This species was long confused with *Rinodina willeyi* because both species have sorediate areolate thalli that react P+ orange-red. *Rinodina buckii* differs from *R. willeyi*, however, in having a thallus composed of bumpy, convex areoles that erupt into soralia from the centers, rather than a thallus composed of thin, tile-like areolae that form soralia along the margins. Thalli of *R. buckii* also often have a distinct purple-gray or purple-brown hue, that sets it apart from other sorediate members of the genus. *Rinodina efflorescens* is a northern species that is rare in the Smokies and also sorediate and P+ orange-red. It can be distinguished from *R. buckii* by the presence of a dull yellow pigment in the soralia, among other, less easily observed, characters.

CHEMISTRY: Pannarin and zeorin. Spot tests. K-, C-, KC-, P+ orange-red, UV-.

NICHE: Buck's Pepperpot has a classic disjunct distribution between the Appalachian Mountains in North America and the mountains of eastern Asia. Although it occurs throughout the Appalachians from Maine south to Alabama, it is only common and abundant in the Smokies where it is found throughout the Park on the bark of hardwoods and conifers.

KEY FEATURES: Light-blue-gray-to-brown-gray areolate crustose thallus, laminal soredia, P+ orange-red, on hardwoods and rarely conifers, at all elevations.

Rinodina bullata

Bubblin' Pepperpot

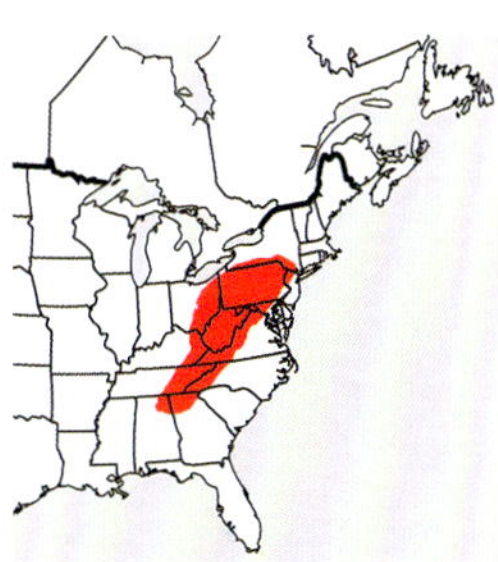

Lendemer 48559 (photo: Tripp)

NOTES: Most of the propagules that lichens make are easy to understand: by this point of the book you have already encountered isidia and soredia and lobules numerous times. You might have noticed the term "blastidia" back in *Baeomyces rufus*, but after so many species, this might not stick out any more. Well, *Rinodina bullata* is one of two pepperpot lichens in the Smokies that makes distinctive blastidia, the term for small rounded propagules that bud directly from the surface of the thallus. They look like an areole giving rise to another, smaller areole. Thankfully, the thallus is not the only character you have to go on for identification, as the species produces atranorin. It is also often fertile, and so one can usually find the characteristic 2-celled brown spores with thickened walls to confirm that it is a *Rinodina*. The only similar species *R. excrescens*, which differs in having a P+ orange-red thallus due to the production of pannarin.

CHEMISTRY: Atranorin. Spot tests. K+ yellow, C-, KC-, P-, UV-.

NICHE: This species is another eastern North American endemic that is known only from the Appalachian Mountains. It is most common in the central Appalachians of Pennsylvania but occurs infrequently in high-elevation hardwood forests farther south into the Smokies. In the Smokies, it is known from a small area along the Cataloochee Divide near Purchase Knob and the S.W.A.G.

KEY FEATURES: Light-blue-gray-to-brown gray areolate crustose thallus, esorediate, minute blastidia budding from convex areoles, K+ yellow and P-, on hardwoods at high elevations.

Rinodina chrysidiata

Yellow-finger Pepperpot

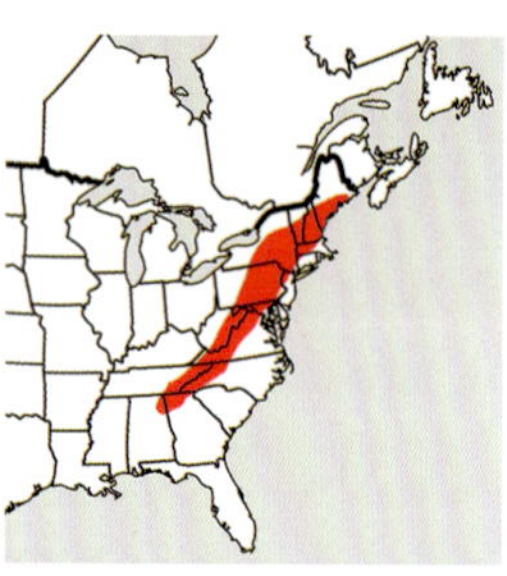

Tripp 5292 (photo: Lendemer)

NOTES: We've said it before that isidiate crustose lichens are not particularly common, but the only thing less common may be isidiate crustose lichens that are yellow in color. These two features make *Rinodina chrysidiata* one of the most easily recognized, and in some ways most attractive, crustose lichens in the Smokies. The epithet "chrysidiata" is derived from the Greek "*chrys-*" for golden yellow and "*-idiata*," which is abbreviation of the Latin for isidium. Under the rules, it should have been spelled "chrysisidiata," but it was decided that the current spelling sounded better and was less prone to errors in typing. This just goes to show that when you get to name a species you can bend the rules sometimes.

CHEMISTRY: Secalonic acid A. Spot tests. K-, C-, KC-, P-, UV-.

NICHE: This is another example of a species that occurs in the Appalachian Mountains and in eastern Asia. It is widely distributed throughout the Appalachians from Maine to Alabama, but is common only in a few middle elevation areas of the Smokies, such as Hyatt Ridge. The species grows on hardwood trees and typically occurs in mature hardwood forests, especially those on ridgetops dominated by oaks.

KEY FEATURES: Yellow areolate crustose thallus, yellow isidia, P-, on hardwoods at middle-to-high elevations.

Rinodina chrysomelaena

Lemon Pepperpot

Lendemer 26912 (photo: Moroz & Tripp)

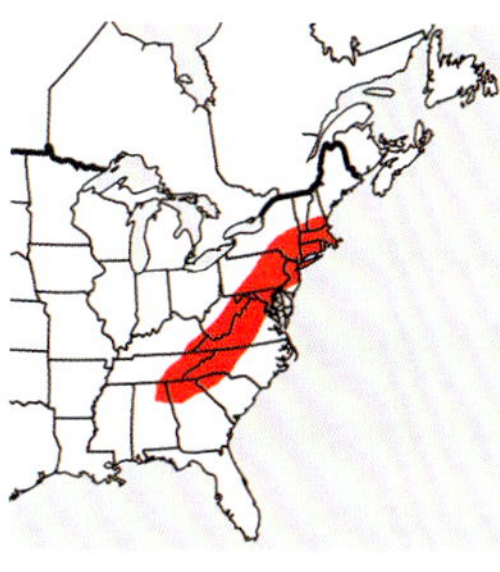

NOTES: Here is a prime example of a species that illustrates how we know that some lichens are endangered. *Rinodina chrysomelaena* was collected in scattered locations in the Appalachians and Piedmont during the 19th century, with the last known collection made at Cherokee Orchard in Tennessee by Gunnar Degelius in the 1930s. It was not seen again until we found it and photographed it for this field guide. It seems unlikely that a yellow crustose lichen with brown spores would have been overlooked for nearly a century. It's more likely that it disappeared from its former haunts. Indeed, this is an extremely distinctive crustose lichen on account of the yellow areolate thallus, lecanorine apothecia, 2-celled brown spores with thickened walls, and occurrence on rocks that at first appearance to be non-calcareous (e.g., granite) but, a closer look, usually reveal hints of a higher pH.

CHEMISTRY: Secalonic acid. Spot tests. K-, C-, KC-, P-, UV-.

NICHE: *Rinodina chrysomelaena* is an eastern North American endemic that was once widespread but is now known from four locations in the southern Appalachians. It grows on shaded non-calcareous rocks, particularly those associated with humid microhabitats. In the Smokies it is known from a single undisclosed location in North Carolina.

KEY FEATURES: Yellow areolate crustose thallus, esorediate, 2-celled spores with thickened walls, P-, on rocks at an undisclosed location.

Rinodina colobinoides

Mite Nibbles

Lendemer 53146 (photo: Tripp)

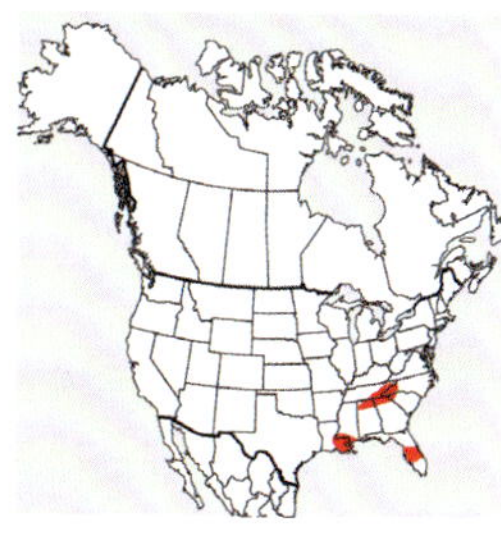

NOTES: Unless you find it fertile, there is nothing about this species that would tip you off down the *Rinodina* path. Instead, its minute but prominent isidia remind one of *Micarea pycnidiophora* or *Pseudosagedia isidiata*. When we first set off on our journey to write this field guide, we considered publishing two volumes, one being *Lichens of the Smokies* and the other being *Lichens of the Smokies You*

Will Never See. Alas, we opted for a combined volume, but that doesn't change the fact that few will ever see *R. colobinoides*. *Rinodina colobinoides* typically has small, granular blastidia rather than isidia. That is true, which is why the individual photographed ultimately may represent a different, as yet undescribed species. We hope that doesn't bother you. After all, the Smokies are full of secrets, and some secrets are best kept in the forest.

CHEMISTRY: No substances. Spot tests. K-, C-, KC-, P-, UV-.

NICHE: Its primary center of distribution is tropical. In United States, *Rinodina colobinoides* really only occurs in the extreme coastal plain, except for disjunct populations in the Smokies where all great things grow.

KEY FEATURES: Inconspicuous, brown-gray thallus, minute isidia or blastidia, coccoid photobiont, all spot tests negative, on bark throughout the Park.

Rinodina degeliana

Swept Under

Lendemer 30270 (photo: Lendemer)

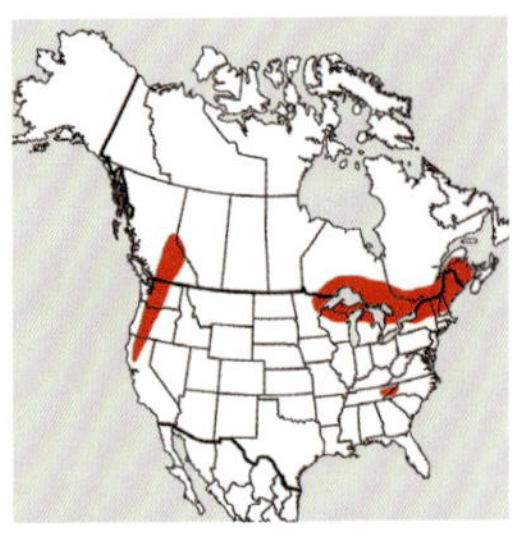

NOTES: Do you ever sweep the dust or dirt under the rug instead of picking it up? We would never admit to doing so, but this lichen might. The thallus is composed of flat, blue-white areoles with soredia that spill out from the margins. The overall appearance is one of tiny carpets that just can't hide dirt anymore. There are a number of other sorediate species in the Smokies that have areolate thalli, but none quite look like *Rinodina degeliana*. *Rinodina buckii* and *R. willeyi* both have grayer thalli that react P+ orange-red due to pannarin. *Rinodina efflorescens*: same story. *Rinodina brodoana* lacks atranorin and is K-, but also has much larger areoles and an overall very different appearance.

CHEMISTRY: Atranorin and zeorin. Spot tests. K+ yellow, C-, KC-, P-, UV-.

NICHE: *Rinodina degeliana* is a common species of northern temperate and boreal eastern North America. So, it shouldn't come as a surprise to learn that in the Smokies, and in the southern Appalachians in general, it is restricted to high elevations. Thankfully Swept Under isn't picky, and will grow on the bark of both hardwoods and conifers.

KEY FEATURES: Blue-white, areolate thallus, flat areoles, marginal soralia, K+ yellow and P-, on bark at high elevations.

Rinodina destituta

The Beggar Lichen

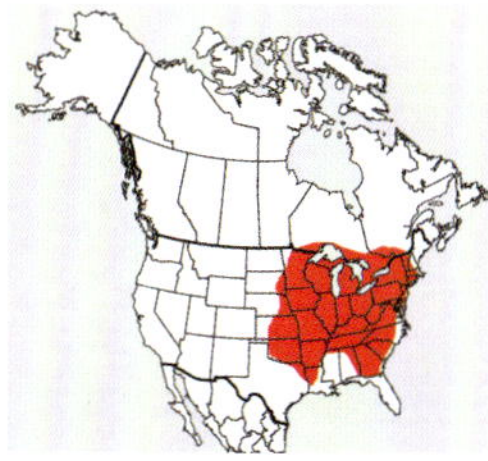

Lendemer 29337 (photo: Lendemer)

NOTES: Species of *Rinodina* that grow on rocks can be difficult to identify, particularly in central and western North America. Thankfully, that is not the case in the Smokies, where there are relatively few such things to confuse you. Of the crustose lichens with brown, 2-celled spores and thickened walls in the Park, this species is only likely to be confused with *R. oxydata*, which differs in having slightly smaller spores. Since both species can look quite similar externally, you'll just have to get out your microscope and measure. If you're looking for a trick to help in the field, then you can go by the relatively thick, greenish-gray thallus and small, reddish-brown apothecia of the Beggar Lichen. In the Smokies, the apothecia tend to lack thalline margins at various stages of development, which is atypical for members of the genus.

CHEMISTRY: Atranorin. Spot tests. K+ yellow, C-, KC-, P-, UV-.

NICHE. Common and widespread in eastern North America, this species is often found on exposed or somewhat shaded, non-calcareous rocks. It is infrequent in the Smokies, likely occurs throughout the Park at middle and low elevations.

KEY FEATURES: Relatively thick, greenish-gray crustose thallus, 2-celled, brown spores with thickened walls, on non-calcareous rock.

Rinodina dolichospora

Mountain Fortress

Lendemer 30270 (photo: Lendemer)

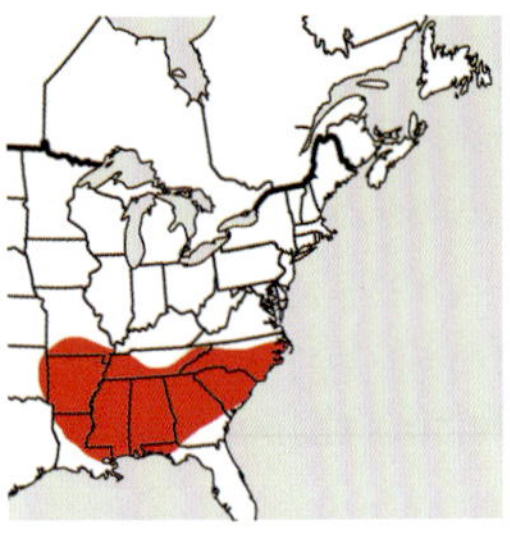

NOTES: *Rinodina dolichospora* is a bit of a mystery. It has bigger spores than most *Rinodina* species, but then again, not nearly as large as those of the other species it looks like. The thallus is thick and verruculous, often greenish-brown, sometimes even appearing torn or ragged. The most similar species is *R. ascociscana*, which differs in having much larger spores and a darker colored thallus. *Rinodina adirondackii* also has large spores, but its thallus is smooth and thin, and it is P+ orange-red due to the presence of pannarin. *Rinodina subminuta* can look similar, but it has erumpent apothecia, a smooth and thin thallus, and smaller spores.

CHEMISTRY: No substances. Spot tests. K-, C-, KC-, P-, UV-.

NICHE: This is a rare species of middle and high elevations in the Smokies, where it grows on the bark of hardwoods and occasionally on logs. The southern Appalachians are its stronghold in North America, although it is known from scattered locations elsewhere in the southeastern United States.

KEY FEATURES: Thick, greenish-brown, verrucuolous thallus, large (but not too large), brown, 2-celled spores with thickened walls, P- thallus, apothecia that are not erumpent, on bark at middle and high elevations.

Rinodina excrescens

Spicy Pepperpot

Tripp 4935 (photo: Lendemer)

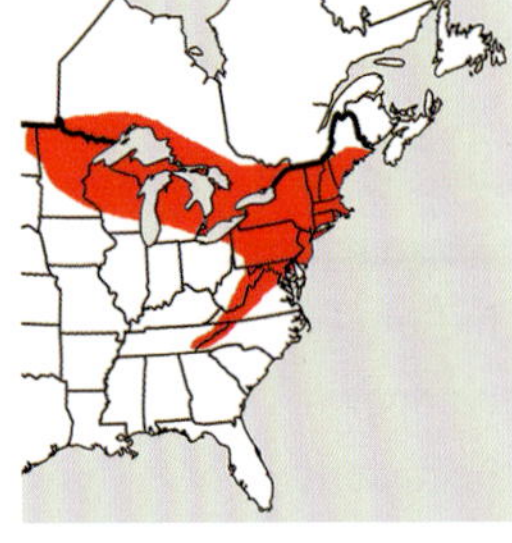

NOTES: This is the second pepperpot lichen that has a thallus composed of convex areoles that give rise to minute, bud-like blastidia from their surfaces. The other species, *Rinodina bullata*, differs chemically in the production of atranorin and thus has a

P- thallus. It might be surprising to learn that there are not really any other P+ orange-red crustose lichens that might be confused with *R. excrescens* in the Smokies, but so it seems. The common name refers to the spicy P+ reaction of the thallus. The most similar species might be *R. buckii*, but that is sorediate rather than blastidiate.

CHEMISTRY: Pannarin. Spot tests. K-, C-, KC-, P+ orange-red, UV-.

NICHE: *Rinodina excrescens* occurs throughout the Appalachian Mountains and Great Lakes Region in North America and its distribution has been particularly well documented in the northern parts of its range. It appears to be rare in the Smokies, where it has been found on the bark of hardwoods in high-elevation northern hardwood forests along the Cataloochee Divide in North Carolina.

KEY FEATURES: Light-blue-gray-to-brown-gray areolate crustose thallus, esorediate, minute blastidia budding from convex areoles, K- and P+ orange-red, on hardwoods at high elevations.

Rinodina subminuta

Itty-bitty Pepperpot

Lendemer 44703 (photo: Tripp)

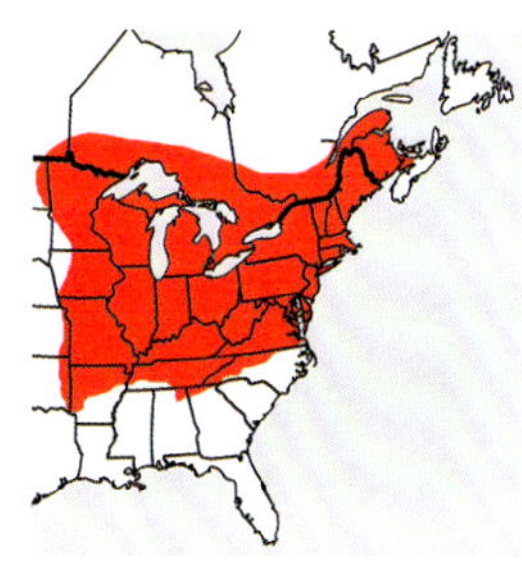

NOTES: Adorable might actually be an appropriate word to describe *Rinodina subminuta*. The species looks like a smaller version of *R. adirondackii*, since both species have smooth, gray thalli. However, in *R. subminuta*, the apothecia are erumpent and thus appear to burst through the thallus surface in the early stages of their development. In addition to having smaller spores, the species differs from *R. adirondackii* in having a P- thallus. It could also be confused with *R. dolichospora* and *R. ascociscana*, but both of those species have larger spores and the apothecia are not erumpent. Unusually fertile thalli of *R. buckii* can also be confused with *R. subminuta*, but that species is always at least sparsely sorediate and has a P+ orange-red thallus.

CHEMISTRY: Zeorin. Spot tests. K-, C-, KC-, P-, UV-.

NICHE: This species is widespread but infrequent throughout temperate eastern North America. In the Smokies, as is the case elsewhere in the southern Appalachians, it occurs primarily at middle-to-high elevations on the bark of hardwood trees.

KEY FEATURES: Thin, blue-gray crustose thallus, erumpent apothecia, esorediate, P-, relatively small brown 2-celled spores with thickened walls, on bark of hardwoods at all elevations.

Rinodina tephraspis

Lumpy Pepperpot

Tripp 3718 (photo: Lendemer)

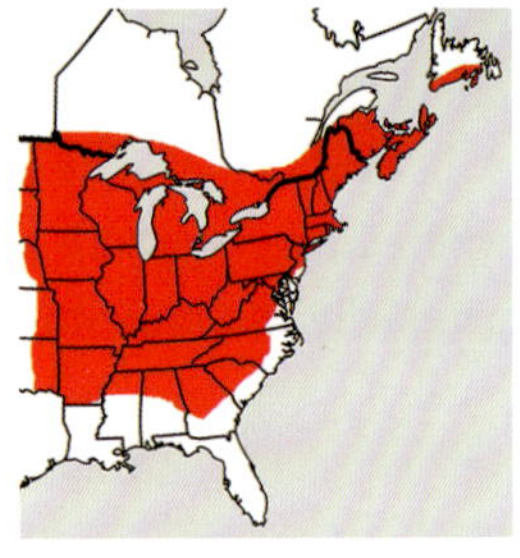

NOTES: *Rinodina tephraspis* is the most common pepperpot lichen in the Smokies and one of the most common crustose lichens on shaded non-calcareous rocks. It can easily be recognized by its lumpy, gray-to-brown-gray areolate thallus, lecanorine apothecia with dark red-brown discs, 2-celled brown spores and frequently C+ pink reaction. The other species of *Rinodina* that occur on rocks in the Smokies have flat, gray or green-gray thalli, and differ in microscopic characters of the spores as well as in chemistry. Species of *Buellia* that occur on rock, such as *B. spuria*, are very different even superficially and have spores that lack thickened walls.

CHEMISTRY: Zeorin with or without '5-o-methylhiascic acid. Spot tests. K-, C- or C+ pink, KC- or KC+ pink, P-, UV-.

NICHE: This species is widespread throughout temperate eastern North America and can be found just about everywhere that there are shaded non-calcareous rocks, especially sandstone. It is common in the Smokies and has been found at all elevations.

KEY FEATURES: Brown-gray areolate thallus, lecanorine apothecia, brown 2-celled spores with thickened walls, on shaded non-calcareous rocks throughout.

Rockefellera crossophylla

Old Gray Crosslobes

Tripp 3963 (photo: Lendemer)

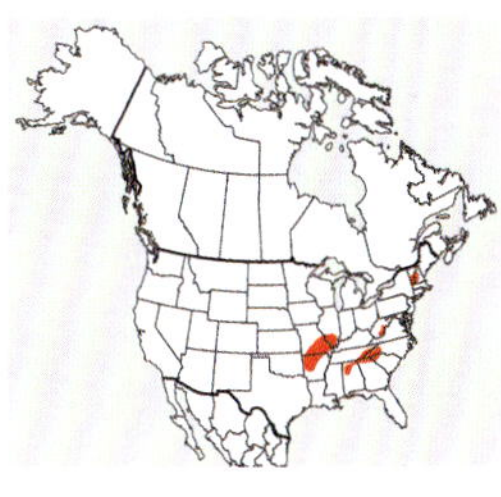

NOTES: *Rockefellera crossophylla* (formerly: *Santessoniella crossophylla*) is a remarkable lichen that, unfortunately, tells the story of habitat destruction and other forms of habitat decline in eastern North America, perhaps better than any other (see below). It is easily distinguished from all other species in the National Park by its cyanobacterial photobiont, small, gray squamules that pile onto one another and comprise the entire thallus, its striking apothecia with bright reddish-orange discs that lack a thalline margin and its distinctive ecology (see Niche). The only species with which this might be confused is *Vahliella leucophaea*, and admittedly these two taxa can be difficult to distinguish at times. However, *V. leucophaea* has flat squamules that are non-overlapping, unlike the digitate, overlapping squamules of *R. crossophylla*.

CHEMISTRY: No substances. Spot tests. K-, C-, KC-, P-, UV-.

NICHE: Old Gray Crosslobes is a species that formerly had a broader geographical distribution in eastern North America but has declined steadily over the last half century, as a result of habitat destruction and increased air pollution. Today, the species is represented in very few localities including the Ozarks and the northern, central, and southern Appalachians. Among these, Great Smoky Mountains National Park hosts the largest remaining populations, yet still fewer than 10 have been documented by lichenologists here. This species is restricted to non-calcareous rocks (we actually once saw it growing at the base of a tree in a very remote, very mature, riparian hardwood forest), where it grows almost exclusively in deeply shaded overhangs, particularly middle-elevation habitats.

KEY FEATURES: Cyanolichen, thallus of small, gray, overlapping squamules, reddish-orange apothecia, on non-calcareous rocks in deeply shaded overhangs, extremely rare (please do not collect this if found).

Ropalospora chlorantha

Plain Spiral-Spores

Tripp 5558 (photo: Lendemer)

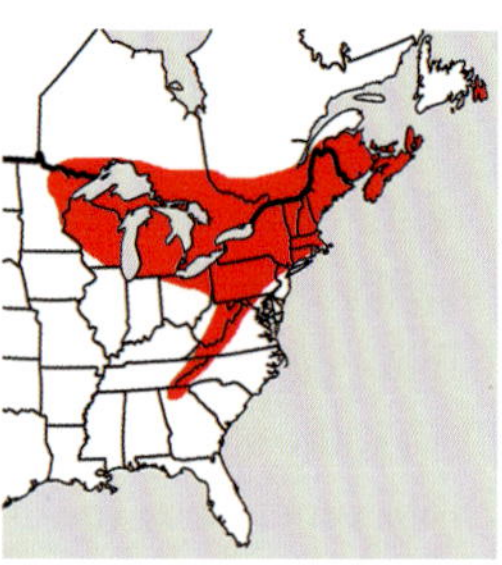

NOTES: There are surprisingly few crustose lichens in the Smokies that have small, circular thalli that are brownish-green-to-green-gray in color as well as black lecideine apothecia. If those features weren't enough to set *Ropalospora chlorantha* apart, you could confirm its identification by the UV+ blue-white medulla and the presence of long, bent, clavate colorless spores that are transversely septate. The only truly similar species is *R. viridis*, which differs in having a sorediate thallus.

CHEMISTRY: Perlatolic acid. Spot tests. K-, C-, KC-, P-, UV+ blue-white.

NICHE: This species has a classic Appalachian-Great Lakes distribution. It is common in northern temperate and boreal regions of eastern North America with a range that extends down the Appalachians to the Smokies where it occurs at middle-to-high elevations. Although it occurs on the bark and branches of both conifers and hardwoods, the species is particularly common on canopy branches and the stems of woody shrubs.

KEY FEATURES: Rosette-forming brownish-green crustose thallus, no soralia, black apothecia, transversely septate colorless spores, green soredia, P- and UV+ blue-white, on bark and branches at middle-to-high elevations.

Ropalospora viridis

Mealy Spiral-spores

Tripp 2667 (photo: Lendemer)

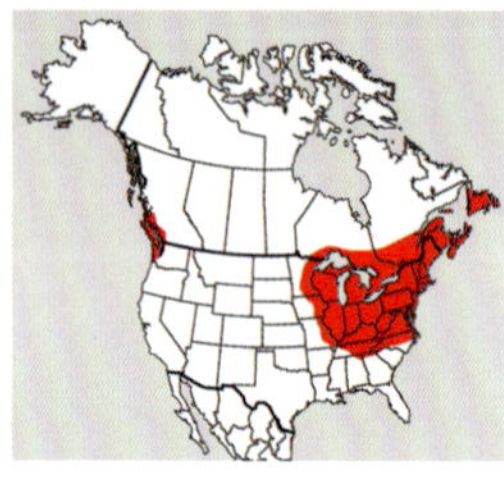

NOTES: *Ropalospora viridis* is most similar to *Ropalospora chlorantha*, from which it differs by the presence of soralia. It is interesting to note that the range of *R. chlorantha* is more restricted than that of *R. viridis* in North America, which suggests that perhaps it is easier to reproduce and spread through asexual propagules than with sexual spores. The small, rosette-forming thalli that are sorediate and brownish-green in color can also be confused with *Fuscidea arboricola* and *Japewiella dollypartoniana*. Both of those species differ chemically, however, as *F. arboricola* is P+ red due to the presence of fumarprotocetraric acid and *J. dollypartoniana* is P+ yellow due to the presence of norstictic acid.

CHEMISTRY: Perlatolic acid. Spot tests. K-, C-, KC-, P-, UV+ blue-white.

NICHE: *Ropalospora viridis* is common and widespread throughout temperate and boreal eastern North America, with a disjunct population in the Pacific Northwest. It is quite common in the Smokies and can be found at all elevations on the bark and branches of hardwoods and conifers.

KEY FEATURES: Rosette forming brownish-green crustose thallus, discrete soralia, green soredia, P- and UV+ blue-white, on bark and branches throughout.

Sarea resinae

Pine Pumpkins

Tripp 2261 (photo: Lendemer)

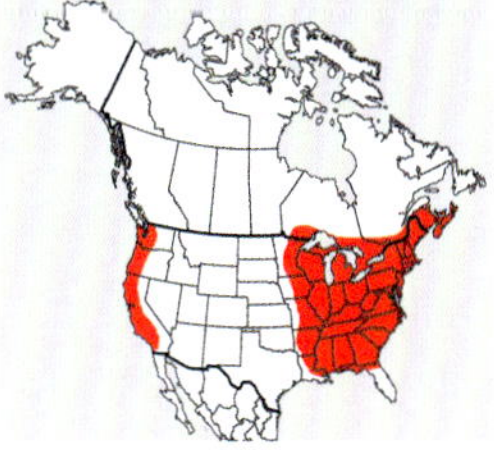

NOTES: Just when you thought that lichens couldn't get more specialized, you come to this one. *Sarea resinae*, pronounced "Sorry-a," incidentally, is a lichen-forming fungus that gave up its photobiont. It is related to *Placynthiella* and *Trapelia*, but you wouldn't guess it from the appearance of its orange apothecia, its asci simple contain large numbers of very small, simple, colorless spores, or by the fact it grows on conifer resin. Anytime you see a conifer with white stains of old resin on the trunk, go over and take a look for this species. While you're at it, look for *S. difformis*, which is an even smaller species with black apothecia that often grows with *S. resinae*.

CHEMISTRY: No substances. Spot tests. K-, C-, KC-, P-, UV-.

NICHE: This species occurs just about everywhere there are conifers that become damaged and exude their resin. It is common throughout eastern North America and parts of western North America. In the Smokies, it is most commonly found on pine resin, but also occasionally on the resin of fir and spruce.

KEY FEATURES: Non-lichenized thallus immersed in old, hardened conifer resin, orange apothecia, many tiny colorless simple spores per ascus, throughout the Smokies.

Schismatomma glaucescens

Hidden Snow Caps

Tripp 2631 (photo: Lendemer)

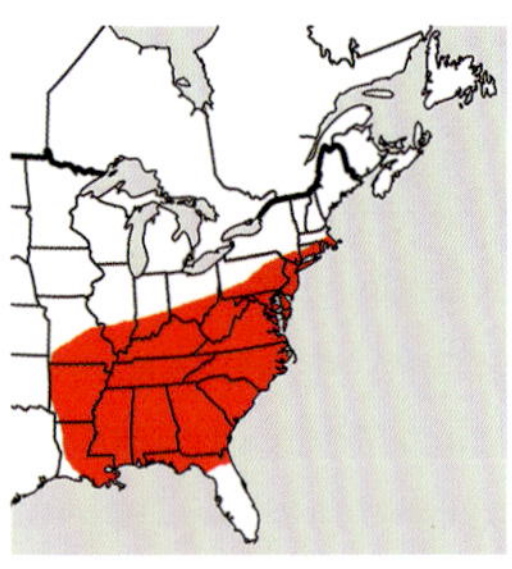

NOTES: No other crustose lichen in the Smokies can be confused with Hidden Snow Caps. The indistinct thallus, densely pruinose apothecia, 4-celled colorless spores, and occurrence in oak bark grooves is distinctive. The species often grows with *Scoliciosporum pensylvanicum*, which while similar in niche differs in having a leprose, UV+ blue-white thallus and epruinose apothecia with needle-like spores. The apothecia and spores of *S. pericleum* are similar internally, but that species produces a distinct albeit unsightly thallus, has pruinose apothecia, and grows on the bark of hemlocks. Like *S. glaucescens*, *Arthonia byssacea* grows on hardwoods and has pruinose apothecia, but that species is restricted to high elevations, has a well-developed white or blue-gray thallus, and does not grow exclusively in the bark grooves of old oak trees.

CHEMISTRY: Unidentified xanthone(?). Spot tests. K-, C-, KC-, UV+ dull orange.

NICHE: This species is endemic to temperate eastern North America, where it is widely distributed and occurs on bark in the grooves of mature oak trees. It is rare in the Smokies and occurs primarily in at middle and low elevations in oak-dominated hardwood forests on ridge tops.

KEY FEATURES: Indistinct crustose thallus, densely white pruinose apothecia, fusiform 4-celled colorless spores, on bark in groves of old oak trunks, at middle-to-low elevations.

Schismatomma pericleum

Bark Wax

Tripp 2684 (photo: Tripp)

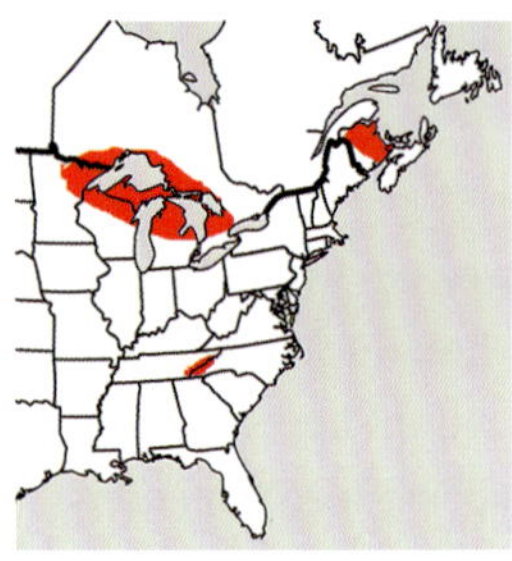

NOTES: *Schismatomma pericleum* is an easily overlooked species due to the small size of the apothecia and the tendency for the thallus to resemble a sickly or poorly developed crustose lichen such as *Graphis scripta*. Nonetheless, it is a very rare species in the southern Appalachians and can be recognized by its palid, gray-brown crustose thallus, dark reddish-brown apothecia that look like they had a failed attempt at being rounded, and by the narrowly fusiform spores that are colorless and 4-celled. *Schismatomma graphidioides* is a similar species known from Alabama that differs in having narrow and branched apothecia, resembling a *Graphis*. No other species in the Smokies closely resembles *S. pericleum*, except perhaps some species of

Arthonia. All such *Arthonia* species, however, differ in having broader spores that are either muriform or transversely celled and clavate.
CHEMISTRY: No substances. Spot tests. K-, C-, KC-, P-, UV-.
NICHE: This is a very rare species with an unusual distribution. It was originally described from Europe and is known from the Canadian Maritimes and Great Lakes in eastern North America. There are, however, a series of disjunct populations in the southern Appalachians at middle-to-low elevations, all found on the bark of hemlocks. It is known from one location in the Smokies near Rabbit Creek in Tennessee.
KEY FEATURES: Poorly developed brownish-gray thallus, rounded to irregularly shaped dark reddish-brown immersed apothecia, colorless fusiform 4-celled spores, on bark of hemlocks at low elevations.

Scoliciosporum chlorococcum

The Grinch

Lendemer 53179 (photo: Tripp)

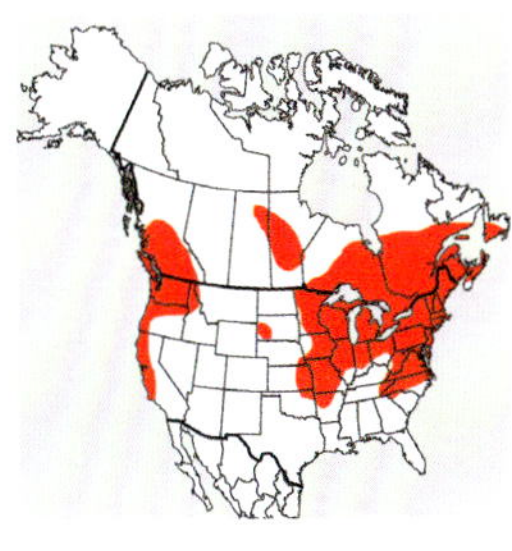

NOTES: There are very few instances in this world where we would argue that it is acceptable to say "really, that's a lichen?" But *Scoliciosporum chlorococcum* is just such an exception. Far from the glamorous world of *Arthonia kermesina* and *Usnea angulata* lives this species. It is an inconspicuous crustose lichen with a dirty, greenish-brown thallus that resembles algal colonies that are long past their prime. The apothecia are small, convex, and dark-reddish-brown-to-black in color. But, if you take the time to cut into the apothecia, you will be rewarded by colorless, transversely septate, snake-like spores. Sounds delightful, no?
CHEMISTRY: No substances. Spot tests. K-, C-, KC-, P-, UV-.
NICHE: A rarity in the Smokies, this species typically occurs on the branches of trees along forest edges and in disturbed habitats. It is widespread in temperate eastern North America and appears to be much more common in the areas surrounding the Park.
KEY FEATURES: Grungy, greenish-brown crustose thallus, small, dark colored, convex apothecia, colorless, snake-like spores, on branches of trees and middle and high elevations.

Scoliciosporum pensylvanicum

Groovy Spiral Spores

Collector Lendemer 33114 (photo: Tripp)

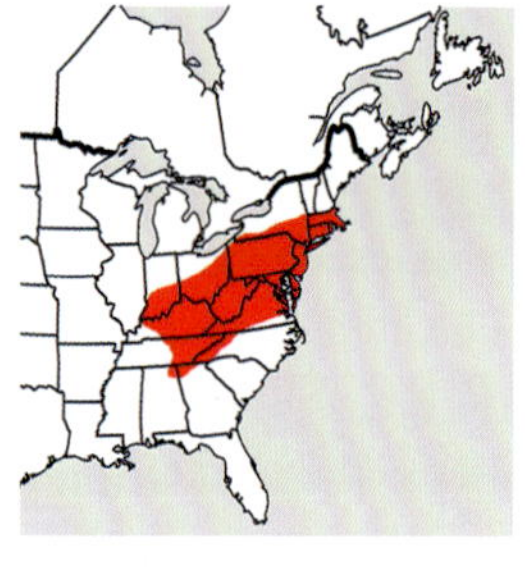

NOTES: *Scoliciosporum pensylvanicum* can easily be recognized by its blue-gray-to-green-gray leprose thallus, small tan apothecia, needle-shaped spores, and UV+ blue-white reaction. Its niche is similar to that of *Schismatomma glaucescens*, which also grows in the deep grooves of oak bark, but differs in having an indistinct, UV+ dull orange thallus and densely white pruinose apothecia. Even though this species is treated in *Scoliciosporum*, it is most likely to be confused with *Bacidia lobarica*, which differs in having longer spores, larger apothecia, and in growing on the smooth surfaces of bark rather than in the deep grooves of oaks. *Scoliciosporum pruinosum* is also similar, but restricted to high elevations of the Smokies, where it usually occurs on sheltered sides of trunks of mature Yellow Birches. It differs from *S. pensylvanicum* in having convex, white apothecia and spores that are spirally arranged within the ascus.

CHEMISTRY: Lobaric acid. Spot tests. K-, C-, KC+ pink, UV+ bright blue-white.

NICHE: Groovy Spiral Spores was originally described from the Appalachian Mountains but is likely widespread in other areas of temperate eastern North America where suitable substrates and habitats occur. It is typically found in the bark grooves of mature oaks, especially Chestnut Oak. It is infrequent in the Smokies, where it occurs mainly in old-growth oak forests, especially on ridge tops at middle and low elevations.

KEY FEATURES: Blue-gray-to-green-gray granular leprose thallus, minute biatorine tan apothecia, needle-like colorless spores, UV+ blue-white, on bark, especially in groves of old oak trunks, at middle-to-low elevations.

Scoliciosporum pruinosum

White Lightening

Lendemer 53143 (photo: Tripp)

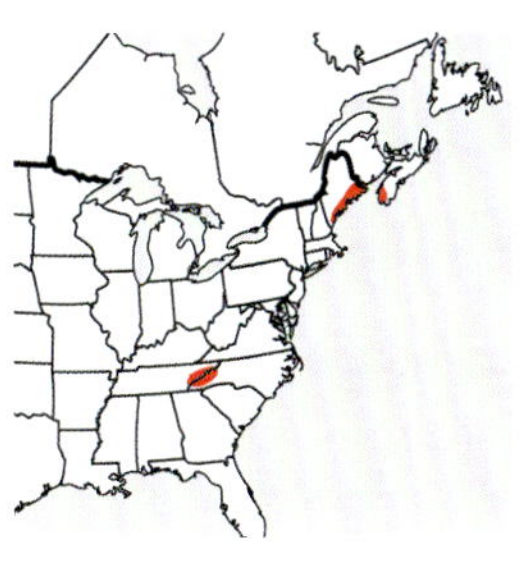

NOTES: This is really a lovely species, not just because it grows on beautiful trees in beautiful places, but because it also yields a flashy reaction under UV. The thin, green thallus contrasts with the pale, convex apothecia and make *Scoliciosporum pruinosum* stand out against the otherwise unornamented bark. It could be confused with *Micarea peliocarpa*, however that species has smaller apothecia and ellipsoid spores that are 4-celled. *Scoliciosporum pensylvanicum* and *Bacidia lobarica* are chemically similar and also have pale apothecia, however the former species has a leprose thallus while the latter species is sorediate.
CHEMISTRY: Lobaric acid. Spot tests. K-, C-, KC+ pink, UV+ bright blue-white.
NICHE: In the Smokies, White Lightning occurs only in high-elevation old growth forests on the bark of mature spruce and Yellow Birch. On the latter substrate, it usually grows on the dry, sheltered side of the trunk where large bark plates peel away. Outside of the Smokies, it is found in the oceanic forests of the Pacific Northwest and northeastern North America.
KEY FEATURES: Green, crustose thallus, small, pale, convex apothecia, spindle-like spores, UV+ bright blue-white, on mature trees at high elevations.

Scoliciosporum umbrinum

Lesser Spiral Spores

Tripp 4949 (photo: Lendemer)

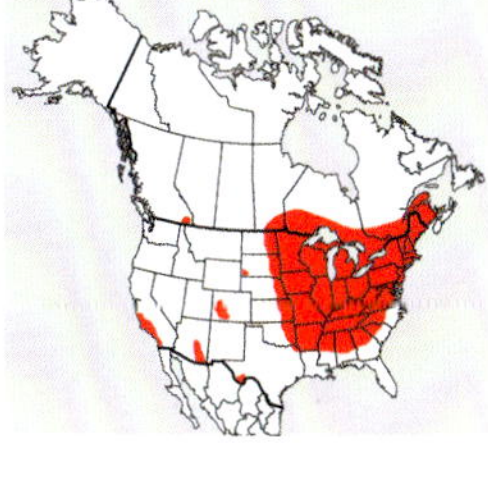

NOTES: The most distinctive features of Lesser Spiral Spores are the poorly developed thalli, reddish-brown apothecia that lack margins, and the colorless, narrowly clavate spores that are transversely septate and often coiled or bent. Indeed, the name of the genus refers to twisted or spiraled spores. *Scoliciosporum umbrinum* is most likely to be confused with species of *Bacidia* or *Bacidina*, but the members of those genera that occur on rocks in the Smokies have pale or lighter-colored apothecia, and spores that are extremely narrow and needle-like. Some species of *Micarea* are also superficially similar, with similarly size spores have spores that are straight rather than bent and twisted.
CHEMISTRY: No substances. Spot tests. K-, C-, KC-, P-, UV-.

NICHE: This species is common and widespread on non-calcareous rocks in eastern North America, with disjunct populations in mountainous areas of the western United States. It often occurs on exposed rocks, or in disturbed habitats, and is rare in the Smokies.
KEY FEATURES: Indistinct crustose thallus, small reddish-brown biatorine apothecia, coiled bent colorless spores that are clavate and transversely septate, on non-calcareous rocks at all elevations.

Stereocaulon dactylophyllum

Fudgy Fingers

Lendemer 33280 (photo: Tripp)

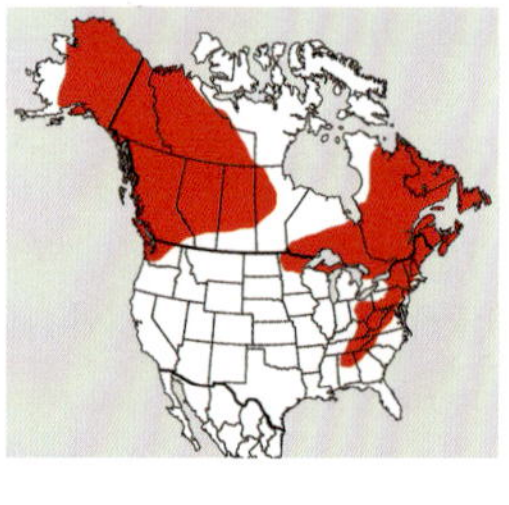

NOTES: *Stereocaulon* is one of the easiest to recognize genera of lichens in the world. Here are the facts: gray-to-white fruticose thalli, distinct stalks upon which apothecia sit (these termed *pseudopodetia*), and a thallus covered by corticate to less commonly ecorticate granules or other types of verrucae, warts, or bumps (these termed phyllocladia). Also, species of *Stereocaulon* usually grow on rocks (or, out West, commonly occur on alpine soils). There are four species of *Stereocaulon* currently known from the Smokies. It just isn't enough, and we want more, but for better or worse, we will have to make do. *Stereocaulon dactylophyllum* is distinguished from *S. pileatum* by having corticate pseudopodetia, elongate, finger-like phyllocladia and by lacking soralia. It differs from both *S. tennesseense* and *S. glaucescens* by having stictic acid in the phyllocladia, which yields a P+ orange reaction. *Stereocaulon dactylophyllum* is usually found with abundant, beautiful brown apothecia, as seen in this photograph.
CHEMISTRY: Atranorin and stictic acid. Spot tests. K+ yellow, C-, KC-, P+ orange, UV-.
NICHE: This species is restricted to—but relatively common in—high elevation forests dominated by Red Spruce, Fraser Fir, American Mountain Ash, and Yellow Birch. It grows on rock, primarily of the Anakeesta Formation. To see a nice population of *Stereocaulon dactylophyllum*, head to the summit of Mt. Guyot, but be prepared to go off-trail! This species is widespread in northern-boreal regions of North America. It reaches its southern limit of distribution in the southern Appalachians, like so many other boreal species.
KEY FEATURES: Grayish-white fruticose thallus, corticate pseudopodetia, brown apothecia, P+ orange thallus, high elevations, on Anakeesta Rock, relatively common.

Stereocaulon pileatum

Pileated Rock Pucker

Tripp 3723 (photo: Lendemer)

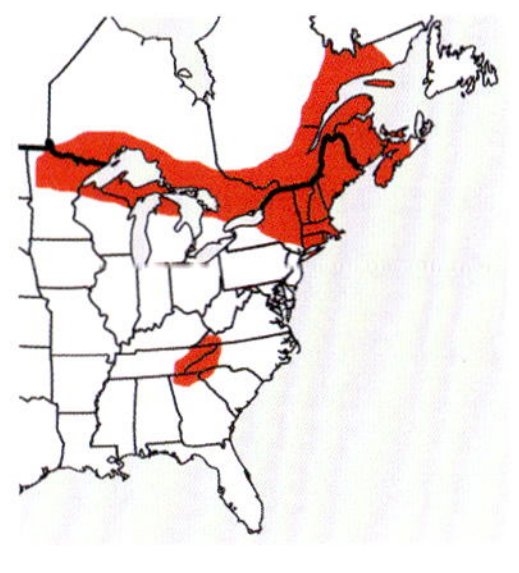

NOTES: *Stereocaulon pileatum* can be differentiated from all other species of *Stereocaulon* in the Smokies by a number of highly distinctive features. Among these, it is relatively small in size, is whitish-gray in color, has sorediate pseudopodetia, is UV+ blue-white owing to the presence of lobaric acid, and has very cool cephalodia scattered about the thallus. So many features packaged into a tiny organism!

CHEMISTRY: Atranorin and lobaric acid. Spot tests. K+ yellow, C-, KC+ pink, P-, UV+ blue-white.

NICHE: *Stereocaulon pileatum* occurs in both middle and high elevation environments in the Smokies, always on rock, especially mineral-rich rock. It is neither rare nor common in the Smokies. Look for it along Forney Creek Trail, not far from Campsite 69. Populations in the southern Appalachians are disjunct from the primary portion of the range of this species, which is northern Appalachia to the Great Lakes.

KEY FEATURES: Small, whitish-gray fruticose thalli, ecorticate pseudopodetia, UV+ blue-white, with cephalodia, occasional at middle and high elevations, on mineral-rich rock.

Stereocaulon tennesseense

Storm Watch

Tripp 3724 (photo: Lendemer)

NOTES: We live and work and hike to find lichens like *Stereocaulon tennesseense*. What a great species! This one is characterized by its gray, fruticose thallus, esorediate pseudopodetia, and brown apothecia. It is similar to *S. dactylophyllum*, but that species has P+ orange phyllocladia. It is perhaps more similar to another species, *S. glaucescens*, but differs in having erect phyllocladia versus flattened phyllocladia in the latter species. Admire that nice chunk of Anakeesta Rock in the photograph. We can't help but wonder how much this one geological addition in the park contributes to the exceptionally high lichen endemism.

CHEMISTRY: Atranorin and lobaric acid. Spot tests. K+ yellow, C-, KC+ pink, P-, UV+ blue-white.

NICHE: This species is restricted to two regions of the Appalachian Mountains: the extreme southern and extreme northern portion. We don't know which came first, but research could probably answer that question. In the Smokies, this species is restricted to the highest elevations, where it occurs in spruce-fir forest on Anakeesta and other mineral-rich rocks.

KEY FEATURES: Small, gray fruticose thallus, corticate pseudopodetia, brown apothecia, erect phyllocladia, uncommon at highest elevations, on Anakeesta Rock.

Sticta beauvoisii

Stinck-ta

Tripp 3549 (photo: Lendemer)

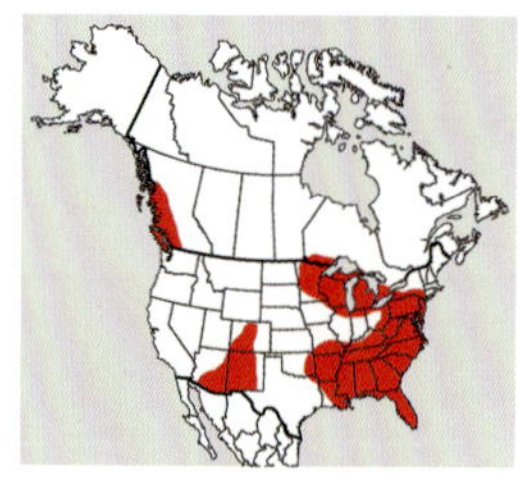

NOTES: Welcome to *Sticta*-ville. This is an important group to learn not only because they are ecologically important, but also because they are extremely abundant throughout the Smokies. First: the genus *Sticta* as a whole is extremely easy to recognize because it smells like fish, has a cyanobacterial photobiont, has relatively large thalli that are almost universally chocolate brown in color, and has very conspicuously cyphellae present on the lower surface. Second: there are six species in the Smokies that you may encounter on occasion, some of them more frequently than others. Of these, three are isidiate (*S. beauvoisii*, *S. fuliginosa*, *S. sylvatica*), two are phyllidiate (*S. fragilinata*, *S. carolinensis*), and one is sorediate (*S. limbata*). *Sticta beauvoisii* can be differentiated from the other two isidiate species by having marginal (only), coralloid isidia. Make this the first *Sticta* you learn in the Smokies.

CHEMISTRY: No substances. Spot tests. K-, C-, KC-, P-, UV-.

NICHE: *Sticta beauvoisii* is the most common member of the genus in the Smokies. It is known from numerous collections and can be found at all elevations growing both on rocks and trees, particularly bases of large hardwoods.

KEY FEATURES: Smells like fish, cyanobacterial photobiont, large, chocolate brown thallus, cyphellae underneath, marginal, coralloid isidia, very common on rocks and trees throughout Park.

Sticta carolinensis
Carolina Moon Lichen

Tripp 3960 (photo: Lendemer)

NICHE: *Sticta carolinensis* is one of two phyllidiate species in the genus in the Park. It is readily differentiated from the other, *S. fragilinata*, by its white medulla that is K-. It is otherwise characterized by its relatively large, brown thalli, but then again, so are most of the other species in the genus. See the *S. beauvoisii* entry for a short crash course in *Sticta* identification.
CHEMISTRY: No substances. Spot tests. K-, C-, KC-, P-, UV-.
NICHE: *Sticta carolinensis* is a relatively common species in the Smokies. It occurs primarily on non-calcareous rock at low-to-middle elevations, particularly in streamside forests with Eastern Hemlock, Sweet Birch, Hickory, and Sugar Maple.

KEY FEATURES: Smells like fish, cyanobacterial photobiont, large, chocolate brown thallus, cyphellae underneath, phyllidiate, white K-medulla, relatively common on rocks at low-to-middle elevations

Sticta fragilinata

Tammy's Pumpkin Pails

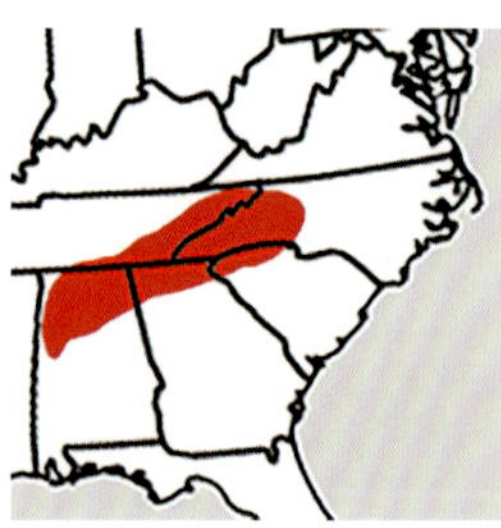

Tripp 5541 (photo: Lendemer)

NICHE: *Sticta fragilinata* is one of two phyllidiate species in the Smokies. It is easily differentiated from the other, *S. carolinensis*, by having a conspicuously yellow-to-orange medulla that reacts K+ purple. Remarkably, this species went unnamed for decades despite its morphological uniqueness. See the *S. beauvoisii* entry for a short crash course in *Sticta* identification.
CHEMISTRY: Fragilin and related pigments. Spot tests. Cortex: K-, C-, KC-, P-, UV-; medulla: K+ purple, C-, KC-, P-, UV-.
NICHE: *Sticta fragilinata* is neither rare nor common in the Smokies. It can be found on occasion, usually at middle elevations, in rich hardwood forests of Black Cherry, Buckeye, and Yellow Birch. This species grows on both hardwoods and on non-calcareous rocks.
KEY FEATURES: Smells like fish, cyanobacterial photobiont, large, chocolate brown thallus, cyphellae underneath, phyllidiate, orange K+ purple medulla, occasional on rocks and trees, middle elevations.

Sticta fuliginosa

Hobble Lobes

Tripp 3441 (photo: Lendemer)

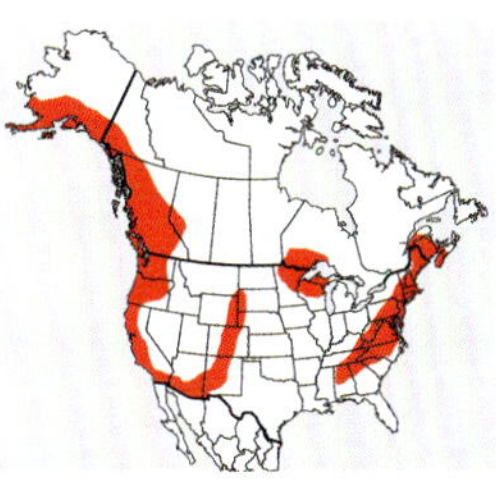

NICHE: *Sticta fuliginosa* is one of three isidiate species in the Smokies. Both this species and *S. sylvatica* have in common isidia that occur on the surfaces of the lobes (vs. marginal isidia in *S. beauvoisii*). At times, these laminal isidia can appear somewhat granulose or lobulate but are still to be considered isidia (vs. true phyllidia, as in *S. fragilinata* and *S. carolinensis*). *Sticta fuliginosa* can be distinguished from *S. sylvatica* by having a thallus that consists of a single, large rounded lobe (vs. consisting of multiple, divided lobes in *S. sylvatica*). See the *S. beauvoisii* entry for a short crash course in *Sticta* identification.

CHEMISTRY: No substances. Spot tests. K-, C-, KC-, P-, UV-.

NICHE: *Sticta fuliginosa* is an uncommon species in the Smokies, where it occurs on bark. We suspect it was once more widespread but may be in decline in recent decades. The Swedish lichenologist Gunnar Degelius collected this species in 1939 on top of Mt. Kephart. Jon Dey made several other early collections of it in the 1970s. We have seen it since only a handful of times. If found, it will be at high elevations in northern hardwood forests such as those of Yellow Birch and Mountain Maple. A great place to see this species is on Welch Ridge, near the spur trail to High Rocks.

KEY FEATURES: Smells like fish, cyanobacterial photobiont, large, chocolate brown thallus consisting of a single lobe, cyphellae underneath, laminal isidia, uncommon, on bark.

Sticta sylvatica

Forest Doilies

Lendemer 32993 (photo: Tripp)

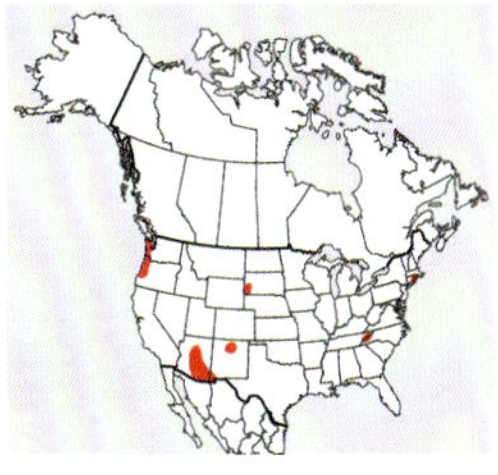

NICHE: *Sticta sylvatica* is one of three isidiate species in the genus in the Smokies. Both this species and *S. fuliginosa* have in common isidia that occur on the surfaces of the lobes (vs. marginal isidia in *S. beauvoisii*). At times, these laminal isidia can appear somewhat granulose or lobulate but are still to be considered isidia (vs. true phyllidia, as in *S. fragilinata* and *S. carolinensis*). *Sticta sylvatica* can be distinguished from *S. fuliginosa* by having a thallus that consists of multiple, divided lobes (vs. consisting of a single, large rounded lobe in *S. fuliginosa*). See the *S. beauvoisii* entry for a short crash course in *Sticta* identification.

CHEMISTRY: No substances. Spot tests. K-, C-, KC-, P-, UV-.

NICHE: This species occurs only in high-elevation habitats, where it grows primarily on vertical rock faces in habitats dominated by Red Spruce, Balsam Fir, Mountain Maple, and Yellow Birch. *Sticta sylvatica* is one of the rarest macrolichens in the Smokies and should not be collected if found (but should be way-pointed).

It is known from only seven collections, five of which date to collections of Jon Dey and A.J. Sharp, pre-1975.

KEY FEATURES: Smells like fish, cyanobacterial photobiont, large, chocolate brown thallus consisting of multiple, divided lobes, cyphellae underneath, on rocks at high elevations, exceptionally rare (do not collect if found).

Strigula americana

Coal Eyes

Lendemer 29565A (photo: Lendemer)

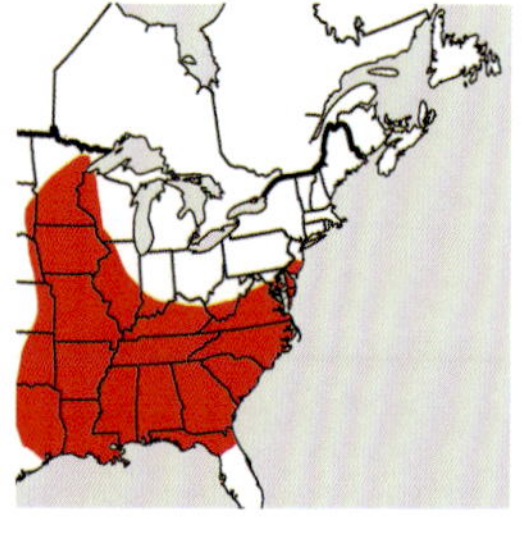

NOTES: *Strigula*, oh *Strigula*. To be a black flask hidden in a forest, dashing about unnoticed from tree to tree. To have a simple presence, whose paraphyses be unbranching and whose spores be parted only in two. To have been born a peritheciate lichen, replete with algae, living in harmony with my tree. Only the most perceptive visitor takes time to notice. She sees my spores, which are longer and wider than those of *S. viridiseda*. And, for a moment, I am finally in the spotlight, or better yet a smoky moonlight, and I am the only lichen that matters.

CHEMISTRY: No substances. Spot tests. K-, C-, KC-, P-, UV-.

NICHE: Coal Eyes occurs throughout eastern North America where, to an observant visitor, it can be found on a wide variety of substrates from low-to-middle elevations including hickories, tulip poplars, black gums, and ironwood. Look for it in areas that might be slightly alkaline, such as near Chillhowee Lake.

KEY FEATURES: Black perithecia on trees, colorless, 2-celled spores, unbranched (and non-anastamosing) paraphyses, on bark of hardwoods, low-to-middle elevations.

Strigula stigmatella

Internal Beauty

Tripp 5288 (photo: Lendemer)

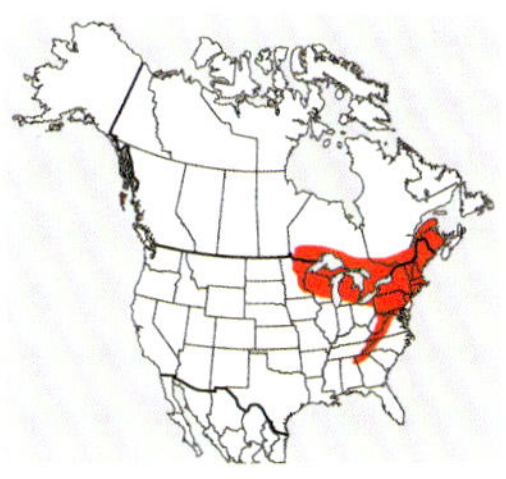

NOTES: From a distance, *Strigula stigmatella* is not particularly striking; the shiny brownish-green crustose thallus tends to blend in with its substrates; and because the perithecia lack carbonized walls and are immersed in the thallus, they are only apparent as brownish ostioles poking ever so slightly above the thallus surface. Given this, the perithecia are often assumed to be pycnidia in the field, and the species is often overlooked as an immature version of something else. The species is nonetheless quite distinctive on account of its large, clavate, colorless spores that are 8–12 celled. The species might be confused with *Porina heterospora*, but that species has much larger spores, a distinctly bronze-colored thallus with superficial rather than immersed perithecia and is restricted to low elevations in the Smokies.

CHEMISTRY: No substances. Spot tests. K-, C-, KC-, P-, UV-.

NICHE: This species has an Appalachian-Great Lakes distribution in North America with a disjunct population in the Pacific Northwest. It is most common in the northern portions of its range but does occur throughout the Appalachians and is widespread in the in Smokies at all elevation. Although *Strigula stigmatella* typically occurs on bark at the mossy bases of mature trees, it can also be found on shaded non-calcareous rocks along streams.

KEY FEATURES: Shiny, brownish-green crustose thallus, immersed perithecia with non-carbonized walls, large 8–12 celled, clavate, colorless spores, on bark and rock at all elevations.

Tephromela atra

Straight Shooter

Tripp 3714 (photo: Lendemer)

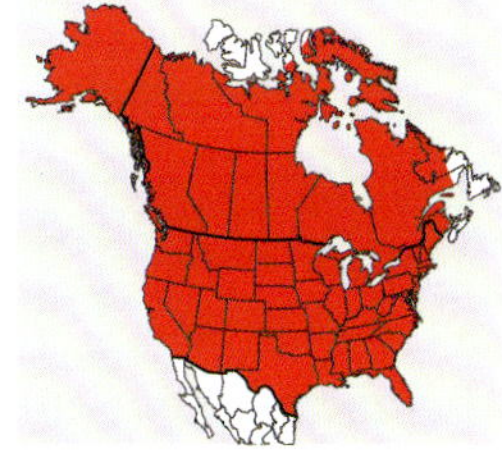

NOTES: *Tephromela atra* is a highly variable yet distinctive crustose lichen first and foremost by jet black apothecial discs that are surrounded by a thick, grayish-white margin concolorous with the thallus. The thallus of *T. atra* ranges from thin to thick, and irregularly areolate to continuous. The photograph here shows an example of thick, areolate morphology, which is typical of the species when it is found on rock (see Niche). On account of its typically (but not always–gasp!) lecanorine apothecia margins and its simple, small, colorless spores, this species is most likely to be confused with members of *Lecanora* that can have very dark discs. In particular, *Lecanora glabrata* is superficially similar but differs in having dark reddish-brown versus black discs ("*tephr*" means ash-colored and "atra" means black, if it helps) as well as in chemistry. As much as we had hoped, the two genera are unfortunately not related.

CHEMISTRY: Atranorin, alectoronic acid and related substances. Spot tests. Cortex: K+ yellow, C-, KC-, P-, UV-; medulla: K-, C-, KC-, P-, UV+ bright blue-white.

NICHE: *Tephromela atra* is an interesting lichen that has an exceptionally broad range geographically but, in several areas where it occurs including the Smokies, it is very uncommon to rare. Over the course of a decade of fieldwork that has included a couple thousand miles of hiking, we have seen this species only once within the Park's borders. We collected it on rock at middle elevations, but elsewhere in its range, the species commonly grows on bark, too.

KEY FEATURES: Irregularly areolate white thallus, jet black apothecia with whitish margins (always concolorous with the thallus), chemistry, on mineral-rich rock (elsewhere on bark), rare in the Smokies.

Thelidium decipiens

Deceptive Dots

Lendemer 53607 (photo: Lendemer)

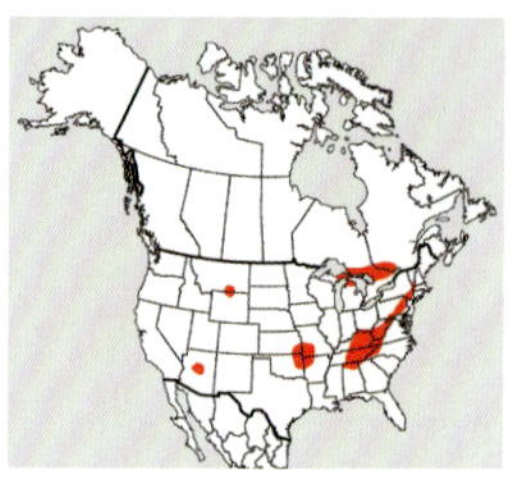

NOTES: Once you have gotten over the excitement of finally finding calcareous rocks in a mountain chain otherwise dominated by other rock types, take a break and a few deep breaths. Before you will be a veritable plethora of tiny black-dotted crusts, many with their fruiting bodies immersed in the substrate. Why does this happen so much on calcareous rocks? Maybe the lichens find this substrate easier to dissolve than jackhammering through granite or schist. *Thelidium decipiens* can be recognized by its immersed thallus and immersed black perithecia, the latter of which have 2-celled colorless spores that are eight-per ascus. Externally, you might confuse *T. decipiens* with species of *Bagliettoa* on account of the fact that members of that genus also have immersed perithecia. But joy, those species have simple spores, although they are rare and difficult to observe in the most common species, *B. baldensis*.

CHEMISTRY: No substances. Spot tests. K-, C-, KC-, P-, UV-.

NICHE: Deceptive Dots is one of the many species restricted to calcareous rocks that is rare in the Smokies, but widespread in temperate eastern North America. Look for the it next time you stop at Ace Gap.

KEY FEATURES: Immersed crustose thallus, black perithecia immersed in the substrate, 2-celled colorless spores, on calcareous rocks and rare.

Thelidium minutulum

Tiny Twin-Spores

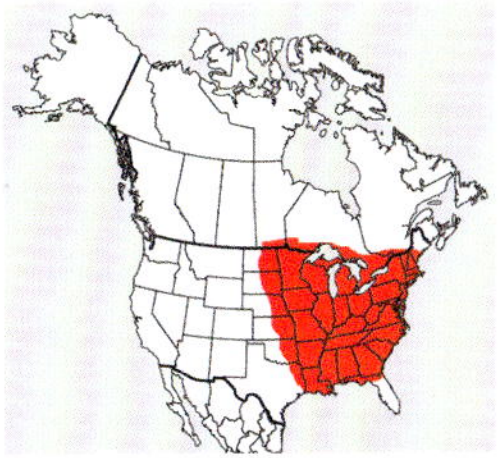

Lendemer 44759 (photo: Tripp)

NOTES: Normally lichenologists train people to avoid peritheciate lichens on rock, especially if they have a poorly developed thallus or one that is immersed in the substrate. The reason for this is that most of the time such lichens turn out to have simple spores and belong to the genus *Verrucaria*, which is hopeless pit of despair in most of North America. However completely avoiding them would mean that you would miss the pleasure of occasionally finding a lichen with septate spores that belongs to another genus and is possible to identify. Such is the case for *Thelidium minutulum*, which grows on shaded rocks, has a poorly developed thallus, black perithecia, and broadly ellipsoid 2-celled colorless spores.

CHEMISTRY: No substances. Spot tests. K-, C-, KC-, P-, UV-.

NICHE: This is a widely distributed, if overlooked, species that occurs throughout temperate eastern North America on shaded non-calcareous or calcareous rocks. In the Smokies, it is known from a single rock outcrop along the Jenkins Ridge Trail near the Tennessee Line at Thunderhead Mountain.

KEY FEATURES: Poorly developed crustose thallus, black perithecia, 2-celled colorless spores, on non-calcareous or calcareous rocks.

Thelocarpon laureri

Lemon Drops

Tripp 3713 (photo: Lendemer)

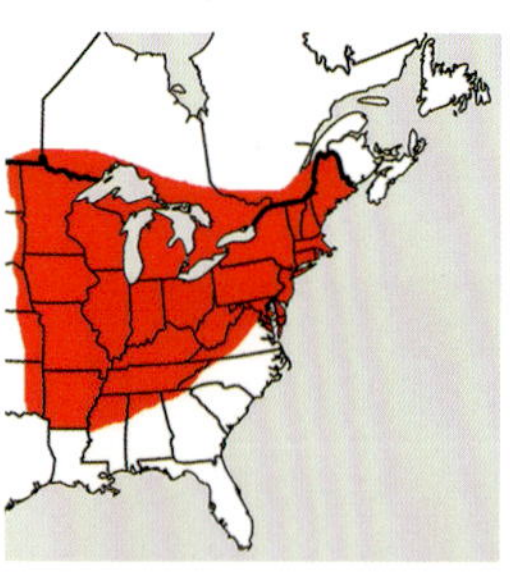

NOTES: A more appropriate common name might be Little Lemon Drops because this species is so small and easily overlooked. In the field, *Thelocarpon laureri* is visible only as tiny yellow dots on the rocks or wood it grows on, and careful examination with a hand lens or microscope is needed to discern the pale, tan-colored ostiole of the perithecia that stick up just above the tops of the yellow areoles that surround them. The yellow thallus and pale perithecia, coupled with the presence of many small ellipsoid colorless spores in each ascus is distinctive and readily separates this species from others. It is most likely to be confused with species of *Candelariella*, although that genus produces true apothecia and not perithecia. It is possible that the pycnidia of *Cyphelium* species with yellow areolae thalli could be confused with *T. laureri*, but in such cases the tiny structures resembling spores would not be contained in bag-like asci.

CHEMISTRY: Pulvinic acid. Spot tests. K-, C-, KC-, P-, UV-.

NICHE: This species is widespread in temperate eastern North America, but is known mostly from scattered occurrences throughout the region. It grows on old wood, especially of wooden bridges and fence posts, and on noncalcareous rocks, often in disturbed habitats. In the Smokies, it is known from a single large rock outcrop near Lonesome Pine.

KEY FEATURES: Small, yellow areolate thallus, pale immersed perithecia, many tiny colorless simple ellipsoid spores, on old wood and noncalcareous rocks.

Thelopsis rubella

Curiouser and Curiouser

Lendemer 30279 (photo: Lendemer)

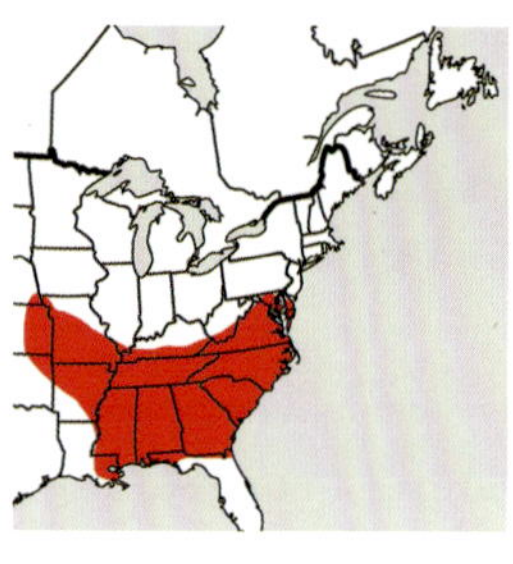

NOTES: Here is a curious little crust, and when we say little crust, we mean it! *Thelopsis rubella* is characterized by its tiny apothecia with a particular curvature that gives them a perithecium-like appearance. They are typically pale beige in color and bear polysporous asci—making this lichen relatively easy to identify through microscopy, if found! Moreover, the spores are 4-celled or 6-celled. There are a limited number of other lichens with polysporous asci in the Smokies, and none of them look like this. Externally the species is very similar to *Topelia aperiens*, but that species has 8-spored asci and globose, muriform spores. *Thelopsis inordinata*, which is not known from the Smokies, shares the polyspo-

rous asci with *T. rubella*, but differs in having muriform globose spores as in *T. aperiens*.
CHEMISTRY: No substances. Spot tests: K-, C-, KC-, P-, UV-.
NICHE: This species occurs throughout the Smokies and southern Appalachians at middle-to-low elevations, but there are few known collections presumably because it is overlooked due to its small size. Although the data are a little spotty, it appears to be widespread throughout the southeastern United States and tends to be found on the bark of hardwoods in moist or humid habitats.
KEY FEATURES: Indistinct, pale brown crustose thallus, partially immersed tan-to-reddish-brown perithecioid-apothecia, polysporous asci containing many 4-celled (or six-celled) colorless spores, on hardwood bark in moist habitats.

Thelotrema subtile

To Live and Die in Dixie

Lendemer 44588 (photo: Tripp)

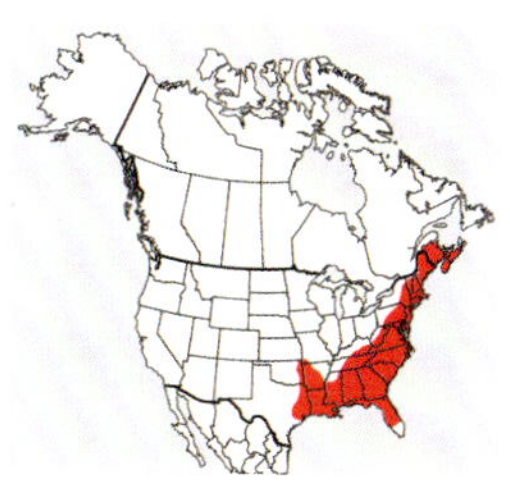

NOTES: *Thelotrema subtile* is a blast from the Coastal Plain. It and other members of the Thelotremataceae are warm-loving species that make their way to the Smokies at low elevations in humid forests. This species is characterized by dirty white to whitish-gray (or sometimes bluish-gray) crustose, continuous thalli with apothecia that are immersed in pores within the thallus and substrate. This gives an appearance of buried flasks and no other group of lichens besides Thelotremataceae look like this. *Thelotrema subtile* is a sexual species with > 2-celled colorless spores. Oh, and it also has an attractive *Trentepohlia* photobiont. Do not confuse this species with *Leucodecton subcompunctum*, which is another Coastal Plain element but differs in having brown, muriform spores. The generic name ("Thelo-") means nipple in Greek.
CHEMISTRY: No substances. Spot tests. K-, C-, KC-, P-, UV-.
NICHE: To Live and Die in Dixie occurs exclusively on bark of hardwoods such as Sweet Birch, Tulip Poplar, and Sugar Maple. It grows in low-elevation humid, rich cove forests and is very common in such areas. Look for it on Schoolhouse Gap Trail between the junction of Laurel Creek Road and Turkeypen Ridge Trail.
KEY FEATURES: Crustose lichen with grayish- or bluish-white, continuous thallus, apothecia immersed in pores within substrate, colorless, transversely (>2-celled) septate spores, very common in low elevations, on hardwoods.

Trapelia coarctata

An Emergent Crisis

Tripp 2426 (photo: Lendemer)

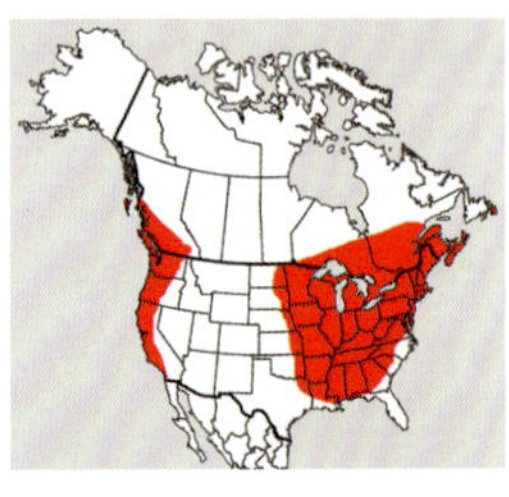

NOTES: *Trapelia coarctata* can be recognized by its continuous blue-gray thallus that is esorediate, C+ pink, and has erumpent apothecia. In case you are wondering, erumpent apothecia are those that erupt out of the thallus, a process so traumatizing that it usually results in ragged bits of thallus around the margins of the apothecia. Often, the apothecia of *T. coarctata* erupt in such a way as to first appear as white bumps on the thallus surface, with some of the apothecia eventually emerging to shed the veil of dead tissue on top of them. *Trapelia placodioides* is similar, but produces soralia along the cracks of the thallus. Likewise, while *T. stipitata* also has erumpent apothecia, these develop a thick stipe and regenerate centrally, causing the buildup of multiple exciples at the margins of the apothecia. Although *T. glebulosa* is also esorediate, it differs in having a dispersed areolate thallus and apothecia that are not strongly erumpent.

CHEMISTRY: Gyrophoric acid. Spot tests. K-, C+ pink, KC+ pink, P-, UV-.

NICHE: This species is widespread in temperate and mountains regions of North America where it occurs on non-calcareous rocks, often in shaded humid habitats. It occurs throughout the Smokies, but is especially common in the wetter northern parts of the Park.

KEY FEATURES: Continuous blue-gray-to-pinkish thallus, C+ pink, esorediate, erumpent apothecia, on non-calcareous rocks especially in humid habitats.

Trapelia glebulosa

Hardscrabble

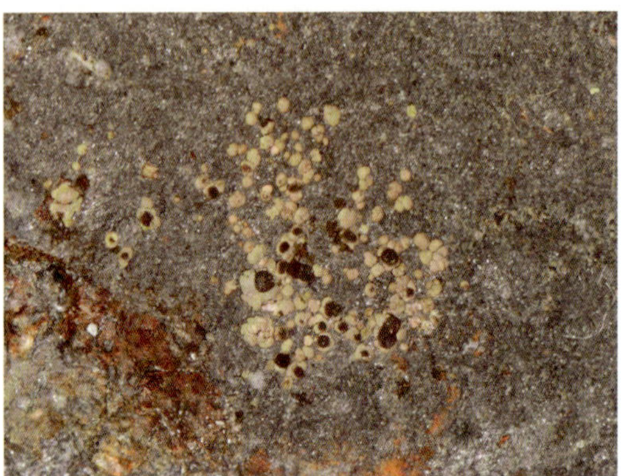

Lendemer 33290 (photo: Tripp)

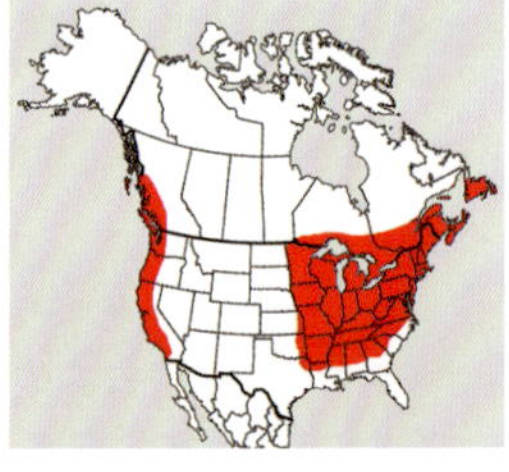

NOTES: In the past, just about every *Trapelia* with an areolate thallus and non-erumpent apothecia was referred to as *T. glebulosa* in North America. However, many such records from shaded, wet forests of the Appalachians were found to have large apothecia with stipes rooted deep into the thallus, and multiple carbonized margins that built up from the regeneration of the hymenium over time. These were eventually named *T. stipitata*, a species that is similar to *T. glebulosa* and also occurs in the Smokies. In many respects, confusion with *T. coarctata* is unlikely because that species has a continuous thallus and the apothecia erupt through the thallus surface.

CHEMISTRY: Gyrophoric acid. Spot tests. K-, C+ pink, KC+ pink, P-, UV-.

NICHE: This species is widespread in many parts of North America, where it occurs on exposed rock outcrops and even on pebbles in disturbed habitats. It is infrequent in the Smokies, but occurs on sun-exposed rocks such as at Lonesome Pine.

KEY FEATURES: Areolate blue-gray to pinkish thallus, C+ pink, esorediate, apothecia without stipes or multiple margins, on non-calcareous rocks especially in open habitats.

Trapelia placodioides

Crumblin' Crusty

Tripp 2395 (photo: Lendemer)

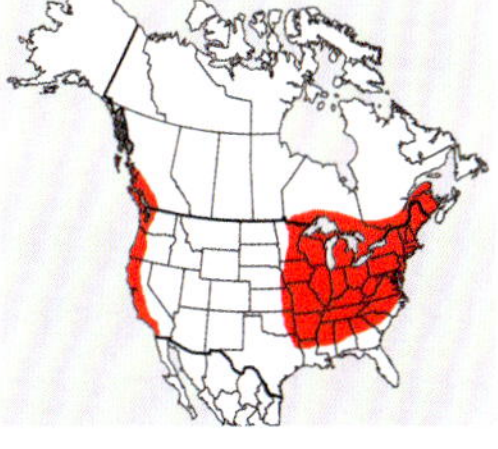

NOTES: There are no other species like *Trapelia placodioides* in eastern North America. It can easily be recognized by its continuous, placodioid thallus that is C+ pink and produces soralia along the cracks of the thallus. *Trapelia coarctata* and *T. stipitata* are similar, but both are esorediate. In some respects, *Herteliana schuyleriana* might be confused with *T. placodioides*, and both species do occur on shaded non-calcareous rocks. However, *H. schuyleriana* produces bud-like blastidia instead of soralia and is C- because it produces atranorin together with roccellic acid.

CHEMISTRY: Gyrophoric acid. Spot tests. K-, C+ pink, KC+ pink, P-, UV-.

NICHE: *Trapelia placodioides* is a typical species of non-calcareous rocks throughout temperate eastern North America and occurs in both shaded and exposed habitats at nearly all elevations. It is extremely common in the southern Appalachians and is widespread in the Smokies.

KEY FEATURES: Continuous blue-gray to pinkish thallus, C+ pink, sorediate along cracks in the thallus, on non-calcareous rocks at all elevations.

Trapeliopsis flexuosa

Lead Shot

Lendemer 33278 (photo: Tripp)

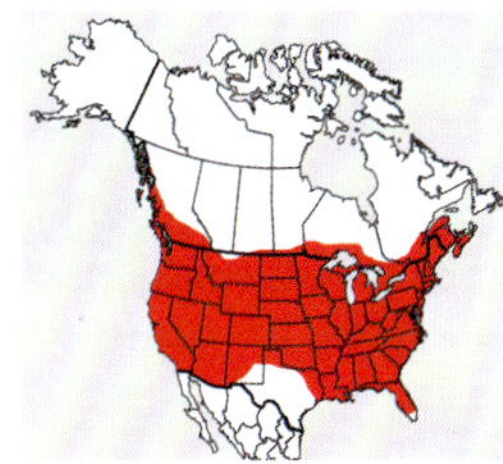

NOTES: *Trapeliopsis* is an attractive and easy-to-learn group of soil-loving and wet-habitat-loving crustose lichens that, in the Smokies, are always sorediate. We have four species in the Park–all of which are common–and they can be differentiated easily by their pigmentation, substrate, and chemistry. When you think you have found a *Trapeliopsis*, the first action to take is to do a C test because nearly all species in the genus react C+ bright pink (this applies to all Smokies species except *T. gelatinosa*). *Trapeliopsis flexuosa* can be identified by its lead gray (except green when wet, as in the photograph), areolate thallus that bears similarly gray and sometimes darker gray soralia, its lack of orange pigment spots, and its substrate (see Niche). This species is most likely to be confused with *T. granulosa*, which differs in having a thallus usually dotted by orange-to-orangish-brown pigments and in having soredia that are frequently lighter in color than the thallus. *Trapeliopsis flexuosa* is sometimes found with apothecia (always in addition to soredia), as in the photograph shown here. For the inquiring mind: spores of this species are smaller than those of *T. granulosa*.

CHEMISTRY: Gyrophoric acid. Spot tests. K-, C+ pink, KC+ pink, P-, UV-.

NICHE: *Trapeliopsis flexuosa* is widespread at middle-to-high elevations throughout the Park. It grows on a variety of substrates that includes humus and bryophytes, but its preference is for old, dry, weathered hardwood. Think of the trail signs that you study for mileage and direction. Next time you see one, look for *T. flexuosa*. There is a high probability it will be there. This species is also widespread on similar substrates throughout the United States.

KEY FEATURES: Areolate, gray crustose lichen with darker gray soredia, no orange pigments in thallus, C+ pink medulla, old dry wood and stumps, middle-to-high elevations, very common.

Trapeliopsis gelatinosa

Through the Looking Glass

Lendemer 32973 (photo: Tripp)

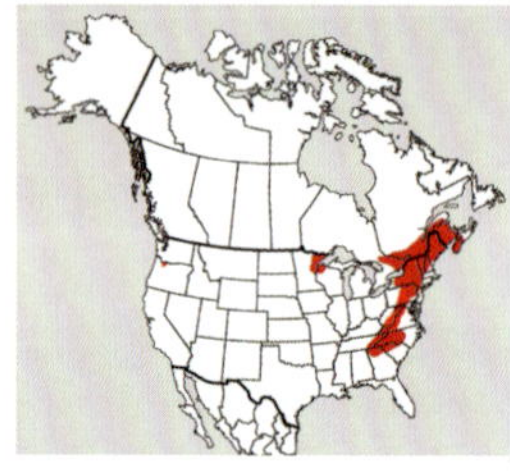

NOTES: *Trapeliopsis gelatinosa* is an attractive species characterized by a shiny, thin olive green thallus that is not areolate (as in *T. flexuosa*) but is rather continuous and bears small, pale green soralia. In addition to soralia, this species is sometimes found with relatively large, brownish-black apothecia, as in the photograph. This species is distinguished from all other *Trapeliopsis* in the Smokies because it is the only one that lacks gyrophoric acid and thus does not react C+ red, which is otherwise very typical of the genus. This lichen also has a distinctive ecology (see Niche).

CHEMISTRY: No substances. Spot tests. K-, C-, KC-, P-, UV-.

NICHE: The center of distribution of *Trapeliopsis gelatinosa* is in boreal forests of the northern Appalachian Mountains. This species is disjunct in the Smokies, where it is occasional and restricted to high elevations in spruce-fir forest. It grows almost exclusively on humus. Look for a population of this species near the summit of Mt. Love and, if found, give it some affection.

KEY FEATURES: Thin, shiny, continuous, olive-green thallus, with soralia sometimes in addition to black apothecia, C- medulla, on humus at highest elevations in Park.

Trapeliopsis granulosa

Tardigrade Jamboree

Tripp 3741 (photo: Lendemer)

NOTES: *Trapeliopsis granulosa* is characterized by its grayish-white thallus that ranges from continuous to areolate, but always with distinctive warts or bumps. It usually bears a brownish-orange pigment, and so, too, does this pigment occur among rather erratically arranged soredia that cover the surface of the thallus. This orange pigment readily differentiates this species from *T. flexuosa*, which lacks it. Additionally, the ecology of *T. granulosa* is different from that of *T. flexuosa* (see Niche). From *T. viridescens*, this species differs in its thallus color (green in the latter species). Finally *T. gelatinosa* is C- and thus not confusable with *T. granulosa*.

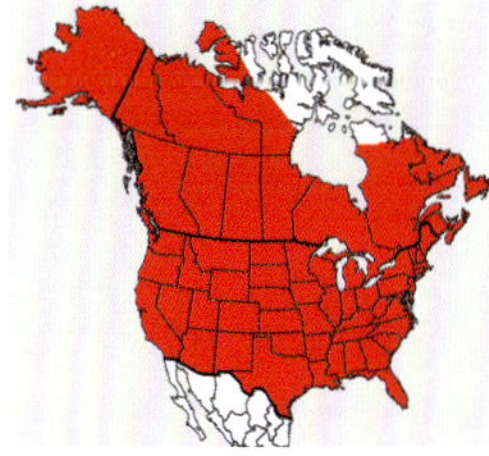

CHEMISTRY: Gyrophoric acid. Spot tests. K-, C+ pink, KC+ pink, P-, UV-.

NICHE: Tardigrade Jamboree is exceptionally widespread throughout all parts of North America. In the Smokies, it occurs primarily at middle elevations (with some "noise" around the elevational edges) and grows on rock, soil, or humus. Look for this species in mixed hardwood forest, such as the Sugar Maple/Sweet Birch/Black Gum stands on Snake Den Trail not far from its junction with Maddron Bald Trail.

KEY FEATURES: Crustose lichen, continuous to areolate gray-to-dirty-white thallus, with warts or bumps, sorediate (sometimes in addition to pale apothecia), with dull orange pigment in thallus and soredia, C+ pink medulla, middle elevations, extremely common on soil, humus, and rock.

Trapeliopsis viridescens

De Composition

Lendemer 33288 (photo: Tripp)

NOTES: *Trapeliopsis viridescens* is a beautiful member of the genus that is quickly recognizable by its distinctive seafoam-green-colored thallus

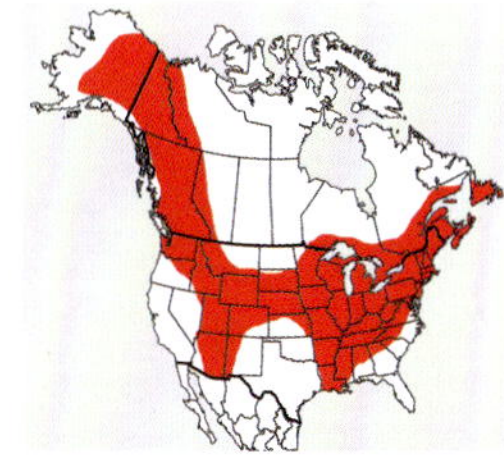

that is composed of minute areoles that dissolve into soredia. These soredia glom together to form a thick, well-developed, continuous crust over the substrate (see Niche). No other member of the genus has a thallus like this. This species can be found with apothecia (always in addition to soredia) on occasion, as in the photo. It seems that *Trapeliopsis* is good at undergoing multiple kinds of reproduction all at once, no? The most likely species to be confused with *T. viridescens* are members of the *Micarea prasina* group, but those have C- thalli and pale apothecia with 2-celled spores.

CHEMISTRY: Gyrophoric acid. Spot tests. K-, C+ pink, KC+ pink, P-, UV-.

NICHE: *Trapeliopsis viridescens* occurs in middle and high elevations, where it grows primarily on logs and stumps. You can find this species in abundance in both rich cove hardwood forest (perhaps in and among old homesteads) as well as in spruce-fir forests on occasion. Look for a nice population of this species near the start of the trail that goes to the summit of Purchase Knob. This species is widespread in similar habitats throughout mesic or sub-mesic regions of North America.

KEY FEATURES: Evenly thick, continuous sea-foam-green thallus that is composed of minute, sorediate granules, sometimes also bearing pale brown apothecia, C+ pink medulla, common old, rotting wood and stumps at middle and high elevations.

Tricharia santessonii

Cassini

Hollinger 4649 (photo: Lendemer)

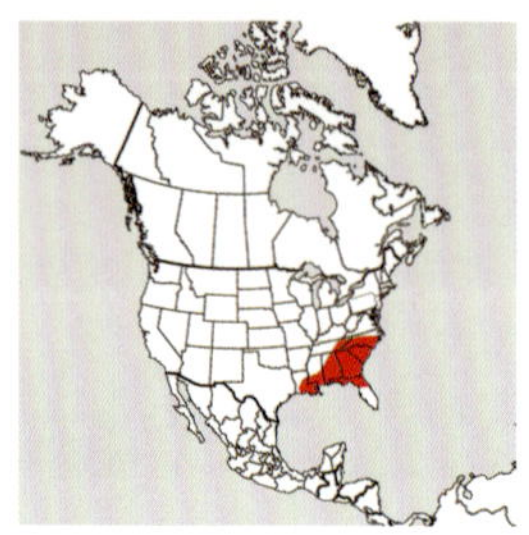

NOTES: Oh *Tricharia*, how does one know thee? This lichen is related to *Gyalideopsis* and would most likely be confused with species of that genus if one were to miss the fact that many of the lashes sticking up from the thallus are sterile black hair-like setae rather than hyphophores that host packets of conidia. Thankfully, most of the species of *Gyalideopsis* that occur in the Smokies do not grow on leaves, and the one that does (*G. pusilla*) produces abundant hyphophores that are bent and slightly widened apically. Once one recognizes the abundance of sterile black setae, then you're golden, because *T. santessonii* is the only member of the genus in the Smokies. So far as known . . .

CHEMISTRY: No substances. Spot tests. K-, C-, KC-, P-, UV-.

NICHE: A species of evergreen leaves in the southern Appalachians and Coastal Plain of southeastern North America. *Tricharia santessonii* is found in the nicest, wettest forests of the Smokies. Look for a waterfall or a swift-moving creek and then avert your gaze to the leaves of the *Rhododendrons*.

KEY FEATURES: Small shiny green thalli on leaves, sterile black lash-like setae, small brown apothecia, submuriform spores, on leaves in the wettest of places near waterfalls and streams.

Trypethelium virens

Beech Sucker

Tripp 2635 (photo: Lendemer)

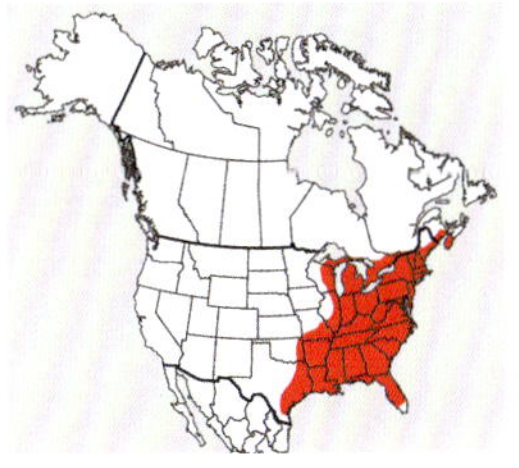

NOTES: *Trypethelium virens* is a lovely species that is exceptionally common throughout eastern North America and, as such, is one you will want to learn without delay! It is very easily identified by its smooth, shiny, continuous thallus that is yellow-to-yellow-brown to yellowish-green in color and its huge, brown perithecia that are clustered into pseudostroma and generally raised above the thallus. This species is only potentially confusable with *Bathelium carolinianum*, which also has aggregated perithecia arranged in a raised pseudostroma, but that species differs by having 4-celled spores, whereas spores of *T. virens* are >4-celled. Additionally, the pseudostroma in *T. virens* are the same color as the thallus and lacks a yellow pigment that can be found in the pseudostroma of *B. carolinianum*.

CHEMISTRY: No substances. Spot tests. K-, C-, KC-, P-, UV-.

NICHE: *Trypethelium virens* is widespread throughout eastern North American but definitely has an affinity for warm, subtropical places. In the Smokies, it is similarly common but most abundant at low elevations in south-facing habitats. Look for this species in riparian acid cove forest on a variety of hardwood substrates, but especially growing on *Ilex opaca* (American Holly).

KEY FEATURES: Continuous, yellowish-green thallus, large, clustered perithecia, these aggregated into pseudostroma that inside are concolorous with the thallus, spores transversely septate and >4-celled, warm, south-facing acid forests at low elevations, common.

Tuckermanella fendleri

Little Edwards

Dey 31420 (photo: Tripp)

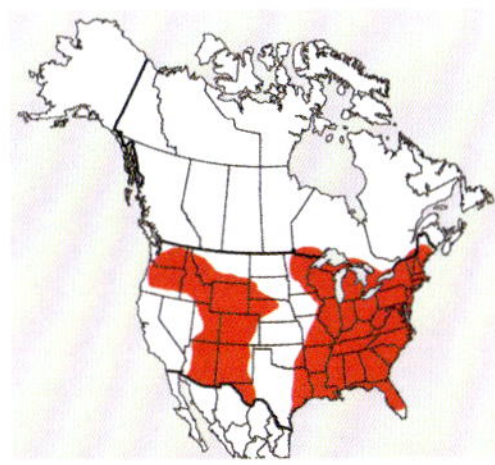

NOTES: *Tuckermanella fendleri* is one of several brown foliose lichens in the Park that all have "Tuckerman" somewhere in its name.

(Well, what can we say? He was an important fellow. . . .) This species is characterized by its small, olive-green-to-brown thalli that are usually dull in appearance (vs. shiny, as in some other brown foliose lichens), adnate to the substrate, and K-, C-, P-, and UV-, pale-to-nearly-pure white lower surfaces, absence of asexual propagules but often presence of apothecia, and its niche (see below). It is most likely to be confused with various species of *Tuckermanopsis* present in the Park that also lack asexual propagules (and usually have apothecia). *Tuckermanopsis ciliaris* differs in having a C+ red medulla and *Tuckermanopsis orbata* differs in having distinctly ruffled lobes that are not adnate to the substrate.

CHEMISTRY: Fatty acids. Spot tests. K-, C-, KC-, P-, UV-.

NICHE: *Tuckermanella fendleri* is an uncommon species that, if found, will likely be in a mid-elevation habitat dominated by acid environment plant species including White Pine, Virginia Pine, Southern Red Oak, Scarlet Oak, Chestnut Oak, and Sourwood. This species, like most brown foliose lichens of the "Tuckerman group," prefers to grow on pines, and in particular pine branches. Look for it fallen from the canopy on such a substrate, perhaps during your next hike up Beauregard Ridge. Elsewhere, it grows throughout eastern and western North America but is conspicuously absent from the Great Plains.

KEY FEATURES: Petite, brown foliose lichen, pale underneath, no asexual propagules, generally with apothecia.

Tuckermanopsis ciliaris

Trailrunner

Tripp 3739 (photo: Lendemer)

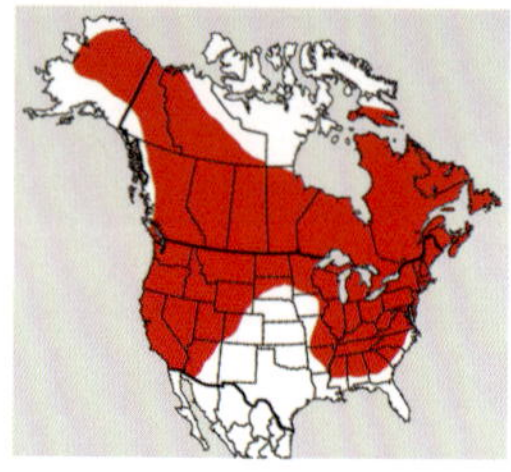

NOTES: *Tuckermanopsis ciliaris* is characterized by comparatively large, olive-green-to-greenish-brown foliose thalli that are pale underneath, generally bear apothecia, and have a medulla that reacts C+ red. It is most likely to be confused with either *Tuckermanopsis orbata* or *Tuckermanella fendleri*, both of which differ by having a C- medulla. It might also be confused for *Melanohalea exasperata*, but that species has pseudocyphellae present on the tips of small bumps that arise from the cortex.

CHEMISTRY: Olivetoric acid. Spot tests. Cortex: K-, C-, KC-, P-, UV-; medulla: K-, C+ red, KC+ red, P-, UV-.

NICHE: This species grows in acid cove forests at all elevations and is quite common there. Look for it on fallen Hickory or Hemlock branches, or alternatively growing on living branches of Mountain Laurel or *Rhododendron*. Elsewhere, this species is similarly common in comparable habitats throughout North America, excepting large portions of the Great Plains.

KEY FEATURES: Relatively large, olive-brown foliose thalli, pale underneath, usually with apothecia, C+ red medulla, common at middle elevations in acidic forests.

Tuckermanopsis orbata

Spruce Globes

Tripp 2280 (photo: Lendemer)

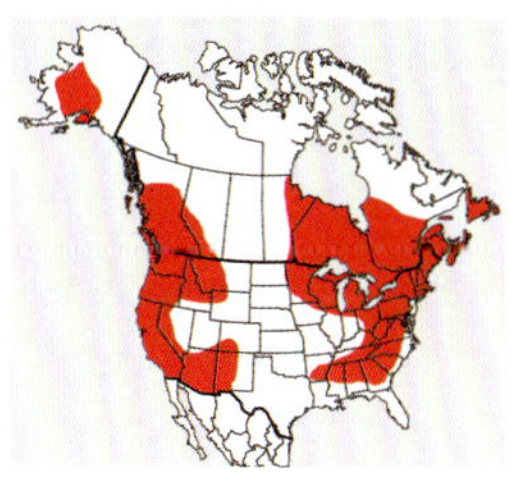

NOTES: *Tuckermanopsis orbata* is an attractive species characterized by small, olive-green-to-greenish-brown foliose thalli that have lobes distinctly ruffled and ascending away from the substrate. It has a C- medulla and therefore is most likely to be confused with *Tuckermanella fendleri*, which is also C- but differs in having smaller lobes that are adnate to the substrate. Another close relative, *Tuckermanopsis ciliaris*, differs in having a C+ red medulla.

CHEMISTRY: Fatty acids. Spot tests. K-, C-, KC-, P-, UV-.

NICHE: *Tuckermanopsis orbata* is relatively common in the Park, assuming you are looking for it in its preferred habitat. This species occurs at high elevations in acidic environments, usually on branches. Look for it on both hardwood branches and conifer branches, including species such as Eastern Hemlock, Red Spruce, Yellow Birch, or high elevation species of *Rhododendron*. Or maybe take a walk to Andrew's Bald or Black Camp Gap, where it has been collected in the past.

KEY FEATURES: Small, greenish-brown thalli with ruffled, ascending lobes, usually with apothecia, C- medulla, common at high elevations.

Tylophoron americanum

Decisions and Revisions

Tripp 3590 (photo: Lendemer)

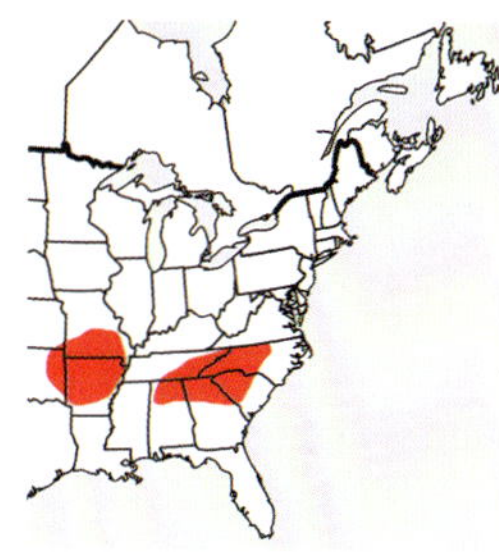

NOTES: *Tylophoron americanum* is one of the many species that we have discovered and named as a result of our work in the Smokies during the last decade. It has a distinctive blue-gray thallus with convex, white sporodochia that produce colorless conidia composed of either one or several cells in number. *Cheiromycina flabelliformis* is somewhat similar but is much smaller and has multicelled conidia joined at the base to resemble a hand with articulated fingers.

CHEMISTRY: Unknown substance related to leprapinic acid. Spot tests. K+ yellow, C-, KC-, P-, UV-.

NICHE: This species is endemic to the southeastern United States where it occurs in old growth forests on bark in grooves of mature hardwood trees. It is infrequent in the Smokies, but often found at middle and low elevations in very high quality forests.
KEY FEATURES: Crustose blue-gray thallus, *Trentepohlia* photobiont, white convex sporodochia, colorless conidia, UV- thallus, on bark of mature hardwoods at middle-to-low elevations.

Umbilicaria mammulata

American Rock Tripe

Tripp 2266 (photo: Lendemer)

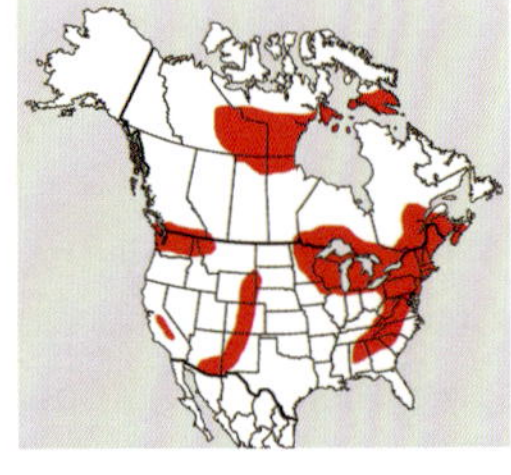

NOTES: Rock Tripe, need we say more? If you have spent time traipsing about eastern forests, you are probably already familiar with this species. This is the huge foliose lichen characterized by having a brown-to-brownish-gray upper surface, a dark black lower surface with abundant rhizines that give it a furry look, and a single point of attachment to its substrate (hence: "umbilicate" and *Umbilicaria*). There are a lot of other features one could write about here, but we don't think you will need them to identify this species. American Rock Tripe is most likely to be confused with another large, brown, foliose lichen that is also very common in the Park and has a single point of attachment to the substrate, *Lasallia papulosa*, but that species has very distinctive bumpy pustules covering the upper surface, in contrast to the smooth upper surfaces of *Umbilicaria mammulata*. American Rocky Tripe might also be confused for *U. muhlenbergii*, but that species lacks rhizines on the lower cortex and is also very rare.
CHEMISTRY: Gyrophoric acid. Spot tests. Cortex: K-, C-, KC-, P-, UV-; medulla: K-, C+ pink, KC+ pink, P-, UV-.
NICHE: American Rock Tripe is very common in the Smokies, particularly in rich hardwood forests with ample non-calcareous rocks dominated by species such as Sugar maple, Silverbell, Buckeye, and American Beech. This species always occurs on rocks. A story goes that this species helped keep George Washington's troops alive when they ran out of food during the freezing winter of 1777 in Valley Forge, Pennsylvania. We can't confirm the extent to which this species sacrificed itself for such a patriotic cause, or how nutritious it was upon boiling, but maybe when mixed with honey and corn meal, it could be sold today as a delicacy in New York City markets?
KEY FEATURES: Very, very large brown-to-gray foliose thalli with smooth upper surfaces a single point of attachment, dense black rhizines covering undersurfaces, sometimes but usually lacking apothecia (and never with asexual propagules), always on rocks, very common in rich forests with abundant rocks.

Usnea angulata

Fisherman's Dream

Tripp 2610 (photo: Lendemer)

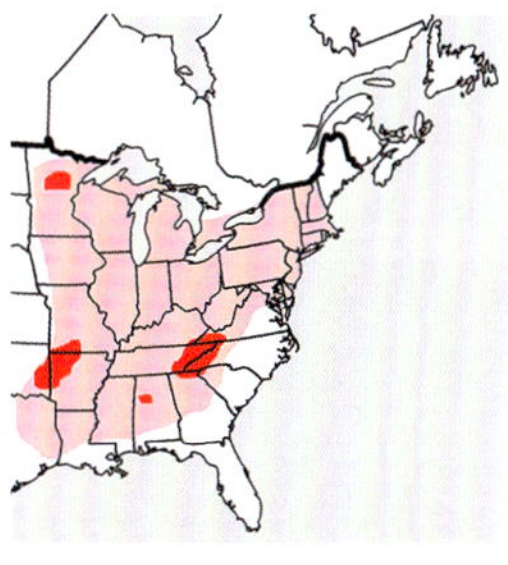

NOTES: *Usnea angulata* is just about as conspicuous and charismatic of a macrolichen as you can possibly get. It forms extensive thalli, often multiple in feet, that dangle down from the canopy branches of trees. Although other species of *Usnea* have such "pendant" thalli, this is the only one with ridged branches. These characteristic ridges resemble the backbone of a fish, hence the common name. The species was once widespread throughout temperate eastern North America (shaded light red in the map) but is now know only from a few areas, most commonly in the southern Appalachians. *Usnea angulata* illustrates that, in many cases, we know enough about lichens to understand whether a given species is threatened or endangered.

CHEMISTRY: Usnic acid and norstictic acid. Spot tests. Cortex: K-, C-, KC+ yellow, P-, UV-; medulla: K+ yellow turning red, C-, KC-, P+ yellow, UV-.

NICHE: Once widespread in temperate eastern North America, but now much more narrowly restricted. In the southern Appalachians, it hangs from the branches of trees and shrubs, particularly mature hemlocks and pines along rivers, and on junipers on mountain bluffs. In the Smokies, it was previously widespread along low- and middle-elevation streams and rivers, but has likely become much rarer with the decline of the hemlocks.

KEY FEATURES: Large, pendant fruticose thallus, strongly ridged branches, K+ red medulla, hanging from branches along streams and rivers.

Usnea ceratina

Pendant Wonder

Lendemer 33247 (photo: Tripp)

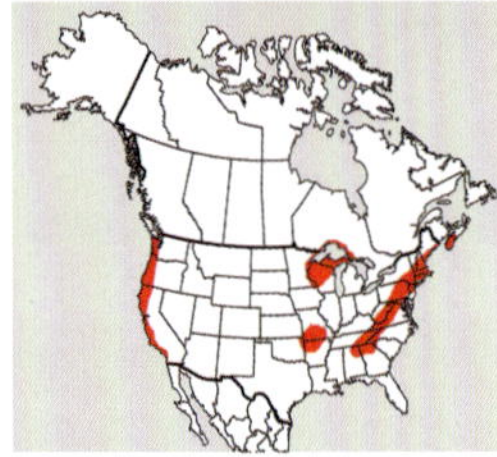

NOTES: Much like *Usnea angulata*, this pendant *Usnea* species was historically more widespread in eastern North America and now is common only in the southern Appalachians. When you see thalli of *U. ceratina* several feet in length and hanging down from the canopies of trees, you can understand why *Usnea* is called Old Man's Beard, but we prefer Pendant Wonder. This species is most likely to be confused with *U. subgracilis*, which differs in having a P+ red medulla, and a soft, lax, pliable thallus with regular, perpendicular side branches. The thallus of *U. ceratina* is always stiff and hard when it is dry. *Usnea entoviolata* is chemically similar, but differs in having smaller, shrubby thalli, and branches covered with excavate soralia rather than mound-like raised tubercles.

CHEMISTRY: Usnic acid and diffractaic acid. Spot tests. Cortex: K-, C-, KC+ yellow, P-, UV-; medulla: K-, C-, KC-, CK+ yellow, P-, UV-.

NICHE: *Usnea ceratina* has a disjunct distribution North America, occurring in the Appalachian Mountains and Great Lakes, the Ozarks, and coastal western North America. In eastern North America, it is only truly common in the southern Appalachians, where it is widespread and commonly found on the bark and branches of both hardwoods and conifers. In the Smokies, it occurs at all elevations; however it is particularly abundant in northern hardwood forests, where it often drips from canopies of trees. You can see this species on your next visit to the Appalachian Highlands Learning Center.

KEY FEATURES: Pendant, often large, fruticose thallus, smooth branches with white tubercles, K-, CK+ yellow medulla, on bark and branches of trees throughout.

Usnea cornuta

Vienna Sausage

Lendemer 32950 (photo: Tripp)

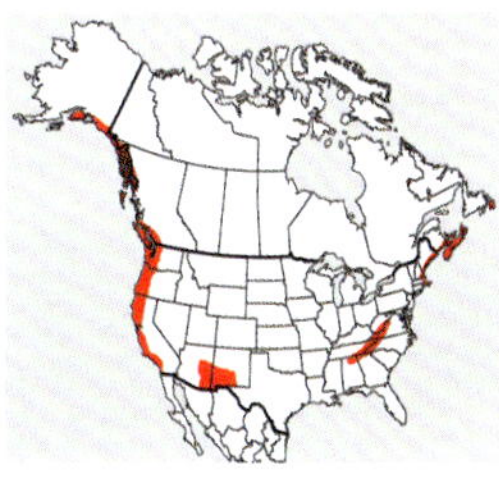

NOTES: One can be sure that *Usnea cornuta* is unmistakable because of its inflated branches and soralia. The branches are always inflated, although the degree of inflation can vary from moderate to extreme. If you aren't sure whether the branches of an *Usnea* are inflated, look at the bases where the branch attaches to the main branch; it will always be distinctly constricted. If you happen to have a razor, you can also carefully slice open the branch to reveal a hollow cavity surrounding the central cord of the branch. *Usnea cornuta* is quite variable chemically and has long been considered to compose a group of closely related species.

CHEMISTRY: Usnic acid and salazinic acid, occasionally with other substances. Spot tests. Cortex: K-, C-, KC+ yellow, P-, UV-; medulla: K+ yellow turning red, C-, KC-, P+ orange, UV-.

NICHE: Like many other species of *Usnea*, *U. cornuta* has a disjunct distribution in North America, occurring in the oceanic forests of the northeast, the middle-to-high elevations of the central and southern Appalachians, coastal western North America, and the high mountains of the desert southwest. In the southern Appalachians, the species is common and widespread at middle-to-high elevations, where it is commonly found on the branches and trunks of both conifers and hardwoods. You should have no trouble finding it on a hike anywhere in the Smokies, especially along a creek or even on a high mountain ridge.

KEY FEATURES: Short, shrubby, fruticose thallus, inflated branches, soralia, on bark and branches of trees throughout.

Usnea dasaea

Proper Prickles

Collector Tripp 1519 (photo: Moroz)

NOTES: Perhaps, dear reader, you might sense a theme with *Usnea*. This species is rare in the Smokies, and in eastern North America is largely known only from historical records collected more than a century ago. Unlike *U. angulata*, however, *Usnea dasaea* appears to have always been rare. This species could easily be confused with other short shrubby-to-subpendant species of the genus, but has a unique appearance due to the production of perpendicular, short, stubby fibrils at regular intervals along the branches. This feature, coupled with the chemistry and the soralia are distinctive. Other species, such as *U. amblyoclada*, are chemically similar, but they always occur on rock and differ in producing copious isidiomorphs (fragile isidia inside soralia) that are often black tipped. *Usnea subgracilis* could also be confused with *U. dasaea*, but has a K- medulla.

CHEMISTRY: Usnic acid, norstictic acid, galbinic acid. Spot tests. Cortex: K-, C-, KC+ yellow, P-, UV-; medulla: K+ yellow turning red, C-, KC-, P+ orange, UV-.

NICHE: Another *Usnea* and another disjunct distribution pattern. In North America, Proper Prickles occurs in coastal southern California, the Ozarks, and the central and southern Appalachians. Clearly, it has an affinity for scenic biodiversity hotspots. As is the case throughout eastern North America, it is very rare in the Smokies and has been found only a small number of times on bark at middle-to-low elevations.

KEY FEATURES: Shrubby to subpendant, fruticose thallus, smooth branches with regularly sized perpendicular fibrils, K+ yellow turning red and P+ orange medulla, on bark and branches of trees at middle-to-low elevations.

Usnea halei

Gneiss Whiskers

Tripp 3708 (photo: Lendemer)

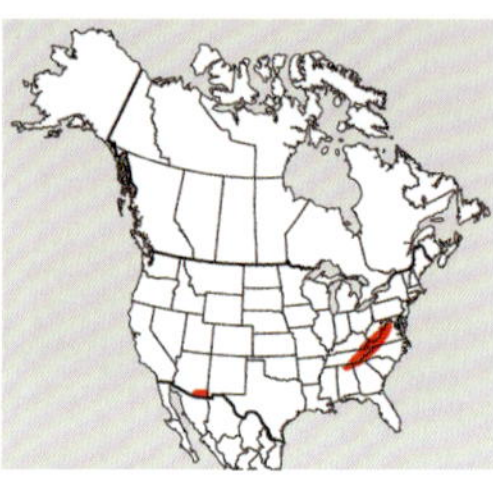

NOTES: For some reason, *Usnea halei* has been a trouble maker from the start—despite the fact that if you see a short, shrubby *Usnea* on a rock in the Smokies, it is almost certainly this one. Historically it was confused with *U. subfusca*, which is a very different species that lacks soralia and grows on bark. Eventually, the name *U. halei* became widely used for all the short shrubby *Usnea* on rocks in eastern North America; however the majority of specimens actually corresponded to *U. amblyoclada*. Although the two species both have shrubby thalli, occur on rocks, and produce abundant isidiomorphs in the soralia, the isidiomorphs of *U. amblyoclada* are often black tipped while those of *U. halei* are not. Unfortunately, while their chemistries are quite different, the spot tests for the two species are essentially the same, unless you were to use the K test to check for norstictic acid crystals in a water mount under the compound microscope. The base of *U. halei* is also often slightly, but distinctly, red in color, and thus it is important not to confuse the species with *U. pensylvanica*. *Usnea pensylvanica* has a K+ dirty yellow medulla and tends to be extensively red pigmented throughout the lower portions of the thallus (see the photograph of that species here).

CHEMISTRY: Usnic acid, norstictic acid, and salazinic acid. Spot tests. Cortex: K-, C-, KC+ yellow, P-, UV-; medulla: K+ yellow turning red, C-, KC-, P+ orange, UV-.

NICHE: *Usnea helei* is common and widespread in the central and southern Appalachian Mountains, where it grows on shaded non-calcareous rocks or rock faces. It is widespread in the Smokies and can be found at nearly all elevations, but only in places where the rock outcrops are not entirely covered by vegetation. There are luxuriant displays of this species on the trail up to the lookout tower on Mt. Cammerer.

KEY FEATURES: Shrubby, fruticose thallus, slightly reddish base, smooth branches with small soralia and isidiomorphs, K+ yellow turning red medulla, on non-calcareous rocks throughout.

Usnea merrillii

Snuggle-me Lichen

Tripp 3813 (photo: Lendemer)

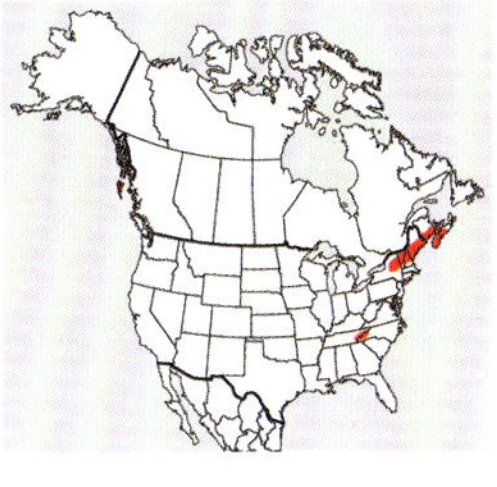

NOTES: *Usnea merrillii* is an extremely elegant species, and it is fitting that it honors George Knox Merrill, an early 20th-century lichenologist who worked in Maine, where the species also commonly occurs. If you look up in a tree in the spruce-fir forests of the Smokies, you will doubtless see a curtain of Old Man's Beard hanging in thin wisps that seem to sway in the breeze. That would be *U. merrillii*, which can easily be recognized by very thin, lax branches that only rarely have side branches and are spaced with narrow white rings all along their lengths. *Usnea ceratina* has thicker branches that lack white rings, and the medulla in that species is P-. *Usnea subgracilis* could be confused with *U. merrillii*, but that species has a P+ red medulla and produces regularly sized fibrils that jut out from the branches at perpendicular angles.

CHEMISTRY: Usnic acid and salazinic acid. Spot tests. Cortex: K-, C-, KC+ yellow, P-, UV-; medulla: K+ yellow turning red, C-, KC-, P+ orange, UV-.

NICHE: The distribution of this species in North America screams "cool and wet all the time." It occurs in some of the wettest forests along the Pacific Coast of British Columbia, and in similarly wet forests along the Atlantic Coast in Maine and adjacent Canada. In the Appalachians, *Usnea merrillii* is restricted to high elevations, particularly spruce-fir forests, where it occurs in Vermont and New Hampshire and then hundreds of miles south in the Smokies, Blacks, and Balsams. If you want to see this species, you need to go no further than the observation tower of Clingman's Dome, where you can look down on it in the canopies of the fir trees while you wait for the clouds to pass so you can see Tennessee.

KEY FEATURES: Pendant, often large, fruticose thallus, smooth branches with white banded rings, K+ yellow turning red medulla, on bark and branches of trees at high elevations, especially in spruce-fir forests.

Usnea mutabilis

Pretty in Pink

Lendemer 33166 (photo: Tripp)

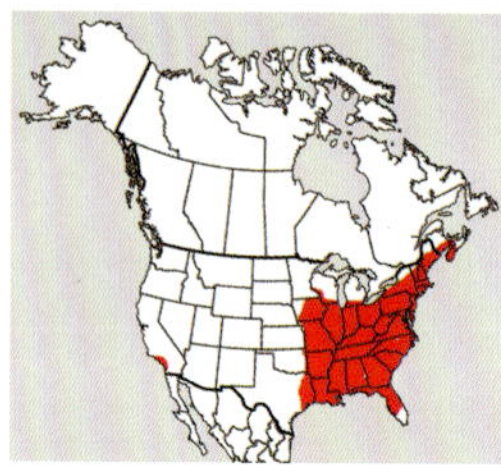

NOTES: This is widely presumed to be the sorediate counterpart to *Usnea strigosa*, which differs from *Usnea mutabilis* in the absence of soralia and the frequent presence of apothecia. The two species are not likely to be confused; however, small thalli of *U. strigosa* that lack apothecia are often misidentified as *U. mutabilis* because the small fibrils produced along the branches are misinterpreted as isidiomorphs forming inside soralia. *Usnea mutabilis* is also externally similar to *U. subscabrosa* because both species have shrubby sorediate thalli. However, *U. subscabrosa* differs in having a white medulla that is P+ red due to the presence of protocetraric acid.

CHEMISTRY: Usnic acid and red/pink pigments. Spot tests. Cortex: K-, C-, KC+ yellow, P-, UV-; medulla: K-, C-, KC-, P-, UV-.

NICHE: This species is widespread throughout temperate eastern North America, with disjunct populations along the Pacific Coast in southern California. It grows on the bark and branches of both conifers and hardwoods, but is most commonly found on conifers and hardwoods with acidic bark. In the Smokies, the species is widespread at all elevations below the spruce-fir zone.

KEY FEATURES: Shrubby, fruticose thallus, pink medulla, branches with small soralia and isidiomorphs, on bark and branches at middle-to-low elevations.

Usnea pensylvanica

A Wine-Dark Sea

Tripp 3687 (photo: Lendemer)

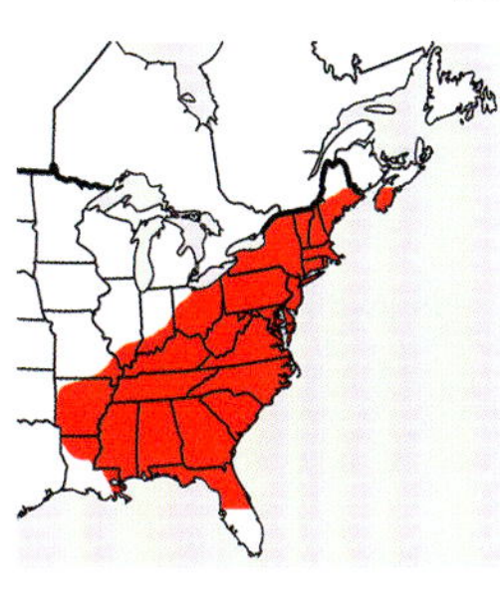

NOTES: *Usnea pensylvanica* was originally described based on a collection made by Henry Muhlenberg in Pennsylvania not long after the Revolutionary War. It belongs to the group of sorediate species with a red pigmented cortex that includes *U. rubicunda*, the name widely used for *U. pensylvanica* in eastern North America. In the Smokies, the species is only likely to be confused with *U. subrubicunda*, a very rare species of high elevations that differs in having a lightly pigmented thallus, appearing light yellow-orange rather than red, and chemically in the production of fumarprotocetraric acid, which is P+ red. *Usnea halei* could also be confused with *U. pensylvanica* because it tends to have a lightly reddish pigmented base, but *U. halei* occurs on rocks and has a K+ red medulla due to the presence of norstictic and salazinic acids.

CHEMISTRY: Usnic acid and stictic acid. Spot tests. Cortex: K-, C-, KC+ yellow, P-, UV-; medulla: K+ yellow, C-, KC-, P+ orange, UV-.

NICHE: A Wine-Dark Sea is widespread and common throughout the Smokies and southern Appalachians, where it occurs extensivelyon the bark and branches of both conifers and hardwoods. Historically, it was common throughout temperate eastern North America, but now is only common in the southeastern United States.

KEY FEATURES: Shrubby, fruticose thallus, strongly red pigmented lower portions, branches with small soralia and isidiomorphs, on bark and branches at middle-to-low elevations.

Usnea silesiaca

Ear Muffs

Tripp 5404A (photo: Lendemer)

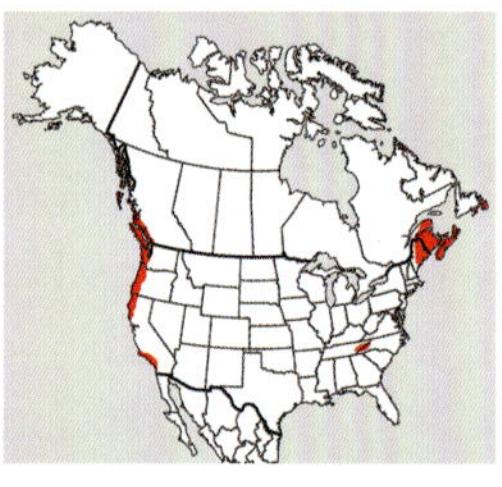

NOTES: *Usnea silesiaca* is easy to overlook as it is likely to blend in with the profusion of *Usnea* that grow in the spruce-fir forests. The subpendant thallus with soralia and a distinctly blackened base are similar to *U. dasopoga*, as is the chemistry. However, Ear Muffs differs from *U. dasopoga* in having large soralia that are wider than half the diameter of the branches whereas those of *U. dasopoga* are small and punctiform, almost never reaching half the diameter of the branches. *Usnea subfloridana* is another species with a distinctly blackened base and soralia, however the thallus in that species is always shrubby and differs chemically in the production of squamatic acid (P-) or thamnolic acid (K+ yellow).

CHEMISTRY: Usnic acid and salazinic acid. Spot tests. Cortex: K-, C-, KC+ yellow, P-, UV-; medulla: K+ yellow turning red, C-, KC-, P+ orange, UV-.

NICHE: *Usnea silesiaca* is a species of the cool, moist spruce-fir forests in the Smokies, where it grows on the bark and branches of trees and shrubs as well as on old standing tree stumps. Outside of the Smokies it is restricted to the high elevations of the southern Appalachians, with disjunct populations in the Canadian Maritimes and Pacific Northwest.

KEY FEATURES: Shrubby to subpendant, fruticose thallus, black pigmented basal portions, branches covered with papillae, soralia large and wider than half the width of the branch, on bark and branches in spruce-fir forests.

Usnea strigosa

Chigger Thicket

Lendemer 33803 (photo: Tripp)

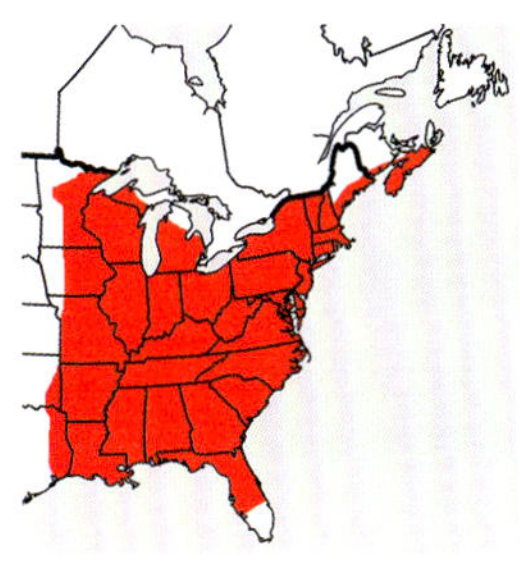

NOTES: If you find an *Usnea* with a shrubby thallus and abundant apothecia then it is almost certainly *U. strigosa* if you are below 5,000 feet in elevation. The identification can be confirmed by the absence of soralia, the lack of black pigment at the base of the thallus, and the frequent presence of a red or pink pigment in the medulla. The only similar species is *U. subfusca*, which replaces *U. strigosa* once you travel to higher elevations in the Smokies. It can easily be distinguished by the blackened base and the production of salazinic acid.

CHEMISTRY: Usnic acid and norstictic acid or psoromic acid, usually with pink-red pigments. Spot tests. Cortex: K-, C-, KC+ yellow, P-, UV-; medulla: K+ yellow turning red or K-, C-, KC-, P+ yellow, UV-.

NICHE: Chigger Thicket is a characteristic species of temperate eastern North America that is widely distributed everywhere except the highest elevations and the northern boreal forests. It is extremely common in the Smokies and can often be found on fallen branches at middle-to-low elevations.

KEY FEATURES: Shrubby, fruticose thallus, unpigmented basal portions, no soralia, apothecia often common, medulla often pink, frequent on bark and branches below 5,000 ft.

Usnea subfloridana

Forest Supervisor

Lendemer 53165 (photo: Tripp)

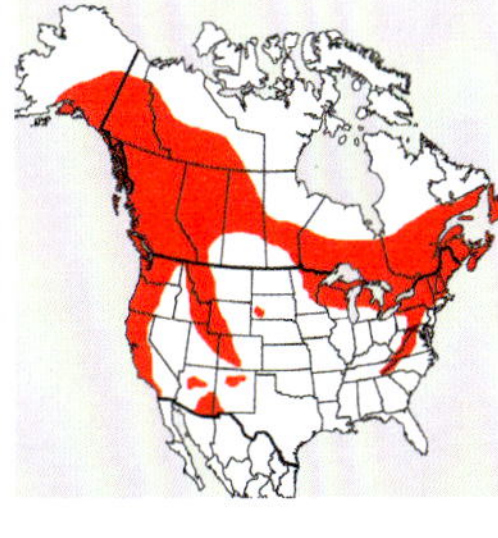

NOTES: *Usnea subfloridana* is one of the most common members of the genus at high elevations in the Smokies. You can easily recognize it by short, shrubby thalli that have abundant soralia on the branches and a distinctly blackened base. *Usnea fulvoreagens* is similar, but has a medulla that reacts K+ yellow turning red and P+ yellow due to the presence of norstictic acid. Thalli of *U. subfusca* that lack their typical apothecia could be confused with *U. subfloridana*, but that species is esorediate and has a K+ yellow turning red medulla due to the presence of salazinic acid.

CHEMISTRY: Usnic acid and squamatic acid or thamnolic acid. Spot tests. Cortex: K-, C-, KC+ yellow, P-, UV-; medulla: K- or K+ yellow, C-, KC-, P- or P+ orange, UV+ blue-white or UV-.

NICHE: *Usnea subfloridana* is a species of northern boreal and northern temperate areas of North America that also occurs at high elevations in the central and southern Appalachians. In the Smokies, it is one of the most common lichens in northern hardwood and spruce-fir forests, where it grows on the bark and branches of hardwoods and conifers.

KEY FEATURES: Shrubby, fruticose thallus, black basal portions, sorediate, medulla white, on bark and branches at high elevations.

Usnea subfusca

Monster Eyes

Tripp 5275 (photo: Lendemer)

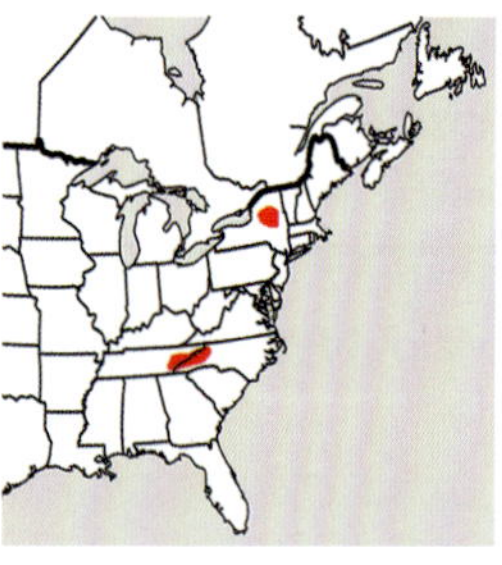

NOTES: Like a hundred eyes watching as you walk down the trail, this species is often lurking above you in the high elevations of the Smokies. The typically abundant apothecia, blackened thallus base, and presence of salazinic acid are distinctive. *Usnea strigosa* is the only other esorediate and commonly fertile member of the genus that occurs in the Smokies and can be distinguished by the lack of a blackened base, its chemistry, and the frequent presence of a pink or red pigment in the medulla. Occasionally, one will find *U. strigosa* and *U. subfusca* growing together at the same location, and in such cases *U. subfusca* typically has more divergent branching and fewer fibrils.

CHEMISTRY: Usnic acid and salazinic acid. Spot tests. Cortex: K-, C-, KC+ yellow, P-, UV-; medulla: K+ yellow turning red, C-, KC-, P+ orange, UV-.

NICHE: The southern Appalachians are the center of the distribution of this eastern North America endemic, which also occurs in the Adirondack Mountains of New York. It is restricted to high elevation forests (above 5,000 ft.) in the Smokies, where it is one of the most common fruticose lichens and is often found growing abundantly on the branches of hardwood trees and shrubs.

KEY FEATURES: Shrubby, fruticose thallus, black basal portions, no soralia, apothecia often common, medulla white, on bark and branches above 5,000 ft.

Usnea subgracilis

First Spring

Tripp 3656 (photo: Lendemer)

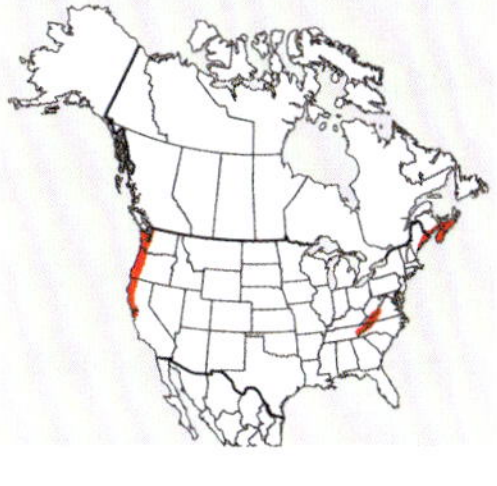

NOTES: Together with *Usnea ceratina* and *U. merrillii*, this species forms the bulk of the biomass of Old Man's Beard that you will see hanging from trees and branches throughout the Smokies. The most distinctive features of *U. subgracilis* are the long, pendant thallus that feels lax and soft if clenched in your hand when it is dry, the regularly spaced perpendicular fibrils on the branches, the occasional white rings that indistinctly divide the branches into segments, and the P+ red medulla due to the presence of protocetraric acid. *Usnea merrillii* is also soft to the touch and has white rings on the branches, but the branches are very thin and brittle in comparison, and it produces salazinic acid instead protocetraric acid. *Usnea ceratina* is hard and stiff in comparison, has conspicuous white tubercles and soralia, and is P-. Small individuals of *U. subgracilis* can be confused with *U. subscabrosa*, but that species has shrubby thalli that are distinctly hard and stiff to the touch, and produces abundant soralia that are often covered with isidiomorphs.

CHEMISTRY: Usnic acid and protocetraric acid. Spot tests. Cortex: K-, C-, KC+ yellow, P-, UV-; medulla: K+ yellow, C-, KC-, P+ red, UV-.

NICHE: Here you have it, another species of cool, moist southern Appalachian forests with disjunct populations along the northern temperate Atlantic and Pacific Coasts. It is common and widespread in the southern Appalachians, particularly in the Smokies, where it can be found on the bark and branches of hardwoods and conifers at all elevations.

KEY FEATURES: Pendant, often large, fruticose thallus, lax or soft and pliable when dry, smooth branches with occasional white rings, P+ red medulla, on bark and branches of trees throughout.

Usnea subrubicunda

Dip of Delight

Lendemer 53177 (photo: Tripp)

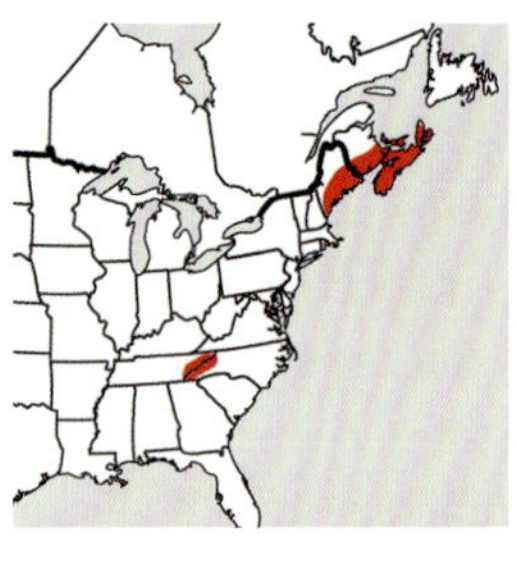

NOTES: As the common name implies, this is one of those species that will make your hike—and maybe even your day. The most distinctive characters of the lichen are a shrubby, isidio-sorediate thallus that just happens to be almost uniformly pigmented a beautiful light yellow-orange color. This pigmentation of the thallus is most likely to cause confusion with the much more common species *U. pensylvanica*. However, the pigment in that species is mottled red and tends to be restricted to the basal portions of the thallus, leaving the rest of the thallus a normal greenish color of *Usnea*. The two species also differ chemically; the medulla of *U. pensylvanica* is P+ orange due to the presence of stictic acid whereas the medulla of *U. subrubicunda* is P+ red due to the presence of fumarprotocetraric acid.

CHEMISTRY: Usnic acid and fumarprotocetraric acid. Spot tests. Cortex: K-, C-, KC+ yellow, P-, UV-; medulla: K+ yellow, C-, KC-, P+ red, UV-.

NICHE: *Usnea subrubicunda* was originally thought to be restricted to the wet, oceanic forests along the Atlantic Coast of northeastern North America; however, we were overjoyed to find it disjunct and equally at home in the fog-cloaked high elevations of the Smokies.

KEY FEATURES: Shrubby, fruticose thallus, isidio-sorediate, medulla white, cortex reddish-orange throughout, on bark and branches at high elevations.

Usnea subscabrosa

Bark Scratch

Tripp 3608 (photo: Lendemer)

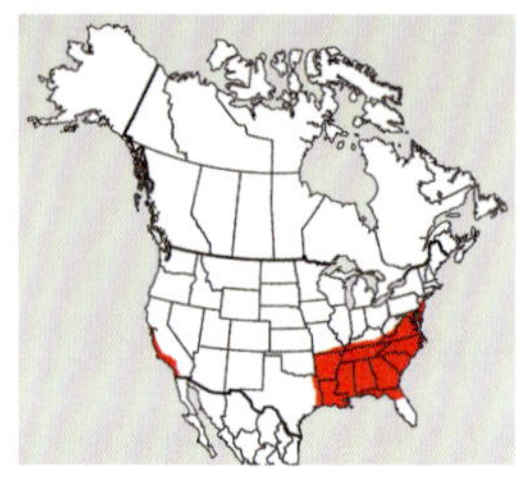

NOTES: Imagine yourself out in the forest, cupping your hand around an *Usnea*, trying to decide if it is soft and lax or hard and stiff. That is the world of lichenology. *Usnea subscabrosa* is a common species of low elevation forests that can be recognized by a shrubby thallus that is distinctly stiff and hard when dry, has abundant soralia with isidiomorphs, and a P+ red medulla due to the presence of fumarprotocetraric acid. *Usnea mutabilis* is the most similar species that occurs with *U. subscabrosa*, but differs in having a P- medulla that is always pink pigmented. Small individuals of *U. subgracilis* can be confused with *U. sub-*

scabrosa, but that species has subpendant-to-pendant thalli that are lax and soft when dry, regularly spaced perpendicular fibrils along the branches, and small, inconspicuous soralia.
CHEMISTRY: Usnic acid and fumarprotocetraric acid. Spot tests. Cortex: K-, C-, KC+ yellow, P-, UV; medulla: K+ yellow, C-, KC-, P+ red, UV-.
NICHE: This is one of the most common and widespread species in southeastern North America. It occurs inland to the southern Appalachian Mountains but only at the middle-to-low elevations, where it is found on the bark and branches of both hardwood and conifers. Like many other *Usnea* species, there is also a disjunct population in coastal southern California.
KEY FEATURES: Shrubby, fruticose thallus that is hard and stiff when dry, non-blackened basal portions, sorediate with isidiomorphs, medulla white, on bark and branches at high elevations.

Usnocetraria oakesiana

The Constitutionalist

Tripp 3822 (photo: Lendemer)

NOTES: *Usnocetraria oakesiana* is a common species in the Smokies that can be recognized on sight because of its diminutive foliose, yellow-green, thallus with crisped, ruffled lobes and marginal soralia. *Ahtiana aurescens* could be confused with *U. oakesiana* because it also has a similarly colored, small rosette-forming thallus with ruffled lobes. However, *A. aurescens* is esorediate and frequently produces apothecia whereas *U. oakesiana* is sorediate and only rarely produces apothecia. *Vulpicida pinastri* might be similar from a distance, but differs markedly in having a bright yellow pigmented medulla. The same is true for *V. viridis*, which in addition to the bright yellow medulla, differs from *U. oakesiana* in being esorediate.

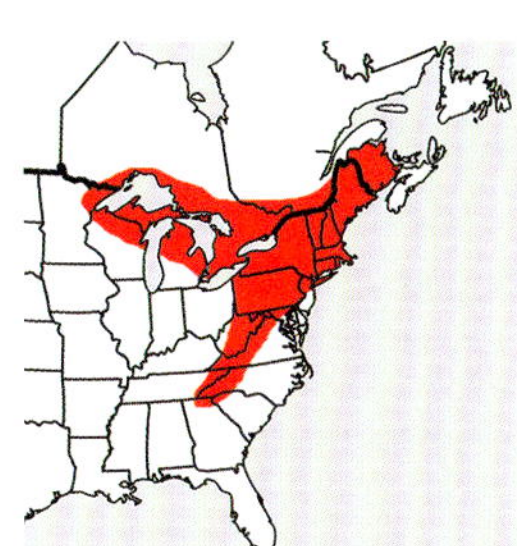

CHEMISTRY: Usnic acid, fatty acids, secalonic acid. Spot tests. Cortex: K-, C-, KC+ yellow, P-, UV-; medulla: K-, C-, KC+ orange, P-, UV-.
NICHE: The Constitutionalist is a very common and widely distributed foliose lichen that occurs throughout northern temperate and boreal North America. In the southeastern United States, it is restricted to the southern Appalachians, where it is found at all elevations and most frequently grows on the bark of conifers, or on conifer logs.
KEY FEATURES: Yellow-green foliose thallus, ruffled lobes, marginal soralia, pale lower surface, on bark at all elevations.

Vahliella leucophaea

Paradise Found

Tripp 5333 (photo: Lendemer)

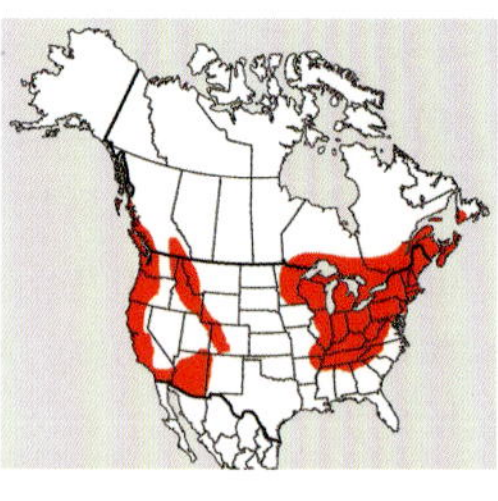

NOTES: This is a small cyanolichen that can be recognized by its flattened brown-gray squamulose thallus and reddish-brown apothecia. Since it occurs on rocks, it is most likely to be confused with *Rockefellera crossophylla*, which differs in having a thallus composed of minutely digitate, overlapping lobes, and apothecia that are strongly convex, and light tan in color. *Vahliella leucophaea* can also be confused with *Fuscopannaria leucosticta* in the rare instances in which that species occurs on rock, as both species have squamulose thalli. Indeed, *V. leucophaea* used to be classified in the genus *Fuscopannaria*. Nonetheless *F. leucosticta* differs markedly from *V. leucophaea* in its distinct white rim on the apothecial margins and the squamules margins.

CHEMISTRY: No substances. Spot tests. K-, C-, KC-, P-, UV-.

NICHE: Paradise Found is widely distributed in mountainous areas throughout North America, where it grows on shaded or sheltered non-calcareous rocks in humid microhabitats. It is rare in the Smokies, but occurs throughout the Park, especially on rock outcrops along creeks and rivers.

KEY FEATURES: Brown-gray squamulose thallus, cyanobacterial photobiont, reddish-brown apothecia, on non-calcareous rocks throughout.

Vainionora americana

Daughters of Liberty

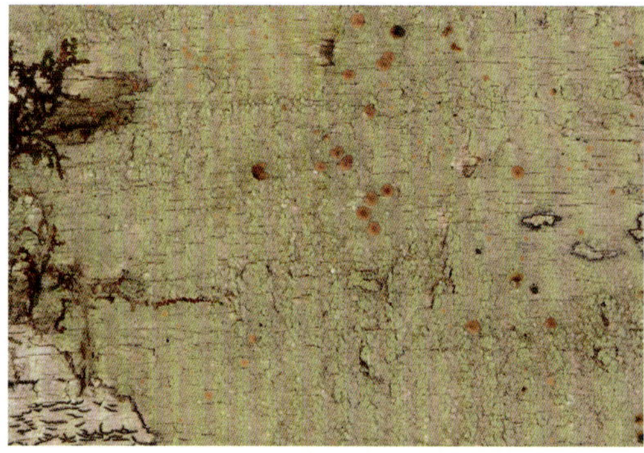

Lendemer 30288 (photo: Tripp)

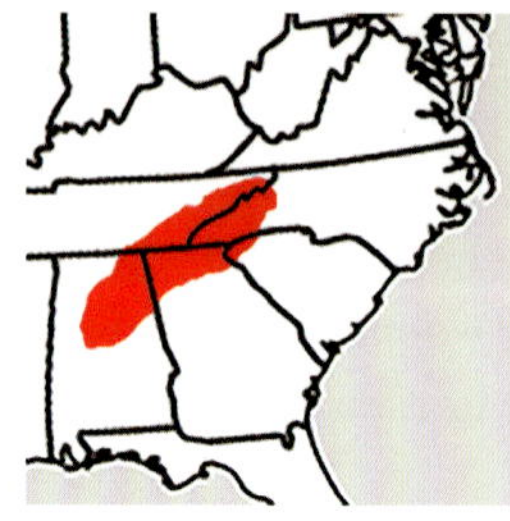

NOTES: *Vainionora* species are superficially similar to members of the *Lecanora subfusca* group, but differ in their dark pigmented hypothecia in addition to other, even more obscure anatomical characters. Most of the species occur in tropical regions, but *V. americana* reaches northward into the Smokies. In the field, *V. americana* resembles species such as *L. rugosella* because of its large lecanorine apothecia with reddish-brown discs, but differs from those species in producing soralia and xanthones. The soralia are often light yellow in color due to the high concentration of xanthones in the medulla of the thallus.

CHEMISTRY: Atranorin, thiophanic acid, arthothelin. Spot tests. Cortex: K+ yellow, C-, KC-, P-, UV; medulla: K-, C+ yellow, KC+ orange, P-, UV+ bright orange.

NICHE: This species is narrowly endemic to the southern Appalachian Mountains where it is widespread at all elevations below the spruce-fir zone. It occurs on the bark and branches of hardwood and conifer trees as well as on old wood of picnic tables, and can be found throughout the Smokies.

KEY FEATURES: Greenish-gray crustose thallus, white prothallus, light yellow soralia that are C+ orange, on bark throughout except spruce-fir forests.

Varicellaria velata

Cowgirl's Blues

Tripp 2681 (photo: Lendemer)

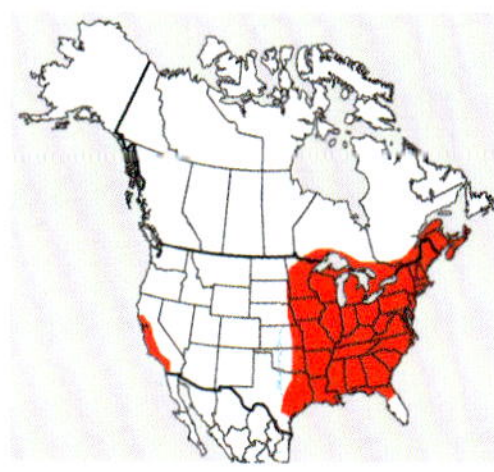

NOTES: *Varicellaria velata* used to be treated in the genus *Pertusaria* but the chemistry, discoid apothecia, and several other anatomical characters are unlike members of that genus. Nonetheless, it shares with *Lepra* and *Pertusaria* a tendency to be rare outside of the southern Appalachians. It is not likely to be confused with other species on account of the thick blue-gray thallus, densely pruinose lecanorine apothecia, and C+ red medulla. *Ochrolechia* species such as *O. pseudopallescens* and *O. trochophora* are somewhat similar, but the apothecial discs in those species are yellow or pink rather than densely white pruinose. In the Smokies, the chemical variant of *V. velata* with lichexanthone is restricted to high elevations, and in such habitats it could be confused with *Ochrolechia arborea* if the densely pruinose apothecia were misinterpreted as soralia with soredia.

CHEMISTRY: Lecanoric acid with or without lichexanthone. Spot tests. Cortex: K-, C-, KC-, P-, UV- or UV+ bright yellow; medulla: K-, C+ red, KC+ red, P-, UV-.

NICHE: This species is common and widespread throughout the Smokies, where it occurs primarily on the bark and branches of hardwoods—more rarely on conifers. Outside of the Smokies, it is common in the southern Appalachians, but elsewhere throughout its distribution in temperate eastern North America, it tends to be infrequent and restricted to mature hardwood trees or high-quality forests.

KEY FEATURES: Blue-gray crustose thallus, densely pruinose discoid lecanorine apothecia, C+ red medulla, on bark throughout.

Verrucaria calkinsiana

Williams' Half Flasks

Lendemer 32798 (photo: Lendemer)

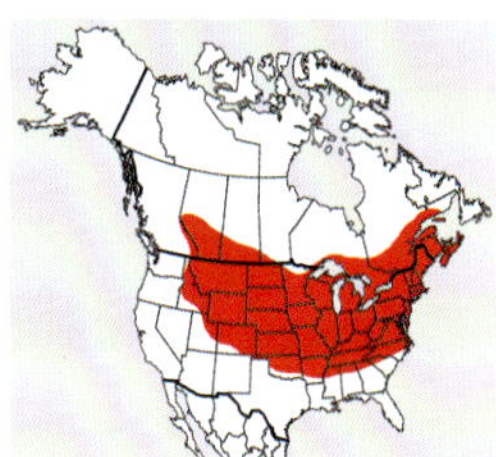

NOTES: What is there to say about *Verrucaria calkinsiana*? It was named after William Wirt

Calkins, a 19th-century lichenologist from Chicago who avidly collected lichens when he vacationed in Florida and Chattanooga. Unfortunately, he did not have a knack for detail or consistency, and so most collections have little data and a confusing array of numbers. Let that be a lesson to budding scientists: if you go through the trouble to collect a specimen, make sure it is clear where you got it and when. Perhaps it is fitting that this *Verrucaria* would carry his name. It can be recognized by its white thallus and relatively large perithecia that are about a third immersed in the rock substrate. The spores are simple and colorless.

CHEMISTRY: No substances. Spot tests. K-, C-, KC-, P-, UV-.

NICHE: *Verrucaria calkinsiana* is a species of calcareous rocks, and thus is relatively rare in the Smokies and the southern Appalachians. It grows on both natural rock outcrops and on concrete. Look for it pretty much everywhere in eastern North America.

KEY FEATURES: Pale, white thallus, relatively large black perithecia ~1/3 immersed in the substrate, simple spores, on calcareous rocks.

Vulpicida viridis

Poke o' Sunshine

Lendemer 33108 (photo: Tripp)

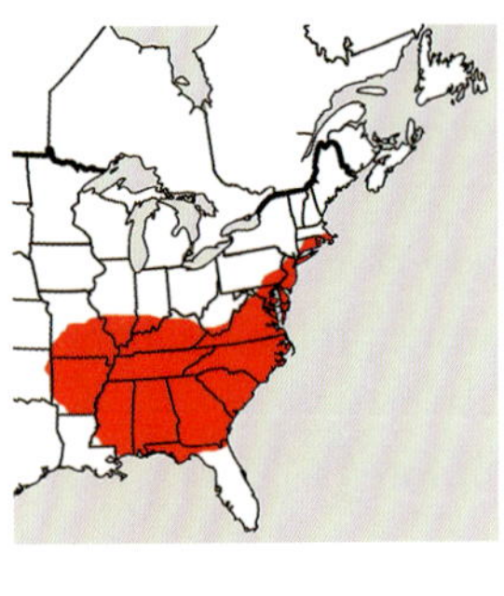

NOTES: *Vulpicida viridis* is a striking foliose lichen of the Smokies. Even though it may be small, the yellow-green thallus with its bright yellow medulla is a thing of beauty, particularly when the thallus is abundantly fertile with apothecia having dark reddish-brown discs. Although *Vulpicida* is mostly a genus of boreal and arctic regions, this species is one of the exceptions. *Vulpicida pinastri*, reportedly occurring in the Smokies, is not likely to be confused with *V. viridis* because it is sorediate instead of esorediate, and occurs only in high-elevation forests. The most similar species is *Ahtiana aurescens*, another relatively small foliose lichen with rosette-forming thalli that are yellow-green in color. That species can easily be distinguished by its white rather than yellow medulla.

CHEMISTRY: Usnic acid, pinastric acid and vulpinic acid. Spot tests. Cortex: K-, C-, KC+ yellow, P-, UV-; medulla: K-, C-, KC-, P-, UV-.

NICHE: Poke o' Sunshine is a relatively rare species known from many scattered occurrences in temperate eastern North America. It typically grows on the canopy branches of conifers, and thus is only visible when the branches fall. In the Smokies, it is known from a small number of locations at low elevations, especially along streams and rivers.

KEY FEATURES: Yellow-green rosette-forming foliose thallus, bright yellow medulla, absence of soredia, frequent apothecia, on branches at low elevations.

Willeya diffractella

Minuteman

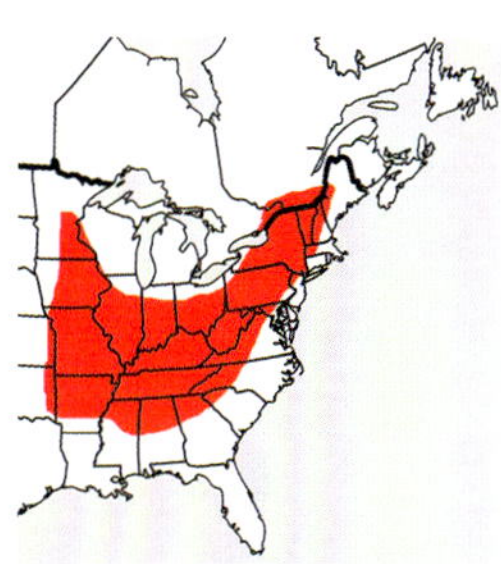

Lendemer 44549 (photo: Tripp)

NOTES: *Willeya diffractella* is one of the several crustose lichens that is characteristic of calcareous rocks in eastern North America. It can be recognized in the field by its thick brown thallus and large black perithecia that stick up ever so slightly above the thallus surface. The species is further characterized by the presence of algae in the hymenium and by colorless muriform spores that are 8-per ascus. The presence of algae in the hymenia of the perithecia is an unusual feature of this and several other members of the Verrucariaceae such as *Endocarpon* and *Staurothele*. Presumably, the algae are dispersed together with the spores and aid in formation of a new lichen when the spores germinate. *Verrucaria fayettensis* is a similar species that also grows on limestone and has a thick brown thallus. However, *V. fayettensis* differs in having a thick carbonized basal layer of thallus tissue that erupts through the upper surface and causes the thallus to appear black speckled.

CHEMISTRY: No substances. Spot tests. K-, C-, KC-, P-, UV-.

NICHE: Minuteman is widespread throughout temperate eastern North America, where it is one of the most common crustose lichens on calcareous rocks such as limestone and dolomite. It is also occasionally found on sandstones and shales that are weakly calcareous. In the Smokies, it is common in the few places where limestone outcrops, such as at Ace Gap.

KEY FEATURES: Thick brown crustose thallus, large black perithecia, algae in hymenium submuriform colorless spores, on calcareous rocks.

Xanthomendoza hasseana

Cock of the Rock Lichen

Tripp 3899 (photo: Lendemer)

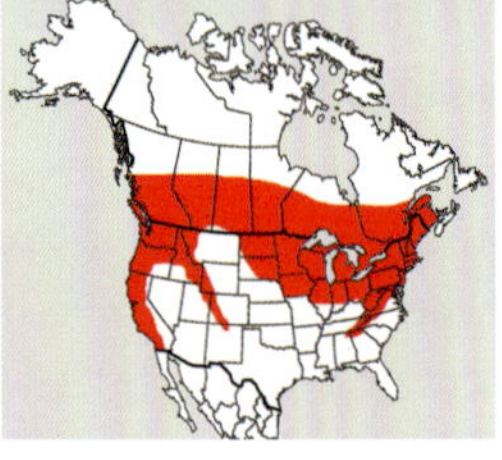

NOTES: This is easy! Shocking, theatric, pyrotechnic, bright orange foliose lichen? . . . Then it can only be one of a few species! *Xanthomendoza hasseana* is readily identifiable by its bright orange-to-orangish-green (to yellowish-green) foliose thallus that has a K+ dark purple thallus that lacks soredia. The only species that is truly comparable is *Xanthomendoza weberi*, but that species is asexual and bears soredia (*X. weberi* sometimes, but not always, has apothecia). This species could also be confused for various, fertile species of *Caloplaca*, but those lichens are crustose and lack a lower cortex. One might also be tempted to confuse *X. hasseana* for species of *Candelaria*, which are similarly foliose and have a lower cortex, but the latter are usually yellow in color and K+ weak red at best, never K+ deep purple. Species of *Candelaria* also have tiny thalli made up of very thin lobes, unlike species of *Xanthomendoza,* which are big and voluptuous.

CHEMISTRY: Parietin and related anthraquinones. Spot tests. Cortex: K+ purple, C-, KC-, P-, UV-; medulla: K-, C-, KC-, P-, UV-.

NICHE: *Xanthomendoza hasseana* occurs primarily and low-to-middle elevations in the Smokies, in rich cove forests. We have collected this species on only a couple of occasions in the Park, but suspect it is more common than current data indicate and is simply under-collected because it tends to hang out in the canopy. Look for it on fallen branches. We have, in fact, seen it abundantly on such substrates along Baxter Creek Trail, between Waterville and the summit of Mt. Sterling.

KEY FEATURES: Relatively large, bright orange foliose lichen, often with apothecia, always without soredia, K+ deep purple, on fallen branches from canopy, low-to-middle elevations

---

Xanthoparmelia angustiphylla

Squiggle Lobes

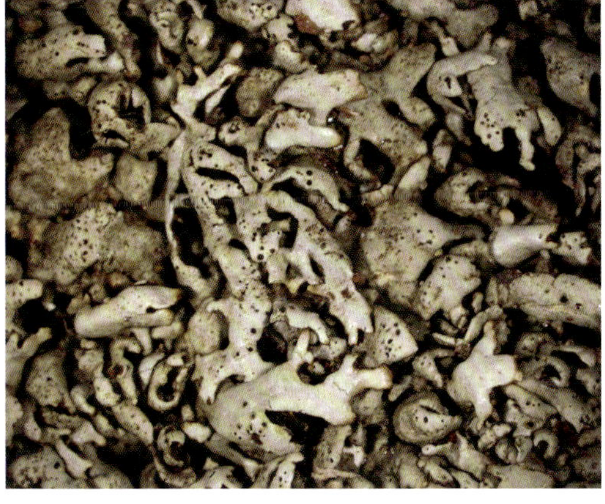

Dey 33000 (photo: Lendemer)

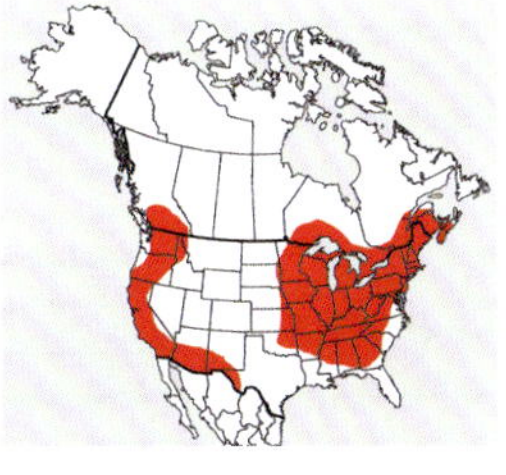

NOTES: This is one of the rarer members of the genus *Xanthoparmelia* in eastern North America. While you commonly find *X. conspersa* and *X. plittii* dotting the rocks with their isidia-laden thalli, the species without isidia—such as *X. angustiphylla*, *X. cumberlandia*, *X. hypofusca*, and *X. viriduloumbrina*—seem to be more reclusive. Maybe it is because asexual propagules like isidia are more likely to become established follwoing dispersal, or maybe some species just like to hide. *Xanthoparmelia angustiphylla* is most likely to be confused with forms of *X. conspersa*, which have sparse isidia, given that both species have a black lower surface and produce stictic acid. *Xanthoparmelia hypofusca* also has a black lower surface and lacks isidia, but the medulla is K+ yellow turning red due to the presence of salazinic acid. *Xanthoparmelia cumberlandia* is similar in the production of stictic acid but differs in having a pale, brown lower surface.

CHEMISTRY: Usnic acid and stictic acid aggregate. Spot tests. Cortex: K-, C-, KC+ yellow, P-, UV-; medulla: K+ yellow, C-, KC-, P+ orange, UV-.

NICHE: A rare species in the Smokies and the southern Appalachians in general, Squiqqle Lobes occurs on non-calcareous rocks in sunny, exposed habitats at middle-to-high elevations. It is similarly widespread throughout eastern North America but is never very common.

KEY FEATURES: Yellow-green foliose thallus, no isidia or soredia, black lower surface, medulla K+ yellow and P+ orange, rare on exposed non-calcareous rocks.

Xanthoparmelia conspersa

Soldier Swords

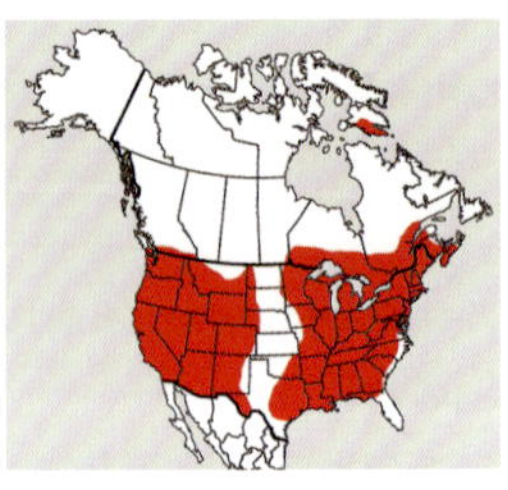

Tripp 3705 (photo: Lendemer)

NOTES: *Xanthoparmelia conspersa* is one of five currently recognized species in the genus known from the Smokies. This is a do-able diversity to learn, especially when one stops to appreciate just how much more diverse, abundant, and taxonomically difficult the genus is in western North America compared to the east. So relax! As a whole, the genus is characterized by a KC+ golden yellow cortex reaction owing to the presence of usnic acid, which gives species of *Xanthoparmelia* their characteristic, unwavering, yellowish-green color. Actually, there ought to be a crayon in the Crayola box called *Xanthoparmelia*, but we digress . . . The genus as a whole universally grows on rock, although some species in western North America are "vagrant'"and apparently grow on nothing, but instead blow around with the Chinook winds. Of this diversity, two species are asexual (*X. conspersa*, *X. plittii*), and three lack asexual propagules and can sometimes (but not always) be found with apothecia: *X. angustiphylla*, *X. cumberlandia*, *X. viriduloumbrina*. *Xanthoparmelia conspersa* is chemically identical to *X. plittii* but can easily be distinguished because it has a black lower cortex, in contrast to the pale lower surface of *X. plittii*.

CHEMISTRY: Usnic acid and stictic acid aggregate. Spot tests. Cortex: K-, C-, KC+ yellow, P-, UV-; medulla: K+ yellow, C-, KC-, P+ orange, UV-.

NICHE: *Xanthoparmelia conspersa* is relatively common in the Park, where it can be found growing at middle-to-high elevations, particularly in Anakeesta or other mineral-rich rock. Look for it on large rock outcrops such as those near Shuckstack Fire Tower or near Chimney Tops. This species is widespread throughout many other portions of the United States, as the map indicates.

KEY FEATURES: Large yellowish-green foliose thallus, black undersides, K+ yellow, P+ orange medulla, relatively common on rocks in the right kind of exposed habitat.

Xanthoparmelia plittii

Desert Birds

Tripp 3943 (photo: Lendemer)

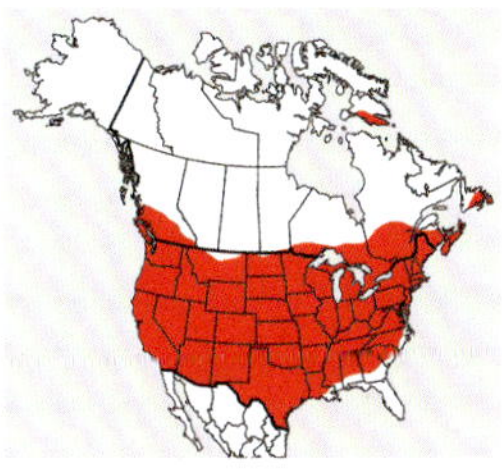

NOTES: *Xanthoparmelia plittii* is one of two isidiate species in the genus present in the Smokies. It is easily distinguished from the other, *X. conspersa*, by its thalli that have pale undersides (vs. black undersides in *X. conspersa*). The two species are otherwise chemically identical. See the entry under *X. conspersa* for a short crash course in Smokies' Xanthoparmeliology.

CHEMISTRY: Usnic acid and stictic acid aggregate. Spot tests. Cortex: K-, C-, KC+ yellow, P-, UV-; medulla: K+ yellow, C-, KC-, P+ orange, UV-.

NICHE: *Xanthoparmelia plittii* is a relatively common species in middle-to-low elevation habitats, where it occurs in acidic forests with exposed rock outcrops. There is one lovely example of such a habitat, dominated by Black Gum, *Rhododendron*, Mountain Laurel, and various species of oaks not far from the Bullhead Trail parking area on Cherokee Orchard Road. Alternatively, instead of watching elk, look for this species in similar habitats near the Cataloochee Valley Campground. Elsewhere, this species occurs in just about every state in the US where there exists exposed rock outcrops. It even visits Canada, on occasion.

KEY FEATURES: Large yellowish-green foliose thallus, pale undersides, K+ yellow, P+ orange medulla, relatively common on rock outcrops at middle-to-low elevations.

Xyleborus sporodochifer

White Caps

Lendemer 53241 (photo: Lendemer)

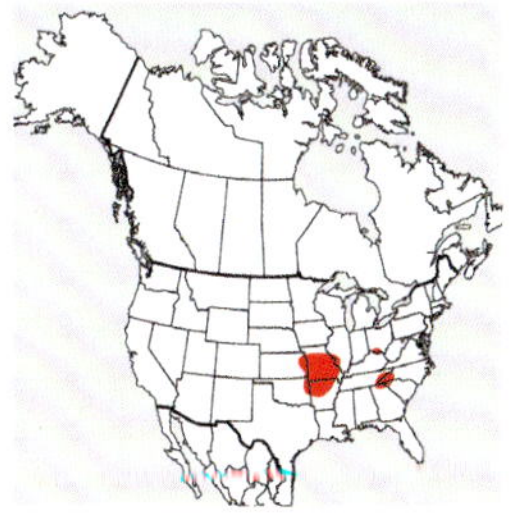

NOTES: *Xyleoborus* is an unusual genus of lichen, mainly because it produces sporodochia instead of pycnidia. Imagine that a pycnidium turned itself inside out so that you could see the fuzzy conidia budding right off of their specialized conidiogenous cells. The small, white sporodochia make the species easy to spot in the field, especially when combined with the fact that it only occurs on the old logs of trees belonging to the Fagaceae. In the southern Appalachians, this species doesn't typically make apothecia, and even when it does, they tend to lack spores. Then again, if you had sporodochia maybe you wouldn't try to have sex, either.

CHEMISTRY: No substances. Spot tests. K-, C-, KC-, P-, UV-.

NICHE: This species is infrequently collected, but nonetheless appears to be widespread in the Smokies, where it grows on large, old, dry logs of oaks and chestnuts. It was originally described from the Ozarks, but has subsequently been found throughout the southern Appalachians. The best way to view the species is by taking a gentle stroll down the Cataloochee Divide Trail from Purchase Knob.

KEY FEATURES: Crustose, brownish thallus, small, convex, white sporodochia, on dry, old oak and chestnut logs at all elevations.

Xylographa disseminata

Gothic Scripts

Lendemer 32941 (photo: Tripp)

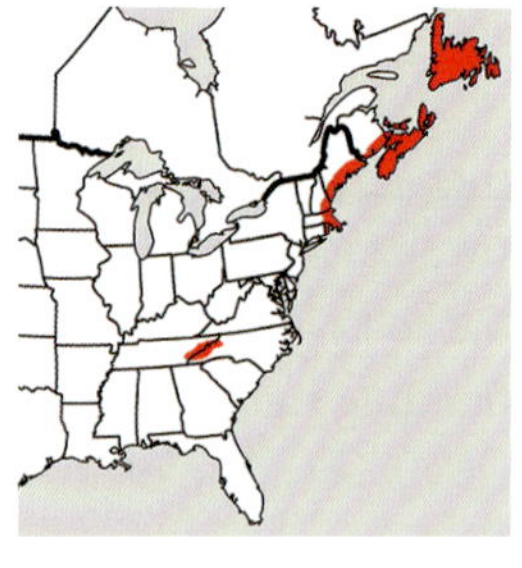

NOTES: *Xylographa disseminata* was originally described by the 19th-century lichenologist Henry Willey from the area near his home in coastal Massachusetts. Like other members of the genus, it was only recognized to occur in the southern Appalachians in modern times when lichenologists began to carefully examine the crustose lichens in places like the Smokies. The species can be recognized by its niche, combined with the bumpy gray thallus and short, dark blue-black apothecia that are elongate and resemble commas. Non-lichenized fungi like *Hysterium pulicare* might be confused with *X. disseminata*, but they do not have a clearly lichenized thallus, and instead grow right out of the wood.

CHEMISTRY: Confriesiic acid. Spot tests. K-, C-, KC-, P-, UV+ blue-white.

NICHE: This species is endemic to eastern North America, where it occurs in oceanic habitats along the Atlantic Coast from Massachusetts north into Canada, and in the southern Appalachians at highest, cloud-covered elevations. In the Smokies, it is known from a small number of locations at high elevations, mostly in the spruce-fir forests, where it was found on the hardened wood of old conifer logs.

KEY FEATURES: Dark, blue-black elongate, lip-like apothecia, bumpy gray thallus, simple spores, on old logs in high-elevation spruce-fir forests.

Xylographa trunciseda

Pallid Palette

Lendemer 33277 (photo: Tripp)

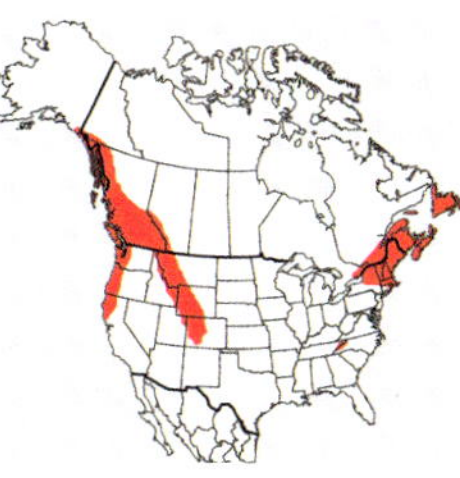

NOTES: This species is one that you would easily overlook as not being a lichen because its thin crustose thallus is largely immersed in the substrate, and the apothecia are pale pink to tan, and relatively inconspicuous. Nonetheless, it is the only species with those features, and the presence of simple colorless spores will confirmation the identification. *Xylograph disseminata* is similar, but differs in chemistry and in having dark blue-black apothecia that resemble little licorice chews when wet. The apothecia of *X. vitiligo* are similar to those of *X. trunciseda*, but that species produces conspicuous soralia and differs chemically in the production of stictic acid.

CHEMISTRY: No substances. Spot tests. K-, C-, KC-, P-, UV-.

NICHE: *Xylographa trunciseda* is a boreal species in North America and populations in the southern Appalachians are disjunct hundreds of miles to the south from their nearest relatives. In the Smokies, it occurs only in the highest-elevation spruce-fir forests, especially along the North Carolina/Tennessee State Line, where it grows on the hardened wood of old conifer longs.

KEY FEATURES: Pale, pink to tan elongate, lip-like apothecia, simple spores, on old logs in high-elevation spruce-fir forests.

PART 5

KEYS TO IDENTIFY THE MORE THAN 900 LICHENS AND ALLIED FUNGI OF GREAT SMOKY MOUNTAINS NATIONAL PARK

KEY TO KEYS

1. Fungus lichenicolous, lacking a photobiont, parasitic on lichens **Key I: Lichenicolous Fungi (Page 473)**

1. Fungus not lichenicolous, typically with a photobiont, not parasitic on lichens 2

2. Thallus (not the fruiting bodes) foliose, squamulose or fruticose 3

3. Thallus entirely fruticose, or dimorphic with a fruticose secondary thallus (e.g., Cladonia which has a foliose primary thallus and fruticose secondary thallus) **Key II: Fruticose Lichens (incl. Cladonia) (Page 477)**

3. Thallus foliose or squamulose, lacking a secondary fruticose component **Key III: Foliose and Squamulose Lichens (Page 485)**

2. Thallus crustose, although the fruiting bodies sometimes appearing fruticose 4

4. Thallus sterile, with asexual propagules (both lichen propagules and fungal propagules) **Key IV: Sterile Asexually Reproducing Lichens (Page 508)**

4. Thallus fertile, with sexual fruiting bodies 5

5. Fruiting bodies not club-like or pin-like (note that species hyphophores key out elsewhere) 6

6. Fruiting bodies discoid or wart-like, apothecia **Key V: Crustose Apotheciate Lichens (Page 521)**

6. Fruiting bodes flask like; perithecia **Key VI: Crustose Peritheciate Lichens (Page 543)**

5. Fruiting bodies club-like or pin-like 7

7. Fruiting bodes pin-like; thallus not a gelatinous algal film; on bark, bracket-fungi, rocks or standing dry wood **Key VII: Calicioid Fungi (Page 547)**

7. Fruiting bodes club-like; thallus a gelatinous algal film; on wet, rotting logs or soil **Multiclavula mucida**

KEY I: NON-LICHENIZED LICHENICOLOUS FUNGI

1. Fruiting body an apothecium or basidium; sexual spores present and produced in asci or on basidia 2

2. Ascospores or basidiospores hyaline 3

3. Fruiting body a basidium; sexual spores basidiospores 4

4. Fruiting bodies bright pink **Marchandiomyces corallinus**

4. Fruiting bodes brown, yellow-brown or concolorous with the thallus 5

5. On *Cladonia* **Tremella cladoniae**

5. On Physciaceae or Parmeliaceae 6

6. On Physciaceae, especially Heterodermia **Syzygospora physciacearum**

6. On Parmeliaceae 7

7. On *Usnea* **Biatoropsis usnearum**

7. On foliose Parmeliaceae (e.g., *Hypogymnia* and *Parmotrema*) 8

8. Galls brown, contrasting in color with the cost thallus; on *Parmotrema* especially at middle and low elevations **Tremella parmeliarum**

8. Galls gray, + concolorous with the host thallus; on *Hypogymnia*, especially at high elevations **Cystobasidium hypogymniicola**

3. Fruiting body an apothecium or perithecium; sexual spores ascospores 9

9. Fruiting body an apothecium 10

10. Ascospores simple 11

11. On *Lepra pustulata*; apothecia black, with radially cracked or striate margins **Skyttea radiatilis**

11. On *Cladonia, Dibaeis* or foliose Parmeliaceae; apothecia dark reddish-brown to black, but usually with smooth margins or lacking distinct margins 12

12. On foliose Parmeliaceae, especially *Punctelia rudecta*; ascospores large, 16-20 x 5-6.5 µm **Phacopsis oxyspora var. defecta**

12. On *Cladonia* or *Dibaeis*; ascospores smaller, 7-14 x 2-3.5 µm 13

13. On *Dibaeis*; ascospores uniseriate, 10.5-14 x 3-3.5 µm **Rhymbocarpus ericetorum**

13. On *Cladonia*; ascospores irregularly arranged, 7-12 x 2-3 µm **Phaeopyxis punctum**

10. Ascospores transversely septate 14

14. Ascospores 2-celled 15

15. On *Peltigera*; apothecia pale to light reddish-brown; ascospores 10-12 x 3-4 µm **Scutula dedicata**

15. On *Stereocaulon*; apothecia dark reddish-brown; ascospores 11-20 x 3-7 µm **Catillaria stereocaulorum**

14. Ascospores 3-5-celled 16

16. Apothecia pale, pallid to tan; on *Cladonia*; ascospores (3-)4(-5)-celled, narrow, 16-22 x 2-3 µm **Lettauia cladoniicola**

16. Apothecia dark reddish-brown to black; on other hosts; ascospores 4-celled, broader, 16-26 x 5-8 µm 17

17. On *Lobaria pulmonaria*; apothecia superficially resembling those of *Lobaria* except much darker in color **Plectocarpon lichenum**

17. On crustose lichens (*Bagliettoa, Ochrolechia, Pertusaria* or *Verrucaria*); apothecia not resembling the fruiting bodies of the host 18

18. Apothecia elongate, lirelliform; on *Bagliettoa* and *Verrucaria* **Opegrapha rupestris**

18. Apothecia circular in outline; on *Ochrolechia* and *Pertusaria* **Opegrapha anomea**

9. Fruiting body a perithecium 19

19. Perithecia pink or orange 20

20. Perithecia pink, covered in cotton-like tomentum; ascospores of two distinct size classes mixed in the same hymenium; on many different lichens but especially foliose Parmeliaceae **Ovicuculispora parmeliae**

20. Perithecia orange, smooth, not covered in tomentum; ascospores uniformly sized; on *Aspicilia* **Nectriopsis rubefaciens**

19. Perithecia black 21

21. Ascospores simple, cylindrical, 33-39 x 2.5-3.0 µm; parasite initially infecting the host, but eventually killing it and the host thallus mostly disappearing **Sarcopyrenia calcarea**

21. Ascospores 2-celled, ellipsoid, 10-19 x 4-7 µm; not entirely killing the host, such that the host thallus remains intact and is readily visible 22

22. On *Phaeophyscia*; internal portions of the perithecium filled with oil droplets **Lichenochora obscuroides**

22. On crustose lichens (*Acarospora, Aspicilia* or *Varicellaria*) 23

23. Ascospores small, 10-12 x 4.0-5.0 µm; on *Acarospora* **Stigmidium fuscatae**

23. Ascospores larger, 14-19 x 4.5-6.5 µm; on *Aspicilia* or *Varicellaria* 24

24. On *Aspicilia*; ascospores 15-19 x 5-6.5 µm **Stigmidium lendemeri**
24. On *Varicellaria*; ascospores 14-16 x 4.5-6.0 µm **Stigmidium eucline**
2. Ascospores brown (all taxa are ascomycetes in this section) 25
25. Fruiting body an apothecium, sometimes resembling a pin 26
26. Fruiting body pin-like 27
27. On *Protoparmelia hypotremella*; ascospores 7-10 µm long **Sphinctrina anglica**
27. On *Pertusaria*; ascospores 4-7 µm long 28
28. Stalk short; exciple reddish-brown, pigment turning K+ red **Sphinctrina turbinata**
28. Stalk relatively tall; exciple brown, pigment K- **Sphinctrina leucopoda**
26. Fruiting body a sessile or immersed apothecium 29
29. Ascospores 2-celled 30
30. Asci polysporous; ascospores up to 30-per ascus; on *Pertusaria*, especially *P. plittiana* **Dactylospora pertusariicola**
30. Asci not polysporous; ascospores 8-per ascus or fewer; on other hosts 31
31. On *Lobaria* **Dactylospora lobariella**
31. On other hosts 32
32. On saxicolous crustose lichens and rarely *Xanthoparmelia*; apothecia flat, plane **Buellia badia**
32. On foliose and fruticose Parmeliaceae, typically corticolous taxa; apothecia convex, without margins 33
33. On *Usnea* **Abrothallus usneae**
33. On foliose Parmeliaceae 34
34. On *Hypotrachyna* **Abrothallus hypotrachynae**
34. On *Flavoparmelia* and *Punctelia* **Abrothallus microspermus**
29. Ascospores 4-celled to submuriform 35
35. On crustose lichens (*Ochrolechia* and *Pertusaria*) 36
36. Ascospores brown early in development, small, 12-15 x 5-7.5 µm; typically on *Ochrolechia yasudae* **Dactylospora glaucomarioides**
36. Ascospores initially hyaline, turning brown with age, larger, 17-26 x 6.5-9.0 µm; on typically on corticolous *Ochrolechia* and *Pertusaria* **Opegrapha anomea**
35. On foliose lichens (Lobaria and Parmeliaceae) 37
37. On *Lobaria*; fungus forming fruiting bodies that resemble deformed apothecia of the host, but much darker in color; ascospores 16-23 x 5-7 µm **Plectocarpon lichenum**
37. On *Flavoparmelia* and *Parmelia*; fungus forming large black patches on the host, these with conspicuous black bumps; ascospores larger, 18-25 x 7-10 µm 38
38. On *Parmelia*; ascospores 18-25 x 7-10 µm **Homostegia piggotii**
38. On *Flavoparmelia*; ascospores 25-32 x 8-9 µm **Homostegia hertelii**
25. Fruiting body a perithecium 39
39. Ascospores simple, uniseriate; on *Trapelia* **Roselliniella microthelia**
39. Ascospores 2-4-celled, mostly irregularly arranged in the ascus; on many different hosts including *Trapelia* 40
40. Ascospores 4-celled; on *Phaeophyscia* **Pyrenidium aggregatum**
40. Ascospores 2-celled; on other hosts 41
41. Asci polysporous (typically more than 16 ascospores per ascus) 42
42. Perithecia <150 µm in diameter; asci with more than 64 ascospores; ascospores 2-4 µm wide **Muellerella lichenicola**
42. Perithecia >150 µm in diameter; asci with fewer than 64 ascospores; ascospores 4-6 µm wide **Muellerella ventosicola**
41. Asci not polysporous; ascospores 8-per ascus or fewer 43

43. Ascospores initially hyaline, becoming brownish late in development 44
44. Ascospores small, 10-12 x 4.0-5.0 µm; on *Acarospora* **Stigmidium fuscatae**
44. Ascospores larger, 14-19 x 4.5-6.5 µm; on *Aspicilia* or *Varicellaria* 45
45. On *Aspicilia*; ascospores15-19 x 5-6.5 µm **Stigmidium lendemeri**
45. On *Varicellaria*; ascospores 14-16 x 4.5-6.0 µm **Stigmidium eucline**
43. Ascospores strong brown from the start 46
46. On *Trapelia* 47
47. Ascospores small, 6-8 x 1.5-1.7 µm **Bellemerella trapeliae**
47. Ascospores larger, 13-16 x 5-6 µm **Polycoccum minutulum**
46. On Hypocenomyce or Porpidia 48
48. On *Hypocenomyce scalaris*; fungus strongly discoloring the host thallus; ascospores 9-13 x 4-7 µm **Clypeococcum hypocenomycis**
48. On *Porpidia albocaerulescens*; fungus not strongly discoloring the host thallus; ascospores 13-25 x 5-9 µm **Endococcus perpusillus**
1. Fruiting body a pycnidium or sporodochium; asexual conidia present and produced from conidiophores 49
49. Conidia hyaline (all species have simple conidia) 50
50. On *Peltigera*; pycnidia superficial on the host thallus; conidia 8-11 x 2-3 µm **Scutula dedicata**
50. On *Cladonia*; pycnidia immersed in the host thallus; conidia 7-11 x 2.5-6 µm 51
51. Fungus inducing the formation of galls; conidia pyriform, pointed at the ends, 8-11 x 4-6 µm; on *Cladonia uncialis* **Bachmanniomyces uncialicola**
51. Fungus not inducing the formation of galls; conidia ellipsoid, not pointed at the ends, 7-9 x 2.5-3.5 µm; on various *Cladonia* species **Epicladonia stenospora**
49. Conidia brown 52
52. Distinct pycnidia or sporodochia absent, the entire fungus consisting of brown, septate fungal hyphae laced through the host hymenium **Intralichen lichenum**
52. Distinct pycnidia or sporodochia present 53
53. Conidia simple, either solitary or forming chains 54
54. Conidia brown, solitary, not forming chains, globose, 2-4 µm in diameter **Lichenoconium erodens**
54. Conidia olive, with a distinct blue-green hue, often forming chains, ellipsoid, 5-9 x 3-4 µm **Vouauxiella lichenicola**
53. Conidia 2-celled or muriform and composed of aggregated cells 55
55. Conidia 2-celled 56
56. Sporodochia present, the conidia produced externally on the surface and visible as a black mealy dust; conidia 6-8 x 3-4 µm; on *Pertusaria texana* **Minutoexcipula tuckerae**
56. Pycnidia present, the conidia produced internally and released through an ostiole; conidia 5-7 x 2-3 µm; on various hosts but not *Pertusaria* **Lichenodiplis lecanorae**
sporodochia 2-celled
55. Conidia muriform, composed of aggregated cells 57
57. On *Fellhanera*; conidia 11-12 x 7.5-8.5 µm, composed of ca. 20 cells **Phaeosporobolus fellhanerae**
57. On other lichens of many different genera; conidia 9-25 µm in diameter, composed of 6-15 cells 58
58. On crustose lichens, especially *Ochrolechia* and *Pertusaria*; conidiomata 20-70 µm in diameter; conidia 9-17 µm in diameter, composed of 10-15 cells **Phaeosporobolus alpinus**
58. On foliose and fruticose lichens; conidiomata 30-90 µm in diameter; conidia 15-25 µm in diameter, composed of 6-12 cells **Phaeosporobolus usneae**

KEY II: FRUTICOSE LICHENS

1. Branches with a central cord (*Usnea*) 2

2. Branches segmented, the segments separated by conspicuous white ring-like annual pseudocyphellae 3

3. Branches with abundant, conspicuous, evenly sized and spaced perpendicular fibrils; thallus typically with a distinct point of attachment to the substrate; common at all elevations **Usnea subgracilis**

3. Branches essentially lacking fibrils; thallus typically without a discernable attachment point, usually draping in masses; uncommon at low elevations or common at high elevations 4

4. Medulla K+ yellow turning red (salazinic acid present) **Usnea merrillii**

4. Medulla K- (diffractaic acid present) or K+ dirty yellow (constictic acid or protocetraric acid present) 5

5. Medulla P+ orange (constictic acid present) **Usnea trichodea**

5. Medulla P- (diffractaic acid present) **Usnea trichodea**

2. Branches not distinctly segmented by conspicuous white ring-like structures 6

6. Thallus without soredia or isidiomorphs, typically with abundant apothecia 7

7. Thallus pendant to subpendant; medulla CK+ yellow (diffractaic acid present); rare **Usnea cristatula**

7. Thallus shrubby, though sometimes becoming surprisingly large and superficially subpendant; medulla CK- (diffractaic acid absent); common 8

8. Thallus distinctly blackened at the base near the point of attachment; medulla white, with salazinic acid (TLC required); common >4000 ft. elevation **Usnea subfusca**

8. Thallus not distinctly blackened at the base, area near the point of attachment mostly some shade of brown or green; medulla often pink or red pigmented, without salazinic acid (TLC required); common <4000 ft. elevation 9

9. Ascospores averaging 7.0-9.0 um long **Usnea strigosa**

9. Ascospores averaging 9.0-12.0 um long **Usnea endochrysea**

6. Thallus with soredia and/or isidiomorphs, typically without apothecia 10

10. Branches strongly ridged and resembling vertebra or collapsing and with deep furrows 11

11. Branches strongly ridged, resembling vertebra; medulla P+ yellow (norstictic acid present); infrequent to rare at middle and low elevations **Usnea angulata**

11. Branches collapsing, with deep furrows; medulla P+ orange (salazinic acid present); not known with certainty from the southern Appalachians **Usnea cavernosa**

10. Branches terete, not with distinct ridges of furrows 12

12. Thallus cortex distinctly red or orange pigmented 13

13. Pigment mostly confined to the basal portions of the thallus, typically mottled red; medulla P+ orange (stictic acid present); very common throughout except at the highest elevations **Usnea pensylvanica**

13. Pigment evenly distributed throughout the thallus, typically light orange; medulla P+ red (fumarprotocetraric acid present); rare mostly in northern hardwood forests at high elevations **Usnea subrubicunda**

12. Thallus cortex uniformly yellow-green, not pigmented 14

14. Medulla pink or red pigmented 15

15. Medulla strongly pink-red pigmented, CK- (diffractaic acid absent); thallus shrubby to subpendant **Usnea mutabilis**

15. Medulla lightly pink pigmented, CK+ strong yellow (diffractaic acid present); thallus subpendant to pendant 16

16. Soralia convex, distinctly raised above the surface of the branches; common **Usnea ceratina**

16. Soralia excavate, initially plane with the of the branches, but then quickly eroding and becoming concave **Usnea entoviolata**

14. Medulla white 17

17. Medulla CK+ strong yellow (diffractaic acid present) 18

18. Soralia convex, distinctly raised above the surface of the branches; common **Usnea ceratina**

18. Soralia excavate, initially plane with the of the branches, but then quickly eroding and becoming concave **Usnea entoviolata**

17. Medulla CK- (diffractaic acid absent) 19

19. Branches inflated, distinctly narrowed at the point of attachment to the previous branch 20

20. Soralia commonly produced on the main branches and the secondary branches; rare **Usnea fragilescens**

20. Soralia mostly absent from the main branches, primarily distributed on the smaller, secondary branches; common 21

21. Medulla P+ orange (salazinic acid present, TLC required for confirmation); common **Usnea cornuta**

21. Medulla P+ red (protocetraric acid present, TLC required for confirmation); rare **Usnea brasiliensis**

19. Branches uniform in thickness, not distinctly narrowed at the point of attachment to the previous branch 22

22. Basal portions of thallus distinctly blackened, especially just adjacent to the point of attachment; middle to high elevations 23

23. Thallus short, shrubby; salazinic acid absent (TLC required) 24

24. Soralia excavate, initially plane with the of the branches, but then quickly eroding and becoming concave, large, often encircling the branches; medulla P+ yellow (norstictic acid present); rare **Usnea fulvoreagens**

24. Soralia plane to convex, small, not encircling the branches; medulla P- or P+ orange (norstictic acid absent); common 25

25. Medulla P+ orange, UV- (thamnolic acid present) **Usnea subfloridana**

25. Medulla P-, UV+ blue-white (squamatic acid present) **Usnea subfloridana**

23. Thallus longer, subpendant to pendant; salazinic acid present (TLC required) 26

26. Basal portions of the thallus without circular cracks; soralia small, punctiform, less than half the diameter of the branch; common **Usnea dasopoga**

26. Basal portions of the thallus often with circular cracks; soralia large, broad, often more than half the diameter of the branch; rare **Usnea silesiaca**

22. Basal portions of thallus not distinctly blackened; throughout at all elevations 27

27. On bark 28

28. Medulla K+ orange-red (galbinic acid present, TLC required for confirmation); branches with numerous, regularly spaced, evenly sized, thorn-like fibrils **Usnea dasaea**

28. Medulla K- or K+ dirty yellow (protocetraric acid present, TLC required for confirmation) 29

29. Thallus coarse, stiff; cortex thick, shiny and glassy when cut with a razor blade; fibrils irregular in size and appearance; soralia large, with numerous isidiomorphs; low to middle elevations **Usnea subscabrosa**

29. Thallus soft, lax; cortex thin, dull with cut with a razor blade; fibrils regular in size, perpendicular to the main branch, often curved up at the ends; soralia minute, often with a single isidiomorph; throughout at all elevations **Usnea subgracilis**

27. On rocks 30

30. Medulla K+ orange-red (galbinic acid or norstictic acid present, TLC required for confirmation); common on rocks 31

31. Medulla with norstictic acid and salazinic acid (TLC required); common throughout **Usnea halei**

31. Medulla with galbinic acid (TLC required); rare, at low elevations **Usnea amblyoclada**

30. Medulla K- or K+ dirty yellow (protocetraric acid present, TLC required for confirmation); infrequent on rocks 32

32. Thallus coarse, stiff; cortex thick, shiny and glassy when cut with a razor blade; fibrils irregular in size and appearance; soralia large, with numerous isidiomorphs; low to middle elevations **Usnea subscabrosa**

32. Thallus soft, lax; cortex thin, dull with cut with a razor blade; fibrils regular in size, perpendicular to the main branch, often curved up at the ends; soralia minute, often with a single isidiomorph; throughout at all elevations **Usnea subgracilis**

1. Branches without a central cord 33

33. Thallus entirely fruticose, not dimorphic and consisting of a crustose or squamulose primary t hallus that produces secondary fruticose structures .34

34. Branches actually lobes, with a distinct upper and lower surface, the two surfaces typically differentiated with distinct coloration 35

35. Thallus without isidia or soredia; restricted to spruce-fir forests **Pseudevernia cladonia**

35. Thallus with isidia or soredia; throughout at all elevations 36

36. Thallus isidiate; common at all elevations **Pseudevernia consocians**

36. Thallus sorediate; mostly at middle and high elevations **Everniastrum catawbiense**

34. True branches present, not actually lobes, either round or flattened, uniformly colored and without differentiated surfaces 37

37. On bark 38

38. Thallus yellow-green (usnic acid present) 39

39. Branches terete, round in section 40

40. Medulla KC- and UV- (unidentified fatty acid present); rare in spruce-fir **Alectoria fallacina**

40. Medulla KC+ pink and UV+ blue-white (alectoronic acid present); not known with certainty from the southern Appalachians **Alectoria sarmentosa**

39. Branches flattened, not round in section 41

41. Thallus sorediate **Ramalina labiosorediata**

41. Thallus esorediate 42

42. Medulla with all spot tests negative (no substances present); middle to high elevations **Ramalina americana**

42. Medulla with at least one spot test positive (lichen substances present, often norbarbatic acid in Smokies populations); middle to low elevations **Ramalina culbersoniorum**

38. Thallus brown, black or gray (usnic acid absent) 43

43. Photobiont a green alga 44

44. Thallus ashy gray; medulla P+ yellow (alectorialic acid present) **Bryoria nadvornikiana**

44. Thallus brown to brown-black; medulla P- or P+ red (alectorialic acid absent) 45

45. Thallus sorediate, the soralia with abundant spinulose isidiomorphs **Bryoria furcellata**

45. Thallus esorediate 46

46. Branches uniformly brown **Bryoria americana**

46. Branches distinctly two-toned, some black and some lighter brown 47

47. Branchlets perpendicular to the main branches, short, regular; thallus dense, shrubby **Bryoria bicolor**

47. Branchlets arising at acute angles from the main branches, longer, irregular in length; thallus lax, whispy **Bryoria tenuis**

43. Photobiont a cyanobacterium 48

48. Corticolous on the trunks of trees **Dendriscocaulon intricatulum**

48. Saxicolous, muscicolous, lignicolous or growing on the damp bases of trees 49

49. Thallus minute fruticose, forming cushions of spine-like branches **Leptogium teretiusculum**

49. Thallus minutely foliose, forming small rosettes around the apothecia **Leptogium subtile**

37. On rock or humus 50

50. Thalli with filamentous branches, mostly resembling tangled mats of hair or minute s piky cushions 51

51. Thallus fuzzy, cotton-like, composed of branches forming loose tufts or masses; in sheltered rock overhangs 52

52. On non-calcareous rocks, especially sandstone; photobiont Trentepohlia; uncommon **Cystocoleus ebeneus**

52. On calcareous rocks; photobiont a cyanobacterium; not known with certainty from the Smokies **Thermutis velutina**

51. Thallus not fuzzy and cotton-like, instead composed of branches forming minute cushions or matted wisp-like locks; on exposed rocks, sometimes inundated with water 53

53. Thallus forming small, spike-like cushions, without long, clearly defined branches **Spilonema revertens**

53. Thallus forming matted locks or rosettes, with clearly defined branches 54

54. Thalli forming small rosettes; apothecia common **Ephebe americana**

54. Thalli forming wisp-like matts; apothecia rare 55

55. Branches coarse, broad, 130-260 µm wide **Ephebe solida**

55. Branches slender, narrow, 70-140 µm wide **Ephebe lanata**

50. Thalli not filamentous, the branches large and broad, resembling blades or erect tubes 56

56. Thallus shrubby, tufted, growing directly attached to the rock; on sheltered rock faces and overhangs 57

57. Thallus yellow-green, K- (usnic acid present), sorediate 58

58. Medulla P+ red and UV- (protocetraric acid present) **Ramalina petrina**

58. Medulla P- and UV+ blue-white (sekekaic acid present) **Ramalina intermedia**

57. Thallus gray to blue-gray, K+ yellow (atranorin present), esorediate 59

59. Medulla of phyllocladia P+ orange and UV- (stictic acid present) **Stereocaulon dactylophyllum**

59. Medulla of phyllocladia P- and UV+ blue-white (lobaric acid present) 60

60. Phyllocladia plate-like, not distinctly digitate and resembling fingers; pseudopodetia dark brown and ligneous in the lower portions **Stereocaulon glaucescens**

60. Phyllocladia digitate, resembling fingers; pseudopodetia white to tan, not ligneous in the lower portions **Stereocaulon tennesseense**

56. Thallus erect and cushion-like, typically growing on humus or bryophytes and not directly attached to the rock; on the ground or on rocks in exposed habitats 61

61. Thallus brown to black, branches with pseudocyphellae; rare 62

62. Pseudocyphellae round, circular in outline; branches terete, round in section; not known with certainty from the southern Appalachians **Cetraria aculeata**

62. Pseudocyphellae irregular or linear in shape; branches flattened, not round in section; known from the southern Appalachians but rare 63

63. Medulla P+ red (fumarprotocetraric acid present) 64

64. Pseudocyphellae linear, distributed along the margins of the branches; very rare and restricted to high elevations **Cetraria laevigata**

64. Pseudocyphellae irregular in shape, distributed both on the margins and the surfaces of the branches; not known from the southern Appalachians **Cetraria islandica**

63. Medulla P- (fumarprotocetraric acid absent) 65

65. Middle to low elevations; margins of branches with long hair-like projections, 0.5-1.0 mm long **Cetraria arenaria**

65. Very high elevations; margins of branches with short hair-like projections, <0.5 mm long **Cetraria ericetorum**

61. Thallus gray or yellow-green, branches without pseudocyphellae; common 66

66. Thallus gray to blue-gray, sometimes tinged brownish (usnic acid absent) 67

67. Branches pseudopodetia, solid and lacking a central cavity; pseudopodetia covered with plate-like phyllocladia 68

68. Medulla of phyllocladia P+ orange and UV- (stictic acid present) **Stereocaulon dactylophyllum**

68. Medulla of phyllocladia P- and UV+ blue-white (lobaric acid present) 69

69. Phyllocladia plate-like, not distinctly digitate and resembling fingers; pseudopodetia dark brown and ligneous in the lower portions **Stereocaulon glaucescens**

69. Phyllocladia digitate, resembling fingers; pseudopodetia white to tan, not ligneous in the lower portions **Stereocaulon tennesseense**

67. Branches podetia, with a central cavity, not solid; either with a continuous cortex or ecorticate and with photobiont colonies +/- visible, never with plate-like phyllocladia 70

70. Podetia corticate; thallus K- and UV+ blue-white (squamatic acid present); rare at high elevations **Cladonia appalachensis**

70. Podetia ecorticate; thallus K+ yellow and UV- (atranorin present); common at middle to high elevations 71

71. Basal portions of the podetia pale to light brown; common at middle to high elevations **Cladonia rangiferina**

71. Basal portions of the podetia distinctly blackened; not known from the southern Appalachians **[Cladonia stygia]**

66. Thallus yellow-green (usnic acid present) 72

72. Podetia corticate, inflated, thorn-like; medulla UV+ blue-white (squamatic acid present) 73

73. Upper portions of podetia with conspicuous white maculae **Cladonia uncialis**

73. Upper portions of podetia without maculae **Cladonia caroliniana**

72. Podetia ecorticate, slender, not distinctly inflated and thorn-like; thallus UV- (squamatic acid absent) 74

74. Thallus P- (fumarprotocetraric acid absent) **Cladonia mitis**

74. Thallus P+ red (fumarprotocetraric acid present) 75

75. Podetia broad, thick, coarse; pycnidial gel hyaline; middle to high elevations **Cladonia arbuscula**

75. Podetia narrow, slender, fragile; pycnidial gel red; low to middle elevations **Cladonia subtenuis**

33. Thallus dimorphic, consisting of a crustose or squamulose primary thallus and a secondary fruticose thallus 76

76. Fruticose portion of thallus composed of pseudopodetia; pseudopodetia solid, lacking a central cavity 77

77. Pseudopodetia sorediate or covered in ecorticate granules 78

78. Primary thallus an ecorticate, leprose crust 79

79. Granules P+ red (protocetraric acid present); thallus gray to blue-gray (atranorin present) **Lepraria arbuscula**

79. Granules P- (protocetraric acid absent); thallus yellow-green (usnic acid present) **Lecanora anakeestiicola**

78. Primary thallus a corticate crust, either areolate or continuous 80

80. Medulla and soralia UV+ blue-white (lobaric acid present); common **Stereocaulon pileatum**

80. Medulla and soralia UV- (lobaric acid absent); very rare **Pilophorus cereolus**

77. Pseudopodetia esorediate, covered with corticate areoles or phyllocladia 81

81. Pseudopodetia very short, <1 mm tall, covered with flattened areoles, resembling tiny matchsticks capped with a globose black apothecium; on rocks associated with waterways at middle to low elevations **Pilophorus fibula**

81. Pseudopodetia much larger, 1-8 cm tall, covered with dispersed phyllocladia, not resembling matchsticks; on rocks at high elevations 82

82. Medulla of phyllocladia P+ orange and UV- (stictic acid present) **Stereocaulon dactylophyllum**

82. Medulla of phyllocladia P- and UV+ blue-white (lobaric acid present) 83

83. Phyllocladia plate-like, not distinctly digitate and resembling fingers; pseudopodetia dark brown and ligneous in the lower portions **Stereocaulon glaucescens**

83. Phyllocladia digitate, resembling fingers; pseudopodetia white to tan, not ligneous in the lower portions **Stereocaulon tennesseense**

76. Fruticose portion of the thallus composed of podetia; podetia hollow, with a central cavity 84
84. Primary thallus crustose; podetia short, thorn-like **Pycnothelia papillaria**
84. Primary thallus squamulose; podetia resembling many things, but not thorns 85
85. Apothecia and pycnidia red, at least one of them 86
86. Podetia sorediate, at least in the upper portions 87
87. Podetia cup-like 88
88. Medulla and soralia K+ instant intense yellow and P+ orange (thamnolic acid present); restricted to high elevations, usually on rotting wood **Cladonia digitata**
88. Medulla and soralia K- and P- (thamnolic acid absent); at all elevations, usually on soil or humus **Cladonia pleurota**
87. Podetia blunt, lacking cups 89
89. Medulla and soralia P- and UV+ blue-white (squamatic acid or barbatic acid present) 90
90. Primary squamules large, broad; podetia short, broad, mostly esorediate and covered with a continuous cortex **Cladonia incrassata**
90. Primary squamules small, narrow; podetia tall, slender, abundantly sorediate and lacking a continuous cortex 91
91. Podetia finely sorediate, soredia often remaining attached to the podetia and if becoming detached, then revealing a whitish, opaque stereome **Cladonia macilenta var. bacillaris**
91. Podetia with a mixture of coarse soredia and fine microsquamules, these readily becoming detached to reveal a brownish, translucent stereome **Cladonia didyma var. didyma**
89. Medulla and soralia P+ orange and UV- (thamnolic acid present) 92
92. Soredia extremely coarse and granular; very rare at high elevations **Cladonia floerkeana**
92. Soredia fine, powdery; very common throughout 93
93. Podetia finely sorediate, soredia often remaining attached to the podetia and if becoming detached, then revealing a whitish, opaque stereome **Cladonia macilenta var. macilenta**
93. Podetia with a mixture of coarse soredia and fine microsquamules, these readily becoming detached to reveal a brownish, translucent stereome **Cladonia didyma var. vulcanica**
86. Podetia esorediate, covered with coarse areoles or a continuous cortex 94
94. Podetia cup-like, at least some of the time; medulla UV- and P- (zeorin present) **Cladonia coccifera**
94. Podetia blunt, lacking cups; medulla UV+ blue-white or P+ orange (zeorin absent) 95
95. Medulla and soralia P+ orange and UV- (thamnolic acid present) 96
96. Podetia covered with fine microsquamules, these readily becoming detached to reveal a brownish, translucent stereome; common at middle and low elevations **Cladonia didyma var. vulcanica**
96. Podetia covered in a continuous, lumpy cortex that does readily detach to reveal the stereome; rare at high elevations **Cladonia floerkeana**
95. Medulla and soralia P- and UV+ blue-white (squamatic acid or barbatic acid present) 97
97. Primary squamules large, broad, marginally sorediate; podetia short, broad; medulla and soredia UV+ bright blue-white (squamatic acid present) **Cladonia incrassata**
97. Primary squamules small, narrow, esorediate; podetia reaching variably heights, slender; medulla UV+ dull blue-white (barbatic acid present) 98
98. Podetia covered with fine microsquamules, these readily becoming detached to reveal a brownish, translucent stereome **Cladonia didyma var. didyma**
98. Podetia covered in a continuous, cortex that does readily detach to reveal the stereome **Cladonia cristatella**

85. Apothecia and pycnidia uniformly brown 99
99. Podetia with cups or cup-like expansions 100
100. Cups and cup-like expansions flared, lacerated, disorganized 101
101. Thallus UV+ blue-white (squamatic acid present) **Cladonia squamosa**
101. Thallus P+ orange (thamnolic acid present) or P+ red (fumarprotocetraric acid present) 102

102. Thallus K+ instant intense yellow and P+ orange (thamnolic acid present) **Cladonia crispata**

102. Thallus K- or K+ dull yellow and P+ red (fumarprotocetraric acid present) 103

103. True cups present, the bottoms completely closed; cups becoming strongly fissured and flared; not known from the southern Appalachians, all specimens examined to date are misidentifications of *C. furcata* **Cladonia multiformis**

103. Cup-like expansions present, these not strongly fissured **Cladonia furcata**

100. Cups and cup-like expansions not as above, normal and resembling cups or funnels 104

104. Podetia with false cups, the bottoms open with a hole 105

105. Podetia with a continuous cortex; very rare **Cladonia crispata**

105. Podetia with a discontinuous cortex, typically covered in squamules that readily detach and reveal the stereome beneath; extremely common **Cladonia squamosa**

104. Podetia with true cups, the bottom completely closed 106

106. Podetia with cups forming tiers, the tiers developing from the centers of the cups **Cladonia verticillata**

106. Podetia typically with a single cup, if forming tiers then the tiers developing from the margins of the cups 107

107. Podetia with a smooth and continuous cortex 108

108. Cups poorly formed, dumpy, appearing melted and deformed; very rare at low and middle elevations **Cladonia mateocyatha**

108. Cups normal, well-formed, stately, not appearing melted and deformed; very rare at high elevations **Cladonia gracilis subsp. turbinata**

107. Podetia sorediate or areolate, not with a smooth and continuous cortex 109

109. Podetia tall, slender; cups small, shallow, goblet-like 110

110. Medulla and soredia UV+ dull blue-white (homosekekaic acid present) **Cladonia rei**

110. Medulla and soredia UV- (homosekekaic acid absent) **Cladonia fimbriata**

109. Podetia short, stout; cups larger, deep, trumpet-let 111

111. Bourgeanic acid present, detectable as long, needle-like crystals deposited on a glass slide after several drops of acetone are added to a fragment of podetium and allowed to dry **Cladonia conista**

111. Bourgeanic acid absent, no long-needle-like crystals deposited on a glass side after application of acetone to a fragment of podetium 112

112. Grayanic acid present (TLC required) **Cladonia grayi**

112. Grayanic acid absent (TLC required) 113

113. Cryptochlorophaeic acid present (TLC required) **Cladonia crypotchlorophaea**

113. Cryptochlorophaea acid absent (TLC required) 114

114. Merochlorophaeic acid present (TLC required) **Cladonia merochlorophaea**

114. Merochlorophaeic acid absent (TLC required) 115

115. Podetia sorediate, the soredia coarse to fine **Cladonia chlorophaea**

115. Podetia areolate, the areoles, large and plate-like **Cladonia pyxidata**

99. Podetia blunt, lacking cups or cup-like expansions 116

116. Thallus yellow-green, KC+ yellow (usnic acid present) 117

118. Primary thallus persistent, composed of large squamules, forming cushion-like thalli **Cladonia robbinsii**

118. Primary thallus ephemeral, composed of small squamules dispersed on the substrate or the basal portions of the podetia 119

119. Upper portions of podetia with conspicuous white maculae **Cladonia uncialis**

119. Upper portions of podetia without maculae **Cladonia caroliniana**

116. Thallus gray, blue-gray, greenish or brownish, KC- or KC+ some color other than yellow (usnic acid absent) 120

120. Medulla C+ green (strepsilin present) **Cladonia strepsilis**

120. Medulla C- (strepsilin absent) 121

121. Medulla and soredia UV+ blue-white (grayanic acid, homosekekaic acid or squamatic acid present) 122

122. Podetia sorediate 123

124. Podetia tall, slender, often with weakly formed cups; homosekekaic acid present (TLC required); typically on soil or organic matter **Cladonia rei**

124. Podetia short, blunt, without cups; grayanic acid present (TLC required); typically corticolous **Cladonia cylindrica**

122. Podetia esorediate, either with a continuous cortex or with squamules 125

125. Podetia slender, abundantly branching; thallus cushion forming; very rare at high elevations **Cladonia appalachensis**

125. Podetia broader, not abundantly branching; thallus not forming cushions 126

126. Podetia with a continuous cortex; very rare **Cladonia crispata**

126. Podetia with a discontinuous cortex, typically covered in squamules that readily detach and reveal the **Cladonia crispata**

121. Medulla and soredia UV- (grayanic acid, homosekekaic acid or squamatic acid present) 127

127. Medulla P+ yellow (norstictic acid or psoromic acid present)

128. Medulla K+ yellow turning red (norstictic acid present); common **Cladonia polycarpoides**

128. Medulla K- (psoromic acid present); rare **Cladonia brevis**

127. Medulla P+ orange or P+ red (psoromic acid absent) 129

129. Thallus K+ instant intense yellow and P+ orange (thamnolic acid present) 130

130. Primary squamules abundant, persistent, dainty, small, crumbling into piles of microsquamules; podetia short, covered in microsquamules; common at all elevations **Cladonia parasitica**

130. Primary squamules infrequent, ephemeral, robust, becoming divided but not crumbling into piles; podetia tall, with a continuous cortex, sometimes covered with large coarse squamules; rare at high elevations **Cladonia crispata**

129. Thallus K- or K+ dull orange and P+ red (fumarprotocetraric acid present) 131

131. Podetia sorediate 132

132. Podetia tall, slender 133

133. Podetia simple, little branching; very rare at high elevations **Cladonia cornuta**

133. Podetia abundantly branching; not known with certainty from the southern Appalachians **Cladonia scabriuscula**

132. Podetia short, blunt 134

134. Primary squamules small, readily dissolving into piles of soredia; podetia often deformed and ragged in appearance **Cladonia ramulosa**

134. Primary squamules large, persistent, not dissolving into piles of soredia; podetia typically well-formed and elegant 135

135. At least one podetium with a continuous cortex on the lower portions, the soredia forming on the upper portions of the podetium and eventually eroding away the cortex in irregular patches down the length of the podetium; common **Cladonia ochrochlora**

135. All podetia, every single one of them, entirely sorediate to the base where the podetium connects to the primary squamule; very rare **Cladonia coniocraea**

131. Podetia esorediate 136

136. Podetia tall, abundantly branching **Cladonia furcata**

136. Podetia short, simple, little branching if at all 137

137. Podetia slender, capped with a large apothecium that is incongruously broad given the stalk that supports it **Cladonia peziziformis**

137. Podetia broad, with apothecia appropriate to the width of the stalks 138

138. Squamules largely small, narrow, with a white lower surface; podetia blunt and not appearing melted and deformed **Cladonia sobolescens**

138. Squamules large, broad, with a brownish-purple lower surface; podetia often with poorly executed attempts at cups that appear melted and deformed **Cladonia mateocyatha**

KEY III: FOLIOSE AND SQUAMULOSE LICHENS

Note that *Tuckermanopsis chlorophylla* is not keyed out here because we consider reported occurrence of the species to be highly doubtful.

1. Thallus black, composed of minute hair-like filaments or minute cushions; always on rock 2

2. Thallus fuzzy, cotton-like, composed of branches forming loose tufts or masses; in sheltered rock overhangs 3

3. On non-calcareous rocks, especially sandstone; photobiont Trentepohlia; uncommon **Cystocoleus ebeneus**

3. On calcareous rocks; photobiont a cyanobacterium; not known with certainty from the Smokies **Thermutis velutina**

2. Thallus not fuzzy and cotton-like, instead composed of branches forming minute cushions or matted wisp-like locks; on exposed rocks 4

4. Thallus forming small, spike-like cushions, without long, clear branches **Spilonema revertens**

4. Thallus forming matted locks or rosettes, with clearly defined branches 5

5. Thalli forming small rosettes; apothecia common **Ephebe americana**

5. Thalli forming wisp-like matts; apothecia rare 6

6. Branches coarse, broad, 130-260 μm wide **Ephebe solida**

6. Branches slender, narrow, 70-140 μm wide **Ephebe lanata**

1. Thallus various colors, squamulose, umbilicate or foliose; on a diverse variety of substrates 7

7. Thallus squamulose or umbilicate 8

8. Thallus umbilicate 9

9. Upper surface with raised bumps and depressions 10

10. Lower surface pale; common **Lasallia papulosa**

10. Lower surface black; rare 11

11. Bumps conspicuous and strongly raised; thallus typically composed of one large, solitary lobe... **Lasallia pensylvanica**

11. Bumps shallow, relatively inconspicuous; thallus composed of many overlapping lobes that group together and form tight colonies **Lasallia caroliniana**

9. Upper surface smooth, without raised bumps and depressions 12

12. Lower surface pale, tan to brown 13

13. Thallus composed of numerous overlapping lobes that group together and form extensive colonies; on rocks associated with water, often directly in streams **Dermatocarpon luridum**

13. Thallus typically composed of one large, solitary lobe; on rocks in dry habitats and never growing directly in streams 14

14. On non-calcareous rocks; perithecia absent, fruiting bodies apothecia if present **Umbilicaria muhlenbergii**

14. On calcareous and weakly calcareous rocks; perithecia present **Dermatocarpon muhlenbergii**

12. Lower surface black 15

15. Thallus typically composed of one large, solitary lobe, although this can often become ragged and appear to be a cluster of lobes; lower surface with abundant rhizines; very common **Umbilicaria mammulata**

Thallus composed of many overlapping lobes that group together and form tight colonies; lower surface with sparse rhizines; rare **Lasallia caroliniana**

8. Thallus squamulose 16

16. Photobiont a cyanobacterium 17

17. Thallus sorediate; apothecia rare **Fuscopannaria sorediata**

17. Thallus esorediate; apothecia common 18

18. Thallus isidiate or lobulate 19

19. Thallus lobulate, consisting of contiguous lobes; prothallus typically fibrous and carpet-like; at middle and high elevations **.Parmeliella appalachensis**

19. Thallus isidiate, consisting flattened squamules; prothallus crustose, smooth; at all elevations **Parmeliella triptophylla**

18. Thallus without isidia or lobules 20

20. Margins of lobes and apothecia with a frosted appearance due to the presence of white tomentum; apothecia with thalline margins **Fuscopannaria leucosticta**

20. Margins of lobes and apothecia concolorous with the thallus, without a white tomentum; apothecia without thalline margins 21

21. Thallus squamulose; squamules tightly adnate, with weakly divided margins; apothecia dull reddish-brown to brown, not contrasting in color with the thallus **Vahliella leucophaea**

21. Thallus lobate; lobes + raised, strongly divided and elongate; apothecia light tan to orange-brown, distinctly lighter in color than the thallus **Rockefellera crossophylla**

16. Photobiont a green alga 22

22. Thallus jade greenish-blue in color; squamules scallop shaped, with a distinct raised lip **Normandina pulchella**

22. Thallus gray, blue-gray or brown in color; squamules lacking a raised lip and not scallop shaped 23

23. Thallus sorediate 24

24. Thallus C+ pink and P- (lecanoric acid present); common **Hypocenomyce scalaris**

24. Thallus C- and P+ red (fumarprotocetraric acid absent); rare **Hypocenomyce anthracophila**

22. Thallus esorediate 25

25. On bark or wood 26

26. Medulla UV+ blue-white (friesiic acid present); fruiting bodies black, sessile apothecia **Hypocenomyce friesii**

26. Medulla UV- (friesiic acid absent); fruiting bodies immersed perithecia or reddish-brown and stipitate apothecia 27

27. Squamules large, adnate, round, margins entire; fruiting bodies perithecia; medulla P- (fumarprotocetraric acid absent) **Placidium arboreum**

27. Squamules small, ascending, elongate, margins finely divided; fruiting bodes reddish-brown, stipitate apothecia; medulla P+ red (fumarprotocetraric acid present) **Cladonia caespiticia**

25. On rock, soil or humus 28

28. Squamules adnate to tightly adnate; lower surface dark or difficult to examine because the squamule cannot be easily removed from the substrate; all spot tests negative; fruiting bodies perithecia 29

29. Ascospores simple; squamules large, round **Placidium arboreum**

29. Ascospores transversely septate or muriform; squamules small, irregular in shape or elongate 30

30. Ascospores 2-celled; squamules minute, often white pruinose; on non-calcareous rocks **Placidiopsis minor**

30. Ascospores muriform; squamules larger, epruinose; on calcareous and weakly calcareous rocks **Endocarpon pallidulum**

28. Squamules ascending; lower surface pale, readily examined; at least one spot test positive; fruiting bodies apothecia, if present 31

31. Medulla C+ green (strepsilin present) **Cladonia strepsilis**

31. Medulla C- (strepsilin absent) 32

32. Medulla UV+ blue-white (sphaerophorin present) **Cladonia petrophila**

32. Medulla UV- (sphaerophorin absent) 33

33. Medulla P- (fatty acids presents) **Cetradonia linearis**

33. Medulla P+ yellow (psoromic acid present) or P+ red (fumarprotocetraric acid) 34

34. Medulla P+ yellow (psoromic acid present) **Cladonia brevis**

34. Medulla P+ red (fumarprotocetraric acid present) 35

35. Squamules with finely divided margins; cortex K- (atranorin absent) **Cladonia caespiticia**

35. Squamules entire, not with finely divided margins; cortex K+ yellow (atranorin present) 36

36. Thallus forming mats; squamules strap-like, long, strongly ascending; on soil, often in disturbed habitats **Cladonia apodocarpa**

36. Thallus forming cushions; squamules short, not strongly ascending; on non-calcareous rock in exposed habitats **Cladonia stipitata**

7. Thallus foliose 37

37. Thallus cortex or medulla strong yellow or orange 38

38. Thallus cortex bright yellow or orange; medulla white 39

39. Cortex yellow, K- (anthraquinones absent) 40

40. Medulla white (pinastric acid and vulpinic acid absent) 41

41. Thallus sorediate; apothecia rare **Candelaria concolor**

41. Thallus esorediate; apothecia common **Candelaria fibrosa**

40. Medulla bright yellow (pinastric acid and vulpinic acid absent)...42

42. Thallus sorediate; apothecia rare; not known with certainty from the southern Appalachians **Vulpicida pinastri**

42. Thallus esorediate; apothecia common; infrequent at middle and low elevations **Vulpicida viridis**

39. Cortex orange, K+ magenta (anthraquinones present) 43

43. Thallus sorediate; apothecia rare **Xanthomendoza weberi**

43. Thallus esorediate; apothecia common **Xanthomendoza hasseana**

38. Thallus cortex yellow-green, gray, blue-gray, or brown; medulla strong yellow or orange 44

44. Thallus without lichenized diaspores; apothecia common 45

45. Medulla yellow, the pigment K- (secanolic acid or vulpinic acid present) 46

46. Upper surface blue-gray to gray (atranorin present); medulla P+ orange-red (galbinic acid present) **Myelochroa galbina**

46. Upper surface yellow-green (usnic acid present); medulla P- (vulpinic acid present) **Vulpicida viridis**

45. Medulla orange-red, the pigment instantly K+ purple (skyrin present) 47

47. Ascospores *Pachysporaria*-type; on shaded rocks at middle and low elevations; rare **Phaeophyscia endococcinodes**

47. Ascospores *Physcia*-type; not known with certainty from the southern Appalachians 48

48. Lobes narrow, <0.5 mm wide **Phaeophyscia endococcinea**

48. Lobes broad, >0.5-1.0 mm wide **Phaeophyscia erythrocardia**

44. Thallus with lichenized diaspores (isidia, phyllidia, pustules or soredia); apothecia rare 49

49. Thallus with isidia or phyllidia 50

50. Photobiont a cyanobacterium; lower surface with conspicuous cyphellae; thallus with abundant marginal phyllidia **Sticta fragilinata**

50. Photobiont a green alga; lower surface without cyphellae; thallus with laminal isidia 51

51. Lower surface ecorticate, orange pigmented, the pigment K+ purple; typically on bark **Heterodermia crocea**

51. Lower surface corticate, black, K-; typically on non-calcareous rocks **Myelochroa obsessa**

49. Thallus with soredia or pustules 52

52. Medulla orange, the pigment K+ purple 53

53. Lower surface corticate, dark back to black 54

54. Medulla uniformly orange; lobes narrow, 0.5-1.0 mm wide; upper cortex K- (atranorin absent); very common at all elevations **Phaeophyscia rubropulchra**

54. Medulla patchily pigmented orange, with the pigment associated with the soralia or pustules; lobes broad, 1.0-3.0 mm wide; upper cortex K+ yellow (atranorin present); common at high elevations 55

55. Thallus with laminal pustules, the pigment unevenly distributed throughout the medulla; common at high elevations **Hypotrachyna croceopustulata**

55. Thallus with capitate soralia, the pigment mostly restricted to the medulla below the soralia; infrequent and mostly in spruce-fir forests **Hypotrachyna gondylophora**

53. Lower surface ecorticate, white but with orange pigment 56

56. Lower surface with orange pigment confined to small spots near the lobe tips **Heterodermia neglecta**

56. Lower surface with orange pigment more-or-less uniformly distributed throughout the thallus **Heterodermia obscurata**

52. Medulla yellow, the pigment K- 57

57. Upper surface yellow-green when dry (usnic acid present); not known with certainty from the southern Appalachians **Vulpicida pinastri**

57. Upper surface blue-gray, gray or brown when dry (usnic acid absent); known from the southern Appalachians 58

58. Lower surface with conspicuous pseudocyphellae 59

59. Photobiont a green alga.. **Pseudocyphellaria aurata**

59. Photobiont a cyanobacterium 60

60. Medulla uniformly yellow **Pseudocyphellaria perpetua**

60. Medulla patchily yellow, the pigment mostly associated with the soralia **Pseudocyphellaria crocata**

58. Lower surface without pseudocyphellae 61

61. Thallus with laminal pustules 62

62. Medulla C- (gyrophoric acid absent); common at all elevations **Myelochroa aurulenta**

62. Medulla C+ pink (gyrophoric acid present); infrequent at low elevations **Hypotrachyna spumosa**

61. Thallus with laminal, marginal or terminal soralia 63

63. Soralia capitate on the tips of the lobes; medulla P+ orange-red (galbinic acid present) **Myelochroa metarevoluta**

63. Soralia laminal or marginal, but not capitate on the tips of the lobes; medulla P- or P+ yellow (galbinic acid absent) 64

64. Lower surface ecorticate, the yellow pigment confined to spots near the lobe tips **Heterodermia casarettiana**

64. Lower surface corticate, brown to black 65

65. Medulla mostly white, the yellow pigment dull and restricted to the soralia **Physconia leucoleiptes**

65. Medulla strongly yellow pigmented throughout 66

66. Upper surface UV- (lichexanthone absent) **Pyxine sorediata**

66. Upper surface UV+ bright yellow (lichexanthone present) **Pyxine subcinerea**

37. Thallus cortex or medulla gray, blue-gray, yellow-green, green, brown or white 67

67. Thallus without lichenized diaspores (e.g., isidia, lobules, phyllidia, pustules or soredia); apothecia typically present 68

68. Upper surface yellow-green (usnic acid present) 69

69. On bark, typically the branches of trees 70

70. Medulla bright yellow (pulvinic acid and vulpinic acid present) **Vulpicida viridis**

70. Medulla white (pulvinic acid and vulpinic acid absent) **Ahtiana aurescens**

69. On rock 71

71. Medulla K+ yellow turning red (salazinic acid present) **Xanthoparmelia viriduloumbrina**
71. Medulla K+ yellow turning dingy brown (stictic acid present) 72
72. Lower surface pale, tan to brown **Xanthoparmelia cumberlandia**
72. Lower surface black, at least in the central portions of the thallus **Xanthoparmelia angustiphylla**
68. Upper surface gray, blue-gray, green or brown (usnic acid absent) 73
73. Primary photobiont a cyanobacterium 74
74. On pebbles and boulders entirely submerged in streams or rivers **Peltigera hydrothyria**
74. On bark, rock or soil in other habitats, never entirely submerged in streams or rivers 75
75. Thallus becoming gelatinous when wet (jelly lichens, *Collema* and *Leptogium*) 76
76. Thallus minute, superficially appearing crustose, composed of cushions or mats of very small lobes 77
77. Ascospores transversely septate, 2(-4) celled **Collema coccophorum**
77. Ascospores submuriform 78
78. Thallus forming cushions of spine-like branches 79
79. Thallus mico-fruticose; lobes 0.1-0.2 mm wide, flattened, becoming regularly finely divided **Leptogium teretiusculum**
79. Thallus meso-fruticose; lobes 1.0-4.0 mm wide, erect to ascending, the margins becoming irregularly finely divided to resemble isidia of varying sizes; ascospores 18-45 x 11-16 µm **Leptogium lichenoides**
78. Thallus minutely foliose, forming small rosettes around the apothecia 80
80. Thallus composed of a distinct upper and lower cortex visible in cross section under the compound microscope; ascospores 20-30 x 10-12 µm **Leptogium subtile**
80. Thallus not stratified, composed of a single layer when viewed in cross section under the compound microscope; ascospores 13-22 x 9-15 µm **Collema occultatum**
76. Thallus larger, not appearing crustose, always with distinct lobes 81
81. Thallus slate gray when dry 82
82. Thallus swelling excessively when wet and becoming strongly shriveled when dry 83
83. Thallus surface covered with numerous tiny squamules **Leptogium phyllocarpum**
83. Thallus surface not covered with tiny squamules **Leptogium chloromelum**
82. Thallus not swelling excessively when wet, mostly retaining non-shriveled and recognizable lobes when dry 84
84. Lobe margins entire, 2-10 mm wide; lobe surface somewhat wrinkled **Leptogium corticola**
84. Lobe margins becoming finely divided, 1-4 mm wide; lobe surface smooth 85
85. Lobes plane, flattened, becoming finely divided, but not in an exaggerated manner; apothecia common; ascospores 12-30 x 5-8 µm **Leptogium dactylinum**
85. Lobes erect to ascending, the margins becoming irregularly finely divided to resemble exaggerated isidia of varying sizes; apothecia rare; ascospores 18-45 x 11-16 µm **Leptogium lichenoides**
81. Thallus olive-brown to black when dry 86
86. Lobe margins becoming finely divided; thallus typically slate-gray to olive-brown in color, composed of a distinct upper and lower cortex visible in cross section under the compound microscope 87
87. Lobes plane, flattened, becoming finely divided, but not in an exaggerated manner; apothecia common; ascospores 12-30 x 5-8 µm **Leptogium dactylinum**
87. Lobes erect to ascending, the margins becoming irregularly finely divided to resemble exaggerated isidia of varying sizes; apothecia rare; ascospores 18-45 x 11-16 µm **Leptogium lichenoides**
86. Lobe margins entire; thallus typically brown-black to black in color, not stratified, composed of a single layer when viewed in cross section under the compound microscope 88
88. Apothecia immersed in the thallus; ascospores submuriform **Collema pustulatum**
88. Apothecia sessile or stipitate; ascospores transversely septate 89
89. Ascospores 2-celled 90
90. On soil and humus over calcareous rocks; ascospores 10-22 x 5-8 µm **Collema coccophorum**

90. On bark; ascospores 15-24 x 3-5 µm **Collema conglomeratum**

89. Ascospores 4-13-celled; on bark and non-calcareous rocks 91

91. Thallus forming tight cushions; lobes small, short, contracted, becoming strongly swollen when wet **Collema leptaleum**

91. Thallus not forming tight cushions; lobes not as above, flat, not becoming strongly swollen when wet 92

92. Apothecia strongly white pruinose 93

93. Proper exciple euthyplectenchymatous, with the hyphae oriented in such a way as to resemble strings in section **Collema pulcellum var. leucopeplum**

93. Proper exciple paraplectenchymatous, with the hyphae oriented in such a way as to resemble circular or angular cubes in section **Collema pulcellum var. pulcellum**

92. Apothecia epruinose 93

93. Ascospores short, 4-6-celled, 25-45 x 5-8 µm long **Collema ryssoleum**

93. Ascospores long, (5-)-6-13-celled, 35-100 x 3-7 µm 94

94. Proper exciple paraplectenchymatous, with the hyphae oriented in such a way as to resemble circular or angular cubes in section **Collema pulcellum var. pulcellum**

94. Proper exciple euthyplectenchymatous, with the hyphae oriented in such a way as to resemble strings in section 95

95. Ascospores narrow, 3-5 µm wide **Collema nigrescens**

95. Ascospores broad, 5-7 µm wide **Collema subnigrescens**

75. Thallus not gelatinous when wet (other cyanolichen genera) 96

96. Lower surface ecorticate, strongly tomentose, often with a network of veins that contrast in color with the rest of the lower surface 97

97. On the boles of trees well above the base; upper surface covered with numerous fine, hyaline hairs; extremely rare **Leioderma cherokeense**

97. On rock, soil, wood, humus or the bases of trees; upper surface glabrous or with a white tomentum, but not covered with numerous fine, hyaline hairs 98

98. Thallus surface with convex cephalodia containing cyanobacteria, the dominant photobiont actually a green alga; rare **Peltigera leucophlebia**

98. Thallus surface without convex cephalodia, the dominant photobiont a cyanobacterium throughout the thallus; common 99

99. Apothecia dark, black 100

100. On mosses over rock and the bases of trees; upper surface smooth, without minute phyllidia **Peltigera neckeri**

100. On bark and occasionally mosses over rocks; upper surface covered with abundant phyllidia **Peltigera phyllidiata**

99. Apothecia reddish-brown 101

101. Upper surface, at least towards the tips of the lobes, distinctly tomentose 102

102. Lobes ascending and erect; thalli small, 1-4 cm in diameter, abundantly fertile with one large apothecium on every lobe **Peltigera didactyla**

102. Lobes mostly plane and not erect; thalli large, 5-20 cm in diameter, typically not abundantly fertile, and if so, then not typically with an apothecium on every lobe 103

103. Lobes very broad, (15-)20-30 mm wide, typically with downturned margins **Peltigera canina**

103. Lobes narrower, 5-10(-20) mm wide; typically with upturned margins 104

104. Upper surface tomentose only near the tips of the lobes; rhizines simple, little branched **Peltigera praetextata**

104. Upper surface tomentose throughout, often with visible as irregular white patches on the thallus; rhizines abundantly branched **Peltigera rufescens**

101. Upper surface uniformly lacking tomentum 105

105. Apothecia flat 106

106. Lobe margins ruffled, lobulate; lower surface without distinct veins, instead with a network of regularly spaced pale spots amid a uniformly blackish tomentum **Peltigera elisabethae**

106. Lobe margins entire, not lobulate; lower surface with a network of distinct veins **Peltigera horizontalis**

105. Apothecia saddle-shaped, curled 107

107. Lower surface with indistinct veins; not known with certainty from the southern Appalachians **Peltigera hymenina**

107. Lower surface with a network of distinct veins; common 108

108. Lower surface pale, with brownish veins; not known with certainty from the southern Appalachians **Peltigera degenii**

108. Lower surface dark brown with black veins 109

109. Lobes narrow, <20 mm wide; rhizines short, <5 mm long **Peltigera seneca**

109. Lobes broad, >20 mm wide; rhizines long, 7-11 mm long **Peltigera neopolydactyla**

96. Lower surface corticate and lacking veins, or the thallus so tightly adnate that the lower surface cannot be readily examined 110

110. On calcareous rocks 111

111. On weakly calcareous rocks; apothecia reddish-brown **Vahliella leucophaea**

111. On strongly calcareous rocks; apothecia black 112

112. Conspicuous prothallus absent; thallus foliose, with marginal lobes, typically forming concentric rings of lobes **Placynthium nigrum**

112. Conspicuous blue-black prothallus present; thallus squamulose without marginal lobes, typically forming a continuous crust **Placynthium petersii**

110. On bark or non-calcareous rocks 113

113. Apothecia produced on the lower surface of the lobe tips, best viewed from below 114

114. Lower surface tomentose 115

115. Lower surface uniformly brown; common **Nephroma helveticum**

115. Lower surface speckled with white, punctiform spots; very rare at high elevations **Nephroma resupinatum**

114. Lower surface smooth, glabrous, not tomentose 116

116. Thallus surface and apothecial margins ragged, with abundant lobules; common **Nephroma helveticum**

116. Thallus surface and apothecial margins smooth, lacking lobules; not known with certainty from the southern Appalachians **Nephroma bellum**

113. Apothecia produced on the upper surface of the lobes, best viewed from above 117

117. On rock 118

118. Margins of lobes and apothecia with a frosted appearance due to the presence of white tomentum; apothecia with thalline margins **Fuscopannaria leucosticta**

118. Margins of lobes and apothecia concolorous with the thallus, without a white tomentum; apothecia without thalline margins 119

119. Thallus squamulose; squamules tightly adnate, with weakly divided margins; apothecia dull reddish-brown to brown, not contrasting in color with the thallus **Vahliella leucophaea**

119. Thallus lobate; lobes + raised, strongly divided and elongate; apothecia light tan to orange-brown, distinctly lighter in color than the thallus **Rockefellera crossophylla**

117. On bark 120

120. Thallus squamulose **Fuscopannaria leucosticta**

120. Thallus foliose 121

121. Thallus, including the lobe tips, P- (pannarin absent) 122

122. Upper surface wrinkled; thallus large, spreading; rare **Pannaria lurida subsp. russellii**

122. Upper surface smooth; thallus small, rosette-forming; infrequent **Pannaria subfusca**

121. Thallus, especially the lobe tips, P+ orange-red (pannarin present) 123

123. Upper surface smooth; lobes and apothecia with a frosted appearance due to the presence of blue-white or white pruina **Pannaria rubiginosa**

123. Upper surface wrinkled, lobes and apothecia epruinose; rare **Pannaria lurida (including subsp. lurida, subsp. russellii, subsp. quercicola)**

73. Primary photobiont a green alga 124

124. On soil or humus; rare 125

125. Thallus surface with convex cephalodia containing cyanobacteria; lobes plane; lower surface ecorticate and with a network of raised veins **Peltigera leucophlebia**

125. Thallus surface without convex cephalodia; lobes strongly ascending, erect; lower surface corticate 126

126. Pseudocyphellae round, circular in outline; branches terete, round in section; not known with certainty from the southern Appalachians **Cetraria aculeata**

126. Pseudocyphellae irregular or linear in shape; branches flattened, not round in section; known from the southern Appalachians but rare 127

127. Medulla P+ red (fumarprotocetraric acid present) 128

128. Pseudocyphellae linear, distributed along the margins of the branches; very rare and restricted to high elevations **Cetraria laevigata**

128. Pseudocyphellae irregular in shape, distributed both on the margins and the surfaces of the branches; not known from the southern Appalachians [**Cetraria islandica**]

127. Medulla P- (fumarprotocetraric acid absent) 129

129. Middle to low elevations; margins of branches with long hair-like projections, 0.5-1.0 mm long [**Cetraria arenaria**]

129. Very high elevations; margins of branches with short hair-like projections, <0.5 mm long **Cetraria ericetorum**

124. On bark, wood or rock; common 130

130. Upper surface dark uniformly dark brown to black 131

131. On rock 132

132. Medulla UV+ blue-white and P- (perlatolic acid present) **Melanelia panniformis**

132. Medulla UV- and P+ orange (stictic acid present) or P+ red (fumarprotocetraric acid present) 133

133. Lobe margins with sessile pycnidia and conspicuous white pseudocyphellae; upper surface more-or-less lacking both pseudocyphellae and conspicuous immersed pycnidia; medulla P+ orange (stictic acid present) **Melanelia hepatizon**

133. Lobe margins without pycnidia and conspicuous white pseudocyphellae; upper surface with both pseudocyphellae and conspicuous immersed pycnidia; medulla P+ red (fumarprotocetraric acid) **Melanelia stygia**

131. On bark or wood 134

134. Upper surface white pruinose 135

135. Upper surface conspicuously white pruinose throughout the thallus; upper cortex scleroplectenchymatous; ascospores 28-37 x 15-20 µm; very rare **Physconia subpallida**

135. Upper surface mostly epruinose, with pruina restricted to the lobe tips; upper cortex prosoplectenchymatous; ascospores 35-45 x 16-23 µm **Anaptychia palmulata**

134. Upper surface epruinose 136

136. Lobe margins with abundant black, sessile pycnidia, often creating a black fringe along the margins 137

137. Thallus small, 1-2 cm in diameter 138

138. Lobes very narrow, 0.5 mm wide; middle to low elevations, typically on bark and branches of pines **Tuckermanella fendleri**

138. Lobes broader, 1-3 mm wide; high elevations, on bark and branches of many substrates **Tuckermanopsis orbata**

137. Thallus larger, 2-7 cm in diameter 139

139. Medulla C- and UV- (fatty acids present); middle to high elevations **Tuckermanopsis orbata**

139. Medulla C+ red (olivetoric acid present) or UV+ blue-white (alectoronic acid present); at all elevations 140

140. Medulla C+ red and UV- (olivetoric acid present); extremely common at all elevations **Tuckermanopsis ciliaris**

140. Medulla C- and UV+ blue-white (alectoronic acid present); rare at middle to high elevations **Tuckermanopsis americana**

136. Lobe margins without sessile pycnidia, the margins lacking a black fringe 141

141. Medulla P+ red (fumarprotocetraric acid present) 142

142. Ascospores 15-20 x 8-12 µm; upper surface largely lacking white pseudocyphellae; thallus often lobulate; common at high elevations **Melanohalea halei**

142. Ascospores 9-16 x 5-9 µm; upper surface with abundant white pseudocyphellae; thallus never lobulate; not known with certainty from the southern Appalachians **Melanohalea olivacea**

141. Medulla P- (fumarprotocetraric acid absent) 143

143. Ascospores brown, 2-celled; upper surface smooth, lacking papillae; very common at nearly all elevations **Anaptychia palmulata**

143. Ascospores hyaline, simple; upper surface conspicuously papillose; very rare at high elevations **Melanohalea exasperata**

130. Upper surface gray, blue-gray, green, or brownish-gray 144

144. On rock 145

145. Thallus microfoliose; upper surface intense green, K- (atranorin absent); in dark, cool, rock overhangs **Flakea papillata**

145. Thallus macrofoliose; upper surface gray to blue-gray, K+ yellow (atranorin present); on rocks in shaded forests or exposed outcrops 146

146. Medulla K+ yellow turning red (salazinic acid present); upper surface with pseudocyphellae **Parmelia omphalodes**

146. Medulla K- or K+ yellow (salazinic acid absent); upper surface without pseudocyphellae 147

147. On shaded rocks in forests; medulla K+ yellow (atranorin present) **Physcia phaea**

147. On exposed rocks; medulla K- (atranorin absent) **Physcia halei**

144. On bark 148

148. Thallus with a conspicuous, black, sponge-like hypothallus **Anzia colpodes**

148. Thallus without a sponge-like hypothallus, although rhizines sometimes becoming dense and forming a conspicuous tangled fringe 149

149. Lower surface ecorticate 150

150. Lobes adnate; apothecial laminal on the surface of the lobes; on the trunks of trees **Heterodermia hypoleuca**

150. Lobes strongly ascending, appearing fruticose; apothecia terminal on the lobe tips; on branches of shrubs and canopy trees **Heterodermia echinata**

149. Lower surface corticate or so closely attached to the substrate that it is impossible to observe 151

151. Lobes inflated, hollow, with a distinct central cavity **Hypogymnia krogiae**

151. Lobes solid, not inflated and hollow 152

152. Upper surface with white pseudocyphellae 153

153. Lower surface uniformly pale, tan to brown; low elevations, rare **Punctelia bolliana**

153. Lower surface black, at least in the central portions of the thallus; middle to high elevations, common **Punctelia appalachensis**

152. Upper surface lacking pseudocyphellae 154

154. Medulla C+ pink (gyrophoric acid present) or C+ red (lecanoric acid or olivetoric acid present) 155
155. Lobes narrow and elongate, resembling a fruticose lichen **Pseudevernia cladonia**
155. Lobes broad, plane, ascending and undulating, but not resembling a fruticose lichen 156
156. Lobes plane, adnate; lobe margins lacking a black fringe of pycnidia; lower surface tomentose; medulla containing cephalodia with cyanobacteria **Lobaria quercizans**
156. Lobes undulating, ascending; lobe margins with abundant black, sessile pycnidia, often creating a black fringe along the margins lower surface; medulla without cephalodia containing cyanobacteria **Tuckermanopsis ciliaris**
154. Medulla C- (gyrophoric acid, lecanoric acid and olivetoric acid absent) 157
157. Medulla K+ yellow turning red (galbinic acid, norstictic acid or salazinic acid present) 158
158. Lobes narrow, 1-3 mm wide; medulla light yellow pigmented, especially in the margins of the apothecia; thallus tightly adnate **Myelochroa galbina**
158. Lobes broad, 10-30 mm wide; medulla uniformly white; thallus loosely adnate 159
159. Medulla K+ yellow turning red, producing red crystals in a water mount when viewed under the compound microscope, P+ yellow (norstictic acid present); lower surface with large blotches of white distributed across the lower surface, especially near the margins; lobes ascending **Parmotrema perforatum**
159. Medulla K+ yellow turning red, not producing red crystals in a water mount when viewed under the compound microscope, P+ orange (salazinic acid present); lower surface centrally black and marginally brown, but not with large blotches of white near the margins; lobes adnate 160
160. Upper cortex with a pattern of reticulate white cracks or maculae; lower surface with rhizines produced up the margins of the lobes **Parmotrema cetratum**
160. Upper cortex without conspicuous white cracks or maculae; lower surface with rhizines absent from a broad, brown zone near the lobe margins **Parmotrema despectum**
157. Medulla K-, K+ yellow or K+ dirty purple-brown (galbinic acid, norstictic acid and salazinic acid absent) 161
161. Medulla P+ orange or red (protocetraric acid or thamnolic acid present) 162
162. Lower surface pale; lobes narrow, 0.5-1.5 mm wide; medulla K+ instant intense yellow and P+ orange (thamnolic acid present) **Imshaugia placorodia**
162. Lower surface centrally black and marginally brown; lobes broad, 3-10 mm wide; medulla K+ yellow turning dirty brownish and P+ red (protocetraric acid present) **Parmotrema submarginale**
161. Medulla P- or P+ yellow (protocetraric acid and thamnolic acid absent) 163
163. Lobe margins with abundant black, sessile pycnidia, often creating a black fringe along the margins 164
164. Medulla UV+ blue-white (alectoronic acid present); very rare **Tuckermanopsis americana**
164. Medulla UV- (fatty acids present); common at high elevations **Tuckermanopsis orbata**
163. Lobe margins lacking a black fringe of pycnidia 165
165. Upper surface with a conspicuous network of ridges and depressions **Platismatia tuckermanii**
165. Upper surface smooth 166
166. Lobes broad, 5-10 mm wide; lower surface with a pale brown tomentum **Lobaria quercizans**
166. Lobes narrower, 1-5 mm wide; lower surface pale or black, often with rhizines, but lacking a tomentum 167
167. Upper cortex K+ yellow (atranorin present, at least in detectable quantities) 168
168. Lower surface black; rhizines abundantly branched, black; medulla K+ purple-brown (lividic acid present) **Hypotrachyna livida**
168. Lower surface pale, white to tan; rhizines mostly simple, pale; medulla K- or K+ yellow (lividic acid absent) 169
169. Medulla K- (zeorin absent); middle to high elevations **Physcia stellaris**
169. Medulla K+ yellow (zeorin present, although spot test reaction due to the presence of atranorin in the medulla); middle to low elevations 170

170. Ascospores 15-17 µm long; common **Physcia pumilior**

170. Ascospores 18-25 µm long; not known with certainty from the southern Appalachians **Physcia aipolia**

167. Upper cortex K- (atranorin absent) 171

171. Upper surface epruinose 172

172. Thallus superficially crustose, extremely adnate to the substrate; rare **Hyperphyscia syncolla**

172. Thallus clearly foliose, loosely adnate to the substrate; common **Anaptychia palmulata**

171. Upper surface white pruinose 173

173. Upper surface conspicuously white pruinose throughout the thallus; upper cortex scleroplectenchymatous; ascospores 28-37 x 15-20 µm; extremely rare **Physconia subpallida**

173. Upper surface mostly epruinose, with pruina restricted to the lobe tips; upper cortex prosoplectenchymatous; ascospores 35-45 x 16-23 µm **Anaptychia palmulata**

67. Thallus with lichenized diaspores (e.g., isidia, lobules, phyllidia, pustules or soredia); apothecia uncommon 174

174. Thallus with lobules or phyllidia (isidia, pustules and soredia absent) 175

175. Dominant photobiont a cyanobacterium 176

176. Lower surface with cyphellae 177

177. Medulla rusty orange-brown, K+ purple (fragilin present) **Sticta fragilinata**

177. Medulla white, K- (fragilin absent) **Sticta carolinensis**

176. Lower surface lacking cyphellae, or the lower surface so tightly attached that it cannot be readily examined 178

178. Thallus not gelatinous when wet (other cyanolichen genera than *Collema* and *Leptogium*) 179

179. Lower surface corticate and lacking veins, or the thallus so tightly adnate that the lower surface cannot be readily examined 180

180. On calcareous rocks; thallus forming concentric rings of lobes **Placynthium petersii**

180. On bark or non-calcareous rocks; thallus mostly consisting of a single organized unit 181

181. Lobes adnate; prothallus conspicuous, typically fibrous, carpet-like; at middle to high elevations **Parmeliella appalachensis**

181. Lobes ascending; prothallus indistinct; at all elevations **Nephroma helveticum**

179. Lower surface ecorticate, strongly tomentose, often with a network of veins that contrast in color with the rest of the lower surface 182

182. Lobules small, flattened, laminal on the thallus surface **Peltigera evansiana**

182. Lobules relatively large, erect, produced on the lobe margins and along cracks on the thallus surface 183

183. Lower surface without distinct veins, instead with a network of regularly spaced pale spots amid a uniformly blackish tomentum **Peltigera elisabethae**

183. Lower surface with a network of distinct veins 184

184. Upper surface glabrous, without tomentum **Peltigera phyllidiosa**

184. Upper surface with patches of white tomentum concentrated near the lobe tips **Peltigera praetextata**

178. Thallus becoming gelatinous when wet (jelly lichens, *Collema* and *Leptogium*) 179

179. Thallus black when dry 180

180. Thallus surface with distinct ridges and bumps; lobules intergrading with cylindrical isidia **Collema furfuraceum**

180. Thallus surface more-or-less smooth; lobules squamiform isidia **Collema flaccidum**

179. Thallus slate fray to olive brown dry 181

181. Lower surface with a continuous white tomentum 182

182. Thallus becoming distinctly swollen and thickened when wet, 200-400 µm thick when hydrated **Leptogium hibernicum**

182. Thallus not distinctly swollen and thickened when wet, reaching a maximum of 150 μm when hydrated **Leptogium laceroides**

181. Lower surface glabrous, occasionally with a few patches of white hairs 183

183. Upper surface wrinkled when dry 184

184. Thallus swelling excessively when wet and becoming strongly shriveled when dry **Leptogium phyllocarpum**

184. Thallus not swelling excessively when wet, mostly retaining non-shriveled and recognizable lobes when dry **Leptogium austroamericanum**

183. Upper surface smooth when dry 185

185. Lobe margins becoming finely divided 186

186. Lobes plane, flattened, becoming finely divided, but not in an exaggerated manner; apothecia common; ascospores 12-30 x 5-8 μm **Leptogium dactylinum**

186. Lobes erect to ascending, the margins becoming irregularly finely divided to resemble exaggerated isidia of varying sizes; apothecia rare; ascospores 18-45 x 11-16 μm **Leptogium lichenoides**

185. Lobe margins entire 187

187. Lobules intergrading with cylindrical isidia; common **Leptogium cyanescens**

187. Lobules uniformly present; not known with certainty from the southern Appalachians **Leptogium denticulatum**

175. Dominant photobiont a green alga 188

188. Upper surface yellow-green (usnic acid present) 189

189. Lower surface black, at least in the central portions of the thallus **Xanthoparmelia angustiphylla**

189. Lower surface pale to dark brown throughout 190

190. Medulla K+ yellow turning red (salazinic acid present) **Xanthoparmelia viriduloumbrina**

190. Medulla K+ yellow turning dingy brownish (stictic acid present) **Xanthoparmelia cumberlandia**

188. Upper surface gray, blue-gray, green-gray, brown or black (usnic acid absent) 191

191. Upper surface uniformly brown to black 192

192. On rock 193

193. Medulla UV+ blue-white and P- (perlatolic acid present) **Melanelia panniformis**

193. Medulla UV- and P+ orange (stictic acid present) or P+ red (fumarprotocetraric acid present) 194

194. Lobe margins with sessile pycnidia and conspicuous white pseudocyphellae; upper surface more-or-less lacking both pseudocyphellae and conspicuous immersed pycnidia; medulla P+ orange (stictic acid present) **Melanelia hepatizon**

194. Lobe margins without pycnidia and conspicuous white pseudocyphellae; upper surface with both pseudocyphellae and conspicuous immersed pycnidia; medulla P+ red (fumarprotocetraric acid) **Melanelia stygia**

192. On bark or wood 195

195. Upper surface white pruinose 196

196. Upper surface conspicuously white pruinose throughout the thallus; upper cortex scleroplectenchymatous; ascospores 28-37 x 15-20 μm; known from one location **Physconia subpallida**

196. Upper surface mostly epruinose, with pruina restricted to the lobe tips; upper cortex prosoplectenchymatous; ascospores 35-45 x 16-23 μm; common **Anaptychia palmulata**

195. Upper surface epruinose 197

197. Lobe margins with abundant black, sessile pycnidia, often creating a black fringe along the margins 198

198. Thallus small, 1-2 cm in diameter 199

199. Lobes very narrow, 0.5 mm wide; middle to low elevations, typically on bark and branches of pines **Tuckermanella fendleri**

199. Lobes broader, 1-3 mm wide; high elevations, on bark and branches of many substrates **Tuckermanopsis orbata**

198. Thallus larger, 2-7 cm in diameter 200

200. Medulla C- and UV- (fatty acids present); middle to high elevations **Tuckermanopsis orbata**

200. Medulla C+ red (olivetoric acid present) or UV+ blue-white (alectoronic acid present); at all elevations 201

201. Medulla C+ red and UV- (olivetoric acid present); extremely common at all elevations **Tuckermanopsis ciliaris**

201. Medulla C- and UV+ blue-white (alectoronic acid present); rare at middle to high elevations **Tuckermanopsis americana**

197. Lobe margins without sessile pycnidia, the margins lacking a black fringe 202

202. Medulla P+ red (fumarprotocetraric acid present) **Melanohalea halei**

202. Medulla P- (fumarprotocetraric acid absent) 203

203. Ascospores brown, 2-celled; upper surface smooth, lacking papillae; very common at nearly all elevations **Anaptychia palmulata**

203. Ascospores hyaline, simple; upper surface conspicuously papillose; very rare at high elevations **Melanohalea exasperata**

191. Upper surface gray, blue-gray, green, or brownish-gray 204

204. Thallus with a conspicuous, black, sponge-like hypothallus **Anzia ornata**

204. Thallus without a sponge-like hypothallus, although rhizines sometimes becoming dense and forming a conspicuous tangled fringe 205

205. Lower surface ecorticate 206

206. Upper cortex K- (atranorin absent) **Phaeophyscia squarrosa**

206. Upper cortex K+ yellow (atranorin present) 207

207. Thallus with abundant marginal phyllidia; phyllidia fragile, often fragmenting **Heterodermia squamulosa**

207. Thallus with coarse laminal and marginal lobules; lobules robust, not fragmenting, often developing in damaged areas of the thallus where the cortex has been consumed by invertebrates **Heterodermia hypoleuca**

205. Lower surface corticate or so closely attached to the substrate that it is impossible to observe 208

208. Upper surface with white pseudocyphellae 209

209. Medulla K+ yellow turning red (salazinic acid present); rare at high elevations **Parmelia omphalodes**

209. Medulla K- (salazinic acid absent); common at all elevations 210

210. Medulla C+ red (lecanoric acid present); common at all elevations **Punctelia rudecta [forms with lobule-like isidia]**

210. Medulla C- (fatty acids present); common only at middle and high elevations 211

211. Lower surface uniformly pale, tan to brown; low elevations, rare **Punctelia bolliana**

211. Lower surface black, at least in the central portions of the thallus; middle to high elevations, common **Punctelia appalachensis**

208. Upper surface lacking pseudocyphellae 212

212. Medulla C+ red (anziaic acid, lecanoric acid or olivetoric acid present) 213

213. Lobes narrow and elongate, resembling a fruticose lichen **Pseudevernia cladonia**

213. Lobes broad, plane, ascending and undulating, but not resembling a fruticose lichen 214

214. Lobe margins lacking a black fringe of pycnidia; lower surface black **Hypotrachyna prolongata**

214. Lobe margins with abundant black, sessile pycnidia, often creating a black fringe along the margins; lower surface pale tan to brown **Tuckermanopsis ciliaris**

212. Medulla C- (anziaic acid, lecanoric acid and olivetoric acid absent) 215

215. Lobe margins with abundant black, sessile pycnidia, often creating a black fringe along the margins 216

216. Medulla UV+ blue-white (alectoronic acid present) **Tuckermannopsis americana**

216. Medulla UV- (fatty acids present) **Tuckermannopsis orbata**

215. Lobe margins without sessile pycnidia, the margins lacking a black fringe 217

217. Medulla P+ orange (salazinic acid present) or P+ red (protocetraric acid present) 218

218. Medulla K+ yellow turning dingy brownish (salazinic acid absent) **Parmotrema submarginale**

218. Medulla K+ yellow turning red (salazinic acid present) 219

219. Upper cortex with a pattern of reticulate white cracks or maculae; lower surface with rhizines produced up the margins of the lobes **Parmotrema cetratum**

219. Upper cortex without conspicuous white cracks or maculae; lower surface with rhizines absent from a broad, brown zone near the lobe margins **Parmotrema despectum**

217. Medulla P- (protocetraric acid and salazinic acid absent) 220

220. Upper cortex K+ yellow (atranorin present in detectable quantities); medulla UV+ blue-white and KC+ yellow-orange (barbatic acid present) **Hypotrachyna virginica**

220. Upper cortex K- (atranorin absent); medulla UV- and KC- (barbatic acid absent) 221

221. Upper surface white pruinose 222

222. Upper surface conspicuously white pruinose throughout the thallus; upper cortex scleroplectenchymatous; ascospores 28-37 x 15-20 µm; known from a single location **Physconia subpallida**

222. Upper surface mostly epruinose, with pruina restricted to the lobe tips; upper cortex prosoplectenchymatous; ascospores 35-45 x 16-23 µm; common **Anaptychia palmulata**

221. Upper surface epruinose 223

223. Lower surface partially ecorticate; phyllidia present, fragile, easily broken; infrequent **Phaeophyscia squarrosa**

223. Lower surface corticate; lobules present, robust, not easily broken; common **Anaptychia palmulata**

174. Thallus with isidia, pustules or soredia (lobules and phyllidia absent) 224

224. Thallus with isidia (pustules and soredia absent) 225

225. Dominant photobiont a cyanobacterium 226

226. Lower surface with cyphellae 227

227. Thallus consisting of a single large lobe, attached to the substrate at a single point and growing outward **Sticta fuliginosa**

227. Thallus consisting of many lobes and resembling a typical foliose lichens, attached to the substrate evenly below and growing parallel to the substrate 228

228. Isidia marginal; thallus surface smooth; common **Sticta beauvoisii**

228. Isidia laminal; thallus ridged or scrobiculate; very rare **Sticta sylvatica**

226. Lower surface without cyphellae 229

229. Thallus not gelatinous when wet (other cyanolichen genera) 230

230. Lower surface ecorticate, strongly tomentose, often with a network of veins that contrast in color with the rest of the lower surface **Peltigera evansiana**

230. Lower surface corticate and lacking veins, or the thallus so tightly adnate that the lower surface cannot be readily examined 231

231. Thallus squamulose, the squamules often somewhat dispersed; prothallus film-like, conspicuous 232

232. On calcareous rocks **Placynthium nigrum**

232. On bark or non-calcareous rocks **Parmeliella triptophylla**

231. Thallus foliose, composed of connected and contiguous lobes; prothallus fibrous, carpet-like 233

233. Thallus P+ orange-red (pannarin present) **Pannaria tavaresii**

233. Thallus P- (pannarin absent) **Coccocarpia palmicola**

229. Thallus becoming gelatinous when wet (jelly lichens, *Collema* and *Leptogium*) 234

234. Lower surface with a conspicuous, continuous white tomentum 235

235. Thallus gray; isidia solitary, cylindrical; at middle to low elevations **Leptogium hirsutum**

235. Thallus dark, blackish; isidia clustered, tree-like; at middle to high elevations **Leptogium acadiense**

234. Lower surface not tomentose, though sometime with scattered tufts of white hairs 236

236. Thallus typically slate-gray to olive-brown in color, composed of a distinct upper and lower cortex visible in cross section under the compound microscope 237

237. Upper surface wrinkled when dry 238

238. Thallus olive-brown, swelling excessively when wet and becoming strongly shriveled when dry **Leptogium millegranum**

238. Thallus slate-gray, not swelling excessively when wet, mostly retaining non-shriveled and recognizable lobes when dry **Leptogium austroamericanum**

237. Upper surface smooth when dry 239

239. Lobe margins becoming finely divided 240

240. Lobes plane, flattened, becoming finely divided, but not in an exaggerated manner; apothecia common; ascospores 12-30 x 5-8 µm **Leptogium dactylinum**

240. Lobes erect to ascending, the margins becoming irregularly finely divided to resemble exaggerated isidia of varying sizes; apothecia rare; ascospores 18-45 x 11-16 µm **Leptogium lichenoides**

239. Lobe margins entire 241

241. Isidia distributed evenly across the surface of the thallus; common **Leptogium cyanescens**

241. Isidia concentrated on prominent ridges of the thallus; rare **Leptogium isidiosellum**

236. Thallus typically brown-black to black in color, not stratified, composed of a single layer when viewed in cross section under the compound microscope 242

242. On rock 243

243. Lobes with conspicuous ridges and bumps; isidia mostly cylindrical **Collema furfuraceum**

243. Lobes more-or-less smooth; isidia squamiform or granular 244

244. Isidia squamiform **Collema flaccidum**

244. Isidia globose or granular 245

245. Thallus small; lobes >5 mm wide **Collema fuscovirens**

245. Thallus larger; lobes 5-10 mm wide **Collema subflaccidum**

242. On bark 246

246. Isidia restricted to conspicuous ridges on the thallus; thallus composed of a distinct upper and lower cortex visible in cross section under the compound microscope **Leptogium isidiosellum**

246. Isidia evenly distributed across the thallus surface; thallus not stratified, composed of a single layer when viewed in cross section under the compound microscope 247

247. Lobes with conspicuous ridges and bumps; isidia mostly cylindrical **Collema furfuraceum**

247. Lobes more-or-less smooth; isidia squamiform or granular 248

248. Isidia squamiform **Collema flaccidum**

248. Isidia globose or granular **Collema subflaccidum**

225. Dominant photobiont a green alga .249

249. Thallus with a conspicuous, black, sponge-like hypothallus **Anzia ornata**

249. Thallus without a sponge-like hypothallus, although rhizines sometimes becoming dense and forming a conspicuous tangled fringe 250

250. Lower surface ecorticate 251

251. Lobes broad, 5-20 mm wide, with a network of ridges and depressions present on both the upper and lower surfaces; lower surface actually corticate, but covered in a tomentum and appearing ecorticate **Lobaria pulmonaria**

251. Lobes narrow, 1-3 mm wide, smooth; lower surface truly ecorticate 252

252. Medulla K+ yellow (atranorin present); lower surface orange-pigmented, the pigment K+ purple **Heterodermia crocea**

252. Medulla K+ yellow turning red (salazinic acid present); lower surface white to brownish, K- **Heterodermia granulifera**

250. Lower surface corticate, or the thallus so tightly adnate that the lower surface cannot be readily examined 253

253. Upper surface yellow-green (usnic acid present) 254
254. Lobes broad, 5-10 mm wide; margins often with conspicuous cilia **Parmotrema xanthinum**
254. Lobes narrower, 1-3(-5) mm wide; margins without cilia 255
255. Lower surface light to dark brown **Xanthoparmelia plittii**
255. Lower surface black, at least in the central portions of the thallus **Xanthoparmelia conspersa**
253. Upper surface gray, blue-gray, green-gray, green, brown or black (usnic acid absent) 256
256. Upper surface brown, black or dark army-green 257
257. Medulla C+ red (lecanoric acid present) 258
258. Isidia granular, fragile, easily abraded and leaving conspicuous white circular scars on the upper surface **Melanelixia subaurifera**
258. Isidia cylindrical, robust, not easily abraded such that the upper surface typically lacks white circular scars **Melanelixia glabratula**
257. Medulla C- (lecanoric acid absent) 259
259. Medulla P+ red (protocetraric acid present); isidia actually flattened lobules; common **Melanohalea halei**
259. Medulla P- (protocetraric acid absent); either short papillae present or true isidia present; rare 260
260. Upper surface brown, papillose, the papillae associated with pseudocyphellae **Melanohalea exasperata**
260. Upper surface gray-brown, isidiate, the isidia laminal and not associated with pseudocyphellae **Phaeophyscia sciastra**
256. Upper surface gray, blue-gray, green, or brownish-gray 261
261. Medulla C+ pink (gyrophoric acid present) or C+ red (anziaic acid or lecanoric acid present) 262
262. Lobes narrow and elongate, resembling a fruticose lichen **Pseudevernia consocians**
262. Lobes broad, plane, ascending and undulating, but not resembling a fruticose lichen 263
263. Upper surface with white pseudocyphellae 264
264. Isidia cylindrical to coralloid, with brown tips, fully corticate; common **Punctelia rudecta**
264. Isidia short, squamiform, uniformly blue-gray, partially ecorticate and often confused with soredia; rare **Punctelia missouriensis**
263. Upper surface without pseudocyphellae 265
265. Lobe margins with a fringe of bulbate cilia 266
266. Lower surface pale, tan to brown; low elevations **Bulbothrix scortella**
266. Lower surface black, not known with certainty from the southern Appalachians **Bulbothrix laevigatula**
265. Lobe margins with or without cilia, but the cilia never with a bulb-like base 267
267. Lobes broad, (5-)10-20 mm wide 268
268. Isidia laminal, never intergrading with lobules; very rare at low elevations **Parmotrema tinctorum**
268. Isidia marginal and laminal, readily intergrading with lobules; rare in spruce-fir forests **Hypotrachyna prolongata**
267. Lobes narrower, 1-5 mm wide 269
269. Isidia marginal and laminal, readily intergrading with lobules; rare in spruce fir **Hypotrachyna prolongata**
269. Isidia laminal, never intergrading with lobules; common, especially at middle and low elevations 270
270. Medulla P- (echinocarpic acid absent); common **Hypotrachyna minarum**
270. Medulla P+ orange-red (echinocarpic acid present); very rare **Hypotrachyna dentella**
261. Medulla C- (anziaic acid, gyrophoric acid and lecanoric acid absent) 271
271. Medulla K+ yellow turning red (galbinic acid, norstictic acid or salazinic acid resent) 272
272. Lobes with a network of ridges and depressions present on both the upper and lower surfaces; lower surface tomentose **Lobaria pulmonaria**

272. Lobes smooth; lower surface smooth or with rhizines, but not tomentose 273

273. Upper surface with white pseudocyphellae 274

274. At least one rhizine squarrose (typically most are squarrose), with many small perpendicular branchlets; common at all elevations **Parmelia squarrosa**

274. Every last rhizine furcate, never with small perpendicular branchlets; rare at high elevations **Parmelia saxatilis**

273. Upper surface without pseudocyphellae 275

275. Thallus tightly adnate; lobes narrow, 1-3(-5) mm wide 276

276. On rock; lobe margins without cilia **Myelochroa obsessa**

276. On bark, especially canopy branches; lobe margins with bulbate cilia **Bulbothrix isidiza**

275. Thallus loosely adnate; lobes broad, (5-)10-20 mm wide 277

277. Lower surface uniformly brown **Parmotrema neotropicum**

277. Lower surface black, at least in the central portions of the thallus, often with a brown zone near the margins 278

278. Medulla UV+ bright yellow (lichexanthone present) **Parmotrema ultralucens**

278. Medulla UV- (lichexanthone absent) **Parmotrema subisidiosum**

271. Medulla K- or K+ yellow turning dingy brown (norstictic acid and salazinic acid absent) 279

279. Medulla P+ orange (echinocarpic acid or stictic acid present) 280

280. Lobes with a network of ridges and depressions present on both the upper and lower surfaces; lower surface tomentose **Lobaria pulmonaria**

280. Lobes smooth; lower surface smooth or with rhizines, but not tomentose 281

281. Thallus adnate; lobes narrow, 1-3(-5) mm wide 282

282. Medulla K- (echinocarpic acid present); lower surface black, at least in the central portions of the thallus; very rare **Hypotrachyna dentella**

282. Medulla K+ instant intense yellow (thamnolic acid present); lower surface pale; common **Imshaugia aleurites**

281. Thallus loosely adnate; lobes broad, (5-)10-20 mm wide 283

283. Lower surface black, at least in the central portions of the thallus; common **Parmotrema crinitum**

283. Lower surface uniformly brown; rare **Parmotrema subtinctorium**

279. Medulla P- (echinocarpic acid and stictic acid absent) 284

284. Medulla UV+ blue-white (alectoronic acid or perlatolic acid present) 285

285. Lobes ruffled, with abundant marginal cilia; isidia mealy, crumbling, resembling soredia **Parmotrema mellissii**

285. Lobes plane, mostly without marginal cilia; isidia remaining intact, not mealy and crumbling 286

286. Thallus adnate to strongly adnate; upper surface white maculate; medulla KC- or KC+ fleeting purple (perlatolic acid present); at nearly every elevation except spruce-fir, especially common at middle and low elevations **Canoparmelia caroliniana**

286. Thallus loosely adnate; upper surface without maculae; medulla KC+ yellow-orange (barbatic acid present) or KC+ pink (alectoronic acid present); lower surface with abundant, richly branched, black rhizines; restricted to spruce-fir 287

287. Medulla KC+ yellow-orange and UV+ dull blue-white (barbatic acid present); rare **Hypotrachyna imbricatula**

287. Medulla KC+ pink and UV+ bright blue-white (alectoronic acid present); presumed extirpated **Hypotrachyna ensifolia**

284. Medulla UV- (alectoronic acid and perlatolic acid absent) 288

288. Lobes broad, (5-)10-20 mm wide; upper surface with a network of ridges and depressions **Platismatia glauca**

288. Lobes narrow, 1-3(-5) mm wide; upper surface smooth 289

289. Upper cortex K+ yellow (atranorin present); at least one isidium with an apical black cilium; common on bark and rocks **Hypotrachyna horrescens**

289. Upper cortex K- (atranorin absent); all isidia without black cilia; rare on rocks **Phaeophyscia sciastra**

224. Thallus with pustules or soredia (isidia absent) 290

290. Primary photobiont a cyanobacterium 291

291. Lower surface with cyphellae **Sticta limbata**

291. Lower surface without cyphellae 292

292. Lower surface ecorticate, strongly tomentose, often with a network of veins that contrast in color with the rest of the lower surface 293

293. Thallus small; lobes scallop-like, concave with upturned margins **Peltigera didactyla**

293. Thallus large; lobes round, but usually plane and not concave [**Peltigera extenuata**]

292. Lower surface corticate and lacking veins, or the thallus so tightly adnate that the lower surface cannot be readily examined 294

294. Upper surface covered with numerous fine, hyaline hairs; presumed extinct **Erioderma mollissimum**

294. Upper surface not covered with numerous fine, color hairs; extant 295

295. Upper cortex or medulla P+ orange (pannarin or stictic acid present) 296

296. Thallus blue-gray when dry, typically with a fibrous, carpet-like prothallus **Pannaria conoplea**

296. Thallus green when dry, lacking a prothallus but the lower surface tomentose **Lobaria scrobiculata**

295. Upper cortex and medulla P- (pannarin and stictic acid absent) 297

297. Thallus brown when dry; soredia white **Nephroma parile**

297. Thallus dark gray when dry; soredia blue-gray **Fuscopannaria sorediata**

290. Primary photobiont a green alga 298

298. Thallus with a conspicuous, black, sponge-like hypothallus 299

299. Lobes adnate, plane; soredia marginal, actually coarse isidioid-lobules **Anzia ornata**

299. Lobes ascending; soredia terminal on the undersides of the lobe tips **Anzia americana**

298. Thallus without a sponge-like hypothallus, although rhizines sometimes becoming dense and forming a conspicuous tangled fringe 300

300. Lobes inflated, hollow, with a distinct central cavity 301

301. Lobes perforated, with distinct circular holes in the upper cortex **Menegazzia subsimilis**

301. Lobes not perforated, the upper cortex continuous and unbroken 302

302. Soralia on the upper surface of the lobe tips **Hypogymnia tubulosa**

302. Soralia on the lower surface of the lobe tips 303

303. Lobes elongate, strap-like; thallus forming a curtain draping down from the substrate; rare at high elevations **Hypogymnia vittata**

303. Lobes not elongate and strap-like, resembling a typical foliose lichen; thallus forming circular colonies; common at all elevations 304

304. Lobe margins black; lobes strongly adnate, contiguous and overlapping; rare at high elevations **Hypogymnia incurvoides**

304. Lobe margins usually not conspicuously black; lobes adnate to ascending, spreading; common at middle and high elevations **Hypogymnia physodes**

300. Lobes not inflated, solid, without a distinct central cavity 305

305. Upper surface yellow-green (usnic acid present) 306

306. Medulla K+ yellow turning red (salazinic acid present); presumed extirpated **Hypotrachyna sinuosa**

306. Medulla K- or K+ yellow turning dingy brown (salazinic acid absent) 307

307. Upper surface with white pseudocyphellae; medulla C+ red (lecanoric acid present) **Flavopunctelia flaventior**

307. Upper surface without pseudocyphellae; medulla C- (lecanoric acid absent) 308
308. Soralia capitate on the lobe tips; medulla UV+ blue-white (divaricatic acid present) **Parmeliopsis capitatea**
308. Soralia marginal or laminal, but not capitate on the lobe tips; medulla UV- (divaricatic acid absent) 309
309. Lobes with upturned, ruffled margins; soralia marginal; medulla P- (protocetraric acid absent) **Usnocetraria oakesiana**
309. Lobes plane; soralia or pustules laminal; medulla P+ red (protocetraric acid present) 310
310. Soredia present, typically fine; typically on bark, rarely on rock **Flavoparmelia caperata**
310. Coarse pustules present; typically on rock **Flavoparmelia baltimorensis**
305. Upper surface gray, blue-gray, green-gray, green, brown or black (usnic acid absent) 311
311. Upper surface brown to black 312
312. Medulla C+ pink (gyrophoric acid present); soredia actually areas of the exposed medulla where granular isidia have worn away from the upper surface leaving scars **Melanelixia subaurifera**
312. Medulla C- (gyrophoric acid absent); soredia not as above 313
313. Medulla P+ orange (stictic acid present) or UV+ dull blue-white (perlatolic acid present) 314
314. Soralia marginal, conspicuous, white; medulla P+ orange (stictic acid present) **Melanelia culbersonii**
314. Soralia laminal on the surface of the thallus, especially near the lobe tips; medulla UV+ dull blue-white (perlatolic acid present) **Melanelia disjuncta**
313. Medulla P- and UV- (perlatolic acid and stictic acid absent) 315
315, Upper surface white pruinose, usually conspicuously so near the lobe tips **Physconia leucoleiptes**
315. Upper surface epruinose 316
316. Soralia capitate, convex, formed on to tops of short secondary lobes **Phaeophyscia pusilloides**
316. Soralia laminal and marginal, but not capitate 317
317. Soralia uniformly laminal; not known with certainty from the southern Appalachians **Phaeophyscia orbicularis**
317. Soralia marginal, in some cases also becoming lamina; common 318
318. Lobes broad, 2-6 mm wide; thallus always with a conspicuous carpet-like fringe of rhizines around the lobes; rare at high elevations **Phaeophyscia hispidula**
318. Lobes narrow, 1-1.5 mm wide; thallus often with rhizines extending outward from the lower surface, but not usually forming a continuous and conspicuous carpet-like fringe; common at all elevations **Phaeophyscia adiastola**
311. Upper surface gray, blue-gray, green, or brownish-gray 319
319. Upper surface with white pseudocyphellae 320
320. Medulla K+ yellow turning red (salazinic acid present) **Parmelia sulcata**
320. Medulla K- (salazinic acid absent) 321
321. Medulla C+ pink (gyrophoric acid) or C+ pink (lecanoric acid or olivetoric acid present) 322
322. Lower surface pale, tan to brown 323
323. Soredia fine, although sometimes becoming coarse and forming aggregation in the central portions of the thallus, fully ecorticate, not easily confused with isidia; common **Punctelia caseana**
323. Soredia very coarse, squamiform, partially corticate and often confused with isidia; rare **Punctelia missouriensis**
322, Lower surface black, at least in the central portions of the thallus 324
324. Medulla C+ red (olivetoric acid present); common at all elevations **Cetrelia olivetorum**
324. Medulla C+ pink (gyrophoric acid present); vey rare and known only from high elevations **Punctelia borreri**
321. Medulla C- (gyrophoric acid, lecanoric acid and olivetoric acid absent) 325

325. Medulla UV- (fatty acids present); rare and restricted to northern hardwood forests **Punctelia reddenda**

325. Medulla UV+ blue-white (alectoronic acid or perlatolic acid present); common at all elevations 326

326. Medulla UV+ bright blue-white (alectoronic acid present); common at all elevations **Cetrelia chicitae**

326. Medulla UV+ dull blue-white (perlatolic acid present); infrequent at high elevations **Cetrelia cetrarioides**

319. Upper surface without pseudocyphellae 327

327. Medulla C+ pink (gyrophoric acid present) or C+ red (anziaic acid or lecanoric acid present) 328

328. Thallus sorediate 329

329. Medulla P+ orange (echinocarpic acid present); restricted to spruce-fir **Hypotrachyna thysanota**

329. Medulla P- (echinocarpic acid absent); common at all elevations 330

330. Lobes narrow, 0.5-1.5 mm wide, stap-like **Everniastrum catawbiense**

330. Lobes broader, 3-5 mm wide, short and regular, not strap-like 331

331. Secondary lobes strongly ascending; lobes revolute, with distinctly undulating margins **Hypotrachyna revoluta**

331. Secondary lobes not strongly ascending; lobes mostly plan, not revolute 332

332. Soralia capitate, strongly convex, on upper surface of terminal portions of the secondary lobes; infrequent at middle and low elevations, often on the branches of shrubs and trees **Hypotrachyna lividescens**

332. Soralia laminal, not strongly capitate; rare at high elevations, on the trunks of trees and rarely on rocks 333

333. Medulla with lecanoric acid and evernic acid (TLC required); rare **Hypotrachyna rockii**

333. Medulla with anziaic acid (TLC required); presumed extirpated and not known from the southern Appalachians in modern times **Hypotrachyna producta**

328. Thallus pustulose, the pustules sometimes exposing the medulla and eventually causing the formation of soredia 334

334. Secondary lobes strongly ascending, jutting out from the thallus; pustules mostly restricted to the secondary lobes; common at middle and high elevations **Hypotrachyna kauffmaniana**

334. Secondary lobes not strongly ascending, usually growing parallel to the thallus; pustules evenly distributed across the thallus; common at all elevations 335

335. Thallus small, tightly adnate to the substrate; uncommon 336

336. Medulla often faintly yellow pigmented; spumosa UV+ blue-white unknown present (TLC required); low elevations **Hypotrachyna spumosa**

336. Medulla uniformly white; two unidentified substances related to hiascic acid present (TLC required); middle and high elevations **Hypotrachyna showmanii**

335. Thallus large, loosely adnate to the substrate and easily removed; common 337

337. Lobes revolute, with undulating margins; medulla C+ pink (gyrophoric acid present); very common at high elevations **Hypotrachyna afrorevoluta**

337. Lobes plane, with even margins; medulla C+ red (lecanoric acid and evernic acid present); infrequent at high elevations **Hypotrachyna taylorensis**

327. Medulla C- (anziaic acid, gyrophoric acid and lecanoric acid absent) 339

339. Lower cortex ecorticate 340

340. Lobes narrow, elongate, strap-like 341

341. Lower surface light yellow pigmented; lobes adnate; rare at middle and low elevations **Heterodermia appalachensis**

341. Lower surface white; lobes ascending; common at all elevations **Heterodermia leucomela**

340. Lobes short, regular, resembling a typical foliose lichen 342

342. Lower surface with yellow or orange or reddish-brown pigmentation (check carefully!) 343

343. Pigment on the lower surface orange or reddish-brown, K+ purple 344

344. Pigment confined to small spots near the lobe tips **Heterodermia neglecta**

344. Pigment more-or-less uniformly distributed across the lower surface, not restricted to tiny spots near the lobe tips **Heterodermia obscurata**

343. Pigment on the lower surface yellow or reddish-brown, K- 345

345. Pigment yellow, confined to small spots near the lobe tips **Heterodermia casarettiana**

345, Pigment reddish-brown, forming a zone between the purple-black central portions and white outer portions of the thallus **Heterodermia langdoniana**

342. Lower surface white, sometimes with a purple-brown pigment toward the central portions of the thallus, entirely lacking yellow, orange or reddish-brown pigments 346

346. Lower surface weakly corticate, uniformly pale tan, without a purple-black pigmented area near the central portions of the thallus; common **Heterodermia speciosa**

346. Lower surface completely ecorticate, with a purple-black pigmented area near the central portions of the thallus; infrequent 347

347. Lobes short, distinctly abbreviated, erect, ascending **Heterodermia erecta**

347. Lobes longer, plane and not strongly erect or ascending **Heterodermia japonica**

339. Lower surface corticate or so closely attached to the substrate that it is impossible to observe 348

348. Medulla K+ yellow turning red (galbinic acid, norstictic acid or salazinic acid present) 349

349. Lobes with a network of ridges and depressions present on both the upper and lower surfaces; lower surface tomentose **Lobaria pulmonaria**

349. Lobes smooth; lower surface smooth or with rhizines, but not tomentose 350

350. Lobes narrow, 1-3(-5) mm wide 351

351. Soralia capitate, convex, produced on short secondary lobes; medulla lightly yellow pigmented; lower surface black, at least in the central portions of the thallus **Myelochroa metarevoluta**

351. Soralia not capitate, plane, produced on the margins of terminal portions of the lobes, occasionally als on the lobe surface; medulla white; lower surface pale, sordid white or tan to brown 352

352. Medulla K+ yellow turning red, producing red crystals in a water mount when viewed under the compound microscope, P+ yellow (norstictic acid present); on rocks, often associated with waterways at middle and low elevations **Heterodermia pseudospeciosa**

352. Medulla K+ yellow turning red, not producing red crystals in a water mount when viewed under the compound microscope, P+ orange (salazinic acid present); on bark, rarely on rocks, at low elevations **Heterodermia albicans**

350. Lobes broad, (5-)10-20 mm wide 353

353. Lower surface uniformly brown 354

354. Norlobaridone present (TLC required) **Parmotrema conferendum**

354. Norlobaridone absent (TLC required) **Parmotrema subsumptum**

353. Lower black in the central portions, often with a brown zone near the margins, sometimes with large white blotches near the lobe margins 355

355. Medulla K+ yellow turning red, producing red crystals in a water mount when viewed under the compound microscope, P+ yellow (norstictic acid present); lower surface typically with numerous white blotches interrupting the brown and black coloration **Parmotrema hypotropum**

355. Medulla K+ yellow turning red, not producing red crystals in a water mount when viewed under the compound microscope, P+ orange (salazinic acid present); lower surface without white blotches 356

356. Upper surface with a regular pattern of white cracks and maculae; lower surface with the rhizines produced right up to the edge of the thallus 357

357. Soredia fine, powdery; soralia conspicuous, produced on upper surface of the tips of the secondary lobes **Parmotrema reticulatum**

357. Soredia coarse, mealy, actually arising from the breakdown of isidiza, typically at least some isidia still intact and with distinct brown tips, produced both marginally and laminally on the thallus surface **Parmotrema subisidiosum**

356. Upper surface without a regular pattern of white cracks and maculae, though maculae sometimes present; lower surface with a zone near the margins that is naked and free of rhizines 358

358. Soralia marginal, produced all along the edges of ruffled secondary lobes; common at high elevations **Parmotrema stuppeum**

358. Soralia terminal, produced on the tips of the secondary lobes, which are usually plane and not ruffled; rare **Parmotrema margaritatum**

348. Medulla K- or K+ yellow turning dingy brown (galbinic acid, norstictic and salazinic acid absent) 359

359. Medulla P+ orange (echinocarpic acid or stictic acid present) or P+ red (fumarprotocetraric acid or protocetraric acid present) 360

360. Lobes with a network of ridges and depressions present on both the upper and lower surfaces; lower surface tomentose **Lobaria pulmonaria**

360. Lobes smooth; lower surface smooth or with rhizines, but not tomentose 361

361. Thallus scrobiculate, the surface with a pattern of ridges and depressions **Crespoa crozalsiana**

361. Thallus not scrobiculate, the surface smooth 362

362. Medulla K- (echinocarpic acid present) **Hypotrachyna thysanota**

362. Medulla K+ yellow turning dingy brown (fumarprotocetraric acid, protocetraric acid or stictic acid present) 363

363. Lobes simple, entire, rounded 364

364. Medulla P+ orange (stictic acid present); middle and high elevations; soralia produced on the upper surface of second secondary lobes **Parmotrema perlatum**

364. Medulla P+ red (protocetraric acid present); middle and low elevations; soralia produced along the margins of ruffled secondary lobes 365

365. Echinocarpic acid present; not known with certainty from the southern Appalachians **Parmotrema dilatatum**

365. Echinocarpic acid absent; rare at middle and low elevations **Parmotrema gardneri**

363. Lobes branching, typically ending in an abbreviated pair resembling a pair of shorts 366

366. Pustules present, laminal **Hypotrachyna croceopustulata**

366. Soralia present, capitate on the tips of secondary lobes 367

367. Middle and low elevations; protocetraric acid present (TLC required); medulla uniformly white **Hypotrachyna pseudosinuosa**

367. High elevations; fumarprotocetraric acid present (TLC required); medulla often orange pigmented just below the soralia **Hypotrachyna gondylophora**

359. Medulla P- and P+ yellow (echinocarpic acid, fumarprotocetraric acid, protocetraric acid and stictic acid absent) 368

368. Medulla UV+ blue-white (alectoronic acid present, barbatic acid, diffractaic acid, divaricatic acid or lividic acid present) 369

369. On rock in protected overhangs; thallus tightly adnate **Dirinaria frostii**

369. On bark or wood, very rarely on exposed rocks; thallus adnate to loosely adnate 370

370. Pustules present 371

371, Upper cortex UV+ bright yellow (lichexanthone present) **Hypotrachyna osseoalba**

371. Upper cortex UV- (lichexanthone absent) 372

372. Pustules scale-like, causing the upper cortex to peel away in large areas and reveal both the medulla and lower cortex

372. Pustules bubble-like, typically dissolving into soredia and not causing large patches of the cortex to peel away 373

373. Upper surface conspicuously white maculate; lobes simple, entire, rounded; divaricatic acid present (TLC required) **Canoparmelia texana**

373. Upper surface not conspicuously maculate; lobes branching, typically ending in an abbreviated pair resembling a pair of shorts; lividic acid present (TLC required) **Hypotrachyna pustulifera**

370. Soredia present 374

374. Lobes simple, entire, rounded 375

375. Thallus without marginal cilia, tightly adnate and not easily removed from the substrate; lobes plane, not ruffled **Canoparmelia texana**

375. Thallus with abundant marginal cilia, loosely adnate and easily removed from the substrate; lobes ruffled, undulating 376

376. Medulla CK+ yellow and UV+ dull blue-white (diffractaic acid present) **Parmotrema diffractaicum**

376. Medulla CK- and UV+ bright blue-white (alectoronic acid present) 377

377. Soralia terminal on the tips of the secondary lobes; soredia fine, powdery; high elevations **Parmotrema arnoldii**

377. Soredia coarse, mealy, actually arising from the breakdown of isidiza, typically at least some isidia still intact and with distinct brown tips, produced both marginally and laminally on the thallus surface; at all elevations, but most common at middle and low elevations **Parmotrema mellissii**

374. Lobes branching, typically ending in an abbreviated pair resembling a pair of shorts 378

378. Thallus forming small rosettes; soralia capitate on the tips of the secondary lobes; medulla KC+ yellow-orange and UV+ dull blue-white (barbatic acid present) **Hypotrachyna laevigata**

376. Thallus forming large colonies; soralia not capitate, terminal and marginal on the secondary lobes; medulla KC+ fleeting pink and UV+ bright blue-white (alectoronic acid present) 377

377. Gyrophoric acid present (TLC required) **Hypotrachyna oostingii**

377. Gyrophoric acid absent (TLC required) **Hypotrachyna densirhizinata**

368. Medulla UV- (alectoronic acid present, barbatic acid, diffractaic acid, divaricatic acid and lividic acid present absent) 378

378. Upper cortex K- (atranorin absent) 379

379. Upper surface white pruinose, usually conspicuously so near the lobe tips **Physconia leucoleiptes**

379. Upper surface epruinose 380

380. Lower surface pale, tan to brown **Physciella chloantha**

380. Lower surface black, at least in the central portions of the thallus 381

381. Soralia capitate, convex, formed on to tops of short secondary lobes **Phaeophyscia pusilloides**

381. Soralia laminal and marginal, but not capitate 382

382. Soralia uniformly laminal; not known with certainty from the southern Appalachians **Phaeophyscia orbicularis**

382. Soralia marginal, in some cases also becoming lamina; common 383

383. Lobes broad, 2-6 mm wide; thallus always with a conspicuous carpet-like fringe of rhizines around the lobes; rare at high elevations **Phaeophyscia hispidula**

383. Lobes narrow, 1-1.5 mm wide; thallus often with rhizines extending outward from the lower surface, but not usually forming a continuous and conspicuous carpet-like fringe; common at all elevations **Phaeophyscia adiastola**

378. Upper cortex K+ yellow (atranorin present) 382

382. Lobes broad, (5-)10 mm wide 383

383. Lower surface uniformly brown; norlobaridone present (TLC required) **Parmotrema commensuratum**

383. Lower surface black, at least in the central portions of the thallus; fatty acids present (TLC required) **Parmotrema simulans**

382. Lobes narrow, 0.5-3(-5) mm wide 384

384. Pustules present, laminal; medulla lightly yellow pigmented **Myelochroa aurulenta**

384. Soredia present, either contained in soralia or produced singularly along the margins; medulla uniformly white 385

385. Soralia laminal, on the lobe surface 386

386. Lower surface pale **Physcia americana**

386. Lower surface black 387

387. On rock; very rare at high elevations **Physcia caesia**

387. On bark; very rare at middle and low elevations [**Physcia sorediosa**]

385. Soralia marginal or terminal 388

388. Soralia hood-shaped, terminal on the tips of the lobes **Physcia adscendens**

388. Soralia not hood-shaped, marginal or terminal on the lobes 389

389. On bark 390

390. Soralia crescent shaped, well delimited, with fine soredia; lobe margins with pale, white cilia; common **Heterodermia speciosa**

390. Soralia not crescent shaped, poorly delimited, with coarse, granular soredia; lobe margins lacking cilia; rare (but common outside of the Smokies in disturbed habitats) **Physcia millegrana**

389. On rock 391

391. Soralia crescent shaped, well delimited, with fine soredia 392

392. Lower surface pale, tan to brownish; common at all elevations **Heterodermia speciosa**

392. Lower surface black, at least in the central portions of the thallus; very rare at high elevations **Physcia subalbinea**

391. Soralia not crescent shaped, poorly delimited, with coarse, granular soredia 393

393. Lobes fairly broad, 0.5-1.0 mm wide; coarse granular soredia produced primarily on the tips and margins of the secondary lobes; very rare, on shaded rocks in humid habitats **Physcia pseudospeciosa**

393. Lobes quite narrow, 0.2-0.5 mm wide; coarse blastidia and soredia produced all along the margins of the lobes; very common, on exposed and shaded rocks 394

394. Thallus tightly adnate, not becoming raised from the surface and easily removed **Physcia subtilis**

394. Thallus adnate to loosely adnate, becoming raised from the surface and easily removed **Physcia thomsoniana**

KEY IV: CRUSTOSE LICHENS WITH ASEXUAL PROPAGULES

1. Thallus bright yellow or orange in color 2

2. On bark or wood 3

3. Thallus isidiate **Rinodina chrysidiata**

3. Thallus sorediate or leprose 4

4. Thallus K+ intense reddish-purple **Caloplaca chrysophthalma**

4. Thallus K- or K+ underwhelming dull reddish-brown 5

5. Thallus entirely leprose, composed of fuzzy, dull, ecorticate granules 6

6. Granules minute, 5-15 μm in diameter; thallus dull yellow; restricted to conifers at low and middle elevations **Chrysothrix chamaecyparicola**

6. Granules larger, 20-50 μm in diameter; thallus intense yellow; on hardwoods and conifers throughout **Chrysothrix xanthina**

5. Thallus areolate or composed of shiny, compact, corticate granules 7

7. Thallus esorediate, composed of large, corticate granules, 50-100 μm in diameter; typically on the bark and wood of mature trees **Candelariella xanthostigma**

7. Thallus sorediate, composed of areoles that erupt with soralia and dissolve into soredia 8

8. Ascospores 8-per ascus **Candelariella xanthostigmoides**

8. Ascospores 16-32 per ascus **Candelariella efflorescens**

2. On rock or humus over rock 9

9. On calcareous rocks; thallus K+ intense reddish-purple10

10. Thallus leprose, composed of fuzzy, dull, ecorticate granules **Caloplaca chrysodeta**

10. Thallus areolate, composed of areoles that produce marginal soralia and dissolve into soredia **Caloplaca flavocitrina**

9. On non-calcareous rocks; thallus K- or K+ underwhelming dull reddish-brown 11

11. Thallus areolate, composed of small, flat areoles that erupt with soralia; very rare... **Lecanora epanora**

11. Thallus leprose, composed of fuzzy, dull, ecorticate granules; common 12

12. Thallus green-yellow 13

13. Thallus UV+ dull orange (rhizocarpic acid present), without patches of rusty orange-yellow pigment; common at all elevations **Psilolechia lucida**

13. Thallus UV+ dull blue-white (obtusatic acid present), with patches of rusty orange-yellow pigment; rare at high elevations **Andreiomyces morozianus**

12. Thallus bright yellow 14

14. Thallus C+ pink (gyrophoric acid present); typically forming small, dispersed thallus on small accumulations of organic matter clinging to vertical rock faces; rare **Chrysothrix susquehannensis**

14. Thallus C- (gyrophoric acid absent); typically forming extensive colonies on sheltered or protected rock faces and overhangs; common 15

15. Thallus thin, forming a dispersed to +/- continuous crust that is not easily detached from the substrate; zeorin present in low concentrations (TLC required); very common **Chrysothrix insulizans**

15. Thallus thick, forming continuous and lumpy crust that is easily detached from the substrate; zeorin absent (TLC required); very rare **Chrysothrix onokoensis**

1. Thallus gray, blue-gray, brown, green, or yellow green in color 16

16. Thallus leprose, composed entirely of granules; lacking sexual fruiting bodies (i.e., apothecia, perithecia) and pycnidia or other specialized structures that disperse conidia 17

17. Thallus P+ orange or P+ red (fumarprotocetraric acid, pannarin, protocetraric acid, salazinic acid or stictic acid present) 18

18. Thallus KC+ pink/red (alectorialic acid present) 19

19. Thallus forming small rosettes; granules shiny, compact, superficially corticate; fumarprotocetraric acid present (TLC required); common, in exposed microhabitats.. **Lepraria neglecta s.l. (alectorialic acid and fumarprotocetraric acid chemotype)**

19. Thallus forming extensive colonies; granules dull, fuzzy, ecorticate; protocetraric acid present (TLC required); rare, in sheltered and protected microhabitats **Lepraria eburnea**

18. Thallus KC- (alectorialic acid absent) 20

20. Thallus K+ yellow turning red (norstictic acid or salazinic acid present) 21

21. Thallus entirely leprose, composed of dispersed piles of granules that eventually overlap to form a continuous crust, K+ yellow turning red, but not producing red crystals in a water mount under the compound microscope (salazinic acid present); atranorin and zeorin present (TLC required); very rare **Lepraria elobata s.l. (salazinic acid chemotype)**

21. Thallus not entirely leprose, with at least some portions smooth and not broken down into granular fragments, K+ yellow turning red, producing red crystals in a water mount under the compound microscope (norstictic acid present); stictic acid present (TLC required); common throughout **Phlyctis petraea s.l. (stictic acid chemotype)**

20. Thallus K-, K+ yellow or K+ brown (norstictic acid and salazinic acid absent) 22

22. Thallus both K+ instant lemon yellow and P+ orange (thamnolic acid present), not entirely leprose, with at least some portions smooth and not broken down into granular fragments **Loxospora elatina**

22. Thallus not both K+ instant lemon yellow and P+ orange (thamnolic acid absent), entirely leprose 23

23. Thallus dimorphic, with a secondary fruticose portion composed of erect pseudopodetia covered with ecorticate granules; very rare, at high elevations **Lepraria arbuscula**

23. Thallus entirely crustose and lacking secondary fruticose portions, not dimorphic; common throughout 24

24. Thallus aggregate-type, composed of dispersed piles of granules (best observed at the thallus edge) that eventually overlap and form a continuous crust 25

25. Granules forming large, airy, conspicuous aggregations that resemble cotton balls; protocetraric acid and roccellic/angardianic acid present (TLC required); in sheltered rock overhangs at high elevations **Lepraria lanata**

25. Granules not forming large and conspicuous aggregations; other substances present; in various habitats throughout 26

26. Granules shiny, compact, superficially corticate; thallus often forming small rosettes; atranorin, roccellic/angardianic acid and fumarprotocetraric acid present (TLC required); in exposed microhabitats **Lepraria neglecta s.l. (fumarprotocetraric acid chemotype)**

26. Granules dull, fuzzy, ecorticate; thallus not forming rosettes; other combinations of substances present; in sheltered and protected microhabitats 27

27. Thallus distinctly lead-gray in color; K- (pannarin and zeorin present, TLC required); throughout the Park **Leprocaulon adhaerens**

27. Thallus light blue-gray in color; K+ yellow (atranorin, zeorin and stictic acid present, TLC required); rare, restricted to high elevations **Lepraria elobata s.str. (stictic acid chemotype)**

24. Thallus placodioid-type, composed of thick, continuous hypothallus (best observed as a raised, delimited thallus edge) upon which the granules rest 28

28. Thallus blue-white, typically with a raised, crisped margin resembling *Normandina*; lower surface with dark rhizohyphae 29

29. Fumarprotocetraric acid present, in addition to atranorin and roccellic/angardianic acid (TLC required); mostly at middle to high elevations **Lepraria oxybapha**

29. Protocetraric acid present, in addition to atranorin and roccellic/angardianic acid (TLC required); mostly at low to middle elevations **Lepraria normandinoides**

28. Thallus greenish-blue to yellow-green, without a raised and crisped margin; lower surface typically with pale rhizohyphae or lacking rhizohyphae 30

30. Thallus greenish-blue, usually without a yellowish tinge; atranorin, zeorin, and stictic acid present (TLC required); common and abundant throughout (when in doubt, pick this species) **Lepraria finkii**

30. Thallus distinctly yellowish; pannaric acid 6-methylester present (TLC required); infrequent but widespread **Lepraria vouauxii**

17. P- or P+ yellow (fumarprotocetraric acid, pannarin, protocetraric acid, salazinic acid or stictic acid absent) 31

31. Photobiont a cyanobacterium; rare at high elevations **Fuscopannaria frullaniae**

31. Photobiont a green alga 32

32. Thallus UV+ blue-white (decarboxysquamatic acid, divaricatic acid or obtusatic acid present) 33

33. Thallus not entirely leprose, with at least some portions smooth and not broken down into granular fragments; usnic acid and decarboxysquamatic acid present, with traces of zeorin (TLC required) **Lecanora strobilina**

33. Thallus entirely leprose; other combinations of substances present 34

34. Thallus yellow-green, with patches of orange-yellow pigments; UV+ dull blue-white (obtusatic acid present, together with isousnic acid; TLC required); high elevations **Andreiomyces morozianus**

34. Thallus blue-gray or white to greenish-gray, without pigmented patches; UV+ bright blue-white (divaricatic acid present; TLC required); throughout 35

35. On bark; thallus aggregate-type, composed of dispersed piles of granules (best observed at the thallus edge) that eventually overlap and form a continuous crust; divaricatic acid and zeorin present (TLC required) **Lepraria hodkinsoniana**

35. On rock in overhangs; thallus placodioid-type, composed of thick, continuous hypothallus (best observed as a raised, delimited thallus edge) upon which the granules rest; divaricatic acid and nordivaricatic acid present (TLC required) **Lepraria cryophila**

32. Thallus UV- or UV+ orange (decarboxysquamatic acid, divaricatic acid and obtusatic acid absent) 36

36. Thallus distinctly dark greenish-brown in color, resembling colonies of green algae, off-putting in appearance, composed of minute goniocysts 37

37. Micareic acid present (TLC required) **Micarea prasina**

37. Methoxymicareic acid present (TLC required) **Micarea micrococca**

36. Thallus gray, blue-gray, yellow-green or light green in color, not as above and distinctly more attractive; composed of granules and not minute goniocysts 38

38. On bark or wood 39

39. Thallus KC+ pink/red and P+ yellow (alectorialic acid present) 40

40. Thallus aggregate-type, composed of dispersed piles of granules (best observed at the thallus edge) that eventually overlap and form a continuous crust; granules shiny, compact, superficially corticate; in exposed microhabitats, common **Lepraria neglecta s. str. (alectorialic acid chemotype)**

40. Thallus placodioid-type, composed of thick, continuous hypothallus (best observed as a raised, delimited thallus edge) upon which the granules rest; granules dull, fuzzy, ecorticate; in sheltered and protected microhabitats, very rare **Lepraria eburnea**

39. Thallus KC- or KC+ yellow, P- or P+ yellow (alectorialic acid absent) 41

41. Thallus P+ intense yellow (psoromic acid present) 42

42. Granules shiny, compact, superficially corticate; thallus without a thick, well-developed white prothallus; atranorin and roccellic/angardianic acid present (TLC required) **Lepraria neglecta s.l. (psoromic acid chemotype)**

42. Granules dull, fuzzy, ecorticate; thallus with a thick, well-developed white prothallus; atranorin and roccellic/angardianic acid absent (TLC required) **Phlyctis boliviensis**

41. Thallus P- or P+ weak yellow (psoromic acid absent) 43

43. Thallus K+ yellow (atranorin present) 44

44. Thallus placodioid-type, composed of thick, continuous hypothallus (best observed as a raised, delimited thallus edge) upon which the granules rest **Lepraria harrisiana**

44. Thallus aggregate-type, composed of dispersed piles of granules (best observed at the thallus edge) that eventually overlap and form a continuous crust 45

45. Granules with a distinctly shaggy appearance due to the presence of numerous, well-developed hyphae extending outward from the surface; nephrosterinic acid present (TLC required); rare **Lepraria rigidula**

45. Granules not appearing shaggy, not with numerous well-developed hyphae extending outward from the surface; nephrosterinic acid absent (TLC required); common 46

46. Typically on boles; common at all elevations; zeorin present (TLC required) **Lepraria caesiella**

46. Typically on tree bases and stumps; rare, at high elevations; zeorin absent (TLC required) **Lepraria jackii**

43. Thallus K- (usnic acid present) 47

47. Thallus not entirely leprose, with at least some portions smooth and not broken down into granular fragments **Lecanora strobilina**

47. Thallus entirely leprose 48

48. Thallus continuous, typically with a well-developed white, fibrous prothallus (best observed as a fringe around the thallus margin); xanthones absent (TLC required); common throughout **Lecanora thysanophora**

48. Thallus composed of dispersed piles of granules (best observed at the thallus edge) that eventually overlap and form a continuous crust, lacking a white fibrous prothallus; xanthones present (TLC required); infrequent at low and middle elevations **Lepraria xanthonica**

38. On rock or humus over rock 49

49. Thallus KC+ pink/red and P+ yellow (alectorialic acid present) **Lepraria neglecta s. str. (alectorialic acid chemotype)**

49. Thallus KC- or KC+ yellow, P- or P+ yellow (alectorialic acid absent) 50

50. Thallus P+ intense yellow (norstictic acid or psoromic acid present) 51

51. Thallus K+ yellow turning red, producing red crystals in a water mount under the compound microscope (norstictic acid present); common **Phlyctis petraea**

51. Thallus K- (psoromic acid absent); rarely on rock, typically on bark **Phlyctis boliviensis**

50. Thallus P- or P+ weak yellow (norstictic acid and psoromic acid absent) 52

52. Thallus green or yellow-green, K- (atranorin absent except in *L. thysanophora*) 53

53. Thallus KC- (usnic acid absent) 54

54. Thallus thin, dispersed, UV+ orange (rhizocarpic acid present) **Psilolechia lucida**

54. Thallus thick, continuous, felt-like, UV- (no substances present) **Botryolepraria lesdainii**

53. Thallus KC+ yellow (usnic acid present) 55

55. Thallus dimorphic, with a secondary fruticose portion composed of flaccid, dangling pseudopodetia covered with ecorticate granules; very rare, at high elevations **Lecanora anakeestiicola**

55. Thallus entirely crustose and lacking secondary fruticose portions, not dimorphic; common throughout 56

56. Thallus aggregate-type, composed of dispersed piles of granules (best observed at the thallus edge) that eventually overlap and form a continuous crust 57

57. Thallus UV+ dull orange (xanthones present, TLC required) **Lepraria xanthonica**

57. Thallus UV- (xanthones absent, TLC required) **Leprocaulon nicholsiae**

56. Thallus placodioid-type, composed of thick, continuous hypothallus (best observed as a raised, delimited thallus edge) upon which the granules rest 58

58. Thallus UV+ dull orange (xanthones present, TLC required); on calcareous rocks **Lepraria disjuncta**

58. Thallus UV- (xanthones absent, TLC required); on non-calcareous rocks **Lecanora thysanophora**

52. Thallus gray to blue-gray, K+ yellow (atranorin present) 59

59. Thallus continuous, typically with a well-developed white, fibrous prothallus (best observed as a fringe around the thallus margin); rare on rocks, typically on bark **Lecanora thysanophora**

59. Thallus composed of dispersed piles of granules (best observed at the thallus edge) that eventually overlap and form a continuous crust, lacking a white fibrous prothallus; common on rocks 60

60. Thallus blue-gray, composed of piles of granules that typically overlap and form a continuous crust; zeorin present (TLC required); common throughout **Lepraria caesiella**

60. Thallus white-gray, composed of piles of granules that often remain separate and form distinct islands; zeorin absent (TLC required); infrequent at middle and high elevations **Lepraria humida**

16. Thallus not both leprose and composed entirely of granules; sometimes with sexual fruiting bodies (i.e., apothecia, perithecia) or pycnidia or other specialized structures that disperse conidia 61

61. Thallus black or very dark brown and forming tufts branches or hair-like masses 62

62. Photobiont a cyanobacterium; on exposed rocks **Spilonema revertens**

62. Photobiont Trentepohlia; on sheltered rocks in overhangs 63

63. On calcareous rocks **Thermutis velutina**

63. On non-calcareous rocks **Cystocoleus ebeneus**

61. Thallus not black or brown and forming tufts or hair-like masses 64

64. Thallus without lichenized diaspores (e.g., blastidia, isidia, soredia) 65

65. Thallus without conspicuous pycnidia and/or highly specialized structures that disperse conidia (e.g., hyphophores, sporodochia, synnemata) 66

66. Thallus K+ yellow (atranorin present), crustose, thick, continuous; on shaded non-calcareous rocks with *Porpidia albocaerulescens* and *Rhizocarpon infernulum* f. *sylvaticum* **Herteliana schuyleriana**

66. Thallus K- (atranorin absent), areolate to squamulose or microfoliose; typically on other substrates in other microhabitats 67

67. Thallus composed of minute, emerald-green, scallop-like squamules **Normandina pulchella**

67. Thallus light green to green-brown, squamulose or micro-foliose, but not scallop-like 68

68. On non-calcareous rocks deep in overhangs; thallus microfoliose, light green in color, lacking short hyaline spines on the upper surface **Flakea papillata**

68. On calcareous exposed rocks, humus or bark; thallus areolate to squamulose, green to green-brown, with short-hyaline spines on the upper surface **Agonimia opuntiella**

65. Thallus with conspicuous pycnidia and/or highly specialized structures that disperse conidia (e.g., hyphophores, sporodochia, synnemata) 69

69. Thallus with conspicuous pycnidia 70

70. Thallus KC+ pink and P+ yellow (alectorialic acid present); on bark and wood at high elevations **Pycnora praestabilis**

70. Thallus KC- and P- (alectorialic acid absent); substrate and elevation various 71

71. Thallus blue-gray to white; pycnidia pruinose, C+ red (lecanoric acid and schizopeltic acid present); on bark at high elevations **Lecanactis abietina**

71. Thallus green to gray to brownish; pycnidia epruinose, C- (lecanoric acid and schizopeltic acid absent); substrate and elevation various 72

72. Pycnidia stipitate; conidia ellipsoid 73

73. Thallus and pycnidia P+ red (fumarprotocetraric acid present; best viewed in a mount under the compound microscope); mostly a low to middle elevations **Micarea neostipitata**

73. Thallus and pycnidia P- (fumarprotocetraric acid absent); middle to high elevations **Micarea pycnidiophora**

72. Pycnidia sessile; conidia bacilliform (bowling-pin shaped) 74

74. Pycnidia pale; on leaves **Fellhanera bouteillei**

74. Pycnidia dark, brown-black; on wood, bark or rocks 75

75. Thallus without blastidia; conidia 6-8 µm long **Fellhanera montesfumosi**

75. Thallus with minute blastidia 76

76. On rocks and bases of trees; conidia 4-5 µm long; common **Fellhanera granulosa**

76. On bark and branches of trees; conidia 5-6 µm long; very rare **Fellhanera eriniae**

69. Thallus with sporodochia, synnemata or hyphophores 77

77. Thallus with sporodochia or synnemata 78

78. Conidia flabellate, resembling hands with fingers; very rare **Cheiromycina flabelliformis**

78. Conidia not flabellate, either globose or multi-cellular; common 79

79. On old logs of oak and chestnut; photobiont coccoid **Xyleborus sporodochifer**

79. On the bark of living trees; photobiont Trentepohlia 80

80. Thallus thin, indistinct; synnemata (stalked sporodochia) present; conidia globose, multi-cellular; typically on the bases of hardwoods at middle to high elevations **Dictyocatenulata alba**

80. Thallus thick, well-developed; sessile sporodochia present; conidia globose to ellipsoid, 1-2 celled; on the boles of mature hardwoods at low to middle elevations **Tylophoron americanum**

77. Thallus with hyphophores 81

81. Hyphophores on tall stalks, with a starburst **Gomphillus americanus**

81. Hyphophores not on tall stalks with an apical starburst 82

82. Hyphophores cylindrical, apically flared, resembling a barrel exploding from the top; very rare **Gyalectidium yahriae**

82. Hyphophores hair-like, lash-like, hook-like or sail-like; common 83

83. On soil or organic matter... **Gyalideopsis moodyae**

83. On leaves, bark or wood 84

84. On leaves; hyphophores pale **Gyalectidium appendiculatum**

84. On bark wood; hyphophores dark, at least in the lower portions 85

85. Hyphophores tall, erect; on spruce and fir branches at high elevations **Gyalideopsis epicorticis**

85. Hyphophores short, bent, hook-like or sail-like; on various substrates at different elevations 86

86. On spruce or fir at high elevations **Gyalideopsis piceicola**

86. On other substrates at low and middle elevations **Gyalideopsis bartramiorum**

64. Thallus with lichenized diaspores (e.g., blastidia, isidia, soredia) 87

87. Thallus with isidia or phyllidia 88

88. On rock or organic matter over rocks 89

89. On calcareous rocks; photobiont a cyanobacterium **Placynthium nigrum**

89. On non-calcareous rocks and organic matter over non-calcareous rocks; photobiont a green alga 90

90. Thallus shiny and bronze in color; photobiont Trentepohlia; on rocks in overhangs; very rare **Clathroporina isidiifera**

90. Thallus white-gray, blue-gray or brown in color; photobiont coccoid; on exposed rocks or organic matter over rocks; common 91

91. Thallus green-brown, C+ pink (gyrophoric acid, best viewed in a water mount under the compound microscope) **Placynthiella icmalea**

91. Thallus white-gray to blue-gray, C- or C+ pink (gyrophoric acid present or absent) 92

92. Isidia tall, thick, coarse; thallus C+ pink (gyrophoric acid present) **Ochrolechia yasudae**

92. Isidia short, thin, fragile; thallus C- (gyrophoric acid absent) **Pertusaria globularis**

88. On bark or wood 93

93. Photobiont *Trentepohlia* 94

94. Thallus thick, blue-gray; rare, on large Yellow Birches at high elevations **Graphis sterlingiana**

94. Thallus thin, bronze to slate-gray; common, on many different substrates, especially at middle and low elevations 95

95. Thallus bronze **Porina scabrida**

95. Thallus slate-gray **Pseudosagedia isidiata**

93. Photobiont coccoid 96

96. True isidia present 97

97. Thallus yellow **Rinodina chrysidiata**

97. Thallus gray, blue-gray, green or brown 98

98. Thallus C+ pink (gyrophoric acid, best viewed in a water mount under the compound microscope) 99

99. Thallus brown to green-brown, isidia minute, fragile **Placynthiella icmalea**

99. Thallus gray to blue-gray; isidia larger, robust **Ochrolechia yasudae**

98. Thallus C- (gyrophoric acid absent) 100

100. Thallus dark green-brown, composed of minute, flattened areoles **Rinodina colobinoides s.l. (isidiate morphotype)**

100. Thallus blue-gray, gray, or light green, continuous and film-like or composed of convex areoles 101

101. Thallus blue-gray to gray, with large, globose warts; isidia short, often with crumbling tips **Pertusaria globularis**

101. Thallus light green, without large, globose warts; isidia variable but not crumbling 102

102. Thallus continuous, film-like; isidia short, minute, +/- translucent; particularly common on ericaceous shrubs at middle and high elevations along streams **Jamesiella anastomosans**

102. Thallus areolate; isidia variable in shape, opaque; typically a middle and low elevations on a variety of substrates 103

103. Thallus with black pycnidia; conidia filiform; atranorin present in low concentrations (TLC required) **Bacidia schweinitzii (forms without apothecia)**

103. Thallus without black pycnidia; conidia ellipsoid if present; atranorin absent (TLC required) 104
104. Isidia short, knob-like, globose **Phyllopsora kalbii**
104. Isidia cylindrical-coralloid or flattened-phyllidiate 105
105. Isidia cylindrical-coralloid **Phyllopsora corallina**
105. Isidia flattened-phyllidiate **Phyllopsora confusa**
96. Isidia actually stipitate pycnidia, often with an apical pale mass of exuded conidia 106
106. Thallus and pycnidia P+ red (fumarprotocetraric acid present; best viewed in a mount under the compound microscope); mostly a low to middle elevations **Micarea neostipitata**
106. Thallus and pycnidia P- (fumarprotocetraric acid absent); middle to high elevations **Micarea pycnidiophora**
87. Thallus with blastidia, pustules, schizidia or soredia 107
107. Thallus with blastidia or schizidia 108
108. Thallus areolate to squamulose, the upper surface with minute, hyaline spine-like hairs; blastidia marginal/terminal **Agonimia opuntiella**
108. Thallus not with the above combination of features, always lacking hyaline, spine-like hairs 109
109. On bark or wood 110
110. Thallus thick, continuous, blue-gray, P+ intense yellow (psoromic acid present) **Phlyctis boliviensis**
110. Thallus thin, areolate, brown-gray to green-gray or green-brown, P- or P+ orange-red (psoromic acid absent) 111
111. Thallus P+ orange-red (pannarin present) **Rinodina excrescens**
111. Thallus P- or P+ yellow (pannarin absent) 112
112. Thallus K+ yellow, P+ weak yellow (atranorin present) **Rinodina bullata**
112. Thallus K- and P- (atranorin absent) 113
113. Pycnidia absent or inconspicuous; blastidia dark, brown to gray-brown **Rinodina colobinoides**
113. Pycnidia present, usually conspicuous; blastidia bright green 114
114. Rarely on bases of trees; conidia 4-5 µm long **Fellhanera granulosa**
114. On boles and branches of trees; conidia 5-6 µm long **Fellhanera eriniae**
109. On rock, soil or humus over rock 115
115. Thallus K+ yellow turning red, producing red crystals in a water mount under the compound microscope (norstictic acid present) 116
116. Thallus P+ orange (stictic acid present) **Phlyctis petraea s.l. (stictic acid chemotype)**
116. Thallus P+ yellow (stictic acid absent) **Phlyctis petraea s. str.**
115. Thallus K- or K+ yellow (norstictic acid absent) 117
117. Thallus P+ orange-red (argopsin or stictic acid present) 118
118. Thallus thick, greenish-gray, rimose-areolate, K+ yellow turning dirty brownish (stictic acid present); schizidia present, short, peg-like; on shaded rocks, often in forests **Baeomyces rufus**
118. Thallus a scurfy, black crust, P+ orange-red (argopsin present); blastidia present, dark, granular, blackish; on sun-exposed rocks **Halecania pepegospora**
117. Thallus P- or P+ yellow (argopsin and stictic acid absent) 119
119. Thallus P+ yellow (atranorin, baeomycesic acid or psoromic acid present) 120
120. On soil or humus over rocks; thallus white, UV+ orange (baeomycesic acid present); schizidia present, large, globose, often becoming detached and leaving scars **Dibaeis baeomyces**
120. On rock; thallus blue-gray to greenish-gray, UV- (baeomycesic acid absent); blastidia or granules present, small, knob-like, not often becoming detached and leaving scars 121
121. Thallus dull, fuzzy, ecorticate, K- (atranorin absent); on sheltered rocks in overhangs **Phlyctis boliviensis**
121. Thallus shiny, compact, corticate, K+ yellow (atranorin present); on exposed rocks in shaded forests 122

122. Thallus green-blue to green-gray, typically lacking a well-developed conspicuous white prothallus; blastidia produced along the cracks in the thallus; zeorin absent (TLC required); on shaded rocks in forests with *Porpidia albocaerulescens* and *Rhizocarpon infernulum* f. *sylvaticum* **Herteliana schuyleriana**

122. Thallus blue-gray to gray, typically with a well-developed, conspicuous, sometimes fibrous, white prothallus; granular soredia produced on the thallus surface; zeorin present (TLC required); on shaded rocks along waterways **Megalaria beechingii**

119. Thallus P- (atranorin, baeomycesic acid and psoromic acid absent) 123

123. Thallus blue-gray to greenish-blue, K+ yellow (atranorin present) 124

124. Thallus green-blue to green-gray, typically lacking a well-developed conspicuous white prothallus; blastidia produced along the cracks in the thallus; zeorin absent (TLC required); on shaded rocks in forests with *Porpidia albocaerulescens* and *Rhizocarpon infernulum* f. *sylvaticum* **Herteliana schuyleriana**

124. Thallus blue-gray to gray, typically with a well-developed, conspicuous, sometimes fibrous, white prothallus; granular soredia produced on the thallus surface; zeorin present (TLC required); on shaded rocks along waterways **Megalaria beechingii**

123. Thallus dark to light green, K- (atranorin absent) 125

125. Prothallus dark, conspicuous; thallus composed of convex areoles that erupt into soredia that contrast strongly in color with the thallus; sphaerophorin present; typically on vertical rock faces associated with waterways or seepages **Micareopsis irriguata**

125. Prothallus white or indistinct; thallus continuous or composed of minute areoles with blastidia that do not contrast strongly with the color of the thallus; no substances present; typically on shaded rocks in forests **Fellhanera granulosa**

107. Thallus with pustules or soredia 126

126. Thallus composed of minute, emerald-green, scallop-like squamules **Normandina pulchella**

126. Thallus light green to green-brown, crustose to squamulose, but not scallop-like 127

127. Thallus areolate to squamulose, the upper surface with minute, hyaline, spine-like hairs; blastidia marginal/terminal **Agonimia opuntiella**

127. Thallus not with the above combination of features, always lacking hyaline, spine-like hairs 128

128. Thallus squamulose 129

129. Thallus C- and P+ red (fumarprotocetraric acid present); squamules usually not strongly separating from the substrate, resembling drops of melted way, not strongly flaring, dark greenish-brown in color **Hypocenomyce anthracophila**

129. Thallus C+ red and P- (lecanoric acid present); squamules usually separating from the substrate, resembling overlapping roof shingles, flaring, light brown in color **Hypocenomyce scalaris**

128. Thallus crustose 130

130. Thallus K+ yellow turning red, producing red crystals in a water mount under the compound microscope (norstictic acid present) 131

131. On sheltered rocks, often in overhangs 132

132. Thallus P+ orange (stictic acid present) **Phlyctis petraea s.l. (stictic acid chemotype)**

132. Thallus P+ yellow (stictic acid absent) **Phlyctis petraea s. str.**

131. On bark 133

133. Thallus thin, green-gray to brownish-green; soralia small, not distinctly raised; soredia green to brown or reddish-brown pigmented **Japewiella dollypartoniana**

133. Thallus thick, blue-gray to white-gray; soralia large, often +/- raised; soredia white to blue-gray 134

134. "Soralia" actually densely pruinose apothecia, strongly raised and with a lecanorine rim, typically with hymenia and ascospores when sectioned; common at middle to high elevations **Lepra waghornei**

134. True soralia present, not strongly raised and with a lecanorine rim, sterile and lacking hymenia and ascospores when sectioned; very rare at high elevations **Lepra excludens**

130. Thallus K- or K+ yellow (norstictic acid absent) 135

135. Thallus P+ yellow, P+ orange or P+ red (many different substances potentially present) 136

136. Thallus, especially soralia, K+ instant intense lemon yellow and P+ orange (thamnolic acid present) 137

137. Thallus pustulose; pustules laminal 138

138. Thallus shiny; pustules breaking down into coarse fragments; elatinic acid absent (TLC required) common and widespread except at the highest elevations **Lepra pustulata**

138. Thallus dull; pustules crumbling and breaking down into soredia; elatinic acid present (TLC required); common at high elevations **Loxospora elatina**

137. Thallus soraliate, with discrete soralia 139

139. Soralia plane, without a rim, often becoming diffuse and confluent **Loxospora elatina**

139. Soralia raised, often with a distinct rim, remaining discrete 140

140. Thallus UV+ yellow (lichexanthone present) **Lepra trachythallina s.l. (lichexanthone chemotype)**

140. Thallus UV- (lichexanthone absent) **Lepra trachythallina s. str.**

136. Thallus not both K+ instant intense lemon yellow and P+ orange (thamnolic acid absent) 141

141. Thallus, especially soralia, P+ orange or P+ red (many different substances potentially present) 142

142. On rock 143

143. On sheltered rocks in overhangs; thallus often orange-tinted; soralia light gray to slate-gray; medulla K+ yellow and P+ orange (stictic acid present) **Porpidia degelii**

143. On sun exposed rocks; thallus black; soralia black, actually minute, granular blastidia; medulla K- and P+ orange-red (argopsin present) **Halecania pepegospora**

142. On bark or wood 144

144. Photobiont *Trentepohlia*; thallus K+ yellow and P+ orange (stictic acid present) **Nadvornikia sorediata**

144. Photobiont coccoid; thallus with various spot tests (stictic acid present or absent) 145

145. Thallus green, C+ pink-orange (gyrophoric acid present) **Biatora printzenii**

145. Thallus gray to blue-gray or brownish, C- (gyrophoric acid absent) 146

146. Thallus (especially soralia) K+ yellow (atranorin or stictic acid present) 147

147. On bark 148

148. Soralia P+ orange (stictic acid present); atranorin and zeorin present (TLC required) **Lecanora layana**

148. Soralia P+ red (fumarprotocetraric or succinoprotocetraric acid present); zeorin absent (TLC required) 149

149. "Soralia" white, actually densely pruinose apothecia that are often sterile and lack hymenia or ascospores; succinoprotocetraric acid present (TLC required); throughout at all elevations **Lepra multipunctoides**

149. True soralia present; soredia typically dark gray and contrasting in color against the white thallus; atranorin and fumarprotocetraric acid present (TLC required); mostly at higher elevations **Violella fucata**

147. On wood 150

150. Soralia P+ orange (stictic acid present); thallus mostly immersed in the substrate; soredia greenish-white, concolorous with the thallus; rare and only at the highest elevations **Xylographa vitiligo**

150. Soralia P+ red (fumarprotocetraric acid present), thallus not immersed in the substrate; soredia dark gray, contrasting with the thallus; common throughout the higher elevations **Violella fucata**

146. Thallus (including soralia) K- (atranorin and stictic acid absent) 151

151. Thallus continuous, with pustulose soralia **Megalospora porphyritis**

151. Thallus areolate, with discrete, non-pustulose soralia 152

152. Soralia laminal and marginal, often completely dissolving the areoles and leaving overlapping heaps of soredia; common **Rinodina buckii**

152. Soralia marginal, rarely completely dissolving the areoles; rare 153

153. Soralia yellow pigmented (secalonic acid present); zeorin present only as a trace compound (TLC required) **Rinodina efflorescens**

153. Soralia white-gray, not yellow pigmented (secalonic acid absent); zeorin present as a major compound (TLC required) **Rinodina willeyi**

141. Thallus, especially soralia, P+ yellow (atranorin or psoromic acid present) 154

154. Thallus K- and P+ intense yellow (psoromic acid present) **Phlyctis boliviensis**

154. Thallus K+ yellow and often P+ yellow (atranorin present) 155

155. On rocks 156

156. Thallus green-blue to green-gray, typically lacking a well-developed conspicuous white prothallus; blastidia produced along the cracks in the thallus; zeorin absent (TLC required); on shaded rocks in forests with *Porpidia albocaerulescens* and *Rhizocarpon infernulum* f. *sylvaticum* **Herteliana schuyleriana**

156. Thallus blue-gray to gray, typically with a well-developed, conspicuous, sometimes fibrous, white prothallus; granular soredia produced on the thallus surface; zeorin present (TLC required); on shaded rocks along waterways **Megalaria beechingii**

155. On bark or wood 157

157. Thallus composed relatively large, flat, dispersed areoles with marginal soralia **Rinodina degeliana**

157. Thallus immersed, poorly-developed, or composed of small, +/- convex, contiguous areoles with laminal soralia 158

158. Thallus dull green, immersed to granular; soralia large, diffuse, irregularly shaped; soredia light green, contrasting against the thallus; typically K+ weak yellow (atranorin present in low concentrations); often overgrowing mosses over bark **Bacidia sorediata**

158. Thallus blue-gray to gray, immersed or areolate; soralia smaller, discrete or diffuse; soredia gray to blue-gray, sometimes yellowish-gray, more-or-less concolorous with the thallus; K+ yellow (atranorin present in high concentrations); typically directly on bark and not overgrowing mosses 159

159. Thallus well-developed, areolate; soralia erupting from the areoles; stictic acid absent (TLC required); uncommon and mostly on *Juglans* or *Juniperus* near waterways **Lecanora appalachensis**

159. Thallus poorly-developed, often immersed; soralia typically erose; stictic acid present or absent; common throughout 160

160. Stictic acid present (TLC required) **Lecanora layana**

160. Stictic acid absent (TLC required) **Lecanora nothocaesiella**

135. Thallus P- or P+ weak yellow (mostly species without secondary compounds except for atranorin) 161

161. Thallus (especially soralia) C+ pink/red (erythrin, gyrophoric acid, or lecanoric acid present) 162

162. On rock, humus over rock, or bryophytes over rock 163

163. Photobiont *Trentepohlia*; on rocks in overhangs; rare 164

164. Low elevations; soralia C+ red (erythrin present) **Opegrapha gyrocarpa**

164. High elevations; soralia C+ pink (gyrophoric acid present) **Dirina massiliensis f. sorediata**

163. Photobiont coccoid; on exposed rocks, humus over rock, or bryophytes over rock; common 165

165. On rock; thallus placodioid; soralia forming along cracks in the thallus **Trapelia placodioides**

165. On humus or bryophytes over rock; thallus areolate; soralia laminal and not forming along cracks in the thallus 166

166. Thallus lead-gray to blue-gray or white; soredia dark gray or blue-gray 167

167. Thallus lead-gray in color; soredia darker in color than the thallus; rarely on soil or bryophytes, typically on bark or wood **Trapeliopsis flexuosa**

167. Thallus blue-gray to white; soredia blue-gray, not darker in color than the thallus; commonly on humus and bryophytes **Trapeliopsis granulosa**

166. Thallus brown to green; soredia brownish-green to green 168

168. Soralia large, convex, conspicuous; typically forming large colonies on bryophytes over rocks in sheltered microhabitats **Biatora chrysantha**

168. Soralia minute, plane, inconspicuous; typically on humus in exposed microhabitats **Placynthiella dasaea**

162. On bark or wood 169

169. Soralia bitter tasting (picrolichenic acid present) **Lepra amara**

169. Soralia not bitter tasting (picrolichenic acid absent) 170

170. Thallus and soralia gray, blue-gray or white 171

171. Thallus areolate, lead-gray; soredia darker in color than the thallus **Trapeliopsis flexuosa**

171. Thallus continuous, blue-gray to white; soralia concolorous with, or lighter than, the thallus 172

172. Thallus UV+ yellow (lichexanthone present) 173

173. "Soralia" actually densely pruinose apothecia, typically raised and with a distinct thalline rim **Varicellaria velata s.l. (lichexanthone chemotype)**

173. True soralia present, typically sessile and without a distinct thalline rim **Ochrolechia arborea**

172. Thallus UV- (lichexanthone absent) 174

174. True soralia present, typically sessile and plane, without a distinct thalline rim **Ochrolechia androgyna**

174. "Soralia" actually densely pruinose apothecia, either convex and hemispherical or raised and with a distinct thalline rim 175

175. Ascospores simple; common throughout **Varicellaria velata s.str.**

175. Ascospores 2-celled, fragmenting into parts; not known with certainty from the southern Appalachians **Varicellaria rhodocarpa**

170. Thallus and soralia brown-green to green 176

176. On bark 177

177. Common on the trunks and branches of trees at high elevations; soralia large, conspicuous **Biatora appalachensis**

177. Rare, on the bases and trunks of trees at middle and high elevations; soralia minute, inconspicuous **Trapelia corticola**

176. On wood 178

178. Thallus continuous, thick; soralia diffuse, irregularly shaped; C+ pink reaction often very difficult to observe **Trapeliopsis viridescens**

178. Thallus areolate, thin; soralia discrete, regularly shaped, C+ pink reaction easily observed 179

179. Soralia well-defined, convex, button-like **Trapelia corticola**

179. Soralia poorly defined, plane, not button-like **Placynthiella dasaea**

161. Thallus C- or C+ yellow/orange (erythrin, gyrophoric acid and lecanoric acid absent) 180

180. Thallus (especially soralia) C+ yellow-orange (xanthones present) 181

181. Thallus gray to blue-gray; soralia large, convex, often lightly yellow pigmented; K+ yellow (atranorin present) **Vainionora americana**

181. Thallus green; soralia smaller, plane, light green in color; K- (atranorin absent) 182

182. Thallus UV+ dull orange, P- (xanthones present); soralia minute, punctiform, irregular in shape and often becoming confluent to resemble a leprose crust **Biatora pontica**

182. Thallus UV- and P+ orange-red (argopsin present); soralia larger, plane, discrete, not becoming confluent **Biatora printzenii**

180. Thallus C- 183

183. Thallus (especially soralia) UV+ blue-white (confluentic acid, divaricatic acid, lobaric acid, perlatolic acid, sphaerophorin present) 184

184. On rock, in overhangs or protected microhabitats 185

185. Photobiont *Trentepohlia* **Opegrapha zonata**

185. Photobiont coccoid 186

186. On sheltered rock faces associated with waterways at middle and low elevations; thallus distinctly greenish; sphaerophorin present (TLC required) **Micareopsis irriguata**

186. On rocks in overhangs at middle to high elevations; thallus distinctly gray to blue-gray; divaricatic acid or confluentic acid present (TLC required) 187

187. Medulla producing an efflux of small bubbles when K is added to a water mount under the compound microscope, confluentic acid present (TLC required) **Porpidia tuberculosa**

187. Medulla not producing an efflux of small bubbles when K is added to a water mount under the compound microscope, divaricatic acid present (TLC required) 188

188. Thallus composed of small, flat areoles dispersed on a dark prothallus; apothecia very rare; ascospores kidney-bean shaped, often some two-celled **Fuscidea recensa**

188. Thallus rimose-areolate; apothecia apparently relatively common; ascospores globose to subglobose, simple **Fuscidea appalachensis**

184. On bark or wood 189

189. Thallus composed of dispersed areoles 190

190. Prothallus dark, blackish, distinct; areoles small, often entirely dissolving into piles and resembling a leprose crust; divaricatic acid present (TLC required); common on spruce at high elevations **Lecidea nylanderi**

190. Prothallus pale, indistinct; areoles relatively large, not entirely dissolving into piles of soredia; perlatolic acid present (TLC required) not known with certainty from the southern Appalachians **Hertelidea botryosa**

188. Thallus continuous, or composed of contiguous areoles 191

191. Thallus dark brownish-green, typically forming small rosettes, often with a conspicuous shiny, pale or brown prothallus **Ropalospora viridis**

191. Thallus lighter in color, yellowish-green, gray or blue-gray, not forming small rosettes, with or without a conspicuous prothallus 192

192. Thallus indistinct, immersed in the substrate; soralia becoming confluent and resemble a leprose crust; soralia KC+ fleeting purple (lobaric acid present) **Bacidia lobarica**

192. Thallus continuous, well-developed, soralia discrete and remaining so; soralia KC- (lobaric acid absent) 193

193. Low elevations; thallus green-gray; prothallus pale; atranorin and sphaerophorin present (TLC required); very rare **Haematomma americanum**

193. High elevations; thallus blue-gray; prothallus dark, blackish; 2'-o-methylerpatolic acid or perlatolic acid present (TLC required); common 194

194. Thallus K+ yellow (atranorin present); 2'-o-methylerpatolic present (TLC required); infrequent **Lecanora darlingiae**

194. Thallus K- (atranorin absent); perlatolic acid present (TLC required); common **Mycoblastus caesius**

183. Thallus UV- (divaricatic acid, lobaric acid, perlatolic acid and sphaerophorin absent) 195

195. Soralia KC+ red/purple, the reaction often fleeting (picrolichenic acid present) **Lepra amara**

195. Soralia KC- (picrolichenic acid absent) 196

196. On rock, soil, humus over rock, or bryophytes over rock 197

197. On soil or humus over rock; thallus thin, film-like **Trapeliopsis gelatinosa**

197. On rock 198

198. On rocks in sheltered overhangs 199

199. Photobiont Trentepohlia; thallus brown, poorly developed; confluentic acid present (TLC required) **Opegrapha zonata**

199. Photobiont coccoid; thallus green or gray, well-developed; confluentic acid absent (TLC required) 200

200. Thallus composed of convex, green areoles; soralia erumpent; prothallus dark, conspicuous; sphaerophorin present (TLC required); typically on vertical rock faces associated with waterways **Micareopsis irriguata**

200. Thallus composed of flat, gray areoles, occasionally with divided margins that appear lobate; soralia excavate; prothallus indistinct; no substances present (TLC required); on rocks in overhangs, not usually associated with waterways **Caloplaca reptans**

198. On exposed rocks, often in forests 201

201. Thallus green-blue to green-gray, typically lacking a well-developed conspicuous white prothallus; blastidia produced along the cracks in the thallus; zeorin absent (TLC required); on shaded rocks in forests with *Porpidia albocaerulescens* and *Rhizocarpon infernulum* f. *sylvaticum* **Herteliana schuyleriana**

201. Thallus blue-gray to gray, typically with a well-developed, conspicuous, sometimes fibrous, white prothallus; granular soredia produced on the thallus surface; zeorin present (TLC required); on shaded rocks along waterways **Megalaria beechingii**

196. On bark and wood 202

202. Thallus distinctly dark greenish-brown in color, resembling colonies of green algae, off-putting in appearance, composed of minute goniocysts 203

203. Micareic acid present (TLC required) **Micarea prasina**

203. Methoxymicareic acid present (TLC required) 204

204. Goniocysts remaining intact **Micarea micrococca**

204. Goniocysts erupting into soralia **Micarea soralifera**

202. Thallus gray, blue-gray, yellow-green or light green in color, not as above and distinctly more attractive; continuous or areolate but composed of not minute goniocysts 205

205. Photobiont *Trentepohlia* **Opegrapha corticola**

205. Photobiont coccoid 206

206. On wood, often rotten wood **Trapeliopsis viridescens**

206. On bark 207

207. Thallus yellow-green, KC+ yellow (usnic acid preset) **Lecanora strobilina**

207. Thallus gray or green, KC- (usnic acid absent) 208

208. Thallus and soredia green 209

209. Thallus minutely areolate, the areoles flat; soralia small, discrete, marginal; always on bark **Lecania croatica**

209. Thallus continuous to granular; soralia large, irregular in shape, diffuse; often overgrowing bryophytes on bark **Bacidia sorediata**

208. Thallus and soredia blue-gray to gray or brownish 210

210. Thallus continuous; "soralia" actually densely pruinose apothecia, discrete, often with a raised thalline rim **Lepra ophthalmiza**

210. Thallus areolate; true soralia present, these not raised and with a thalline rim 211

211. Thallus blue-gray, sorediate **Rinodina brodoana**

211. Thallus brown-gray, blastidiate **Rinodina colobinoides**

KEY V: CRUSTOSE APOTHECIATE LICHENS WITHOUT ASEXUAL PROPAGULES

1. Apothecia borne on stalks, brown or bright pink 2

2. Thallus green to brown, K+ yellow and P+ orange (stictic acid present); apothecia brown; on shaded rocks **Baeomyces rufus**

2. Thallus white, K- and P+ yellow (baeomycesic acid present); apothecia pink; on exposed soil **Dibaeis baeomyces**

1. Apothecia immersed, sessile or somewhat raised, but never brown on stalks, color various 3

3. Thallus and/or apothecia bright yellow, orange, red or pink 4

4. Apothecia brightly colored, yellow, orange, red or pink 5

5. Thallus brightly colored yellow or orange 6

6. Thallus and apothecia K+ magenta/purple (anthraquinones present); ascospores polarilocular 7

7. On bark **Caloplaca flavorubescens**

7. On rock 8

8. Thallus continuous **Caloplaca flavovirescens**

8. Thallus areolate 9

9. Areoles yellow, typically reduced to a small area just around the apothecial margins **Caloplaca feracissima**

9. Areoles orange, typically remaining intact and not becoming reduced **Caloplaca subsoluta**

6. Thallus and apothecia K- (anthraquinones absent); ascospores simple 10

10. On rock in overhangs; thallus green, leprose **Psilolechia lucida**

10. On bark or wood; thallus yellow, areolate...11

11. Thallus esorediate, composed entirely of large, globose areoles, 50-100 μm in diameter **Candelariella xanthostigma**

11. Thallus sorediate, composed of flattened areoles that erupt into soralia and often dissolve entirely into soredia 12

12. Ascospores 8-per ascus **Candelariella efflorescens**

12. Ascospores 16-32-per ascus **Candelariella xanthostigmoides**

5. Thallus not brightly colored, gray, blue-gray or brown 13

13. Apothecia yellow 14

14. On rock in overhangs; thallus green, leprose, UV+ orange (rhizocarpic acid present); ascospores simple **Psilolechia lucida**

14. On bark of mature spruce; thallus yellow-green, indistinct, UV- (rhizocarpic acid absent); ascospores 4-celled **Arthonia cupressina**

13. Apothecia orange, red or pink 15

15. On rock 16

16. On non-calcareous rocks in overhangs and on seepy vertical faces; thallus a slimy algal colony; apothecia pink, K- (anthraquinones absent) **Dibaeis absoluta**

16. On exposed calcareous rocks; thallus a continuous crust; apothecia orange, K+ red (anthraquinones present) **Protoblastenia rupestris**

15. On bark, resin, wood or humus 17

17. Thallus thick, blue-gray; apothecia pink; on wood and humus at high elevations **Icmadophila ericetorum**

17. Thallus thin or indistinct, gray to brownish if present; on bark or resin at various elevations 18

18. Ascospores simple 19

19. Ascospores 8-per ascus; apothecia red **Ramboldia russula**

19. Ascospores >100-per ascus; apothecia orange or orange-red 20

20. On conifer resin; apothecia orange, K- **Sarea resinae**

20. On bark; apothecia orange-red, K+ purple/red **Piccolia ochrophora**

18. Ascospores transversely septate, 8-per ascus 21

21. Ascospores brown **Buellia elizae**

21. Ascospores hyaline 22

22. Ascospores polarilocular, with lumina pushed to the poles **Caloplaca cerina**

22. Ascospores not polarilocular, with normal central lumina 23

23. Apothecia circular, convex, bright scarlet; on mature spruce **Arthonia kermesina**

23. Apothecia irregularly shaped, blotch-like, flattened, dark reddish-brown; on Yellow Birch and other hardwoods 24

24. Ascospores 3-celled, 8-12 x 3-5 µm; on bases and roots of hardwoods **Arthonia helvola**

24. Ascospores 2-celled, 11-15 x 4-5 µm; on boles of mature hardwoods, especially Yellow Birch **Arthonia vinosa**

4. Apothecia not brightly colored, typically brown or black 25

25. On bark or wood; ascospores in a mazaedium 26

26. Apothecia immersed in the thallus; exciple thin throughout **Cyphelium tigillare**

26. Apothecia sessile on the thallus; exciple basally thickened **Cyphelium lucidum**

25. On rock; ascospores remaining in asci 27

27. Thallus rusty orange-red; ascospores hyaline **Rhizocarpon oederi**

27. Thallus yellow; ascospores brown 28

28. Ascospores muriform **Rhizocarpon geographicum**

28. Ascospores 2-celled 29

29. On Anakeesta rocks at high elevations; ascospores 12-14 x 7.5-8.5 µm, lumina not thickened to form angular locules **Buellia sharpiana**

29. On other rocks types, at middle and low elevations; ascospores 20-30 x 10-13 µm, lumina thickened to form angular locules **Rinodina chrysomelaena**

3. Thallus and/or apothecia not brightly colored, typically shades of gray, blue-gray, greenish, brown or pale 30

30. Ascospores brown 31

31. Ascospores submuriform or muriform 32

32. On bark 33

33. On mature Yellow Birch at high elevations; apothecia lirellate; ascospores pale brown, very large, 100-150 x 30-48 µm **Graphis sterlingiana**

33. On mature beech and maple at low and middle elevation; apothecia immersed in the thallus and mostly visible a slit or circular pore on the thallus surface; ascospores dark brown, smaller, 45-55 x 10-13 µm **Leucodecton subcompunctum**

33. On rock 34

34. Apothecia circular, with persistent margins or immersed in the thallus and opening through a small pore 35

35. Apothecia circular and with visible discs, not immersed and opening through a small pore 36

36. Thallus white-gray, typically on soil, misbehaved, parasitic on other lichens, especially *Cladonia* **Diploschistes muscorum**

36. Thallus tan, always on rock and well behaved, never parasitic on other lichens **Diploschistes scruposus**

35. Apothecia with hidden discs, immersed and opening through a small pore 37

37. Thallus dark, chocolate brown; very rare **Diploschistes aeneus**

37. Thallus white, gray or tan, but never dark chocolate brown; common Diploschistes actinostomus

34. Apothecia +/- angular in shape, without persistent margins 38

38. Thallus yellow **Rhizocarpon geographicum**

38. Thallus gray or brown 39

39. Medulla C+ pink (gyrophoric acid present) **Rhizocarpon grande**

39. Medulla C- (gyrophoric acid absent) 40

40. Medulla K+ yellow turning red (norstictic acid present), producing red crystals in a water mount (best observed under the compound microscope); ascospores 8-per ascus; not known with certainty from the southern Appalachians **Rhizocarpon eupetraeum**

40. Medulla K- (norstictic acid absent); ascospores 1-2-per ascus; infrequent at high elevations **Rhizocarpon subgeminatum**

31. Ascospores transversely septate 41

41. Ascospores 4-celled 42
42. Apothecia circular, not lirellate **Buellia vernicoma**
42. Apothecia lirellate 43
43. Hymenium not inspersed with oil droplets; thallus typically P+ yellow or P+ orange (norstictic acid and/or stictic acid present) **Phaeographis brasiliensis**
43. Hymenium densely inspersed with oil droplets; thallus P- (norstictic acid and stictic acid absent) 44
44. Exciple apically to laterally carbonized, the portion below the hymenium not carbonized **Phaeographis inusta**
44. Exciple completely carbonized, including the area below the hymenium **[Leiorreuma sericeum]**
41. Ascospores 2-celled 45
45. On rock 46
46. Thallus yellow or yellow-green 47
47. Very common; thallus yellow-green, effigurate, with elongate marginal areoles that resemble lobes; cortex KC+ yellow (usnic acid present) 48
48. Medulla C+ pink and P- (gyrophoric acid present) **Dimelaena oreina s.str.**
48. Medulla C- and P+ red (fumarprotocetraric acid present) **Dimelaena oreina s.l. (fumarprotocetraric acid chemotype)**
47. Very rare; thallus yellow, not effigurate, the marginal areoles not distinctly elongate and lobe-like; cortex KC- or KC+ orange-red (usnic acid absent) 49
49. On Anakeesta rocks at high elevations; ascospores 12-14 x 7.5-8.5 μm, lumina not thickened to form angular locules **Buellia sharpiana**
49. On other rocks types, at middle and low elevations; ascospores 20-30 x 10-13 μm, lumina thickened to form angular locules **Rinodina chrysomelaena**
46. Thallus gray, blue-gray or brown 50
50. Apothecia with thalline margins, lecanorine or pseudolecanorine; ascospores with thickened walls creating angular lumina 51
51. Thallus K- (atranorin absent); common **Rinodina tephraspis**
51. Thallus K+ yellow (atranorin present), although the reaction often weak and only visible when performed on a water mount under the compound microscope; infrequent 52
52. Ascospores (19-)20-22.5(-26) × (9.5-)12.0-13.0(-15.0) μm, averaging >19 μm long **Rinodina destituta**
52. Ascospores (15.5-)18.5-19.5(-22) × (8.5-)10.5-11.5(-13.5) μm, averaging <19 μm long **Rinodina oxydata**
50. Apothecia without thalline margins, lecideine; ascospores walls not thickened and with angular lumina 53
53. Thallus well developed, areolate, K+ yellow and P+ orange (stictic acid present) or K+ yellow turning red and P+ yellow (norstictic acid present) **Buellia spuria**
53. Thallus poorly-developed and indistinct, K- (norstictic acid absent) 54
54. Ascospores 10-13 × 5-8 μm; conidia bacilliform, 3-5 × 1-1.5 μm; on exposed rocks; typically initially lichenicolous on other saxicolous lichens **Buellia badia**
54. Ascospores 9-13 × 4-6 μm; conidia filiform, 12-18 × 0.5-1.0 μm; on sheltered or protected rocks; never lichenicolous on other lichens **Amandinea punctata**
45. On bark or wood 55
55. Thallus yellow; ascospores forming mazaedium, a loose mass of spores formed through the disintegration of asci 56
56. Apothecia immersed in the thallus; exciple thin throughout **Cyphelium tigillare**
56. Apothecia sessile on the thallus; exciple basally thickened **Cyphelium lucidum**
55. Thallus gray, blue-gray or brown; ascospores not forming a mazaedium, remaining inside the hymenium and asci 57
57. Ascospores 12-32-per ascus 58
58. Thallus well-developed, blue-white; medulla K+ yellow turning red (norstictic acid present),

producing red crystals in a water mount (best observed under the compound microscope); ascospores medium, 13-17 x 6-8 µm **Hafellia subnexa**

58. Thallus poorly-developed and often an indistinct gray stain; medulla K- (norstictic acid absent); ascospores small, 8-11 x 3-5 µm **Amandinea polyspora**

57. Ascospores 8 or fewer per ascus 59

59. Apothecia elongate, lirellae; thallus immersed in the substrate and indistinct; photobiont *Trentepohlia*; ascospores pale brown at maturity **Melaspilea demissa**

59. Apothecia circular; thallus distinct, not immersed in the substrate; photobiont coccoid; ascospores dark brown at maturity 60

60. Thallus truly foliose, tightly appressed and only appearing crustose **Hyperphyscia syncolla**

60. Thallus truly crustose, areolate or continuous, but never appearing foliose 61

61. Apothecia lecanorine, thalline margin present; apothecial margins contrasting in color with the discs and concolorous with the thallus 62

62. Thallus P+ orange-red (pannarin present) 63

63. Thallus composed of flattened, tile-like areoles; apothecia immersed; ascospores large, 21-36 x 9-20 µm; blastidia absent **Rinodina adirondackii**

63. Thallus composed of bullate, rounded areoles; apothecia sessile; ascospores smaller, 15-20 x 7-12 µm; blastidia present **Rinodina excrescens**

62. Thallus P- or P+ yellow (pannarin absent) 64

64. Ascospores large, mostly 20-40 x 11-17 µm 65

65. Ascospores *Physcia*-type, lumina with sharply angular edges and sides at maturity, 23-40 x 11-17 µm **Rinodina ascociscana**

65. Ascospores *Pachysporaria*-type, lumina often appearing rounded and elongate at maturity, 19-33 x 12-16 µm **Rinodina dolichospora**

64. Ascospore smaller, 12-27 x 7-16 66

66. Ascospores *Physcia*-type, lumina with sharply angular edges and sides at maturity 16-22 x 8-11 **Rinodina subminuta**

66. Ascospores *Pachysporaria*-type, lumina often appearing rounded and elongate at maturity 67

67. Thallus composed of bullate, rounded areoles, P+ yellowish (atranorin present) **Rinodina bullata**

67. Thallus composed of flattened areoles or continuous, P- (atranorin absent) **[Rinodina maculans]**

61. Apothecia lecideine, thalline margin absent; apothecial margins concolorous with the discs and contrasting in color with the thallus 68

68. Epihymenium K+ fleeting purple; apothecial discs red-pruinose, but the pruina sometimes sparse and difficult to observe under the dissecting microscope **Buellia elizae**

68. Epihymenium K-; apothecial discs epruinose or white-pruinose 69

69. Thallus and exciple K+ yellow turning red (norstictic acid present), producing red crystals in a water mount (best observed under the compound microscope) 70

70. Ascospores ellipsoid to ovoid, blunt with rounded ends, 11-17 x 6-8 µm; common **Buellia stillingiana**

70. Ascospores football shaped, with pointed ends, 16-23 x 6-10 µm; rare **Buellia curtisii**

69. Thallus and exciple K- or K+ yellow (norstictic acid absent) 71

71. Apothecia convex, often with a bumpy or roughened appearance; thallus P+ red (fumarprotocetraric acid present) **Buellia dialyta**

71. Apothecia plane, smooth; thallus P- or P+ yellowish (fumarprotocetraric acid absent) 72

72. Thallus K+ yellow and P+ yellowish (atranorin present); ascospores large, 18-25 x 6-11 µm **Hafellia disciformis**

72. Thallus K- and P- (atranorin absent); ascospores small, 9-13 x 4-6 µm **Amandinea punctata**

30. Ascospores hyaline 73

73. Ascospores transversely septate or muriform 74

74. Ascospores submuriform or muriform 75

75. On rock 76

76. Thallus pale, immersed in the substrate; apothecia tan or pinkish; on calcareous rock, rare **Gyalecta farlowii**

76. Thallus gray to brown or rusty-orange, areolate, not immersed in the substrate; apothecia black, superficial; on non-calcareous rocks, common 77

77. Thallus K+ yellow turning red (norstictic acid present), producing red crystals in a water mount (best observed under the compound microscope) **Rhizocarpon rubescens**

77. Thallus K- or K+ yellow, not producing red crystals in a water mount (norstictic acid absent) 78

78. Ascospores 1-2-per ascus, 30-50 x 20-25 μm; rare **Rhizocarpon subgeminatum**

78. Ascospores 8-per ascus; common 79

79. Thallus gray; ascospores muriform, 20-30 x 10-13 μm **Rhizocarpon reductum**

79. Thallus rusty colored, with a distinct red or orange tinge; ascospores submuriform, with only one or two longitudinal septa, 15-20 x 5-8 μm **Rhizocarpon oederi**

75. On bark, wood, humus or leaves 80

80. On leaves; thalli island like, composed of small patches of light green-gray areoles, each with a large central black fruiting body **Aulaxina quadrangula**

80. On bark or wood; thalli not as above 81

81. Thallus K+ yellow turning red (norstictic acid present), producing red crystals in a water mount (best observed under the compound microscope) **Phlyctis speirea**

81. Thallus K- or K+ yellow, not producing red crystals in a water mount (norstictic acid absent) 82

82. Apothecia circular, typically with persistent margins 83

83. Thallus with eyelash-like or hook-like hyphophores in addition to the apothecia (note that if apothecia are present, ascospores are very rare in these species, and thus the species are keyed based on characters of hyphophores) 84

84. Hyphophores tall, eyelash-like, erect; on spruce and fir branches at high elevations **Gyalideopsis epicorticis**

84. Hyphophores short, bent, hook-like or sail-like; on various substrates at different elevations 85

85. On spruce or fir at high elevations **Gyalideopsis piceicola**

85. On other substrates at low and middle elevations **Gyalideopsis bartramiorum**

83. Thallus without hyphophores, only apothecia and pycnidia present 86

86. Low elevations; apothecia pale; ascospores sub-globose, 13-15 x 10-12 μm **Gyalecta obesispora**

86. High elevations; apothecia black; ascospores ellipsoid, 70-115 x 20-50 μm 87

87. On bark; known from the southern Appalachians **Lopadium disciforme**

87. On humus or soil; not known with certainty from the southern Appalachians **Lopadium pezizoideum**

82. Apothecia lirellate or irregular in shape and blotch-like 88

88. Apothecia elongate, lirellate; thallus thick, gray, shiny; ascospores 1-per ascus, 110-150 x 30-50 μm **Graphis sterlingiana**

88. Apothecia irregular in shape or blotch-like; thallus thin, dull; ascospores 8-per ascus, 1
5-35 x 7-15 μm 89

89. Photobiont coccoid **Arthonia susa**

89. Photobiont *Trentepohlia* 90

90. Ascospores 25-35 x 12-15 μm; thallus white **Arthothelium spectabile**

90. Ascospores 15-30 x 7-10 μm; thallus brown or indistinct 91

91. Ascospores ellipsoid, not bent, turning brown with age, 15-25 x 7-10 μm; widespread at middle to low elevations **Arthonia ruana**

91. Ascospores distinctly shoe-like, bent, remaining hyaline, 22-30 x 8-10 μm; known from a single collection in the southern Appalachians **Arthonia interveniens**

74. Ascospores transversely septate 92
92. On rock or mosses over rock 93
93. Ascospores 2-celled 94
94. On mosses over rock 95
95. Thallus thin, dark, greenish-gray; apothecia small, pale, white or light yellowish; ascospores 10-15 x 3-3.5 µm; at all elevations **Coenogonium pineti**
95. Thallus thick, light, blue-gray; apothecia large, pink; ascospores 13-27 x 4-6 µm; restricted to high elevations **Icmadophila ericetorum**
94. On rock 96
96. On calcareous rocks 97
97. Thallus gray to greenish-gray; prothallus inconspicuous, pale; photobiont a coccoid green alga; ascospores small, 7-12 x 2-4 µm **Catillaria lenticularis**
97. Thallus dark, black; prothallus conspicuous, blue-black; photobiont a cyanobacterium; ascospores larger, 8-20 x 4-6 µm **Placynthium nigrum**
96. On non-calcareous rocks 98
98. Apothecia lecanorine; thallus black, covered with abundant dark, granular blastidia **Halecania pepegospora**
98. Apothecia lecideine; thallus blue-gray to gray or brown, without dark granular blastidia 99
99. Ascospores narrowly ellipsoid, 4-6 µm wide, often bent and kidney bean shaped 100
100. Medulla K- and UV+ blue-white (divaricatic acid present); ascospores 9-12 x 4-5 µm **Fuscidea arcuatula**
100. Medulla K+ yellow and UV- (atranorin present); ascospores 13-15 x 4.5-6 µm **Megalaria beechingii**
99. Ascospores broadly ellipsoid, 6-11 µm wide, not bent and kidney bean shaped 101
101. Thallus brown; K- (no substances present); ascospores 17-20 x 8-11 µm; very common **Rhizocarpon infernulum f. sylvaticum**
101. Thallus white to gray; K+ yellow (stictic acid present) or K+ yellow turning red (norstictic acid present); ascospores 12-18 x 6-8 µm; uncommon **Rhizocarpon cinereovirens**
93. Ascospores >2-celled 102
102. Thallus dark, black; prothallus conspicuous, blue-black; photobiont a cyanobacterium **Placynthium nigrum**
102. Thallus blue-gray, gray, green, or brown; prothallus continuous or not, with a variable color; photobiont *Trentepohlia* or coccoid green alga 103
103. Photobiont *Trentepohlia* 104
104. Ascospores 4-celled; thallus C+ pink (gyrophoric acid present), typically with at least some soralia **Opegrapha gyrocarpa**
104. Ascospores 6-8-celled; thallus C- (confluentic acid present); with or without soralia 105
105. Thallus typically with at least some soralia; apothecia sessile, rounded to somewhat elongate **Opegrapha zonata**
105. Thallus esorediate; apothecia immersed, elongate **Enterographa hutchinsiae**
103. Photobiont coccoid 106
106. Thallus and apothecia C+ pink (gyrophoric acid present) **Micarea peliocarpa**
106. Thallus and apothecia C- (gyrophoric acid absent) 107
107. Thallus P+ yellow (psoromic acid present) or P+ orange (thamnolic acid present) 108
108. Ascospores 2-4-celled, 13-27 x 4-6 µm; thallus P+ orange (thamnolic acid present); restricted to the highest elevations **Icmadophila ericetorum**
108. Ascospores up to 16-celled, 105-205 x 27-47 µm; thallus P+ yellow (psoromic acid present); mostly at low to middle elevations **Phlyctis boliviensis**
107. Thallus P- (psoromic acid and thamnolic acid absent) 108
108. Thallus rusty orange or red 109

109. Hypothecium dark brown pigmented; ascospores ellipsoid, not bent and curved, 15-20 x 5-8 µm **Rhizocarpon oederi**

109. Hypothecium hyaline; ascospores acicular to elongate-fusiform, bent and curved, 20-40 x 2-3.5 µm **Scoliciosporum umbrinum**

108. Thallus blue-gray, gray or green 110

110. Thallus composed of minute, green goniocysts; apothecia pale, entirely lacking pigments; ascospores needle-like, <1.5 µm wide **Bacidina delicata**

110. Thallus continuous or areolate, not composed of goniocysts; apothecia reddish-brown to black; ascospores fusiform, 3-7 µm wide 111

111. Hypothecium hyaline 112

112. Conidia 10-19 µm long; paraphyses with unpigmented tips **Lecania cuprea (A. Massal.) van den Boom & Coppins**

112. Conidia 31-55 µm long; paraphyses with pigmented tips **Lecania subfuscula (Nyl.) S. Ekman**

111. Hypothecium reddish-brown to black pigmented 113

113. Ascospores >4-celled 114

114. Apothecia black; ascospores 14-18 x 4-5 µm **Fellhanera montesfumosi**

114. Apothecia reddish-brown; ascospores 18-27 × 5-7 µm **Bilimbia sabuletorum**

113. Ascospores 4-celled 115

115. On calcareous rocks; thallus without minute, green blastidia; conidia filiform **Bacidia coprodes**

115. On on-calcareous rocks; thallus with minute, green blastidia; conidia bacilliform **Fellhanera granulosa**

92. On bark, wood, soil, humus or leaves 116

116. Ascospores 2-celled 117

117. On rock 118

118. Apothecia lecanorine, margins contrasting with the disc and concolorous with the thallus **Lecania erysibe**

118. Apothecia lecideine, margins concolorous with the disc and contrasting with the thallus 119

119. Ascospores 12-16 x 3.5-5 µm; thallus green, K- (atranorin absent) **Micareopsis irriguata**

119. Ascospores 13-15 x 4.5-6 µm; thallus blue-gray, K+ yellow (atranorin present) **Megalaria beechingii**

117. On bark, wood, leaves or humus 120

120. Apothecia elongate, lirellae; thallus immersed in the substrate and indistinct **Melaspilea demissa**

120. Apothecia circular; thallus distinct, not immersed in the substrate 121

121. Thallus and apothecia C+ pink/red (gyrophoric acid or lecanoric acid present) 122

122. Ascospores large, >100 µm long, fragmenting into part-spores; thallus and apothecia C+ red (lecanoric acid present); not known with certainty from the southern Appalachians **Varicellaria rhodocarpa**

122. Ascospores small, 7-15 x 2-3.5 µm, not fragmenting into part-spores; thallus and apothecia C+ pink (gyrophoric acid present); rare on wood at middle and high elevations **Micarea denigrata**

121. Thallus and apothecia C- (gyrophoric acid and lecanoric acid absent) 123

123. Thallus thick, blue-gray; apothecia large, pink; restricted to high elevations **Icmadophila ericetorum**

123. Thallus thin, variable in color; apothecia small, reddish-brown, pale or black; at all elevations 124

124. Apothecia black 125

125. Hypothecium dark, reddish-brown; ascospores large, 12-18 x 5-7 µm **Megalaria laureri**

125. Hypothecium hyaline to light brown; ascospores small, 8-13 x 2-5 µm 126

126. Ascospores broadly ellipsoid, with the upper cell distinctly enlarged, 10-13 x 4-5 µm **Arthonia apatetica**

126. Ascospores ellipsoid, with both cells equal in size, 8-18 x 2-4 µm 127

127. Paraphyses capitate, with brown pigmented caps; apothecia plane; ascospores 8-12 x 2-4 µm **Catillaria nigroclavata**

127. Paraphyses not capitate, without brown pigmented caps; apothecia convex; ascospores 9-18 x 2-3.5 µm **Micarea elachista**

124. Apothecia pale to orange-brown or reddish-brown 128

128. Photobiont *Trentepohlia* 129

129. Apothecia pale white to light orange; ascospores 10-15 x 3-3.5 µm; common **Coenogonium pineti**

129. Apothecia waxy yellow-orange; ascospores 7-10 x 2.5-3.5 µm; rare **Coenogonium luteum**

128. Photobiont coccoid 130

130. Ascospores polarilocular, with lumina pushed to the poles **Caloplaca camptidia**

130. Ascospores not polarilocular, with normal central lumina 131

131. On leaves **Fellhanera bouteillei**

131. On bark, wood or humus 132

132. Ascospores small, 7-11 x 2.5-4 µm 133

133. Thallus pale, blue-gray, K+ yellow (atranorin present); apothecia plane, at least some with persistent margins **Cliostomum griffithii**

133. Thallus green, K- (atranorin absent); apothecia convex, without a margin 134

134. Thallus composed of areoles, not resembling an algal colony; very rare **Micarea elachista**

134. Thallus composed of green goniocysts, resembling an algal colony; very common 135

135. Methoxymicareic acid present (TLC required) **Micarea micrococca**

135. Micareic acid present (TLC required) **Micarea prasina**

132. Ascospores larger, 10-23 x 3.5-8.5 136

136. Thallus P+ orange or P+ red (argopsin or fumarprotocetraric acid present) 137

137. Apothecia pale; ascospores consistently 2-celled, 14–17 × 6.5–8.5 µm; thallus K+ yellow (atranorin present); known from a single historical collection in the Smokies **Megalaria albocincta**

137. Apothecia reddish-brown to brown; ascospores inconsistently 2-celled, 10-18 x 3.5-5 µm; thallus K- (atranorin absent); uncommon at high elevations **Biatora pycnidiata**

136. Thallus P- or P+ yellow (argopsin and fumarprotocetraric acid absent) 138

138. Thallus K+ yellow (atranorin present); very rare and restricted to high elevations **Cliostomum griffithii**

138. Thallus K- (atranorin absent); common throughout 139

139. Ascospores ellipsoid, 10-15 x 4-7.5 µm **Catinaria atropurpurea**

139. Ascospores elongate ellipsoid, 9-23 x 2-6 µm 140

140. Ascospores relatively short and narrow, mostly 2-celled, 9-18 x 2-3.5 µm; rare **Micarea elachista**

140. Ascospores relatively long and broader, a mixed of simple and 2-celled, 10-23 x 3.5-6 µm; common 141

141. On bark; apothecia reddish-brown to orange-brown; ascospores 17-22 x 3.5-4 µm **Biatora longispora**

141. On mosses, often at the bases of trees; apothecia pale brown; ascospores 10-23 x 4-6 **Biatora vernalis**

116. Ascospores >2-celled 142

142. Apothecia lirellate or irregular in shape and blotch-like 143

143. Ascospores large, up to 16-celled, 105-205 x 27-47 µm; thallus thick, white, ecorticate, P+ intense yellow (psoromic acid present) **Phlyctis boliviensis**

143. Ascospores smaller, <50 µm long; thallus thin, variable in color, corticate or ecorticate, P- (psoromic acid absent) 144

144. Apothecia lirellate, aggregated into raised, carbonized pseudostroma **Glyphis cicatricosa**

144. Apothecia lirellate, irregularly shaped or blotch like, but not aggregated into raised, carbonized pseudostroma 145

145. Apothecia elongate, lirellae; exciple carbonized at least in the apical portions 146
146. Exciple completely carbonized, including below the hymenium 147
147. Ascospores 8-14-celled, 30–50 × 3–5 µm **Opegrapha viridis**
147. Ascospores 6-celled, 18-30 x 3-6 µm 148
148. Apothecial disc exposed, often yellow-green pruinose; ascospores broad, 18-20 x 5-6 µm **Opegrapha varia**
148. Apothecial disc concealed, epruinose; ascospores narrow, 25–30 × 3–5 µm **Opegrapha vulgata**
146. Exciple apically or laterally carbonized, not carbonized below the hymenium 149
149. Exciple striate, apically carbonized **Graphis endoxantha**
149. Exciple simple, laterally carbonized to the base of the hymenium 150
150. Hymenium inspersed with oil droplets; apothecial discs epruinose; infrequent at low elevations **Graphis lineola**
150. Hymenium clear, not inspersed with oil droplets; apothecial discs typically white-pruinose; common throughout **Graphis scripta**
145. Apothecia irregularly shaped or lirelliform; exciple not carbonized 151
151. Thallus thick, green, shiny, corticate; apothecial lirellate, fissurine, buried in the thallus and opening through a long, narrow slit **Fissurina insidiosa**
151. Thallus thin, sordid white to brown or gray, dull, mostly ecorticate or with a poorly developed cortex; apothecia irregularly shaped or blotch-like, flat, not immersed and opening through a slit 152
152. Photobiont absent; thallus a white stain 153
153. Ascospores long, slender, clavate; on *Pinus* **Arthonia caudata**
153. Ascospores short, broad, ellipsoid; mostly on hardwoods, very rarely on *Pinus* 154
154. Ascospores 4-(6)-celled, 15-26 x 6-8 µm; not known with certainty from the southern Appalachians **Arthonia punctiformis**
154. Ascospores 6-celled, 17-19 x 6-8 µm; common at low and middle elevations **Arthonia quintaria**
152. Photobiont present; thallus brown to gray, not stain-like 155
155. Apothecia with red pigments turning K+ purple in section under the compound microscope 156
156. On the bases of hardwoods at middle and high elevations; apothecia blotch-like; ascospores small, 3-celled, 8-12 x 3-5 µm **Arthonia helvola**
156. On boles and branches of trees at low elevations; apothecia irregular in shape, often +/- elongate; ascospores larger, 5-8-celled, 20-33 x 5.5-11 µm **Arthonia cinnabarina**
155. Apothecia with pigments that are K- or K+ intensifying, but not K+ purple 157
157. Ascospores 6-celled, with both end cells distinctly enlarged, those two cells sandwiching the remaining narrow cells **Arthonia rubella**
157. Ascospores 4-celled, with only one end cell distinctly enlarged 158
158. Thallus and exciple C+ fleeting pink (gyrophoric acid present), best observed in a water mound under the compound microscope) **Arthonia anglica**
158. Thallus and exciple C- (gyrophoric acid absent) **Arthonia mediella**
142. Apothecia circular, typically with persistent margins 159
159. Ascospores needle-shaped, long and narrow 160
160. Ascospores exceedingly long, 100-200 µm or longer 161
161. Apothecia black, urn-like; ascospores lumbricoid, worm-like **Conotrema urceolatum**
161. Apothecia reddish-brown, sessile to short-stipitate; ascospores needle-like or filiform 162
162. Hyphophores present, tall, slender, with starburst-like opening; common **Gomphillus americanus**
162. Hyphophores absent; rare **Gomphillus calycioides**
160. Ascospores up to 100 µm, mostly <80 µm .163
163. Thallus and apothecia UV+ blue-white (lobaric acid or perlatolic acid present) 164
164. Only the thallus medulla UV+ blue-white (perlatolic acid present) **Ropalospora chlorantha**
164. Entire thallus and apothecia conspicuously UV+ blue-white (lobaric acid present) 165

165. Pycnidia stipitate, conspicuous 166

166. Thallus and basal portions of pycnidia P+ red (fumarprotocetraric acid present) **Micarea neostipitata**

166. Thallus and basal portions of pycnidia P- (fumarprotocetraric acid absent) **Micarea pycnidiophora**

165. Pycnidia sessile or immersed, mostly unremarkable 167

167. Thallus sorediate; ascospores 40-50 µm long **Bacidia lobarica**

167. Thallus granular, but not sorediate; ascospores 20-30 µm long 168

168. On hardwoods in humid habitats, especially at low and middle elevations in the grooves of *Quercus prinus*; apothecia light brown varying to whitish; ascospores straight, not spirally arranged within the ascus **Scoliciosporum pensylvanicum**

168. On mature Yellow Birch and conifers at high elevations; apothecia white; ascospores curved, spirally arranged within the ascus **Scoliciosporum pruinosum**

163. Thallus and apothecia UV- (lobaric acid and perlatolic acid absent) 169

169. On leaves **Bacidia apiahica**

169. On bark, rock or wood 170

170. Pycnidia stipitate, conspicuous 171

171. Pycnidia dark; ascospores 15-55 x 1.5-3.0 µm **Ropalospora chlorantha**

171. Pycnidia pale; ascospores 15-25 x 1.5-2.5 µm 172

172. Thallus and basal portions of pycnidia P+ red (fumarprotocetraric acid present) **Micarea neostipitata**

172. Thallus and basal portions of pycnidia P- (fumarprotocetraric acid absent) **Micarea pycnidiophora**

170. Pycnidia sessile or immersed, mostly unremarkable 173

173. Apothecia and pycnidia white, entirely without pigments; thallus composed of goniocysts **Bacidina delicata**

173. Apothecia and pycnidia pigmented, reddish-brown to brown or black; thallus continuous, areolate or granular, but not composed of goniocysts 174

174. On rock; not known with certainty from the southern Appalachians **Ropalospora lugubris**

174. On bark or wood; common in the southern Appalachians 175

175. Apothecia black 176

176. Apothecia mostly plane, with persistent margins 177

177. Epihymenium blue-green to green 178

178. Hypothecium reddish-brown to black; ascospores 30-90 x 2-4 µm; common **Bacidia schweinitzii**

178. Hypothecium hyaline; ascospores 11-30 x 2-3 µm; rare **Bacidia circumspecta**

177. Epihymenium reddish-brown to brown or hyaline 179

179. Hymenium inspersed with oil droplets; thallus dark greenish-brown, areolate, often forming circular rosettes; medulla UV+ blue-white (perlatolic acid present); common **Ropalospora chlorantha**

179. Hymenium clear, not inspersed with oil droplets; thallus pale gray to brownish, not conspicuously areolate and forming circular rosettes; medulla UV- (perlatolic acid absent); rare 180

180. Epihymenium K+ fleeting purple; ascospores 30-80 x 2-4 µm **Bacidia helicospora**

180. Epihymenium K- ; ascospores 11-30 x 2-3 µm **Bacidia circumspecta**

176. Apothecia convex, with margins rapidly excluded 181

181. Hypothecium hyaline; ascospores noodle-like, 20-40 x 4-5 µm **Scoliciosporum chlorococcum**

181. Hypothecium dark reddish-brown to black; ascospores needle-like, 15-25 x 1.5-2.5 µm 182

182. Thallus well-developed, often bullate; unidentified substance present (TLC required); common, especially at middle and low elevations on *Tsuga* **Micarea chlorosticta**

182. Thallus poorly-developed, indistinct; no substances present (TLC required); infrequent at middle and high elevations, on hardwoods and conifers **Micarea endocyanea**

175. Apothecia reddish-brown to tan 183

183. Hypothecium dark brown to black 184

184. Brown pigment in the exciple and hypothecium K- or K+ intensifying, but not K+ rose and bleeding pink-red; apothecia epruinose **Bacidia ekmaniana**

184. Brown pigment in the exciple and hypothecium K+ rose, bleeding pink-red; apothecia often pruinose 185

185. Thallus granulose **Bacidia diffracta**

185. Thallus smooth **Bacidia polychroa**

183. Hypothecium hyaline to pale yellowish 186

186. Apothecia small, convex; ascospores noodle-like, 4-5 μm wide **Scoliciosporum chlorococcum**

186. Apothecia larger, plane; ascospores needle-like, 1.2-3 μm wide 187

187. Ascospores short, 30-40 μm long **Bacidina assulata**

187. Ascospores longer, (30-)40-90 μm long 188

188. Thallus granulose; not known with certainty from the southern Appalachians **Bacidia rubella**

188. Thallus smooth; rare in the southern Appalachians 189

189. Apothecia pruinose; epihymenium K- **Bacidia suffusa**

189. Apothecia epruinose; epihymenium K+ fleeting purple **Bacidia helicospora**

159. Ascospores ellipsoid, fusiform, globose, or lumbricoid 190

190. On leaves 191

191. Apothecia with lecideine margins that are compact and concolorous with the disc **Fellhaneropsis cf. vezdae**

191. Apothecia with byssoid margins that are fuzzy, white and contrast against the disc 192

192. Apothecial disc dark blue-black **Byssoloma subdiscordans**

192. Apothecial disc pale to dark brown **Byssoloma leucoblepharum**

190. On bark, wood, soil, humus or rock 193

193. Apothecia black and urn-like; ascospores lumbricoid, worm-like, >100 μm long **Conotrema urceolatum**

193. Apothecia various colors including black, sessile to immersed; ascospores ellipsoid or fusiform 194

194. Apothecia immersed in the thallus, the disc visible through a pore-like opening **Thelotrema subtile**

194. Apothecia not immersed, without a pore-like opening 195

195. Apothecia buried in thalline warts, resembling perithecia 196

196. Ascospores >100 per ascus, 4-6-celled, small, 12-16 x 4-8 μm **Thelopsis rubella**

196. Ascospores 8-per ascus, 10-14-celled, large, 85-110 x 12-14 μm **Porina heterospora**

195. Apothecia discoid, not buried in thalline warts and resembling perithecia 197

197. Thallus P+ yellow (psoromic acid present) or P+ orange-red (argopsin or thamnolic acid present) 198

198. Ascospores large, up to 16-celled, 105-205 x 27-47 μm; thallus P+ intense yellow (psoromic acid present) **Phlyctis boliviensis**

198. Ascospores smaller, <50 μm long; thallus P+ orange-red (argopsin or thamnolic acid present) 199

199. Thallus thick, continuous, blue-gray, K+ instant lemon yellow and P+ orange (thamnolic acid present); apothecia large, plane, pink; ascospores 2-4-celled, 13-27 x 4-6 μm; restricted to high elevations **Icmadophila ericetorum**

199. Thallus thin, areolate, greenish to gray, K- and P+ red (argopsin present); apothecia small, convex, black; ascospores 15-30 x 4-5 μm, 4-8-celled; not known with certainty from the southern Appalachians **Micarea lignaria**

197. Thallus P- (argopsin, psoromic acid and thamnolic acid absent) 200

200. On organic matter or dead tufts of grass; thallus with short, bent, hyphophores **Gyalideopsis moodyae**

200. On bark, wood, rock or mosses over one of those substrates; thallus without hyphophores 201

201. Hypothecium hyaline 202

202. Thallus and apothecia C- (gyrophoric acid absent) 203

203. Apothecia plane, with persistent margins; ascospores 9-18 x 2-3.5 µm, 4-8-celled **Bacidia circumspecta**

203. Apothecia convex, with rapidly excluded margins; ascospores 11-30 x 2-3 µm, (1)-2(-4) celled **Micarea elachista**

202. Thallus and apothecia C+ pink (gyrophoric acid present) 204

204. Ascospores (4)-6-8-celled, (20-)26-40 x (3.5-)4-5(-5.5) µm; conidia long, 50-100 x 1-1.2 µm; not known with certainty from the southern Appalachians **Micarea cinerea**

204. Ascospores 4-(5)-celled, 12-20 x (3.5-)4-5(-5.5) µm; conidia short, <50 µm long; common 205

205. At least some apothecia gray-blue or gray-brown pigmented; common and widespread **Micarea peliocarpa**

205. All apothecia white, entirely lacking pigment; not known with certainty from the southern Appalachians **Micarea alabastrites**

201. Hypothecium dark, reddish-brown to black 206

206. Apothecia pruinose 207

207. Apothecial discs yellow-green pruinose; apothecial margins often radially cracked **Cresponea flava**

207. Apothecial discs white or blue-gray pruinose; apothecial margins not radially cracked 208

208. Thallus with conspicuous, pruinose pycnidia 209

209. Pruina on the pycnidia C+ red (lecanoric acid present) **Lecanactis abietina**

209. Pruina on the pycnidia C- (lecanoric acid absent) **Arthonia byssacea**

208. Thallus without conspicuous pycnidia 210

210. Thallus green, granular, well-developed; apothecia convex, blue-gray pruinose **Arthonia caesia**

210. Thallus blue-gray or distinct; apothecia flat, white pruinose 211

211. Thallus blue-gray, continuous; ascospores 12-18 x 4-6 µm **Arthonia byssacea**

211. Thallus pale, brownish, indistinct; ascospores 22-26 x 6-7 µm **Schismatomma glaucescens**

206. Apothecia epruinose 212

212. Thallus with abundant bright-green blastidia or soredia 213

213. Ascospores 6-10-celled, 25-30 x 3-5 µm; on bark; very rare **Fellhanera eriniae**

213. Ascospores 4-celled, 12-16 x 3-5 µm; typically on rock, but rarely also on the bases of trees; common 214

214. On sheltered and vertical rock faces associated with waterways or seeps; prothallus dark, conspicuous; areoles large, convex, sorediate **Micareopsis irriguata**

214. On exposed rocks in shaded forests; prothallus pale or indistinct; areoles small, with minute blastidia **Fellhanera granulosa**

212. Thallus without blastidia 215

215. Apothecia convex, with rapidly excluded margins that are concolorous with the disc **Micarea melaena**

215. Apothecia plane, with persistent margins that contrast in color with the disc 216

216. Photobiont *Trentepohlia*; apothecia with thalline margins; very rare **Schismatomma pericleum**

216. Photobiont coccoid; apothecia with non-thalline margins that lack algae; not uncommon 217

217. Apothecial margins compact, not byssoid and fuzzy **Byssoloma marginatum**

217. Apothecial margins byssoid, fuzzy 218

218. Apothecial disc dark blue-black **Byssoloma subdiscordans**

218. Apothecial disc pale to dark brown **Byssoloma leucoblepharum**

73. Ascospores simple 219

219. On rock or soil over rock 220

220. Ascospores >100-per ascus 221

221. Apothecia immersed in thalline areoles, the areoles remaining intact or becoming reduced to a lecanorine margin 222

222. Thallus cortex or medulla C+ pink (gyrophoric acid present) 223

223. Thallus areoles becoming reduced to a lecanorine margin around the apothecia; apothecia 1-per areole **Acarospora obpallens**

223. Thallus areoles remaining intact and well-developed; apothecia often multiple per areole 224

224. Areoles light to medium brown, adnate, not becoming squamulose; lower surface pale; common **Acarospora fuscata**

224. Areoles dark brown, becoming squamulose; lower surface black; rare [**Acarospora thamnina**]

222. Thallus cortex and medulla C- (gyrophoric acid absent) 225

225. Thallus rusty orange-brown or reddish-brown; areoles mostly contiguous; apothecia often multiple per areole **Acarospora sinopica**

225. Thallus brown, not rusty-orange or red in color; areoles dispersed; apothecia 1-per areole **Trimmatothelopsis dispersa**

221. Apothecia naked, not immersed in thalline areoles, never with a thalline margin 226

226. Apothecia deformed and crinkled, the margins becoming cracked, the discs often becoming carbonized **Polysporina simplex**

226. Apothecia circular, not deformed and crinkled, the margins smooth and discs not carbonized **Sarcogyne similis**

220. Ascospores 1-8-per ascus 227

227. Photobiont a cyanobacterium 228 [the correct identity of *Pyrenopsis* material from the southern Appalachians is unclear, the ascospores sizes for the species given below was taken from the protologues]

228. Ascospores 11 x 6 μm **Pyrenopsis sanguinea**

228. Ascospores 6-8 x 5-6 μm **Pyrenopsis subfuliginea**

227. Photobiont a green alga 229

229. On strongly calcareous rocks 230

230. Apothecia lecanorine, with white margins contrasting strongly against the reddish-brown discs **Lecanora dispersa**

230. Apothecia lecideine, without margins, either convex and orange or purple-brown and immersed in the substrate 231

231. Apothecia convex, sessile, orange, K+ purple **Protoblastenia rupestris**

231. Apothecia flat, immersed in the substrate, dark purple-brown, K- 232

232. Hypothecium orange, K+ dull purple **Kephartia crystalligera**

232. Hypothecium light to dark brown, K- **Clauzadea chondrodes**

229. On non-calcareous rocks and weakly calcareous rocks 233

233. Thallus C+ pink (gyrophoric acid present), best viewed in a water mount under the compound microscope 234

234. Hypothecium dark brown 235

235. Ascospores narrow, 10-20 x 5-7 μm; restricted to high elevations **Lecidea fuscoatra**

235. Ascospores broad, 10-16 x 7-10 μm; at middle to high elevations **Rimularia badioatra**

234. Hypothecium hyaline 236

236. Thallus thin, continuous; apothecia erumpent, first appearing as a pale circular area below the thallus surface, then breaking through **Trapelia coarctata**

236. Thallus thicker, areolate; apothecia not erumpent, not punching through the surface of the thallus 237

237. Exciple striate, composed of multiple brown layers; apothecia with a distinct hyaline stipe; on shaded rocks, often in forested habitats **Trapelia stipitata**

237. Exciple simple, composed of a single brown layer; apothecia without a hyaline stipe; on exposed rocks, often in disturbed habitats **Trapelia glebulosa**

233. Thallus C- (gyrophoric acid absent) 238

238. Thallus K+ yellow turning red (norstictic acid present), producing red crystals in a water mount (best observed under the compound microscope) 239

239. Apothecia buried in thalline warts and resembling perithecia **Pertusaria plittiana**

239. Apothecia sessile, easily recognized, not resembling perithecia 240

240. Apothecial discs densely white pruinose; hypothecium dark, brown **Porpidia albocaerulescens**

240. Apothecia discs epruinose or rarely lightly white pruinose; hypothecium hyaline **Aspicilia cinerea**

238. Thallus K- or K+ yellow not turning red (norstictic acid absent) 241

241. Apothecia lecanorine, with a distinct thalline margin that usually contrasts with the disc and is concolorous with the margin 242

242. Thallus yellow-green, K- and KC+ yellow (usnic acid present) 243

243. Thallus very thick, with bullate areoles, often not appearing crustose; apothecial discs orange, often pruinose; ascospores narrowly ellipsoid, 7-12 x 3-5 µm **Rhizoplaca subdiscrepans**

243. Thallus thinner, with mostly flattened areoles, always clearly crustose; apothecial discs waxy yellow to dark blue-gray, epruinose; ascospores ellipsoid, 9-14 x 5-7 µm 244

244. Thallus thin, poorly developed, +/- dispersed areolate; apothecia sessile; common on exposed rocks at middle and high elevations **Lecanora polytropa**

244. Thallus thicker, well-developed, rimose-areolate; apothecia immersed in the thallus; rare at the highest **Lecanora intricata**

242. Thallus green-gray, gray, or brown, not both K- and KC+ yellow (usnic acid absent) 245

245. Apothecial discs black to dark gray 246

246. Medulla UV+ blue-white (alectoronic acid present); hymenium stained dark purple **Tephromela atra**

246. Medulla UV- (alectoronic acid absent); hymenium hyaline, not stained purple 247

247. Medulla K+ yellow and P+ orange (stictic acid present) **Aspicilia laevata**

247. Medulla K- and P- (stictic acid absent) **Aspicilia caesiocinerea/Aspicilia olivaceopallida** [note that the delimitation of these species requires further study]

245. Apothecial discs reddish-brown 248

248. Apothecial margins white, ecorticate; thallus areolate, often poorly developed and indistinct; on weakly calcareous rocks or rocks with calcareous influence **Lecanora dispersa**

248. Apothecial margins gray to green-gray, corticate; thallus continuous to rimose-areolate, well-developed; on non-calcareous rocks 249

249. Apothecial discs and margins P+ orange-red (pannarin present) **Lecanora saxigena**

249. Apothecia discs and margins P- or P+ weak yellow (pannarin absent) 250

250. Thallus areolate to rimose areolate; 2'-o-merthylperlatolic acid present (TLC required) **Lecanora pseudistera**

250. Thallus continuous; zeorin present (TLC required) **Lecanora subimmergens**

241. Apothecia not lecanorine, either without a distinct margin, or with a margin that is concolorous with the disc and contrasts with the thallus 251

251. On soil or organic matter over rock 252

252. Thallus thin, greenish, film-like; apothecia pale; hypothecium hyaline **Trapeliopsis gelatinosa**

252. Thallus minutely areolate, brown to black; apothecia black; hypothecium reddish-brown **Placynthiella uliginosa**

251. Directly on rock 253

253. Apothecia light in color, white, pink, tan, yellow or bronze 254

254. Thallus a slimy algal colony; apothecia pink **Dibaeis absoluta**

254. Thallus leprose, areolate, continuous or continuous, but not resembling a slimy algal colony; apothecia white, tan, yellow or bronze 255

255. Thallus leprose, green, UV+ dull orange (rhizocarpic acid present); apothecia yellow **Psilolechia lucida**

255. Thallus areolate or continuous, yellow-green, gray or brownish, UV- (rhizocarpic acid absent); apothecia whit, tan or bronze 256

256. Thallus yellow-green, K- and KC+ yellow (usnic acid present) 257

257. Thallus thin, poorly developed, +/- dispersed areolate; apothecia sessile; common on exposed rocks at middle and high elevations **Lecanora polytropa**

257. Thallus thicker, well-developed, rimose-areolate; apothecia immersed in the thallus; rare at the highest **Lecanora intricata**

256. Thallus gray, or brown, not both K- and KC+ yellow (usnic acid absent) 258

258. Apothecia convex, sessile; ascospores narrow, 7-11 x 3.5-4.5 µm **Micarea bauschiana**

258. Apothecia plane, immersed in the thallus; ascospores broad, 12-17 x 7-9 µm 259

259. Apothecia pale, white; epihymenium hyaline, lacking pigment or granules **Ionaspis alba**

259. Apothecia bronze to brownish; epihymenium brown pigmented and granular **Ionaspis lacustris**

253. Apothecia dark reddish-brown to black 260

260. Hypothecium hyaline 261

261. Thallus yellow-green, K- and KC+ yellow (usnic acid present) **Lecanora intricata**

261. Thallus white, blue-gray or gray or brown, not both K- and KC+ yellow (usnic acid absent) 262

262. Apothecia immersed in the thallus 263

263. Apothecia bronze to light brown; ascospores ovoid, 12-17 x 7-9 µm **Ionaspis lacustris**

263. Apothecia black; ascospores ellipsoid to narrowly ellipsoid, 6-14 x 3-7 µm 264

264. Thallus K+ yellow (atranorin present) **Lecanora oreinoides**

264. Thallus K- (atranorin absent) 265

265. Apothecia often white pruinose; confluentic acid present (TLC required); on exposed rocks at middle and high elevations **Lecidea tessellata**

265. Apothecia epruinose; planaic acid present (TLC required); not known with certainty from the southern Appalachians **Lecidea plana**

262. Apothecia sessile, not immersed in the thallus 266

266. Medulla UV+ blue-white (divaricatic acid present) 267

267. Thallus sorediate; ascospores globose to subglobose, 8-9 x 6-7 µm, always simple; rare **Fuscidea appalachensis**

267. Thallus esorediate; ascospores ellipsoid to kidney bean shaped, 9-12 x 4-5 µm, often some 2-celled; common **Fuscidea arcuatula**

266. Medulla UV- or UV+ dull orange (divaricatic acid absent) 268

268. Epihymenium intense blue-green; thallus poorly developed; rare **Micarea polycarpella**

268. Epihymenium olive-brown to brown or pale; common 269

269. Thallus UV+ dull orange and KC+ orange (xanthones present) **Lecidea deminutula**

269. Thallus UV- and KC- (xanthones absent) 270

270. Apothecia convex, brown; ascospores small, 7-11 x 3.5-4.5 µm; thallus poorly developed, dull greenish-gray, continuous **Micarea bauschiana**

270. Apothecia plane, dark reddish-brown to pale; ascospores larger, 5-15 x 3-7 µm; thallus well-developed, brown or gray, areolate 271

271. Thallus brown; apothecia black; ascospores narrow, 5-12 x 3-4.5 µm **Lecidea atrobrunnea**

271. Thallus gray; apothecia reddish-brown; ascospores broader, 8-15 x 4-7 µm **Miriquidica leucophaea**

260. Hypothecium dark, brown to black 272

272. Ascospores small, narrowly ellipsoid, mostly 5-10 x 3-4(-5) µm 273

273. Apothecia small, convex, mostly with excluded margins; common 274

274. Thallus white to gray, areolate; epihymenium blue-green **Leimonis erratica**

274. Thallus brown, indistinct; epihymenium olive-brown to brown or hyaline 275

275. Ascospores 5-8 x 2-3 µm; not known with certainty from the southern Appalachians **Micarea lutulata**

275. Ascospores 7-11 x 2-4 µm; infrequent but widespread **Lecidea cyrtidia**

273. Apothecia relatively large, plane, with persistent margins; rare 276

276. Thallus brown; no substances present (TLC required); rare at high elevations **Lecidea atrobrunnea**

276. Thallus gray; planaic acid present (TLC required); not known with certainty from the southern Appalachians **Lecidea plana**

272. Ascospores relatively large, broadly ellipsoid 277

277. Apothecial discs strongly white pruinose 278

278. Thallus K+ yellow and P+ orange (stictic acid present); common at all elevations **Porpidia albocaerulescens**

278. Thallus K- and P- (confluentic acid present); rare and restricted to high elevations **Porpidia cinereoatra**

277. Apothecial discs epruinose or occasionally weakly pruinose 279

279. Thallus poorly developed and indistinct, mostly immersed in the substrate 280

280. Ascospores narrow, 10-18 x 4-6 µm; apothecia convex, aggregated, typically with excluded margins; rare and restricted to high elevations **Miriquidica pycnocarpa**

280. Ascospores broad, 12-18 x 6-8 µm; apothecia plane, solitary, with persistent margins that frequently become radially cracked; common throughout **Porpidia subsimplex**

279. Thallus well-developed and distinct, not immersed in the substrate 281

281. Apothecia aggregated in groups, convex, with rapidly excluded margins; ascospores narrowly ellipsoid, 10-18 x 4-6 µm; rare and restricted to the highest elevations

281. Apothecia solitary, plane to weakly convex, with persistent margins, ascospores broader, 8-24 x 5-10 µm 283

282. Epihymenium green to blue-green283

283. Thallus K+ yellow (atranorin present); ascospores 8-16 x 5-9 µm; rare **Lecidella carpathica**

283. Thallus K- (atranorin absent); ascospores 15-24 x 6-10 µm; common at high elevations **Porpidia contraponenda**

282. Epihymenium brown or hyaline 284

284. Apothecia reddish-brown, with excluded margins; exciple pale, light brown pigmented **Lecidea ahlesii**

284. Apothecia black, with persistent margins; exciple dark brown 285

285. Apothecia large, mostly 1.0-3.5 mm in diameter; typically on rocks associated with waterways or seeps **Porpidia macrocarpa**

285. Apothecia small, mostly 0.3-1.0 mm in diameter; typically on rocks in areas away from water 286

286. Ascospores 10-17 x 5-9 µm; stictic acid present or absent (TLC required); common at all elevations **Porpidia crustulata**

286. Ascospores 15-24 x 6-10 µm; methyl 2'-*O*-methylmicrophyllinate and related substances present (TLC required); restricted to high elevations **Porpidia contraponenda**

219. On bark, wood, leaves or bryophytes over bark 287

287. Apothecia buried in thalline warts, often resembling perithecia 288

288. Ascospores 8-per ascus 289

289. Thallus UV+ bright yellow (lichexanthone present) **Pertusaria paratuberculifera**

289. Thallus UV-, UV+ dull pink/orange or UV+ bright orange 290

290. Medulla K+ yellow turning red (norstictic acid present), producing red crystals in a water mount (best observed under the compound microscope) 291

291. Thallus yellowish, UV+ orange **Pertusaria rubefacta**

291. Thallus gray, UV- **Pertusaria propinqua**

290. Medulla K-, K+ yellow or K+ yellow turning orange/brown, but not producing red crystals in a water mount (norstictic acid absent) 292

292. Thallus UV+ bright orange (thiophanic acid present) 293

293. Ostiole pale to yellow, plane to depressed, not nipple-like; variolaric acid present (TLC required) **Pertusaria epixantha**

293. Ostiole typically yellow, raised, nipple-like; variolaric acid absent (TLC required) 294

294. Medulla K+ yellow and P+ orange (stictic acid present) **Pertusaria texana s. str.**

294. Medulla K- and P- (stictic acid absent) **Pertusaria texana s.l.**

292. Thallus UV- or UV+ dull orange (thiophanic acid absent) 295

295. Thalline warts small, globose; medulla C- (arthothelin absent); low to middle elevations **Pertusaria ostiolata**

295. Thalline warts large, conical; medulla C+ orange (arthothelin present); high elevation northern hardwood forests **Pertusaria appalachensis**

288. Ascospores 1-6-per ascus 296.

296. Thallus UV+ bright yellow (lichexanthone present) **Pertusaria valliculata**

296. Thallus UV-, UV+ dull pink or orange, or UV+ bright orange (lichexanthone absent) 297

297. Medulla K+ yellow turning red (norstictic acid present), producing red crystals in a water mount (best observed under the compound microscope) 298

298. Thalline warts small, conical; thallus yellowish, UV+ bright orange (thiophanic acid present) **Pertusaria neolecania**

298. Thalline warts larger, flattened; thallus gray to blue-gray, UV- (thiophanic acid absent) **Pertusaria neoscotica**

297. Medulla K-, K+ yellow or K+ yellow turning orange/brown, but not producing red crystals in a water mount (norstictic acid absent) 299

299. Medulla K+ yellow and P+ orange or red (stictic acid or succinoprotocetraric acid present) 300

300. Ascospores mostly 4-per ascus **Pertusaria tetrathalmia**

300. Ascospores mostly 2-per ascus 301

301. Thallus yellowish, UV+ bright orange (thiophanic acid absent) **Pertusaria pustulata**

301. Thallus gray to blue-gray, UV- or UV+ dull pink or orange (thiophanic acid absent) 302

302. Medulla P+ red (succinoprotocetraric acid present) **Pertusaria subpertusa**

302. Medulla P+ orange (stictic acid present) 303

303. Epihymenium K+ purple **Pertusaria consocians**

303. Epihymenium K- **Pertusaria macounii**

299. Medulla K- and P- (stictic acid and succinoprotocetraric acid absent) 304

304. Ascospores 2-per ascus; thallus greenish, often with white striations; thalline warts often shallow and mostly immersed in the thallus; low elevations **Pertusaria obruta**

304. Ascospores 4-per ascus; thallus gray to blue-gray, without white striations; thalline warts rounded and sessile; at all elevations 305

305. Thallus isidiate **Pertusaria globularis**

305. Thallus without isidiate **Pertusaria superiana**

287. Apothecia not buried in thalline warts, not resembling perithecia 306

306. Ascospores large, 30-200 μm long 307

307. Ascospores very large, 70-200 μm long 308

308. Medulla and pruina C+ red (lecanoric acid present) **Varicellaria velata**

308. Medulla and pruina C- (lecanoric acid absent) **309**

309. Apothecia densely pruinose, the pruina thick and often causing the apothecia to resemble soralia; basal portions of apothecia not blood-red pigmented 310

310. Medulla K+ yellow turning red (norstictic acid present), producing red crystals in a water mount (best observed under the compound microscope) 311

311. Apothecia, strongly raised and with a lecanorine rim, typically with hymenia and ascospores when sectioned; common at middle to high elevations **Lepra waghornei**

311. Apothecia not strongly raised and with a lecanorine rim, typically sterile and lacking hymenia and ascospores when sectioned; very rare at high elevations **Lepra excludens**

310. Medulla K-, K+ yellow or K+ yellow turning orange/brown, but not producing red crystals in a water mount (norstictic acid absent) 312

312. Medulla P+ orange (thamnolic acid present) or P+ red (succinoprotocetraric acid present) 313

313. Medulla K+ yellowish turning dingy brown, P+ red (succinoprotocetraric acid present) **Lepra multipunctoides**

313. Medulla K+ instant strong lemon yellow then turning dark reddish-brown, P+ orange (thamnolic acid present) 314

314. Thallus UV- (lichexanthone absent); common throughout **Lepra trachythallina s.str.**

314. Thallus UV+ yellow (lichexanthone present); restricted to low elevations **Lepra trachythallina s.l. (lichexanthone chemotype)**

312. Medulla P- (succinoprotocetraric acid and thamnolic acid absent) 315

315. Pruina and medulla bitter tasting, KC+ purple/lavender (picrolichenic acid present) **Lepra amara**

315. Pruina and medulla not bitter tasting, KC- (fatty acids present) **Lepra ophthalmiza**

309. Apothecia epruinose; basal portions of apothecia blood-red pigmented 316

316. Epihymenium and hymenium inspersed with POL+ crystals **Mycoblastus sanguinarioides**

316. Epihymenium and hymenium POL-, lacking crystals **Mycoblastus sanguinarius**

307. Ascospores smaller, 30-60 µm long 317

317. Epihymenium C- (gyrophoric acid absent); Thallus K+ instant lemon yellow and P+ orange (thamnolic acid present); apothecia erumpent, often with ragged thalline margins; ascospores elongate, 30-50 x 4-6 µm, with multiple transverse septa but the septa so faint and indistinct that they appear simple 318

318. Apothecial disc purplish, densely pruinose; infrequent throughout **Loxospora cismonica**

318. Apothecial disc reddish-brown, epruinose to lightly pruinose; common but restricted to s pruce-fir forests **Loxospora ochrophaea**

317. Epihymenium C+ pink (gyrophoric acid present); thallus K- and P- (thamnolic acid absent); apothecia not erumpent, with smooth thalline margins; ascospores ovoid, 30-60 x 20-30 µm, simple 319

319. Cortex of apothecial margin C+ pink 320

320. Medulla of apothecial margin C-; thallus UV- (lichexanthone absent); common throughout **Ochrolechia trochophora**

320. Medulla of apothecial margin C+ pink; thallus often UV+ yellow (lichexanthone present); rare and typically on canopy branches at high elevations **Ochrolechia mexicana**

319. Cortex of apothecial margin C- 321

321. Medulla of apothecial margin C-; thallus UV- (lichexanthone absent); rare and usually on conifers **Ochrolechia pseudopallescens**

321. Medulla of apothecial margin C+ pink; thallus often UV+ yellow (lichexanthone present); rare and usually on canopy branches at low elevations **Ochrolechia africana**

306. Ascospores small, 2-30 µm long 322

322. Ascospores minute, 2-3 µm long, >100 per ascus .323

323. Apothecia lecanorine, with thalline margins; typically on canopy branches 324

324. Hymenium clear, not inspersed with oil droplets; low and middle elevations **Maronea polyphaea**

324. Hymenium inspersed with oil droplets; high elevations **Maronea constans**

323. Apothecia not lecanorine, without thalline margins; typically on boles or bases of trees 325

325. Apothecia pallid, white, immersed in the substrate, with radially cracked margins **Ramonia microspora**

325. Apothecia orange or orange-red, sessile, with smooth margins if present 326

326. Apothecia K-; on conifer resin **Sarea resinae**

326. Apothecia K+ purple; on bark or wood **Piccolia ochrophora**

322. Ascospores larger, 5-30 μm long, 8-32-per ascus 327

327. Apothecia lecanorine, with a distinct thalline margin that usually contrasts with the disc and is concolorous with the margin 328

328. Apothecial discs strongly pruinose, the pruina obscuring the pinkish or reddish coloration of the disc 329

329. Apothecial discs UV+ bright yellow (lichexanthone present); rare **Lecanora miculata**

329. Apothecial discs UV- (lichexanthone absent); common 330

330. Epihymenium C+ yellow (xanthones present); apothecial discs with a pinkish hue; common at low and middle elevations, becoming rare and absent at the highest elevations 331

331. Apothecial and thallus medulla K+ yellow turning red (norstictic acid present); common, especially at middle and low elevations **Lecanora subpallens**

331. Apothecial and thallus medulla K- (norstictic acid absent); not known with certainty from the southern Appalachians **Lecanora carpinea**

330. Epihymenium C- (xanthones absent); apothecial discs bone white; at high elevations 332

332. Apothecial and thallus medulla K+ yellow turning red (norstictic acid present) **Lecanora albella var. rubescens**

332. Apothecial and thallus medulla K- (norstictic acid absent) 333

333. Thallus containing virensic acid (TLC required) **Lecanora caesiorubella var. caesiorubella**

333. Thallus containing protocetraric acid (TLC required) **Lecanora albella var. albella**

328. Apothecial discs epruinose or weakly pruinose, the color of the disc easily observed 334

334. Apothecial discs UV+ bright yellow (lichexanthone present) **Lecanora miculata**

334. Apothecia discs UV- or UV+ dull orange (lichexanthone absent) 335

335. Epihymenium C+ yellow (xanthones present) 336

336. Apothecia large, 0.8-1.0 mm in diameter; common at high elevations **Lecanora masana**

336. Apothecia small, 0.2-0.6 mm in diameter; not known with certainty from the southern Appalachians **Lecanora louisianae**

335. Epihymenium C- (xanthones absent) 337

337. Thallus yellow-green, K- and KC+ yellow (usnic acid present) 338

338. Apothecia large, 0.8-1.0 mm in diameter; ascospores broadly ellipsoid, 14-16 x 7-10 μm **Lecanora masana**

338. Apothecia small, 0.2-0.5 mm in diameter; ascospores narrowly ellipsoid, 10-15 x 3-5 μm 339

339. All apothecia lacking a thalline margin; rare at high elevations **Lecanora symmicta**

339. At least one apothecium with a thalline margin; common at all elevations 340

340. Apothecial margins fuzzy, ecorticate; common at all elevations **Lecanora strobilina**

340. Apothecial margins shiny, corticate; not known with certainty from the southern Appalachians **Lecanora albellula**

337. Thallus gray to blue-gray or brown, not both K- and KC+ yellow (usnic acid absent) 341

341. Apothecial margins ecorticate, fuzzy **Lecanora imshaugii**

341. Apothecial margins corticate, shiny 342

342. Apothecial discs dark, purple-black; hymenium stained purple; medulla UV+ blue-white (alectoronic acid present) **Tephromela atra**

342. Apothecial discs reddish-brown to brown or yellowish; hymenium hyaline; medulla UV- (alectoronic acid absent) 343

343. Apothecial discs or margins P+ orange-red (fumarprotocetraric acid or pannarin present) 344

344. Epihymenium P-; apothecial margin P+ red (fumarprotocetraric acid present), smooth, even; infrequent at high elevations **Lecanora pulicaris**

344. Epihymenium P+ orange-red (pannarin present); apothecial margin P-, apothecial margins uneven, beaded; very common 345

345. Ascospores small, 10-15 x 7-10 µm; very common at all elevations **Lecanora cinereofusca**
345. Ascospores large, 14-21 x 8-13 µm; very rare and restricted to high elevations **Lecanora insignis**
343. Apothecial discs and margins P- or P+ yellow (fumarprotocetraric acid and pannarin absent) 346
346. Apothecia with thalline margins only in early stages of developed, without thalline margins when mature **Lecanora sachsiana**
346. Apothecia with persistent thalline margins 347
347. Ascospores very small, 5-8 x 3-3.5 µm; on the bark and cones of pines **Lecanora minutella**
347. Ascospores larger, 10-20 x 5-12 µm; on various substrates, but typically not on pines 348
348. Epihymenium POL-, not inspersed with granules 349
349. Medulla of the apothecial margin containing clumps of large, POL+ crystals 350
350. Apothecia small, 0.4-0.8(-1.) mm, not basally constricted **Lecanora argentata**
350. Apothecia large, 0.2-2.0(-2.7) mm, basally constricted **Lecanora subrugosa**
349. Medulla of the apothecial margin containing many small, POL+ crystals 351
351. Apothecia small, 0.3-0.6(-1.2) mm in diameter; not basally constricted; ascospores 9-13 x 6-8 µm **Lecanora glabrata**
351. Apothecia large, 0.7-2.0 mm in diameter, basally constricted; ascospores 15-19 x 7-11 µm **Lecanora allophana**
348. Epihymenium POL+, inspersed with granules 352
352. Granules in the epihymenium fine, becoming inspersed into the upper layer of the hymenium 353
353. Ascospores 10-14 x 5-8 µm; common at all elevations **Lecanora hybocarpa**
353. Ascospores 11-18 x 7-12 µm; rare at high elevations **Lecanora circumborealis**
352. Granules in the epihymenium coarse, restricted to the uppermost layer of the epihymenium 354
354. Apothecia small, 0.3-1.0 mm in diameter, not basally constricted **Lecanora chlarotera**
354. Apothecia larger, 0.6-2.0(-3.) mm in diameter, basally constricted 355
355. Apothecial discs reddish brown **Lecanora rugosella**
355. Apothecial discs yellow 356
356. Apothecia discs epruinose; rare at high elevations **Lecanora wisconsinensis**
356. Apothecial discs pruinose; common at high elevations **Lecanora masana**
327. Apothecia not lecanorine, either without a distinct margin, or with a margin that is concolorous with the disc and contrasts with the thallus 357
357. Thallus squamulose **Hypocenomyce friesii**
357. Thallus crustose 358
358. Apothecia elongate, short-lirelliform 359
359. Apothecia pale, reddish-brown; thallus indistinct; confriesiic acid absent (TLC required) **Xylographa trunciseda**
359. Apothecia dark gray to blue-black; thallus distinct, granular, gray; confriesiic acid present (TLC required) **Xylographa disseminata**
358. Apothecia rounded, circular 360
360. On bark 361
361. Hypothecium dark, reddish-brown to black 362
362. Apothecia reddish-brown; ascospores large, broadly ellipsoid to subglobose, 17-20 x 12-15 µm **Japewia tornoensis**
362. Apothecia black; ascospores smaller, ellipsoid, 7-16 x 4-8 363
363. Thallus P+ strong yellow (psoromic acid present); asci polysporous **Lecidea roseotincta**
363. Thallus P- (psoromic acid absent); asci with 8 ascospores 364
364. Ascospores relatively narrow, 9-16 x 4-6 µm; hypothecium dark brown; thallus thick, green, verruculous, often overgrowing bryophytes **Lecidea berengeriana**
364. Ascospores relatively broad, 7-16 x 6-8 µm; hypothecium golden-brown; thallus thinner, gray, +/- smooth, directly on bark 365

365. Hymenium inspersed with oil droplets **Lecidella elaeochroma**

365. Hymenium clear, not inspersed with oil droplets; Thallus KC- (xanthones absent) **Lecidella euphorea**

361. Hypothecium hyaline to pale yellow 366

366. Thallus dirty green, resembling an algal colony, composed of goniocysts, distinctly not attractive 367

367. Methoxymicareic acid present (TLC required) **Micarea micrococca**

367. Micareic acid present (TLC required) **Micarea prasina**

366. Thallus various colors, but not resembling an algal colony, either continuous, areolate or indistinct, overall prettier 368

368. Thallus P+ orange-red (argopsin present) **Biatora pycnidiata**

368. Thallus P- or P+ strong yellow (argopsin absent) 369

369. Thallus P+ strong yellow (psoromic acid present); asci polysporous **Lecidea roseotincta**

369. Thallus P- (psoromic acid absent); asci with 8 ascospores 370

370. Ascospores large, broadly ellipsoid to subglobose, 17-20 x 12-15 μm **Japewia tornoensis**

370. Ascospores smaller, ellipsoid to elongate ellipsoid, 5-23 x 3-7 μm 371

371. Ascospores broader, 5-7 μm long 372

372. Ascospores short, ellipsoid, 8-12 x 5-7 μm; apothecia often highly variable in color and degree of pruinosity **Pyrrhospora varians**

372. Ascospores elongate ellipsoid, 10-23 x 3.5-6 μm; apothecia uniform in color, consistently epruinose 373

373. On bark; apothecia reddish-brown to orange-brown; ascospores 17-22 x 3.5-4 μm **Biatora longispora**

373. On mosses, often at the bases of trees; apothecia pale brown; ascospores 10-23 x 4-6 **Biatora vernalis**

371. Ascospores narrow, 2-5 μm long 372

372. Ascospores consistently simple, short 5-12 x 2.5-4 μm; apothecia plane, reddish-brown to brown; 373

373. On the bark and cones of pines; ascospores small, 5-8 x 3-3.5 μm **Lecanora minutella**

373. On the bark and wood of mature cherry and oaks; ascospores larger, 9–12 × 2.5–4 μm **Lecanora sachsiana**

372. Ascospores predominately 2-celled, long, 7-18 x 2-3.5 μm; apothecia convex, dark brownish 374

374. Thallus C- (gyrophoric acid absent) **Micarea elachista**

374. Thallus C+ pink (gyrophoric acid present) **Micarea denigrata**

360. On wood 375

375. Hypothecium dark, reddish-brown to black .376

376. Thallus sorediate, UV+ blue-white (perlatolic acid present) **Hertelidea botryosa**

376. Thallus esorediate, UV- (perlatolic acid absent) 377

377. Thallus P+ yellow (alectorialic acid present) **Pycnora praestabilis**

377. Thallus P- (alectorialic acid absent) 378

378. Thallus thick, green, granular, C+ pink (gyrophoric acid present), but the reaction often very difficult to observe; ascospores 9-12 x 4-6 μm; on wet, rotting wood **Trapeliopsis viridescens**

378. Thallus thin, indistinct, mostly minutely areolate, C- (gyrophoric acid absent); ascospores 7-11 x 2-4.5 μm; on dry, rot resistant wood 379

379. Apothecia dark brownish-black; lower layers of the hymenium brown; on conifer lignum **Lecidea plebeja**

379. Apothecia light, reddish-brown; lower layers of the hymenium blue-black; on oak and chestnut lignum **Xyleborus sporodochifer**

375. Hypothecium hyaline to pale yellow 380

380. Thallus dirty green, resembling an algal colony, composed of goniocysts, distinctly not attractive 381

381. Methoxymicareic acid present (TLC required) **Micarea micrococca**

381. Micareic acid present (TLC required) **Micarea prasina**

380. Thallus various colors, but not resembling an algal colony, either continuous, areolate or indistinct, overall prettier 382

382. Ascospores a mix of simple and 2-celled 383

383. Thallus C- (gyrophoric acid absent) **Micarea elachista**

383. Thallus C+ pink (gyrophoric acid present) **Micarea denigrata**

382. Ascospores consistently simple 384

384. Apothecia yellowish to reddish-brown 385

385. Apothecia yellowish; thallus yellow-green, K- and KC+ yellow (usnic acid present) **Lecanora symmicta**

385. Apothecia reddish-brown; thallus indistinct or gray, K- and KC- (usnic acid absent) 386

386. Thallus areolate, lichenized; ascospores 9–12 × 2.5–4 µm; on the bark and wood of mature cherry and oaks **Lecanora sachsiana**

386. Thallus indistinct, not lichenized; ascospores 10-15 x 6-8 µm; on conifer wood **Agyrium rufum**

384. Apothecia dark, deep reddish-brown to black 387

387. Thallus thick, green, granular, C+ pink (gyrophoric acid present), but the reaction often very difficult to observe; on wet, rotting wood **Trapeliopsis viridescens**

387. Thallus indistinct or thick, gray and areolate C- (gyrophoric acid absent); on dry, not rotting wood 388

388. Thallus thick, white-gray, areolate **Ramboldia elabens**

388. Thallus thin and indistinct 389

389. Apothecia dark reddish-brown, epruinose; ascospores 7-11 x 2-3.5 µm **Lecanora hypoptoides**

389. Apothecia black, pruinose; ascospores 6-10 x 3-5 µm **Lecidea turgidula**

KEY VI: CRUSTOSE PERITHECIATE LICHENS

Note that *Naetrocymbe punctiformis* is not included in the key below because the report is likely based on a misidentification.

1. Ascospores brown at maturity, remaining brown when post-mature 2

2. Ascospores submuriform or muriform 3

3. On bark 4

4. Apothecioid perithecia present; thallus K+ yellow and P+ orange (stictic acid present); ascospores small, 45-55 x 10-13 µm **Leucodecton subcompunctum**

4. True perithecia present; thallus K- and P- (stictic acid absent); ascospores large, 45-65 x 28-27 µm 5

5. Perithecia +/- solitary, with mostly apical ostioles **Pyrenula leucostoma**

5. Perithecia aggregated, many with lateral ostioles **Pyrenula ravenelii**

3. On rock, soil or other lichens 6

6. Thallus K+ yellow and P+ orange (stictic acid present) **Leucodecton subcompunctum**

6. Thallus K- and P- (stictic acid absent) 7

7. Thallus dark, chocolate brown; very rare **Diploschistes aeneus**

7. Thallus white, gray or tan, but never dark chocolate brown; common **Diploschistes actinostomus**

2. Ascospores transversely septate 8

8. Ascospores 2-celled; photobiont absent; on bark and branches of Yellow Birch **Distopyrenis americana**

8. Ascospores 4-7-celled; photobiont present or absent; on the bark of various trees 9
9. Thallus thick, glossy, corticate 10
10. Ascospores +/- citriform (lemon-shaped), with the terminal lumina pushed against the exospore 11
11. Thallus UV+ bright yellow (lichexanthone present) **Pyrenula pseudobufonia**
11. Thallus UV- (lichexanthone absent) **Pyrenula citriformis**
10. Ascospores ellipsoid, with the terminal lumina positionally normally in the centers of the end cells 12
12. Ascospores small, 15-19 × 6-7.5 µm **Pyrenula confoederata**
12. Ascospores larger, 22-40 x 8-18 13
13. Ascospores with elliptical lumina **Pyrenula subelliptica**
13. Ascospores with angular lumina 14
14. Hymenial gel IKI+ orange **Pyrenula micheneri**
14. Hymenial gel IKI- **Pyrenula punctella**
9. Thallus thin or lacking, dull, ecorticate 15
15. Photobiont absent; ascospores 25-29 × 8-10 µm **Requienella subcollapsa**
15. Photobiont Trentepohlia; ascospores larger or smaller than the above 16
16. Ascospores 4-8 celled, 18-24 x 5-9 µm; thallus white, ecorticate **Eopyrenula intermedia**
16. Ascospores 4-celled, 30-45 x 12-16 µm; thallus brown, indistinct **Lithothelium phaeosporum**
1. Ascospores hyaline at maturity, sometimes turning brownish when post-mature and collapsing 17
17. Ascospores simple 18
18. On bark or wood 19
19. Thallus brown, composed of large squamules; ascospores 8-per ascus; on bark of living trees, especially *Quercus alba* **Placidium arboreum**
19. Thallus bright yellow, composed of minute areoles; ascospores many per ascus; on wood **Thelocarpon laureri**
18. On rock, soil, or mosses over rock or soil 20
20. Thallus large, composed of large, overlapping, brown squamules, loosely adnate on the substrate **Placidium arboreum**
20. Thallus smaller, either continuous, areolate or absent, tightly adnate on the substrate 21
21. Ascospores minute, many per ascus 22
22. Thallus and perithecia bright yellow 23
23. Perithecia not immersed thalline areoles; thallus an indistinct film; on soil **Thelocarpon intermediellum**
23. Perithecia immersed in thalline areoles; thallus composed of minute areoles; on rock and wood **Thelocarpon laureri**
22. Thallus brown to rusty orange, composed of larger, flat areoles 24
24. Thallus brown; apothecia immersed in the areoles, with the visible portion of the disc resembling a pore and the widest part of the apothecium buried in the apothecium **Trimmatothelopsis dispersa**
24. Thallus rusty orange; apothecia immersed in the areoles, but with the disc broad and visible, not resembling a pore **Acarospora sinopica**
21. Ascospores larger, 8-per ascus 25
25. Perithecia immersed in pits in the substrate 26
26. Thallus gray-white, thin or immersed; perithecia with a distinct disk-like structure at the apex **Bagliettoa baldensis**
26. Thallus concolorous with the substrate, immersed; perithecia without a distinct disk-like structure at the apex **Bagliettoa calciseda**
25. Perithecia not immersed in pits in the substrate 27
27. Ascospores rod-shaped, very strange; perithecia conical; parasitic on *Protoblastenia rupestris* **Sarcopyrenia calcarea**

27. Ascospores ellipsoid, normal looking; perithecia globose; not parasitic on other lichens 28
28. On calcareous rocks, not associated with waterways **Verrucaria calkinsiana**
28. On non-calcareous rocks, closely associated with water **Verrucaria aethiobola**
17. Ascospores transversely septate, submuriform or muriform 29
29. Ascospores transversely septate 30
30. On bark or leaves 31
31. Ascospores 2-celled 32
32. Ascospores long and narrow, 20-30 x 2.5-3 µm; photobiont absent **Leptorhaphis parameca**
32. Ascospores ellipsoid to fusiform, never as above; photobiont present or absent 33
33. Ascospores quite large, 33-60 x 15-23 µm **Acrocordia megalospora**
33. Ascospores smaller, 12-30 x 4-10 µm 34
34. Ascospores large, 20-30 x 6-10 µm 35
35. Ascospores broad, 7-10 µm wide; on smooth barked trees, primarily at low elevations in disturbed habitats **Arthopyrenia cinchonae**
35. Ascospores narrow, 6-8 µm wide; on rough, dry bark of mature *Betula alleghaniensis* at high elevations **Arthopyrenia betulicola**
34. Ascospores smaller, 10-25 x 4-7 µm 36
36. Photobiont absent; on branches of *Hamamelis* **Arthopyrenia degelii**
36. Photobiont *Trentepohlia*; on boles and bases of trees 37
37. Paraphyses branched and anastomosing; ascospores remaining intact, not breaking apart 38
38. Ascospores clavate, with a submedian septum, resembling a loafer; common **Anisomeridium polypori**
38. Ascospores ovoid, with median septum, not resembling a loafer; rare **Anisomeridium biforme**
37. Paraphyses unbranched, not anastomosing; ascospores often becoming +/- fragmented at the septum 39
39. Ascospores 12-17 x 4-5 µm; macroconidia 7-9 x 2-3 µm **Strigula viridiseda**
39. Ascospores 15-25 x 4-5 µm; macroconidia 14-19 x 3-5 µm **Strigula americana**
31. Ascospores 3-to-many-celled 40
40. Ascospores 3-celled, 19–26 × 6.5–8.5 µm; on fast growing trees and shrubs (e.g., *Prunus, Viburnum*) at high elevations **Mycoporum biseptatum**
40. Ascospores 4-many-celled, of various sizes; occurring on a diversity of substrates in many different habitats 41
41. Ascospores 4-celled 42
42. Perithecioid apothecia present, tan; ascospores many per ascus **Thelopsis rubella**
42. True perithecia present, reddish-black to black; ascospores 8-per ascus 43
43. Perithecial walls red pigmented **Segestria leptalea**
43. Perithecial walls black 44
44. Thallus brown, thick, shiny; perithecia aggregated and forming large raised pseudostroma; pseudostroma filled with yellow pigment **Bathelium carolinianum**
44. Thallus gray to pale brown, indistinct, dull; perithecia not forming raised pseudostroma 45
45. Perithecia large; photobiont present, *Trentepohlia* **Lithothelium hyalosporum**
45. Perithecia small; photobiont absent...46
46. Perithecia simple, with a single chamber; 28-40 x 8-12 **Mycoporum antecellens**
46. Perithecia compound, with multiple chambers; 17-22 x 5-7 **Mycoporum eschweileri**
41. Ascospores 5-to-many-celled 47
47. Ascospores many per ascus, relatively small, 12-16 x 4-8 µm; perithecial walls not dark pigmented **Thelopsis rubella**
47. Ascospores 8-ascus, larger, 25-140 x 3-10 (or larger); perithecial walls dark, at least in the apical portions 48

48. Perithecial wall dark pigmented only in the apical portions; ascospores 5-8-celled, 24-42 x 5-7 µm **Strigula stigmatella**

48. Perithecial wall dark pigmented throughout; ascospores 8-21-celled (or more), 25-140 x 3-10 µm (or larger) 49

49. Perithecia urn-like, with an open apical pore 50

50. Ascospores very long, many celled, >100 µm long, lumbricoid, worm-like; high elevations **Conotrema urceolatum**

50. Ascospores shorter, 9-15-celled, 25-40 × 6-8 µm, clavate; low elevations **Thelotrema subtile**

49. Perithecia flask-like, with a closed pore 51

51. Thallus thick, shiny, light brown or bronze; perithecia aggregated into pseudostroma; ascospores broad, 7-10 µm wide **Trypethelium virens**

51. Thallus thin, dull, dark brown to purple-brown; perithecia solitary; ascospores narrower, 3-7 µm wide 52

52. Ascospores clavate, 8-13-celled 28-65 x 4-7 µm **Pseudosagedia cestrensis**

52. Ascospores filiform, 14-21-celled, 100-140 x 3-5 µm **Pseudosagedia rhaphidosperma**

30. On rock, organic matter, or mosses over rocks and organic matter 53

53. Ascospores 2-celled 54

54. Thallus composed of small, brown, +/- pruinose areoles; perithecia immersed in the areoles; ascospores small, 8-10 x 4-5 µm **Placidiopsis minor**

54. Thallus continuous or immersed in the substrate; perithecia superficial or immersed in the substrate; ascospores larger, 11-18 x 5-7 µm 55

55. Paraphyses present, branched and anastomosing; ascospores narrowly ellipsoid 56

56. On rocks in waterways; thallus shiny, well developed **Anisomeridium carinthiacum**

56. On rocks not associated with waterways; thallus dull, indistinct **Anisomeridium distans**

55. Paraphyses absent; ascospores broadly ellipsoid 57

57. Perithecia not immersed in pits in the rock; on various substrates; involucrellum absent; ascospores 25-32 x 11-14 µm **Thelidium decipiens**

57. Perithecia immersed in pits in the rock; on calcareous rock; involucrellum present; ascospores 19-25 x 8-11 µm **Thelidium minutulum**

53. Ascospores 4-10-celled 58

58. Ascospores 6-10-celled 59

59. Perithecial walls dark pigmented only the upper portions; thallus shiny, thick, green-brown **Strigula stigmatella**

59. Perithecial walls dark pigmented throughout; thallus dull, thin, gray-brown **Pseudosagedia guentheri**

58. Ascospores 4-celled 60

60. Perithecial walls dark red in section 61

61. Ascospores 20-30 x 5-8 µm **[Segestria lectissima]**

61. Ascospores 16-23 x 3-5 µm **Segestria leptalea**

60. Perithecial walls black, purple-brown, or tan in section 62

62. Ascospores uniformly 4-celled, fusiform, 18-25 x 4-5 µm; thallus thin, brown to gray-brown, with a distinct black prothallus **Pseudosagedia chlorotica**

62. Ascospores a mixture of 4-celled and 2-celled, broadly ellipsoid, 16-22 x 6-8 µm; thallus indistinct **Anisomeridium distans**

29. Ascospores submuriform or muriform 63

63. Foliicolous on leaves **Aulaxina quadrangula**

63. On bark or rocks 64

64. On bark 65

65. Perithecia solitary, horizontal, laid on their sides, appearing blown over, with lateral ostioles; thallus typically UV+ yellow (lichexanthone present); ascospores I+ purple **Polymeridium proponens**

65. Perithecia compound, vertical, not laid on their sides and appearing blown over, with apical ostioles; thallus UV- (lichexanthone absent); ascospores I- 66

66. Ascospores 30-38(-43) x 12-17(-18) μm **Mycoporum compositum**

66. Ascospores (36-)39-48(-55) x (14-)16-21(-24) μm **Mycoporum pycnocarpoides**

64. On rocks 67

67. Ascospores 2-ascus; on non-calcareous rocks; rare **Staurothele fissa**

67. Ascospores 8-ascus; on calcareous rocks; common 68

68. Thallus brown, areolate, well developed; hymenium inspersed with algal cells **Willeya diffractella**

68. Thallus pale gray to white, continuous, poorly developed; hymenium not inspersed with algal cells69

69. Perithecioid apothecia present, pale pinkish-brown, immersed in the substrate **Gyalecta farlowii**

69. True perithecia present, black, superficial **Polyblastia cupularis**

KEY VII: CALICIOID FUNGI

1. Pins lichenicolous on lichen thalli 2

2. Ascospores 2-celled; lichenicolous on *Arthonia* **Chaenothecopsis brevipes**

2. Ascospores simple; lichenicolous on *Protoparmelia hypotremella* or *Pertusaria* 3

3. On *Protoparmelia hypotremella*; ascospores 7-10 μm long **Sphinctrina anglica**

3. On *Pertusaria*; ascospores 4-7 μm long 4

4. Stalk short; exciple reddish-brown, pigment turning K+ red **Sphinctrina turbinata**

4. Stalk relatively tall; exciple brown, pigment K- **Sphinctrina leucopoda**

1. Pins not lichenicolous on other lichens 5

5. Ascospores simple 6

6. Mass of ascospores (i.e., top of the pin viewed from above) black 7

7. Pins associated with *Trentepohlia* 8

8. Stalk K+ fleeting red, reaction must be observed under the compound microscope; ascospores 5.5-9.0 x 2.5-3.5 μm **Chaenothecopsis rubescens**

8. Stalks K- or K+ darker in color, reaction must be observed under the compound microscope; ascospores 5.0-7.5 x 2.5 μm **Chaenothecopsis australis**

7. Pins not associated with a photobiont 9

9. Ascospores large, 8.5-16 x 3-7.0 μm; stalk and lower surface of the capitulum with a whitish tinge **Mycocalicium fuscipes**

9. Ascospores smaller, 5-8.5 x 2.5-3.5 μm; stalk and lower surface of the capitulum uniformly brown 10

10. Stalks slender, appearing fragile and easily broken; capitulum rounded; at least some asci with a narrow apical tube, but this almost impossible to observe; not terribly common 11

11. On resin **Chaenothecopsis resinicola**

11. On bark or wood 12

12. Stalk short, <0.6 mm tall **Chaenothecopsis nana**

12. Stalk larger, 0.5-1.0 mm tall **Chaenothecopsis savonica**

10. Stalks stocky, appearing robust and not easily broken, simple; capitulum flattened and button-like; asci without a narrow apical tube; extremely common 13

13. Exciple with cells that are angular in outline **Mycocalicium subtile**

13. Exciple with cells that are rounded in outline **Mycocalicium albonigrum**

6. Mass of ascospores (i.e., top of the pin viewed from above) light to medium brown or yellow 14

14. On the bracket fungus *Trichaptum abietinum*, on dead conifers **Chaenotheca balsamconensis**

14. On bark or wood, but not on bracket fungi 15

15. Stalks and/or lower surface of the capitulum with yellow pruina 16
16. Photobiont *Trentepohlia* 17
17. Ascospores large, 4.5-7.5 µm in diameter **Chaenotheca hispidula**
17. Ascospores small, 3.5-4.5 µm in diameter **Chaenotheca olivaceorufa**
16. Photobiont either a unicellular green with globose cells (Trebouxioid) or chain-forming with the cells resembling tiny sausages (*Stichococcus*) 18
18. Photobiont a unicellular green with globose cells (Trebouxioid) 19
19. Thallus poorly-developed, dull greenish; ascospores ellipsoid, varying from simply to multi-septate **Chaenotheca laevigata**
19. Thallus well-developed, bright yellow; ascospores globose to subglobose, uniformly simple **Chaenotheca chrysocephala**
18. Photobiont chain-forming with the cells resembling tiny sausages (*Stichococcus*) 20
20. Thallus bright yellow-green, matching the pruina on the stalks and lower surface of the capitulum **Chaenotheca furfuracea**
20. Thallus dull greenish-gray, not matching the bright pruina on the stalks and capitulum **Chaenotheca chlorella**
15. Stalks and/or lower surface of the capitulum either white pruinose, brown pruinose, red pruinose or entirely lacking pruina 21
21. Photobiont Trentepohlia **Chaenotheca olivaceorufa**
21. Photobiont either a unicellular green with globose cells (Trebouxioid) or chain-forming with the cells resembling tiny sausages (*Stichococcus*) 22
22. Stalk and lower surface of the capitulum red-pruinose **Chaenotheca gracillima**
22. Stalk and lower surface of the capitulum white pruinose, brown-pruinose or entirely epruinose 23
23. Photobiont chain-forming with the cells resembling tiny sausages (*Stichococcus*) 24
24. Lower surface of the capitulum brown pruinose **Chaenotheca stemonea**
24. Lower surface of the capitulum white pruinose or epruinose 25
25. Lower surface of the capitulum epruinose **Chaenotheca cinerea**
25. Lower surface of the capitulum white pruinose 26
26. Thallus immersed and indistinct **Chaenotheca xyloxena**
26. Thallus thick, well-developed and verruculose **Chaenotheca trichialis**
23. Photobiont a unicellular green with globose cells (Trebouxioid) 27
27. Thallus well-developed and with scattered patches of orange-pigment, the pigment reacting K+ red **Chaenotheca ferruginea**
27. Thallus either poorly-developed, or if well-developed then talking orange-pigmented patches 28
28. Photobiont cells large, 15-20 µm in diameter (*Dictyochloropsis*) **Chaenotheca brunneola**
28. Photobiont cells smaller, 8.0-15 µm in diameter (other Trebouxioid photobionts) 29
29. Asci catenulate, forming chains; exciple poorly-developed **Chaenotheca sphaerocephala**
29. Asci solitary and not forming chains; exciple well-developed **Chaenotheca hygrophila**
5. Ascospores transversely septate, 2-5-celled 30
30. Mass of ascospores (i.e., top of the pin viewed from above) light to medium brown or yellow **Chaenotheca laevigata**
30. Mass of ascospores (i.e., top of the pin viewed from above) black or greenish-blue 31
31. Apothecia immersed or sessile; thallus bright yellow 32
32. Apothecia immersed in the thallus; exciple thin throughout **Cyphelium tigillare**
32. Apothecia sessile on the thallus; exciple basally thickened **Cyphelium lucidum**
31. Apothecia clearly staked; thallus not bright yellow 33
33. Asci readily dissolving such that the mature ascospores form a readily visible mass 34
34. Mass of ascospores (i.e., top of the pin viewed from above) distinctly greenish-blue in color **Microcalicium ahlneri**

34. Mass of ascospores (i.e., top of the pin viewed from above) black 35
35. Stalk strongly I+ blue when viewed under the compound microscope and mounted in iodine **Calicium lenticulare**
35. Stalk I- when viewed under the compound microscope and mounted in iodine 36
36. Lower portion of the capitulum yellow pruinose 37
37. Ascospores smooth, unornamented **Calicium trabinellum**
37. Ascospores with spiral ornamentation **Calicium adspersum**
36. Lower portion of the capitulum either white pruinose, brown pruinose, or entirely lacking pruina 38
38. Lower portion of the capitulum brown pruinose 39
39. Thallus poorly-developed, immersed in the substrate; ascospores <12 μm long **Calicium salicinum**
39. Thallus well-developed, green to yellow-green; ascospores >12 μm long **Calicium viride**
38. Lower portion of the capitulum white pruinose or epruinose 40
40. Pins minute, <0.5 mm tall **Calicium pinastri**
40. Pins larger, 0.5-1.5 mm tall 41
41. Lower surface of the capitulum epruinose **Calicium abietinum**
41. Lower surface of the capitulum white pruinose 42
42. Thallus well-developed; ascospores with spiral ornamentation **Calicium quercinum**
42. Thallus poorly-developed, mostly immersed and indistinct; ascospores with non-spiral ornamentation 43
43. Asci clavate **Calicium parvum**
43. Asci cylindrical **Calicium glaucellum**
33. Asci not readily dissolving such that the mature ascospores mostly remain in the asci and do not form a readily visible mass 44
44. Ascospores 4-celled **Phaeocalicium matthewsianum**
44. Ascospores 2-celled 45
45. On bracket fungi, especially *Trichaptum biforme* on dead hardwoods **Phaeocalicium polyporaeum**
45. On bark or wood, but not on bracket fungi 46
46. On the twigs and branches of trees 47
47. On Betula; capitulum compressed laterally, flattened **Phaeocalicium flabelliforme**
47. On Alnus; capitulum not compressed and flattened **Stenocybe pullatula**
46. On resin, boles, bases and lignum of trees 48
48. On resin 49
49. Stalks gray in color; pins short, 0.2-0.4 mm tall **Chaenothecopsis marcineae**
49. Stalks black; pins taller, 0.5-1.5 mm tall 50
50. Stalk and lower portion of the capitulum epruinose **Chaenothecopsis dolichocephala**
50. Stalk and lower portion of the capitulum greenish to yellow-green pruinose 51
51. Stalk and lower surface of the capitulum green pruinose **Chaenothecopsis eugenia**
51. Stalk and lower surface of the capitulum yellow-green pruinose **Chaenothecopsis edbergii**
48. On bark or wood 52
52. Stalk K+ intense green reaction must be observed under the compound microscope **Chaenothecopsis viridireagens**
52. Stalk K+ yellow turning orange-red, K+ fleeting red or K-, reaction must be observed under the compound microscope 53
53. Stalk K+ yellow turning orange-red or K+ fleeting red 54
54. Stalk K+ fleeting red (norstictic acid absent), reaction must be observed under the compound microscope **Chaenothecopsis pusiola**
54. Stalk K+ yellow turning orange-red, with the precipitation of red crystals (norstictic acid present), reaction must be observed under the compound microscope **Chaenothecopsis norstictica**
53. Stalks K- or K+ darker in color, reaction must be observed under the compound microscope 55

55. Stalk and lower portion of the capitulum white pruinose **Chaenothecopsis fennica**
55. Stalk and lower portion of the capitulum epruinose 56
56. Tissue of stalk with a distinct reddish tinge **Chaenothecopsis debilis**
56. Tissue of stalk brown 57
57. Ascospores light brown, with a strongly pigmented septum **Chaenothecopsis nigra**
57. Ascospores normal brown, with a weakly pigmented septum that is difficult to discern **Chaenothecopsis pusilla**

Epilogue

FIGURE 48. SMOKY MOUNTAINS

Great Smoky Mountains National Park derives its name from "Shaconage" or "a place of blue smoke" as named by its early human inhabitants: the Eastern Band of the Cherokee. It is a fitting name, as seen in this photograph, but so too would have been Great Lichen Mountains National Park. Photo by Erin Tripp.

Great Smoky Mountains National Park lies near the southern terminus of the Appalachian Mountains. The name "Appalachia" likely arose from cartographic uncertainties by early French and Spanish explorers. Historically, the Native American Apalachee Tribe lived in the Florida Panhandle but now resides primarily in Louisiana. There is no evidence that they ever occupied the mountainous regions of eastern North America. The place name "Apalatchi" was variously plotted on early maps and was subsequently adopted for the name of the mountain chain.

The Eastern Band of the Cherokee call their homeland "Shaconage" (Figure 48), or "a place of blue smoke." Congress established GSMNP on 15 June 1934 and on 2 September 1940, President Franklin D. Roosevelt dedicated it and turned its stewardship over to the National Park Service. Amid a massive financial crisis, budgetary woes shrouded the early years of the Park. Many donors eventually breathed life into the nation's nascent Park, notably the Rockefeller family, who contributed five million dollars to complete its infrastructure development and opening (their generosity is today honored by the lichen *Rockefellera crossophylla*). Today, Newfound Gap stands as a memorial, and millions of visitors, volunteers, donors, and tax dollars continue to contribute to the maintenance of GSMNP. We highly recommend two books that provide excellent history and context about the early years of the Great Smoky Mountains and its transition to a National Park. The first is *Our Southern Highlanders*, written by the legendary Horace Kephart and first published by the Outing Publishing Company in 1913 and revised in 1922. The second is *The Birth of a National Park*, written by the influential Carlos Campbell and first published by the University of Tennessee Press in 1960.

People from all backgrounds visit the Smokies to marvel at fields of Fringed Phacelia in Walnut Bottoms, the endless cascades of Gunter Fork, black bear sightings on Snake Den, magnificent old growth of Albright Grove, April wildflower spectacles of Porter's Flats, elegant trees of Pretty Hollow Gap, towering hemlocks on Rough Fork, or swimming holes on Big Creek. Some come to photograph the famous "Rhododendron Gardens" atop Gregory Bald while the off-trail explorer experiences them on hands and knees, "Rhodo surfing" through the impenetrable thickets of Middle Prong east of Ramsey Cascades. Our chief incentive in composing the *Field Guide to the Lichens of Great Smoky Mountains National Park* is that the wonderful diversity of lichens in the Park might soon be added to list of "great things to see" throughout this remarkable landscape. Time will determine whether we were successful.

APPENDIX 1

Tables

TABLE 1. ABRIDGED TIMELINE OF SALIENT LICHENOLOGICAL WORKS IN EASTERN NORTH AMERICA, PRIOR TO 1941, WHEN DEGELIUS'S REPORT ON GSMNP LICHENS WAS PUBLISHED.

1741—Dillenius	Recognized 16 taxa from collections of Bartram and others in the Historia Muscorum
1743—Gronovius	Recognized 10 taxa from Clayton's collections in the Flora Virginica
1803, 1810, 1814—Acharius	Described many new species from collections of G.H.E. Muhlenberg in multiple works
1803—Michaux	Based on his collections from the late 1700's, named 21 species in Flora Boreali-Americana; 14 taxa from "Carolina," including four from Grandfather Mountain
1803—Michaux	14 from Carolina (incl. 4 from Grandfather) from 1780s/1790s travel
1813—Muhlenberg	Published first checklist in Catalogus Plantarum Americae Septentrionalis, based largely on specimens he collected and sent to Acharius.
1823—Halsey	First checklist of lichens of New York City published in Annales of the New York Lyceum
1845—Tuckerman	Published An Enumeration of North American Lichens, a checklist much expanded from Muhlenberg
1848—Tuckerman	A Synopsis of the Lichens of New England, the other Northern States, and British America, first checklist of the northeast enumerated 300 taxa
1867—Curtis	Published first checklist of North Carolina lichens, reported 217 taxa
1872, 1882, 1888—Tuckerman	Three major works published for the continent: Genera Lichenum and A Synopsis of North American Lichens in two parts
1887—Eckfeldt & Calkins	First checklist of lichens of Florida published in Journal of Mycology
1890—Calkins	200 species stated to occur in eastern Tennessee, but only 38 reported by name
1896—Calkins	First checklist of lichens of Chicago published, reported 120 species
1896—Millspaugh & Nuttall	First checklist of lichens of West Virginia published, reported 115 species
1898—Macoun	First checklist of lichens of Ottawa published in The Ottawa Naturalist
1902—Macoun	First checklist of Canadian lichens, reported 614 taxa
1910—Fink	Flora of lichens from Minnesota published in Contributions from the US National Herbarium
1926—Evans & Meyrowitz	First checklist of lichens of Connecticut published
1930—Fink	Lichens of North America, first and only flora published for the continent, albeit incomplete

TABLE 2. GSMNP LOCALITIES VISITED AND COLLECTED BY THE AUTHORS OF THIS BOOK, GUNNAR DEGELIUS (1941), AND JONATHAN DEY (1978).

State	County	Locality	This Book	Degelius (1941)	Dey (1978)
TN	Blount	Abrams Creek/Abrams Falls	X		
TN	Blount	Ace Gap	X		
TN	Sevier	Alum Cave	X	X	X
NC	Swain	Andrews Bald	X		X
NC	Haywood	Balsam High Top	X		X
NC	Haywood	Baxter Creek	X		
NC	Swain	Beaureguard Ridge	X		
NC	Haywood	Big Cataloochee (= "Big Cataloochie")	X		X
NC	Swain	Blockhouse Mountain	X		
NC	Swain	Bone Valley	X		
NC	Haywood	Boogerman Loop	X		
TN	Sevier	Boulevard Trail	X		X
NC	Haywood	Caldwell Fork	X		
TN	Sevier	Charlie's Bunion	X		X
TN	Sevier	Cherokee Orchard		X	
NC	Haywood/Swain	Chiltoes Mountain	X		X
TN	Sevier	Chimney Tops (= "The Chimneys")	X	X	
NC/TN	Swain/Sevier	Clingmans Dome	X	X	X
NC/TN	Swain/Sevier	Collins Gap	X		X
NC/TN	Haywood/Cocke	Deer Creek Gap	X		X
NC	Haywood	Double Gap	X		
NC/TN	Swain/Sevier	Eagle Rocks	X		
TN	Blount	Flint Gap	X		
NC	Swain	Forney Creek	X		
NC	Swain	Forney Ridge	X	X	X
TN	Sevier	Greenbrier Pinnacle	X		
NC/TN	Swain/Blount	Gregory Bald	X		
TN	Blount	Hannah Mountain	X		
NC	Swain	Hazel Creek	X		
NC	Haywood	Hemphill Bald	X		
NC	Swain	High Rocks	X		
NC	Swain	Hyatt Ridge	X		
NC/TN	Haywood/Cocke	Inadu Knob	X		X
NC/TN	Swain/Sevier	Jenkins Knob			X
NC	Swain	Jenkins Ridge	X		

State	County	Locality	This Book	Degelius (1941)	Dey (1978)
NC	Swain	Lakeshore Trail (Fontana Dam to Flint Gap)	X		
NC	Swain	Lakeshore Trail (Forney Creek to Tunnel Ridge)	X		
NC	Swain	Long Hungry Ridge	X		
TN	Sevier	Laurel Falls		X	
TN	Sevier	Laurel Top	X		X
NC	Haywood/Swain	Luftee Knob	X		X
NC/TN	Swain/Sevier	Masa Knob	X		X
NC	Swain	Mingus Mill/ Mingus Creek	X		
NC/TN	Swain/Sevier	Mount Ambler (= "Mount Amber")	X		X
NC/TN	Swain/Sevier	Mount Buckley	X		X
TN	Cocke	Mount Cammerer	X		
NC/TN	Swain/Sevier	Mount Chapman	X		X
NC/TN	Swain/Sevier	Mount Guyot	X		X
NC	Swain	Mount Hardison	X		X
NC/TN	Swain/Sevier	Mount Kephart	X	X	X
TN	Sevier	Mount LeConte	X	X	X
NC/TN	Swain/Sevier	Mount Love	X		X
NC/TN	Swain/Sevier	Mount Sequoyah	X		X
NC	Haywood	Mount Sterling	X		X
NC	Haywood/Swain	Mount Yonaguska	X		X
TN	Sevier	"near park office"		X	
NC/TN	Swain/Sevier	Newfound Gap	X	X	X
NC	Swain	Noland Creek	X		
NC/TN	Haywood/Cocke/ Sevier	Old Black	X		X
NC/TN	Swain/Sevier	Pecks Corner	X		X
NC	Swain	Pole Road Creek	X		
NC	Haywood	Polls Gap	X		
NC/TN	Swain/Sevier	Porters Gap			X
NC	Haywood	Pretty Hollow	X		
NC	Haywood	Purchase Knob	X		
TN	Blount	Rabbit Creek			
TN	Blount	Rich Mountain	X		
NC	Swain	Shuckstack	X		

State	County	Locality	This Book	Degelius (1941)	Dey (1978)
NC/TN	Swain/Sevier	Silers Bald			X
TN	Cocke	Snake Den Mountain	X		
NC	Haywood	Spruce Mountain			X
TN	Sevier	Sugarland Mountain			X
NC	Swain	Sunkota Ridge	X		
NC	Haywood/Swain	Thermo Knob	X		X
NC/TN	Haywood/Sevier	Tricorner Knob	X		X
NC	Swain	Welch Ridge	X		
NC	Swain	Whiteoak Brank			
TN	Blount	Whiteoak Sink	X		
NC/TN	Haywood/Sevier	Yellow Creek Gap	X		X

APPENDIX 2

Revised Checklist for the More Than 900 Lichens, Lichenicolous, and Allied Fungi of Great Smoky Mountains National Park

Since we published the first checklist of GSMNP lichens (Lendemer et al. 2013) many new records have accumulated through our own fieldwork while preparing this *Field Guide* and through the examination of specimens collected by others. Likewise, a substantial number of new species has been described from GRSMP, and in some cases the names for certain species have changed. Therefore, we here provide an updated checklist. As was the case with the earlier checklist edition, the list is arranged alphabetically, and names listed in bold face text are species that we have confirmed as occurring in the Park. Names of lichenicolous fungi are denoted by an asterisk (*); species whose type specimens were collected in GSMNP are denoted by a double dagger ("‡"); and species whose occurrence in GSMNP we consider to be doubtful are denoted by a single dagger "†". In addition to species we have newly reported in publications after 2013, and species reported by others (e.g., Selva 2016), we here include 78 newly reported species and confirm the presence of 30 more species that had been previously reported but for which we had not seen or collected vouchers from GSMNP ourselves.

Abrothallus hypotrachynae Etayo & Diederich*
Abrothallus microspermus Tul.*
Abrothallus usneae Rabenh.*
Acarospora fuscata (Nyl.) Arnold
Acarospora obpallens (Nyl. *ex* Hasse) Zahlbr.
Acarospora sinopica (Wahlenb.) Korb.
Acrocordia megalospora (Fink) R.C. Harris
Agonimia opuntiella (Buschardt & Poelt) Vězda
Agyrium rufum (Pers.) Fr.
Ahtiana aurescens (Tuck.) A. Thell & Randlane
Alectoria fallacina Motyka‡
Alectoria sarmentosa Ach. subsp. *sarmentosa*
Amandinea polyspora (Willey) E. Lay & P. May
Amandinea punctata (Hoffm.) Coppins & Scheid.
Anaptychia palmulata (Michx.) Vain.
Andreiomyces morozianus (Lendemer) B.P. Hodkin. & Lendemer‡
Anisomeridium biforme (Borrer) R.C. Harris
Anisomeridium carinthiacum (J. Steiner) R.C. Harris
Anisomeridium distans (Willey) R.C. Harris
Anisomeridium polypori (Ellis & Everh.) M.E. Barr
Anzia americana Yoshim. & Sharp
Anzia colpodes (Anzi) Yoshim.
Arthonia anglica Coppins
Arthonia byssacea (Weigel) Almq.
Arthonia caesia (Flot.) Körb.
Arthonia caudata Willey
Arthonia cinnabarina (DC.) Wallr.
Arthonia cupressina Tuck.
Arthonia helvola (Nyl.) Nyl.
Arthonia interveniens Nyl.
Arthonia kermesina R.C. Harris, E. Tripp & Lendemer‡
Arthonia mediella Nyl.
Arthonia punctiformis Ach.†
Arthonia quintaria Nyl.
Arthonia ruana A. Massal.

Arthonia rubella (Fée) Nyl.
Arthonia susa R.C. Harris & Lendemer
Arthonia vinosa Leight.
Arthopyrenia betulicola R.C.Harris, E. Tripp & Lendemer‡
Arthopyrenia cinchonae (Ach.) Müll. Arg.
Arthopyrenia degelii R.C. Harris
Arthothelium spectabile A. Massal.
Aspicilia caesiocinerea (Nyl. *ex* Malbr.) Arnold
Aspicilia cinerea (L.) Körb.
Aspicilia laevata (Ach.) Arnold
Aspicilia olivaceopallida (H. Magn.) Lendemer‡
Aulaxina quadrangula (Stirt.) R. Sant.
Bachmanniomyces uncialicola (Zopf) D. Hawksw.*
Bacidia apiahica (Müll.Arg.) Zahlbr.
Bacidia circumspecta (Norrl. & Nyl.) Malme
Bacidia coprodes (Körb.) Lettau
Bacidia diffracta S. Ekman
Bacidia ekmaniana R.C. Harris, Lendemer & Ladd
Bacidia helicospora S. Ekman
Bacidia lobarica Printzen & Tønsberg
Bacidia polychroa (Th. Fr.) Körb.
Bacidia rubella (Hoffm.) A. Massal.
Bacidia schweinitzii (Fr. *ex* Tuck.) A. Schneid.
Bacidia sorediata Lendemer & R.C. Harris
Bacidia suffusa (Fr.) A. Schneid.
Bacidina assulata (Körb.) S. Ekman
Bacidina delicata (Leight.) V. Wirth & Vězda
Baeomyces rufus (Huds.) Rebent.
Bagliettoa baldensis (A. Massal.) Vězda
Bagliettoa calciseda (DC.) Gueidan & Cl. Roux
Bathelium carolinianum (Tuck.) R.C. Harris
Bellemerella trapeliae Nav.-Ros. & Cl. Roux*
Biatora appalachensis Printzen & Tønsberg‡
Biatora chrysantha (Zahlbr.) Printzen
Biatora longispora (Degel.) Lendemer & Printzen
Biatora pontica Printzen & Tønsberg
Biatora printzenii Tønsberg
Biatora pycnidiata Printzen & Tønsberg
Biatora vernalis (L.) Fr.
Biatoropsis usnearum Räsänen*
Bilimbia sabuletorum (Schreb.) Arnold
Botryolepraria lesdainii (Hue) Canals, Hern.-Mar., Gómez-Bolea & Llimona
Brigantiaea leucoxantha (Spreng.) R. Sant. & Hafellner
Bryoria americana Motyka
Bryoria bicolor (Ehrh.) Brodo & D. Hawksw.
Bryoria furcellata (Fr.) Brodo & D. Hawksw.
Bryoria nadvornikiana (Gyeln.) Brodo & D. Hawksw.
Bryoria tenuis (Å.E. Dahl) Brodo & D. Hawksw.
Buellia badia (Fr.) A. Massal.
Buellia curtisii (Tuck.) Imsh.
Buellia dialyta (Nyl.) Tuck.
Buellia elizae Tuck.
Buellia sharpiana Lendemer & R.C. Harris‡
Buellia spuria (Schaer.) Anzi
Buellia stillingiana J. Steiner
Buellia vernicoma (Tuck.) Tuck.
Bulbothrix isidiza (Nyl.) Hale
Bulbothrix laevigatula (Nyl.) Hale†
Bulbothrix scortella (Nyl.) Hale
Byssoloma leucoblepharum (Nyl.) Vain.
Byssoloma marginatum (Arnold) Sérus.
Byssoloma subdiscordans (Nyl.) P. James
Calicium abietinum Pers.
Calicium adspersum Pers.
Calicium glaucellum Ach.
Calicium lenticulare Ach.
Calicium parvum Tibell
Calicium pinastri Tibell (Reported by Selva 2016)
Calicium quercinum Pers. (Reported by Selva 2016)
Calicium salicinum Pers.
Calicium trabinellum (Ach.) Ach.
Calicium viride Pers. (Confirmed by Selva 2016)
Caloplaca camptidia (Tuck.) Zahlbr.
Caloplaca cerina (Hedw.) Th. Fr.
Caloplaca chrysodeta (Vain. *ex* Räsänen) Dombr.
Caloplaca chrysophthalma Degel.
Caloplaca feracissima H. Magn.
Caloplaca flavocitrina (Nyl.) H. Olivier
Caloplaca flavorubescens (With.) J.R. Laundon
Caloplaca flavovirescens (Wulfen) Dalla Torre & Sarnth.
Caloplaca reptans Lendemer & B.P. Hodkin.
Caloplaca subsoluta (Nyl.) Zahlbr.
Candelaria concolor (Dicks.) Stein

Candelaria fibrosa (Fr.) Müll. Arg.
Candelariella efflorescens R.C. Harris & W.R. Buck
Candelariella xanthostigma (Ach.) Lettau
Candelariella xanthostigmoides (Müll. Arg.) Rogers
Canoparmelia caroliniana (Nyl.) Elix & Hale
Canoparmelia texana (Tuck.) Elix & Hale
Catillaria lenticularis (Ach.) Th. Fr.
Catillaria stereocaulorum (Th. Fr.) H. Olivier*
Catinaria atropurpurea (Schaer.) Vězda & Poelt
Cetradonia linearis (A. Evans) J.C. Wei & Ahti‡
Cetraria aculeata (Schreb.) Fr.
Cetraria ericetorum Opiz
Cetraria laevigata Rass.
Cetrelia cetrarioides (Delise) W.L. Culb. & C.F. Culb. *s.l.*
Cetrelia chicitae (W.L. Culb.) W.L. Culb. & C.F. Culb.
Cetrelia olivetorum (Nyl.) W.L. Culb. & C.F. Culb.
Chaenotheca balsamconensis J.L. Allen & McMullin
Chaenotheca brunneola (Ach.) Müll. Arg.
Chaenotheca chlorella (Ach.) Müll. Arg.
Chaenotheca chrysocephala (Turner *ex* Ach.) Th. Fr.
Chaenotheca cinerea (Pers.) Tibell
Chaenotheca ferruginea (Turner *ex* Sm.) Mig.
Chaenotheca furfuracea (L.) Tibell
Chaenotheca gracillima (Vain.) Tibell
Chaenotheca hispidula (Ach.) Zahlbr.
Chaenotheca hygrophila Tibell
Chaenotheca laevigata Nádv.
Chaenotheca olivaceorufa Rikkinen (Reported by Selva 2016)
Chaenotheca sphaerocephala Nádv. (Confirmed by Selva 2016)
Chaenotheca stemonea (Ach.) Müll. Arg.
Chaenotheca trichialis (Ach.) Th. Fr.
Chaenotheca xyloxena Nádv.
Chaenothecopsis australis Tibell (Reported by Selva 2016)
Chaenothecopsis brevipes Tibell (Confirmed by Selva 2016)
Chaenothecopsis debilis (Turner & Borrer *ex* Sm.) Tibell
Chaenothecopsis dolichocephala Titov
Chaenothecopsis edbergii Selva & Tibell
Chaenothecopsis eugenia Titov (Reported by Selva 2016)
Chaenothecopsis fennica (Laurila) Tibell (Reported by Selva 2016)
Chaenothecopsis marcineae Selva (Confirmed by Selva 2016)
Chaenothecopsis nana Tibell (Reported by Selva 2016)
Chaenothecopsis nigra Tibell
Chaenothecopsis norstictica R.C. Harris (Reported by Selva 2016)
Chaenothecopsis pusilla (A. Massal.) A.F.W. Schmidt
Chaenothecopsis pusiola (Ach.) Vain.
Chaenothecopsis resinicola Tibell & Titov
Chaenothecopsis rubescens Vain.
Chaenothecopsis savonica (Räsänen) Tibell
Chaenothecopsis viridireagens (Nádv.) A.F.W. Schmidt
Cheiromycina flabelliformis B. Sutton
Chrysothrix chamaecyparicola Lendemer
Chrysothrix insulizans R.C. Harris & Ladd
Chrysothrix onokoensis (F. Wolle) R.C. Harris & Ladd
Chrysothrix susquehannensis Lendemer & Elix
Chrysothrix xanthina (Vain.) Kalb
Cladonia apodocarpa Robbins
Cladonia appalachensis Yoshim. & Sharp *ex* Lendemer & R.C. Harris‡
Cladonia arbuscula (Wallr.) Flot.
Cladonia brevis (Sandst.) Sandst.
Cladonia caespiticia (Pers.) Flörke
Cladonia caroliniana Schwein. *ex* Tuck.
Cladonia chlorophaea (Flörke *ex* Sommerf.) Spreng.
Cladonia coccifera (L.) Willd.
Cladonia coniocraea (Flörke) Spreng.†
Cladonia conista (Ach.) Robbins
Cladonia cornuta (L.) Hoffm.
Cladonia crispata (Ach.) Flot.
Cladonia cristatella Tuck.
Cladonia crypotchlorophaea Aasah.
Cladonia cylindrica A. Evans
Cladonia didyma (Fée) Vain. var. *didyma*
Cladonia didyma var. *vulcanica* (Zoll. & Moritzi) Vain.

Cladonia digitata (L.) Hoffm.
Cladonia fimbriata (L.) Fr.
Cladonia floerkeana (Fr.) Flörke
Cladonia furcata (Huds.) Schrad.
Cladonia gracilis (L.) Hoffm. subsp. *turbinata* (Ach.) Ahti
Cladonia grayi G. Merr. *ex* Sandst.
Cladonia incrassata Flörke
Cladonia macilenta Hoffm. var. *macilenta*
Cladonia macilenta var. *bacillaris* (Genth.) Schaer.
Cladonia mateocyatha Robbins
Cladonia merochlorophaea Asahina
Cladonia mitis Sandst.
Cladonia multiformis G. Merr.
Cladonia ochrochlora Flörke
Cladonia parasitica (Hoffm.) Hoffm.
Cladonia petrophila R.C. Harris
Cladonia peziziformis (With.) J.R. Laundon
Cladonia pleurota (Flörke) Schaer.
Cladonia polycarpoides Nyl.
Cladonia pyxidata (L.) Fr.
Cladonia ramulosa (With.) J.R. Laundon
Cladonia rangiferina (L.) F.H. Wigg.
Cladonia rei Schaer.
Cladonia robbinsii A. Evans
Cladonia scabriuscula Delise *ex* Nyl.†
Cladonia sobolescens Nyl. *ex* Vain.
Cladonia squamosa Hoffm.
Cladonia stipitata Lendemer & Hodkinson
Cladonia strepsilis (Ach.) Grognot
Cladonia subtenuis (Abbayes) Mattick
Cladonia uncialis (L.) F.H. Wigg.
Cladonia verticillata (Hoffm.) Schaer.
Clathroporina isidiifera R.C. Harris
Clauzadea chondrodes (A. Massal.) Clauzade & Cl. Roux *ex* Hafellner & Türk
Cliostomum griffithii (Sm.) Coppins
Clypeococcum hypocenomycis D. Hawksw.*
Coccocarpia palmicola (Spreng.) Arv. & Galloway
Coenogonium luteum (Dicks.) Kalb & Lücking
Coenogonium pineti (Ach.) Lücking & Lumbsch
Collema coccophorum Tuck.
Collema conglomeratum Hoffm.
Collema flaccidum (Ach.) Ach.
Collema furfuraceum Du Rietz
Collema fuscovirens (With.) J.R. Laundon
Collema leptaleum Tuck.
Collema nigrescens (Huds.) DC.
Collema occultatum Bagl.
Collema pulcellum Ach. var. *leucopeplum* (Tuck.) Degel.
Collema pulcellum Ach. var. *pulcellum*
Collema pustulatum Ach.
Collema ryssoleum (Tuck.) A. Schneid.
Collema subflaccidum Degel.
Collema subnigrescens Degel.†
Conotrema urceolatum (Ach.) Tuck.
Crespoa crozalsiana (B de Lesd.) Lendemer & B.P. Hodkin.
Cresponea flava (Vain.) Egea & Torrente
Cyphelium lucidum (Th. Fr.) Th. Fr.
Cyphelium tigillare (Ach.) Ach.
Cystobasidium hypogymniicola Diederich & Ahti*
Cystocoleus ebeneus (Dillwyn) Thwaites
Dactylospora glaucomarioides (Willey ex Tuck.) Hafellner*
Dactylospora lobariella (Nyl.) Hafellner*
Dactylospora pertusariicola (Willey) Hafellner*
Dendriscocaulon intricatulum (Nyl.) Henssen
Dermatocarpon luridum (With.) J.R. Laundon
Dermatocarpon muhlenbergii (Ach.) Müll. Arg.
Dibaeis absoluta (Tuck.) Kalb & Gierl
Dibaeis baeomyces (L.f.) Rambold & Hertel
Dictyocatenulata alba Finley & E.F. Morris
Dimelaena oreina (Ach.) Norman
Diploschistes actinostomus (Ach.) Zahlbr.
Diploschistes aeneus (Müll. Arg.) Lumbsch
Diploschistes muscorum (Scop.) R. Sant.
Diploschistes scruposus (Schreb.) Norman
Dirina massiliensis Durieu & Mont. f. *sorediata* (Müll. Arg.) Tehler
Dirinaria frostii (Tuck.) Hale & W.L. Culb.
Distopyrenis americana Aptroot
Endocarpon pallidulum (Nyl.) Nyl.
Endococcus perpusillus Nyl.*
Enterographa hutchinsiae (Leight.) A. Massal.
Eopyrenula intermedia Coppins
Ephebe americana Henssen

Ephebe lanata (L.) Vain.
Ephebe solida Bornet
Epicladonia stenospora (Harm.) D. Hawksw.*
Erioderma mollissimum (Samp.) Du Rietz
Everniastrum catawbiense (Degel.) Hale *ex* Sipman‡
Fellhanera bouteillei (Desm.) Vězda
Fellhanera eriniae R.C. Harris & Lendemer‡
Fellhanera granulosa R.C. Harris & Lendemer
Fellhanera montesfumosi R.C. Harris & Lendemer‡
Fellhaneropsis cf. *vezdae* (Coppins & P. James) Sérus. & Coppins
Fissurina insidiosa C. Knight & Mitt.
Flakea papillata O.E. Erikss.
Flavoparmelia baltimorensis (Gyeln. & Fóriss.) Hale
Flavoparmelia caperata (L.) Hale
Flavopunctelia flaventior (Stirt.) Hale
Fuscidea appalachensis Fryday
Fuscidea arcuatula (Arnold) V. Wirth & Vězda
Fuscidea recensa (Stirt.) Hertel, V. Wirth & Vězda
Fuscopannaria darlingiae E. Tripp & Lendemer
Fuscopannaria leucosticta (Tuck.) P.M. Jørg.
Fuscopannaria sorediata P.M. Jørg.
Glyphis cicatricosa Ach.
Gomphillus americanus Essl.
Gomphillus calycioides (Delise *ex* Duby) Nyl.
Graphis endoxantha Nyl.
Graphis lineola Ach.
Graphis scripta (L.) Ach.
Graphis sterlingiana E. Tripp & Lendemer
Gyalecta farlowii Nyl.
Gyalecta obesispora R.C. Harris & Lendemer‡
Gyalectidium appendiculatum Lücking, Lendemer & Tripp
Gyalectidium yahriae W.R. Buck & Sérus.
Gyalideopsis bartramiorum Lendemer
Gyalideopsis epicorticis (A. Funk) Tønsberg & Vězda
Gyalideopsis moodyae Lendemer & Lücking
Gyalideopsis piceicola (Nyl.) Vězda & Poelt
Haematomma americanum Kalb & Staiger
Hafellia disciformis (Fr.) Marbach & H. Mayrhofer
Hafellia subnexa (Nyl.) Marbach
Halecania pepegospora (H. Magn.) v.d. Boom
Herteliana schuyleriana Lendemer
Hertelidea botryosa (Fr.) Printzen & Kantvilas
Heterodermia albicans (Pers.) Swinsc. & Krog
Heterodermia appalachensis (Kurok.) W.L. Culb.
Heterodermia casarettiana (A. Massal.) Trevis.
Heterodermia crocea R.C. Harris
Heterodermia echinata (Taylor) W.L. Culb.
Heterodermia erecta Lendemer
Heterodermia granulifera (Ach.) W.L. Culb.
Heterodermia hypoleuca (Ach.) Trevis.
Heterodermia japonica (M. Satô) Swinsc. & Krog
Heterodermia langdoniana Lendemer & E. Tripp
Heterodermia leucomela (L.) Poelt
Heterodermia neglecta Lendemer, R.C. Harris & E. Tripp
Heterodermia obscurata (Nyl.) Trevis.
Heterodermia pseudospeciosa Kurok.
Heterodermia speciosa (Wulfen) Trevis.
Heterodermia squamulosa (Degel.) W.L. Culb.‡
Homostegia hertelii D. Hawksw., C. Atienza & M.S. Cole*
Homostegia piggotii (Berk. & Broome) P. Karst.*
Hyperphyscia syncolla (Tuck. *ex* Nyl.) Kalb
Hypocenomyce anthracophila (Nyl.) P. James & Gotth. Schneid.
Hypocenomyce friesii (Ach.) P. James & Gotth. Schneid.
Hypocenomyce scalaris (Ach. *ex* Lilij.) M. Choisy
Hypogymnia incurvoides Rass.
Hypogymnia krogiae Ohlsson
Hypogymnia physodes (L.) Nyl.
Hypogymnia tubulosa (Schaer.) Hav.
Hypogymnia vittata (Ach.) Parrique
Hypotrachyna afrorevoluta (Swinsc. & Krog) Swinsc. & Krog
Hypotrachyna croceopustulata (Kurok.) Hale
Hypotrachyna densirhizinata (Kurok.) Hale
Hypotrachyna dentella (Hale & Kurok.) Hale
Hypotrachyna ensifolia (Kurok.) Hale
Hypotrachyna gondylophora (Hale) Hale‡
Hypotrachyna horrescens (Taylor) Swinsc. & Krog
Hypotrachyna imbricatula (Zahlbr.) Hale
Hypotrachyna kauffmaniana ined.
Hypotrachyna laevigata (Sm.) Hale
Hypotrachyna livida (Taylor) Hale

Hypotrachyna lividescens (Kurok.) Hale
Hypotrachyna minarum (Vain.) Swinsc. & Krog
Hypotrachyna oostingii (J.P. Dey) Hale
Hypotrachyna osseoalba (Vain.) Y.S. Park & Hale
Hypotrachyna producta Hale‡
Hypotrachyna prolongata (Kurok.) Hale
Hypotrachyna pseudosinuosa (Asahina) Hale
Hypotrachyna pustulifera (Hale) Skorepa
Hypotrachyna revoluta (Flörke) Hale
Hypotrachyna rockii (Zahlbr.) Hale
Hypotrachyna showmanii Hale
Hypotrachyna sinuosa (Sm.) Hale
Hypotrachyna spumosa (Asah.) Swinsc. & Krog
Hypotrachyna taylorensis (M.E. Mitch.) Hale
Hypotrachyna thysanota (Kurok.) Hale
Hypotrachyna virginica (Hale) Hale
Icmadophila ericetorum (L.) Zahlbr.
Imshaugia aleurites (Ach.) S.L.F. Mey.
Imshaugia placorodia (Ach.) S.L.F. Mey.
Intralichen lichenum (Diederich) D. Hawksw. & M.S. Christ.*
Ionaspis alba Lutzoni
Ionaspis lacustris (With.) Lutzoni
Jamesiella anastomosans (P. James & Vězda) Lücking, Sérus. & Vězda
Japewia tornoensis (Nyl.) Tønsberg
Japewiella dollypartoniana J.L. Allen & Lendemer
Kephartia crystalligera R.C. Harris & Lendemer‡
Lasallia caroliniana (Tuck.) Davydov, Peršoh & Rambold
Lasallia papulosa (Ach.) Llano
Lasallia pensylvanica (Hoffm.) Llano
Lecanactis abietina (Ehrh. *ex* Ach.) Körb.
Lecania croatica (Zahlbr.) Kotlov
Lecania erysibe (Ach.) Mudd
Lecania subfuscula (Nyl.) S. Ekman
Lecanora albella (Pers.) Ach. var. *albella*
Lecanora albella var. *rubescens* (Imshaug & Brodo) Lumbsch
Lecanora albellula (Nyl.) Th. Fr.†
Lecanora allophana (Ach.) Nyl.
Lecanora anakeestiicola Lendemer & E. Tripp‡
Lecanora appalachensis Lendemer & R.C. Harris‡
Lecanora argentata (Ach.) Malme
Lecanora caesiorubella Ach. var. *caesiorubella*
Lecanora carpinea (L.) Vain.†
Lecanora chlarotera Nyl.
Lecanora cinereofusca H. Magn. var. *cinereofusca*
Lecanora circumborealis Vitik. & Brodo
Lecanora dispersa (Pers.) Rohl.
Lecanora epanora (Ach.) Ach.
Lecanora glabrata Ach.
Lecanora hybocarpa (Tuck.) Brodo
Lecanora hypoptoides (Nyl.) Nyl.
Lecanora imshaugii Brodo
Lecanora insignis Degel.‡
Lecanora intricata (Ach.) Ach.
Lecanora layana Lendemer
Lecanora louisianae B. de Lesd.
Lecanora masana Lendemer & R.C. Harris‡
Lecanora miculata Ach.
Lecanora minutella Nyl.
Lecanora nothocaesiella Lendemer & R.C. Harris
Lecanora oreinoides (Körb.) Hertel & Rambold
Lecanora polytropa (Ehrh.) Rabenh.
Lecanora pseudistera Vain.
Lecanora pulicaris (Pers.) Ach.
Lecanora rugosella Zahlbr.
Lecanora sachsiana E. Tripp & Lendemer
Lecanora saxigena Lendemer & R.C. Harris
Lecanora strobilina (Spreng.) Kieff.
Lecanora subimmergens Vain.
Lecanora subpallens Zahlbr.
Lecanora subrugosa Nyl.
Lecanora symmicta (Ach.) Ach.
Lecanora thysanophora R.C. Harris
Lecanora wisconsinensis H. Magn.
Lecidea ahlesii (Hepp) Nyl.
Lecidea atrobrunnea (DC.) Schaer.
Lecidea berengeriana (A. Massal.) Nyl.
Lecidea cyrtidia Tuck.
Lecidea deminutula H. Magn.‡
Lecidea fuscoatra (L.) Ach.
Lecidea nylanderi (Anzi) Th. Fr.
Lecidea plana (J. Lahm) Nyl.
Lecidea plebeja Nyl.
Lecidea roseotincta Coppins & Tønsberg
Lecidea tessellata Flörke

Lecidea turgidula Th. Fr.
Lecidella carpathica Körb.
Lecidella elaeochroma (Ach.) M. Choisy
Lecidella euphorea (Flörke) Hertel
Leimonis erratica (Korber) R.C. Harris & Lendemer
Leioderma cherokeense P.M. Jørg.‡
Lepra amara (Ach.) Hafellner
Lepra excludens (Nyl.) Hafellner
Lepra multipunctoides (Dibben) Lendemer & R.C. Harris
Lepra ophthalmiza (Nyl.) Hafellner
Lepra pustulata (Brodo & W.L. Culb.) Lendemer & R.C. Harris
Lepra trachythallina (Erichsen) Lendemer & R.C. Harris
Lepra waghornei (Hult.) Lendemer & R.C. Harris
Lepraria arbuscula (Nyl.) Lendemer & B.P. Hodkin.
Lepraria caesiella R.C. Harris
Lepraria cryophila Lendemer‡
Lepraria disjuncta Lendemer
Lepraria eburnea J.R. Laundon
Lepraria elobata Tønsberg
Lepraria finkii (Hue) R.C. Harris
Lepraria harrisiana Lendemer
Lepraria hodkinsoniana Lendemer
Lepraria humida Slav.-Bayr.
Lepraria jackii Tønsberg
Lepraria lanata Tønsberg‡
Lepraria neglecta (Nyl.) Erichsen
Lepraria normandinoides Lendemer & R.C. Harris
Lepraria oxybapha Lendemer
Lepraria rigidula (B. de Lesd.) Tønsberg
Lepraria vouauxii (Hue) R.C. Harris
Lepraria xanthonica Lendemer
Leprocaulon adhaerens Lendemer & B.P. Hodkin.
Leprocaulon nicholsiae Lendemer & E. Tripp
Leptogium acadiense J.W.Hinds, F.L.Anderson & Lendemer
Leptogium austroamericanum Malme
Leptogium chloromelum (Ach.) Nyl.
Leptogium corticola (Taylor) Tuck.
Leptogium cyanescens (Rabenh.) Körb.
Leptogium dactylinum Tuck.
Leptogium denticulatum Tuck.
Leptogium hibernicum M.E. Mitch.
Leptogium hirsutum Sierk
Leptogium isidiosellum (Riddle) Sierk
Leptogium laceroides B. de Lesd.
Leptogium lichenoides (L.) Zahlbr.
Leptogium millegranum Sierk
Leptogium phyllocarpum (Pers.) Mont.
Leptogium subtile (Schrad.) Torss.
Leptogium teretiusculum (Flörke) Arnold
Leptorhaphis parameca (A. Massal.) Körb.
Lettauia cladoniicola D. Hawksw. & R. Sant.*
Leucodecton subcompunctum (Nyl.) A. Frisch
Lichenochora obscuroides (Lindsay) Triebel & Rambold*
Lichenoconium erodens M. S. Christ. & D. Hawksw.*
Lichenodiplis lecanorae (Vouaux) Dyko & D. Hawksw.*
Lithothelium hyalosporum (Nyl.) Aptroot
Lithothelium phaeosporum (R.C. Harris) Aptroot
Lobaria pulmonaria (L.) Hoffm.
Lobaria quercizans Michx.
Lobaria scrobiculata (Scop.) P. Gaertn.
Lopadium disciforme (Flot.) Kullh.
Lopadium pezizoideum (Ach.) Körb.†
Loxospora cismonica (Beltr.) Hafellner
Loxospora elatina (Ach.) A. Massal.
Loxospora ochrophaea (Tuck.) R.C. Harris
Marchandiomyces corallinus (Roberge) Diederich & D. Hawskw.*
Maronea constans (Nyl.) Hepp
Maronea polyphaea H. Magn.
Megalaria albocincta (Degel.) Tønsberg‡
Megalaria beechingii Lendemer
Megalaria laureri (Hepp *ex* Th. Fr.) Hafellner
Megalospora porphyritis (Tuck.) R.C. Harris
Melanelia culbersonii Hale
Melanelia disjuncta (Erichsen) Essl.
Melanelia hepatizon (Ach.) A. Thell
Melanelia panniformis (Nyl.) Essl.
Melanelia stygia (L.) Essl.
Melanelixia glabratula (Lamy) Sandler & Arup
Melanelixia subaurifera (Nyl.) O. Blanco, A. Crespo, Divakar, Essl., D. Hawksw. & Lumbsch
Melanohalea exasperata (De Not.) O. Blanco, A. Crespo, Divakar, Essl., D. Hawksw. & Lumbsch

Melanohalea halei (Ahti) O. Blanco, A. Crespo, Divakar, Essl., D. Hawksw. & Lumbsch
Melanohalea olivacea (L.) O. Blanco, A. Crespo, Divakar, Essl., D. Hawksw. & Lumbsch†
Melaspilea demissa (Tuck.) Zahlbr.
Menegazzia subsimilis H. Magn.
Micarea alabastrites (Nyl.) Coppins†
Micarea bauschiana (Körb.) V. Wirth & Vězda
Micarea chlorosticta (Willey) R.C. Harris
Micarea cinerea (Schaer.) Hedl.†
Micarea denigrata (Fr.) Hedl.
Micarea elachista (Körb.) Coppins & R. Sant.
Micarea endocyanea (Willey) R.C. Harris
Micarea lignaria (Ach.) Hedl.
Micarea lutulata (Nyl.) Coppins
Micarea melaena (Nyl.) Hedl.
Micarea micrococca (Körb.) Gams
Micarea neostipitata Coppins & P. May
Micarea peliocarpa (Anzi) Coppins & R. Sant.
Micarea polycarpella (Erichsen) Coppins & Palice
Micarea prasina Fr.
Micarea pycnidiophora Coppins & P. James
Micareopsis irriguata R.C. Harris & Lendemer
Microcalicium ahlneri Tibell
Minutoexcipula tuckerae V. Atienza & D. Hawksw.*
Miriquidica leucophaea (Flörke *ex* Rabenh.) Hertel & Rambold
Miriquidica pycnocarpa (Korb.) M.P. Andreev
Muellerella lichenicola (Sommerf.) D. Hawksw.*
Muellerella ventosicola (Mudd.) D. Hawksw.*
Multiclavula mucida (Fr.) R.H. Petersen
Mycoblastus caesius (Coppins & P. James) Tønsberg
Mycoblastus sanguinarioides Kantvilas
Mycoblastus sanguinarius (L.) Norman
Mycocalicium albonigrum (Nyl.) Tibell (Confirmed by Selva 2016)
Mycocalicium fuscipes (Tuck.) Fink (Reported by Selva 2016)
Mycocalicium subtile (Pers.) Szatala
Mycoporum antecellens (Nyl.) R.C. Harris
Mycoporum biseptatum R.C. Harris & Lendemer
Mycoporum compositum (A. Massal.) R.C. Harris
Mycoporum eschweileri (Müll. Arg.) R.C. Harris
Mycoporum pycnocarpoides Müll. Arg.
Myelochroa aurulenta (Tuck.) Elix & Hale
Myelochroa galbina (Ach.) Elix & Hale
Myelochroa metarevoluta (Asahina) Elix & Hale
Myelochroa obsessa (Ach.) Elix & Hale
Nadvornikia sorediata R.C. Harris
Naetrocymbe punctiformis (Pers.) R.C. Harris†
Nectriopsis rubefaciens (Ellis & Everh.) M.E. Barr
Nephroma bellum (Spreng.) Tuck.
Nephroma helveticum Ach.
Nephroma parile (Ach.) Ach.
Nephroma resupinatum (L.) Ach.
Normandina pulchella (Borrer) Nyl.
Ochrolechia africana Vain.
Ochrolechia androgyna (Hoffm.) Arnold
Ochrolechia arborea (Kreyer) Almb.
Ochrolechia mexicana Vain.
Ochrolechia pseudopallescens Brodo
Ochrolechia trochophora (Vain.) Oshio
Ochrolechia yasudae (Vain.) Oshio
Opegrapha anomea Nyl.*
Opegrapha corticola Coppins & P. James
Opegrapha gyrocarpa Flot.
Opegrapha rupestris Pers.*
Opegrapha varia Pers.
Opegrapha viridis (Ach.) Behlen & Desberger
Opegrapha vulgata Ach.
Opegrapha zonata Körb.
Pannaria conoplea (Ach.) Bory
Pannaria lurida (Mont.) Nyl. subsp. *lurida*
Pannaria lurida subsp. *quercicola* P.M. Jørg.†
Pannaria lurida subsp. *russellii* (Tuck.) P.M. Jørg
Pannaria rubiginosa (Ach.) Bory
Pannaria subfusca P.M. Jørg.
Pannaria tavaresii P.M. Jørg.
Parmelia omphalodes (L.) Ach.
Parmelia saxatilis (L.) Ah.
Parmelia squarrosa Hale
Parmelia sulcata Taylor
Parmeliella appalachensis P.M. Jørg.
Parmeliella triptophylla (Ach.) Müll. Arg.
Parmeliopsis capitata R.C. Harris *ex* J.W. Hinds & P. Hinds
Parmotrema arnoldii (Du Rietz) Hale
Parmotrema cetratum (Ach.) Hale

Parmotrema commensuratum (Hale) Hale
Parmotrema conferendum Hale
Parmotrema crinitum (Ach.) M. Choisy
Parmotrema despectum Kurok.
Parmotrema diffractaicum (Essl.) Hale
Parmotrema dilatatum (Vain.) Hale†
Parmotrema gardneri (C.W. Dodge) Hale
Parmotrema hypotropum (Nyl.) Hale
Parmotrema margaritatum (Hue) Hale
Parmotrema mellissii (C.W. Dodge) Hale
Parmotrema neotropicum (Kurok.) Hale
Parmotrema perforatum (Jacq.) A. Massal.
Parmotrema perlatum (Huds.) M. Choisy
Parmotrema reticulatum (Ach.) M. Choisy
Parmotrema simulans (Hale) Hale
Parmotrema stuppeum (Taylor) Hale
Parmotrema subisidiosum (Müll. Arg.) Hale
Parmotrema submarginale (Michx.) DePriest & B. Hale
Parmotrema subsumptum (Nyl.) Hale
Parmotrema subtinctorium (Zahlbr.) Hale
Parmotrema tinctorum (Despr. *ex* Nyl.) Hale
Parmotrema ultralucens (Krog) Hale
Parmotrema xanthinum (Müll. Arg.) Hale
Peltigera canina (L.) Willd.
Peltigera degenii Gyeln.†
Peltigera didactyla (With.) J.R. Laundon
Peltigera elisabethae Gyeln.
Peltigera evansiana Gyeln.
Peltigera horizontalis (Huds.) Baumg.
Peltigera hydrothyria Miadl. & Lutzoni
Peltigera hymenina (Ach.) Delise
Peltigera leucophlebia (Nyl.) Gyeln.
Peltigera neckeri Hepp. *ex* Müll. Arg.
Peltigera neopolydactyla (Gyeln.) Gyeln.
Peltigera phyllidiosa Miadl. & Goffinet
Peltigera praetextata (Flörke *ex* Sommerf.) Zopf
Peltigera rufescens (Weiss) Humb.
Peltigera seneca Magain, Miądl. & Sérus.
Pertusaria appalachensis Lendemer, R.C. Harris & Elix
Pertusaria consocians Dibben
Pertusaria epixantha R.C. Harris
Pertusaria globularis (Ach.) Tuck.
Pertusaria macounii (I.M. Lamb) Dibben
Pertusaria neolecania Lumbsch & T.H. Nash
Pertusaria neoscotica I.M. Lamb
Pertusaria obruta R.C. Harris
Pertusaria ostiolata Dibben
Pertusaria paratuberculifera Dibben
Pertusaria plittiana Erichsen
Pertusaria propinqua Müll. Arg.
Pertusaria pustulata (Ach.) Duby
Pertusaria rubefacta Erichsen
Pertusaria subpertusa Brodo
Pertusaria superiana Lendemer & E. Tripp
Pertusaria tetrathalmia (Fèe) Nyl.
Pertusaria texana Müll. Arg.
Phacopsis oxyspora (Tul.) Triebel & Rambold var. *defecta* Triebel & Rambold*
Phaeocalicium flabelliforme Tibell (Confirmed by Selva 2016)
Phaeocalicium matthewsianum Selva & Tibell
Phaeocalicium polyporaeum (Nyl.) Tibell
Phaeographis brasiliensis (A. Massal.) Kalb & Matthes-Leicht
Phaeographis inusta (Ach.) Müll. Arg.
Phaeophyscia adiastola (Essl.) Essl.
Phaeophyscia ciliata (Hoffm.) Moberg
Phaeophyscia endococcinea (Körb.) Moberg†
Phaeophyscia endococcinodes (Poelt) Essl.
Phaeophyscia erythrocardia (Tuck.) Essl.†
Phaeophyscia hispidula (Ach.) Essl.
Phaeophyscia orbicularis (Neck.) Moberg†
Phaeophyscia pusilloides (Zahlbr.) Essl.
Phaeophyscia rubropulchra (Degel.) Essl.
Phaeophyscia squarrosa Kashiw.
Phaeopyxis punctum (A. Massal.) Rambold, Triebel & Coppins*
Phaeosporobolus alpinus R. Sant., Alstrup & D. Hawksw.*
Phaeosporobolus fellhanerae R.C. Harris & Lendemer*‡
Phaeosporobolus usneae D. Hawksw. & Hafellner*
Phlyctis boliviensis Nyl.
Phlyctis petraea R.C. Harris, Muscavitch, Ladd & Lendemer
Phlyctis speirea G. Merr.

Phyllopsora confusa Swinsc. & Krog
Phyllopsora corallina (Eschw.) Müll. Arg.
Phyllopsora kalbii Brako
Physcia adscendens (Fr.) H. Olivier
Physcia aipolia (Ehrh. *ex* Humb.) Fürnr.†
Physcia americana G. Merr.
Physcia caesia (Hoffm.) Fürnr.
Physcia halei J.W. Thomson
Physcia millegrana Degel.
Physcia phaea (Tuck.) J.W. Thomson
Physcia pseudospeciosa J.W. Thomson
Physcia pumilior R.C. Harris
Physcia stellaris (L.) Nyl.
Physcia subalbinea Nyl.
Physcia subtilis Degel.‡
Physcia thomsoniana Essl.
Physciella chloantha (Ach.) Essl.
Physconia leucoleiptes (Tuck.) Essl.
Physconia subpallida Essl.
Piccolia ochrophora (Nyl.) Hafellner
Pilophorus cereolus (Ach.) Tuck.
Pilophorus fibula (Tuck.) Th. Fr.
Placidiopsis minor R.C. Harris
Placidium arboreum (Schwein. *ex* Tuck.) Lendemer
Placynthiella dasaea (Stirt.) Tønsberg
Placynthiella icmalea (Ach.) Coppins & P. James
Placynthiella uliginosa (Schrad.) Coppins & P. James
Placynthium nigrum (Huds.) Gray
Placynthium petersii (Tuck.) Burnham
Platismatia glauca (L.) W.L. Culb. & C.F. Culb.
Platismatia tuckermanii (Oakes) W.L. Culb. & C.F. Culb.
Plectocarpon lichenum (Sommerf.) D. Hawksw.*
Polyblastia cupularis A. Massal.
Polycoccum minutulum Kocourk. & F. Berger*
Polymeridium proponens (Nyl.) R.C. Harris
Polysporina simplex (Davies) Vězda
Porina heterospora (Fink) R.C. Harris
Porina scabrida R.C. Harris
Porpidia albocaerulescens (Wulfen) Hertel & Knoph
Porpidia cinereoatra (Ach.) Hertel & Knoph
Porpidia contraponenda (Arnold) Knoph & Hertel
Porpidia crustulata (Ach.) Hertel & Knoph
Porpidia degelii (H. Magn.) Lendemer & R.C. Harris
Porpidia macrocarpa (DC.) Hertel & Schwab
Porpidia subsimplex (H. Magn.) Fryday
Porpidia tuberculosa (Sm.) Hertel & Knoph
Protoblastenia rupestris (Scop.) J. Steiner
Pseudevernia cladonia (Tuck.) Hale & W.L. Culb.
Pseudevernia consocians (Vain.) Hale & W.L. Culb.
Pseudocyphellaria aurata (Ach.) Vain.
Pseudocyphellaria crocata (L.) Vain.†
Pseudocyphellaria perpetua McCune & Miadl.
Pseudosagedia cestrensis (Tuck.) R.C. Harris
Pseudosagedia chlorotica (Ach.) Hafellner & Kalb
Pseudosagedia guentheri (Flot.) Hafellner & Kalb
Pseudosagedia isidiata (R.C. Harris) R.C. Harris
Pseudosagedia rhaphidosperma (Müll. Arg.) R.C. Harris
Psilolechia lucida (Ach.) M. Choisy
Punctelia appalachensis (W.L. Culb.) Krog
Punctelia bolliana (Müll. Arg.) Krog
Punctelia borreri (Turner) Krog†
Punctelia caseana Lendemer & Hodkinson
Punctelia missouriensis G. Wilh. & Ladd
Punctelia reddenda (Stirt.) Krog
Punctelia rudecta (Ach.) Krog
Pycnora praestabilis (Nyl.) Hafellner
Pycnothelia papillaria Dufour
Pyrenidium aggregatum K. Knudsen & Kocourk.*
Pyrenopsis sanguinea Anzi†
Pyrenopsis subfuliginea Nyl.†
Pyrenula citriformis R.C. Harris
Pyrenula confoederata R.C. Harris
Pyrenula leucostoma Ach.
Pyrenula macounii R.C. Harris
Pyrenula pseudobufonia (Rehm) R.C. Harris
Pyrenula punctella (Nyl.) Trevis.
Pyrenula ravenelii (Tuck.) R.C. Harris
Pyrenula subelliptica (Tuck.) R.C. Harris
Pyrrhospora varians (Ach.) R.C. Harris
Pyxine sorediata (Ach.) Mont.
Pyxine subcinerea Stirt.
Ramalina americana Hale
Ramalina culbersoniorum LaGreca
Ramalina intermedia Delise
Ramalina labiosorediata Gasparyan, Sipman & Lücking

Ramalina petrina Bowler & Rundel
Ramboldia elabens (Fr.) Kantvilas & Elix
Ramboldia russula (Ach.) Kalb, Lumbsch & Elix
Ramonia microspora Vězda
Requienella subcollapsa (Ellis & Everh.) R.C. Harris
Rhizocarpon cinereovirens (Müll. Arg.) Vain.
Rhizocarpon eupetraeum (Nyl.) Arnold
Rhizocarpon geographicum (L.) DC.
Rhizocarpon grande (Flörke *ex* Flot.) Arnold
Rhizocarpon infernulum (Nyl.) Lynge f. *sylvaticum* Fryday
Rhizocarpon oederi (Weber) Körber
Rhizocarpon reductum Th. Fr.
Rhizocarpon rubescens Th. Fr.
Rhizocarpon subgeminatum Eitner
Rhizoplaca subdiscrepans (Nyl.) R. Sant.
Rhymbocarpus ericetorum (Körb.) Etayo, Diederich & Ertz*
Rimularia badioatra (Kremp.) Hertel & Rambold
Rinodina adirondackii H. Magn.
Rinodina ascociscana (Tuck.) Tuck.
Rinodina brodoana Sheard, Lendemer & E. Tripp‡
Rinodina buckii Sheard‡
Rinodina bullata Sheard & Lendemer
Rinodina chrysidiata Sheard
Rinodina chrysomelaena (Ach.) Tuck.
Rinodina colobinoides (Nyl.) Zahlbr.
Rinodina degeliana Coppins
Rinodina destituta (Nyl.) Zahlbr.
Rinodina dolichospora Malme
Rinodina efflorescens Malme
Rinodina excrescens H. Magn.
Rinodina oxydata (A. Massal.) A. Massal.
Rinodina subminuta H. Magn.
Rinodina tephraspis (Tuck.) Herre
Rinodina willeyi Sheard & Giralt
Rockefellera crossophylla (Tuck.) Lendemer & E. Tripp
Ropalospora chlorantha (Tuck.) S. Ekman
Ropalospora lugubris (Sommerf.) Poelt
Ropalospora viridis (Tønsberg) Tønsberg
Roselliniella microthelia (Wallr.) N. Hoffm. & Hafellner*
Sarcogyne similis H. Magn.
Sarcopyrenia calcarea Lendemer & R.C. Harris*‡
Sarea resinae (Fr.) Kuntze
Schismatomma glaucescens (Nyl. *ex* Willey) R.C. Harris
Schismatomma pericleum (Ach.) Branth & Rostr.
Scoliciosporum chlorococcum (Stenh.) Vězda
Scoliciosporum pensylvanicum R.C. Harris
Scoliciosporum pruinosum (P. James) Vězda
Scoliciosporum umbrinum (Ach.) Arnold
Scutula dedicata Triebel, Wedin & Rambold*
Segestria leptalea (Durieu & Mont.) R.C. Harris
Skyttea radiatilis (Tuck.) R. Sant., Etayo & Diederich*
Sphinctrina anglica Nyl.*
Sphinctrina leucopoda Nyl.* (Confirmed by Selva 2016)
Sphinctrina turbinata (Pers.: Fr.) De Not.*
Spilonema revertens Nyl.
Staurothele fissa (Taylor) Zwackh
Stenocybe pullatula (Ach.) Stein (Confirmdd by Selva 2016)
Stereocaulon dactylophyllum Flörke
Stereocaulon glaucescens Tuck.
Stereocaulon pileatum Ach.
Stereocaulon tennesseense H. Magn.‡
Sticta beauvoisii Delise
Sticta carolinensis T. McDonald
Sticta fragilinata T. McDonald
Sticta fuliginosa (Hoffm.) Ach.
Sticta limbata (Sm.) Ach.
Sticta sylvatica (Huds.) Ach.
Stigmidium eucline (Nyl.) Vězda*
Stigmidium fuscatae (Arn.) R. Sant.*
Stigmidium lendemeri Kocourk. & K. Knudsen*
Strigula americana R.C. Harris
Strigula stigmatella (Ach.) R.C. Harris
Strigula viridiseda (Nyl.) R.C. Harris
Syzygospora physciacearum Diederich*
Tephromela atra (Huds.) Hafellner
Thelidium decipiens (Nyl.) Kremp.
Thelidium minutulum Körber
Thelocarpon intermediellum Nyl.
Thelocarpon laureri (Flot.) Nyl.
Thelopsis rubella Nyl.

Thelotrema subtile Tuck.
Thermutis velutina (Ach.) Flot.†
Trapelia coarctata (Turner *ex* Sm. & Sowerby) M. Choisy
Trapelia corticola Coppins & P. James
Trapelia glebulosa (Sm.) J.R. Laundon
Trapelia placodioides Coppins & P. James
Trapelia stipitata Brodo & Lendemer
Trapeliopsis flexuosa (Fr.) Coppins & P. James
Trapeliopsis gelatinosa (Flörke) Coppins & P. James
Trapeliopsis granulosa (Hoffm.) Lumbsch
Trapeliopsis viridescens (Schrad.) Coppins & P. James
Tremella cladoniae Diederich & M.S. Christ.*
Tremella parmeliarum Diederich*
Trimmatothelopsis dispersa (H. Magn.) K. Knudsen & Lendemer
Trypethelium virens Tuck.
Tuckermanella fendleri (Nyl.) Essl.
Tuckermannopsis americana (Spreng.) Hale
Tuckermannopsis chlorophylla (Willd.) Hale
Tuckermannopsis ciliaris (Ach.) Gyelnik
Tuckermannopsis orbata (Nyl.) M.J. Lai
Tylophoron americanum Lendemer, E. Tripp & R.C. Harris‡
Umbilicaria mammulata (Ach.) Tuck.
Umbilicaria muhlenbergii (Ach.) Tuck.
Usnea amblyoclada (Müll. Arg.) Zahlbr.
Usnea angulata Ach.
Usnea brasiliensis (Zahlbr.) Motyka
Usnea cavernosa Tuck.†
Usnea ceratina Ach.
Usnea cornuta Körb.
Usnea cristatula Motyka
Usnea dasaea Stirt.
Usnea dasopoga (Ach.) Nyl.
Usnea endochrysea Stirt.
Usnea entoviolata Motyka
Usnea fragilescens Hav. *ex* Lynge
Usnea fulvoreagens (Räsänen) Räsänen
Usnea halei P. Clerc
Usnea merrillii Motyka
Usnea mutabilis Stirt.
Usnea pensylvanica Motyka
Usnea silesiaca Motyka
Usnea strigosa (Ach.) A. Eaton
Usnea subfloridana Stirt.
Usnea subfusca Stirt.
Usnea subgracilis Vain.
Usnea subrubicunda P. Clerc
Usnea subscabrosa Nyl. *ex* Motyka
Usnea trichodea Ach.
Usnocetraria oakesiana (Tuck.) M.J. Lai & J.C. Wei
Vahliella leucophaea (Vahl.) P.M. Jørg.
Vainionora americana Kalb, Tønsberg & Elix
Varicellaria rhodocarpa (Körb.) Th. Fr.
Varicellaria velata (Turner) Lendemer, Hodkinson & R.C. Harris
Verrucaria aethiobola Wahlenb.†
Verrucaria calkinsiana Servít
Violella fucata (Stirt.) T. Sprib.
Vouauxiella lichenicola (Linds.) Petrak & Syd.*
Vulpicida pinastri (Scop.) J.E. Mattsson & M.J. Lai
Vulpicida viridis (Schwein.) J.E. Mattsson & M.J. Lai
Willeya diffractella (Nyl.) Müll. Arg.
Xanthomendoza hasseana (Räsänen) Søchting, Kärnef. & S. Kondr.
Xanthomendoza weberi (S.Y. Kondr. & Kärnef.) L. Lindblom
Xanthoparmelia angustiphylla (Gyeln.) Hale
Xanthoparmelia conspersa (Ehrh. *ex* Ach.) Hale
Xanthoparmelia cumberlandia (Gyeln.) Hale
Xanthoparmelia plittii (Gyeln.) Hale
Xanthoparmelia viriduloumbrina (Gyeln.) Hale
Xyleborus sporodochifer R.C. Harris & Ladd
Xylographa disseminata Willey
Xylographa trunciseda (Th. Fr.) Redinger
Xylographa vitiligo (Ach.) J.R. Laundon

Literature Cited

Ahmadjian V, Jocobs JB. 1981. Relationship between fungus and alga in the lichen *Cladonia cristatella* Tuck. *Nature* 289: 169–72.

Allen, JL, Lendemer JC. 2016. Quantifying the impacts of sea-level rise on coastal biodiversity: a case study on lichens in the mid-Atlantic Coast of eastern North America. *Biol Conserv* 202: 119–26.

Arnold AE, Miadlikowska J, Higgins KL, Sarvate SD, Gugger P, Way A, Hofstetter V, Kauff F, Lutzoni F. 2009. A phylogenetic estimation of trophic transition networks for ascomycetous fungi: are lichens cradles of symbiotrophic fungal diversification? *Syst Biol* 58: 283–97.

Becker VE. 1980. Nitrogen-fixing lichens in forests of the southern Appalachian Mountains of North Carolina. *Bryologist* 83: 29–39.

Bergamini A, Stofer S, Bolliger J, Scheidegger C. 2007. Evaluating macrolichens and environmental variables as predictors of the diversity of epiphytic microlichens. *Lichenologist* 39: 475–89.

Baldwin BG, Goldman DH, Keil DJ, Patterson R, Rosatti TJ, Wilken DH (eds.). 2012. *The Jepson Manual: Vascular Plants of California, 2nd Edition.* Berkeley (CA): University of California Press.

Braun EL. 1950. *Deciduous Forest of Eastern North America.* Philadelphia (PA): Blakiston.

Brodo IM. 1968. The lichens of Long Island, New York: a vegetational and floristic analysis. *New York State Museum Science & Service Bulletin* 410: 1–330.

Brodo IM. 1973. Substrate ecology. In: Ahmadjian V, editor. *The Lichens.* New York (NY): Academic. p. 401–441.

Brodo IM, Sharnoff SD, Sharnoff S. 2001. *Lichens of North America.* New Haven (CT): Yale University Press.

Brunialti G, Frati L, Aleffi, Marignani M, Rosati L, Burrascano S, Ravera S. 2010. Lichens and bryophytes as indicators of old-growth features in Mediterranean forests. *Plant Biosyst* 144: 221–33.

Campbell C. 1960. *Birth of a national park in the Great Smoky Mountains.* Knoxville (TN): University of Tennessee Press.

Ciegler A, Eliasson UH, Keller HW. 2003. Tree-canopy lichens of the Great Smoky Mountains National Park. *Evansia* 20: 114–31.

Culberson WL. 1955. The corticolous communities of lichens and bryophytes in the upland forests of northern Wisconsin. *Ecol Monogr* 25: 215–31.

Culberson WL. 1972. Disjunct distributions in the lichen-forming fungi. *Ann Mo Bot Gard* 59: 165–73.

Degelius, G. 1941. Contributions to the lichen flora of North America II: The lichen flora of the Great Smoky Mountains. *Arkiv fur Botanik* 30 (A:3): 1–84.

Delcourt PA, Delcourt, HR, Morse, DF, Morse, PA. 1993. History, evolution, and organization of vegetation and human culture. In: Martin WH, Boyce, SG, Echternacht AC, editors. *Biodiversity of the Southeastern United States: lowland terrestrial communities.* New York (NY): Wiley. p. 47–79.

Denevan WM. 1992. The pristine myth: the landscape of the Americas in 1492. *Ann Am Assoc Geogr* 82: 369–85.

DePriest P. 1984. *Southern Appalachian lichens: an indexed bibliography.* Atlanta (GA): National Park Service, Southeast Regional Office, Atlanta.

De Wit T. 1983. Lichens as indicators for air quality. *Environ Monit Assess* 3: 273–82.

Dey JP. 1978. Fruticose and foliose lichens of the high-mountain areas of the southern Appalachians. *Bryologist* 81: 1–93.

Dey JP. 1998. *IAR: A report prepared for the Resources Management and Science Division, Great Smoky Mountains National Park.* Gatlinburg, TN.

Dey JP. 2004. *IAR: A report prepared for the Resources Management and Science Division, Great Smoky Mountains National Park.* Gatlinburg, TN.

Dyer LA, Letourneau DK. 2007. Determinants of lichen diversity in a rain forest understory. *Biotropica* 39: 525–29.

Esslinger TL. 2016. A cumulative checklist for the lichen-forming, lichenicolous and allied fungi of the continental United States and Canada, version 21. *Opuscula Philolichenum* 15: 136–390.

Evans AW. 1947. A study of certain North American Cladoniae. *Bryologist* 50: 14–51.

Fanning E, Ely JS, Lumbsch HT, Keller HW. 2007. Vertical distribution of lichen growth forms in tree canopies of Great Smoky Mountains National Park. *Southeast Nat* 6: 83–88.

Flora of North America Editorial Committee, eds. 1993+. *Flora of North American North of Mexico.* 20+ vols. New York and Oxford.

Fridley JD. 2009. Downscaling climate over complex terrain: high finescale (<1000 m) spatial variation of near-ground temperatures in a montane forested landscape (Great Smoky Mountains). *J Appl Meteorol Climatol* 48: 1033–49.

Friedmann EI. 1982. Endolithic microorganisms in the Antarctic cold desert. *Science* 215: 1045–53.

Gaffin D, Hotz, DG, Getz TI. 2001. An evaluation of temperature variations around the Great Smoky Mountains National Park and their associated synoptic weather patterns. NOAA Technical Memorandum, Weather Forecast Office (WFO) Morristown, Tennessee. Available from http://www.weather.gov/mrx/smokytemp

Geiger R, Aron RH, Todhunter P. 2003. *The climate near the ground.* Lanham (MD): Rowman and Littlefield.

Hale ME. 1965. Vertical distributions of cryptogams in a red maple swamp in Connecticut. *Bryologist* 69: 193–97.

Hawksworth DL. 2002. Bioindication: calibrated scales and their utility. In: Nimis PL, Scheidegger C, Wolseley PA , editors. *Monitoring with lichens.* Dordrecht (NETH): Springer Science & Business Media. p. 11–20.

Hitch CJB, Stewart WDP. 1973. Nitrogen fixation by lichens in Scotland. *New Phytol* 72: 509–24.

Hodkinson BP. 2010. A first assessment of lichen diversity for one of North America's "Biodiversity Hotspots" in the Southern Appalachians of Virginia. *Castanea* 75: 126–33.

Hoffman G, Kazmierski R. 1969. An ecological study of epiphytic bryophytes and lichens on *Pseudotsuga menziesii* on the Olympic Peninsula, Washington. I. A description of the vegetation. *Bryologist* 72: 1–19.

Honegger R, Zippler U. 2007. Mating systems in representatives of *Parmeliaceae, Ramalinaceae* and *Physciaceae* (*Lecanoromycetes*, lichen-forming ascomycetes). *Mycol Res* 111: 424–32.

Jägerbrand AK, Alatalo JM. 2015. Effects of human trampling on abundance and diversity of vascular plants, bryophytes and lichens in alpine heath vegetation, northern Sweden. *SpringerPlus* 4: 95.

Jenkins MA. 2007. Vegetation communities of Great Smoky Mountains National Park. *Southeast Nat* 6: 35–56.

Jones D, Wilson MJ, McHardy WJ. 1981. Lichen weathering of rock-forming minerals: application of scanning electron microscopy and microprobe analysis. *J Microsc* 124: 95–104.

Jüriado I, Paal J, Liira J. 2003. Epiphytic and epixylic lichen species diversity in Estonian natural forests. *Biodivers Conserv* 12: 1587–1607.

Kantvilas, G. 1990. Succession in rainforest lichens. *Tasforests* 2: 167–71.

Keller HW, Skrabal M, Eliasson UH, Gaither TW. 2003. Tree canopy biodiversity in the Great Smoky Mountains National Park: ecological and developmental observations of a new myxomycete species of *Diachea. Mycologia* 96: 537–47.

Keller HW, Ely JS, Lumbsch HT, Selva SB. 2007. Great Smoky Mountains National Park's first lichen bio-quest. *Southeast Nat* 6: 89–98.

Kephart H. 1922. *Our southern highlanders: a narrative of adventure in the Southern Appalachians and a study of the life among the mountaineers.* New York (NY): Macmillan Company.

Knudsen K, Lendemer J, Schultz M, Kocourková J, Sheard JW, Pigniolo A, Wheeler T. 2017. Lichen biodiversity and ecology in the San Bernardino and San Jacinto Mountains in southern California (U.S.A.). *Opuscula Philolichenum* 16: 15–138.

Ladd D. 1996. *Lichen assessment and monitoring in two oak woodlands, Mark Twain National Forest, Missouri.* Final Report, USDA Forest Service Contract 40-64R7-3-50.

Ladd D. 1998. Looking at lichens. *Missouri Conservationist* 59: 8–13.

Lang GE, Reiners WA, Pike LH. 1980. Structure and biomass dynamics of epiphytic lichen communities of balsam fir forests in New Hampshire. *Ecol* 61: 541–50.

Lendemer JC, Tripp EA. 2008. Contributions to the lichen flora of North Carolina: a preliminary checklist of the lichens of Gorges State Park. *Bryologist* 111: 57–67.

Lendemer JC, Harris RC, Tripp EA. 2013. The lichens and allied fungi of Great Smoky Mountains National Park: an annotated checklist with comprehensive keys. *Mem N Y Bot Gard* 104: 1–152.

Lendemer JC, Harris RC. 2014. Studies in lichens and lichenicolous fungi-No. 18: resolution of three names introduced by Degelius and Magnusson based on material from the Great Smoky Mountains. *Castanea* 79: 106–17.

Lendemer JC, Tripp EA, Sheard J. 2014. A review of *Rinodina* (Physciaceae) in Great Smoky Mountains National Park highlights the growing significance of this "island of biodiversity" in eastern North America. *Bryologist* 117: 259–81.

Lendemer, JC, Harris RC, Ruiz AM. 2016. A review of the lichens of the Dare Regional Biodiversity Hotspot in the Mid-Atlantic Coastal Plain of North Carolina, eastern North America. *Castanea* 81: 1–77.

Lücking R. 1998. Foliicolous lichens and their lichenicolous fungi collected during the «Smithsonian International Cryptogamic Expedition 1996» to Guyana. *Trop Bryol* 15: 45–76.

Lücking R, Seavey F, Common R, Beeching SQ, Breuss O, Buck WR, Crane L, Hodges M, Hodkinson BP, Lay E, Lendemer JC, McMullin RT, Mercado-Díaz JA, Nelsen M, Rivas Plata E, Safranek W, Sanders WB, Schaefer Jr. H, Seavey J. 2011. The lichens of Fakahatchee Strand Preserve State Park, Florida: proceedings from the 18th Tuckerman Workshop. *Bull Fla Mus Nat Hist* 49(4): 127–86.

Manos PS, Meireles JE. 2015. Biogeographic analysis of the woody plants of the southern Appalachians: implications for the origins of a regional flora. *Am J Bot* 102: 1–25.

Matmon A, Bierman PR, Larsen J, Southworth S, Pavich M, Caffee M. 2003. Temporally and spatially uniform rates of erosion in the southern Appalachian Great Smoky Mountains. *Geology* 31: 155–58.

McCune B, Dey J, Peck J, Heiman K, Will-Wolf S. 1997. Regional gradients in lichen communities of the Southeast United States. *Bryologist* 100: 145–58.

McCune B. 1993. Gradients in epiphyte biomass in three *Pseudotsuga-Tsuga* forests of different ages in western Oregon and Washington. *Bryologist* 96: 405–11.

McCune B, Rosentreter R, Ponzetti JM, Shaw DC. 2000. Epiphyte habitats in an old conifer forest in western Washington, U.S.A. *Bryologist* 103: 417–27.

McDonald T, Miadlikowska J, Lutzoni F. 2003. The lichen genus *Sticta* in the Great Smoky Mountains: a phylogenetic study of morphological, chemical, and molecular data. *Bryologist* 106: 61–79.

Mozingo HN. 1954. Two notable additions to the lichen flora of the Great Smoky Mountains National Park. *Bryologist* 57: 31–33.

Naczi RFC, Abbott JR, and Collaborators. 2017[in press]. *New manual of vascular plants of Northeastern United States and Adjacent Canada*. Bronx (NY): New York Botanical Garden Press.

Nilsson S-G, Arup U, Baranowki R, Ekman S. 1995. Tree-dependent lichens and beetles as indicators in conservation forests. *Conserv Biol* 9: 1208–15.

Nordén B, Paltto H, Götmark F, Wallin K. 2007. Indicators of biodiversity, what do they indicate? –lessons for conservation of cryptogams in oak-rich forest. *Biol Conserv* 135: 369–97.

Paoletti M, Seymour FA, Alcocer MJC, Kaur N, Calvo AM, Archer DB, Dyer PS. 2007. Mating type and the genetic basis of self-fertility in the model fungus *Aspergillus nidulans*. *Curr Biol* 17: 1384–89.

Peck J, Grabner J, Ladd, D. 2002. *The lichen communities of the southeastern Missouri Ozarks: lessons in species associates, habitat partitioning, and distribution from the MOFEP study.* Report to the Missouri Department of Conservation – Missouri Ozark Forest Ecosystem Project.

Perlmutter GB, Blank GB, Wentworth TR, Lowman MD, Neufeld HS, Rivas Plata E. 2017. Effects of highway pollution on forest lichen community structure in western Wake County, North Carolina, USA. *Bryologist* 120: 1–10.

Perry LB, Konrad CE, Hotz DG, Lee LG. 2007. Synoptic classification of snowfall events in the Great Smoky Mountains, USA. *64th Eastern Snow Conference*, St. Johns, Newfoundland. Canadian Meteorological Society.

Pettersson RK, Esseen P-A, Sjöbert K. 1995. Invertebrate communities in boreal forest canopies as influenced by forestry and lichens with implications for passerine birds. *Biol Conserv* 74: 57–63.

Pike LH. 1978. The importance of epiphytic lichens in mineral cycling. *Bryologist* 81: 247–57.

Pogoda C, Keepers K, Lendemer JC, Kane N, Tripp EA. 2018. Reductions in complexity of mitochondria of lichen-forming fungi shed light on genome architecture of obligate symbioses. *Mol Ecol*. DOI: 10.1111/mec.14519

Printzen C, Tønsberg T. 2007. *Bacidia lobarica* (Bacidiaceae, Lecanorales) sp. nov., a sorediate lichen from the southeastern U.S.A. *Bryologist* 110: 487–89.

Sanders WB. 2014. Complete life cycle of the lichen fungus *Calopadia puiggarii* (Pilocarpaceae, Ascomycetes) documented in situ: Propagule dispersal, establishment of symbiosis, thallus development, and formation of sexual and asexual reproductive structures. *Am J Bot* 101: 1836–48.

Schmitt CK, Slack NG. 1990. Host specificity of epiphytic lichens and bryophytes: A comparison of the Adirondack Mountains (New York) and the Southern Blue Ridge Mountains (North Carolina). *Bryologist* 93: 257–74.

Selva S. 2016. Calicioid lichens and fungi of Great Smoky Mountains National Park: not a healthy population. *Evansia* 33: 106–22.

Sharnoff S. 1994. Use of lichens by wildlife in North America. *Research and Exploration* 10: 370–71.

Sheard, JW. 2010. *The Lichen Genus Rinodina (Lecanoromycetidae, Physciaceae) in North America, North of Mexico*. Ottowa: National Research Council of Canada, NRC Research Press. 246 p.

Sherrer S, Zippler U, Honegger R. 2005. Characterisation of the mating-type locus in the genus *Xanthoria* (lichen-forming ascomycetes, Lecanoromycetes). *Fungal Genet Biol* 42: 976–88.

Singh G, Dal Grande F, Cornejo C, Schmitt I, Scheidegger C. 2012. Genetic basis of self-incompatibility in the lichen-forming fungus *Lobaria pulmonaria* and skewed frequency distribution of mating-type idiomorphs: implications for conservation. *PLoS ONE*. https://doi.org/10.1371/journal.pone.0051402.

Skorepa AC. 1972. A catalog of the lichens reported from Tennessee. *Bryologist* 75: 481–500.

Smith MT. 1987. *Archaeology of aboriginal culture change in the interior southeast: depopulation during the early historic period*. Gainesville (FL): University of Florida Press.

Søderstrom L. 1988. Sequence of bryophytes and lichens in relation to substrate variables of decaying coniferous wood in northern Sweden. *Nord J Bot* 8: 89–97.

Southworth S, Schultz A, Denenny D. 2005. *Geologic map of the Great Smoky Mountains national park region, Tennessee and North Carolina*. U.S. Geological Survey Open-File Report 2005–1225.

Spribille T, Pérez-Ortega S, Tønsberg T, Schirokauer D. 2010. Lichens and lichenicolous fungi of the Klondike Gold Rush National Historic Park, Alaska, in a global biodiversity context. *Bryologist* 113: 439–515.

Spribille, T., Resl P, Ahti T, Pérez-Ortega S., Tønsberg T, Mayerhofer H, Lumbsch HT. 2014. Molecular systematics of the wood-inhabiting, lichen-forming genus *Xylographa* (Baeomycetales, Ostropomycetidae) with eight new species. *Acta Universitatis Upsaliensis, Symbolae Botanicae Upsalienses* 37(1): 1–87.

Thomson JW. 1967. *The lichen genus Cladonia in North America*. Toronto (ONT): University of Toronto Press.

Tønsberg T. 2004. Additions to the lichen flora of Great Smoky Mountains National Park. *All Taxa Biodiversity Newsletter (ATBI Quarterly)* 5(1): 6.

Tønsberg T. 2005. Notes on the lichen flora of the Great Smoky Mountains National Park. *Southeastern Biol* 52(1): 26–28.

Tønsberg, T. 2007. Notes on the lichen genus *Lepraria* in Great Smoky Mountains National Park, southeastern North America: *Lepraria lanata* and *L. salazinica* spp nov. *Opuscula Philolichenum* 4: 51–54.

Tripp EA, Lendemer JC. 2010. The genus *Platygramme* in North America. *Castanea* 75: 388–93.

Tripp EA, Lendemer JC, Harris RC. 2010. Resolving the genus *Graphina* Müll. Arg. in North America: new species, new combinations, and treatments for *Acanthothecis*, *Carbacanthographis*, and *Diorygma*. *Lichenologist* 42: 55–71.

Tripp EA. 2016. 2016: Is asexual reproduction an evolutionary dead end in lichens? *Lichenologist* 48: 559–80.

Tripp EA, Lendemer JC. 2017. Twenty-seven modes of reproduction in the obligate lichen symbiosis. *Brittonia*: 10.1007/s12228-017-9500-6.

Tripp EA, Zhuang Y, Lendemer JC. 2017. A review of existing whole genome data suggests lichen mycelia may be haploid or diploid. Bryologist 120: 302–10.

Tripp EA, Zhang N, Schneider H, Huang Y, Mueller GM, Hu Z, Häggblom M, Bhattacharya D. 2017. Reshaping Darwin's Tree: Impact of the Symbiome. *Trends Ecol Evol* 32: 552–55.

Tripp EA, Lendemer JC, McCain CM. In review. Impacts of disturbance on lichenz-obligate symbiotic organismz-in a temperate biodiversity hotspot.

Tripp EA, Lendemer JC. In Press. Highlights from 10 years of lichenological research in Great Smoky Mountains National Park: celebrating the NPS centennial. *Syst Bot.*

Wei JC., Wang, XY, Wu, JL, Wu, JN, Chen, XL, Hou, JL. 1982. *Lichenes Officinales Sinensis*. Beijing (China): Science Press.

Wetmore, CM. 1967. Lichens of the Black Hills of South Dakota and Wyoming. *Publications of the Museum of Michigan State University, Biological Series* 3: 209–464.

Whittaker RH. 1956. Vegetation of the Great Smoky Mountains. *Ecol Monogr* 26: 1–80.

Wolf JHD. 2005. Diversity patterns and biomass of epiphytic bryophytes and lichens along an altitudinal gradient in the North Andes. *Ann Mo Bot Gard* 80(4): 928–60.

Wuerthner G. 2003. *Great Smoky Mountains: a visitors companion*. Mechanicsburg (PA): Stackpole Books. 240 p.

Yoshimura I, Sharp AJ. 1968. Some lichens from the southern Appalachians and Mexico. *Bryologist* 71: 108–13.